Periodic Table of the Elements with the Gmelin System Numbers

1	2	3	4	5	6	7	8	9	10	11	12	13	14	15	16	17	18
1 H 2																	2 He 1
3 Li 20	4 Be 26											5 B 13	6 C 14	7 N 4	8 O 3	9 F 5	10 Ne 1
11 Na 21	12 Mg 27											13 Al 35	14 Si 15	15 P 16	16 S 9	17 Cl 6	18 Ar 1
19 * K 22	20 Ca 28	21 Sc 39	22 Ti 41	23 V 48	24 Cr 52	25 Mn 56	26 Fe 59	27 Co 58	28 Ni 57	29 Cu 60	30 Zn 32	31 Ga 36	32 Ge 45	33 As 17	34 Se 10	35 Br 7	36 Kr 1
37 Rb 24	38 Sr 29	39 Y 39	40 Zr 42	41 Nb 49	42 Mo 53	43 Tc 69	44 Ru 63	45 Rh 64	46 Pd 65	47 Ag 61	48 Cd 33	49 In 37	50 Sn 46	51 Sb 18	52 Te 11	53 I 8	54 Xe 1
55 Cs 25	56 Ba 30	57** La 39	72 Hf 43	73 Ta 50	74 W 54	75 Re 70	76 Os 66	77 Ir 67	78 Pt 68	79 Au 62	80 Hg 34	81 Tl 38	82 Pb 47	83 Bi 19	84 Po 12	85 At	86 Rn 1
87 Fr	88 Ra 31	89*** Ac 40	104 71	105 71													

* NH₄ 23 → NH_4 23

**Lanthanides 39	58 Ce	59 Pr	60 Nd	61 Pm	62 Sm	63 Eu	64 Gd	65 Tb	66 Dy	67 Ho	68 Er	69 Tm	70 Yb	71 Lu
***Actinides	90 Th 44	91 Pa 51	92 U 55	93 Np 71	94 Pu 71	95 Am 71	96 Cm 71	97 Bk 71	98 Cf 71	99 Es 71	100 Fm 71	101 Md 71	102 No 71	103 Lr 71

A Key to the Gmelin System is given on the Inside Back Cover

Gmelin Handbook of Inorganic Chemistry

8th Edition

Organometallic Compounds in the Gmelin Handbook

The following listing indicates in which volumes these compounds are discussed or are referred to:

Ag Silber B5 (1975)

Au Organogold Compounds (1980)

Bi Bismut-Organische Verbindungen (Erg.-Werk, Bd. 47, 1977)

Co Kobalt-Organische Verbindungen 1 (Erg.-Werk, Bd. 5, 1973), 2 (Erg.-Werk, Bd. 6, 1973), Kobalt Erg.-Bd. A (1961), B1 (1963), B2 (1964)

Cr Chrom-Organische Verbindungen (Erg.-Werk, Bd. 3, 1971)

Fe Eisen-Organische Verbindungen A1 (Erg.-Werk, Bd. 14, 1974), A2 (Erg.-Werk, Bd. 49, 1977), A3 (Erg.-Werk, Bd. 50, 1978), A4 (1980), A5 (1981), A6 (Erg.-Werk, Bd. 41, 1977), A7 (1980), B1 (partly in English; Erg.-Werk, Bd. 36, 1976), Organoiron Compounds B2 (1978), Eisen-Organische Verbindungen B3 (partly in English; 1979), B4 (1978), B5 (1978), Organoiron Compounds B6 (1981), B7 (1981), B11 (1983; present volume), Eisen-Organische Verbindungen C1 (1979), C2 (1979), Organoiron Compounds C3 (1980), C4 (1981), C5 (1981), and Eisen B (1929–1932)

Hf Organohafnium Compounds (Erg.-Werk, Bd. 11, 1973)

Nb Niob B4 (1973)

Ni Nickel-Organische Verbindungen 1 (Erg.-Werk, Bd. 16, 1975), 2 (Erg.-Werk, Bd. 17, 1974), Register (Erg.-Werk, Bd. 18, 1975), Nickel B3 (1966), and C (1968–1969)

Np, Pu Transurane C (partly in English; Erg.-Werk, Bd. 4, 1972)

Pt Platin C (1939) and D (1957)

Ru Ruthenium Erg.-Bd. (1970)

Sb Organoantimony Compounds 1 (1981), 2 (1981), 3 (1982)

Sn Zinn-Organische Verbindungen 1 (Erg.-Werk, Bd. 26, 1975), 2 (Erg.-Werk, Bd. 29, 1975), 3 (Erg.-Werk, Bd. 30, 1976), 4 (Erg.-Werk, Bd. 35, 1976), 5 (1978), 6 (1979), Organotin Compounds 7 (1980), 8 (1981), 9 (1982), 10 (1983)

Ta Tantal B2 (1971)

Ti Titan-Organische Verbindungen 1 (Erg.-Werk, Bd. 40, 1977), 2 (1980)

U Uranium Suppl. Vol. E2 (1980)

V Vanadium-Organische Verbindungen (Erg.-Werk, Bd. 2, 1971), Vanadium B (1967)

Zr Organozirconium Compounds (Erg.-Werk, Bd. 10, 1973)

Gmelin Handbuch
der Anorganischen Chemie

Achte völlig neu bearbeitete Auflage

BEGRÜNDET VON   Leopold Gmelin

ACHTE AUFLAGE BEGONNEN   im Auftrag der Deutschen Chemischen Gesellschaft
von R.J. Meyer

FORTGEFÜHRT VON   E.H.E. Pietsch und A. Kotowski
Margot Becke-Goehring

HERAUSGEGEBEN VOM   Gmelin-Institut für Anorganische Chemie
der Max-Planck-Gesellschaft zur Förderung der Wissenschaften
Direktor: Ekkehard Fluck

Springer-Verlag Berlin Heidelberg GmbH 1983

Gmelin Handbook
of Inorganic Chemistry

8th Edition

Fe
Organoiron Compounds

Part B11

Mononuclear Compounds 11

With 29 illustrations

AUTHORS	Konrad Holzapfel, Wolfgang Petz, Christa Siebert (Maintal), Bernd Wöbke
EDITORS	Johannes Füssel, Ulrich Krüerke, Adolf Slawisch
FORMULA INDEX	Alfred Drechsler, Edgar Rudolph
CHIEF EDITOR	Ulrich Krüerke

Springer-Verlag Berlin Heidelberg GmbH 1983

LITERATURE CLOSING DATE: END OF 1979

IN MANY CASES MORE RECENT DATA HAVE BEEN CONSIDERED

Library of Congress Catalog Card Number: Agr 25-1383

ISBN 978-3-662-06917-2   ISBN 978-3-662-06915-8 (eBook)
DOI 10.1007/978-3-662-06915-8

© by Springer-Verlag Berlin Heidelberg 1982
Originally published by Springer-Verlag, Berlin · Heidelberg · New York in 1982
Softcover reprint of the hardcover 8th edition 1982

Preface

The present volume is a continuation of Series B on the mononuclear organoiron compounds. It covers the literature completely to the end of 1979 and includes occasional references to the literature up to 1982.

This volume begins the description of compounds with ^5L ligands, i.e., ligands bonded to the iron atom by five carbon atoms. ^5LFe compounds are numerous, particularly those where ^5L is cyclopentadienyl. The stability of a ^5LFe unit permits many different ligands at the other three coordination sites of the iron atom. Thus the present volume can only cover a part of the ^5L compounds. Main sections of the volume deal with ^5LFe compounds containing other ligands of the D and X type and ^5LFe compounds with an additional ^1L ligand, which include compounds with $^5LFe(CO)$ units (complete in this volume) and $^5LFe(CO)_2$ units (to be continued in B 12).

Series B so far comprises volumes B 1 to B 7, and a survey of this series has been given in the preface to B 7 (1981). The missing parts, B 8 to B 10, describe the remaining ^4LFe compounds and will appear in the near future.

Formulas and symbols have been explained in the prefaces to "Kobalt-Organische Verbindungen" 1, New Suppl. Ser., Vol. 5, and "Nickel-Organische Verbindungen" 1, New Suppl. Ser., Vol. 16. Much of the data, particularly in tables, is given in abbreviated form without dimensions; for explanations see p. 375. Additional remarks, if necessary, are given in the texts heading the tables. The location of substances in other organoiron volumes is given in the form "B 4, 1.1.6.1", i.e., Series B (mononuclear compounds), Volume 4, Chapter 1.1.6.1, or "C 3, 2.4.1.1.1", i.e., Series C (polynuclear compounds), Volume 3, Chapter 2.4.1.1.1.

The volume contains an empirical formula index on p. 376 and a ligand formula index on p. 398.

Frankfurt am Main
December 1982

Ulrich Krüerke

Table of Contents

Organoiron Compounds, Part B

Mononuclear Compounds 11

1.5 Compounds with Ligands Bonded by Five C Atoms

1.5.1 Compounds with One ^{5}L Ligand

$[C_5H_5Fe]^+C_6H_5COO^-$ forms in solution by photolysis ($\lambda = 359$ and 466 nm) of benzoylferrocene and an excess of water in $(CH_3)_2SO$, pyridine, or dimethylformamide. The ^1H NMR spectrum of a C_5D_5N solution shows a C_5H_5 resonance at $\delta = 3.8$ ppm along with signals of $C_6H_5COO^-$ and cyclopentadiene. Characteristic IR absorptions indicate the presence of a carboxylate group. Admission of oxygen produces a purple solution ($\lambda_{max} = 550$ nm), which displays the ESR spectrum of high-spin Fe^{III}. Benzoic acid was isolated after acidification [1].

$[C_5H_5Fe]^+$ can be produced in the gas phase from ferrocene by electron impact, and generally it is an abundant fragment ion in the mass spectra of cyclopentadienyliron compounds. An ion cyclotron resonance investigation of the alkylation of $[C_5H_5Fe]^+$ by CH_3X ($X = F$, Cl, Br, I) revealed sequential alkylation steps, $[C_5H_5Fe]^+ + nCH_3X \rightarrow [(CH_3)_nC_5H_{5-n}Fe]^+ + nHX$ with $n = 1$ to 4, but significant product yields were obtained only in the case of CH_3Br [2].

$[C_5H_5FeH]^+$ was recently observed along with $[C_5H_8Fe]^+$ ($\sim 1:3$ mole ratio) in ion–molecule reactions of Fe^+ with cyclopentane ($< 10^{-6}$ Torr). It presumably forms by initial insertion of Fe^+ into a C–H bond followed by the stepwise loss of $2H_2$. The five-membered carbon ring remains intact [3].

References:

[1] L.H. Ali, A. Cox, T.J. Kemp (J. Chem. Soc. Chem. Commun. **1972** 265/6). – [2] R.R. Corderman, J.L. Beauchamp (Inorg. Chem. **17** [1978] 68/70). – [3] G.D. Byrd, R.C. Burnier, B.S. Freiser (J. Am. Chem. Soc. **104** [1982] 3565/9).

1.5.1.1 Compounds with $[^5LFe(^2D)_2]^-$ Anions

The two following salts represent singular cases since $[^5LFe(^2D)_2]^-$ anions with ^2D ligands other than PF_3 have so far not been reported. The preparation of the salts below is based on the proton abstraction from $C_5H_5Fe(PF_3)_2H$, see 1.5.1.3, p. 16.

$K[C_5H_5Fe(PF_3)_2]$ was formed on slow addition of $KOC(CH_3)_3$ in THF to an excess of $C_5H_5Fe(PF_3)_2H$ in THF. The excess hydride prevented the easy exchange of $OC(CH_3)_3$ for F on the PF_3 ligand. Concentration of the solution and addition of pentane gave a white precipitate, which was purified by repeated washing with pentane, 86% yield.

IR spectrum (KBr): ν(P–F) bands at 798 (sh), 820 (vs, br), 838 (vs, br), 852 (vs), and 878 (s) cm^{-1}; δ(PF) bands at 537 (vs) and 554 (sh) cm^{-1}. The ν(P–F) frequencies are shifted to lower wave numbers relative to the hydride because the negative charge enhances the $Fe-PF_3$ backbonding and thus reduces the P-F double-bond fraction.

$[(C_2H_5)_3NH][C_5H_5Fe(PF_3)_2]$ was obtained as a by-product in the synthesis of $C_5H_5Fe(PF_3)_2H$ from $C_5H_6Fe(PF_3)_3$ and $N(C_2H_5)_3$. After vacuum distillation of the hydride, repeated recrystallization of the residue from THF/pentane gave colorless crystals of the salt, 1 to 5% yield.

IR spectrum (KBr): 529 (s), 550 (m), 792 (vs, sh), 809 (vs, br), 817 (vs, sh), 833 (vs, sh), and 882 (m, br) cm^{-1}.

Reference:

T. Kruck, L. Knoll (Chem. Ber. **105** [1972] 3783/8).

1.5.1.2 Compounds with $[^5LFe(^2D)_3]^{n+}$, $[^5LFe(^2D)_2{}^2D']^+$, $[^5LFe(^2D-^2D)^2D']^+$, and $[^5LFe^2D-^2D-^2D]^+$ Cations

This chapter describes cationic compounds of the types represented by Formulas I to III. Much of the material is summarized in Table 1. ^5L is always C_5H_5 and the only two compounds with a substituted C_5H_5 ring are described outside the table. With one exception (on p. 3) all cations are singly charged, i.e., $n=1$. ^2D represents phosphine, arsine, stibine, or another nitrogen- or oxygen-containing ligand. $^2D-^2D$ stands for diphosphine, o-phenanthroline, or 2,2'-bipyridyl; and $^2D-^2D-^2D$ stands for polyphosphine ligands.

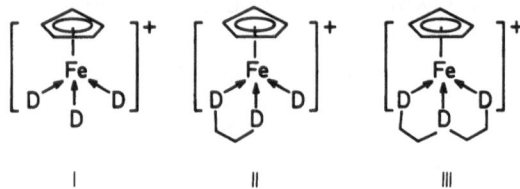

I	II	III

The ionic nature of these compounds is responsible for their conductivity in solution which shows that there are two ions per molecule. Molar conductivities have been determined in acetone, nitromethane, and methanol but only for a few complexes. For the compounds No. 9, 16, and 17 to 24 the molar conductivities were reported to be 82 to 89 $\Omega^{-1}\cdot cm^2\cdot mol^{-1}$ but neither solvent nor temperature was given [5]. The complexes No. 25 to 30 have Λ values of 95 to 99 (in methanol at 33 °C) [11], and the polyphosphine-containing compounds No. 41 to 49 have conductivities of 90 to 140 (in acetone) and 60 to 95 $\Omega^{-1}\cdot cm^2\cdot mol^{-1}$ (in nitromethane) [1 to 3, 8].

$[CH_3-C_5H_4Fe(P(OC_2H_5)_3)_3]PF_6$ is obtained in a 36% yield when the cationic arene complex $[CH_3-C_5H_4FeC_6H_5NO_2]PF_6$ is heated under reflux at 130 to 140 °C with 4 equivalents of $P(OC_2H_5)_3$ for 18 h. A 60% yield is obtained when the same reaction is carried out with $[CH_3-C_5H_4FeC_6H_5NH_2]PF_6$ as the starting material at 145 to 150° for 20 h. The excess phosphine is removed, the solid residue taken up in acetone, and the solution chromatographed on Al_2O_3 with CH_3COCH_3/CH_2Cl_2 (1:1).

The ^1H NMR spectrum (in CD_3COCD_3) exhibits signals at $\delta = 1.17$ (t, CH_3 in C_2H_5), 1.93 (s, CH_3 on C_5H_4), 3.97 to 4.45 (m, CH_2), and 4.52 to 4.82 ppm (m, C_5H_4) [19].

$[C_6H_5-C_5H_4Fe(P(OC_6H_5)_3)_3]I$ is prepared by the action of iodine on $C_6H_5-C_5H_5Fe(P(OC_6H_5)_3)_3$ (see Formula IV) in CH_2Cl_2. After 3 h the mixture is concentrated and the solution chromatographed on Al_2O_3 with petroleum ether. Two bands can be eluted, the first with ether/CH_2Cl_2 to give the neutral $C_6H_5-C_5H_4-Fe(P(OC_6H_5)_3)_2I$ (p. 16), the second with acetone/methanol, followed by recrystallization from CH_2Cl_2/ether, to give yellow crystals of the compound in a 30% yield. The crystals are the monoetherate $[C_6H_5-C_5H_4Fe(P(OC_6H_5)_3)_3]I \cdot (C_2H_5)_2O$ [5].

The ^1H NMR spectrum exhibits the signals $\delta = 4.83$, 5.54 (m, C_5H_4) and 6.6 to 7.6 (m, C_6H_5), which are consistent with a substituted cyclopentadienyl ring [5].

$$Fe(P(OC_6H_5)_3)_3$$

IV

$[C_5H_5Fe(P(CH_3)_3)_3][PF_6]_2$ (Formula I, n=2) is prepared by addition of an equimolar amount of $NOPF_6$ to a solution of $[C_5H_5Fe(P(CH_3)_3)_3]PF_6$ (Table 1, No. 4) in acetone. Gas evolves immediately, and the color of the solution changes from orange to red. After 1 h the volume is reduced, and ether added to the mixture. This precipitates an orange powder, which is then dissolved in CH_3CN and filtered. The pure product is obtained from this solution, in a 68% yield, upon addition of ether [17].

The product melts at 180 °C (dec.) and is paramagnetic, μ_{eff}: 2.12 B.M. [17].

The complexes described in Table 1 can be prepared by the following methods.

Method I: The neutral complex of the $C_5H_5Fe(CO)_2X$ type is irradiated for several hours in CH_3CN or benzene in the presence of excess ligand ^2D, $^2D-^2D$, or $^2D-^2D-^2D$ [1, 15, 17].

Method II: The monocarbonyl complex of the $[C_5H_5Fe(CO)(^2D')_2]X$ type is irradiated in the presence of the ligand ^2D in acetone, THF, or CH_3CN or in ^2D as the solvent [13, 17].

Method III: An excellent leaving group, such as $^2D'=CH_3CN$ or CH_3COCH_3, in the cationic complex $[C_5H_5Fe(^2D'')_2{}^2D']X$ is replaced by the new ligand ^2D [13, 17].

Method IV: Addition of the ligand ^2D to the neutral complex $C_5H_5Fe(^2D')_2X$. The ligands $^2D'$ of the starting material also may be partially replaced during the reaction [5].

Method V: Metathesis between the cationic complex $[C_5H_5Fe(^2D)_3]X$ and NH_4BF_4, NH_4PF_6, $AgBF_4$, or $AgPF_6$ in a polar solvent such as methanol or water [1, 3].

References on p. 15

Table 1
Compounds with $[C_5H_5Fe(^2D)_3]^+$, $[C_5H_5Fe(^2D)_2{}^2D']^+$, $[C_5H_5Fe(^2D-^2D)^2D']^+$, and $[C_5H_5Fe{}^2D-^2D-^2D]^+$ Cations.
Further information for all compounds (except for No. 16) is given at the end of the table.
For abbreviations and dimensions, see p. 375

No.	donor ligands method of preparation (yield in %)	anion	properties and remarks	Ref.
$[C_5H_5Fe(^2D)_3]^+$ type				
1	$P(CH_3)_3$ I (87)	Br	yellow, m.p. 250° ¹H NMR (CD_3SOCD_3): 1.45 (m, PCH_3), 4.40 (q, C_5H_5, $J(P,H)=1.9$)	[17]
2	$P(CH_3)_3$ II (29)	I	no properties reported	[17]
3	$P(CH_3)_3$ III	BF_4	no properties reported	[17]
4	$P(CH_3)_3$ V (93)	PF_6	m.p. 181° (dec.)	[17]
5	$P(OCH_3)_3$ see further information	PF_6	light yellow powder, m.p. 265 to 268° (dec.) ¹H NMR (CD_3COCD_3): 3.79 (m, CH_3, $J(P,H)=10.8$), 4.84 (q, C_5H_5, $J(P,H)=1.2$)	[18]
6	$P(OC_2H_5)_3$ see further information	PF_6	yellow powder, m.p. 254 to 255° (dec.) ¹H NMR (CD_3COCD_3): 1.33 (t, CH_3, $J(H,H)=7$), 4.18 (m, CH_2, $J(H,H)=7$, $^3J(P,H)=4.1$), 4.73 (q, C_5H_5, $J(P,H)=1.3$)	[18, 19]
7	$P(OC_6H_5)_3$ IV, V (98)	BF_4	bright yellow, with 0.5 mol ether ¹H NMR (CD_3COCD_3): 5.03 (q, C_5H_5, $J(P,H)=1.6$), 7.1 (m, C_6H_5)	[5, 7]
8	$P(OC_6H_5)_3$ IV, V	PF_6	yellow, with 1 mol ether ¹H NMR: 4.78 (q, C_5H_5, $J(P,H)=1.5$), 6.81 to 7.17	[5]

No.	2D	2D'		properties	Ref.
9	P(OC$_6$H$_5$)$_3$ see further information		SCN	yellow, with 1 mol ether, m.p. 146 to 147° IR (Nujol): ν(SCN) 2050	[5]

[C$_5$H$_5$Fe(^{2}D)$_2$ ^{2}D']$^+$ type

No.	2D	2D'		properties	Ref.
10	P(CH$_3$)$_3$ see further information	OS(CD$_3$)$_2$	BF$_4$	^{1}H NMR (CD$_3$SOCD$_3$): 1.52 (m, PCH$_3$), 4.54 (t, C$_5$H$_5$)	[17]
11	P(CH$_3$)$_3$ IV, V (74)	OS(CH$_3$)$_2$	PF$_6$	yellow, m.p. 283 to 286° (dec.) ^{1}H NMR: 1.52 (m, PCH$_3$), 3.23 (s, SCH$_3$), 4.54 (t, C$_5$H$_5$, J(P,H)=2)	[17]
12	P(CH$_3$)$_3$ IV	NCCH$_3$	I	no properties reported	[17]
13	P(CH$_3$)$_3$ II (94)	NCCH$_3$	BF$_4$	orange, m.p. 168 to 175° (dec.)	[17]
14	P(CH$_3$)$_3$ II (96)	NCCH$_3$	PF$_6$	m.p. 300° (dec.) ^{1}H NMR (CD$_3$SOCD$_3$): 1.55 (m, PCH$_3$), 2.05 (s, CH$_3$CN), 4.60 (t, C$_5$H$_5$, J(P,H)=1.5) IR: ν(CN) 2260	[17]
15	P(CH$_3$)$_3$ III (67)	P(C$_6$H$_5$)$_3$	BF$_4$	orange, m.p. 185 to 190° ^{1}H NMR (CD$_2$Cl$_2$): 1.37 (m, PCH$_3$), 4.25 (q, C$_5$H$_5$, J(P,H)=1.8), 7.42 (m, C$_6$H$_5$)	[17]
16	P(C$_2$H$_5$)$_3$ IV, V	P(OC$_6$H$_5$)$_3$	PF$_6$	orange ^{1}H NMR: 1.21 (t, CH$_3$, J(H,H)=6), 1.93 (m, CH$_2$), 4.56 (s, C$_5$H$_5$), 6.98 to 7.10 (m, C$_6$H$_5$)	[5]
17	P(OC$_6$H$_5$)$_3$ IV, V	SO$_2$	PF$_6$	yellow, with 1 mol ether ^{1}H NMR: 4.94 (s, C$_5$H$_5$), 7.33 (m, C$_6$H$_5$) IR (Nujol): ν(SO) 1298	[5]
18	P(OC$_6$H$_5$)$_3$ IV, V	NCCH$_3$	PF$_6$	orange, m.p. 126 to 130° ^{1}H NMR: 2.11 (s, CH$_3$), 4.21 (s, C$_5$H$_5$), 7.04 to 7.16 (m, C$_6$H$_5$)	[5]

References on p. 15

2

Table 1 [continued]

No.	donor ligands method of preparation (yield in %)	anion	properties and remarks	Ref.
19	$P(OC_6H_5)_3$ IV, V	$NCCH_2CH_3$	PF_6	orange, m.p. 124 to 125° ^1H NMR: 1.15 (t, CH_3, $J(H,H)=6$), 2.61 (q, CH_2, $J(H,H)=6$), 4.23 (s, C_5H_5), 7.06 to 7.34 (m, C_6H_5) IR (Nujol): $\nu(CN)$ 2267	[5]
20	$P(OC_6H_5)_3$ IV, V	$NCCH_2Cl$	PF_6	orange, m.p. 148 to 151° ^1H NMR: 4.38 (t, C_5H_5, $J(P,H)=1.5$), 4.90 (t, CH_2, $J(P,H)=1$), 7.36 to 7.43 (m, C_6H_5) IR: $\nu(CN)$ 2271	[5]
21	$P(OC_6H_5)_3$ IV, V	$NCCH_2CH_2Cl$	BF_4	orange, m.p. 110 to 113° ^1H NMR: 3.09 (t, $NCCH_2$, $J(H,H)=5$), 3.68 (t, CH_2Cl, $J(H,H)=5.6$), 4.23 (s, C_5H_5), 7.09 to 7.29 (m, C_6H_5) IR: $\nu(CN)$ 2278	[5]
22	$P(OC_6H_5)_3$ IV, V	$NCN(CH_3)_2$	BF_4	orange, m.p. 111 to 112° ^1H NMR: 2.83 (s, CH_3), 4.19 (s, C_5H_5), 7.12 to 7.23 (m, C_6H_5) IR: $\nu(CN)$ 2292	[5]
23	$P(OC_6H_5)_3$ IV, V	$NCCH_2CONH_2$	BF_4	orange, m.p. 123 to 125° ^1H NMR: 3.74 (s, CH_3), 4.16 (t, C_5H_5, $J(P, H) \sim 1$), 5.67 (s, NH_2), 7.13 to 7.28 (m, C_6H_5) IR (Nujol): $\nu(C=O)$ 1716	[5]
24	$P(OC_6H_5)_3$ IV, V	$NCCH_2COOH$	BF_4	orange, m.p. 128 to 131° IR (Nujol): $\nu(CN)$ 2280	[5]

$[C_5H_5Fe(^2D-^2D)^2D']^+$ type

No.	donor ligands method of preparation (yield in %)	anion	properties and remarks	Ref.
25	$P(C_6H_5)_3$	—	red violet	[11]

No.	Ligand		Anion	Properties	Ref.
26	(bipyridine)	As(C$_6$H$_5$)$_3$	–	red violet	[11]
	see further information				
27	(bipyridine)	Sb(C$_6$H$_5$)$_3$	–	red violet	[11]
	see further information				
28	(phenanthroline)	P(C$_6$H$_5$)$_3$	–	orange red	[11]
	see further information				
29	(phenanthroline)	As(C$_6$H$_5$)$_3$	–	orange red	[11]
	see further information				
30	(phenanthroline)	Sb(C$_6$H$_5$)$_3$	–	orange red	[11]
	see further information				
31	(P(CH$_3$)$_2$CH$_2$–)$_2$	(CH$_3$)$_2$CO	BF$_4$	brown crystals IR (Nujol): ν(C=O) 1650	[4]
32	(P(CH$_3$)$_2$CH$_2$–)$_2$ IV (51)	P(OC$_6$H$_5$)$_3$	–	yellow ^{1}H NMR (CDCl$_3$): 4.38 (C$_5$H$_5$)	[12]
33	(P(CH$_3$)$_2$CH$_2$–)$_2$ V (82)	P(OC$_6$H$_5$)$_3$	BF$_4$	yellow	[12]
34	(P(C$_6$H$_5$)$_2$CH$_2$–)$_2$	(CH$_3$)$_2$CO	Cl	see further information	[13]
35	(P(C$_6$H$_5$)$_2$CH$_2$–)$_2$ II (70)	(CH$_3$)$_2$CO	PF$_6$	graphite, dec. p. 80 to 85° ^{1}H NMR (CD$_3$COCD$_3$): 2.20 (s, CH$_3$), 2.63, 2.73 (CH$_2$), 4.56 (t, C$_5$H$_5$), 7.70 (m, C$_6$H$_5$) IR (KBr): ν(C=O) 1710	[10, 13]

Table 1 [continued]

No.	donor ligands method of preparation (yield in %)	anion	properties and remarks	Ref.
36	$(P(C_6H_5)_2CH_{2^-})_2$　C_4H_8O (THF) II (15)	PF_6	blue, dec. p. 80 to 85° not sufficiently soluble for ^1H NMR	[13]
37	$(P(C_6H_5)_2CH_{2^-})_2$　NH_3 III (10)	PF_6	red crystals, with 1 mol THF, dec. p. 200 to 205° ^1H NMR (CD_3COCD_3): 0.75 (NH_3), 2.81, 2.92 (CH_2), 4.53 (t, C_5H_5), 7.75 (C_6H_5) IR (KBr): $\nu(NH)$ 3285, 3350	[13]
38	$(P(C_6H_5)_2CH_{2^-})_2$　N_2H_4 III (55)	PF_6	red crystals, with 1 mol THF, dec. p. 160 to 165° ^1H NMR (CD_3COCD_3): 2.66, 2.90 (CH_2), 4.63 (t, C_5H_5), 7.82 (m, C_6H_5) IR (KBr): $\nu(NH)$ 3185, 3270	[13]
39	$(P(C_6H_5)_2CH_{2^-})_2$　$NCCH_3$ I (87)	Br	red, dec. p. ~118 to 125° ^1H NMR $(CDCl_3)$: 1.77 (s, CH_3), 2.45 to 2.55 (m, CH_2), 4.36 (t, C_5H_5, J(P,H) = 1.2), 7.0 to 7.8 (m, C_6H_5)	[15]
40	$(P(C_6H_5)_2CH_{2^-})_2$　$NCCH_3$ see further information	$B(C_6H_5)_4$	orange red, m.p. 190° (dec.) ^1H NMR (CD_3COCD_3): 1.40 (m, CH_3), 2.40 (m), 2.62 (m), 4.40 (t, C_5H_5), 6.85 (m, BC_6H_5), 7.90, 7.52 (m, PC_6H_5) ^{13}C NMR (CD_3COCD_3): 79.7 (C_5H_5), 129.9, 131.2, 131.5, 132.3, 132.5, 133.5, 133.7 (C_6H_5) IR (CH_2Cl_2): $\nu(CN)$ 2270	[14]

$[C_5H_5Fe^2D_-^2D_-^2D]^+$ **type, R = C_6H_5**

No.	donor ligands method of preparation (yield in %)	anion	properties and remarks	Ref.
41	$RP\begin{cases} C_2H_4-P(CH_3)_2 \\ C_2H_4-P(CH_3)_2 \end{cases}$ I	—	yellow, m.p. 268 to 270° ^1H NMR $(CDCl_3)$: 1.30 (m, CH_3), 1.4 to 2.7 (br, CH_2), 4.20 (s, C_5H_5), 7.47 (m, C_6H_5) ^{31}P NMR (CH_2Cl_2): 76.1 (d, PCH_3), 120.4 (t, PC_6H_5, J(P,P) = 37)	[8, 9]

References on p. 15

No.	Structure	Anion	Properties	Ref.
42	RP$_b$—C$_2$H$_4$—P$_a$(CH$_3$)$_2$ / —C$_2$H$_4$—P$_c$R$_2$ I	I	^{31}P NMR (CH$_2$Cl$_2$): 68.7 (m, PCH$_3$, J(P$_b$,P$_a$)=39), 104.3 (m, PR$_2$, J(P$_a$,P$_c$)=44, J(P$_b$,P$_c$)=29), 123.1 (m, PR)	[9]
43	RP$\langle$—C$_2$H$_4$—PR$_2$ / —C$_2$H$_4$—PR$_2$ I (52)	Br	yellow, m.p. 240 to 242° (dec.)	[2]
44	RP$\langle$—C$_2$H$_4$—PR$_2$ / —C$_2$H$_4$—PR$_2$ V	PF$_6$	yellow, m.p. 253 to 255° ^{1}H NMR (CDCl$_3$): 2(CH$_2$), 4.23 (C$_5$H$_5$), 7.3 (C$_6$H$_5$)	[2, 16]
45	CH$_3$—C$\langle$—CH$_2$—PR$_2$ / —CH$_2$—PR$_2$ / —CH$_2$—PR$_2$ I (33)	Cl	yellow, 2–H$_2$O solvate, m.p. 139 to 141°	[1]
46	CH$_3$—C—CH$_2$—$\langle$—CH$_2$—PR$_2$ / —CH$_2$—PR$_2$ / —CH$_2$—PR$_2$ I (50)	Br	yellow, m.p. 127 to 130°	[1]
47	CH$_3$—C$\langle$—CH$_2$—PR$_2$ / —CH$_2$—PR$_2$ / —CH$_2$—PR$_2$ V (93)	PF$_6$	yellow, m.p. 298° (dec.) ^{1}H NMR (CD$_3$COCD$_3$): 5.06 (C$_5$H$_5$)	[1]
48	RP—C$_2$H$_4$—PR$_2$ / C$_2$H$_4$ / RP—C$_2$H$_4$—PR$_2$ I (77)	I	green, m.p. 211 to 214°	[3]

Table 1 [continued]

No.	donor ligands method of preparation (yield in %)	anion	properties and remarks	Ref.
49	RP$-$C$_2$H$_4$$-PR_2$ $\mid$ C$_2$H$_4$ $\mid$ RP$-$C$_2$H$_4$$-PR_2$ V	PF$_6$	yellow, m.p. 169 to 173°	[3]

supplement, [C$_5$H$_5$Fe(^{4}D)^{2}D']$^+$ type

| 50 | (P(CH$_3$)$_2$CH$_2-$)$_2$
III (50) | N$_2$Mo-
(P(C$_6$H$_5$)$_3$)$_2-$
C$_6$H$_5$CH$_3$

BF$_4$ | 1$-$C$_6$H$_5$CH$_3$ solvate, golden$-$brown needles
IR: ν(NN) 1930 in Nujol, 1945 in C$_6$H$_5$CH$_3$ | [20] |

References on p. 15

Further information:

[C$_5$H$_5$Fe(P(CH$_3$)$_3$)$_3$]Br (Table 1, No. 1) is prepared by irradiation of a solution of C$_5$H$_5$Fe(CO)$_2$Br and excess P(CH$_3$)$_3$ (mole ratio 1:8) in refluxing CH$_3$CN for 6.5 h. Ether is added to the concentrated solution to precipitate crystals, which are recrystallized from CH$_2$Cl$_2$/petroleum ether [17].

[C$_5$H$_5$Fe(P(CH$_3$)$_3$)$_3$]I (Table 1, No. 2) is obtained along with C$_5$H$_5$(CO)-Fe(P(CH$_3$)$_3$)$_2$I by photolysis of [C$_5$H$_5$Fe(CO)(P(CH$_3$)$_3$)$_2$]I in acetone for 3.5 h as a mixture with the starting material [17].

[C$_5$H$_5$Fe(P(CH$_3$)$_3$)$_3$]BF$_4$ and [C$_5$H$_5$Fe(P(CH$_3$)$_3$)$_3$]PF$_6$ (Table 1, Nos. 3 and 4). Complex No. 3 was mentioned to form rapidly by ligand replacement in [C$_5$H$_5$Fe(P(CH$_3$)$_3$)$_2$NCCH$_3$]BF$_4$ (No. 13) in the presence of P(CH$_3$)$_3$. The hexafluorophosphate is obtained by metathesis of the bromide (No. 1) and NH$_4$PF$_6$ (mole ratio 1:2) in water. It is recrystallized from CH$_2$Cl$_2$/ether [17]. The cation [C$_5$H$_5$Fe(P(CH$_3$)$_3$)$_3$]$^+$ (isolated as the PF$_6^-$ salt) is also formed from C$_5$H$_5$-Fe(CO)$_2$SnCl$_3$ and an excess of P(CH$_3$)$_3$ (mole ratio 1:8) in CH$_3$CN on irradiation for 3 h [17].

Electrochemical oxidation by cyclic voltammetry in CH$_2$Cl$_2$ gives a di-cation at E$_{1/2}$=0.71 V, see [C$_5$H$_5$Fe(P(CH$_3$)$_3$)$_3$][PF$_6$]$_2$ on p. 3.

[C$_5$H$_5$Fe(P(OCH$_3$)$_3$)$_3$]PF$_6$ and [C$_5$H$_5$Fe(P(OC$_2$H$_5$)$_3$)$_3$]PF$_6$ (Table 1, Nos. 5 and 6) are obtained by photolyzing the arene complex [C$_5$H$_5$Fe(C$_6$H$_4$(CH$_3$)$_2$-1,4)]PF$_6$ in CH$_2$Cl$_2$ in the presence of excess P(OCH$_3$)$_3$ for 1 h at 3±3 °C (mole ratio ~1:26) or excess P(OC$_2$H$_5$)$_3$ for 2.5 h at 10±3 °C, respectively, with the help of sunlight. The compounds are precipitated from concentrated CH$_2$Cl$_2$ solutions by addition of hexane, 82 and 96% yield, respectively. No product is obtained in the absence of sunlight even if the reaction mixture is heated. Complex No. 6 is also produced from the cation [C$_5$H$_5$Fe(C$_8$H$_8$)]$^+$ and P(OC$_2$H$_5$)$_3$ in the dark at room temperature after several hours [18].

A replacement reaction giving compound No. 6, with various yields, was also described. The nitroarene complexes [C$_5$H$_5$FeC$_6$H$_5$NO$_2$]PF$_6$; o-, m-, and p-[C$_5$H$_5$Fe(C$_6$H$_4$(NO$_2$)CH$_3$)]PF$_6$ (reaction temperature 130 to 140 °C); the related amino arene complexes (reaction temperature of 145 to 150 °C); or [C$_5$H$_5$Fe(C$_6$H$_4$(CH$_3$-1)Cl-4)]PF$_6$ is refluxed with four equivalents of P(OC$_2$H$_5$)$_3$. After removal of excess phosphine, the solids are taken up in acetone, and the solutions chromatographed on Al$_2$O$_3$ with acetone/CH$_2$Cl$_2$ eluent. Recrystallization from CH$_2$Cl$_2$/ether gives the ring replacement product No. 6 [19].

[C$_5$H$_5$Fe(P(OC$_6$H$_5$)$_3$)$_3$]BF$_4$ and [C$_5$H$_5$Fe(P(OC$_6$H$_5$)$_3$)$_3$]PF$_6$ (Table 1, Nos. 7 and 8). To a concentrated solution of C$_5$H$_5$Fe(P(OC$_6$H$_5$)$_3$)$_2$I in benzene is added a slight excess of AgBF$_4$ and P(OC$_6$H$_5$)$_3$. The resulting suspension is intensively stirred for 3 h and then filtered. The residue is washed with benzene and dissolved in CH$_2$Cl$_2$, and filtered through charcoal. Ether is added to precipitate the voluminous product (No. 7).

A mole of the crystals contains 0.5 mol ether of solvation [7]. Compound No. 8 is obtained like Nos. 17 to 24 by reacting the intermediate [C$_5$H$_5$Fe(P(OC$_6$H$_5$)$_3$)$_2$]$^+$ with P(OC$_6$H$_5$)$_3$. It also forms by slow (12 h) disproportionation of the same intermediate (PF$_6^-$ anion) in benzene [5].

References on p. 15

Nucleophilic attack of $NaBH_4$ in THF containing a little water or of C_6H_5Li in THF between $-78\,^\circ C$ and room temperature takes place at the C_5H_5 ligand of No. 7 to give the $^4LFe(^2D)_3$ complex type $C_5H_5RFe(P(OC_6H_5)_3)_3$ with $R=H$ or C_6H_5 [5]. No addition on the ring ligand is observed with $(CH_3)_3SnLi$ [6]. The reaction with $LiSn_4(CH_3)_9$ in THF forms $C_5H_5Fe(P(OC_6H_5)_3)_2Sn(Sn(CH_3)_3)_3$ [6, 7], see 1.5.1.3, p. 16, No. 43.

$[C_5H_5Fe(P(OC_6H_5)_3)_3]SCN$ (Table 1, No. 9) is obtained along with C_5H_5-$Fe(P(OC_6H_5)_3)_2SCN$ when a benzene solution of $[C_5H_5Fe(P(OC_6H_5)_3)_2]^+$ is treated with LiSCN [5].

$[C_5H_5Fe(P(CH_3)_3)_2OS(CD_3)_2]BF_4$ (Table 1, No. 10) is present in equilibrium with compound No. 15 according to the ^1H NMR spectrum of No. 15 in $(CD_3)_2SO$ solution [17].

$[C_5H_5Fe(P(CH_3)_3)_2OS(CH_3)_2]PF_6$ (Table 1, No. 11) is prepared by adding a few drops of $OS(CH_3)_2$ to a mixture of $C_5H_5Fe(P(CH_3)_3)_2I$ and NH_4PF_6 in CH_2Cl_2. A yellow solid precipitates. It can be recrystallized from CH_2Cl_2/ether [17].

$[C_5H_5Fe(P(CH_3)_3)_2NCCH_3]I$ (Table 1, No. 12) is formed by dissolving $C_5H_5Fe(P(CH_3)_3)_2I$ in CH_3CN. Presumably the large excess of the solvent drives the equilibrium towards the CH_3CN complex [17].

$[C_5H_5Fe(P(CH_3)_3)_2NCCH_3]X$ ($X=BF_4$ and PF_6, Table 1, Nos. 13 and 14) are prepared from the corresponding $[C_5H_5Fe(CO)(P(CH_3)_3)_2]X$ in CH_3CN by irradiation at reflux temperature for 3.5 h. However, a satisfactory analysis was not obtained for the BF_4^- salt [17].

The CH_3CN ligand can be replaced by $P(C_6H_5)_3$.

$[C_5H_5Fe(P(CH_3)_3)_2P(C_6H_5)_3]BF_4$ (Table 1, No. 15) is the product of the 18-h reaction of $[C_5H_5Fe(P(CH_3)_3)_2NCCH_3]BF_4$ (No. 13) with an excess of $P(C_6H_5)_3$ (mole ratio $\sim 1:3$) in THF at room temperature [17].

$[C_5H_5Fe(P(OC_6H_5)_3)_2{}^2D]X$ compounds ($^2D=SO_2$, $NCCH_3$, $NCCH_2CH_3$, $NCCH_2Cl$, $NCCH_2CH_2Cl$, $NCN(CH_3)_2$, $NCCH_2CONH_2$, $NCCH_2COOH$; $X=BF_4$, PF_6; Table 1, Nos. 17 to 24) are prepared from $C_5H_5Fe(P(OC_6H_5)_3)_2I$ and $AgPF_6$ or $AgBF_4$ in benzene at room temperature (30 min) by subsequent treatment with the appropriate ^2D molecule (12 h). After removal of the solvent, repeated precipitation of the residue from CH_2Cl_2 with ether gives the pure salt in about a 30% yield. The salts No. 21 to 23 do not readily crystallize, but crystallization can be induced by very slow addition of ether or ether/hexane mixtures, while shaking and storage overnight at $-5\,^\circ C$ [5].

$[C_5H_5Fe(C_{10}H_8N_2)^2D]I$ and $[C_5H_5Fe(C_{12}H_8N_2)^2D]I$ ($^2D=P(C_6H_5)_3$, $As(C_6H_5)_3$, $Sb(C_6H_5)_3$; Table 1, Nos. 25 to 27 and 28 to 30) are prepared by the reaction of $[C_5H_5Fe(CO)_2{}^2D]I$ and the chelating ligands o–phenanthroline or 2,2'–bi-pyridyl in CH_2Cl_2. Removal of the solvent in vacuo and washing with benzene gives the crystalline compounds. They can also be prepared by refluxing the monocarbonyl complexes $C_5H_5Fe(CO)(^2D)I$ with the chelating ligand in benzene for 2 h. The solids are washed several times with benzene and are recrystallized from CH_3OH.

The complexes do not melt below $200\,^\circ C$. The air-stable solids dissolve in H_2O, CH_2Cl_2, $CHCl_3$, $HCON(CH_3)_2$, CH_3OH, C_2H_5OH, and CH_3COCH_3 [11].

$[C_5H_5Fe(P(CH_3)_2CH_2-)_2OC(CH_3)_2]BF_4$ (Table 1, No. 31) is obtained from $C_5H_5Fe(P(CH_3)_2CH_2-)_2I$ and $TlBF_4$ in acetone at $0\,^\circ C$ [4].

References on p. 15

An acetone solution reacts readily with N_2 to give $[C_5H_5Fe(P(CH_3)_2CH_2-)_2N]_2$-$[BF_4]_2 \cdot 2H_2O$. With CO a high yield of $[C_5H_5Fe(P(CH_3)_2CH_2-)_2(CO)]BF_4$ is obtained. Reduction with $LiAlH_4$ in THF gives the hydride $C_5H_5Fe(P(CH_3)_2CH_2-)_2H$ [4], see 1.5.1.3, p. 16, No. 1.

$[C_5H_5Fe(P(CH_3)_2CH_2-)_2P(OC_6H_5)_3]I$ and $[C_5H_5Fe(P(CH_3)_2CH_2-)_2$-$P(OC_6H_5)_3]BF_4$ (Table 1, Nos. **32** and **33**). Complex No. 32 is obtained along with $C_5H_5Fe(P(CH_3)_2CH_2-)_2I$ by reacting $C_5H_5Fe(P(OC_6H_5)_3)_2I$ with the diphosphine in toluene at 60 °C (6 h) and room temperature (overnight). The neutral compound can be removed from the solid mixture by extraction with CH_2Cl_2. The residue of No. 32 is purified on Al_2O_3 with acetone eluent. The tetrafluoroborate (No. 33) precipitates on addition of saturated aqueous NH_4BF_4 to No. 32 in acetone. It is recrystallized from acetone/water [12].

Irradiation of complex No. 32 in the presence of KI and excess $(P(CH_3)_2CH_2-)_2$ gives $C_5H_5Fe(P(CH_3)_2CH_2-)_2I$ [12], see 1.5.1.3, p. 16, No. 55.

$[C_5H_5Fe(P(C_6H_5)_2CH_2-)_2OC(CH_3)_2]Cl$ (Table 1, No. **34**). When an acetone solution of $[C_5H_5Fe(CO)(P(C_6H_5)_2CH_2-)_2]Cl$ under argon is irradiated, a black solution of No. 34 is obtained. No further details are available [13]. Dissolution in THF/pentane gives the violet complex $[C_5H_5Fe(P(C_6H_5)_2CH_2-)_2]Cl \cdot OC(CH_3)_2$ in a 10% yield [13].

$[C_5H_5Fe(P(C_6H_5)_2CH_2-)_2OC(CH_3)_2]PF_6$ and $[C_5H_5Fe(P(C_6H_5)_2CH_2-)_2$-$(C_4H_8O)]PF_6$ (Table 1, Nos. **35** and **36**) are prepared by irradiation of $[C_5H_5Fe(CO)$-$(P(C_6H_5)_2CH_2-)_2]PF_6$ in acetone at -30 °C or in THF at room temperature for 3 h, respectively, in a stream of argon in order to remove carbon monoxide. Compound No. 36, a fine crystalline solid, is washed with cold THF. The THF complex contains some unreacted starting material. Complex No. 35 is recrystallized from argon-saturated THF on cooling to -78 °C [13].

The solids are stable for a time, even in air. However, in solution the THF complex decomposes immediately, even at -80 °C [13]. Both complexes can be converted into the cation $[C_5H_5Fe(P(C_6H_5)_2CH_2-)_2N]_2^{2+}$ (see C3, 2.5.2.1.3, p. 102) by the action of nitrogen. The reaction is reversible, and compound No. 36 is formed, when the nitrogen complex is treated with THF [10, 13].

$[C_5H_5Fe(P(C_6H_5)_2CH_2-)_2NH_3]PF_6$ and $[C_5H_5Fe(P(C_6H_5)_2CH_2-)_2N_2H_4]PF_6$ (Table 1, Nos. **37** and **38**). To an argon saturated solution of complex No. 35 an excess of N_2H_4 is added which causes the black solution to turn red. Removal of the volatile products in vacuo, dissolution in THF, and crystallization at -78 °C gives pure No. 38. The ammonia complex No. 37 is prepared by passing NH_3 through the solution of No. 35 for 5 min. After removal of the solvent the residue is recrystallized in the same manner from acetone/THF. Both complexes are obtained solvated with 1 mol THF [10, 13].

$[C_5H_5Fe(P(C_6H_5)_2CH_2-)_2NCCH_3]Br$ (Table 1, No. **39**) results when an equimolar mixture of $C_5H_5Fe(CO)_2Br$ and the ^2D-^2D ligand in CH_3CN is photolyzed until the CO evolution stops. Concentration of the filtered solution gives small red crystals.

IR bands were reported for the 400 to 1500 cm^{-1} region. The compound is stable and unreactive towards oxygen. It reacts with a variety of anions (X = CN, SCN, Br, I, SC_6H_5, CH_3 (from CH_3MgI), H (from $LiAlH_4$) in THF to give $C_5H_5Fe(P(C_6H_5)_2$-$CH_2-)_2X$ in good yields (1.5.1.3, p. 16) [15].

References on p. 15

[C₅H₅Fe(P(C₆H₅)₂CH₂-)₂NCCH₃][B(C₆H₅)₄] (Table 1, No. **40**) results from C₅H₅Fe(P(C₆H₅)₂CH₂-)₂CH₂OCH₃ in two steps. The starting material is treated with HBF₄ in acetic anhydride at 0 °C for 11 min. The complex dissolves to give a dark solution, which is poured into anhydrous ether. The resulting precipitate is dissolved in NCCH₃. Attempts to recrystallize the resulting product gave only a red glassy material. The BF₄⁻ anion is exchanged for [B(C₆H₅)₄]⁻ in methanol to produce orange crystals.

Deep red irregular blocky crystals were grown on cooling an acetone solution from 50 °C to room temperature. A structure determination was carried out at −34 °C. The crystals belong to the monoclinic system with a = 23.542(8), b = 11.494(3), c = 8.774(4) Å, and β = 112.26(2)°; space group Cc − C$_s^4$ or C2/c − C$_{2h}^6$. Z = 4 gives D$_c$ = 1.248, while D$_m$ = 1.24 g·cm⁻³. The molecular structure, which is shown in **Fig. 1** consists of well separated cations and anions with no unusual interionic contacts. The chelate ring exhibits an envelope configuration, i.e., the atoms Fe, P(1), P(2), C(1) are essentially coplanar, while atom C(2) − the flap atom of the envelope − lies more than 0.7 Å from the plane. This diminishes the contact between the phenyl groups and the CH₃CN and C₅H₅ ligands. The torsion angles of the chelating ring were compared with those of other metal-(P(C₆H₅)₂CH₂-)₂ complexes. A summary of metal-CH₃CN bond parameters was given in a separate table [14].

Fig. 1

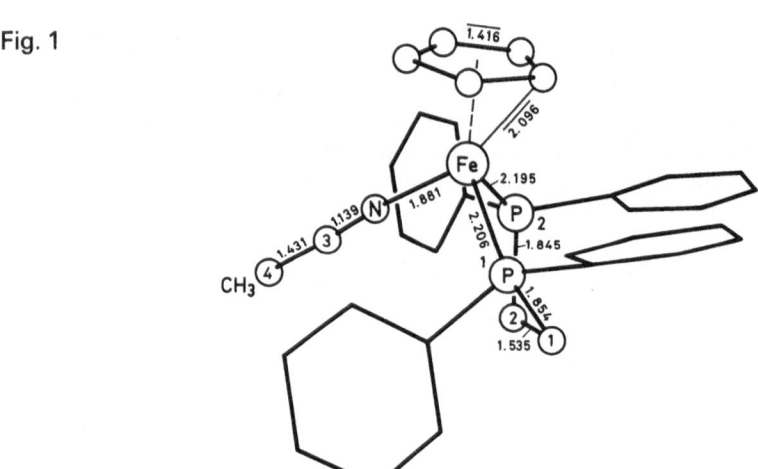

Molecular structure of the [C₅H₅Fe(P(C₆H₅)₂CH₂-)₂NCCH₃]⁺ cation [14].

[C₅H₅Fe(C₁₄H₂₅P₃)]I and **[C₅H₅Fe(C₂₄H₂₉P₃)]I** (Table 1, Nos. **41** and **42**) are obtained by UV irradiation of a mixture of C₅H₅Fe(CO)₂I and an equivalent amount of the phosphine in benzene for 1 h. The crude products are chromatographed on Al₂O₃/CH₂Cl₂ with acetone as the eluent [8, 9].

The complexes contain tridentate ligands, which were identified by their characteristic proton-decoupled ³¹P NMR spectra. In the case of No. 41 the spectrum has a low-field triplet and a high-field doublet. The nonequivalence of the phosphine ends of compound No. 42 makes the three different P,P coupling constants directly observable. These are quite different because the couplings between neighboring P atoms have pathways both through the metal atom and through the ethane bridge whereas the coupling between two terminal P atoms has only the one pathway through the metal atom [9].

$[C_5H_5Fe(C_{34}H_{33}P_3)]Br$ and $[C_5H_5Fe(C_{34}H_{33}P_3)]PF_6$ (Table 1, Nos. **43** and **44**). A mixture of equimolar amounts of the phosphine and $C_5H_5Fe(CO)_2Br$ in benzene is exposed to UV irradiation for 15 h. The bromide (No. 43) is obtained by chromatography on Al_2O_3/CH_2Cl_2 with acetone eluent. The oil obtained crystallizes upon trituration with a mixture of benzene/petroleum ether; recrystallization is from $CH_2Cl_2/$ hexane. No. 44 is formed from No. 43 in acetone with an excess of aqueous NH_4PF_6; it is recrystallized from acetone/benzene [2]. This salt was also obtained from $(C_{34}H_{33}P_3)FeCl_2$ and TlC_5H_5 followed by addition of NH_4PF_6 [16].

$LiAlH_4$ (or $LiAlD_4$) in THF produces the hydride, $C_5H_5Fe(C_{34}H_{33})H$ (or the deuteride). It has an Fe-H bond and one uncoordinated phosphine site [16], see 1.5.1.3, p. 16, Nos. 47 and 48.

$[C_5H_5Fe(C_{41}H_{39}P_3)]X$ (X=Cl, Br, PF_6; Table 1, Nos. **45** to **47**). A mixture of the $C_5H_5Fe(CO)_2X$ (X=Cl, Br) and 1,1,1-tris(diphenylphosphinomethyl)ethane $(C_{41}H_{39}P_3)$ in benzene is exposed to UV irradiation (12 h). The filtered black reaction mixture is evaporated, and the residue chromatographed on Al_2O_3 with CH_2Cl_2 eluent. Further purification can be carried out by a second chromatography or by recrystallization from $CH_2Cl_2/$hexane. An acetone solution of the bromide (No. 46) is treated with excess concentrated aqueous NH_4PF_6 to give compound No. 47. It is recrystallized from acetone/benzene.

The $[C_5H_5Fe(C_{41}H_{39}P_3)]^+$ salts are quite soluble in benzene and CH_2Cl_2 because of the large bulky organic triphosphine ligand [1].

$[C_5H_5Fe(C_{42}H_{42}P_4)]I$ and $[C_5H_5Fe(C_{42}H_{42}P_4)]PF_6$ (Table 1, Nos. **48** and **49**). A mixture of $C_5H_5Fe(CO)_2I$ and hexaphenyl-1,4,7,10-tetraphosphadecane $(C_{42}H_{42}P_4)$ in benzene is exposed to UV irradiation for 29 h. The solvent is removed from the filtered reaction mixture at 25 °C/40 Torr to give No. 48. The hexafluorophosphate, prepared from an acetone solution of the iodide and an excess aqueous NH_4PF_6, can be purified on Al_2O_3 with acetone eluent [3].

The molar conductivity of the iodide shows that it is an 1:1 electrolyte containing a tridentate tetratertiary phosphine $C_{42}H_{42}P_4$ ligand [3].

$[C_5H_5Fe(P(CH_3)_2CH_2-)_2N_2Mo(P(C_6H_5)_3)_2(C_6H_5CH_3)]$ BF_4 (Table 1, No. **50**) is obtained, when a solution of compound No. 31 in pure acetone at -78 °C is stirred with $(C_6H_5CH_3)Mo(P(C_6H_5)_3)_2N_2$ at -78 °C for 15 min and than at -15 °C for 15 min. Light petroleum is added and after filtration the solution is concentrated under reduced pressure at -15 °C to give a finely divided, brown solid. Recrystallization from acetone/toluene at -15 °C gives the solvate [20].

References:

[1] R.B. King, L.W. Houk, K.H. Pannell (Inorg. Chem. **8** [1969] 1042/8). –
[2] R.B. King, P.N. Kapoor, R.N. Kapoor (Inorg. Chem. **10** [1971] 1841/50). –
[3] R.B. King, R.N. Kapoor, M.S. Saran, P.N. Kapoor (Inorg. Chem. **10** [1971] 1851/60). – [4] W.E. Silverthorn (Chem. Commun. **1971** 1310/1). – [5] M.L.H. Green, R.N. Whiteley (J. Chem. Soc. A **1971** 1943/6).

[6] W. Kläui, H. Werner (Chimia [Aarau] **27** [1973] 3869). – [7] W. Kläui, H. Werner (J. Organometal. Chem. **54** [1973] 331/40). – [8] R.B. King, J.A. Zinich, J.C. Cloyd Jr. (Inorg. Chem. **14** [1975] 1554/9). – [9] R.B. King, J.C. Cloyd (Inorg.

Chem. **14** [1975] 1550/4). − [10] D. Sellmann, E. Kleinschmidt (Angew. Chem. **87** [1975] 595/6; Angew. Chem. Intern. Ed. Engl. **14** [1975] 571).

[11] S.C. Tripathi, S.C. Srivastava, V.N. Pandey (Transition Metal Chem. [Weinheim] **1** [1976] 266/8). − [12] G. Balavoine, M.L.H. Green, J.P. Sauvage (J. Organometal. Chem. **128** [1977] 247/52). − [13] D. Sellmann, E. Kleinschmidt (J. Organometal. Chem. **140** [1977] 211/19). − [14] P.E. Riley, C.F. Capshew, R. Pettit, R.E. Davies (Inorg. Chem. **17** [1978] 408/14). − [15] P.M. Treichel, D.C. Molzahn (Syn. Reactiv. Inorg. Metal-Org. Chem. **9** [1979] 21/9).

[16] S.G. Davies, H. Felkin, O. Watts (J. Chem. Soc. Chem. Commun. **1980** 159/60). − [17] P.M. Treichel, D.A. Komar (J. Organometal. Chem. **206** [1981] 77/88). − [18] T.P. Gill, K.R. Mann (J. Organometal. Chem. **216** [1981] 65/71). − [19] C.C. Lee, U.S. Gill, M. Iqbal, C.I. Azogu, R.G. Sutherland (J. Organometal. Chem. **231** [1982] 151/9). − [20] M.L.H. Green, W.E. Silverthorn (J. Chem. Soc. Dalton Trans. **1973** 301/6).

1.5.1.3 Compounds of the $^5LFe(^2D)_2X$, $^5LFe(^2D-^2D)X$, and $^5LFe(^2D-X)^2D$ Type

The complex types described in this section are represented by the general Formulas I to III. The individual compounds are arranged in Table 2 in the same sequence:

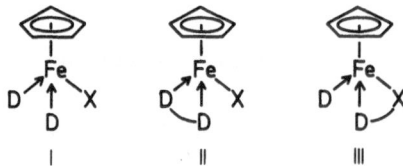

I	II	III

Nos. 1 to 43 belong to Type I, Nos. 44 to 74 belong to Type II, and Nos. 75 to 80 belong to Type III. Within each group the compounds are arranged by the kind of the covalently bonded X ligand, where X stands for hydrogen, halogen atoms, CN, SCN, and a variety of groups bonded to iron through O, S, N, Si, Ge, and Sn atoms. A Grignard-like complex belonging to Type II with X= MgBr is included as No. 74.

There is only one complex which contains a substituted cyclopentadienyl ligand:

C_6H_5-$C_5H_4Fe(P(OC_6H_5)_3)_2I$, formed in an approximately 15% yield along with $[C_6H_5$-$C_5H_4Fe(P(OC_6H_5)_3)_3]I$ when C_6H_5-$C_5H_5Fe(P(OC_6H_5)_3)_3$ ($^4LFe(^2D)_3$ type, Formula IV on p. 3) was treated with iodine in CH_2Cl_2 for 3 h. The complex was eluted from Al_2O_3 with ether/CH_2Cl_2 and formed black crystals on partial evaporation of a solution in CH_2Cl_2/ether/petroleum ether. Only an elemental C, H, Fe analysis was reported [13].

$C_5H_5Fe(P(C_6H_5)_2CH_2-)_2S_2O_3$ (Table **2**, No. **82**) represents another special case because it must be a 17-electron species, which is only formally identical with the compounds in this section.

The preparative methods for the compounds in Table 2 are briefly summarized below. Details and special procedures are given under further information on pp. 26/38.

Method I: The carbonyl groups of the corresponding $C_5H_5Fe(CO)_2X$ compound are substituted by the reaction with an excess of the ^2D molecule in aliphatic solvents or benzene at elevated temperatures for several hours [31, 33].

Method II: Substitution as in Method I but at room temperature under UV irradiation for several hours. Stoichiometric amounts [6, 9, 10] or an excess of the donor molecule [14, 17, 48] are used. Reactions were also carried out in refluxing pentane or acetonitrile [48].

Method III: Salts of the $[C_5H_5Fe(^2D)_2NCCH_3]^+$ or $[C_5H_5Fe(^2D-^2D)NCCH_3]^+$ cations are treated with KX or $[NR_4]I$ salts in polar solvents (acetone, CH_3OH, THF) at reflux temperature, whereby the good leaving group CH_3CN is replaced by the X^- ion [41, 48]. This method was also successful for the reactants $[C_5H_5Fe(P(OC_6H_5)_3)_3]^+/LiSn_4(CH_3)_9$ [21], see No. 43.

Method IV: Halides of the $C_5H_5Fe(^2D)_2X$ or $C_5H_5Fe(^2D-^2D)X$ type (X = Cl, Br, or I) are treated in THF with Na/Hg [1], $NaBH_4$ [4, 19], or $LiAlH_4$ ($LiAlD_4$) [12, 37, 41] to give the corresponding hydride complexes.

Method V: a. $Fe(P(C_6H_5)_3)_2I_2$ is reacted with $Cu(RN_3R)$ and subsequently with TlC_5H_5.
            b. $C_5H_5Fe(^2D)_2I$ compounds are treated with $Cu(RN_3R)$ or $Ag(RN_3R)$ [43].

The electrochemical oxidation of a series of $C_5H_5Fe(P(C_6H_5)_2CH_2-)_2X$ compounds was investigated [26, 42]. The oxidation potentials E_p (in V, vs. SCE) were obtained by cyclic voltammetry in CH_2Cl_2 ($c \sim 5 \times 10^{-3}$ M and 0.1 M in $[N(C_4H_9)_4]ClO_4$); they are arranged below by increasing values of E_p. The numbers in parentheses refer to the number of the compound in Table 2:

X (No.)	S_2O_3 (82)	SC_6H_5 (63)	H (45)	Cl (50)	$Sn(CH_3)_3$ (72)
E_p	−0.366	−0.252	−0.08	0.08	0.065
X (No.) . . .	Br (53)	I (56)	NCS (59)	CN (81)	$SnCl_3$ (69)
E_p	0.11	0.15	0.316	0.535	0.90

Second, irreversible oxidation steps were observed for No. 82 at 0.769 V and for No. 63 at 1.166 V [42]. Chemical oxidation with Ag^+ or NO^+ was used to prepare salts of the $[C_5H_5Fe(^2D)_2X]^+$ and $[C_5H_5Fe(^2D-^2D)X]^+$ cations [26, 42], see 1.5.1.4, p. 40.

Table 2
Compounds of the $C_5H_5Fe(^2D)_2X$, $C_5H_5Fe(^2D-^2D)X$, and $C_5H_5Fe(^2D-X)^2D$ Type. Further information on numbers preceded by an asterisk is given at the end of the table. For abbreviations and dimensions, see p. 375.

No.	group X or ^2D-X method of preparation (yield in %)	donor ^2D or $^2D-^2D$	properties and remarks	Ref.
$C_5H_5Fe(^2D)_2X$ type				
*1	H see further information	PF_3	orange-yellow viscous oil, m.p. $\sim -50°$, b.p. $40°/10^{-3}$ Torr ^1H NMR: -13.99 (m, Fe-H), 5.22 (s, C_5H_5) IR (C_4Cl_6): ν(Fe-H) 1912	[16]
*2	D see further information	PF_3	IR (neat or KBr): ν(Fe-D) 1429	[16]

Table 2 [continued]

No.	group X or ²D-X method of preparation (yield in %)	donor ²D or ²D-²D	properties and remarks	Ref.
3	H	$P(CH_3)_3$ ⎫	prepared from the corresponding $C_5H_5Fe(CO)(^2D)H$ complexes and the ²D ligands under irradiation;	[52]
4	H	$P(OCH_3)_3$ ⎬	no properties reported	[52]
*5	H IV (23)	$P(OC_6H_5)_3$	yellow, m.p. 121 to 122° ¹H NMR: -14.05 (t, Fe-H, J(PFeH)=82), 3.86 (s, C_5H_5) IR (vaseline): ν(Fe-H) 1970	[1, 4]
*6	Cl I (~45)	$PF_2-NCH_3-PF_2$	red, m.p. 62 to 64°	[33]
*7	Br II (12)	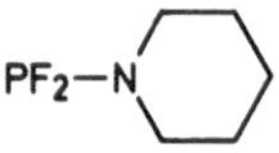 PF₂–N⟨hexagon⟩	red, m.p. 94 to 96° IR (KBr): ν(P-F) 800, 824	[17]
*8	Br II (34)	$P(CH_3)_2C_6H_5$	black, m.p. 115 to 118° (dec.) ¹H NMR (CS_2): 1.58 (CH_3, 5 peaks), 3.52 (t, C_5H_5, J(PFeCH)=1.5), 7.10 to 7.80 (m, C_6H_5)	[36]
*9	Br see further information	$P(OCH_3)_3$	no properties reported	[22]
*10	Br see further information	$P(OC_6H_5)_3$	black, m.p. 117 to 118° (dec.)	[4]
*11	I I (68.1)	$NH_2-C_4H_9-n$ ⎫		[31]
*12	I I (60.3)	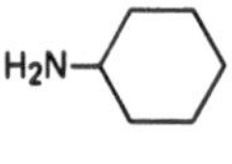 H₂N–⟨hexagon⟩	red brown IR (KBr): ν(N-H) 3320 to 3350, δ(NH) 1560 to 1570, 1620	[31]
*13	I I (61.5)	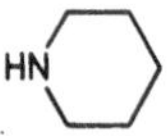 HN⟨hexagon⟩	decompose at higher temperature without melting	[31]
*14	I I (66.8)	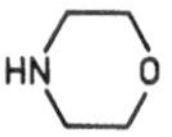 HN⟨ring O⟩		[31]
*15	I I (62.4)	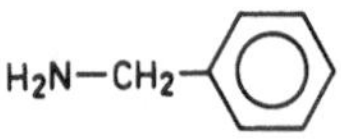 H₂N–CH₂–⟨ring⟩ ⎭		[31]
*16	I II (30)	$PF_2-N(C_2H_5)_2$	dark red, m.p. 99 to 100° ¹H NMR ($CDCl_3$): 1.22 (t, CH_3), ~3.4 (broad, CH_2), 4.45 (t, C_5H_5) IR (KBr): ν(P-F) 796, 826	[17]

References on p. 39

Table 2 [continued]

No.	group X or ^{2}D-X method of preparation (yield in %)	donor ^{2}D or ^{2}D-^{2}D	properties and remarks	Ref.
*17	I II (70)	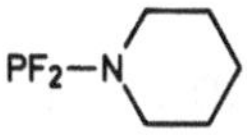 $PF_2-N\bigcirc$	dark red, m.p. 128 to 130° ^{1}H NMR (CDCl$_3$): 1.63 (s, CH$_2$), ~3.6 (broad, CH$_2$), 4.42 (t, C$_5$H$_5$) ^{19}F NMR (CH$_2$Cl$_2$): 28.0 (d, J(P,F)= 1099), 32.2 (d, J(P,F) =1156) IR (KBr): ν(P–F) 788, 833	[17]
*18	I III (66)	P(CH$_3$)$_3$	black, m.p. 128 to 130° (dec.) ^{1}H NMR (CS$_2$): 1.57 (t, CH$_3$, J(PCH)=4), 3.77 (t, C$_5$H$_5$, J(PFeCH)=2)	[48]
*19	I II (90)	P(OCH$_3$)$_3$	no properties reported	[43]
*20	I II (36)	P(OC$_6$H$_5$)$_3$	black crystals, m.p. >130° (dec.) ^{1}H NMR (C$_6$D$_6$): 4.06 (t, C$_5$H$_5$, J(PFeCH)=1.3), 6.7 to 7.7 (m, C$_6$H$_5$)	[1 to 5, 21]
21	SCN VI (−)	P(OC$_6$H$_5$)$_3$	red, m.p. 100° (dec.) IR (Nujol): ν(SCN) 2101 also see 1.5.1.2, No. 9 in Table 1, p. 5	[13]
*22	OOCCF$_3$ II (78)	P(OCH$_3$)$_3$	red brown, m.p. 64 to 65° ^{1}H NMR (CH$_3$COCH$_3$): 3.70 (t, CH$_3$, J(POCH)=5.6), 4.44 (t, C$_5$H$_5$, J(PFeCH)=0.8) ^{19}F NMR (CH$_3$COCH$_3$): 75.8 IR (KBr): ν(C=O) 1686	[9]
*23	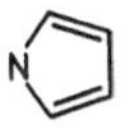 $\begin{smallmatrix}N\end{smallmatrix}$ see further information	PF$_3$	no properties reported	[44]
*24	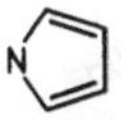 $\begin{smallmatrix}N\end{smallmatrix}$ see further information	PF$_2$-NCH$_3$-PF$_2$	no properties reported	[44]
*25	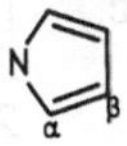 $\begin{smallmatrix}N\end{smallmatrix}$ α β see further information	PF$_2$-N(CH$_3$)$_2$	red, m.p. 86° ^{1}H NMR (CDCl$_3$): 2.66 (b, CH$_3$), 4.59 (s, C$_5$H$_5$), 6.07, 6.30 (m's, β- and α-CH) ^{13}C NMR (CDCl$_3$): 35.9 (m, CH$_3$), 81.1 (s, C$_5$H$_5$), 109.2, 135.8 (s's, β- and α-C) ^{19}F NMR (CDCl$_3$): 22.1, 36.7 (m's, J(P,F)=1231) IR (Nujol): ν(P–F) 710 to 842	[44, 51]

References on p. 39

Table 2 [continued]

No.	group X or ^{2}D–X method of preparation (yield in %)	donor ^{2}D or ^{2}D–^{2}D	properties and remarks	Ref.
*26	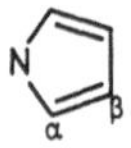 see further information	PF_2–$N(C_2H_5)_2$	m.p. 54° ^{1}H NMR ($CDCl_3$): 1.15 (t, CH_3), 3.17 (b, CH_2), 4.55 (s, C_5H_5), 6.06, 6.27 (m's, β- and α–CH) ^{13}C NMR ($CDCl_3$): 13.9 (s, CH_3), 38.9 (m, CH_2), 80.8 (s, C_5H_5), 109.0, 135.6 (s's, β- and α–C) ^{19}F NMR ($CDCl_3$): 21.6, 36.0 (m's, J(P,F)=1213) IR (Nujol): ν(P–F) 710 to 842	[51]
*27	$Si(CH_3)_3$ II (17)	$P(CH_3)_3$	orange, m.p. 240 to 242° (dec.) ^{1}H NMR (CS_2): 0.08 (s, $SiCH_3$), 1.28 (m, PCH_3), 3.80 (t, C_5H_5, J(PFeCH)=2)	[48]
*28	 CH$_3$Si II (9)	$PCH_3(C_6H_5)_2$	orange, m.p. 176 to 177° ^{1}H NMR (C_6H_6): 0.59 (s, $SiCH_3$), 1.3 to 1.8 (m, CH_2), 1.70 (virtual t), 2.4 to 3.6 (m, CH_2), 4.10 (t, C_5H_5, J(PFeCH)=1.8)	[39]
*29	SnF_3 see further information	$P(OC_6H_5)_3$	yellow crystals, dec. p. 195° ^{1}H NMR ($CDCl_3$/C_6D_6): 4.25/4.28 (t, C_5H_5, J(PFeCH)=1.5) IR (KBr): ν(Sn–F) 500, 545	[29]
*30	$SnCl_3$ II (−)	$P(OCH_3)_3$	red orange ^{1}H NMR ($CDCl_3$): 4.60 (t, C_5H_5, J(PFeCH)=1.2 to 1.4) ^{57}Fe–γ (80 K): δ=0.298(Fe), Δ=1.87 ^{119}Sn–γ (80 K): δ=1.99, Δ=1.91	[27]
*31	$SnCl_3$ II (−)	$P(OC_2H_5)_3$	red orange ^{1}H NMR ($CDCl_3$): 4.30 (t, C_5H_5, J(PFeCH)=1.5) ^{57}Fe–γ(80 K): δ=0.294(Fe), Δ=1.84 ^{119}Sn–γ (80 K): δ=1.97, Δ=1.89	[27]
*32	$SnCl_3$ II (92)	$P(OC_6H_5)_3$	orange crystals, m.p. 172 to 177° (dec.) ^{1}H NMR ($CDCl_3$/C_6D_6): 4.38/4.26 (t, C_5H_5, J(PFeCH)=1.5) ^{57}Fe–γ (298 K): δ=0.44, Δ=1.80 ^{119}Sn–γ (80 K): δ=1.88, Δ=1.92 IR (KBr): ν(Sn–Cl) 308, 330	[24, 27 to 29]
*33	$SnBr_3$ II (90)	$P(OC_6H_5)_3$	orange red, dec. p. 165° ^{1}H NMR ($CDCl_3$/C_6D_6): 4.34/4.27 (t, C_5H_5, J(PFeCH)=1.5) IR (KBr): ν(Sn–Br) 240, 257	[29]

References on p. 39

Table 2 [continued]

No.	group X or ^{2}D–X method of preparation (yield in %)	donor ^{2}D or ^{2}D–^{2}D	properties and remarks	Ref.
*34	SnI_3 see further information	$P(OC_6H_5)_3$	^{1}H NMR (C_6D_6): 4.28 (t, C_5H_5, J(PFeCH) = 1.6), 6.8 (m, C_6H_5)	[20, 29]
*35	$Sn(C_6H_5)Cl_2$ see further information	$P(OC_6H_5)_3$	yellow, m.p. 160 to 163° ^{1}H NMR $(CDCl_3)$: 4.39 (t, C_5H_5) ^{119}Sn-γ (298 K): $\delta = 1.75$, $\Delta = 2.90$	[24, 28]
*36	$Sn(C_6H_5)_2Cl$ see further information	$P(OC_6H_5)_3$	yellow, m.p. 105 to 108° (dec.) ^{1}H NMR $(CDCl_3)$: 4.34 (t, C_5H_5) ^{119}Sn-γ (298 K): $\delta = 1.65$, $\Delta = 2.71$	[24, 28]
*37	$Sn(CH_3)_3$ II (81)	$P(CH_3)_3$	yellow orange, m.p. 261 to 265° (dec.) ^{1}H NMR (C_6D_6): 0.42 (s, $SnCH_3$), 1.00 (t, PCH_3, J(PCH) = 18), 3.75 (t, C_5H_5, J(PFeCH) = 2)	[48]
*38	$Sn(CH_3)_3$ II (23)	$Sb(C_6H_5)_3$	dark red, m.p. 130 to 148° (dec.) ^{57}Fe-γ (78 K): $\delta = 0.80$, $\Delta = 2.09$ ^{119}Sn-γ (78 K): $\delta = 1.47$, $\Delta = 0.73$	[14]
39	$Sn(C_6H_5)_3$ II (66)	$P(CH_3)_3$	orange, m.p. 210 to 215° ^{1}H NMR (C_6D_6): 0.98 (m, PCH_3), 3.96 (t, C_5H_5, J(PFeCH) = 2), 7.22, 7.84 (m's, C_6H_5)	[48]
*40	$Sn(C_6H_5)_3$ II (30)	$P(CH_3)_2C_6H_5$	red brown, dec. p. 170 to 180° ^{1}H NMR (CS_2): 1.22 (t, CH_3, J(PCH) = 4.5), 1.28 (t, CH_3, J(PCH) = 4.5), 3.88 (t, C_5H_5, J(PFeCH) = 1.2), 7.10 (m, C_6H_5) ^{57}Fe-γ (78 K): $\delta = 0.63$, $\Delta = 1.70$ ^{119}Sn-γ (78 K): $\delta = 1.71$, $\Delta = 1.25$	[14]
*41	$Sn(C_6H_5)_3$ II (−)	$PCH_3(C_6H_5)_2$	red brown, dec. p. 110 to 140° ^{1}H NMR (CS_2): 1.72 (t, CH_3, J(PCH) ~4), 4.22 (t, C_5H_5, J(PFeCH) = 1.2), 7.32 (m, C_6H_5) ^{57}Fe-γ (78 K): $\delta = 0.67$, $\Delta = 1.70$ ^{119}Sn-γ (78 K): $\delta = 1.58$, $\Delta = 1.14$	[14]
*42	$Sn(C_6H_5)_3$ II (41)	$P(OC_6H_5)_3$	yellow, m.p. 185 to 187° ^{1}H NMR: 4.31 (s, C_5H_5) ^{119}Sn-γ (80 K): $\delta = 1.49$, $\Delta = 0.78 \pm 0.10$	[24, 28]

Table 2 [continued]

No.	group X or ^{2}D–X method of preparation (yield in %)	donor ^{2}D or ^{2}D–^{2}D	properties and remarks	Ref.

*43	Sn(Sn(CH$_3$)$_3$)$_3$
	III (15)
		P(OC$_6$H$_5$)$_3$	yellow, dec. in vacuum ~150°
			^{1}H NMR (C$_6$D$_6$): 0.56 (s, CH$_3$,
			  J(SnCH) = 40.5, 42.0,
			  J(SnSnCH) = 8.1), 4.38 (t, C$_5$H$_5$,
			  J(PFeCH) = 1.35), 6.9 (m, C$_6$H$_5$)
			^{13}C NMR (CS$_2$): −0.2 (q, CH$_3$,
			  J(H,C) = 125), 80.3 (d, C$_5$H$_5$,
			  J(H,C) ~165)
				[18, 21]

C$_5$H$_5$Fe(^{2}D-^{2}D)X type				
*44	H III (−)	(P(CH$_3$)$_2$CH$_2$–)$_2$	yellow ^{1}H NMR (C$_6$D$_6$): −15.4 (FeH) IR (Nujol): ν(Fe–H) 1830	[12]

*45	H
	III (61),
	IV (50)
		(P(C$_6$H$_5$)$_2$CH$_2$–)$_2$	yellow, m.p. 146 to 156°
			^{1}H NMR (C$_6$D$_6$): −16.12 (t, FeH,
			  J(PFeH) = 72), 4.16 (t, C$_5$H$_5$,
			  J(PFeCH) = 1.5), 7 to 8 (m, C$_6$H$_5$)
			^{31}P NMR (C$_6$H$_5$CH$_3$): 110.8 (d)
			IR: ν(Fe–H) 1830, 1880 in Nujol,
			  ν(Fe–H) 1840 in CH$_2$Cl$_2$
				[19, 25, 37, 40, 41]

*46	D IV (48)	(P(C$_6$H$_5$)$_2$CH$_2$–)$_2$	yellow ^{1}H NMR (C$_6$D$_6$): 4.16 (t, C$_5$H$_5$, J(PFeCH) = 1.5), 7 to 8 (m, C$_6$H$_5$) IR (Nujol): ν(Fe–D) 1340	[37]
*47	H see further information	C$_2$H$_4$–P(C$_6$H$_5$)$_2$ PC$_6$H$_5$ C$_2$H$_4$–P(C$_6$H$_5$)$_2$	^{1}H NMR: −17.0 (double doublet, J(PFeH) = 65 and 75) IR: ν(Fe–H) 1875	[46]
*48	D see further information	C$_2$H$_4$–P(C$_6$H$_5$)$_2$ PC$_6$H$_5$ C$_2$H$_4$–P(C$_6$H$_5$)$_2$	no properties reported	[46]
*49	Cl II (26)	(P(CH$_3$)$_2$CH$_2$–)$_2$	black, m.p. 205° ^{1}H NMR (CDCl$_3$): 4.25 (s, C$_5$H$_5$)	[10]
*50	Cl II (45)	(P(C$_6$H$_5$)$_2$CH$_2$–)$_2$	black, m.p. 177 to 178° ^{1}H NMR (CDCl$_3$): 4.20 (s, C$_5$H$_5$) ^{57}Fe-γ (80 K): δ = 0.70, Δ = 1.97	[10, 19]
*51	Cl II (52)	P(C$_6$H$_5$)$_2$CH=CH– P(C$_6$H$_5$)$_2$	black, m.p. 198 to 202° ^{1}H NMR (CDCl$_3$): 4.11 (s, C$_5$H$_5$)	[10]
*52	Br see further information	(P(CH$_3$)$_2$CH$_2$–)$_2$	black, m.p. 260 to 265° (dec.) ^{1}H NMR (CS$_2$): 1.30 to 1.95 (m, CH$_2$, CH$_3$), 3.95 (broad s, C$_5$H$_5$)	[36]

References on p. 39

Table 2 [continued]

No.	group X or ^{2}D-X method of preparation (yield in %)	donor ^{2}D or ^{2}D-^{2}D	properties and remarks	Ref.
*53	Br II (41), III (32)	$(P(C_6H_5)_2CH_2-)_2$	black, m.p. 191 to 193°, 207 to 215° ^{1}H NMR (CDCl$_3$): 4.20 (s, C$_5$H$_5$)	[10, 41]
*54	Br II (43)	P(C$_6$H$_5$)$_2$CH=CH- P(C$_6$H$_5$)$_2$	black, m.p. 220 to 225° ^{1}H NMR (CDCl$_3$): 4.20 (s, C$_5$H$_5$)	[10]
*55	I see further information	$(P(CH_3)_2CH_2-)_2$	black	[12, 32]
*56	I VI (74)	$(P(C_6H_5)_2CH_2-)_2$	black, m.p. 212 to 215°	[13, 41]
*57	I see further information	(C$_6$H$_5$)$_2$P, (C$_6$H$_5$)$_2$P structure with O, O, CH$_3$, CH$_3$	black red, m.p. 125° ^{1}H NMR (CDCl$_3$): 3.80 (C$_5$H$_5$)	[32, 45]
*58	CN III (92)	$(P(C_6H_5)_2CH_2-)_2$	orange, m.p. 188 to 190° ^{1}H NMR (CDCl$_3$): 2 to 2.8 (m, CH$_2$), 4.28 (s, C$_5$H$_5$), 7 to 8 (m, C$_6$H$_5$) IR (KBr): ν(CN) 2060	[41]
*59	SCN III (80)	$(P(C_6H_5)_2CH_2-)_2$	dark purple, dec. p. 199° ^{1}H NMR (CDCl$_3$): 2.2 to 2.4 (m, CH$_2$), 4.14 (s, C$_5$H$_5$), 7 to 8 (m, C$_6$H$_5$) IR (KBr): ν(CN) 2084 UV (CH$_2$Cl$_2$): $\lambda_{max}(\varepsilon) = 510$ (534)	[41]
*60	CF$_3$COO II (31)	$(P(CH_3)_2CH_2-)_2$	violet, m.p. 200 to 210° (dec.) IR (KBr): ν(C=O) 1690	[9]
*61	CF$_3$COO II (50)	$(P(C_6H_5)_2CH_2-)_2$	violet, m.p. 140 to 142° ^{1}H NMR (CH$_3$COCH$_3$): 4.26 (t, C$_5$H$_5$, J(PFeCH) = 1.3) ^{19}F NMR (CH$_2$Cl$_2$): 76.6 (s, CF$_3$) IR (KBr): ν(C=O) 1680	[9]
*62	CF$_3$COO II (54)	P(C$_6$H$_5$)$_2$CH=CH- P(C$_6$H$_5$)$_2$	violet, m.p. 158 to 160° ^{1}H NMR (CH$_3$COCH$_3$): 4.22 (s, C$_5$H$_5$, J(PFeCH) = 1.5) ^{19}F NMR (CH$_2$Cl$_2$): 76.6 (CF$_3$) IR (KBr): ν(C=O) 1690	[9]
*63	SC$_6$H$_5$ III (43)	$(P(C_6H_5)_2CH_2-)_2$	black, m.p. 177.5 to 179.5° ^{1}H NMR (CS$_2$): 1.9 to 2.7 (m, CH$_2$), 4.0 (s, C$_5$H$_5$), 6.4 to 6.6 (m, SC$_6$H$_5$), 6.8 to 8.0 (m, PC$_6$H$_5$)	[41]

Table 2 [continued]

No.	group X or ^{2}D–X method of preparation (yield in %)	donor ^{2}D or ^{2}D–^{2}D	properties and remarks	Ref.
*64	S–CH=S see further information	$(P(C_6H_5)_2CH_2-)_2$	red brown, dec. p. 105° ^{1}H NMR (C_6D_6): 1.32 to 2.0 (m, CH_2), 4.17 (t, C_5H_5, J(PFeCH) = 1.1), 11.01 (s, CH=S) IR (KBr): ν(S–CH=S) 985, 1240	[30]
*65	SO_2–CH_3 see further information	$(P(C_6H_5)_2CH_2-)_2$	orange, dec. p. 158 to 159° ^{1}H NMR (C_6D_6): 1.7 to 2.0 (m, CH_2), 2.17 (s, S–CH_3), 4.31 (t, C_5H_5), 6.7 to 8.1 (m, C_6H_5) IR (KBr): ν(SO) 1020, 1142, ν(S–CH_3) 930	[30]
*66	$Si(CH_3)_3$ II (16)	$(P(C_6H_5)_2CH_2-)_2$	orange, m.p. 179 to 180° (dec.) ^{1}H NMR $(CDCl_3)$: −0.47 (s, $SiCH_3$), 2.4 (CH_2), 4.15 (C_5H_5), 7.2 to 7.4 (m, C_6H_5) ^{29}Si NMR (C_6D_6): 25.16 (t, J(PFeSi) = 46)	[6, 50]
*67	$Si(CH_3)_3$ II (11)	$P(C_6H_5)_2CH=CH-$ $P(C_6H_5)_2$	orange, m.p. ~176° (dec.) ^{1}H NMR $(CDCl_3)$: −0.57 (s, $SiCH_3$), 4.20 (t, C_5H_5, J(PFeCH) = 1.2), 7.2 to 7.4 (m, C_6H_5)	[6]
*68	$Ge(C_6H_5)_3$ see further information	$(P(C_6H_5)_2CH_2-)_2$	orange needles ^{1}H NMR (C_6D_6): 1.7 to 2.1 (m, CH_2), 4.4 (C_5H_5), 6.9 to 7.7 (m, C_6H_5)	[49]
*69	$SnCl_3$ see further information	$(P(C_6H_5)_2CH_2-)_2$	red ^{57}Fe-γ (80 K): δ = 0.59, Δ = 1.75 ^{119}Sn-γ (80 K): δ = 1.29, Δ = 1.76	[19]
*70	$SnBr_3$ see further information	$(P(C_6H_5)_2CH_2-)_2$	dark red ^{57}Fe-γ (80 K): δ = 0.61, Δ = 1.84 ^{119}Sn-γ (80 K): δ = 1.99, Δ = 1.60	[19]
*71	SnI_3 see further information	$(P(C_6H_5)_2CH_2-)_2$	dark green ^{57}Fe-γ (80 K): δ = 0.63, Δ = 1.80 ^{119}Sn-γ (80 K): δ = 2.26, Δ = 1.53	[19]
*72	$Sn(CH_3)_3$ II (49)	$(P(C_6H_5)_2CH_2-)_2$	orange, m.p. 190 to 200° (dec.) ^{1}H NMR $(CDCl_3)$: −0.55 (s, $SnCH_3$, J(SnCH) = 35), 2.37 (d, CH_2, J(PCH) = 12), 4.18 (t, C_5H_5, J(PFeCH) = 1.3), 7.3, 7.4 (m, C_6H_5) ^{57}Fe-γ (80 K): δ = 0.59, Δ = 1.63 ^{119}Sn-γ (80 K): δ = 1.50, Δ = 0.70	[6, 19]

References on p. 39

Table 2 [continued]

No.	group X or ^{2}D-X method of preparation (yield in %)	donor ^{2}D or ^{2}D-^{2}D	properties and remarks	Ref.
*73	Sn(CH$_3$)$_3$ II (30)	P(C$_6$H$_5$)$_2$CH=CH- P(C$_6$H$_5$)$_2$	red, m.p. 175° (dec.) ^{1}H NMR (CDCl$_3$): −0.67 (s, SnCH$_3$, J(SnCH) = 35), 4.19 (t, C$_5$H$_5$, J(PFeCH) = 1.3), 7.2, 7.4 (m, C$_6$H$_5$)	[6]
*74	MgBr see further information	(P(C$_6$H$_5$)$_2$CH$_2$-)$_2$	3-THF solvate, red ^{1}H NMR (C$_6$D$_6$): 1.3 (CH$_2$), 3.42 (m, OCH$_2$), 4.23 (broad s, C$_5$H$_5$), 7.0 to 8.0 (m, C$_6$H$_5$)	[23, 37]

C$_5$H$_5$Fe(^{2}D-X)^{2}D type

No.	group X or ^{2}D-X method of preparation (yield in %)	donor ^{2}D or ^{2}D-^{2}D	properties and remarks	Ref.
*75	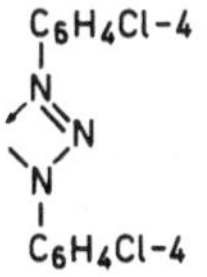 C$_6$H$_4$Cl-4 N N N C$_6$H$_4$Cl-4 Va (30)	P(C$_6$H$_5$)$_3$	purple ^{1}H NMR (CDCl$_3$): 4.10 (d, C$_5$H$_5$, J(PFeCH) = 1.4), 6.8 to 7.6 (m, C$_6$H$_4$) ^{31}P NMR (C$_6$D$_6$): 72.5 IR (KBr): ν(NNN) 1260	[43]
*76	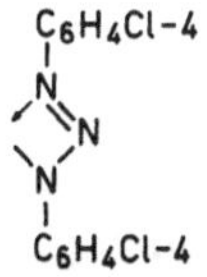 C$_6$H$_4$Cl-4 N N N C$_6$H$_4$Cl-4 Vb (30)	P(OCH$_3$)$_3$	red ^{1}H NMR (CDCl$_3$): 3.55 (d, OCH$_3$, J(POCH) = 11), 4.30 (d, C$_5$H$_5$, J(PFeCH) = 1.4), 7.0 (m, C$_6$H$_4$) ^{31}P NMR (C$_6$D$_6$): 185.0 IR (KBr): ν(NNN) 1258	[43]
*77	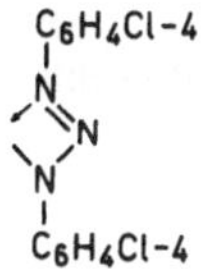 C$_6$H$_4$Cl-4 N N N C$_6$H$_4$Cl-4 Vb (50)	P(OC$_6$H$_5$)$_3$	red ^{1}H NMR (CDCl$_3$): 4.16 (d, C$_5$H$_5$, J(PFeCH) = 1.2), 6.9 (m, C$_6$H$_4$), 7.0 to 7.2 (m, C$_6$H$_5$) ^{31}P NMR (C$_6$D$_6$): 168.4 IR (KBr): ν(NNN) 1262	[43]
*78	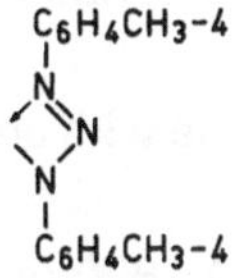 C$_6$H$_4$CH$_3$-4 N N N C$_6$H$_4$CH$_3$-4 Va (30)	P(C$_6$H$_5$)$_3$	purple ^{1}H NMR (CDCl$_3$): 2.20 (s, CH$_3$), 4.09 (d, C$_5$H$_5$, J(PFeCH) = 1.4), 6.8 to 7.6 (m, C$_6$H$_5$) ^{31}P NMR (C$_6$D$_6$): 73.8 IR (KBr): ν(NNN) 1264	[43]
*79	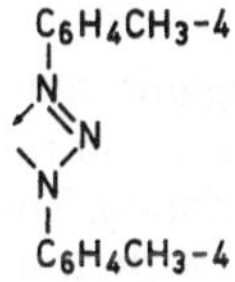 C$_6$H$_4$CH$_3$-4 N N N C$_6$H$_4$CH$_3$-4 Vb (30)	P(OCH$_3$)$_3$	red ^{1}H NMR (CDCl$_3$): 2.25 (s, CH$_3$), 3.55 (d, OCH$_3$, J(POCH) = 11), 4.28 (d, C$_5$H$_5$, J(PFeCH) = 1.4), 7.0 (m, C$_6$H$_4$) ^{31}P NMR (C$_6$D$_6$): 186.9 IR (KBr): ν(NNN) 1274	[43]

References on p. 39

Table 2 [continued]

No.	group X or ^{2}D–X method of preparation (yield in %)	donor ^{2}D or ^{2}D–^{2}D	properties and remarks	Ref.
*80	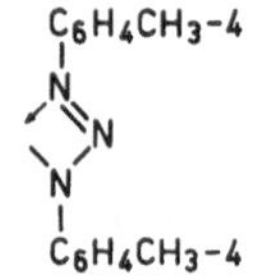 C$_6$H$_4$CH$_3$–4 C$_6$H$_4$CH$_3$–4 Vb (50)	P(OC$_6$H$_5$)$_3$	red ^{1}H NMR (CDCl$_3$): 2.28 (s, CH$_3$), 4.12 (d, C$_5$H$_5$, J(PFeCH) = 1.2), 6.8 (m, C$_6$H$_4$), 7.0 to 7.2 (m, C$_6$H$_5$) ^{31}P NMR (C$_6$D$_6$): 169.4 IR (KBr): v(NNN) 1270	[43]

supplement

*81	CN III (30)	P(CH$_3$)$_3$	yellow, m.p. 178 to 180° ^{1}H NMR (C$_6$D$_6$): 1.08 (m, CH$_3$), 3.95 (t, C$_5$H$_5$, J(PFeCH) = 2) IR (KBr): v(CN) 2042	[48]
*82	S$_2$O$_3$ III (25)	(P(C$_6$H$_5$)$_2$CH$_2$–)$_2$	dark red, m.p. 172.5 to 173° IR (KBr): 520, 600, 665, 690, 738, 785, 825, 870, 1010, 1085, 1155, 1210, 1302, 1430, 1475 UV (CH$_2$Cl$_2$): $\lambda_{max}(\varepsilon) = 570$ (1680)	[42]

*Further information:

$C_5H_5Fe(PF_3)_2H$ and $C_5H_5Fe(PF_3)_2D$ (Table 2, Nos. 1 and 2). The hydride was prepared by irradiating $Fe(PF_3)_5$ and excess cyclopentadiene in refluxing ether for 2 h. The oily reaction product (obtained after evaporation of solvent and excess cyclo-C_5H_6 below 10 °C) was distilled repeatedly at 40 °C/0.001 Torr to give the pure compound in a 27% yield. A yellow solid by–product was not identified. An almost quantitative yield resulted from the proton transfer (1) under the influence of small amounts of $N(C_2H_5)_3$ in ether:

$$C_5H_6Fe(PF_3)_3 \xrightarrow[\text{warming}]{\text{base}} C_5H_5Fe(PF_3)_2H + PF_3 \qquad (1)$$

The starting material of reaction (1) is a $^4LFe(^2D)_3$ type compound described in "Organoiron Compounds" B6, 1.4.1.1.1, p. 3.

The deuterated compound No. 2 was obtained by stirring the hydride with an excess of D_2SO_4 for 6 h.

The strong P–F valence vibrations at 853, 861 (sh), 880 (sh), 890 (sh), and 920 cm^{-1} (gas-phase spectrum) indicate C_s symmetry and thus a localized hydrogen atom.

The disgusting smelling compound dissolves in all common organic solvents. It is thermally stable up to 169 °C. Electron impact (20 and 70 eV) gave the molecular ion [M]$^+$ and the [M − nPF$_3$]$^+$ fragments (n = 1 and 2) in high abundance. The cleavage of hydrogen from the parent molecule is small but increases with the loss of PF$_3$ ligands. Decomposition in air begins at about 40 °C. Bases remove the hydrogen as a proton to give the $[C_5H_5Fe(PF_3)_2]^-$ anion [16], see 1.5.1.1, p. 1.

References on p. 39

C$_5$H$_5$Fe(P(OC$_6$H$_5$)$_3$)$_2$H (Table 2, No. 5) was prepared from C$_5$H$_5$Fe(P(O-C$_6$H$_5$)$_3$)$_2$I (No. 20) and Na/Hg in THF at 25 °C. The reduction with two equivalents of NaBH$_4$ in boiling THF gave a 7:5 mixture of the hydride No. 5 and compound IV, which was initially formulated as the dimer (C$_5$H$_5$Fe(P(OC$_6$H$_5$)$_3$)$_2$)$_2$ [1, 4], see 1.5.2.1.1, p. 50.

IV   V

The complete IR spectrum (from 670 to 3110 cm^{-1}) is reported in [4]. The solid is stable in air. In C$_6$H$_6$ under Ar in a sealed tube, the hydride does not change on heating at 130 °C for 30 min [1]. UV irradiation in boiling C$_6$H$_6$ leads to compound IV by elimination of H$_2$ [1, 4].

C$_5$H$_5$Fe(PF$_2$-NCH$_3$-PF$_2$)$_2$Cl (Table 2, No. 6). Preparative Method II with an excess of the ^{2}D molecule in boiling hexane for several hours [33].

C$_5$H$_5$Fe(PF$_2$-NC$_5$H$_{10}$)$_2$X (X = Br and I, Table 2, Nos. 7 and 17). Preparative Method II with an approximately 3.5-fold excess of PF$_2$-NC$_5$H$_{10}$ in C$_6$H$_6$ at 25 °C for 20 h (No. 7) or 11 h (No. 17). The compounds were purified by chromatography on SiO$_2$ with ether/petroleum ether (No. 7) or by direct recrystallization of the crude product from CH$_2$Cl$_2$/C$_6$H$_{14}$ (No. 17).

The ^{19}F NMR spectrum shows nonequivalent F atoms (F and F′ in Formula V) which must be due to restricted rotation around the Fe-P bonds. The large chemical shift difference (4.2 ppm) relative to the ^{2}J(F,F) coupling constant (<0.6 ppm) produces an AMX pattern consisting of a pair of quartets with all signals of approximately equal intensities [17].

C$_5$H$_5$Fe(P(CH$_3$)$_2$C$_6$H$_5$)$_2$Br (Table 2, No. 8). Preparative Method II with two equivalents of the phosphine in benzene for 17 h, recrystallization from C$_6$H$_6$/C$_6$H$_{14}$.

The five CH$_3$ signals in the ^{1}H NMR spectrum may be attributed to two triplets at δ = 1.51 and 1.65 ppm with ^{2}J(PCH) + ^{4}J(PFePCH) = 9 Hz.

The compound reacts with Mg metal in THF or ether/benzene to give a dark red, very air-sensitive solution containing variable amounts of magnesium (0.6 to 1 Mg per Fe atom). The solutions react with C$_6$H$_5$CH$_2$Cl or C$_6$H$_5$CH$_2$Br to give dibenzyl as the main product; a benzyl-iron compound could not be isolated [36].

C$_5$H$_5$Fe(P(OCH$_3$)$_3$)$_2$Br (Table 2, No. 9) was isolated from a reaction of P(OCH$_3$)$_3$ with C$_5$H$_5$Fe(CO)(Br)-P(C$_6$H$_5$)$_2$CH$_2$CH$_2$P(C$_6$H$_5$)$_2$-(Br)(CO)FeC$_5$H$_5$ [22].

C$_5$H$_5$Fe(P(OC$_6$H$_5$)$_3$)$_2$Br (Table 2, No. 10) was obtained among several other products from the reaction of complex IV with Br$_2$ in CHCl$_3$ between −60 °C and room temperature, recrystallization from CH$_2$Cl$_2$/C$_7$H$_{16}$ [4]. The starting material was formulated in [4] as a dimer, see compound No. 5.

C$_5$H$_5$Fe(^{2}D)$_2$I (^{2}D = C$_4$H$_9$NH$_2$, cyclo-C$_6$H$_{11}$NH$_2$, cyclo-C$_5$H$_{10}$NH, cyclo-C$_4$H$_8$ONH, and C$_6$H$_5$CH$_2$NH$_2$, Table 2, Nos. 11 to 15). The preparation by Method I

References on p. 39

from $C_5H_5Fe(CO)_2I$ and a large excess of the amine in refluxing C_6H_6 (4 h) was described for No. 12. The reaction with cyclo-$C_6H_{11}NH_2$ started even at room temperature. Negligible amounts of $(C_5H_5Fe(CO)_2)_2$ occurred as a by-product and could be removed with CH_2Cl_2.

Other IR bands (KBr) were found at about 890 (s), 1045 (m), 1065 (w), 1085 (s), 1155 (m), 1180 (m), 1225 (m), 1240 (s), 1260 (m), 1310 (s), 1345 (vw), 1395 (m), 1435 (m), 1445 (sh), 2850 (s), and 2920 (s) cm^{-1}.

Complex No. 12 is insoluble in all common organic solvents [31].

$C_5H_5Fe(PF_2-N(C_2H_5)_2)_2I$ (Table **2**, No. **16**) was obtained as a mixture with $C_5H_5Fe(CO)(PF_2-N(C_2H_5)_2)I$ (62% yield) since only ~1.2 equivalents of the ligand molecule were used in Method II. Chromatography on Al_2O_3 with petroleum ether/ether and crystallization from a little pentane gave the pure complex.

A ^{19}F NMR spectrum (in either THF or CH_2Cl_2) could not be obtained. From a reaction with TlC_5H_5 in THF at room temperature only the starting material could be isolated by chromatography [17].

$C_5H_5Fe(P(CH_3)_3)_2I$ (Table **2**, No. **18**). In the preparation by Method V, $[C_5H_5Fe(P(CH_3)_3)_2NCCH_3]BF_4$ and $[N(C_4H_9)_4]I$ (mole ratio ~1:4) were refluxed in THF for 2 h. The compound remained as a crystalline material on complete evaporation of an ether/petroleum ether solution.

^1H NMR spectra in CD_3CN or $(CD_3)_2SO$ (^2D) showed that these solvent molecules displace the iodine atom from the first coordination sphere. Only resonances of the $[C_5H_5Fe(P(CH_3)_3)_2{}^2D]I$ species were observed, see 1.5.1.2, Table 1, Nos. 10 to 14. Cyclic voltammetry in CH_2Cl_2 revealed two one-electron oxidation steps, at $E_p=$ +0.025 and +1.25 V (vs. SCE) [48].

$C_5H_5Fe(P(OCH_3)_3)_2I$ (Table **2**, No. **19**). Prepared by Method II with 2.2 equivalents of $P(OCH_3)_3$ in CH_3CN (6 h). The compound was used as starting material for Nos. 76 and 79 [43].

$C_5H_5Fe(P(OC_6H_5)_3)_2I$ (Table **2**, No. **20**). Prepared by Method II with ~3 equivalents of $P(OC_6H_5)_3$ in refluxing C_6H_6 (10 h). Repeated chromatography on Al_2O_3 with petroleum ether/benzene gave the pure compound [1, 4]. The treatment of the $C_5H_5Fe(^1L-^2D)^2D$ complex IV on p. 27 with an equimolar amount of I_2 in $CHCl_3$ at −60 °C (0.5 h) and room temperature (1 h) resulted in a mixture of No. 20 (17% yield) and compound VI (28% yield). This $^2D-^5LFe(^2D)I$ type compound (see 1.5.1.6, p. 44) was initially assumed to be an isomer of the iodide No. 20 [3, 4] ("isomer II" in [3]).

VI

The ^1H NMR spectrum in CS_2 shows the chemical shifts at $\delta=3.75$ (t, C_5H_5) and 7.1 (m, C_6H_5) ppm [21]. The complete IR spectrum (from 690 to 3120 cm^{-1}) is reported in [4]. The compound crystallizes in the monoclinic system with $a=19.72$,

References on p. 39

$b = 18.18$, $c = 10.99$ (all ± 0.01) Å, and $\beta = 102.9(3)°$; space group $P2_1/a\text{-}C_{2h}^5$. $Z = 4$ gives $D_c = 1.51$ g·cm^{-3}; $D_m = 1.45$ g·cm^{-3} was measured. The structure of the molecule is represented in **Fig. 2**, which shows average values of structurally equivalent spatial parameters. The Fe atom is at a distance of 1.72 Å from the C_5H_5 plane. The distances of the I, P(1), and P(2) atoms from this plane are very similar (3.07, 2.95, and 2.91 Å, respectively). The Fe-I bond length is little longer than the sum of Fe octahedral and I single-bond covalent radii. The rather short Fe-P bonds reflect the $d\pi\,(Fe) - d\pi\,(P)$ back-donation. All other bond lengths coincide with usual values. Steric hindrance between the bulky phosphite ligands results in the P-Fe-P bond angle being larger than the P-Fe-I angles. Significant variations of bond angles at P atoms must be due to various spatial requirements of their substituents [2, 5, 8]. The packing of molecules and shortest intermolecular distances are given in a figure in [5].

Fig. 2

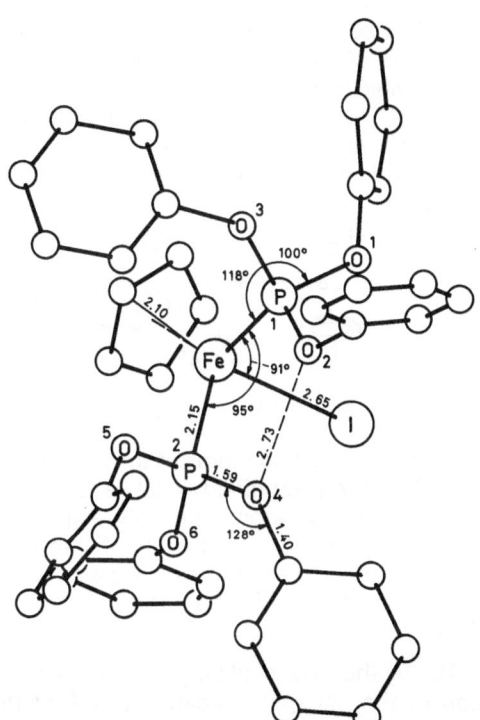

Molecular structure of $C_5H_5Fe(P(OC_6H_5)_3)_2I$ [2, 5, 8].

Electrochemical oxidation of the complex (cyclic voltammetry in CH_2Cl_2) occurred at $E_p \sim 0.57$ V (vs. SCE); the anodic peak was poorly defined [26]. The polarographic reduction in dimethylformamide at $E_{1/2} = 1.33$ V was mentioned without further detail in [3]. Chemical oxidation was achieved with NOPF$_6$ to give [C_5H_5-Fe(P(OC$_6$H$_5$)$_3$)$_2$I]PF$_6$ [26]. Reaction with AgPF$_6$ or AgBF$_4$ in the presence of P(OC$_6$H$_5$)$_3$ leads to the [$C_5H_5Fe(P(OC_6H_5)_3)_3$]$^+$ cation [13, 21]. Cations of the [$C_5H_5Fe(P(OC_6H_5)_3)_2(^2D,L)$]$^+$ type form on such oxidations in the presence of various coordinating molecules [13], see 1.5.1.2, p. 2. Chemical reductions with Na/Hg

References on p. 39

or $NaBH_4$ [49] are described under the hydride No. 5. Ferrocene (65%) forms on treatment with C_5H_5Na in THF at 20 to 65 °C [3, 4]. The reaction with C_6H_5MgBr gave only a low yield of the complex IV on p. 27 [4].

$C_5H_5Fe(P(OCH_3)_3)_2OOCCF_3$ (Table 2, No. **22**). Prepared by Method II with 2 equivalents of $P(OCH_3)_3$ in C_6H_6/C_6H_{14} at room temperature for 3 h, recrystallization from CH_2Cl_2/C_6H_{14}.

The complete IR spectrum (from 712 to 3110 cm^{-1}) is reported. The triplet nature of the CH_3 proton resonance indicates strong coupling with both P atoms, which implies a high $^2J(PFeP)$ coupling. The replacement of CO ligands by the weaker π acceptor $P(OCH_3)_3$ causes a shift of the ^{19}F resonance to higher field, possibly due to an increased negative charge on the Fe atom which is partially transmitted to the CF_3 group [9].

$C_5H_5Fe(^2D)_2NC_4H_4$ ($^2D = PF_3$, $PF_2-NCH_3-PF_2$, $PF_2-N(CH_3)_2$, and $PF_2-N(C_2H_5)_2$, Table 2, Nos. **23** to **26**). These compounds were obtained by π→σ rearrangement of the pyrrolyl ligand of azaferrocene (Formula VII). Thus Nos. 25 and 26 formed when azaferrocene was reacted with the appropriate donor molecule (mole ratio 1:3.3) in C_6H_6 at 70 °C for 2 h. The crude reaction products were extracted with pentane and the complexes were crystallized from the extracts at −78 °C after concentration [51]. Compounds No. 23 and 24 were only mentioned in [44].

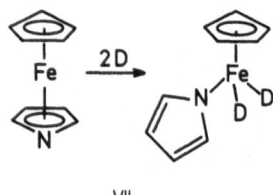

VII

In the ^{13}C NMR spectrum the α carbon atoms of the pyrrolyl ligand show an average chemical shift change of 16.8 ppm (downfield) relative to the corresponding C atoms of free pyrrole, whereas the β carbon atoms are only slightly affected. An extensive pπ(N) − dπ(Fe) interaction may account for the significant deshielding at the α carbons [51]. All C atoms are greatly deshielded as compared with those of the pyrrolyl ring in azaferrocene. A much smaller deshielding effect was observed for the cyclopentadienyl C atoms [44, 51].

In the mass spectra (70 eV) the most facile fragmentation sequence of the molecular ions involves elimination of the phosphine ligands to afford the azaferrocenium ion $[C_5H_5FeC_4H_4N]^+$. The next abundant fragment is $[C_5H_5Fe]^+$, while $[C_4H_4NFe]^+$ occurs only with low intensity [51].

$C_5H_5Fe(P(CH_3)_3)_2Si(CH_3)_3$ (Table 2, No. **27**). Prepared by Method II with an approximately fivefold excess of $P(CH_3)_3$ in petroleum ether at 25 °C/5 h. The compound was isolated by chromatography on Al_2O_3 with ether/petroleum ether and subsequent sublimation at 67 °C/0.03 Torr [48].

$C_5H_5Fe(PCH_3(C_6H_5)_2)_2SiC_4H_9$ (Table 2, No. **28**) formed in Method II along with $C_5H_5Fe(CO)(PCH_3(C_6H_5)_2)SiC_4H_9$ (13% yield) from the dicarbonyl complex and $PCH_3(C_6H_5)_2$ (mole ratio 1:1.5) in hexane (8 h). Most of the product crystallized from the reaction solution; some more was isolated by chromatography on Al_2O_3 with hexane/benzene.

References on p. 39

Prominent ions in the mass spectrum are the molecular ion $[M]^+$, $[M - C_4H_9Si - H]^+$, $[M - C_4H_9Si - CH_3]^+$, and $[M - C_{13}H_{13}P]^+$. The compound shows considerable stability towards O_2 and H_2O. Like other silacyclobutane compounds, it is catalytically polymerized by $(Pt(C_2H_4)Cl_2)_2$ in C_6H_6 at room temperature. Polymer formation apparently also occurred in the presence of $Fe_2(CO)_9$ [39].

$C_5H_5Fe(P(OC_6H_5)_3)_2SnF_3$ (Table 2, No. 29) was prepared from the $SnCl_3$ complex No. 32 and AgF suspended in acetone/water (2:1) at room temperature (1 h). Recrystallization from CH_2Cl_2/C_5H_{12}, 85% yield [29].

$C_5H_5Fe(P(OR)_3)_2SnCl_3$ (R = CH_3, C_2H_5, and C_6H_5, Table 2, Nos. 30 to 32). Prepared by Method II in C_6H_6 [27] or CH_2Cl_2 [29] with two equivalents of $P(OR)_3$ [29], 5 to 20 h. The crude oily products of Nos. 30 and 31 solidified on addition of petroleum ether and crystallized from CH_3OH at $-15\,°C$ to give needles [27]. Recrystallization of No. 32 from CH_2Cl_2/CH_3OH at $-78\,°C$ [27] or from C_6H_6 [29]. According to [29], only small amounts of No. 32 formed when the photochemical reaction was carried out in benzene. No. 32 was also prepared from the $Sn(C_6H_5)_3$ compound No. 42 by Sn–C cleavage with an excess of HCl in benzene/ether at room temperature (3 h), 92% yield [24].

The crystals of No. 32 are stable in air [19].

$C_5H_5Fe(P(OC_6H_5)_3)_2SnBr_3$ (Table 2, No. 33). Prepared by Method II in CH_2Cl_2 (5 h) like No. 32 [29]. The compound forms also from No. 43 by cleavage of the Sn–Sn bonds with Br_2 [20].

$C_5H_5Fe(P(OC_6H_5)_3)_2SnI_3$ (Table 2, No. 34) forms in a 40.3% yield from compound No. 43 and I_2 (mole ratio 1:3.3) in C_6H_6 at room temperature. Intermediate products, $C_5H_5Fe(P(OC_6H_5)_3)_2Sn(Sn(CH_3)_3)_nI_{3-n}$ (n = 1 and 2) were not observed, not even when only one mole of I_2 per mole of No. 43 was used. Excess I_2 cleaves the Fe–Sn bond to give $C_5H_5Fe(P(OC_6H_5)_3)_2I$ [20]. The compound was apparently also prepared by Method II [29].

$C_5H_5Fe(P(OC_6H_5)_3)_2SnCl_n(C_6H_5)_{3-n}$ (n = 2 and 1, Table 2, Nos. 35 and 36) formed from the $Sn(C_6H_5)_3$ compound No. 42 by cleavage of phenyl groups with one or two equivalents of HCl in benzene/ether at room temperature (3 h), 68 and 75% yield, respectively. Recrystallization from C_6H_{12}/C_6H_6 or C_6H_{12} [24].

$C_5H_5Fe(P(CH_3)_3)_2Sn(CH_3)_3$ (Table 2, No. 37). Prepared by Method II with 3.6 equivalents $P(CH_3)_3$ in pentane at $36\,°C$ (10.5 h) and isolated by sublimation at $120\,°C/0.1$ Torr.

The complex was noticeably decomposed after a few days in the air [48].

$C_5H_5Fe(Sb(C_6H_5)_3)_2Sn(CH_3)_3$ (Table 2, No. 38). Prepared by Method II with a 50 to 100% excess of $Sb(C_6H_5)_3$ in benzene or acetone, recrystallized from benzene/ petroleum ether. The $C_5H_5Fe(CO)(Sb(C_6H_5)_3)Sn(CH_3)_3$ by-product was separated from the reaction mixture by fractional crystallization.

The very large quadrupole splitting of the ^{57}Fe-γ resonance could be a result of the bulk of the donor ligands [14].

$C_5H_5Fe(P(CH_3)_n(C_6H_5)_{3-n})_2Sn(C_6H_5)_3$ (n = 2 and 1, Table 2, Nos. 40 and 41). Prepared like No. 38.

In the ^1H NMR spectra, the CH_3 triplets arise from a strong coupling to both P atoms. The CH_3 groups of No. 40 are not equivalent as indicated by a doublet of triplets. From the changes of the isomer shifts of both ^{57}Fe and ^{119}Sn relative to the dicarbonyl complex, it was concluded that upon substitution of ^{2}D for CO, the s-electron density decreases at Fe and increases at Sn [14].

$C_5H_5Fe(P(OC_6H_5)_3)_2Sn(C_6H_5)_3$ (Table 2, No. **42**). No details of the preparation reported. The compound formed along with the monocarbonyl complex and was separated by fractional crystallization. Recrystallization from C_6H_6/C_6H_{12} [24]. For the cleavage of Sn-C_6H_5 bonds, see Nos. 35 and 36.

$C_5H_5Fe(P(OC_6H_5)_3)_2Sn(Sn(CH_3)_3)_3$ (Table 2, No. **43**) was obtained by slow addition of $LiSn_4(CH_3)_9$ to $[C_5H_5Fe(P(OC_6H_5)_3)_3]BF_4$ in THF. The chromatographic isolation on basic Al_2O_3 with C_6H_6 is accompanied by noticeable decomposition. Recrystallization from pentane.

Figures of the ^{13}C NMR spectrum are given. The chemical shifts of the phenyl C atoms (C-1 bonded to oxygen) are $\delta = 124.3$ (C-2), 126.6 (C-4), 131.7 (C-3), and 154.5 (C-1) ppm with $^1J(C,H) = 160$ to 167 Hz, $^3J(^{13}CCC^1H) = 6$ to 8 Hz, and $^3J(^{13}CCO^{31}P) = 8$ Hz. The NMR spectra confirm the nonlinear structure of the $Sn_4(CH_3)_9$ group [21].

The reactions with Br_2 and I_2 are described under the preparation of Nos. 33 and 34.

$C_5H_5Fe(P(CH_3)_2CH_2-)_2H$ (Table 2, No. **44**) was obtained from the iodine compound No. 55 and $LiAlH_4$ in THF. Details were not reported [12].

$C_5H_5Fe(P(C_6H_5)_2CH_2-)_2X$ (X=H and D, Table 2, Nos. **45** and **46**). Method IV was carried out with the Cl compound No. 50 and $NaBH_4$ (mole ratio $\sim$1:18) in THF (30 min). Hexane extracts of the crude oily product gave on cooling at $-20\,°C$ crystals of No. 45 [19]. No. 46 was obtained from the Br compound No. 53 and $LiAlD_4$ (mole ratio $\sim$1:1.3) in THF (overnight) followed by hydrolysis and the usual workup. No H/D exchange occurred on hydrolysis, but exchange was observed during chromatography on Al_2O_3 [39]. In Method III, $[C_5H_5Fe(P(C_6H_5)_2CH_2-)_2NCCH_3]Br$ and $LiAlH_4$ (mole ratio $\sim$1:1.6) were reacted for 2 h. The hydride No. 45 was extracted from the crude product with large amounts of pentane [41]. The hydride also formed in high yields when $C_5H_5Fe(P(C_6H_5)_2CH_2-)_2MgBr$ was reacted in THF ($-78\,°C$ to room temperature) with C_2H_5OH (83% yield), CH_3COCH_3 (87%), or cyclohexene oxide (67%). Solvolysis with D_2O gave a 70 to 75% yield of the deuteride No. 46 [39]. The formation of No. 45 from $[(C_5H_5Fe(P(C_6H_5)_2CH_2-)_2)_2N_2][PF_6]_2$ and $NaBH_4$ was mentioned in [25]. No. 45 also resulted from a one-week reaction of $[(P(C_6H_5)_2CH_2-)_2]_2C_2H_4Fe$ with cyclopentadiene in THF, 50% yield [40].

The Fe-H bond adds to the alkynes $RC\equiv CR$ (R=CF_3 and $COOCH_3$) to give the products $C_5H_5Fe(P(C_6H_5)_2CH_2-)_2CR=CHR$ [30], see 1.5.2.1.1, p. 50. For the addition of CS_2, see compound No. 64.

$C_5H_5Fe(PC_6H_5(CH_2CH_2P(C_6H_5)_2)_2)X$ (X=H and D, Table 2, Nos. **47** and **48**). The two compounds were obtained by brief treatment of the $[C_5H_5FePC_6H_5(CH_2CH_2P(C_6H_5)_2)_2]^+$ cation (see 1.5.1.2, p. 2) with $LiAlH_4$ or $LiAlD_4$, respectively, since H^- or D^- replaces one coordination site of the tridentate ligand of the cation (Formulas VIII and IX).

The ^{1}H NMR spectrum ($C_6D_5CD_3$) of the deuteride No. 48 stored at 20 °C for 18 h showed the typical Fe-H resonance while the IR spectrum contained a new absorption at 2280 cm^{-1} attributable to the C-D stretching vibration of the C_5H_4D

References on p. 39

ligand. Thus the deuteride undergoes smooth intramolecular H–D scrambling at room temperature. The mechanism below was proposed [46]:

VIII   IX

C$_5$H$_5$Fe(P(CH$_3$)$_2$CH$_2$-)$_2$Cl (Table **2**, No. **49**). Prepared by Method II with an equimolar amount of the diphosphine in benzene (3 to 6 h). Recrystallization from CH$_2$Cl$_2$/C$_6$H$_{14}$ by slow evaporation and cooling to −15 °C [10]. For the complete IR spectrum (KBr), see [10].

C$_5$H$_5$Fe(P(C$_6$H$_5$)$_2$CH$_2$-)$_2$X (X=Cl and Br, Table **2**, Nos. **50** and **53**). Prepared by Method II as described under No. 49 [10]. Compound No. 50 was also obtained from Fe(P(C$_6$H$_5$)$_2$CH$_2$-)$_2$Cl$_2$ and TlC$_5$H$_5$ in benzene (30 min). Purification on Al$_2$O$_3$ (with CHCl$_3$ eluent) and crystallization from CHCl$_3$/ether gave a 70% yield [19]. In the preparation of No. 53 by Method III, [C$_5$H$_5$Fe(P(C$_6$H$_5$)$_2$CH$_2$-)$_2$NCCH$_3$]Br and KBr (mole ratio 1:5) were boiled in acetone for 10 to 45 min. The compound was precipitated from CH$_2$Cl$_2$ by slow addition of hexane [41]. − The IR spectrum (KBr) is completely reported for No. 50 [10].

The oxidation of both complexes with AgPF$_6$ leads to paramagnetic cations [26], which are described in 1.5.1.4, p. 40. For the reaction of No. 53 with Mg, see compound No. 74. Treatment with CH$_2$=CH(CH$_2$)$_4$MgBr gave the C$_5$H$_5$Fe(P(C$_6$H$_5$)$_2$-CH$_2$-)$_2$R complex with R = − (CH$_2$)$_4$CH=CH$_2$ [18, 36].

C$_5$H$_5$Fe(P(C$_6$H$_5$)$_2$CH=CHP(C$_6$H$_5$)$_2$)X (X=Cl and Br, Table **2**, Nos. **51** and **54**). Prepared by Method II as given under No. 49 [10]. For the complete IR spectra (KBr), see [10].

C$_5$H$_5$Fe(P(CH$_3$)$_2$CH$_2$-)$_2$Br (Table **2**, No. **52**) cannot be prepared by Method II. Instead, C$_5$H$_5$Fe(CO)$_2$Br in benzene was first irradiated in the presence of two equivalents of P(OC$_6$H$_5$)$_3$ (18 h) to give No. 10, which was not isolated. After addition of one equivalent of P(CH$_3$)$_2$CH$_2$CH$_2$P(CH$_3$)$_2$, further irradiation for 18 h converted the intermediate to No. 52, which crystallized from benzene/heptane on cooling, 51% yield.

The compound reacts with Mg metal in THF or ether/benzene as is described for No. 8 on p. 27 [36].

C$_5$H$_5$Fe(P(CH$_3$)$_2$CH$_2$-)$_2$I (Table **2**, No. **55**) was prepared like No. 56 from C$_5$H$_5$Fe(P(OC$_6$H$_5$)$_3$)$_2$I and P(CH$_3$)$_2$CH$_2$CH$_2$P(CH$_3$)$_2$ under irradiation [12]. Details were not reported. The same components react in toluene at 60 °C (6 h) without irradiation to give No. 55 in a 41% yield along with [C$_5$H$_5$Fe(P(CH$_3$)$_2$CH$_2$-)$_2$-P(OC$_6$H$_5$)$_3$]I (51% yield). No. 55 was extracted from the mixture with CH$_2$Cl$_2$ and was recrystallized from CH$_2$Cl$_2$/petroleum ether. The above iodide was converted to No. 55 with an excess of KI in acetone under irradiation for 24 h [32]. The direct preparation of No. 55 from C$_5$H$_5$Fe(CO)$_2$I by Method II fails because it gives only

References on p. 39

a dinuclear complex with a bridging rather than chelating ligand, C_5H_5-Fe(CO)(I)P(CH$_3$)$_2$CH$_2$CH$_2$P(CH$_3$)$_2$(I)(CO)FeC$_5$H$_5$ [10], see C3, 2.5.2.2.11, p. 161.

[$C_5H_5Fe(P(CH_3)_2CH_2-)_2OC(CH_3)_2$]BF$_4$ formed on treatment of No. 55 with TlBF$_4$ in acetone [12]. In the presence of an activated olefin, this treatment gave, in a slow reaction, cations of the [$C_5H_5Fe(P(CH_3)_2CH_2-)_2{}^2L$]$^+$ type with $^2L=$ CH$_2$=CHCOOCH$_3$ and C_2H_5OOCCH=CHCOOC$_2$H$_5$ [32].

$C_5H_5Fe(P(C_6H_5)_2CH_2-)_2I$ (Table 2, No. 56) was prepared from C_5H_5-Fe(P(OC$_6H_5)_3)_2I$ (No. 20) and an excess of the bisphosphinoethane in C$_6$H$_6$ under irradiation (4 h), whereby the temperature increased to 70 °C. Purification on Al$_2$O$_3$ with ether/CH$_2$Cl$_2$ and recrystallization from CH$_2$Cl$_2$/ether/petroleum ether [13]. Method III was carried out as for No. 53 but with an excess of KI [41].

Reactions of No. 56 with TlX (X=BF$_4$ or PF$_6$) in acetone gave a yellow–orange crystalline product which probably contained the 16-electron cation [C_5H_5-Fe(P(C$_6H_5)_2CH_2-)_2$]$^+$ together with varying amounts of incorporated solvent, either as a weak ligand or in the lattice [32], see also 1.5.1.2, p. 2.

$C_5H_5Fe(C_{36}H_{37}O_2P_2)I$ (Table 2, No. 57) was obtained in a 69% yield by reacting $C_5H_5Fe(P(OC_6H_5)_3)_2I$ (No. 20) with (−)-DIOP ($C_{36}H_{37}O_2P_2$) in toluene at 80 °C for 15 h. Recrystallization from CH$_2$Cl$_2$/petroleum ether [32].

The compound crystallizes in the orthorhombic system with a=16.920(3), b=19.902(4), and c=10.649(6) Å; space group P2$_1$2$_1$2$_1$ − D$_2^4$. Z=4 gives D$_c$=1.38 g·cm^{-3}. The molecular structure is shown in **Fig. 3**. The geometric parameters of the coordination at the Fe atom are normal and comparable to those of analogous complexes. The coordination site of the iodine atom is quite crowded by two phenyl groups and several H atoms on C(1) to C(4), the nonbonding H–I distances being between 2.92 and 3.38 Å. This proximity may explain the failure to prepare olefin compounds of the [$C_5H_5Fe(DIOP)^2L$]$^+$ type. The conformation of the DIOP ligand was compared with conformations in various other DIOP complexes [45].

Fig. 3

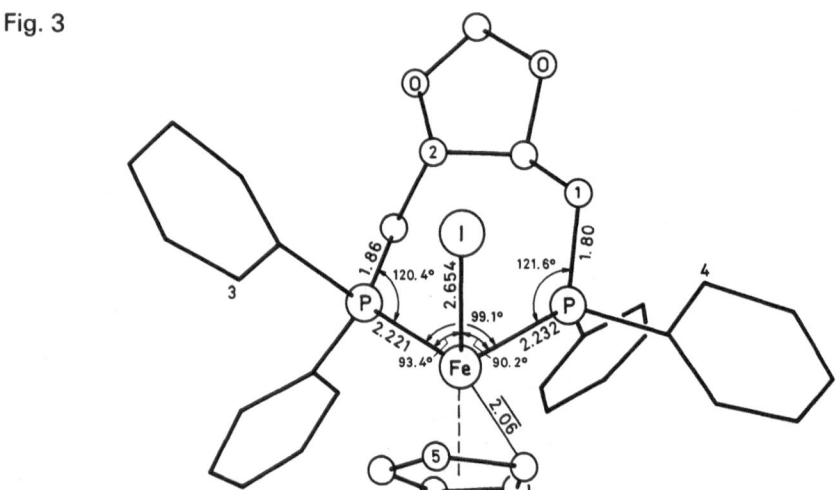

Molecular structure of $C_5H_5Fe(C_{36}H_{37}O_2P_2)I$ [45].

References on p. 39

The optical rotation of the compound is $[\alpha]_D^{25} = -518°$ measured in $CHCl_3$ at $c = 0.27$ [32, 45].

$C_5H_5Fe(P(C_6H_5)_2CH_2-)_2X$ (X = CN, SCN, and SC_6H_5, Table 2, Nos. **58, 59**, and **63**) were prepared by Method III from $[C_5H_5Fe(P(C_6H_5)_2CH_2-)_2NCCH_3]Br$ and an excess of KCN in CH_3OH, KSCN in CH_3CN/CH_3COCH_3 (1:1), or KSC_6H_5 in CH_3OH by refluxing for 10 to 15 min. No. 58 was isolated by extraction of the crude product with C_6H_6 and recrystallization from C_6H_6/CH_3OH. Nos. 59 and 63 were chromatographed on Al_2O_3 with $CHCl_3$ or CH_2Cl_2 and were crystallized from $CHCl_3$ (No. 59) or ether (No. 63) [41].

The IR spectra (KBr) are completely reported for No. 59 (665 to 2084 cm^{-1} range) and No. 63 (420 to 1570 cm^{-1} range) [41]. The $\nu(CN)$ band of No. 59 indicates that the SCN group is bonded to Fe through the N atom [42].

The reaction of No. 63 with $NOPF_6$ yielded $[C_5H_5Fe(P(C_6H_5)_2CH_2-)_2NO][PF_6]_2$ [42], see 1.5.1.8, p. 48.

$C_5H_5Fe(^2D-^2D)OOCCF_3$ ($^2D-^2D = P(CH_3)_2CH_2CH_2P(CH_3)_2$, $P(C_6H_5)_2CH_2-$ $CH_2P(C_6H_5)_2$, and $P(C_6H_5)_2CH=CHP(C_6H_5)_2$, Table 2, Nos. **60** to **62**). Prepared by Method II with an equimolar quantity of the $^2D-^2D$ ligand in hexane (with some benzene) for 1 h (Nos. 60 and 61) or 4 h (No. 62). No. 61 was recrystallized from ether. No. 62 separated as crystals during the irradiation. No. 60 was recrystallized from a minimum of CH_2Cl_2 on addition of hexane [9].

The IR spectra (KBr) were completely reported for all complexes [9].

$C_5H_5Fe(P(C_6H_5)_2CH_2-)_2S-CH=S$ (Table 2, No. **64**) formed on dissolution of $C_5H_5Fe(P(C_6H_5)_2CH_2-)_2H$ in CS_2. During a period of 2 h, well formed, air-stable crystals precipitated, 77% yield [30].

$C_5H_5Fe(P(C_6H_5)_2CH_2-)_2SO_2CH_3$ (Table 2, No. **65**) was prepared by bubbling SO_2 through a solution of $C_5H_5Fe(P(C_6H_5)_2CH_2-)_2CH_3$ in C_6H_{14}/CH_2Cl_2 (8:1) at $-72°C$ for 10 min. The compound was isolated by chromatography on Al_2O_3 with acetone/methanol (9:1). Recrystallization from CH_2Cl_2/ether, 35% yield [30].

$C_5H_5Fe(^2D-^2D)Si(CH_3)_3$ ($^2D-^2D = P(C_6H_5)_2CH_2CH_2P(C_6H_5)_2$ and $P(C_6H_5)_2-$ $CH=CHP(C_6H_5)_2$, Table 2, Nos. **66** and **67**). Prepared by Method II with an equimolar quantity of the bisphosphines in hexane during 15 to 18 h, No. 66 was chromatographed on Al_2O_3 with CH_2Cl_2, while No. 67 was recrystallized from hexane [6].

The IR spectra (KBr) of the two compounds were completely reported (500 to 3075 cm^{-1} range) [6]. The ^{29}Si NMR spectra of various C_5H_5Fe complexes containing silyl groups were discussed with respect to bonding and molecular motions [50].

$C_5H_5Fe(P(C_6H_5)_2CH_2-)_2Ge(C_6H_5)_3$ (Table 2, No. **68**) was prepared from $C_5H_5Fe(P(C_6H_5)_2CH_2-)_2MgBr$ (No. 74) and $(C_6H_5)_3GeBr$ in THF at 20°C/38 h. Workup by hydrolysis and chromatography on Al_2O_3 with toluene gave a solid, which was recrystallized from ether (yield not reported) [49].

$C_5H_5Fe(P(C_6H_5)_2CH_2-)_2SnX_3$ (X = Cl, Br, and I, Table 2, Nos. **69** to **71**). Compound No. 69 was prepared by refluxing $C_5H_5Fe(P(C_6H_5)_2CH_2-)_2Cl$ (No. 50) with a slight excess of anhydrous $SnCl_2$ in THF for 2 h. The crude product was precipitated by addition of petroleum ether and recrystallized from THF/petroleum ether, 80% yield.

The $SnCl_3$ complex was converted to Nos. 70 and 71 by refluxing it with excess anhydrous $SnBr_2$ or SnI_2 in THF/CH_3OH for 30 min. The partially evaporated solutions gave pure products, No. 70 in 60% yield and No. 71 in 30% yield, on cooling [19].

References on p. 39

$C_5H_5Fe(P(C_6H_5)_2CH_2-)_2Sn(CH_3)_3$ (Table **2**, No. **72**) formed in a rapid reaction when the solid $SnCl_3$ complex No. 69 was slowly added to a large excess of CH_3MgI in ether. Hydrolysis and chromatography on Al_2O_3 with ether gave the crystalline complex in a 20% yield [19]. Also prepared by Method II under the conditions given for Nos. 66 and 67 [6].

For No. 72, but also for Nos. 69 to 71, the isomer shifts of the ^{57}Fe and ^{119}Sn Mössbauer spectra are greater than for the corresponding dicarbonyl complexes indicating a higher s-electron density at Fe, which is transmitted to Sn through the Fe–Sn bond [19]. $^{117,119}Sn$ NMR studies were briefly reported. These gave the Taft constant $\sigma^* = -1.32$ for the $C_5H_5Fe(P(C_6H_5)_2CH_2-)_2$ moiety [11].

$C_5H_5Fe(P(C_6H_5)_2CH=CHP(C_6H_5)_2)Sn(CH_3)_3$ (Table **2**, No. **73**). Prepared by Method II under the conditions given for Nos. 66 and 67 [6]. The Taft constant is the same as for No. 72 [11].

$C_5H_5Fe(P(C_6H_5)_2CH_2-)_2MgBr \cdot 3C_4H_8O$ (Table **2**, No. **74**) forms in 40 to 60% yield from $C_5H_5Fe(P(C_6H_5)_2CH_2-)_2Br$ (No. 53) and excess Mg turnings (activated with some CH_2Br-CH_2Br) in refluxing THF (1 h). The compound crystallizes from the filtered solution on repeated cooling to $-30\,°C$ [23, 37]. A deep red solution of the complex was also prepared at 20 °C (16 h) using activated Mg powder [49]. Attempts to prepare the compound by metalation of the hydride No. 45 with $n-C_4H_9MgBr$ were unsuccessful [37]. One THF molecule can easily be displaced by recrystallization from C_6H_6 to give $C_5H_5Fe(P(C_6H_5)_2CH_2-)_2MgBr \cdot 2C_4H_8O \cdot C_6H_6$ [23].

The complex crystallizes in the monoclinic system with $a=12.258(4)$, $b=13.027(4)$, $c=26.577(11)$ Å, and $\beta=102.48(2)°$; space group $P2_1/c - C_{2h}^5$ and $Z=4$. The crystals are built up from the molecule shown in **Fig. 4** and one THF molecule of crystallization. The Fe–Mg distance indicates a bond with a strong covalent charac-

Fig. 4

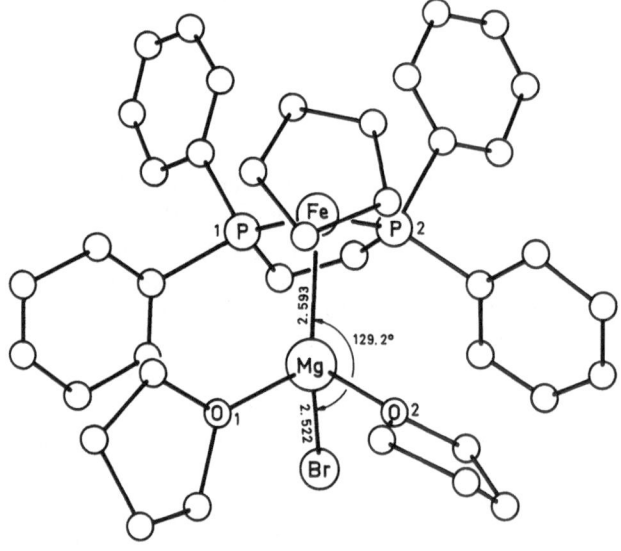

Molecular structure of $C_5H_5Fe(P(C_6H_5)_2CH_2-)_2Mg(OC_4H_8)_2Br$ [23].

References on p. 39

ter. The distorted tetrahedral environment of the Mg atom is similar to that of many Grignard reagents. Other parameters not given in Fig. 4 are Mg-O(1) 2.075(15) and Mg-O(2) 2.098(15) Å and the angles Fe-Mg-O(1) 119.0(5)°, Fe-Mg-O(2) 115.8(6)°, and O(1)-Mg-O(2) 91.7(9)°. The molecular geometry of the C_5H_5-Fe(^{2}D-^{2}D) unit is normal [23].

Solutions of the complex are rapidly oxidized to give ferrocene and the bromide No. 53. The hydride No. 45 forms on solvolysis with H_2O or C_2H_5OH. The reaction with CO_2 in C_6H_6 in the presence of $MgBr_2$ probably gives the intermediate $C_5H_5Fe(P(C_6H_5)_2CH_2-)_2CO-OMgBr$ (IR band at 1970 cm^{-1}), which is converted to [$C_5H_5Fe(P(C_6H_5)_2CH_2-)_2CO]BF_4$ by subsequent treatment with $AgBF_4$ (65% yield). Complex No. 74 is sufficiently basic to abstract a proton from acetone or cyclohexene oxide to form the hydride No. 45. In the second case, cyclohexanol and small amounts of 2-cyclohexene-1-ol were detected as organic products [37].

Scheme 1

$$C_5H_5Fe(^2D-^2D)MgBr + RBr$$

(a) $\Big\downarrow$ THF, 20 °C, 0.5 to 5 h

$$[C_5H_5Fe(^2D-^2D)MgBr]^+ + R^\bullet + Br^-$$

(b) $\Big\downarrow$

$$C_5H_5Fe(^2D-^2D)Br + C_5H_5Fe(^2D-^2D)R + RH + R-R + \text{olefins}$$

X   XI   XII   XIII   XIV

Reactions with organic bromides have been studied [35, 37], see Scheme 1 (^{2}D-^{2}D = $P(C_6H_5)_2CH_2CH_2P(C_6H_5)_2$ and R = n-, s-, and t-C_4H_9, CH_2=CHCH$_2$, $C_6H_5CH_2$, and 4-$CH_3C_6H_4$). The bromide X (No. 53) is formed in all cases, but there was no or only minor formation of the alkylated compound XI (~17% for n-C_4H_9Br and 9% for $C_6H_5CH_2$Br). The distribution of the products XII to XIV varies widely with the kind of the group R, thus indicating that compound No. 74 initially reacts by a one-electron transfer to RBr to give the complex cation in Scheme 1 and an R radical. The radicals presumably abstract hydrogen from the solvent (varying amounts of XII in all cases) or dimerizes (high yields of XIII for allyl and benzyl bromide) or disproportionates to give XIV (1- and 2-butenes and isobutene for the butyl bromides). Step (a) in Scheme 1 was further supported by an ESR spectrum of the reaction mixture with s-C_4H_9Br (at 77 K). It showed an extremely broad signal (centered at g = 2.19) corresponding to the paramagnetic cationic iron species. The reaction of No. 74 with CH_2=CH(CH$_2$)$_4$Br gave methylcyclopentane (40% yield) and the complex type XI with R = $CH_2C_5H_9$-cyclo, consistent with the fact that the 5-hexene-1-yl radical rapidly cyclizes to cyclopentylmethyl. The highest yield of complex XI (R = $C_6H_5CH_2$, 34%) was obtained with $C_6H_5CH_2Cl$, an indication of S_N2 participation since the chlorides are less readily reduced than the bromides [35, 37].

$C_5H_5Fe(P(C_6H_5)_2CH_2-)_2COC_6H_5$ was isolated (45% yield) from the reaction of No. 74 with C_6H_5COCl in THF at room temperature [34, 37]. In contrast, the reaction with CH_3COBr gave only the bromide No. 53 in an 84% yield [37]. Alkylation to complex XI (R = CH_3) was also achieved with 4-$CH_3C_6H_4SO_2CH_3$ in THF at room temperature [37].

Various compounds, which are usually good substrates for nucleophilic reactants, like Na[$C_5H_5Fe(CO)_2$] or $C_5H_5Fe(CO)_2MgBr$, did not react with No. 74 in THF at room temperature, e.g., $C_6H_5CH=CH_2$, C_6H_5CHO, $C_6H_5COC_6H_5$, $(CH_3)_3SiCl$, $(CH_3)_3SnCl$ [37], $(C_6H_5)_3SiBr$, and $C_6H_5(CH_3)_2GeBr$ [49]. Reactions with $CH_3(C_6H_5)_2GeBr$ and $(C_6H_5)_3GeBr$ gave $C_5H_5Fe(P(C_6H_5)_2CH_2-)_2Br$ (No. 53) and $C_5H_5Fe(P(C_6H_5)_2CH_2-)_2Ge(C_6H_5)_3$ (No. 68), respectively. The results were also explained by a primary electron transfer from No. 74 and by differences in the electrophilic character of the organogermyl radicals [49].

The reaction with $(CF_3CO)_2O$ leads probably to an intermediate $C_5H_5Fe(P(C_6H_5)_2CH_2-)_2COCF_3$. Subsequent treatment with $AgBF_4$ in acetone gave some $[C_5H_5Fe(P(C_6H_5)_2CH_2-)_2CO]BF_4$ [37].

$C_5H_5Fe(N_3(C_6H_4Cl-4)_2)^2D$ and $C_5H_5Fe(N_3(C_6H_4CH_3-4)_2)^2D$ ($^2D = P(C_6H_5)_3$, $P(OCH_3)_3$, and $P(OC_6H_5)_3$, Table 2, Nos. 75 to 77 and 78 to 80). In the preparation of Nos. 75 and 78 by Method Va, a clear solution formed on addition of the $Cu(RN_3R)$ compound in THF to an emulsion of $Fe(P(C_6H_5)_3)_2I_2$ in toluene at room temperature. This was treated after 5 min with solid TlC_5H_5. After evaporation of solvent and extraction with hexane the products precipitated from the extracts only at very low temperature. Recrystallizations from hexane/ether (4:1) gave products containing one mole of ether per mole of complex. Solvent-free samples were obtained from hexane. Method Vb was carried out with $Cu(RN_3R)$ for Nos. 76 and 79 and with $Ag(RN_3R)$ for Nos. 77 and 80 in refluxing benzene (4 h). Evaporation of solvent, extraction with hexane, and cooling gave impure products, which were freed from $Cu(P(OCH_3)_3)I$ or $P(OC_6H_5)_3$ by recrystallization from hexane (Nos. 76 and 79) or ether (Nos. 77 and 80).

XV

The aromatic groups of the triazenido ligands are magnetically equivalent in the ^1H NMR spectra. All observations are consistent with the symmetrical bonded triazenido ligand shown by Formula XV. There was no evidence for a dissociation of the ^2D ligands on the NMR time scale.

The complexes are soluble in the common organic solvents. Their sensitivity towards oxidation decreases in the order $P(C_6H_5)_3 > P(OCH_3)_3 > P(OC_6H_5)_3$ [43].

$C_5H_5Fe(P(CH_3)_3)_2CN$ (Table 2, No. 81). Method III with an equimolar amount of KCN in CH_3OH at room temperature for 17 h. The compound was isolated by sublimation at 120 °C/0.02 Torr [48].

$C_5H_5Fe(P(C_6H_5)_2CH_2-)_2S_2O_3$ (Table 2, No. 82). Method III with an excess of $Na_2S_2O_3$ in refluxing CH_3OH (10 min); purification on acidic Al_2O_3 with CH_2Cl_2/CH_3OH (9:1) and recrystallization from CH_2Cl_2.

The field-dependent magnetic moment is $\mu_{eff} = 1.95$ B.M. extrapolated to $1/H = 0$. Thus the compound is a 17-electron species and presumably contains a unidentate S-bonded $S_2O_3^{2-}$ ligand. The IR bands at 600 (m), 1010 (s), and 1210 (s) cm^{-1} are in accord with this formulation. As no oxidant was present in the synthesis, oxidation perhaps occurred by contact with air (or on chromatography?) in the workup. It was noted that the 18-electron precursor, $[C_5H_5Fe(P(C_6H_5)_2CH_2-)_2NCCH_3]Br$, has a very low oxidation potential [42].

References:

[1] A.N. Nesmeyanov, Yu.A. Chapovsky, Yu.A. Ustynyuk (Izv. Akad. Nauk SSSR Ser. Khim. **1966** 1871; Bull. Acad. Sci. USSR Div. Chem. Sci. **1966** 1815). – [2] Yu.A. Chapovsky, B.G. Andrianov, Yu.T. Struchkov, V.A. Semion (Zh. Strukt. Khim. **8** [1967] 559/60; J. Struct. Chem. [USSR] **8** [1967] 501/2). – [3] A.N. Nesmeyanov, Yu.A. Chapovsky (Izv. Akad. Nauk SSSR Ser. Khim. **1967** 223/4; Bull. Acad. Sci. USSR Div. Chem. Sci. **1967** 224). – [4] A.N. Nesmeyanov, Yu.A. Chapovsky, Yu.A. Ustynyuk (J. Organometal. Chem. **9** [1967] 345/53). – [5] V.G. Andrianov, Yu.T. Struchkov (Zh. Strukt. Khim. **9** [1968] 240/9; J. Struct. Chem. [USSR] **9** [1968] 182/90).

[6] R.B. King, K.H. Pannell (Inorg. Chem. **7** [1968] 1510/3). – [7] V.G. Andrianov, Yu.T. Struchkov (Zh. Strukt. Khim. **9** [1968] 503/12; J. Struct. Chem. [USSR] **9** [1968] 426/34). – [8] V.G. Andrianov, Yu.A. Chapovsky, V.A. Semion, Yu.T. Struchkov (Chem. Commun. **1968** 282/4). – [9] R.B. King, R.N. Kapoor (J. Inorg. Nucl. Chem. **31** [1969] 2169/77). – [10] R.B. King, L.W. Houk, K.H. Pannell (Inorg. Chem. **8** [1969] 1042/8).

[11] G. Singh (4th Intern. Conf. Organometal. Chem., Bristol 1969, Paper B10). – [12] W.E. Silverthorn (J. Chem. Soc. Chem. Commun. **1971** 1310/1). – [13] M.L.H. Green, R.N. Whiteley (J. Chem. Soc. A **1971** 1943/6). – [14] W.R. Cullen, J.R. Sams, J.A.J. Thompson (Inorg. Chem. **10** [1971] 843/50). – [15] R.B. King, C.N. Kapoor, R.N. Kapoor (Inorg. Chem. **10** [1971] 1841/50).

[16] T. Kruck, L. Knoll (Chem. Ber. **105** [1972] 3783/8). – [17] R.B. King, W.C. Zipperer, M. Ishaq (Inorg. Chem. **11** [1972] 1361/70). – [18] W. Kläui, M. Werner (Chimia [Aarau] **17** [1973] 386). – [19] M.J. Mays, P.L. Sears (J. Chem. Soc. Dalton Trans. **1973** 1873/5). – [20] W. Kläui, M. Werner (J. Organometal. Chem. **60** [1973] C19/C21).

[21] W. Kläui, M. Werner (J. Organometal. Chem. **54** [1973] 331/40). – [22] Tai-Wing Ng (Diss. Univ. Waterloo, Can., 1973, from Diss. Abstr. Intern. B **34** [1974] 4865/6). – [23] M. Felkin, P.J. Knowler, B. Meunier, A. Mitschler, L. Richards, R. Weiss (J. Chem. Soc. Chem. Commun. **1974** 44). – [24] R.E.J. Bichler, M.C. Clark, B.K. Hunter (J. Organometal. Chem. **69** [1974] 367/76). – [25] D. Sellmann, E. Kleinschmidt (Angew. Chem. **87** [1975] 595/6; Angew. Chem. Intern. Ed. Engl. **14** [1975] 571).

[26] P.M. Treichel, K.P. Wagner, M.J. Mueh (J. Organometal. Chem. **86** [1975] C13/C16). – [27] T.N. Pecoraro (Diss. Rutgers State Univ. 1975, pp. 1/56; Diss. Abstr. Intern. B **36** [1975] 746). – [28] G.M. Bancroft, A.T. Rake (Inorg. Chim. Acta **13** [1975] 175/9). – [29] B. Herber, M. Werner (Syn. Reactiv. Inorg. Metal-Org. Chem. **5** [1975] 381/91). – [30] P.M. Treichel, D.C. Molzahn (Inorg. Chim. Acta **36** [1976] 267/8).

[31] V.N. Pandey (Inorg. Chim. Acta **25** [1977] L37/L38). – [32] G. Balavoine, M.L.H. Green, J.P. Sauvage (J. Organometal. Chem. **128** [1977] 247/52). – [33] R.B. King, M.G. Newton, J. Gimeno, M. Chang (Inorg. Chim. Acta **25** [1977] L35/L37). – [34] M. Felkin, B. Meunier, C. Pascard, T. Prange (J. Organometal. Chem. **135** [1977] 361/72). – [35] M. Felkin, B. Meunier (Nouv. J. Chim. **1** [1977] 281/2).

[36] B. Meunier (Diss. Univ. Paris Sud 1977, No. 1881). – [37] M. Felkin, P.J. Knowles, B. Meunier (J. Organometal. Chem. **146** [1978] 151/67). – [38] H. Felkin, B. Meunier (J. Organometal. Chem. **146** [1978] 169/78). – [39] C.S. Cundy, M.F.

Lappert, C.-K. Yuen (J. Chem. Soc. Dalton Trans. **1978** 427/33). − [40] S.D. Ittel, C.A. Tolman, P.J. Krusic, A.D. English, J.P. Jesson (Inorg. Chem. **17** [1978] 3432/8).

[41] P.M. Treichel, D.C. Molzahn (Syn. Reactiv. Inorg. Metal-Org. Chem. **9** [1979] 21/9). − [42] P.M. Treichel, D.C. Molzahn, K.P. Wagner (J. Organometal. Chem. **174** [1979] 191/7). − [43] E. Pfeiffer, K. Vrieze (Transition Metal Chem. [Weinheim] **4** [1979] 385/8). − [44] A. Efraty, N. Jubran (Inorg. Chim. Acta **44** [1980] L191/ L192). − [45] G. Balavoine, S. Brunie, M.B. Kagan (J. Organometal. Chem. **187** [1980] 125/39).

[46] S.G. Davies, H. Felkin, O. Watts (J. Chem. Soc. Chem. Commun. **1980** 159/60). − [47] J.G.M. van der Linden, A.M. Dix, E. Pfeiffer (Inorg. Chim. Acta **39** [1980] 271/4). − [48] P.M. Treichel, D.A. Komar (J. Organometal. Chem. **206** [1981] 77/8). − [49] N. Aktoğu, S.G. Davies, J. Dubac, P. Mazerolles (J. Organometal. Chem. **212** [1981] C13/C15). − [50] K.M. Pannell, A.R. Basindale (J. Organometal. Chem. **229** [1982] 1/9).

[51] A. Efraty, N. Jubran, A. Goldman (Inorg. Chem. **21** [1982] 868/73). − [52] M.G. Alt, M.E. Eichner (Chemie Dozententagung Kaiserslautern 1982, Abstr. B 28). − [53] D.A. Komar (Diss. Univ. Wisconsin 1981, pp. 1/224; Diss. Abstr. Intern. B **42** [1982] 2828).

1.5.1.4 Compounds with [^{5}LFe(^{2}D)$_2$X]$^+$ and [^{5}LFe(^{2}D-^{2}D)X]$^+$ Cations

The salts in this section, compiled in Table 3, contain paramagnetic 17-electron cations. A neutral 17-electron species has been placed in Table 2 (No. 82, pp. 26 and 38) because of its formal identity with the complexes in Section 1.5.1.3.

The salts were prepared from the respective $C_5H_5Fe(^2D)_2X$ and $C_5H_5Fe(^2D-^2D)X$ compounds by oxidation with AgPF$_6$ or AgBF$_4$. However, the preparation of salt No. 1 required the more powerful oxidant NOPF$_6$. Some typical preparative procedures are described under further information. The oxidation potentials for the formation of several [$C_5H_5Fe(P(C_6H_5)_2CH_2-)_2X$]$^+$ cations are given in the introductory remarks of 1.5.1.3, p. 16.

The magnetic moments of the salts are equivalent to one unpaired electron, excepting the cyano compound (No. 7) which has an anomalously low value. This compound also has an unusually low conductivity for a 1:1 electrolyte in acetonitrile [2]. No explanation has been offered.

Explanations for Table 3: Molar electrical conductivities were measured in CH_2Cl_2 except for Nos. 7 and 10. The conductivity values are given in $\Omega^{-1}\cdot cm^2\cdot mol^{-1}$. (Table 3, see p. 41.)

Further information:

[$C_5H_5Fe(P(OC_6H_5)_3)_2I$]PF$_6$ (Table 3, No. 1) precipitated as a dark solid when a slight excess of solid NOPF$_6$ was added to $C_5H_5Fe(P(OC_6H_5)_3)_2I$ in C_6H_6 and the mixture was stirred for 6 h at room temperature; recrystallization from acetone/ether [2].

[$C_5H_5Fe(P(CH_3)_3)_2X$]BF$_4$ (X=SC$_6H_5$ and Sn(C$_6H_5)_3$, Table 3, Nos. 2 and 3) were prepared by the addition of an equimolar amount of AgBF$_4$ to solutions (THF) of the $C_5H_5Fe(P(CH_3)_3)_2X$ complexes (prepared in situ without isolation for No. 2). This causes silver metal and much of the salt to precipitate immediately. Large volumes

Table 3
Salts with $[C_5H_5Fe(^2D)_2X]^+$ and $[C_5H_5Fe(^2D-^2D)X]^+$ Cations.
Further information on all salts is given on pp. 40 and 42.
For abbreviations and dimensions, see p. 375.

No. group X (yield in %)	donor ^2D or $^2D-^2D$ anion	properties and remarks explanations on p. 40	Ref.
$[C_5H_5Fe(^2D)_2X]^+$ type			
1 I (92)	$P(OC_6H_5)_3$ PF_6	blue–green needles, m.p. 112 to 114° $\mu_{eff}=2.20$ B.M., $\Lambda_M=58.8$	[1, 2]
2 SC_6H_5 (29)	$P(CH_3)_3$ BF_4	deep purple crystals, m.p. 157 to 159° (dec.) $\mu_{eff}=1.88$ B.M.	[3]
3 $Sn(C_6H_5)_3$ (77)	$P(CH_3)_3$ BF_4	yellow powder, m.p. 183 to 183.5° (dec.) $\mu_{eff}=2.05$ B.M.	[3]
$[C_5H_5Fe(^2D-^2D)X]^+$ type			
4 Cl (58)		orange–red prisms, m.p. 175 to 177° $\mu_{eff}=2.13$ B.M., $\Lambda_M=55.0$	[1, 2]
5 Br (37)		red–brown prisms, m.p. 188 to 189° $\mu_{eff}=2.12$ B.M., $\Lambda_M=53.5$	[1, 2]
6 I (70)		m.p. 145° (dec.)	[1, 2]
7 CN (71)	$(P(C_6H_5)_2CH_2-)_2$ PF_6	yellow powder, dec. 200° IR (KBr): $\nu(CN)$ 2055 $\mu_{eff}=0.486$ to 0.566 B.M., $\Lambda_M=66.75$ (in CH_3CN)	[2]
8 NCS (68)		burgundy crystals, dec. $>200°$ IR (KBr): $\nu(CN)$ 2040 UV (CH_2Cl_2): $\lambda_{max}(\varepsilon)=231$ (340000), 260 (~35000, sh), 582 (2300) $\mu_{eff}=1.97$ B.M.	[2]
9 SC_6H_5 (40)		black crystals, m.p. 205 to 207° UV (CH_2Cl_2): $\lambda_{max}(\varepsilon)=582$ (1390)	[2]
10 SC_6H_5 (75)	$(P(C_6H_5)_2CH_2-)_2$ BF_4	black microcrystalline, CH_3OH solvate, softens at 120°, m.p. 200 to 202° $\Lambda_M=138$ (in CH_3CN)	[2]
11 $Sn(CH_3)_3$ (33)	$(P(C_6H_5)_2CH_2-)_2$ PF_6	orange prisms $\Lambda_M=51.4$	[1, 2]

of acetone were used to extract the salts from the precipitates. The crude solids obtained from the solutions were recrystallized from CH_2Cl_2/ether [3].

$[C_5H_5Fe(P(C_6H_5)_2CH_2-)_2X]PF_6$ (X=Cl, Br, I, CN, NCS, SC_6H_5, and $Sn(CH_3)_3$, Table 3, Nos. 4 to 9 and No. 11). The oxidation of the starting materials $C_5H_5Fe(P(C_6H_5)_2CH_2-)_2X$ was carried out in acetone by adding an approximately equivalent amount of $AgPF_6$ and stirring the solutions at room temperature for a few minutes up to 30 min. After removal of the precipitated silver metal and the solvent by evaporation at reduced pressure, the products were recrystallized from mixtures of polar solvents like CH_3CN/ether (No. 8), CH_3OH/ether (No. 9), CH_2Cl_2/ether (Nos. 4 and 5), CH_3OH/CH_3COCH_3 (No. 7), or $CH_3OH/i-C_5H_{12}$ (No. 11).

Complex No. 4 was also obtained (21% yield) in an attempt to oxidize $C_5H_5Fe(P(C_6H_5)_2CH_2-)_2H$ with $AgPF_6$ in a similar fashion.

The iodine complex (No. 6) was not obtained completely pure since it proved very difficult to crystallize.

An excess of $AgPF_6$ caused the solutions of No. 8 to turn red, while the $\nu(CN)$ band shifted from 2040 to 2050 cm^{-1}. Addition of CH_3CN or KBr reversed the color change. This effect was interpreted as a complexation of Ag^+ by the S atom of the NCS ligand.

The reversible one–electron oxidation of No. 9 observed by cyclic voltammetry could not be duplicated by chemical means. Thus the treatment of the starting material for No. 9 with excess $NOPF_6$ led to a replacement of SC_6H_5 by NO^+ to give $[C_5H_5Fe(P(C_6H_5)_2CH_2-)_2NO][PF_6]_2$ [2], see 1.5.1.8, p. 48.

$[C_5H_5Fe(P(C_6H_5)_2CH_2-)_2SC_6H_5]BF_4$ (Table 3, No. 10) was prepared by oxidation with H_3O^+ using a large excess of aqueous HBF_4 and a solution of $C_5H_5Fe(P(C_6H_5)_2CH_2-)_2SC_6H_5$ in acetone. After the volume of the solution was reduced, the salt slowly crystallized on standing. Recrystallization from CH_3OH/ether gave the methanol solvate [2].

References:

[1] P.M. Treichel, K.P. Wagner, H.J. Mueh (J. Organometal. Chem. **86** [1975] C13/C16). – [2] P.M. Treichel, D.C. Molzahn, K.P. Wagner (J. Organometal. Chem. **174** [1979] 191/7). – [3] P.M. Treichel, D.A. Komar (J. Organometal. Chem. **206** [1981] 77/88).

1.5.1.5 Other Compounds with Chelating Ligands Formally of the D-X Type

The compounds in this section contain paramagnetic C_5H_5Fe units, either $C_5H_5Fe^I$ bonded to neutral dithietene (Formula I) or $[C_5H_5Fe^{III}]^{2+}$ combined with anionic ligands of the acetylacetonato or salicylidene–iminato type (Formulas II and III).

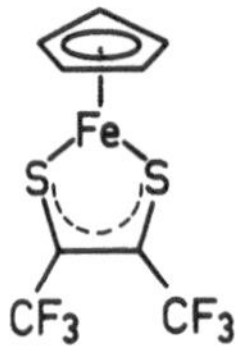

I 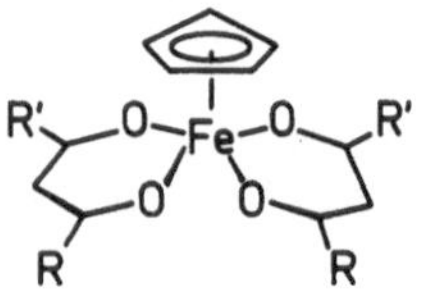   II 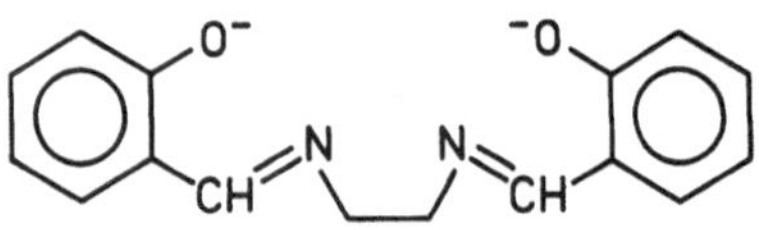   III

C$_5$H$_5$Fe(S(CF$_3$)C-C(CF$_3$)S) (Formula I) appeared to be present in a brown to black mixture of paramagnetic species obtained from (C$_5$H$_5$Fe(CO)$_2$)$_2$ and bis(trifluoromethyl)dithietene (mole ratio ~1:1.7) in C$_6$H$_6$ on lengthy UV irradiation. Pure compounds could not be isolated but the mass spectrum suggested the mixture consisted of complex I and (C$_5$H$_5$FeCO)$_2$S(CF$_3$)C-C(CF$_3$)S [2]. Without further comment on preparation and purification, reference was made to I in [3] (black, m.p. 150 °C) and in [1] (reversible two-electron reduction in dimethoxyethane at E$_p$ = −2.0 V, vs. Ag/AgClO$_4$).

A rational synthesis was later achieved by reacting (C$_5$H$_5$FeCO)$_4$ with the dithietene in refluxing toluene for 12 h and by recrystallizing the product from CH$_2$Cl$_2$/C$_6$H$_{14}$. A melting point of >300 °C was observed. The elemental analysis, the solution molecular weight (in HCON(CH$_3$)$_2$), and the IR spectrum (not reported) were consistent with the formulation above and a monomeric nature. A single one-electron reduction wave (polarography) occurred at −0.44 V (vs. SCE) [7].

C$_5$H$_5$Fe(RCO-CH-COR′)$_2$ compounds (Formula II) were prepared (Method I) by heating a ferricenium salt, [(C$_5$H$_5$)$_2$Fe]X, in C$_2$H$_5$OH with two equivalents of the β-diketone at 60 °C for 3 to 4 h. After evaporation of solvent, the residue was washed with cyclohexane to remove (C$_5$H$_5$)$_2$Fe. Chromatography in C$_6$H$_6$ on deactivated Al$_2$O$_3$ gave the pure compounds. Fe(RCO-CH-COR′)$_3$ was a by-product in all cases [4, 6]. In Method II, the [C$_5$H$_5$FeC$_6$H$_6$]$^+$ cation in aqueous solution is treated with the β-diketone at 20 °C for 3 to 4 h [5, 6]. The two methods can be represented by the following disproportionations (Ch = diketonate anion):

$$2\,[(C_5H_5)_2Fe]Ch \rightarrow C_5H_5Fe(Ch)_2 + (C_5H_5)_2Fe + 0.5\,C_{10}H_{10} \tag{1}$$
$$2\,[C_5H_5FeC_6H_6]Ch \rightarrow C_5H_5Fe(Ch)_2 + C_5H_5FeC_6H_6 + C_6H_6 \tag{2}$$

The unstable C$_5$H$_5$FeC$_6$H$_6$ formed in (2) is rapidly oxidized to the cation [6].

All compounds disproportionate on storage for long periods of time or on heating to give Fe(Ch)$_3$ and [(C$_5$H$_5$)$_2$Fe]Ch. Cyclopentadiene was produced on heating at 250 to 300 °C in evacuated tubes [6].

C$_5$H$_5$Fe(CH$_3$CO-CH-COCH$_3$)$_2$ (Formula II, R=R′=CH$_3$) was obtained by Method I in a 40 to 45% yield [6]. Method II was carried out with an excess of acetylacetone, 49% yield [5, 6].

The red crystalline substance melts at 187 °C (dec.). IR spectrum: 420, 490, 580, 595, 665, 695, 780, 795, 945, 1025 to 1033, 1200, 1280, 1300, 1379, 1425, 1530, and 1590 cm^{-1} [4, 6]. UV spectrum: $\lambda_{max}(\varepsilon)$ = 288 (24000), 350 (3200), and 430 (300) nm [6]. A figure is given in [5].

The compound is readily soluble in benzene, acetone, CH$_2$Cl$_2$, and CCl$_4$ [5].

C$_5$H$_5$Fe(C$_6$H$_5$CO-CH-COC$_6$H$_5$)$_2$ (Formula II, R=R′=C$_6$H$_5$). No yield reported for Method I, 23% yield in Method II [5, 6].

The dark red finely crystalline substance melts at 265 to 270 °C (dec.). IR spectrum: 425, 465, 535, 560, 590 to 595, 625, 635, 690, 705, 730, 760, 790, 820, 950, 1030, 1080, 1100, 1130, 1170, 1180, 1230, 1310, 1330 to 1340, 1360, 1400, 1425, 1500, 1550, 1560, and 1602 cm^{-1} [4]. UV spectrum: $\lambda_{max}(\varepsilon)$ = 255 (43000), 345 (90000), and 410 (6300) nm [6]. A figure is given in [5].

The compound is moderately soluble in acetone and CHCl$_3$ and sparingly soluble in C$_6$H$_6$ and C$_2$H$_5$OH [6]. The decomposition with Br$_2$ in CCl$_4$ affords [(C$_5$H$_5$)$_2$Fe][FeBr$_4$] [4].

$C_5H_5Fe(CF_3CO-CH-COC_4H_3S)_2$ (Formula II, $R=CF_3$, $R'=$3-thienyl). Method II was carried out by shaking aqueous $[C_5H_5FeC_6H_6]^+$ with a hexane solution of the β–diketone. A very unstable blue solid, which formed at the phase boundary, was considered to be $[C_5H_5FeC_6H_6][CF_3CO-CH-COC_4H_3S]$, which was rapidly converted to $C_5H_5Fe(CF_3CO-CH-COC_4H_3S)_2$ in polar solvents [5, 6]. No yields were given.

The red crystalline complex melts at 143 to 145 °C [4, 6]. IR spectrum: 425, 475, 493, 535, 565, 595, 600, 610, 650, 660, 693, 730, 755 (sh), 780, 795, 805, 820, 860, 945, 1005, 1030, 1065, 1110, 1130, 1140, 1162, 1180 (sh), 1200, 1235, 1258, 1310 to 1325, 1355, 1408, 1510 (sh), 1520, 1580 to 1590, and 1633 cm^{-1} [4]. UV spectrum: $\lambda_{max}(\varepsilon) = 280$ (20600), 347 (50000), and 430 (2400) nm [6]. A figure is given in [5].

$C_5H_5Fe(C_{16}H_{14}N_2O_2) \cdot Hg(C_{16}H_{14}N_2O_2)$ ($C_{16}H_{14}N_2O_2 = N,N'$-ethylene-bis(salicylideneiminato), Formula III). A product of the above composition formed when a suspension of $Hg(C_{16}H_{14}N_2O_2)$ in toluene was refluxed with freshly recrystallized $(C_5H_5Fe(CO)_2)_2$ for 2 h.

The light brown, microcrystalline, insoluble powder shows in its IR spectrum bands which are characteristic for a Schiff base. Two bands in the 830 cm^{-1} region can be assigned to the C_5H_5 ligand. The product is paramagnetic, giving $\mu_{eff} = 5.80$ B.M. at room temperature, which corresponds to the high–spin Fe^{III} state. This was confirmed by an ESR spectrum of the solid at -140 °C.

The reaction with CH_3COOH in refluxing $CH_3COOC_2H_5$ resulted in the formation of $[Fe(C_{16}H_{14}N_2O_2)]CH_3COO$ along with metallic Hg. The μ–oxo compound $[Fe(C_{16}H_{14}N_2O_2)]_2O$ was formed in refluxing pyridine [8].

References:

[1] R.E. Dessy, R.B. King, M. Waldrup (J. Am. Chem. Soc. **88** [1966] 5112/7). – [2] R.B. King, M.B. Bisnette (Inorg. Chem. **6** [1967] 469/79). – [3] J.A. McCleverty (Progr. Inorg. Chem. **10** [1968] 49/221, 200/6). – [4] O.N. Suvorova, G.A. Domrachev, G.A. Razuvaev (Dokl. Akad. Nauk SSSR **183** [1968] 850/1; Dokl. Chem. Proc. Acad. Sci. USSR **178/183** [1968] 1057/8). – [5] A.V. Pavlycheva, G.A. Domrachev, G.A. Razuvaev, O.N. Suvorova (Dokl. Akad. Nauk SSSR **184** [1969] 105/7; Dokl. Chem. Proc. Acad. Sci. USSR **184/189** [1969] 10/2).

[6] G.A. Razuvaev, G.A. Domrachev, O.N. Suvorova, L.G. Abakumova (J. Organometal. Chem. **32** [1971] 113/20). – [7] J.S. Miller (Inorg. Chem. **14** [1975] 2011/2). – [8] R. Lancashire, T.D. Smith (J. Chem. Soc. Dalton Trans. **1982** 693/8).

1.5.1.6 Compounds of the $(^2D-^5L)Fe(^2D)X$ Type

There is only one example of the title complex type (Formula II). Its singular formation involves the shift of a $C(C_6H_4)$-Fe bond (Formula I, $^5LFe(^1L-^2D)^2D$ type) to a $C(C_6H_4)-C(C_5H_4)$ bond (Formula II) within an initial $P(OC_6H_5)_3$ ligand:

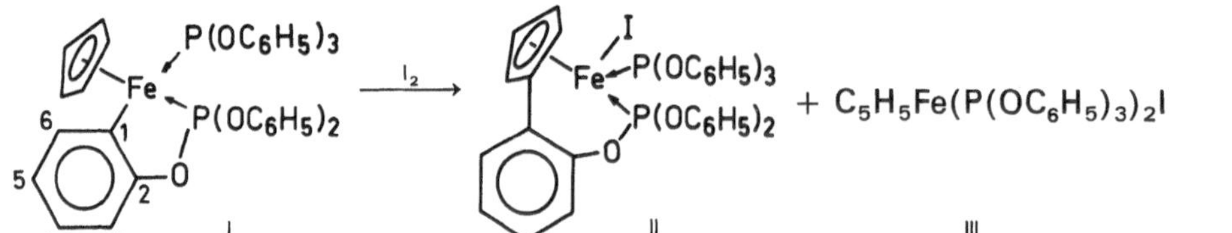

Since the starting material I is distinguished only by one H atom from the composition $C_5H_5Fe(P(OC_6H_5)_3)_2$, it was assumed to be a dimer of this unit [1 to 4]. It is described in 1.5.2.1.1, p. 50. Similarly, complex II is distinguished only by two H atoms from the normal iodide III and was thus supposed to be an isomer ("isomer II") of III [1 to 3].

$(P(OC_6H_5)_2OC_6H_4-C_5H_4-2)Fe(P(OC_6H_5)_3)I$ (Formula II) was obtained in a 28% yield along with III (17% yield) by adding an equimolar amount of I_2 in $CHCl_3$ to compound I in $CHCl_3$ at $-60\,°C$ (30 min) and further stirring at room temperature for 1 h. The two components could be separated by repeated chromatography on Al_2O_3 with benzene/petroleum ether (7:3) as eluent [2].

Complex II is a dark violet substance [3, 5] melting at 180 to 190 °C without decomposition [3, 5]; however, 172 to 174 °C with decomposition had been reported in [1]. The ^1H NMR spectrum (in $CDCl_3$) shows two symmetrical doublets of C_5H_4 protons centered at $\delta = 3.92$ and 4.27 ppm corresponding to two pairs of nonequivalent protons [2 to 5]. The IR spectrum (solid in Vaseline) is given from 690 to 2020 cm^{-1} [2].

Fig. 5

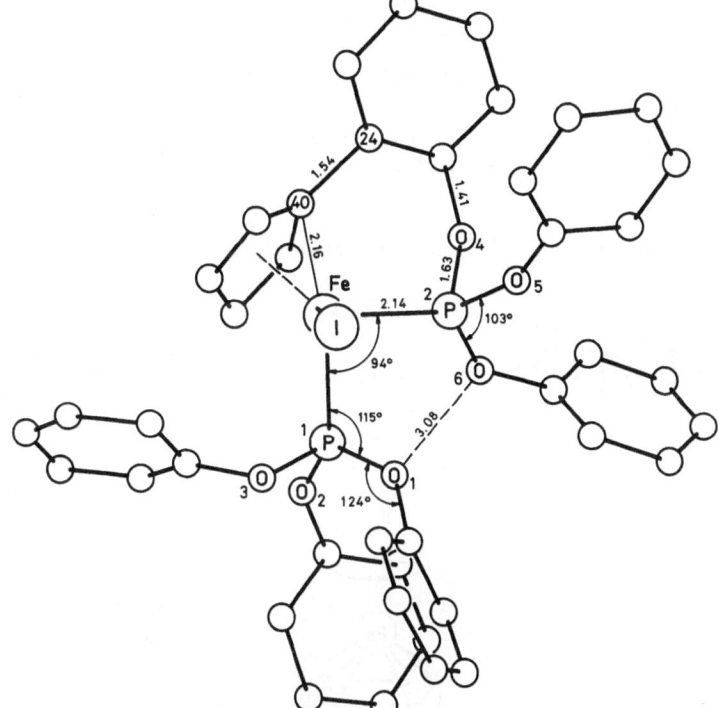

Molecular structure of $(P(OC_6H_5)_2OC_6H_4-C_5H_4-2)Fe(P(OC_6H_5)_3)I$ [4, 5].

The compound crystallizes in the orthorhombic system with $a = 20.43$, $b = 27.15$, and $c = 13.73$ (all ± 0.01) Å, space group Pbca-D_{2h}^{15}. $Z = 8$ gives $D_c = 1.52\ g \cdot cm^{-3}$, while $D_m = 1.45\ g \cdot cm^{-3}$. The molecular structure is shown in **Fig. 5**. The coordination geometry around the Fe atom is very similar to that of the iodide III (see Fig. 2 on

p. 29) in spite of the single bond C(24)–C(40) connecting two rings. This bond is nearly coplanar with the planes of the benzene and cyclopentadienyl rings, which are rotated relative to each other by 48°. The mean Fe–C(C_5H_4) distance is elongated by 0.06 Å as compared with III, indicating a weakened Fe–C_5H_4 interaction due to C–C bond formation. The shortest nonbonding O(1)···O(6) distance is distinctly greater than in the iodide III (2.73 Å). This reflects some decrease of steric hindrance between the phosphite ligands one of them being displaced towards the C_5H_4 ring because of bond formation. Other mean bond lengths and bond angles shown in Fig. 5 largely coincide with those found in III [4, 5]. Nonbonding intramolecular and intermolecular distances are given, and the molecular packing in the crystal is shown in a figure [4].

Polarographic reduction in dimethylformamide occurs at 1.31 V (vs. SCE). The compound was not converted to the iodide III by UV irradiation in boiling benzene. Small amounts of methylferrocene were obtained with $CH_3C_5H_4Na$ in refluxing THF (5 h) [1, 2].

References:

[1] A.N. Nesmeyanov, Yu.A. Chapovsky (Izv. Akad Nauk SSSR Ser. Khim. **1967** 223/4; Bull. Acad. Sci. USSR Div. Chem. Sci. **1967** 224). – [2] A.N. Nesmeyanov, Yu.A. Chapovsky, Yu.A. Ustynyuk (J. Organometal. Chem. **9** [1967] 345/53). – [3] V.G. Andrianov, Yu.T. Struchkov (Zh. Strukt. Khim. **9** [1968] 240/9; J. Struct. Chem. USSR **9** [1968] 182/90). – [4] V.G. Andrianov, Yu.T. Struchkov (Zh. Strukt. Khim. **9** [1968] 503/12; J. Struct. Chem. USSR **9** [1968] 426/34). – [5] V.G. Andrianov, Yu.A. Chapovsky, V.A. Semion, Yu.T. Struchkov (Chem. Commun. **1968** 282/4).

1.5.1.7 Neutral and Ionic Compounds with Various Borane Ligands

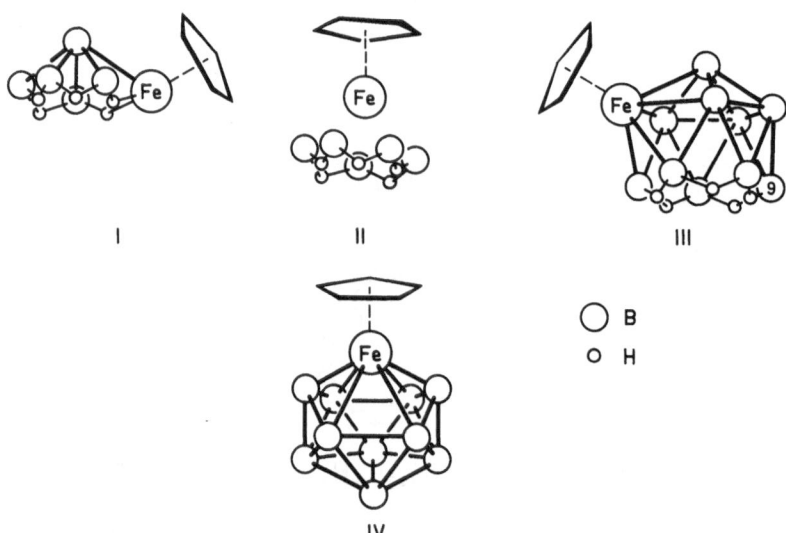

The neutral $C_5H_5FeB_mH_n$ compounds in this section are represented by the schematic structures I to III (only bridging H atoms shown). These are proposed structures based on ^1H and ^{11}B NMR spectra [1, 2]. The ionic compounds contain

$[C_5H_5FeB_{10}H_{10}]^{n-}$ anions with $n = 2$ and 3, for which the closo-1-ferra-undecaborate structure IV was assumed [3].

2-$C_5H_5FeB_5H_{10}$ (Formula I) was obtained along with smaller amounts of 2-$C_5H_5FeB_{10}H_{15}$ and large quantities of ferrocene from the reaction of $FeCl_2$ with pentaborane(9) and C_5H_5Na (approximate mole ratio 12:10:21) in THF at $-78\,°C$ (apparently no reaction) and room temperature (3 h). The usual workup by chromatography on SiO_2 with hexane eluent gave the compound in a 2.5% yield (based on B_5H_9) [1, 2]. Elution with hexane/benzene (4:1) and further development by TLC gave very little 2-$C_5H_5FeB_{10}H_{15}$ ($\sim$7 wt% of the first product). Some paramagnetic $(C_5H_5)_2FeB_8H_8$ and traces of other unidentified species were also detected [2].

The violet crystals are moderately air sensitive. The ^1H NMR spectrum (in $C_6D_5CD_3$, ^{11}B-decoupled) confirms the presence of five bridging units with $\delta(J(B,H)$ in Hz) $= -15.99$ (q, 70, Fe-H-B), -2.33 and -1.70 (broad singlets, B-H-B) and of five terminal protons at $\delta = -0.55$ (q, 141), 3.86 (q, 153), and 6.96 (q, 147) ppm. The C_5H_5 signal occurs at $\delta = 4.04$ (s) ppm. Likewise in agreement with the structure proposal I, the ^{11}B NMR (in $C_6D_5CD_3$) shows one single and two pairs of B atoms as doublets at $\delta(J(B,H)$ in Hz) $= -53.0$ (140), 8.2 (146), and 44.4 (145) ppm, respectively. The mass spectra exhibit the parent ion $[M]^+$ (electron impact) and the $[M+1]^+$ peak in the case of chemical ionization [1, 2].

1-$C_5H_5FeB_5H_{10}$ (Formula II) can be obtained from the isomer I by thermal rearrangement. Complete conversion in $C_6D_5CD_3$ requires 70 h at 175 °C and 5 h at 180 °C; small amounts of insoluble decomposition products and ferrocene were also formed.

The dark violet solid is air sensitive. The ^1H NMR spectrum consists of three resonances at $\delta = -4.52$ (s, broad, bridging H), 3.50 (s, terminal H), and 4.23 (s, C_5H_5) ppm (1:1:1 ratio). The single ^{11}B NMR resonance expected for structure II is at $\delta = 5.1$ ppm (d, $J(H,B) = 145$ Hz). The absence of a H signal in the high-field Fe-H region and of a B signal in the range of apical B-H also strongly points to occupancy of the apex position by the Fe atom [1, 2].

2-$C_5H_5FeB_{10}H_{15}$ (Formula III). The yellow compound is a minor component of the preparation of 2-$C_5H_5FeB_5H_{10}$.

The ^1H NMR spectrum (in $C_6D_5CD_3$, ^{11}B-decoupled) shows three kinds of bridging and terminal protons at $\delta = -3.46$, -2.64, -2.44 ppm (2:1:2 ratio) and $\delta = 3.54$, 4.30, 5.10 ppm (6:2:2 ratio), respectively. The ratio of the terminal protons must be produced by some coincidental superposition. $\delta(C_5H_5) = 4.92$ (s) ppm. ^{11}B NMR spectrum (in $C_6D_5CD_3$, all doublets, $J(H,B)$ in Hz): $\delta = -38.4$ (152, apex BH group), 9.7 (141), 13.4 (120), 24.2 (102), and 27.0 (104) ppm (2:3:2:2 ratio). This was interpreted as being due to four pairs of equivalent B atoms, with the unique B(9) atom (Formula III) superimposed on one of the pairs [2].

$M_2[C_5H_5FeB_{10}H_{10}]$ Compounds with $M = [N(C_2H_5)_4]^+$, K^+, and H_3O^+. Salts of the dianion (Formula IV) were prepared from $C_5H_5Fe(CO)_2I$ and $M_2[B_{10}H_{10}]$ under irradiation at room temperature (30 h) followed by chromatography on SiO_2. The yields were 43 to 45% for the $[N(C_2H_5)_4]^+$ and K^+ salts. The K^+ salt was also obtained by the thermal reaction of $C_5H_5Fe(CO)_2I$ with $K_2[B_{10}H_{10}]$ in refluxing 1,2-dimethoxyethane (8 h). Metathesis with saturated aqueous $[N(C_2H_5)_4]Cl$ gave a 35% yield of the tetraethylammonium salt. More details were not reported in the English abstract of [3].

[N(C₂H₅)₄]₂[C₅H₅FeB₁₀H₁₀] forms platelike reddish brown crystals. Only ethyl resonances were observed in the ¹H NMR spectrum: δ=1.4 (CH₃) and 3.4 (CH₂) ppm. IR spectrum: 820 and 1000 cm⁻¹ (C₅H₅), 1380 and 1460 cm⁻¹ (C₂H₅), and 2480 cm⁻¹ (terminal B–H).

The salt decomposes at 290 °C. It can be reduced with NaBH₄. In the field–desorption mass spectrum, the [N(C₂H₅)₄]⁺ and [C₅H₅FeB₁₀H₁₀]⁺ ions were observed [3].

K₂[C₅H₅FeB₁₀H₁₀] was isolated as a red sticky product. On a strongly acidic ion exchange resin it was converted to [H₃O]₂[C₅H₅FeB₁₀H₁₀], which is a strong dibasic acid, similar to H₂B₁₀H₁₀ [3].

[N(C₂H₅)₄]₃[C₅H₅FeB₁₀H₁₀] formed from [N(C₂H₅)₄]₂[C₅H₅FeB₁₀H₁₀] by re-duction with NaBH₄. ¹H NMR spectrum: δ=1.4 (CH₃), 3.4 (CH₂), and 4.78 (C₅H₅) ppm [3].

References:

[1] R. Weiss, R.N. Grimes (J. Am. Chem. Soc. **99** [1977] 8087/8). – [2] R. Weiss, R.N. Grimes (Inorg. Chem. **18** [1979] 3291/4). – [3] Lu Yi-xin, Huang Xiu-yun, Sun Cui-fang, Ding Hong-xun (Acta Chim. Sinica **40** [1982] 191/3).

1.5.1.8 Compounds Containing Nitrosyl Ligands

Na₂[C₅H₅FeNO] is formed as a red-brown solution when (C₅H₅FeNO)₂ (see C3, 2.5.2.1.1, p. 97) was reduced with Na/Hg in THF at ambient temperature for 15 min. It decomposed on attempted isolation. Attempts to bridge two Fe atoms by reactions with various halogeno compounds (CH₂Br₂, CH₂I₂, (CO)₄FeI₂, and (C₅H₅)₂TiCl₂) were unsuccessful since the redox reaction predominates, even at low temperatures, giving (C₅H₅FeNO)₂. The reaction with CH₃I at 0 °C yielded (C₅H₅Fe(NO)CH₃)₂ [1].

C₅H₅Fe(NO)I₂ was prepared by refluxing (C₅H₅FeNO)₂ and I₂ (mole ratio 1:2) in CH₂Cl₂ for 1 h. Chromatography on SiO₂ with CH₂Cl₂ eluent gave the pure compound in a 77% yield. Small amounts were also formed in the reaction of Na₂[C₅H₅FeNO] with CH₂I₂.

The black-violet crystals decompose at 140 °C. The complex sublimes in a vacuum at 110 to 120 °C with partial decomposition. ¹H NMR spectrum (CD₃COCD₃): δ=5.9 ppm. IR spectrum (KBr): ν(NO) 1835 cm⁻¹ and C₅H₅ bands at 835, 859, 1004, 1124, 1420, and 3067 cm⁻¹.

The most abundant fragments of the mass spectrum are [C₅H₅Fe(NO)I]⁺, [C₅H₅Fe]⁺, and [C₅H₅FeI]⁺. The peak of the molecular ion is very weak. All fragments are listed including several metastable peaks. The compound is stable towards air. It is soluble in CH₂Cl₂ and less soluble in C₆H₆ at low temperature [1].

C₅H₅Fe(²D)(NO)I with ²D=P(C₆H₅)₃, As(C₆H₅)₃, and Sb(C₆H₅)₃. These red complexes are prepared by passing NO into a solution of the C₅H₅Fe(²D)(CO)I com-pounds in C₆H₆ at 25 °C for 1 h (50.6, 52.3, and 58.2% yield, respectively). IR spectrum (KBr): ν(NO) 1660 to 1662 cm⁻¹. The compounds are soluble in CH₂Cl₂, CHCl₃, and CH₃COCH₃ and insoluble in hydrocarbon solvents [3].

[C₅H₅Fe(P(OC₆H₅)₃)(NO)I]PF₆ was formed by oxidation of C₅H₅-Fe(P(OC₆H₅)₃)(CO)I with NOPF₆. More was not reported [2].

[C₅H₅Fe(P(C₆H₅)₂CH₂-)₂NO][PF₆]₂ was obtained in an attempted oxidation of $C_5H_5Fe(P(C_6H_5)_2CH_2-)_2SC_6H_5$ with an excess of $NOPF_6$ in CH_3COCH_3 (5 min). The complex slowly crystallized from a methanol/ether solution of the crude oily reaction product, 55% yield.

The orange crystals decompose at 153 to 154 °C. ¹H NMR spectrum (CD_3COCD_3): $\delta = 4.0$ (m, CH_2), 6.4 (br, C_5H_5), and 7.6 to 8.3 (m, C_6H_5) ppm. IR spectrum (KBr): $\nu(NO)$ 1885 cm⁻¹ [4].

[C₅H₅Fe(²D)(NO)₂]I with ²D = $P(C_6H_5)_3$, $As(C_6H_5)_3$, and $Sb(C_6H_5)_3$ were formed when NO gas was passed for 30 min through solutions of [C₅H₅Fe(CO)₂²D]I compounds in CH_2Cl_2 (42.7, 47.4, and 53.8% yield, respectively). The IR spectra (KBr) of the pink-red complexes show two $\nu(NO)$ bands at 1633 to 1640 (s) and close to 1725 (m) cm⁻¹ [3].

References:

[1] H. Brunner, H. Wachsmann (J. Organometal. Chem. **15** [1968] 409/21). – [2] P.M. Treichel, D.C. Molzahn (J. Organometal. Chem. **86** [1975] C13/C16). – [3] V.N. Pandey (Transition Metal Chem. **2** [1977] 48/50). – [4] P.M. Treichel, D.C. Molzahn (J. Organometal. Chem. **174** [1979] 191/7).

1.5.2 Compounds with One ⁵L Ligand and Additional ¹L Ligands

This second main chapter contains a very large part of the ⁵LFe complexes since it deals with the numerous compounds containing carbonyl groups and the various other ¹L ligands combined with ligands of the X and D type. Therefore, this Chapter 1.5.2 can not be completed within the present volume but will continue at least through the next volume B 12.

The compounds of the first four sections (1.5.2.1 to 1.5.2.4) contain carbonyl groups and/or such ¹L ligands which are bonded to iron through an sp³ carbon atom (generally designated as R). These compounds are arranged by the number of carbonyl groups. The sections from 1.5.2.5 onwards will describe compounds with ¹L ligands such as thiocarbonyl, isonitrile, vinylidene, and carbene. The present volume B 11 ends with Section 1.5.2.3.9.

1.5.2.1 Compounds without Iron Carbonyl Groups

C₅H₅Fe(NO)CH₃ is the heaviest fragment in the mass spectrum of (C_5H_5-$Fe(NO)CH_3)_2$. It is present in the gas phase of the dimer as indicated by an ESR signal of the vapor quenched at liquid nitrogen temperature [1].

C₅H₅FeRu₃(CO)₉C₅H₇ (Formula I) is obtained along with $Ru_2(CO)_6C_5H_6$ and $C_5H_5Ru_3(CO)_7C_5H_7$ from refluxing $HRu_3(CO)_9C_5H_7$ and $(C_5H_5Fe(CO)_2)_2$ in $n-C_8H_{18}$ for 20 h (3% yield). The yield can be increased to 15% when the reaction is carried out in the presence of 1,3-pentadiene. The products are separated by chromatography on SiO_2 with petroleum ether/ether (9:1) as the eluent.

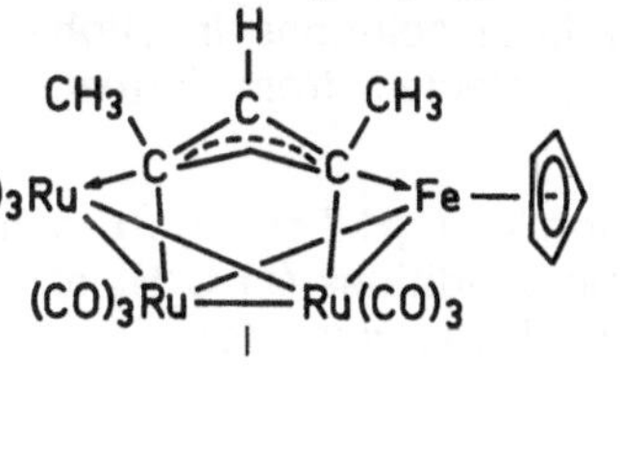

The compound is a dark brown powder. Its structure above is inferred from
^{1}H NMR, IR, and mass spectra. ^{1}H NMR (CDCl$_3$): $\delta = 3.47$ (s, CH$_3$), 4.11 (s, C$_5$H$_5$),
and 6.58 (s, CH) ppm. The IR spectrum shows ν(CO) bands at 1966 (m), 1983 (w),
1993 (m), 2024 (vs), 2034 (vs), and 2071 (s) cm^{-1}. The mass spectrum shows stepwise loss of nine CO groups from the molecular ion. Doubly charged ions corresponding to CO loss are present as intense peaks, indicating a great stability of the organometallic core [2].

References:

[1] H. Brunner, H. Wachsmann (J. Organometal. Chem. **15** [1968] 409/21). —
[2] S. Aime, D. Osella (Inorg. Chim. Acta **57** [1982] 207/10).

1.5.2.1.1 ^{5}LFe^{1}L Compounds with Various Donor Ligands

The compounds in this section are listed in Table 4. They are arranged by the
various types of donor ligands illustrated by the general Formulas I to IV. Most of
the compounds belong to the group II (Nos. 6 to 27) containing predominantly
P(C$_6$H$_5$)$_2$CH$_2$CH$_2$P(C$_6$H$_5$)$_2$ as the bidentate ligand. In the text, this frequent formula
will be abbreviated C$_{26}$H$_{24}$P$_2$. Only compounds No. 16, 19, and 21 contain another
bidentate phosphine. Nitrogen donors occur only in group III where –C=N–C=N–,
–C–C–N=N–, and –C–C–O–P– chains form a five-membered ring with the Fe atom.
The coordination IV is represented by compound No. 33 and can be described as
a 1-ferra-2,5-diphospha-bicyclo[2.1.1]hexane system.

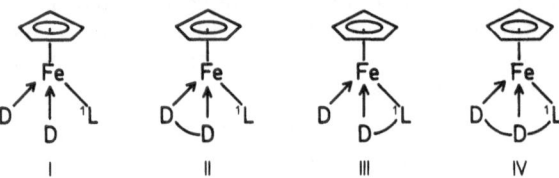

The preparative methods for the compounds in Table 4 are summarized below.
Details and special procedures are given under further information on p. 59/66. All
reactions are carried out at room temperature if not stated otherwise.

Method I: The carbonyl groups of C$_5$H$_5$Fe(CO)$_2$^{1}L and C$_5$H$_5$Fe(CO)^{1}L-^{2}D
                   complexes are replaced by phosphorus donors by irradiating the com-
                   ponents in hydrocarbon solvents.

          a. The substitution of two CO groups is a stepwise process which re-
             quires reaction times from about 6 [41] to 75 h [11, 27, 30]. Shorter
             reactions (<2.5 h) result in the formation of either pure C$_5$H$_5$-
             Fe(CO)(^{2}D)^{1}L compounds (see 1.5.2.2.12, p. 150) [27, 30] or
             mixtures of mono- and disubstituted products [37]. Benzene is a pre-
             ferred solvent for reactions with C$_{26}$H$_{24}$P$_2$ [11, 27, 30] since the
             C$_5$H$_5$Fe(CO)(C$_{26}$H$_{24}$P$_2$)^{1}L intermediates (monodentate C$_{26}$H$_{24}$P$_2$)
             precipitate from solutions in aliphatic hydrocarbons and are thus
             somewhat protected from further photochemical decarbonylation
             [11].

             Irradiation of C$_5$H$_5$Fe(CO)$_2$^{1}L complexes and P(OC$_6$H$_5$)$_3$ under
             more drastic conditions (in C$_6$H$_6$ at 80 °C or in THF at 30 to 40 °C)
             did not yield the expected C$_5$H$_5$Fe(P(OC$_6$H$_5$)$_3$)$_2$^{1}L compounds.

Instead, compounds V (No. 32) and VI were formed by photolytic cleavage of the Fe-^{1}L bond and ortho-metalation at one of the phenyl groups of the phosphite [1, 2, 13, 20]. Details are given under further information for No. 32.

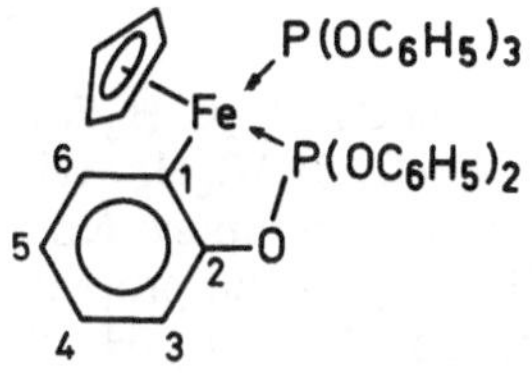

V

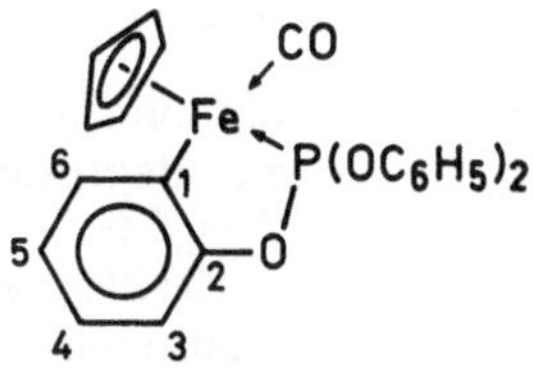

VI

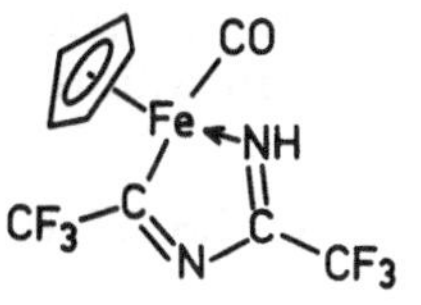

VII

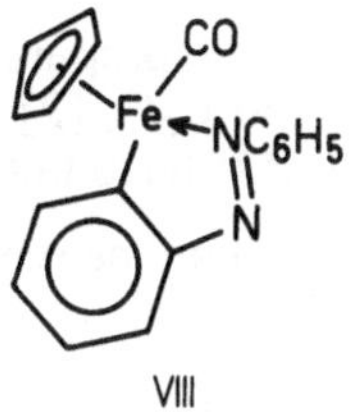

VIII

b. The $C_5H_5Fe(CO)^1L$-^2D complexes VII and VIII are reacted (VII) with $P(CH_3)_2C_6H_5$ [18], $P(C_6H_5)_3$, or $P(OC_6H_5)_3$ [6] in hexane for 12 h [6] or 10 d [18] or (VIII) with $P(C_6H_5)_3$ [12] in cyclohexane for 43 h.

Method II: The Grignard reaction is used to prepare compound type II with $C_{26}H_{24}P_2$ as the bidentate ligand.

a. $C_5H_5Fe(C_{26}H_{24}P_2)Cl$, $C_5H_5Fe(C_{26}H_{24}P_2)Br$, or $[C_5H_5Fe(C_{26}H_{24}P_2)-NCCH_3]Br$ are treated in the usual manner with RMgX reagents (X= Cl, Br, I) in THF/ether [16, 25, 26, 31, 36]. The simple Grignard substitution of $C_5H_5Fe(P(OC_6H_5)_3)_2I$ with C_6H_5MgBr in boiling THF failed but gave small amounts of the ortho-metalated V (No. 32) [2].

b. The Grignard analogue $C_5H_5Fe(C_{26}H_{24}P_2)MgBr$ is reacted with 4-$CH_3C_6H_4SO_2OCH_3$ in THF for 1 h to give compound No. 6 [31] or it is treated in THF with various halides RX (R= Cl or Br) to give compounds No. 7 to 10. The cyclo-pentylmethyl ligand of No. 7 forms in a reaction with 6-bromo-1-hexene by simultaneous cyclization [26, 31]. The benzoyl complex No. 26 was obtained with C_6H_5COCl [27]. Because of side reactions the yields are quite variable depending on the kind of the organic halide. Also see the chemical behavior of $C_5H_5Fe(C_{26}H_{24}P_2)MgBr$ in 1.5.1.3, p. 37.

Method III: This is the insertion of an alkyne into the Fe-H bond of C_5H_5-$Fe(C_{26}H_{24}P_2)H$ to give alkenyl complexes of the $C_5H_5Fe(C_{26}H_{24}P_2)$-CR=CHR type. So far, this reaction was only realized with activated triple bonds (R= CF_3 and $COOCH_3$, Nos. 22 and 27) [25].

Method IV: $[C_5H_5Fe(C_{26}H_{24}P_2)=CR_2]^+$ and $[C_5H_5Fe(C_{26}H_{24}P_2)=C=CR_2]^+$ cations (carbene and vinylidene ligands) are converted to neutral

References on p. 66

$C_5H_5Fe(C_{26}H_{24}P_2)^1L$ compounds (Nos. 11, 12, and 25) by addition of H^- to the α-carbon atom using $NaBH_4$ in THF [39] or in C_2H_5OH [27] and $Na[B(OCH_3)_3H]$ in THF [39]. In one special case (No. 3), the Fe-bonded ethyl group was formed by the addition of H^- ($NaBH_4$) to the π-bonded ethylene (^2L ligand) of the $[C_5H_5Fe(P(OC_6H_5)_3)_2-C_2H_4]^+$ cation [14].

Method V: Carbene and vinylidene ligands of $[C_5H_5Fe(C_{26}H_{24}P_2)^1L]^+$ cations are deprotonated, for $^1L=C(OH)C_6H_5$ with KOH (K_2CO_3?) in C_2H_5OH to give the benzoyl complex No. 26 [27] and for $^1L=C{=}CHCH_3$ with KOH [29] to give the propynyl complex No. 15. Deprotonation of a CH_2 group of the $C_{26}H_{24}P_2$ ligand was observed for $^1L=C{=}C(CH_3)_2$ where no hydrogen is available in α or β position, see compound No. 33.

Attempts to prepare the compounds $C_5H_5Fe(P(OC_6H_5)_3)_2R$ with $R=C_6H_5$, C_6H_4Cl-4, and C_6F_5 [2] and $C_5H_5Fe(C_{26}H_{24}P_2)R$ with $R=t$-C_4H_9, $C_6H_4CH_3$-4, and COOH [31, 33] have failed so far.

Incorrect formulas were initially given for Nos. 29, 30, and 32, see further information.

Explanations for Table 4: Free valencies in the ligand formulas of Nos. 28 to 33 indicate the bonding to the Fe atom.

Table 4
$C_5H_5Fe(^2D)_2{}^1L$ Compounds with ^1L and Donor Ligands or Chelating Ligands Combining the Two Functions.
Further information on numbers preceded by an asterisk is given at the end of the table.
For abbreviations and dimensions, see p. 375.

No.	^1L ligand method of preparation (yield in %)	donor ligand	properties and remarks	Ref.
$C_5H_5Fe(^2D)_2{}^1L$ type				
*1	CH_3 Ia (81)	$P(CH_3)_3$	deep red, m.p. 93 to 100° ^1H NMR (CS_2): −0.82 (t, $FeCH_3$, J=6), 1.20 (t, PCH_3, J=3), 3.56 (t, C_5H_5, J=6)	[41, 43]
*2	CH_3 Ia (33)	$P(OCH_3)_3$	m.p. 82° ^1H NMR (CD_3COCD_3): −0.44 (t, $FeCH_3$, J=5.5), 3.51 (t, OCH_3, J=5.4), 4.13 (t, C_5H_5, J=1.0)	[37]
*3	C_2H_5 IV (~20)	$P(OC_6H_5)_3$	yellow ^1H NMR: 1.75 (m, C_2H_5), 4.18 (s, C_5H_5), 7.1 (m, C_6H_5)	[14]
*4	C_6H_5 IIa (44)	$P(CH_3)_2C_6H_5$	red, m.p. 103 to 106° (dec.) ^1H NMR (C_6D_6): 1.08, 1.32 (t's, CH_3, J(P,H)=7.5), 4.0 (t, C_5H_5, J(P,H)=1.5), 6.90 to 7.80 (m, C_6H_5) IR (Nujol): 1560 (C_6H_5)	[32]
*5	CH_2CH_2CN see further information	$P(OC_6H_5)_3$	yellow ^1H NMR ($CDCl_3$): 1.45 (m, CH_2), 2.48 (m, CH_2), 3.95 (t, C_5H_5, J=1), 6.85 to 7.4 (C_6H_5) IR (Nujol): $\nu(CN)$ 2290	[19]
$C_5H_5Fe(^2D-^2D)^1L$ type, $R=C_6H_5$				
*6	CH_3 IIa (40, 88), IIb (31)	$(PR_2CH_2-)_2$	dark red needles, m.p. 170 to 174° ^1H NMR (CS_2): −1.63 (t, CH_3, J=6.7), 1.95 to 2.35 (m, C_2H_4), 3.90 (t, C_5H_5, J=1.5), 7.00 to 7.90 (m, C_6H_5) ^{57}Fe-γ (80 K): $\delta=0.53$, $\Delta=1.82$	[16, 25, 31, 36]

Table 4 [continued]

No.	^{1}L ligand method of preparation (yield in %)	donor ligand	properties and remarks	Ref.
*7	CH$_2$C$_5$H$_9$–cyclo IIb (27)	(PR$_2$CH$_2$–)$_2$	red; ^{1}H NMR (C$_6$D$_6$): -0.2 to $+0.20$ (m, FeCH$_2$), 0.50 to 2.40 (m, C$_2$H$_4$ and C$_5$H$_9$), 4.30 (t, C$_5$H$_5$, J=1.2), 6.90 to 7.80 (m, C$_6$H$_5$)	[26, 31]
*8	CH$_2$C$_6$H$_5$ IIa (66), IIb (34)	(PR$_2$CH$_2$–)$_2$	red; ^{1}H NMR (CS$_2$): 0.51 (t, FeCH$_2$, J=7.3), 2.15 (d, C$_2$H$_4$, J=10), 3.62 (t, C$_5$H$_5$, J=1.5), 5.85 to 7.90 (m, C$_6$H$_5$)	[26, 31]
*9	C$_4$H$_9$–n IIa ($\sim$70), IIb ($\sim$17)	(PR$_2$CH$_2$–)$_2$	red oil; ^{1}H NMR (CS$_2$): -0.60 (m, FeCH$_2$), 0.20 to 0.60 (m, CH$_2$CH$_2$CH$_3$), 1.80 to 2.30 (m, C$_2$H$_4$), 3.88 (t, C$_5$H$_5$, J=1), 6.90 to 7.70 (m, C$_6$H$_5$)	[26, 31]
*10	CH(CH$_3$)CH$_2$CH$_3$ IIb ($<$2)	(PR$_2$CH$_2$–)$_2$	no data reported	[31]
*11	CH$_2$C$_4$H$_9$–t IV ($\sim$8)	(PR$_2$CH$_2$–)$_2$	^{1}H NMR (C$_6$D$_6$): 0.84 (s, CH$_3$), 1.30 (t, FeCH$_2$, J=10), 2.00 (t, C$_2$H$_4$, J=10), 4.40 (br s, C$_5$H$_5$), 6.7 to 7.4, 7.45 to 7.85 (m's, C$_6$H$_5$)	[39]
*12	CH=C(CH$_3$)$_2$ IV (20)	(PR$_2$CH$_2$–)$_2$	^{1}H NMR (C$_6$D$_6$): 1.08, 2.05 (s's, CH$_3$), 1.7 to 2.3 (br m, C$_2$H$_4$), 4.30 (t, C$_5$H$_5$, J=1.1), 5.33 (t, FeCH, J=8.5), 7.15, 7.60 (m's, C$_6$H$_5$); ^{13}C NMR (C$_6$D$_6$): 21.9, 33.7 (s's, CH$_3$), 27.0 (t, C$_2$H$_4$, J(P,C)=21.5), 78.4 (s, C$_5$H$_5$), 127.7 (t, FeCH, J(P,C)=3.9), 128.6 (br s, C–β), 131.0 to 144.5 (m, C$_6$H$_5$)	[39]

No.	R		Properties	Ref.
*13	$CH_2(CH_2)_3CH=CH_2$ IIa (38)	$(PR_2CH_2-)_2$	red oil ^1H NMR (C_6D_6): −0.1, 0.50 to 2.70 (m, Fe(CH_2)$_4$ and C_2H_4), 4.30 (br s, C_5H_5), 4.70 to 6.30 (m, $CH=CH_2$), 7.00 to 8.10 (m, C_6H_5) IR (Nujol): ν($C=C$) 910, 1640	[26, 31]
*14	$C≡CH$ see further information	$(PR_2CH_2-)_2$	red solid ^1H NMR (C_6D_6): 1.76 (t, $C≡CH$, $J=5.5$), 2.04 (t, C_2H_4, $J=38.0$), 4.28 (t, C_5H_5, $J=1.0$), 7.4 to 8.0 (m, C_6H_5) ^{13}C NMR (C_6D_6): 68.3 (s, CH), 79.7 (s, C_5H_5), 105.7 (s, FeC), 127.4 to 143.1 (m, C_6H_5) IR: ν($C≡C$) 1925, ν(C–H) 3270	[29]
*15	$C≡CCH_3$ V	$(PR_2CH_2-)_2$	red orange, m.p. 178.5 to 180.5° ^1H NMR ($CDCl_3$): 1.60 (t, CH_3, $J=2.15$), 1.9 to 2.9 (m, C_2H_4), 4.17 (t, C_5H_5, $J=1.3$), 7.2 to 8.0 (C_6H_5) ^{13}C NMR ($CDCl_3$): 7.7 (s, CH_3), 28.4 (t, C_2H_4, $J(P,C)=22.5$), 78.8 (s, C_5H_5), 97.5 (s, C-β), 112.6 (s, FeC), 127.1 to 142.7 (m, C_6H_5) IR: ν($C≡C$) 2100 UV (CH_3CN): $\lambda_{max}(\varepsilon)$ = 475 (450), shoulder on intense UV absorption	[29, 34]
*16	C_6H_5 IIa (67)	$(P(CH_3)_2CH_2-)_2$	red, m.p. 65 to 68° ^1H NMR (C_6D_6): 0.80 to 1.35 (m, CH_3, CH_2), 4.0 (t, C_5H_5, $J=1.5$), 6.80 to 7.80 (m, C_6H_5) IR (Nujol): 1555 (C_6H_5)	[32]
*17	C_6H_5 see further information	$(PR_2CH_2-)_2$	no properties reported	[27]
*18	C_2F_5 Ia (37)	$(PR_2CH_2-)_2$	orange, m.p. 159 to 160° (dec.) ^1H NMR ($CDCl_3$): 4.33 (s, C_5H_5) ^{19}F NMR (CH_2Cl_2): 63.5 (CF_2), 79.7 (CF_3)	[11]
19	C_2F_5 Ia (80)	$PR_2CH=CHPR_2$	red brown, m.p. 140 to 141° ^{19}F NMR (CH_2Cl_2): 66.4 (CF_2), 80.3 (CF_3)	[11]

Table 4 [continued]

No.	¹L ligand method of preparation (yield in %)	donor ligand	properties and remarks	Ref.
*20	$CF(CF_3)_2$ Ia (85)	$(PR_2CH_2^-)_2$	red brown, m.p. 136° ¹H NMR (CDCl₃): 4.33 (s, C_5H_5) ¹⁹F NMR (CH_2Cl_2): 65.1 (d, CF_3, J=12), δ(CF) not observed	[11]
21	$CF(CF_3)_2$ Ia (87)	$PR_2CH=CHPR_2$	red purple, m.p. 140° (dec.) ¹⁹F NMR (CH_2Cl_2): 64.0 (br, CF_3), δ(CF) not observed	[11]
*22	$C(CF_3)=CHCF_3$-cis III (48)	$(PR_2CH_2^-)_2$	red purple, m.p. 174 to 177° ¹H NMR (C_6D_6): 1.85 to 2.4 (m, C_2H_4), 4.24 (s, C_5H_5), 4.84 (q, CH, J∼10.3), 6.8 to 7.6 (m, C_6H_5) IR (KBr): ν(C=C) 1545	[25]
*23	CH_2OCH_3 Ia (19)	$(PR_2CH_2^-)_2$	orange red, m.p. 150° (dec.) ¹H NMR (CDCl₃): 2.13 (s), 2.20 (br), 3.80 (m), 3.92 (t), 7.30 (m, C_6H_5)	[30]
*24	$CH_2OCH_2CH_3$ method not described	$(PR_2CH_2^-)_2$	no properties reported	[38, 40]
*25	$CH(C_6H_5)OCH_3$ IV (68)	$(PR_2CH_2^-)_2$	red, m.p. 155 to 158° (dec.) ¹H NMR (C_6D_6): 2.15 (s, CH_3O), 2.20 to 3.20 (m, C_2H_4 and CH), 4.00 (t, C_5H_5, J=1.0), 6.40 to 8.00 (m, C_6H_5) IR (Nujol): δ(C−O−CH_3) 1100	[27]
*26	COC_6H_5 Ia, IIb (45), V (79)	$(PR_2CH_2^-)_2$	yellow, m.p. 194 to 196° (dec.) ¹H NMR (CD_2Cl_2): 2.10 to 3.40 (m, C_2H_4), 4.10 (t, C_5H_5, J=1), 6.10, 6.80 to 7.90 (m's, C_6H_5) ¹³C NMR (CDCl₃): 81.8 (s, C_5H_5), 122.2 to 132.6 (C_6H_5), CO not observed IR (Nujol): ν(C=O) 1510	[27]

References on p. 66

No.	$^1L-^2D$ ligand	^2D type	Properties	Ref.
*27	C(COOCH$_3$)=CHCOOCH$_3$ III (50)	(PR$_2$CH$_2$–)$_2$	dark red, dec.p. 180 to 190° ^{1}H NMR (C$_6$D$_6$): 2.19, 2.63 (m's, C$_2$H$_4$), 3.61 (s, CH$_3$O), 4.23 (s, C$_5$H$_5$), 4.76 (s, CH), 6.8 to 7.8 (m, C$_6$H$_5$) IR (KBr): ν(C=C) 1514, ν(C=O) 1693	[25]

C$_5$H$_5$Fe($^1L-^2D$)^2D type

No.	$^1L-^2D$ ligand	^2D type	Properties	Ref.
*28	–C(CF$_3$)=N–C(CF$_3$)=NH→ Ib (21)	P(CH$_3$)$_2$C$_6$H$_5$	dark green, m.p. 104 to 105° ^{1}H NMR (CDCl$_3$): 0.80, 1.60 (d's, CH$_3$, J=9.0), 4.40 (d, C$_5$H$_5$), 7.4 to 7.5 (m, C$_6$H$_5$), 8.50 (br s, NH) ^{19}F NMR (CHCl$_3$): 63.90, 68.35 (s's, CF$_3$)	[18]
*29	–C(CF$_3$)=N–C(CF$_3$)=NH→ Ib (27)	P(C$_6$H$_5$)$_3$	olive green, m.p. ~173° (dec.) ^{1}H NMR (CH$_3$COCH$_3$): 4.47 (d, C$_5$H$_5$, J=1.8), 7.39 (C$_6$H$_5$), 10.2 (br, NH) IR (KBr): ν(CN) 1479, ν(NH) 3315	[6]
*30	–C(CF$_3$)=N–C(CF$_3$)=NH→ Ib (29)	P(OC$_6$H$_5$)$_3$	green black, m.p. 133 to 134° ^{1}H NMR (CH$_3$COCH$_3$): 4.40 (d, C$_5$H$_5$, J=1.8), 7.1 to 7.4 (m, C$_6$H$_5$), 10.1 (br, NH) IR (KBr): ν(CN) 1491, ν(NH) 3340	[6]
*31	[structure: 2-methylphenyl with N=N–C$_6$H$_5$] Ib (73)	P(C$_6$H$_5$)$_3$	deep purple, m.p. 155° (dec.) ^{1}H NMR (CS$_2$): 4.12 (d, C$_5$H$_5$), 6.54 to 7.18 (C$_6$H$_4$), 7.18 (m, C$_6$H$_5$)	[12]
*32	[structure: 2-methylphenyl with O–P(OC$_6$H$_5$)$_2$]	P(OC$_6$H$_5$)$_3$	yellow to orange, greenish yellow, m.p. 118 to 120°, 132 to 133° ^{1}H NMR (CCl$_4$): 4.08 (t, C$_5$H$_5$, J=1.4), 6.5 to 7.3 (m, C$_6$H$_5$, C$_6$H$_4$) ^{31}P NMR (CHCl$_3$): 200.0 (d, P in the metalated ligand), 166.4 (d, P in the unmetalated ligand), J(P, P)=150	[1, 2, 15, 20, 23, 28]

see further information

Table 4 [continued]

No.	¹L ligand method of preparation (yield in %)	donor ligand	properties and remarks	Ref.
C₅H₅Fe¹L⁻²D⁻²D type				
*33	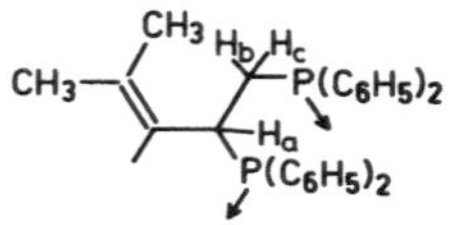 see further information		red, m.p. 190 to 191° (dec.) ¹H NMR (C₆D₆): 1.87 (H_c), 1.98 (CH₃), 2.05 (CH₃), 2.29 (H_b), 4.41 (C₅H₅), 4.47 (H_a), 6.9 to 7.2 (m, C₆H₅), 7.39, 7.56, 7.82 (ortho-H in C₆H₅) ¹³C NMR (C₆D₆): 20.9 (poor t, CH₃, J(P,C) ~J(P',C) = 4.4), 26.3 (d, CH₃, J(P,C) = 2.9), 33.3 (d, CH₂, J(P',C) = 35.2), 56.9 (dd, CH, J(P,C) = 37.4, J(P',C) = 16.8), 74.4 (s, C₅H₅), 126.3 to 147.9 (m, C₆H₅ and C=C) ³¹P NMR (C₆H₆): 42.29, 86.63 (AB, ³J(P,P) = 44.8)	[34]
supplement				
*34	C₂H₅ 1a	P(CH₃)₃	no properties reported	[44]
35	C₂H₅ 1a	P(OCH₃)₃	no properties reported	[44]

References on p. 66

*Further information:

$C_5H_5Fe(P(CH_3)_3)_2CH_3$ (Table **4**, No. **1**). Method I a in petroleum ether for 6.5 h. The complex was isolated from the crude solid by sublimation at 70 to 80 °C/0.05 Torr [41]. Irradiation of the same starting material in pentane for 1 h gave only C_5H_5-$Fe(CO)(P(CH_3)_3)CH_3$ [37]. Heating of $C_5H_5Fe(CO)_2CH_3$ and excess $P(CH_3)_3$ in refluxing toluene (15 h) did not yield No. 1 but $C_5H_5Fe(CO)(P(CH_3)_3)COCH_3$ [41].

The compound is rapidly destroyed in air. A reversible one-electron oxidation (cyclic voltammetry in CH_2Cl_2) was observed at $E_p = -0.47$ V (vs. SCE) [41].

$C_5H_5Fe(P(OCH_3)_3)_2CH_3$ (Table **4**, No. **2**). Method I a in pentane for 20 min gave comparable amounts of No. 2 and $C_5H_5Fe(CO)(P(OCH_3)_3)CH_3$, which could be separated on SiO_2 with benzene eluent. No. 2 was eluted with ether and recrystallized from pentane at -78 °C [37].

$C_5H_5Fe(P(OC_6H_5)_3)_2C_2H_5$ (Table **4**, No. **3**). Method IV with $[C_5H_5Fe$-$(P(OC_6H_5)_3)_2C_2H_4]BF_4 \cdot C_6H_6$ as the starting material and $NaBH_4$ in THF. Hydrolysis, chromatography on Al_2O_3 with petroleum ether/ether (1:1), and recrystallization from the same solvent gave the pure product [14].

$C_5H_5Fe(P(CH_3)_2C_6H_5)_2C_6H_5$ (Table **4**, No. **4**). Method II a with $C_5H_5Fe(P(CH_3)_2$-$C_6H_5)_2Br$ and C_6H_5MgBr in benzene at room temperature for 4 h. A 53% yield was obtained with a tenfold excess of the Grignard reagent and a reaction time of 3 d. Recrystallization from hexane at -30 °C.

In reactions of C_6H_5MgBr and $C_6H_5CH_2Cl$, the compound catalyzed the homo-coupling (in benzene, 4 h) to give a 46% yield of C_6H_5-C_6H_5 and a 32% yield of $C_6H_5CH_2CH_2C_6H_5$ along with small amounts of $C_6H_5CH_2C_6H_5$. Similar behavior was observed for compound No. 16. Without catalyst, this reaction gives only a small amount of $C_6H_5CH_2C_6H_5$ after 72 h [32].

$C_5H_5Fe(P(OC_6H_5)_3)_2CH_2CH_2CN$ (Table **4**, No. **5**) was prepared in a 69% yield by the reaction of $[C_5H_5Fe(P(OC_6H_5)_3)_2C_2H_4]BF_4$ with an excess of $[N(C_2H_5)_4]CN$ in CH_3CN for 1 h. The solid reaction product was extracted with C_6H_6. Addition of hexane caused the complex to precipitate slowly. It was recrystallized from benzene/hexane.

Hydrogenolysis in the presence of Pd/C catalyst gave small amounts of C_2H_5CN [19].

$C_5H_5Fe(C_{26}H_{24}P_2)CH_3$ (Table **4**, No. **6**). Method II a with $C_5H_5Fe(C_{26}H_{24}P_2)Cl \cdot$ $CHCl_3$ and a tenfold amount of CH_3MgI in THF/ether (30 min, 40% [16] or 80 to 90% yield [31]) or with $[C_5H_5Fe(C_{26}H_{24}P_2)NCCH_3]Br$ and a tenfold amount of CH_3MgBr in THF/ether (12 h, 88% yield) [36]. Method II b with C_5H_5-$Fe(C_{26}H_{24}P_2)MgBr$ and a fourfold amount of $4-CH_3C_6H_4SO_2OCH_3$ in THF (1 h). The last reaction gives an approximately equal amount of $C_5H_5Fe(C_{26}H_{24}P_2)Br$, which crystallizes first from concentrated benzene/pentane solutions. Compound No. 6 is eluted from Al_2O_3 with pentane/benzene (3:1) [31].

Cyclic voltammetry in CH_2Cl_2 showed a one-electron oxidation at $E_p = -0.26$ V (vs. SCE). Oxidation with $AgPF_6$ in acetone gave $[C_5H_5Fe(C_{26}H_{24}P_2)CH_3]PF_6$ [21, 35], see 1.5.2.1.2, p. 67. $C_5H_5Fe(C_{26}H_{24}P_2)SO_2CH_3$ was formed by insertion of SO_2 [25], see 1.5.1.3, p. 16, No. 65.

$C_5H_5Fe(C_{26}H_{24}P_2)CH_2C_5H_9$-**cyclo** (Table **4**, No. **7**). Method IIb is carried out in THF for 3 h and gives methylcyclopentane as a major by-product (40% yield) along with small amounts of 1-hexene and 1,5-hexadiene. $C_5H_5Fe(C_{26}H_{24}P_2)Br$ is the other major by-product and can be separated by crystallization from benzene. No. 7 was recrystallized from ether/hexane. Only traces of No. 7 are formed in Method IIa using $C_5H_5Fe(C_{26}H_{24}P_2)Br$ and $CH_2=CH(CH_2)_4MgBr$, also see compound No. 13. For Method IIb this indicates a mechanism via the 1-hexen-6-yl radical, which is known to cyclize rapidly to cyclopentylmethyl [26, 31].

The cleavage of No. 7 with HCl gives exclusively methylcyclopentane [31].

$C_5H_5Fe(C_{26}H_{24}P_2)CH_2C_6H_5$ (Table **4**, No. **8**). Method IIa with C_5H_5-$Fe(C_{26}H_{24}P_2)Br$ and $C_6H_5CH_2MgCl$ (mole ratio 1:1.5) in ether/THF for 2 h [31]. Method IIb with $C_5H_5Fe(C_{26}H_{24}P_2)MgBr$ and $C_6H_5CH_2Cl$ in THF for 1 h also gave appreciable amounts of toluene and dibenzyl as by-products. The yield of No. 8 decreased to 9% when $C_6H_5CH_2Br$ was used, and the amount of the by-product dibenzyl (from a radical reaction, see No. 7) increased sharply (44%) [26, 31]. Compound No. 8 was eluted from Al_2O_3 with pentane and was recrystallized from CH_2Cl_2/hexane [31].

No. 8 reacts with benzyl halides to give dibenzyl, but this reaction results in only ~5% conversion after 4 d at room temperature [31].

$C_5H_5Fe(C_{26}H_{24}P_2)R$ (R=n-C_4H_9 and i-C_4H_9, Table **4**, Nos. **9** and **10**). The yields in Method IIb (in THF, 3 h at room temperature) are low because of radical reactions, which produce much butane and, in the case of No. 10, various butenes. The main product, $C_5H_5Fe(C_{26}H_{24}P_2)Br$, can be separated by crystallization from benzene. Compound No. 9 was obtained by Method IIa (from the bromide and n-C_4H_9MgBr in THF/ether overnight) as an impure oil which could not be crystallized [26, 31].

$C_5H_5Fe(C_{26}H_{24}P_2)CH_2C_4H_9$-**t** (Table **4**, No. **11**) was formed by Method IV along with $C_5H_5Fe(C_{26}H_{24}P_2)H$ (mole ratio ~2:3) from $[C_5H_5Fe(C_{26}H_{24}P_2)=CH-C(CH_3)_3]BF_4$ and excess $NaBH_4$ in THF and was isolated by chromatography on SiO_2 [39].

$C_5H_5Fe(C_{26}H_{24}P_2)CH=C(CH_3)_2$ (Table **4**, No. **12**). An approximately 4:1 mixture of No. 12 and $C_5H_5Fe(C_{26}H_{24}P_2)H$ was obtained by Method IV from $[C_5H_5Fe(C_{26}H_{24}P_2)=C=C(CH_3)_2]SO_3F$ and $Na[B(OCH_3)_3H]$ in THF. No. 12 was isolated by fractional crystallization. Its alkylation with $[(CH_3)_3O]BF_4$ in CH_2Cl_2 gives $[C_5H_5Fe(C_{26}H_{24}P_2)=CH-C(CH_3)_3]BF_4$ in ~60% crude yield [39].

$C_5H_5Fe(C_{26}H_{24}P_2)CH_2(CH_2)_3CH=CH_2$ (Table **4**, No. **13**). After separation of nonreacted $C_5H_5Fe(C_{26}H_{24}P_2)Br$ by crystallization from ether/pentane, No. 13 did not crystallize from pentane at −30 °C but separated as an oil contaminated with 2 to 3% of the cyclopentylmethyl complex No. 7 [26, 31]. The decomposition of the impure product with HCl in dioxane gave 1-hexene and some methylcyclopentane [31].

$C_5H_5Fe(C_{26}H_{24}P_2)C≡CH$ (Table **4**, No. **14**) was obtained from C_5H_5-$Fe(C_{26}H_{24}P_2)BF_4$ and the $LiC≡CH/NH_2CH_2CH_2NH_2$ complex in THF [29].

The propensity of the compound to undergo electrophilic attack at the β-carbon atom is remarkable. Thus protonation with HPF_6 gave $[C_5H_5Fe(C_{26}H_{24}P_2)=C=CH_2]PF_6$ [29] and methylation with CH_3SO_3F in C_6H_6 rapidly gave an equimolar mixture of cations X and XI (Scheme 2) along with small amount of IX. This was explained by nucleophilic attack (1), equilibrium (2), and further nucleophilic attack (3) [39]:

References on p. 66

Scheme 2 $2 C_5H_5Fe(C_{26}H_{24}P_2)C \equiv CH$

$$\Big\downarrow + CH_3SO_3F \qquad\qquad\qquad (1)$$

$[C_5H_5Fe(C_{26}H_{24}P_2)=C=CHCH_3]^+ \; + \; C_5H_5Fe(C_{26}H_{24}P_2)C\equiv CH \; \rightleftharpoons \qquad (2)$

IX

$$C_5H_5Fe(C_{26}H_{24}P_2)C\equiv CCH_3 \; + \; [C_5H_5Fe(C_{26}H_{24}P_2)=C=CH_2]^+$$
$$X$$

$$\Big\downarrow + CH_3SO_3F \qquad\qquad\qquad (3)$$

$$[C_5H_5Fe(C_{26}H_{24}P_2)=C=C(CH_3)_2]^+$$
$$XI$$

$C_5H_5Fe(C_{26}H_{24}P_2)C\equiv CCH_3$ (Table **4**, No. **15**). Method V was carried out in THF/ H_2O above pH 12 [29]. The compound was also obtained from $CH_3C\equiv CLi$ and excess $C_5H_5Fe(C_{26}H_{24}P_2)Cl$ in THF at room temperature (7 h). It was purified on SiO_2 with ether eluent and recrystallization from C_6H_6/C_7H_{16}, 32% yield [29, 34]. A 53% yield was reported for a similar reaction using $C_5H_5Fe(C_{26}H_{24}P_2)BF_4$ as the starting material [29].

The addition of $HPF_6 \cdot O(C_2H_5)_2$ to a benzene solution of the compound rapidly precipitates the PF_6^- salt of cation IX in Scheme 2 [29]. But the propynyl group is less basic than the ethynyl group of No. 14 [39]. For the equilibrium $C_5H_5Fe(C_{26}H_{24}P_2)C\equiv CCH_3 + H_3O^+ \rightleftharpoons [C_5H_5Fe(C_{26}H_{24}P_2)=C=CHCH_3]^+ + H_2O$ in THF/H_2O (2:1) at 20 °C, pK = 7.74±0.05 was determined by titration from either side, with aqueous HCl and KOH [29]. The reaction with CH_3SO_3F in C_6H_6 gives $[C_5H_5Fe(C_{26}H_{24}P_2)=C=C(CH_3)_2]SO_3F$ [29, 34, 39], reaction (3) in Scheme 2.

$C_5H_5Fe(P(CH_3)_2CH_2-)_2C_6H_5$ (Table **4**, No. **16**). Preparative Method IIa with $C_5H_5Fe(P(CH_3)_2CH_2-)_2Br$ and excess C_6H_5MgBr in ether/THF overnight. Purification on SiO_2 with hexane eluent and crystallization from ether/hexane [32]. For a catalytic action, see compound No. 4.

$C_5H_5Fe(C_{26}H_{24}P_2)C_6H_5$ (Table **4**, No. **17**) forms along with the benzoyl complex No. 26 in the second slow substitution step of Method Ia, but its isolation was not described. In contrast to No. 26, it does not decompose during chromatography on Al_2O_3 or SiO_2 [27].

$C_5H_5Fe(C_{26}H_{24}P_2)R$ (R = C_2F_5 and $CF(CF_3)_2$, Table **4**, Nos. **18** and **20**). The yields given in Table 4 resulted from Method Ia carried out in benzene solution. Much lower yields (13% No. 18 and 36% No. 20) were obtained in hexane due to precipitation of intermediate $C_5H_5Fe(CO)(C_{26}H_{24}P_2)R$ complexes [11].

$C_5H_5Fe(C_{26}H_{24}P_2)C(CF_3)=CHCF_3$-cis (Table **4**, No. **22**). In Method III, the compound slowly precipitates during the reaction in pentane. It was recrystallized from ether/hexane. The crystals are stable to air.

^{19}F NMR spectrum (CD_3COCD_3): δ = 101.4 (m, β-CF_3) and 107.6 (pseudo quartet, α-CF_3) ppm downfield from C_6F_6. The coupling constant J(F,F) = 14 Hz ascertaines the cis configuration [25].

$C_5H_5Fe(C_{26}H_{24}P_2)CH_2OCH_3$ (Table **4**, No. **23**) was obtained by Method Ia in C_6H_6 at ~0 °C. Fine crystals resulted from recrystallizations from benzene and Skelly B.

References on p. 66

The IR spectrum (CDCl$_3$) is given from 1060 to 3060 cm^{-1}. The reaction with cyclohexene in the presence of HBF$_4$ produced a 23% yield of norcarane (Formula XII) when it was conducted in acetic anhydride at $-20\,^\circ$C. Attempts to prepare the $[C_5H_5Fe(C_{26}H_{24}P_2)=CH_2]^+$ cation from No. 23 by treatment with HBF$_4$ in acetic anhydride failed and gave instead a light brown solid which evolved CH$_2$=CH$_2$ as solid and in polar solvents. Dissolution of the solid in CH$_3$CN and treatment with $[B(C_6H_5)_4]^-$ in anhydrous CH$_3$OH produced $[C_5H_5Fe(C_{26}H_{24}P_2)NCCH_3][B(C_6H_5)_4]$ [30], see 1.5.1.2, p. 2, No. 40.

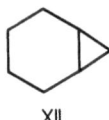

XII

C$_5$H$_5$Fe(C$_{26}$H$_{24}$P$_2$)CH$_2$OCH$_2$CH$_3$ (Table 4, No. 24) produces a dark red solution of the $[C_5H_5Fe(C_{26}H_{24}P_2)=CH_2]^+$ cation when it is reacted with CF$_3$COOH or CF$_3$SO$_3$H in CD$_2$Cl$_2$ or CD$_2$Cl$_2$/SO$_2$ at $-78\,^\circ$C [38, 40].

C$_5$H$_5$Fe(C$_{26}$H$_{24}$P$_2$)CH(C$_6$H$_5$)OCH$_3$ (Table 4, No. 25). A suspension of $[C_5H_5$-Fe(C$_{26}$H$_{24}$P$_2$)=C(C$_6$H$_5$)OCH$_3$]BF$_4$ in C$_2$H$_5$OH was used in Method IV. The product was recrystallized from ether/hexane.

The crystals decompose in air within a few hours. Rapid decomposition on SiO$_2$ or Al$_2$O$_3$ was observed. A yellow–green product which resulted from the treatment with HPF$_6$ did not appear to be a carbene compound (^{1}L=CHC$_6$H$_5$) [27].

C$_5$H$_5$Fe(C$_{26}$H$_{24}$P$_2$)COC$_6$H$_5$ (Table 4, No. 26) was also obtained in a 64% yield by Method Ia by using the C$_5$H$_5$Fe(CO)(C$_{26}$H$_{24}$P$_2$)COC$_6$H$_5$ intermediate (monodentate diphosphine ligand) and irradiating for 18 h in C$_6$H$_6$. Diffusion of ether into a concentrated solution of the complex in CH$_2$Cl$_2$ afforded crystals suitable for an X-ray study.

The monoclinic crystals have the parameters $a=10.801(4)$, $b=22.186(3)$, $c=13.779(2)$ Å, and $\beta=112.13(3)^\circ$; space group P2$_1$/n $-$ C$_{2h}^5$ and Z$=4$. The molecular structure is shown in **Fig. 6**. The angles around the C(1) atom are strongly distorted: Fe-C(1)-C(2) 120.8°, Fe-C(1)-O 126.1°, and C(2)-C(1)-O 112.9°. This distortion along with an Fe-C(1) distance shorter, a C(1)-O distance longer than usual, and the extremely low-energy ν(C=O) band indicate a contribution of a carbene bonding system, Fe$^+$=C(C$_6$H$_5$)-O$^-$. The corresponding decrease of conjugation between the carbonyl group and the phenyl ring is expressed by the dihedral angle C(1)-O/C(2)-C(3) $=61.2^\circ$.

The complex is stable in air for several hours, but it rapidly decomposes on Al$_2$O$_3$ or SiO$_2$ during chromatography. Protonation with HBF$_4$ in CH$_3$OH gives $[C_5H_5Fe(C_{26}H_{24}P_2)=C(OH)C_6H_5]BF_4$, which can be converted back to No. 26 with alkali in alcohol. Alkylation with $[(CH_3)_3O]BF_4$ in CH$_2$Cl$_2$ leads to $[C_5H_5Fe(C_{26}H_{24}P_2)=C(OCH_3)C_6H_5]BF_4$ [27].

C$_5$H$_5$Fe(C$_{26}$H$_{24}$P$_2$)C(COOCH$_3$)=CHCOOCH$_3$ (Table 4, No. 27) was isolated (Method III) by chromatography on Al$_2$O$_3$ and recrystallization from CHCl$_3$/C$_6$H$_{14}$. A single isomer was found, but its configuration at the C=C bond could not be determined.

The complex is air stable. The product of the thermal decomposition melts at 200 to 201.5 °C [25].

References on p. 66

Fig. 6

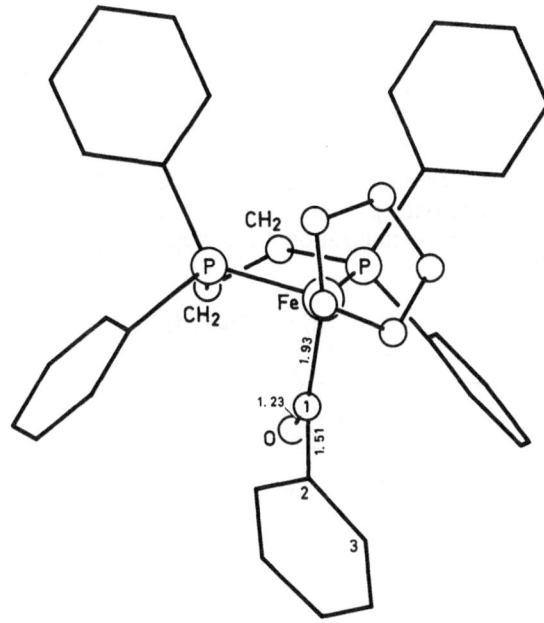

Molecular structure of $C_5H_5Fe(P(C_6H_5)_2CH_2-)_2COC_6H_5$ [27].

$C_5H_5Fe(^2D)C(CF_3)=NC(CF_3)=NH$ ($^2D = P(CH_3)_2C_6H_5$, $P(C_6H_5)_3$,   and
$P(OC_6H_5)_3$, Table **4**, Nos. **28** to **30**). Nos. 29 and 30 were first formulated incorrectly
as having two nonchelating ligands derived from CF_3CN [6]. An X-ray analysis of
the starting material, $C_5H_5Fe(CO)C(CF_3)=NC(CF_3)=NH$, revealed the true nature of
the ligand [18], see Formula VII on p. 51. The IR spectra are completely reported
in [6, 18].

$C_5H_5Fe(P(C_6H_5)_3)C_{12}H_9N_2$ (Table **4**, No. **31**) was purified on SiO_2 with petroleum ether/ether (7:3) eluent. The IR spectrum (CS_2) is completely given [12].

$C_5H_5Fe(C_6H_4OP(OC_6H_5)_2)P(OC_6H_5)_3$ (Table **4**, No. **32**) was originally formulated as a $(C_5H_5Fe(P(OC_6H_5)_3)_2)_2$ dimer [1 to 4, 13, 15, 17, 22], but it is in fact
the orthometalated compound XIII, p. 64 [20]. It forms from a variety of starting
materials under the action of $P(OC_6H_5)_3$, usually by substitution of carbonyl groups
and simultaneous cleavage of Fe–R and Fe–X bonds.

The best yield (73%) was obtained by irradiating $C_5H_5Fe(CO)(P(OC_6H_5)_3)C_6H_5$
and the phosphite in C_6H_6 for 14 h [1, 2]. The readily accessible $(C_5H_5Fe(CO)_2)_2$
also supplied a good yield (63%) when irradiated with the phosphite (mole ratio
1:4) in hexane for 15 h [28]. Other photochemical formations involved about two
molar equivalents or excess phosphite and $C_5H_5Fe(CO)_2R$ compounds (mostly in
C_6H_6) where $R = CH_3$ (THF, 30 to 40 °C, 60 h, 66% yield) [20], C_6H_5 (80 °C, 9 h,
up to 65%) [1, 2, 13], C_6H_4F-4 (80 °C, 6.5 h, 35%) [13], C_6H_4F-3 (20 °C, 15.5 h,
21%) [15], C_6H_4Cl-4 (20 °C, 15 h, 43%) [2], $COCH_3$ (80 °C, 8 h, 40%) [4], 2-thienyl
(15 °C, 4 h, 60%) [22], 2-furfuryl, 2-methyl-2-furfuryl (?), 4,5-benzo-2-furfuryl [22].
Yields ranging from 8 to 60% were obtained from reactions of the phosphite with
$C_5H_5Fe(CO)_2SiR_3$ compounds ($Si(CH_3)_3$ [7], $Si(C_6H_5)_2CH_3$, $Si(C_6H_5)_3$, and
$Si(CH_3)(C_6H_5)C_{10}H_7$-1 [28]) under irradiation in hexane, cyclohexane, aromatic hy-

drocarbons, or CH_3CN for several hours. The silyl groups occur as R_3SiH or $R_3SiOSiR_3$ [28]. Compound No. 32 also forms in the preparation of $C_5H_5Fe(P(OC_6H_5)_3)_2H$ from the corresponding iodide and $NaBH_4$ in THF, by irradiation of the hydride in boiling benzene, or by treatment of the iodide with C_6H_5MgBr in boiling THF/ether (small amounts) [2]. The unstable neutral C_5H_5Fe-arene compounds, $C_5H_5FeC_6H_6$ and $C_5H_5FeC_{10}H_8$, give No. 32 along with ferrocene even at -10 to $+20\,°C$ when treated with the phosphite in THF for a few hours (27% yield) [17]. The complex was separated and purified on Al_2O_3 with petroleum ether/benzene (4:1) [2, 17] or CH_2Cl_2/C_6H_{14} (1:1) eluent [28]; recrystallization from heptane [2], CH_2Cl_2/heptane [20], or ether/hexane [28].

XIII

The ^{1}H-decoupled ^{13}C NMR spectrum ($CHCl_3$) from [23] is given below (δ in ppm, in parentheses assignment, multiplicity, and J(P,C) in Hz):

$P(OC_6H_5)_3$	$P(OC_6H_5)_2$	POC_6H_4Fe (Formula XIII)
121.6 (o, d, 5)	119.9 (o, d, 5)	110.2 (C-3, d, 15)
	121.2 (o, d, 4)	120.7, 122.7 (C-4,5, s)
124.0 (p, s)	123.5 (p, s)	143.0 (C-1, d, 11)
	124.0 (p, s)	146.0 (C-6, s)
129.2 (m, s)	129.2 (m, s)	163.2 (C-2, d, 22)
	129.5 (m, s)	
152.3 (C-O, d, 11)	152.5 (C-O, d, 12)	
	152.7 (C-O, d, 7)	

The C_5H_5 singlet was at $\delta = 80.2$ ppm [23]. The difference of the ^{31}P resonances of POC_6H_5 and POC_6H_4Fe (33.6 ppm for No. 32, see Table 4) is characteristic for orthometalation. Similar differences were found with other orthometalated complexes [24]. The IR spectrum (Vaseline) is completely reported in [2]. Bands at 798 and 1095 cm^{-1} (Nujol) were cited to be characteristic for the POC_6H_4Fe system [20]. The coordinated $P(OC_6H_5)_3$ produces strong bands at 1180, 1200, and 1225 cm^{-1} [2].

In the mass spectrum, phosphite ligands are split off before the C_5H_5 ligand [2]. The most abundant particles are the molecular ion $[M]^+$ (relative intensity 61), $[M - P(OC_6H_5)_3]^+$ (100), and $[P(OC_6H_5)_2]^+$ (74) [20, 28].

Compound No. 32 is unusually stable. Its benzene solution did not decompose on boiling in air. Heating a C_6D_{12} solution in a sealed tube to 140 °C did not markedly affect the ^{1}H NMR spectrum. No distinct wave on polarographic reduction in dimethylformamide was observed. Decomposition occurred on treatment with HCl, excess I_2, Na/Hg, or on heating in CCl_4 [2]. Smaller amounts of I_2 reacted in $CHCl_3$ at $-60\,°C$

to give two products [2, 3], which were later identified as $C_5H_5Fe(P(OC_6H_5)_3)_2I$ [5, 8, 9] and $(P(OC_6H_5)_2OC_6H_4-C_5H_4-2)Fe(P(OC_6H_5)_3)_3)I$ [8, 10], see 1.5.1.6, p. 44. A similar reaction with Br_2 yielded $C_5H_5Fe(P(OC_6H_5)_3)_2Br$ and other unidentified products [2].

$C_5H_5FeP(C_6H_5)_2CH_2CH(P(C_6H_5)_2)C=C(CH_3)_2$ (Table **4**, No. **33**) was obtained from $[C_5H_5Fe(C_{26}H_{24}P_2)=C=C(CH_3)_2]SO_3F$ by deprotonation of one methylene group within the diphosphinoethane ligand. An approximately equimolar amount of $NaN(Si(CH_3)_3)_2$ was added to a vigorously stirred suspension of the salt in ether. Filtration after 5 min, addition of hexane, concentration in vacuo, and cooling to $-15\,°C$ gave the pure compound in a 65.6% yield. The deprotonation was also accomplished with excess KOH in THF in the presence of some water (3 h).

The P,H coupling constants (in Hz, from a 270 MHz ^1H NMR spectrum) are listed below, where the labels refer to the ligand Formula XIV:

		H_a	H_b	H_c	CH_3	C_5H_5
	P	44.8	43.3	16.9	3.2, 2.2	0.7
	P′	7.3	10.1	4.6	3.0, 2.1	1.5

XIV

The well-separated apparent C_6H_5 triplets ($J(P,H) \sim 8.0$ Hz) may be due to the ortho protons of the three phenyl rings, which are held in close proximity to the sterically locked bicyclic ring system. Other coupling constants are $J(H_a,H_b) = 4.2$, $J(H_a,H_c) \sim 0$, $J(H_b,H_c) = 13.0$, and $J(H,H$ phenyl$) \sim 7.2$ Hz. H_a is on a bridgehead C atom and thus exhibits a large down-field shift. The resonance of this C atom in the ^{13}C NMR spectrum also occurs at low field, while the other shifts are in the usual range. Resonances of the olefinic carbons could not be identified. In the ^{31}P NMR spectrum, the P(2) atom (Fig. 7), which is contained in a four-membered chelate ring, probably produces the higher-field signal.

IR spectrum (mulls): 491, 527, 533, 697, 715, 742, 751, 799, 850, 1094 (broad), and 1433 cm^{-1}. All these bands are strong.

Single monoclinic crystals were grown from ether at $-20\,°C$. The crystal parameters (at 23 °C) are $a = 14.738(4)$, $b = 12.471(5)$, $c = 16.982(7)$ Å, and $\beta = 106.17(3)°$; space group $Cc-C_s^4$. $Z = 4$ gives $D_c = 1.268$ g·cm^{-3}. The molecular structure is shown in **Fig. 7**, p. 66. The Fe-P(2)-C(2)-C(3) ring is folded with a dihedral angle of 47.4° between the Fe-P(2)-C(2) and Fe-C(3)-C(2) planes. The C(2)-C(3) bond is abnormally long. The five-membered Fe-P(1)-C(1)-C(2)-P(2) ring shows less sign of strain although the P(1)-C(1)-C(2) angle is rather acute. The exocyclic double bond has a normal length, and the CH_3 groups are nearly coplanar. The Fe atom lies only 0.063 Å from the plane containing the atoms C(2) to C(6).

The mass spectrum displays a fairly strong molecular ion $[M]^+$ and the $[M-P(C_6H_5)_2]^+$ fragment. The complex reacted with $HBF_4 \cdot O(CH_3)_2$ or $CF_3SO_2OCH_3$ at room temperature, but neither product was satisfactorily characterized [34].

$C_5H_5Fe(P(CH_3)_3)_2C_2H_5$ (Table **4**, No. **34**) reacts on SiO_2 with $CHCl_3$ to give $[C_5H_5Fe(P(CH_3)_3)_3]Cl$ [44].

References on p. 66

Fig. 7

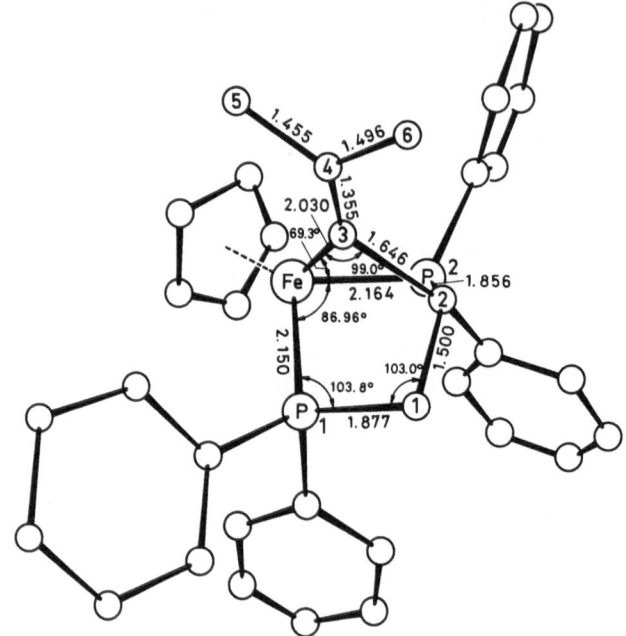

Molecular structure of $C_5H_5FeP(C_6H_5)_2CH_2CH(P(C_6H_5)_2)C=C(CH_3)_2$ [34].

Other bond angles (°):

Fe-P(2)-C(2)	88.2(2)	P(2)-C(2)-C(1)	107.8(5)
Fe-C(3)-C(4)	144.1(6)	P(2)-C(2)-C(3)	85.8(4)
P(1)-Fe-C(3)	78.1(2)		

References:

[1] A.N. Nesmeyanov, Yu.A. Chapovsky, Yu.A. Ustynyuk (Izv. Akad. Nauk SSSR Ser. Khim. **1966** 1870/1; Bull. Acad. Sci. USSR Div. Chem. Sci. **1966** 1814). – [2] A.N. Nesmeyanov, Yu.A. Chapovsky, Yu.A. Ustynyuk (J. Organometal. Chem. **9** [1967] 345/53). – [3] A.N. Nesmeyanov, Yu.A. Chapovsky (Izv. Akad. Nauk SSSR Ser. Khim. **1967** 223/4; Bull. Acad. Sci. USSR Div. Chem. Sci. **1967** 224). – [4] A.N. Nesmeyanov, Yu.A. Chapovsky (Izv. Akad. Nauk SSSR Ser. Khim. **1967** 2075/7; Bull. Acad. Sci. USSR Div. Chem. Sci. **1967** 1988/90). – [5] Yu.A. Chapovsky, V.G. Andrianov, Yu.T. Struchkov, V.A. Semion (Zh. Strukt. Khim. **8** [1967] 559/60; J. Struct. Chem. [USSR] **8** [1967] 501/2).

[6] R.B. King, K.H. Pannell (J. Am. Chem. Soc. **90** [1968] 3984/7). – [7] R.B. King, K.H. Pannell (Inorg. Chem. **7** [1968] 1510/3). – [8] V.G. Andrianov, Yu.A. Chapovsky, V.A. Semion, Yu.T. Struchkov (Chem. Commun. **1968** 282/4). – [9] V.G. Andrianov, Yu.T. Struchkov (Zh. Strukt. Khim. **9** [1968] 240/9; J. Struct. Chem. [USSR] **9** [1968] 182/90). – [10] V.G. Andrianov, Yu.T. Struchkov (Zh. Strukt. Khim. **9** [1968] 503/12; J. Struct. Chem. [USSR] **9** [1968] 426/34).

[11] R.B. King, R.N. Kapoor, K.H. Pannell (J. Organometal. Chem. **20** [1969] 187/93). – [12] M.I. Bruce, M.Z. Iqbal, F.G.A. Stone (J. Chem. Soc. A **1970** 3204/9). – [13] A.N. Nesmeyanov, L.G. Makarova, I.V. Polovyanyuk (J. Organometal.

Chem. **22** [1970] 707/12). − [14] M.L.H. Green, R.N. Whiteley (J. Chem. Soc. A **1971** 1943/6). − [15] A.N. Nesmeyanov, L.G. Makarova, I.V. Polovyanyuk (Izv. Akad. Nauk SSSR Ser. Khim. **1972** 607/9; Bull. Acad. Sci. USSR Div. Chem. Sci. **1972** 567/9).

[16] M.J. Mays, P.L. Sears (J. Chem. Soc. Dalton Trans. **1973** 1873/5). − [17] A.N. Nesmeyanov, N.A. Vol'kenau, L.S. Shilovtseva, V.A. Petrakova (J. Organometal. Chem. **61** [1973] 329/35). − [18] M. Bottrill, R. Goddard, M. Green, R.P. Hughes, M.K. Lloyd, S.H. Taylor, P. Woodward (J. Chem. Soc. Dalton Trans. **1975** 1150/5). − [19] W.H. Knoth (Inorg. Chem. **14** [1975] 1566/72). − [20] R.P. Stewart Jr., J.J. Benedict, L. Isbrandt, R.S. Ampulski (Inorg. Chem. **14** [1975] 2933/6).

[21] P.M. Treichel, K.P. Wagner (J. Organometal. Chem. **86** [1975] C13/C16). − [22] A.N. Nesmeyanov, N.E. Kolobova, L.V. Goncharenko, K.N. Anisimov (Izv. Akad. Nauk SSSR Ser. Khim. **1976** 153/9; Bull. Acad. Sci. USSR Div. Chem. Sci. **1976** 142/6). − [23] R.P. Stewart Jr., L.R. Isbrandt, J.J. Benedict, J.G. Palmer (J. Am. Chem. Soc. **98** [1976] 3215/9). − [24] R.P. Stewart Jr., L.R. Isbrandt, J.J. Benedict (Inorg. Chem. **15** [1976] 2011/3). − [25] P.M. Treichel, D.C. Molzahn (Inorg. Chim. Acta **36** [1976] 267/8).

[26] H. Felkin, B. Meunier (Nouv. J. Chim. **1** [1977] 281/2). − [27] H. Felkin, B. Meunier, C. Pascard, T. Prange (J. Organometal. Chem. **135** [1977] 361/72). − [28] G. Cerveau, E. Colomer, R. Corriu (J. Organometal. Chem. **136** [1977] 349/54). − [29] A. Davison, J.P. Selegue (J. Am. Chem. Soc. **100** [1978] 7763/5). − [30] P.E. Riley, C.E. Capshew, R. Pettit, R.E. Davis (Inorg. Chem. **17** [1978] 408/14).

[31] H. Felkin, P.J. Knowles, B. Meunier (J. Organometal. Chem. **146** [1978] 151/67). − [32] H. Felkin, B. Meunier (J. Organometal. Chem. **146** [1978] 169/78). − [33] N. Grice, S.C. Kao, R. Pettit (J. Am. Chem. Soc. **101** [1979] 1627/8). − [34] R.D. Adams, A. Davison, J.P. Selegue (J. Am. Chem. Soc. **101** [1979] 7232/8). − [35] P.M. Treichel, D.C. Molzahn, K.P. Wagner (J. Organometal. Chem. **174** [1979] 191/7).

[36] P.M. Treichel, D.C. Molzahn (Syn. Reactiv. Inorg. Metal−Org. Chem. **9** [1979] 21/9). − [37] H.G. Alt, M. Herberhold, M.D. Rausch, B.H. Edwards (Z. Naturforsch. **34b** [1979] 1070/7). − [38] M. Brookhart, J.R. Tucker, T.C. Flood, J. Jensen (J. Am. Chem. Soc. **102** [1980] 1203/5). − [39] A. Davison, J.P. Selegue (J. Am. Chem. Soc. **102** [1980] 2455/6). − [40] J.E. Jensen (Diss. Univ. Southern California 1981 from Diss. Abstr. Intern. B **42** [1981] 1019).

[41] P.M. Treichel, D.A. Komar (J. Organometal. Chem. **206** [1981] 77/88). − [42] G. Cerveau, G. Chauviere, E. Colomer, R.J.P. Corriu (J. Organometal. Chem. **210** [1981] 343/51). − [43] D.A. Komar (Diss. Univ. Wisconsin 1981 from Diss. Abstr. Intern. B **42** [1982] 2828). − [44] H.G. Alt, M.E. Eichner (Chemie Dozententagung, Kaiserslautern 1982, Abstr. B28, p. 133).

1.5.2.1.2 Compounds with a $[^5LFe(^2D-^2D)^1L]^+$ Cation

$[C_5H_5Fe(P(C_6H_5)_2CH_2CH_2P(C_6H_5)_2)CH_3]PF_6$ is the only example known at present. The cation forms by electrochemical or chemical oxidation of $C_5H_5Fe(P(C_6H_5)_2CH_2CH_2P(C_6H_5)_2)CH_3$, see 1.5.2.1.1, p. 50, No. 6. For the preparation of the salt, the neutral complex was reacted in acetone with an equimolar amount of $AgPF_6$ at room temperature for 20 min. Removal of precipitated Ag and solvent gave a crude solid, which was recrystallized from $CHCl_3$/ether [1, 2].

The brown–yellow plates melt at 155 °C with decomposition. The paramagnetism, $\mu_{eff} = 2.18$ B.M., is equivalent to one unpaired electron. Molar conductivity in CH_2Cl_2: $\Lambda_M = 62.5 \; \Omega^{-1} \cdot cm^2 \cdot mol^{-1}$ [2].

References:

[1] P.M. Treichel, K.P. Wagner (J. Organometal. Chem. **86** [1975] C13/C16). − [2] P.M. Treichel, D.C. Molzahn, K.P. Wagner (J. Organometal. Chem. **174** [1979] 191/7).

1.5.2.2 Compounds with One CO Group

1.5.2.2.1 Compounds of the $^5LFe(CO)^3D$ Type

The ligand term ^3D stands for the methyleneimino group, which frequently acts as a 3-electron donor bridging group in dinuclear complexes, see C1, 2.1.3.2.3.1, p. 127. However, it can also occur as a terminal ligand and is believed to donate three electrons to the metal atom. The bonding was discussed in earlier studies on the analogous Mo and W compounds [1, 2].

$C_5H_5Fe(CO)N=C(C_4H_9-t)_2$ was obtained in a 55% yield along with $(C_5H_5Fe(CO)_2)_2$ and decomposition products by adding $(t-C_4H_9)_2C=NLi$ to a frozen solution of $C_5H_5Fe(CO)_2Cl$ in ether and by reacting the mixture at room temperature for 30 min. The compound was isolated as a blue oil which was purified by vacuum distillation (20 °C/10^{-3} Torr).

^1H NMR spectrum (in cyclo-$C_6D_{11}CD_3$): $\delta = 1.06$ (s, CH_3) and 4.39 (C_5H_5) ppm. The single CH_3 signal was observed at all temperatures between ± 40 °C, and a linear FeNC skeleton was inferred. IR spectrum (in Nujol): $\nu(CN)$ 1610 (w) and $\nu(CO)$ 1947 (s) cm^{-1} (1953 cm^{-1} in $CHCl_3$). The IR and mass spectra confirm the mononuclear formulation.

The complex is air and light sensitive and decomposes slowly at room temperature, even under nitrogen. It has a camphor-like odor. The rapid reaction with I_2 in hexane produced an inseparable oily mixture of noncarbonyl compounds. No reaction occurred with $P(C_6H_5)_3$ in refluxing hexane over 2 d [3].

$C_5H_5Fe(CO)N=C(C_6H_5)_2$ was possibly generated in a reaction similar to that above, from $C_5H_5Fe(CO)Cl$ and $(C_6H_5)_2C=NLi$ in hexane/ether. This was indicated by a $\nu(CO)$ band at 1946 cm^{-1}, but all attempts to isolate the complex, even at low temperatures, led to its decomposition. Chromatography gave only starting material and $(C_5H_5Fe(CO)_2)_2$ [3].

References:

[1] K. Farmery, M. Kilner, C. Midcalf (J. Chem. Soc. A **1970** 2279/85). − [2] M. Kilner, C. Midcalf (J. Chem. Soc. A **1971** 292/7). − [3] M. Kilner, C. Midcalf (J. Chem. Soc. Dalton Trans. **1974** 1620/4).

1.5.2.2.2 Compounds of the $[C_5H_5Fe(CO)(CN)_2]^-$ Anion and Related Products

There is only fragmentary information on a neutral $C_5H_5Fe(CO)(CN)_2$ (formally Fe^{III}) and the $[C_5H_5Fe(CO)(CN)_2]^{2-}$ anion (formally Fe^I).

$C_5H_5Fe(CO)(CN)_2$ was mentioned as a water-soluble, readily oxidable compound. Its preparation was not described. UV spectra of the aqueous solutions (220 to 300 nm range) in the presence of $Hg(CN)_2$ indicated a rapid reversible formation of a 1:1 adduct (equilibrium constant $K = 46$ mol^{-1}) followed by an irreversible reaction. The molar absorptivity of the $C_5H_5Fe(CO)(CN)_2 \cdot Hg(CN)_2$ adduct was calculated to be $\varepsilon = 1.24 \times 10^3$ mol$^{-1} \cdot$cm^{-1}. The results suggest that the two adjacent CN groups form a cyclic system with $Hg(CN)_2$ [4]. For a discussion of the bonding, also see [5].

$K_2[C_5H_5Fe(CO)(CN)_2]$ [2] was obtained from the reaction of $(C_5H_5Fe(CO)_2)_2$ and an 100% excess of KCN (1:8 mole ratio) in refluxing CH_3OH for 15 min [1]. Its isolation and properties were not described [1, 2], except for the oxidation with Br_2, which gives the Fe^{II} complex anion treated below (reaction 3).

$K[C_5H_5Fe(CO)(CN)_2]$ can be prepared by reactions (1) to (3):

$$C_5H_5Fe(CO)_2Br + 2\,KCN \quad\quad \rightarrow K[C_5H_5Fe(CO)(CN)_2] + CO + KBr \quad\quad (1)$$

$$C_5H_5Fe(CO)_2HgCl + 2\,KCN \quad\quad \rightarrow K[C_5H_5Fe(CO)(CN)_2] + CO + Hg + KCl \quad (2)$$

$$K_2[C_5H_5Fe(CO)(CN)_2] + 0.5\,Br_2 \rightarrow K[C_5H_5Fe(CO)(CN)_2] + KBr \quad\quad (3)$$

Conditions: (1) with excess KCN in CH_3OH/H_2O under reflux for 30 min, (2) with excess KCN in C_2H_5OH/H_2O under reflux for 30 min, 65% crude yield, and (3) from starting material prepared in situ and an equimolar amount of Br_2 in CH_3OH at room temperature for 15 min, 45% crude yield. In all three methods, the reaction mixture is filtered, evaporated to dryness under reduced pressure, and the residue extracted with hot C_2H_5OH. Addition of excess ether precipitates the crude salt. Reaction (1) must be carried out in aqueous alcohol. In pure alcohol, CN$^-$ displaces only the bromine atom and gives $C_5H_5Fe(CO)_2CN$.

The salt forms golden-brown crystals on recrystallization from C_2H_5OH [1]. ^{1}H NMR spectrum (in D_2O): $\delta = 4.4$ (C_5H_5) ppm [7]. ^{13}C NMR spectrum (in H_2O/CH_3OH): $\delta = 82.6$ (C_5H_5), 159.0 (CN), and 219.2 (CO) ppm. The ^{13}CN signal is shifted 10 ppm to high field from the free CN$^-$ value. But the ^{13}CO atom is clearly less shielded than in $C_5H_5Fe(CO)_2CN$, by 8 ppm, and other $C_5H_5Fe(CO)_2X$ compounds (X = halogen) [3]. Solution IR spectra (acetone) show single bands for v(CO) at 1955 and v(CN) at 2095 cm^{-1}. In solid spectra (KBr and Nujol) these bands are split into 1950, 1970 and 2080, 2095 cm^{-1}, respectively [1, 7, 8].

The salt is insoluble in $CHCl_3$, sparingly soluble in polar organic solvents, and very soluble in water. The solid is stable in air for weeks but in hot solution it was noticeably oxidized in a few minutes. Aqueous solutions give insoluble precipitates with $AgNO_3$, $HgCl_2$, and $(CH_3)_2SnCl_2$ [1]. The $Ag[C_5H_5Fe(CO)(CN)_2]$ and $Hg[C_5H_5Fe(CO)(CN)_2]_2$ (?) salts are mentioned in [8] without further characterization. The anion was only slightly decomposed (to give $[Fe(CN)_6]^{3-}$) by aqueous KCN at 200 °C for 2 h. This indicates a remarkable strengthening of the last Fe–CO bond, presumably due to an increased negative charge at the Fe atom. $H[C_5H_5Fe(CO)(CN)_2]$ forms with concentrated hydrochloric acid [1].

The CN groups of the anion can be alkylated to give isonitrile complexes of the $C_5H_5Fe(CO)(CN)CNR$ and $[C_5H_5Fe(CO)(CNR)_2]^+$ types. The extent of alkylation depends critically on the alkylating agent, the mole ratio, and the solvent. Alkyl chlorides react very slowly and aromatic halides do not react at all [1, 6, 7]. Alkylations were carried out in refluxing CH_3CN with equimolar amounts of $CH_2{=}CHCH_2Br$,

$C_6H_5CH_2Br$, $4-XC_6H_4CH_2Br$ (X=Cl, $COOC_2H_5$), and $(C_6H_5)_3CBr$ to give neutral monoisonitrile compounds [1] and with excess CH_3I, C_2H_5I, $CH_2=CHCH_2Br$, and $C_6H_5CH_2Br$ to give diisonitrile cations [1, 6]. The yields were shown to diminish in THF or CH_2Cl_2. In CH_3OH, only $C_5H_5Fe(CO)(CN)CNR$ compounds are formed, even with excess alkylating agent. Alkylation with excess $[(C_2H_5)_3O]BF_4$ in CH_2Cl_2 gave a high yield of $[C_5H_5Fe(CO)(CNC_2H_5)_2]BF_4$ [7]. The use of FSO_3CH_3 was described in [9]. The alkylation by $C_6H_4CH_2Cl$ groups in chloromethylpolystyrene in the presence of NaI in refluxing CH_3CN fixes $C_5H_5Fe(CO)(CN)CN-$ units into the polymer to an extent of ~80% [11].

The reactions of $K[C_5H_5Fe(CO)(CN)_2]$ with various boron compounds were studied. $H[C_5H_5Fe(CO)(CN)_2]$ formed with $BF_3 \cdot O(C_2H_5)_2$ in CH_3OH. No reaction occurred with neat BCl_3 or with $B(OC_6H_5)_3$ and $B(NC_4H_4$-cyclo$)_3$ in ether or THF. Two equivalents of $BF_3 \cdot O(C_2H_5)_2$ in ether eliminated KBF_4 and produced an inseparable product containing 3 Fe and 3 B atoms (mass spectrum) and uncoordinated CN groups. This rapidly hydrolyzed to give $[C_5H_5Fe(CO)(CN)_2]^-$. Adducts were formed with $BH_3 \cdot THF$ and $B(C_6H_5)_3$, see below. Partial hydroboration of the C_5H_5 ligand was observed with more than two equivalents of $BH_3 \cdot THF$ in ether/THF. Structure I (E=B or Al) was proposed for compounds obtained with $BF_3 \cdot O(C_2H_5)_2$, $BCl_3 \cdot CH_3CN$, BBr_3, $(C_6H_5)_2BCl$, and $AlCl_3$ [8]; for details, see C4, 2.5.3.2.8, p. 230.

I

$H[C_5H_5Fe(CO)(CN)_2]$ precipitates on addition of concentrated hydrochloric acid to a cooled aqueous solution of the potassium salt [1]. It also forms from the salt and 1.5 equivalents of $BF_3 \cdot O(C_2H_5)_2$ in CH_3OH [8]. The IR spectrum of the solid (in KBr) showed one $\nu(CO)$ band at 1990 and two $\nu(CN)$ bands at 2100 and 2135 cm^{-1}. The dark yellow needles (from ethanol/ether) do not melt at 250 °C but do darken. The acid is somewhat air sensitive [1]. It forms with NH_3 the NH_4^+ salt.

$NH_4[C_5H_5Fe(CO)(CN)_2]$ precipitates from an NH_3-saturated solution of $H[C_5H_5Fe(CO)(CN)_2]$ in C_2H_5OH when excess ether is added. The salt, yellow crystals from ethanol/ether, is much more soluble in organic solvents than the potassium salt, but it is more air sensitive [1].

$K[C_5H_5Fe(CO)(CNBH_3)_2]$ could only be isolated as an impure resinous material formed from $K[C_5H_5Fe(CO)(CN)_2]$ and two equivalents of $BH_3 \cdot THF$ in THF. IR bands (Nujol) at 1980 (CO) and 2165 (CN) cm^{-1} are consistent with the presence of two $-C\equiv N-BH_3$ ligands [8].

$K[C_5H_5Fe(CO)(CNB(C_6H_5)_3)_2] \cdot 2(CH_3)_2CO$ was prepared from $K[C_5H_5$-$Fe(CO)(CN)_2]$ and $B(C_6H_5)_3$ (mole ratio $\sim 1:1$) in refluxing ether (2 h). Recrystallization of the precipitated solid from acetone/ether gave yellow crystals in an 85% yield. IR spectrum (in KBr): $\nu(CO)$ at 1970 and $\nu(CN)$ at 2160 and 2180 cm^{-1} [8].

References:

[1] C.E. Coffey (J. Inorg. Nucl. Chem. **25** [1963] 179/85). – [2] J.C. Thomas (unpublished work from [1]). – [3] L.F. Farnell, E.W. Randall, E. Rosenberg (Chem. Commun. **1971** 1078/9). – [4] E.C. Porzsolt, M.T. Beck (Proc. 3rd Conf. Coord. Chem., Bratislava, Czech., 1971, pp. 261/6). – [5] M.T. Beck, E.C. Porzsolt (J. Coord. Chem. **1** [1971] 57/66).

[6] J.A. Dineen, P.L. Pauson (J. Organometal. Chem. **43** [1972] 209/12). – [7] J.A. Dineen, P.L. Pauson (J. Organometal. Chem. **71** [1974] 77/85). – [8] J. Emri, B. Györi, A. Bakos, G. Czira (J. Organometal. Chem. **112** [1976] 325/31). – [9] M.E. Grant, J.J. Alexander (J. Coord. Chem. **9** [1979] 205/10). – [10] B.V. Johnson, W.P. Sturtzel, J.E. Shade (Inorg. Chim. Acta **32** [1979] 243/8).

[11] H. Menzel, W.P. Fehlhammer, W. Beck (Z. Naturforsch. **37b** [1982] 201/8).

1.5.2.2.3 Hydrides of the $^5LFe(CO)(Si(X,R)_3)_2H$ Type and Derived Salts

The compounds in this section are collected in Table 5 and arranged by hydrides and salts. Cyclopentadienyl and methylcyclopentadienyl occur as ^5L ligands. The preparative methods are described under further information.

Attempts to prepare analogous compounds with $Si(CH_3)Cl_2$, $Si(CH_3)_2Cl$, and $Si(CH_3)_3$ ligands by the thermal reaction given under No. 1 were not successful [6].

Table 5
Hydrides of the $^5LFe(CO)(Si(X,R)_3)_2H$ Type and Derived Salts.
Further information for numbers preceded by an asterisk is given at the end of the table.
For abbreviations and dimensions, see p. 375.

No. ^5L ligand Si(X,R)$_3$ ligand cation	properties and remarks	Ref.
hydrides		
*1 C_5H_5 $SiCl_3$	pale yellow needles, m.p. 130 to 132° subl. at 60 to 65°/0.01 Torr ^1H NMR: -12.2 (FeH), 3.7 (C_5H_5) in C_6H_6, -11.64 (FeH), 5.43 (s, C_5H_5) in CH_3CN, $^1J(^{57}Fe,H) = 14.5$, $^2J(^{29}SiFeH) = 20$ IR (C_6H_{12}, C_6H_{14}): 1960 (FeH), 2025 (CO), $\nu(^{13}CO)$ 1976	[1, 2, 4, 6]
*2 C_5H_5 $Si(CH_3)F_2$	pale yellow prismatic crystals	[7]

Table 5 [continued]

No.	^5L ligand Si(X,R)$_3$ ligand cation	properties and remarks	Ref.
*3	C$_5$H$_5$ Si(CH$_3$)$_2$C$_6$H$_5$	bond lengths (Å): Fe–Si 2.336(3), Fe–CO 1.71(1), Fe–C(C$_5$H$_5$) 2.10(1)	[8]
*4	C$_5$H$_4$CH$_3$ SiCl$_3$	needle–like white crystals, m.p. 121 to 122°, subl. at 60°/0.01 Torr ^{1}H NMR (CCl$_4$): 2.2 (CH$_3$), 5.0, 5.3 (t's, C$_5$H$_4$) IR (C$_6$H$_{14}$): 2019, 2023 (CO)	[6]

salts

No.			
5	C$_5$H$_5$ SiCl$_3$ [N(C$_2$H$_5$)$_4$]$^+$	formed from No. 1 by displacement of (CO)$_4$Fe(SiCl$_3$)H from [N(C$_2$H$_5$)$_4$]– [(CO)$_4$Fe(SiCl$_3$)] in CH$_2$Cl$_2$, not isolated	[6]
*6	C$_5$H$_5$ SiCl$_3$ [As(C$_6$H$_5$)$_4$]$^+$	yellow crystals, m.p. 142 to 145° ^{1}H NMR (CH$_3$CN): 4.60 (C$_5$H$_5$) IR (CH$_3$COCH$_3$): 1936 (CO)	[6]
*7	C$_5$H$_5$ SiCl$_3$ [C$_5$H$_5$Fe(CO)$_3$]$^+$	yellow crystals, dec. above 220° ^{1}H NMR: 4.63, 5.75 (C$_5$H$_5$) in CH$_3$CN, 4.60, 6.15 (C$_5$H$_5$) in CH$_3$COCH$_3$ IR (CH$_3$COCH$_3$): 1936 (CO, anion), 2074, 2121 (CO, cation)	[6]
*8	C$_5$H$_4$CH$_3$ SiCl$_3$ [C$_5$H$_4$CH$_3$Fe(CO)$_3$]$^+$	needle–like yellow crystals, dec. above 105° ^{1}H NMR (CD$_3$COCD$_3$): 1.35, 1.7 (s's, CH$_3$), 3.75, 4.0 (t's, C$_5$H$_4$), 5.4 (br s, C$_5$H$_4$) IR (CH$_2$Cl$_2$): 1934 (CO, anion), 2070, 2118 (CO, cation)	[6]

*Further information:

   C$_5$H$_5$Fe(CO)(SiCl$_3$)$_2$H (Table 5, No. 1) was obtained in a 35% yield along with
C$_5$H$_5$Fe(CO)$_2$SiCl$_3$ (11%) and [C$_5$H$_5$Fe(CO)$_3$][C$_5$H$_5$Fe(CO)(SiCl$_3$)$_2$] (No. 7, 53%)
by heating (C$_5$H$_5$Fe(CO)$_2$)$_2$ with excess HSiCl$_3$ at 140 °C for 30 min (sealed tube).
The yield ratio depends critically on the temperature in the 120 to 170 °C range. The
highest yield of the salt No. 7 was observed at ~130 °C. At 180 °C, C$_5$H$_5$Fe(CO)$_2$SiCl$_3$
is practically the sole product [6], also see [1]. After removal of excess HSiCl$_3$, No. 1
and C$_5$H$_5$Fe(CO)$_2$SiCl$_3$ could be extracted from the solid with pentane and were
separated by sublimation at 0.01 Torr. No. 7 was isolated by further extraction with
CH$_2$Cl$_2$ and crystallization from hot CH$_2$Cl$_2$ [6]. No. 1 was also prepared in a 37%
yield by irradiation of C$_5$H$_5$Fe(CO)$_2$SiCl$_3$ and excess HSiCl$_3$ in hexane for 4 h (substantial decomposition was observed) [4]. **C$_5$H$_5$Fe(CO)(SiCl$_3$)$_2$D** was obtained from
C$_5$H$_5$Fe(CO)$_2$SiCl$_3$ and DSiCl$_3$ by the irradiation method [6].

   The ^{1}H NMR spectrum of an acetone solution corresponds to the spectrum of
the [C$_5$H$_5$Fe(CO)(SiCl$_3$)$_2$]$^-$ anion: There is no Fe–H signal and the C$_5$H$_5$ resonance

appears at $\delta=4.6$ to 4.7 ppm (see Nos. 6 and 7). In CH_3CN some dissociation is indicated by a second weak C_5H_5 signal at $\delta=4.62$ ppm (intensity $\sim6\%$ of the main signal). The IR spectrum of an acetone solution also reflects the dissociation, $\nu(CO)$ at 1936 cm^{-1} (see Nos. 6 to 8). The $\nu(FeD)$ of the deuteride (in hexane) is at 1414 cm^{-1} [6].

The compound crystallizes in the monoclinic system with $a=7.493(5)$, $b=11.867(9)$, $c=8.651(5)$ Å, and $\beta=99.73(2)°$; space group $P2_1-C_2^2$. $Z=2$ gives $D_c=1.835$, while $D_m=1.88(2)$ g·cm^{-3}. The molecular structure is shown in **Fig. 8**. There are no unusual features. The Fe–Si distances, very similar to those of No. 2, appear to indicate some $\pi(d–d)$ contribution [3]. For a comparison with the structures of Nos. 2 and 3, also see [7].

Fig. 8

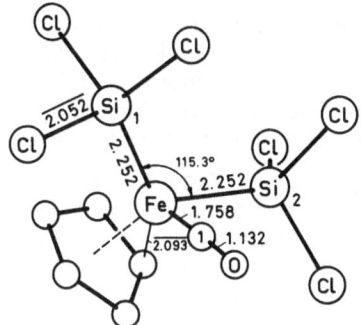

Molecular structure of $C_5H_5Fe(CO)(SiCl_3)_2H$, the H atom on Fe not located [3].

Other bond angles (°):

Si(1)–Fe–C(1)	85.1(3)	C(1)–Fe–C$_5$H$_5$(c)	125.8
Si(2)–Fe–C(1)	84.4(3)	Fe–C(1)–O	177.0(10)
Si(1)–Fe–C$_5$H$_5$(c)	119.4	Fe–Si–Cl	114.7(8) (av)
Si(2)–Fe–C$_5$H$_5$(c)	118.1	Cl–Si–Cl	103.8(3) (av)

The solid is moderately stable in air, whereas solutions are much less stable [2 to 4]. Electron impact gives $[M-Cl]^+$ as the heaviest fragment [2]. The compound dissolves in hydrocarbons. It also dissolves in CH_3CN and CH_3COCH_3 but behaves like a strong acid. In CH_3CN, $pK_a\sim2.6$ was estimated. The dissociation is complete in CH_3COCH_3 (also see ^1H NMR spectra), giving a molar conductivity $\Lambda=150$ $\Omega^{-1}\cdot$cm$^2\cdot$mol^{-1} for 10^{-3} M solutions [1, 2, 6]. Polarography in CH_3CN showed two reduction processes, the first at $E_{1/2}=-1.5$ V (vs. Ag/AgNO$_3$, 0.01 M) associated with the one-electron reduction of H^+ and the second at $E_{1/2}=-2.4$ V associated with a many-electron reduction (about 6 to 7) and decomposition of $[C_5H_5Fe(CO)(SiCl_3)_2]^-$ [5]. In CH_2Cl_2, the compound displaces the weaker $(CO)_4Fe(SiCl_3)H$ acid from its salts [6], also see No. 5.

$C_5H_5Fe(CO)(Si(CH_3)F_2)_2H$ (Table 5, No. 2). The preparation of the extremely air-sensitive compound was not described. It crystallizes in the orthorhombic system with $a=11.821(2)$, $b=14.640(2)$, and $c=7.157(2)$ Å; space group $Pnma-D_{2h}^{16}$. $Z=4$ gives $D_c=1.64$ g·cm^{-3}. The molecular structure is shown in **Fig. 9**, p. 74. The mole-

References on p. 74

Fig. 9

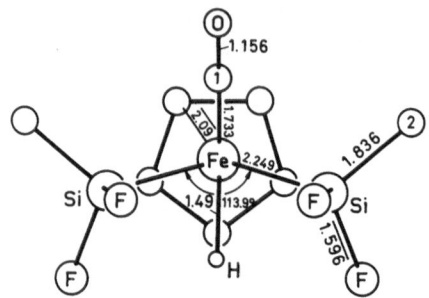

Molecular structure of $C_5H_5Fe(CO)(Si(CH_3)F_2)_2H$ [7].

cule has mirror symmetry. The Fe–bonded H atom was located by direct observation at the point where minimization of intramolecular nonbonded contacts would place it. It is 2.06(7) Å from each of the Si atoms and 2.56 to 2.57(7) Å from each of the F atoms. The Fe–H bond length given in Fig. 9 is a value corrected by 0.1 Å because of the systematic X-ray diffraction error for metal–hydrogen bonds [7].

$C_5H_5Fe(CO)(Si(CH_3)_2C_6H_5)_2H$ (Table 5, No. 3). A view of the molecular structure like that in Fig. 9 is presented in [7] and shows the lack of exact mirror symmetry due to slight rotations of the silyl ligands around the Fe–Si bonds. But the differences in rotations are small and the H atom will approximately be equidistant from two of the substituents on each silyl ligand [7].

$C_5H_4CH_3Fe(CO)(SiCl_3)_2H$　　and　　$[C_5H_4CH_3Fe(CO)_3][C_5H_4CH_3Fe(CO)$-$(SiCl_3)_2]$ (Table 5, Nos. 4 and 8) were obtained from $(C_5H_4CH_3Fe(CO)_2)_2$ and $HSiCl_3$ by the first method (115 °C for 10 h) described for No. 1. Additional CO bands in the IR spectrum of No. 4 may originate from conformational isomers [6].

$[As(C_6H_5)_4][C_5H_5Fe(CO)(SiCl_3)_2]$ (Table 5, No. 6) was prepared in a 90% yield from No. 7 and an equivalent amount of $[As(C_6H_5)_4]Cl$ in acetone, also see No. 7. The compound crystallized on slow diffusion of hexane into a CH_2Cl_2 solution. The molar conductivity is $\Lambda = 119\ \Omega^{-1} \cdot cm^2 \cdot mol^{-1}$ at 10^{-3} M [6].

$[C_5H_5Fe(CO)_3][C_5H_5Fe(CO)(SiCl_3)_2]$ (Table 5, No. 7). The preparation is described under No. 1. The compound is sparingly soluble in cold CH_2Cl_2 and very soluble in acetone. The molar conductivity in acetone, $\Lambda = 139\ \Omega^{-1} \cdot cm^2 \cdot mol^{-1}$ at $\sim 10^{-3}$ M, is consistent with a 1:1 electrolyte. No. 7 reacts with $[As(C_6H_5)_4]Cl$ in acetone to give No. 6; CO and $C_5H_5Fe(CO)_2Cl$ form from the cation [6].

References:

[1] W.A.G. Graham, W. Jetz (Abstr. Papers 155th Natl. Meeting Am. Chem. Soc., San Francisco 1968, M82). − [2] W. Jetz, W.A.G. Graham (J. Am. Chem. Soc. **91** [1969] 3375/6). − [3] L. Manojlović-Muir, K.W. Muir, J.A. Ibers (Inorg. Chem. **9** [1970] 447/52). − [4] W. Jetz, W.A.G. Graham (Inorg. Chem. **10** [1971] 4/9). − [5] J.H. Breckenridge, H.W. van den Born, W.E. Harris (Can. J. Chem. **49** [1971] 398/402).

[6] W. Jetz, W.A.G. Graham (Inorg. Chem. **10** [1971] 1159/65). − [7] R.A. Smith, M.J. Bennett (Acta Cryst. B **33** [1977] 1118/22). − [8] K.A. Simpson (Diss. Univ. Alberta 1973 from [7]).

1.5.2.2.4 Intermediate [C$_5$H$_5$Fe(CO)^{2}D]$^+$ Cations

Removal of I$^-$ from C$_5$H$_5$Fe(CO)(^{2}D)I compounds with AgBF$_4$ in CH$_2$Cl$_2$ at room temperature presumably produces the unstable 16-electron species [C$_5$H$_5$Fe(CO)^{2}D]$^+$. Their free coordination site is readily filled by Lewis bases (^{2}D') or π-electron donors like alkenes or alkynes. No attempts were made to isolate salts of the cations [1 to 3].

[C$_5$H$_5$Fe(CO)P(C$_6$H$_5$)$_3$]BF$_4$ is immediately formed as a bright green solution from the dark green C$_5$H$_5$Fe(CO)(P(C$_6$H$_5$)$_3$)I in CH$_2$Cl$_2$. The solution rapidly decomposes to give [C$_5$H$_5$Fe(CO)$_2$P(C$_6$H$_5$)$_3$]$^+$ if no incoming ligand is present or is added within a few minutes. [C$_5$H$_5$Fe(CO)(P(C$_6$H$_5$)$_3$)^{2}L]BF$_4$ compounds were obtained with 1-butene, cyclooctene, allene, 2-butyne, and 3-hexyne. Acrylonitrile and various cyanoalkenes fill the coordination site with their nitrogen lone-electron pair to give [C$_5$H$_5$Fe(CO)(P(C$_6$H$_5$)$_3$)^{2}D]BF$_4$ compounds [1, 2], see 1.5.2.2.5 below.

[C$_5$H$_5$Fe(CO)P(OC$_6$H$_5$)$_3$]BF$_4$ appears to be somewhat more stable than the previous compound. It was formed as a dark green solution from C$_5$H$_5$Fe(CO)-(P(OC$_6$H$_5$)$_3$)I over a 30-min period in CH$_2$Cl$_2$ and showed only little decomposition after 2 d at − 20 °C in the absence of air. A ν (CO) band at 2005 cm^{-1} was observed. Solutions rapidly turn brown if exposed to air. [C$_5$H$_5$Fe(CO)(P(OC$_6$H$_5$)$_3$)^{2}L]BF$_4$ compounds were prepared with ethylene, propene, 1-butene, (Z)-2-butene, cycloheptene, cyclooctene, allene, 3-hexyne, and diphenylacetylene. Treatment with (C$_2$H$_5$)$_2$NH did not result in a simple addition of the amine but produced the ortho-metalated complex I [3], see 1.5.2.2.14, p. 233.

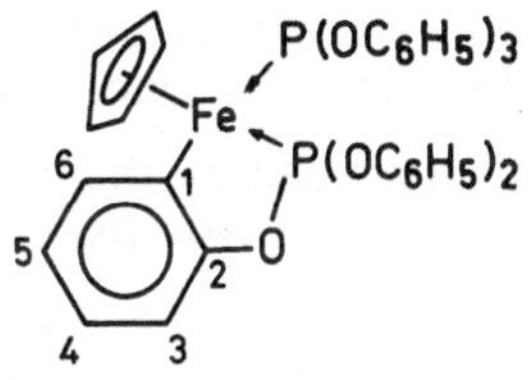

I

References:

[1] C.J. Coleman (Diss. Univ. South Carolina 1978, pp. 1/123; Diss. Abstr. Intern. B **39** [1979] 5909). − [2] D.L. Reger, C.J. Coleman, P.J. McElligott (J. Organometal. Chem. **171** [1979] 73/84). − [3] D.L. Reger, C.J. Coleman (Inorg. Chem. **18** [1979] 3155/60).

1.5.2.2.5 Compounds with [^{5}LFe(CO)(^{2}D)$_2$]$^+$, [^{5}LFe(CO)(^{2}D)^{2}D']$^+$, and [^{5}LFe(CO)^{2}D-^{2}D]$^+$ Cations

The basic structures of the compounds in this section are shown by the cation Formulas I to III. The compounds are listed in Table 6, p. 78, and are arranged by the same sequence. Most compounds contain C$_5$H$_5$ as the ^{5}L ligand (Nos. 1 to 30). Two other compounds of the type I contain C$_6$H$_7$ (cyclohexadienyl, Formula IV) and C$_7$H$_9$ (cycloheptadienyl, Formula VI) as the ^{5}L ligands. They are placed at the end of the table (Nos. 31 and 32).

In the text, the $P(C_6H_5)_2CH_2CH_2P(C_6H_5)_2$ ligand will be abbreviated $C_{26}H_{24}P_2$.

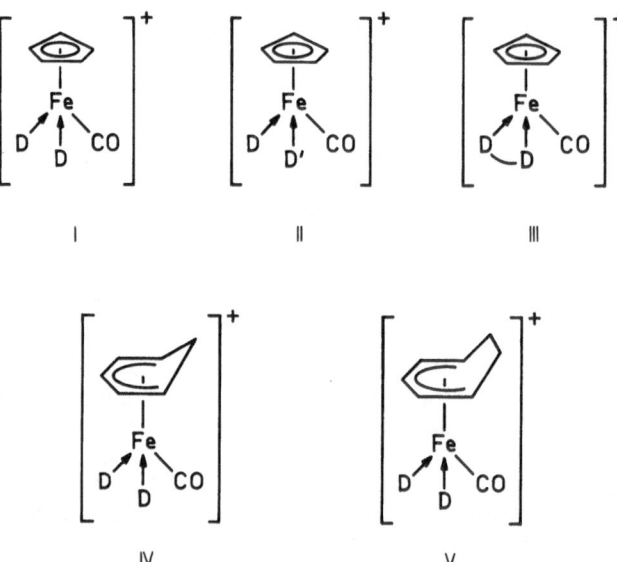

The preparative methods for the compounds in Table 6 are briefly summarized below. Details and special procedures are given under the further information on p. 82.

Method I: $[C_5H_5Fe(CO)_2{}^2D]X$ compounds are used as the starting material.

    a. One CO group is replaced by the donor in refluxing CH_3CN to give the $[C_5H_5Fe(CO)({}^2D)_2]X$ type (No. 1) [29].

    b. If the starting material contains a potential bidentate ligand bonded through only one donor site, $[C_5H_5Fe(CO)_2{}^2D(-{}^2D)]X$, intramolecular CO substitution is achieved by brief UV irradiation in CH_2Cl_2 or CH_3COCH_3 to give $[C_5H_5Fe(CO){}^2D-{}^2D]X$ compounds (Nos. 15, 16, 21, and 23) [4, 8].

    c. If the donor in the starting material is a rather good leaving group, such as THF or CH_3CN, both one CO and ${}^2D$ are replaced by $P(CH_3)_3$ in refluxing acetone to give No. 2 [29, 31] or by a bidentate donor in refluxing toluene to give $[C_5H_5Fe(CO){}^2D-{}^2D]X$ compounds (Nos. 16 and 23) [14].

Method II: Compounds of the $C_5H_5Fe(CO)_2X$ type (X = Cl and I) are used as the starting material. A ${}^2D-{}^2D$ donor replaces one CO group and displaces the halogen atom as X^- from the first coordination sphere to give $[C_5H_5Fe(CO){}^2D-{}^2D]X$ compounds. The anion may then be exchanged by Method V (Nos. 13 and 27).

    a. Reaction of $C_5H_5Fe(CO)_2Cl$ with ${}^2D-{}^2D$ under UV irradiation in THF or acetone at room temperature to give Nos. 20 and 27 [9].

    b. Reaction of a solid mixture of $C_5H_5Fe(CO)_2Cl$ and $C_{26}H_{24}P_2$ on addition of a little THF at room temperature to give No. 20 [11, 19].

References on p. 85

c. Reaction of $C_5H_5Fe(CO)_2Cl$ with bipyridyl in the presence of $AlCl_3$ in refluxing benzene to give a low yield of No. 13 [2].

d. Reaction of $C_5H_5Fe(CO)_2I$ with $^2D-^2D$ in refluxing benzene to give Nos. 28 and 29 [7, 30].

Method III: Compounds of the $C_5H_5Fe(CO)(^2D)X$ type ($^2D=P(CH_3)_3$ or $P(C_6H_5)_3$, $X = Br$ or I) are used as starting material. The displacement of the halogen atom is usually assisted by Al salts or, more conveniently, Ag salts. The anion may then be exchanged by Method V (Nos. 3, 4, and 8).

a. Reaction of $C_5H_5Fe(CO)(P(C_6H_5)_3)I$ with ^2D in the presence of AlX_3 ($X = Cl$ or Br) in benzene at room temperature (No. 4) or with a large excess of the donor in refluxing benzene (No. 8) [2].

b. Reaction of $C_5H_5Fe(CO)(P(C_6H_5)_3)I$ with $AgBF_4$ (1:1 mole ratio) and an RCN donor in CH_2Cl_2 at room temperature to give Nos. 9 to 12 [23]. The THF complex No. 6 forms when the same starting material and excess $AgBF_4$ are stirred in THF at room temperature [17].

c. $C_5H_5Fe(CO)(P(CH_3)_3)Br$ reacts with a slight excess of $P(CH_3)_3$ in refluxing toluene to give No. 3 (after anion exchange), even in the absence of a halogen-trapping agent [29].

Method IV: Elimination of I^- from $C_5H_5Fe(P(OC_6H_5)_3)_2I$ or $C_5H_5Fe(C_{26}H_{24}P_2)I$ with $AgBF_4$ or $AgPF_6$ in benzene at room temperature and reaction of the intermediate complex cation with CO (Nos. 5, 23, and 25) [6]. $TlBF_4$ in acetone was also used to prepare No. 25 [16].

Method V: Exchange of anions in water or aqueous solutions. Noncoordinating complex anions are commonly substituted for the primary halides to obtain readily crystallizable salts. But this method was also used for Nos. 20 and 22 [25], see further information.

Attempts to use Method Ib to prepare a chelated complex with $P(C_6H_5)_2$-$CH_2CH_2CH_2P(C_6H_5)_2$ were unsuccessful [8].

IR spectra are completely reported for Nos. 4, 8, 13 [2], and 17 [3]. The electrical conductivities of Nos. 5 and 25 to 27 are in the range characteristic for 1:1 electrolytes [6, 9]. Compounds No. 9 to 12 were described to be stable in air [23]. The great air-sensitivity of No. 19 is caused by the anion [3]. Irradiation of Nos. 1 to 3, 20, and 23 in CH_3COCH_3, THF, or CH_3CN results in replacement of the CO ligand by the solvent molecule [11, 19, 29, 31].

Explanations for Table 6: The term special in the second column indicates other preparative methods described under further information.

References on p. 85

Table 6

Compounds with $[^5LFe(CO)(^2D)_2]^+$, $[^5LFe(CO)(^2D)^2D']^+$, and $[^5LFe(CO)^2D-^2D]^+$ Cations.

Further information on numbers preceded by an asterisk is given at the end of the table. For abbreviations and dimensions, see p. 375.

No.	donor ligands method of preparation (yield in %)	anion	properties and remarks	Ref.
$[C_5H_5Fe(CO)(^2D)_2]^+$ type				
*1	$P(CH_3)_3$ Ia (87)	I	m.p. 302 to 340° (?)	[29]
*2	$P(CH_3)_3$ Ic (88)	BF_4	yellow, m.p. >300° ^1H NMR (CD_3SOCD_3): 1.55 (m, CH_3), 5.02 (t, C_5H_5, J(P,H)=1.9) IR (KBr): 1960 (CO)	[29, 31]
*3	$P(CH_3)_3$ IIIc (67)	PF_6	m.p. >300°	[29]
*4	$P(C_6H_5)_3$ IIIa (~8)	PF_6	orange powder, darkens gradually on heating and melts at 186° ^1H NMR (CD_3COCD_3): 4.95 (t, C_5H_5, J(P,H)=1.5), 7.42 (m, C_6H_5) ^{57}Fe-γ: δ=0.146 (Fe), Δ=1.85 (295 K), δ=0.234 (steel), Δ=1.91 IR (CH_2Cl_2, Nujol): 1970 (CO), k(CO)=15.67 mdyn·Å^{-1}	[2, 22, 26]
5	$P(OC_6H_5)_3$ IV	PF_6	pale yellow, m.p. 151 to 153° ^1H NMR: 4.54 (s, C_5H_5), 7.08 to 7.42 (m, C_6H_5) IR (mull): 2020 (CO)	[6]
$[C_5H_5Fe(CO)(^2D)^2D']^+$ type				
*6	$P(C_6H_5)_3$ C_4H_8O (THF) IIIb (80)	BF_4	green, dec. >105°, m.p. 108° ^1H NMR (CD_3COCD_3): 1.80 (m, CH_2), 3.64 (m, CH_2O), 5.06 (s, C_5H_5), 7.60 (m, C_6H_5) IR ($CHCl_3$): 1987 (CO)	[17]
7	$P(OC_6H_5)_3$ C_4H_8O (THF) IIIb (?)	BF_4	only mentioned as a solid, the exchange of alkenes (^2L) for THF is slow but gives low yields of $[C_5H_5Fe(CO)(P(OC_6H_5)_3)^2L]^+$	[24]
8	$P(C_6H_5)_3$ CH_3CN IIIa (9)	PF_6	^1H NMR ($CDCl_3$): 2.22 (d, CH_3, J=1.5), 4.98 (d, C_5H_5, J=1.5), 7.55 (m, C_6H_5) IR (Nujol): 1970 (CO), other bands given from 523 to 1480	[2]

References on p. 85

Table 6 [continued]

No.	donor ligands method of preparation (yield in %)	anion	properties and remarks	Ref.
9	P(C$_6$H$_5$)$_3$ CH$_2$=CHCN IIIb (84)	BF$_4$	red, m.p. 168 to 170° ^{1}H NMR (CD$_3$COCD$_3$): 5.10 (d, C$_5$H$_5$, J=1.5), 5.90 (br, CH$_2$), 6.03 (br, CH), 7.59 (br, C$_6$H$_5$) ^{13}C NMR (CH$_2$Cl$_2$): 84.96 (s, C$_5$H$_5$), 106.69 (s, CH), 129.33 (d, C$_6$H$_5$, C-3,5, J=10.3), 130.62 (s, CN), 131.62 (s, C$_6$H$_5$, C-4), 132.42 (d, C$_6$H$_5$, C-1, J=48.4), 133.14 (d, C$_6$H$_5$, C-2,6, J=9.8), 140.79 (s, CH$_2$), 216.08 (d, CO, J=28) IR (CH$_2$Cl$_2$): 1980 (CO)	[23]
10	P(C$_6$H$_5$)$_3$ CH$_2$=CHCH$_2$CN IIIb (96)	BF$_4$	red, m.p. 180 to 182° ^{1}H NMR (CD$_3$COCD$_3$): 3.38 (m, CH$_2$CN), 5.09 (d, C$_5$H$_5$, J=1.5), 5.35 (m, CH), 5.60 (br, CH$_2$=), 7.59 (br, C$_6$H$_5$) ^{13}C NMR (CH$_2$Cl$_2$): 23.71 (s, CH$_2$CN), 85.48 (s, C$_5$H$_5$), 121.71 (s, CH$_2$=), 124.61 (s, CH), 128.26 (d, C$_6$H$_5$, C-3,5, J=10.3), 130.85 (s, CN), 131.50 (s, C$_6$H$_5$, C-4), 133.10 (d, C$_6$H$_5$, C-2,6, J=10.0), 136.46 (d, C$_6$H$_5$, C-1, J=49.0), 218.19 (d, CO, J=28.0) IR (CH$_2$Cl$_2$): 1988 (CO)	[23]
11	P(C$_6$H$_5$)$_3$ CH$_2$=C(CH$_3$)CH$_2$CN IIIb (95) BF$_4$		red, m.p. 176 to 177° ^{1}H NMR (CD$_3$COCD$_3$): 1.54 (s, CH$_3$), 3.39 (s, CH$_2$CN), 4.68, 4.82 (s's, CH$_2$=), 5.02 (d, C$_5$H$_5$, J=1.8), 7.56 (br, C$_6$H$_5$) ^{13}C NMR (CH$_2$Cl$_2$): 21.04 (s, CH$_2$CN), 28.04 (s, CH$_3$), 85.44 (s, C$_5$H$_5$), 100.21 (s, =C), 115.75 (s, CH$_2$=), 129.35 (d, C$_6$H$_5$, C-3,5, J=10.3), 131.28 (s, CN), 131.62 (s, C$_6$H$_5$, C-4), 132.26 (d, C$_6$H$_5$, C-2,6, J= 10.1), 134.78 (d, C$_6$H$_5$, C-1, J= 48.0), 210.00 (d, CO, J=28.0) IR (CH$_2$Cl$_2$): 1988 (CO)	[23]
12	P(C$_6$H$_5$)$_3$ (CH$_3$)$_2$C=CHCH$_2$CN IIIb (87) BF$_4$		red, m.p. 156 to 157.5° ^{1}H NMR (CD$_3$COCD$_3$): 1.42, 1.58 (s's, CH$_3$), 2.70 (t, CH, J=7.2),	[23]

References on p. 85

Table 6 [continued]

No.	donor ligands method of preparation (yield in %)	anion	properties and remarks	Ref.
			3.25 (d, CH_2, J=8.1), 5.03 (d, C_5H_5, J=1.8), 7.60 (br, C_6H_5) ^{13}C NMR (CH_2Cl_2): 17.58 (d, CH_2, J=0.2), 18.66, 25.53 (s's, CH_3), 84.54 (s, C_5H_5), 109.53 (s, =C), 110.58 (s, CH), 121.70 (s, CN), 129.10 (d, C_6H_5, C–3,5, J=9.8), 131.08 (d, C_6H_5, C–4, J=9.4), 134.26 (d, C_6H_5, C–2,6, J=9.6), 138.59 (d, C_6H_5, C–1, J=48.4), 210.00 (d, CO, J=28.0) IR (CH_2Cl_2): 1988 (CO)	

$[C_5H_5Fe(CO)^2D-^2D]^+$ type

No.	donor ligands method of preparation (yield in %)	anion	properties and remarks	Ref.
13	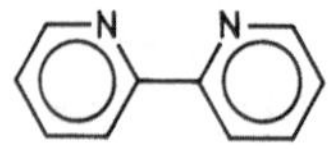 IIc	PF_6	red orange ^{1}H NMR ($CDCl_3$): 5.21 (s, C_5H_5) IR (CH_2Cl_2): 1990 (CO), other bands given from 505 to 1610	[2]
*14	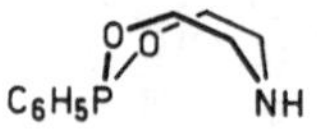 C_6H_5P—NH special	PF_6	^{31}P NMR: 208 (PC_6H_5) IR: 1975 (CO)	[28]
15	$P(C_6H_5)_2CH_2P(C_6H_5)_2$ Ib	ClO_4	^{1}H NMR (CD_3COCD_3): 5.25 (C_5H_5), 7.30 (C_6H_5) IR (CH_2Cl_2): 1982 (CO)	[8]
16	$P(C_6H_5)_2CH_2P(C_6H_5)_2$ Ib, Ic (88)	PF_6	orange yellow ^{1}H NMR (CD_3CN): 4.5, 5.1 (CH_2, AB pattern, J(A,B)=16, J(P,H)= 12), 5.08 (t, C_5H_5, J=1.4) ^{31}P NMR (CD_3CN): −8.9 (cation) IR (CH_2Cl_2): 1983 (CO)	[8, 14]
*17	$P(CH_3)_2CH_2CH_2P(CH_3)_2$ V (43)	PF_6	pale yellow, m.p. >300° IR (KBr): 1970 (CO), ν(PF) 835, other bands given from 547 to 3115	[3]
*18	$P(CH_3)_2CH_2CH_2P(CH_3)_2$ special	BF_4	pale yellow IR ($CHCl_3$): 1960 (CO)	[5]
*19	$P(CH_3)_2CH_2CH_2P(CH_3)_2$ special	$[C_5H_5Fe(CO)_2]$	orange solid IR (KBr): 1770, 1858 (CO, anion), 1970 (CO, cation)	[3]
*20	$P(C_6H_5)_2CH_2CH_2P(C_6H_5)_2$ IIa, IIb (100), V	Cl	yellow IR: 1980 (CO) ^{57}Fe–γ (80 K): δ=0.60, Δ=1.09	[9, 10, 11, 19, 25]

References on p. 85

Table 6 [continued]

No.	donor ligands method of preparation (yield in %)	anion	properties and remarks	Ref.
21	$P(C_6H_5)_2CH_2CH_2P(C_6H_5)_2$ Ib	ClO_4	^1H NMR (CD_3COCD_3): 5.10 (C_5H_5), 7.55 (C_6H_5) IR (CH_2Cl_2): 1980 (CO)	[4, 8]
*22	$P(C_6H_5)_2CH_2CH_2P(C_6H_5)_2$ V	OH	yellow needles, m.p. 89 to 91° IR: 1980 (CO)	[25]
*23	$P(C_6H_5)_2CH_2CH_2P(C_6H_5)_2$ Ib, Ic (58), IV, V	PF_6	yellow, m.p. 191 to 194° ^1H NMR: 2.5 (br, CH_2), 5.08 (t, C_5H_5, J=1.5) in CD_3CN, 2.06, 2.76 (CH_2), 5.08 (t, C_5H_5), 7.51 (m, C_6H_5) in CD_3COCD_3 ^{31}P NMR (CD_2Cl_2): 92.3 (cation) IR: 1981 (CO) in CH_2Cl_2, 1970 or 1990 (CO) in KBr	[8, 11, 14, 19, 30]
*24	$P(C_6H_5)_2CH_2CH_2P(C_6H_5)_2$ V	BH_4	yellow powder, m.p. 132 to 133° IR: 1980 (CO)	[25]
*25	$P(C_6H_5)_2CH_2CH_2P(C_6H_5)_2$ IV (~25), V	BF_4	golden yellow, m.p. 220° ^1H NMR: 4.88 (br, C_5H_5) in $CDCl_3$, 2.58 to 3.07 (m, CH_2), 4.83 (s, C_5H_5), 7.30 to 7.58 (m, C_6H_5), solvent not given IR: 1973 in mull, 1980 in CH_2Cl_2 (CO)	[6, 9, 16]
26	$P(C_6H_5)_2CH_2CH_2P(C_6H_5)_2$ V	$[B(C_6H_5)_4]$	yellow ^1H NMR (CD_3COCD_3): 5.07 (t, C_5H_5, J=1.3) IR (CH_2Cl_2): 1981 (CO)	[9]
*27	$P(C_6H_5)_2CH=CHP(C_6H_5)_2$ IIa	SbF_6	yellow ^1H NMR (CD_3COCD_3): 5.09 (t, C_5H_5, J=1.5) IR (CH_2Cl_2): 1992 (CO)	[9]
*28	$P(CH_2CH_2P(C_6H_5)_2)_3$ IId (66)	I	yellow orange, m.p. 116 to 118° ^{31}P NMR (CH_2Cl_2): −13.3, −12.5 (uncoordinated P), 93.6, 96.5 (coordinated P) IR (CH_2Cl_2): 1970 (CO)	[7, 13]
*29	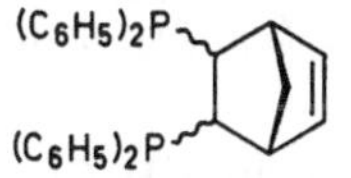 $(C_6H_5)_2P$ $(C_6H_5)_2P$ IId (80 to 86)	I	diastereomers	[30]

References on p. 85

Table 6 [continued]

No.	donor ligands method of preparation (yield in %)	anion	properties and remarks	Ref.
*30	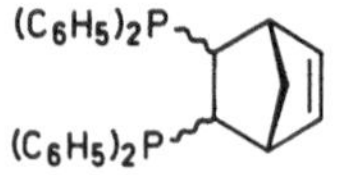 $(C_6H_5)_2P$, $(C_6H_5)_2P$ V (97)	PF_6	diastereomers yellow cubic crystals ^1H NMR (CD_3COCD_3): 4.95, 4.99 (t's, C_5H_5, J = 1.6) IR (KBr): 1975 (CO)	[30]

$[^5LFe(CO)(^2D)_2]^+$ type with $^5L = C_6H_7$ and C_7H_9

No.	donor ligands	anion	properties and remarks	Ref.
*31	$^5L =$ (structure with C_2H_5, positions 1–6) special	BF_4	^1H NMR (CD_3COCD_3): 0.13 (A_2B_3 pattern, C_2H_5), 2.84, 3.93 (H–6), 3.33 (H–1,5), 4.46 (OCH_2), 5.43 (H–2,4), 6.50 (H–3) IR (CH_3COCH_3): 2015 (CO)	[12]
*32	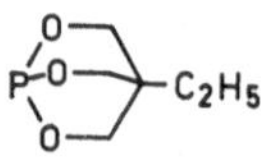 $^5L =$ 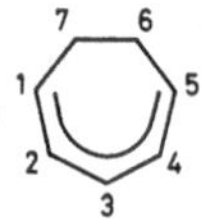 (structure with C_2H_5, positions 1–7) special	BF_4	^1H NMR (CD_3COCD_3): 0.11 (A_2B_3 pattern, C_2H_5), 1.66, 2.48 (H–6,7), 4.17 (H–1,5), 4.46 (OCH_2), 5.55 (H–2,4), 6.50 (H–3) IR (CH_2COCH_3): 2007 (CO)	[12]

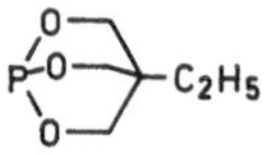

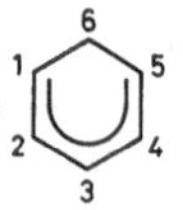

* Further information:

$[C_5H_5Fe(CO)(P(CH_3)_3)_2]X$ (X = I, BF_4, PF_6, Table 6, Nos. 1 to 3). No. 1 was formed in a low yield (1.5%) by refluxing $C_5H_5Fe(CO)_2I$ and $P(CH_3)_3$ in heptane for 1 h. Cyclic voltammetry in CH_2Cl_2 (~5×10^{-3} M) shows the formation of the $[C_5H_5Fe(CO)(P(CH_3)_3)_2]^{2+}$ cation at $E_p = 1.30$ V (vs. SCE). Irradiation of No. 1 in acetone gives approximately equal amounts of $C_5H_5Fe(CO)(P(CH_3)_3)I$ and $[C_5H_5Fe(P(CH_3)_3)_3]I$ [29]. Irradiation of Nos. 2 and 3 in refluxing CH_3CN gave high yields of $[C_5H_5Fe(P(CH_3)_3)_2NCCH_3]X$ [29, 31].

$[C_5H_5Fe(CO)(P(C_6H_5)_3)_2]PF_6$ (Table 6, No. 4). A more rapid and more productive preparation than Method IIIa is the irradiation of $[C_5H_5Fe(CO)_2P(C_6H_5)_3]PF_6$ and excess $P(C_6H_5)_3$ in CH_2Cl_2 for 4 h, 63% yield (nonoptimized) [26]. The IR spectrum (in Nujol) is completely given from 505 to 1435 cm^{-1} [2]. Polarographic one-electron reduction was observed in dimethoxyethane (2×10^{-3} M) at $E_{1/2} = -1.4$ V (vs. Ag/AgClO$_4$) [1] and in acetonitrile (10^{-3} M) at $E_p = -0.88$ V (vs. Ag, estimated value due to the presence of maxima) [22].

$[C_5H_5Fe(CO)(P(C_6H_5)_3)C_4H_8O]BF_4$ (Table 6, No. 6) reacts in CH_2Cl_2 with $CH_2=CH_2$ [17] or $CH_2=C=CH_2$ [23] to give $[C_5H_5Fe(CO)(P(C_6H_5)_3)^2L]BF_4$ com-

References on p. 85

pounds. No stable olefin complexes were obtained with cyclic olefins [17]. Treatment with $CH_2=P(C_6H_5)_3$ in THF/C_6H_6 for 15 h produced $[C_5H_5Fe(CO)(P(C_6H_5)_3)-CH_2P(C_6H_5)_3]BF_4$ [18], see 1.5.2.2.13, p. 232.

VI   VII

$[C_5H_5Fe(CO)NH(C_2H_4O)_2PC_6H_5]PF_6$ (Table **6**, No. **14**) was prepared from $C_5H_5Fe(CO)_2Br$ and the bicyclophosphorane VI in C_2H_5OH at $\sim$60 °C [28]. Details of the procedure were described for the molybdenum analogue in [27].

Unexpectedly, deprotonation with a slight excess of $LiCH_3$ results in P–N ring closure of the bidentate ligand and migration of the C_6H_5 group from P to Fe to give complex VII in a 30% yield. This reaction is even reversible. Bubbling HCl gas through a THF solution of complex VII followed by addition of NH_4PF_6 regenerates No. 14 in a 60% yield [28].

$[C_5H_5Fe(CO)(P(CH_3)_2CH_2-)_2]PF_6$ (Table **6**, No. **17**) was obtained (Method V) by first decomposing the anion of No. 19 in air and then dissolving the product in H_2O/C_2H_5OH and precipitating No. 17 with NH_4PF_6 [3].

$[C_5H_5Fe(CO)(P(CH_3)_2CH_2-)_2]BF_4$ (Table **6**, No. **18**) rapidly forms in high yields when $[C_5H_5Fe(P(CH_3)_2CH_2-)_2OC(CH_3)_2]BF_4$ or $[(C_5H_5Fe(P(CH_3)_2CH_2-)_2)_2-\mu-N_2]BF_4$ (see C3, 2.5.2.1.3, p. 102) are treated in acetone with carbon monoxide [5].

$[C_5H_5Fe(CO)(P(CH_3)_2CH_2-)_2][C_5H_5Fe(CO)_2]$ (Table **6**, No. **19**) was prepared from $(C_5H_5Fe(CO)_2)_2$ and $(P(CH_3)_2CH_2-)_2$ (mole ratio 1:1.5) in benzene at room temperature overnight. The precipitated complex No. 19 was washed with hexane, 72% yield. The compound is very air sensitive due to the instability of the anion [3].

$[C_5H_5Fe(CO)C_{26}H_{24}P_2]Cl$ (Table **6**, No. **20**). UV irradiation in acetone converts the compound to $C_5H_5Fe(C_{26}H_{24}P_2)Cl$ via the isolable $[C_5H_5Fe(C_{26}H_{24}P_2)-OC(CH_3)_2]Cl$ intermediate [19]. No. 20 is not decomposed by aqueous NaOH or $NaBH_4$ [25], also see Nos. 22 and 24.

$[C_5H_5Fe(CO)C_{26}H_{24}P_2]OH$ (Table **6**, No. **22**) was prepared by Method V from the chloride No. 20 in aqueous solution by passage through an hydroxide-loaded anion–exchange resin, followed by addition of concentrated NaOH solution to the eluate and extraction with $CH_3COOC_2H_5$ [25].

$[C_5H_5Fe(CO)C_{26}H_{24}P_2]PF_6$ (Table **6**, No. **23**) is sparingly soluble in THF. Irradiation of a saturated THF solution at room temperature gives $[C_5H_5Fe(C_{26}H_{24}P_2)-OC_4H_8]PF_6$, see 1.5.1.2, p. 2. $[C_5H_5Fe(C_{26}H_{24}P_2)OC(CH_3)_2]PF_6$ (see 1.5.1.2, p. 2) and $[C_5H_5Fe(C_{26}H_{24}P_2)(\mu-N_2)(C_{26}H_{24}P_2)FeC_5H_5][PF_6]_2$ (see C3, 2.5.2.1.3, p. 102) form under irradiation in acetone at -30 °C in N_2 atmosphere [11, 19]. No. 23 is recovered from the all three salts when CO is passed through their solution [19].

References on p. 85

[C$_5$H$_5$Fe(CO)C$_{26}$H$_{24}$P$_2$]BH$_4$ (Table 6, No. 24) was obtained by Method V from the chloride No. 20 and NaBH$_4$ without attack of BH$_4^-$ on the cation. The salt reduces CH$_3$COCH$_3$ and C$_6$H$_5$CHO in CH$_3$OH/H$_2$O to give CH$_3$CH(OH)CH$_3$ and C$_6$H$_5$CH$_2$OH, respectively [25].

[C$_5$H$_5$Fe(CO)C$_{26}$H$_{24}$P$_2$]BF$_4$ (Table 6, No. 25) was prepared in a 65% yield from C$_5$H$_5$Fe(C$_{26}$H$_{24}$P$_2$)MgBr and CO$_2$ in benzene in the presence of MgBr$_2$ etherate, followed by addition of AgBF$_4$. This reaction presumably proceeds via the C$_5$H$_5$Fe(C$_{26}$H$_{24}$P$_2$)COOMgBr intermediate. It does not occur in donor solvents like THF. A 13% yield was also obtained from a reaction of C$_5$H$_5$Fe(C$_{26}$H$_{24}$P$_2$)MgBr with (CF$_3$CO)$_2$O in THF at −78 to 0 °C, followed by addition of AgBF$_4$; C$_5$H$_5$-Fe(C$_{26}$H$_{24}$P$_2$)COCF$_3$ may be an intermediate [20].

Some results of an unpublished structure analysis [15], [21, footnote 15] are reported in [21]: mean distances 2.094(7) for Fe-C(C$_5$H$_5$), 1.717(1) for Fe-center(C$_5$H$_5$), 2.210(1) for Fe-P, 1.408(7) for C-C(C$_5$H$_5$), and 1.833(7) Å for P-C. Torsion angles about bonds in the chelate ring are also given.

[C$_5$H$_5$Fe(CO)P(C$_6$H$_5$)$_2$CH=CHP(C$_6$H$_5$)$_2$]SbF$_6$ (Table 6, No. 27) formed also in a ∼60% yield from C$_5$H$_5$Fe(CO)P(C$_6$H$_5$)$_2$CH=CHP(C$_6$H$_5$)$_2$(CO)FeC$_5$H$_5$ and AgSbF$_6$ in acetone under irradiation for 5 h [9].

[C$_5$H$_5$Fe(CO)P(CH$_2$CH$_2$P(C$_6$H$_5$)$_2$)$_3$]I (Table 6, No. 28). Method IId gave an oily product which was dissolved by addition of CH$_3$OH, filtered, evaporated, and then recrystallized from CH$_3$OH at −78 °C. Attempted purification by chromatography on Al$_2$O$_3$ (?) led to almost complete decomposition [7].

[C$_5$H$_5$Fe(CO)P(C$_6$H$_5$)$_2$C$_7$H$_8$P(C$_6$H$_5$)$_2$]X (X=I and PF$_6$, Table 6, Nos. 29 and 30). Method IId was carried out with both (±)- and (−)-P(C$_6$H$_5$)$_2$C$_7$H$_8$P(C$_6$H$_5$)$_2$ as the starting materials, Formula VIII. This gives the cation as two diastereomeric pairs of enantiomers (from the (±)-donor) or two diastereomers (from the (−)-donor). The isomeric iodides No. 29 precipitated from the reaction mixture on cooling to room temperature. They were purified by addition of benzene to a solution of CH$_2$Cl$_2$. The isomers were not separated.

The conversion of No. 29, obtained from the (−)-donor, to No. 30 with NH$_4$PF$_6$ in aqueous CH$_3$OH gave a mixture of diastereomers which crystallized from CH$_2$Cl$_2$/C$_6$H$_5$CH$_3$ (1:1) at −25 °C without isomer separation.

The optically pure less soluble diastereomer of No. 30 was obtained by repeated precipitation from CH$_2$Cl$_2$/benzene (1:1) with stirring at room temperature for 30 min. It melts at 250 to 252 °C with decomposition. The high-field triplet in the ^{1}H NMR spectrum is assigned to this isomer. Optical rotations (in CH$_2$Cl$_2$, 3×10^{-3} M, 17 °C):

$$[\alpha]_{578} -200°, \quad [\alpha]_{546} -195°, \quad \text{and} \quad [\alpha]_{436} -1280°.$$

The more soluble diastereomer was only enriched to a 3:1 mixture in the mother liquor of the first crystallization. Optical rotations:

$$[\alpha]_{578} -190°, \quad [\alpha]_{546} -210°, \quad \text{and} \quad [\alpha]_{437} -870°.$$

The very similar optical rotations and CD spectra of the diastereomers, unusual for opposite metal configurations in organometallic diastereomers, were explained by a rigid λ conformation for the five-membered chelate ring enforced by the norbornene skeleton [30].

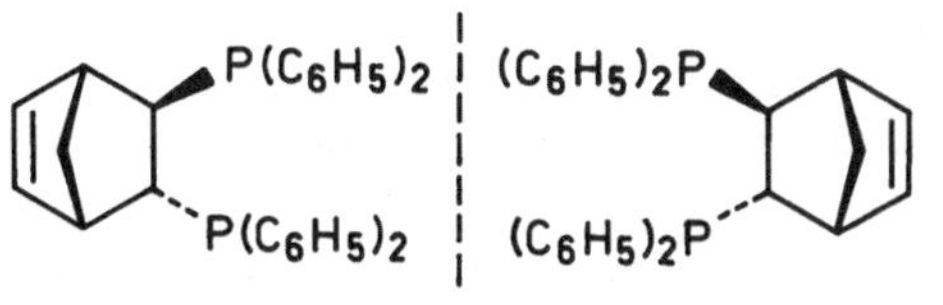

VIII   IX

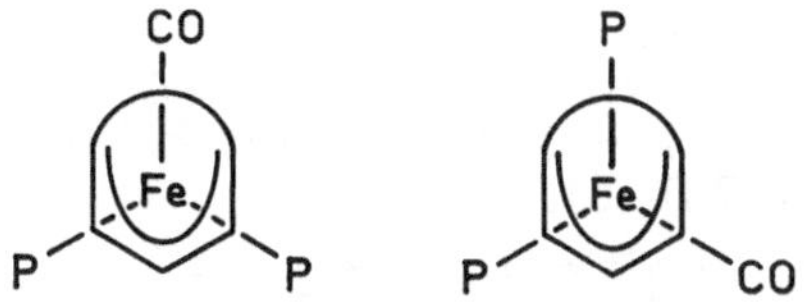

$[C_6H_7Fe(CO)(C_6H_{11}O_3P)_2]BF_4$ and $[C_7H_9Fe(CO)(C_6H_{11}O_3P)_2]BF_4$ (Table 6, Nos. 31 and 32) were prepared from the $^4LFe(CO)(C_6H_{11}O_3P)_2$ compounds with ^4L=cyclohexadiene or cycloheptadiene by deprotonation of the ^4L ligands with $[(C_6H_5)_3C]BF_4$ in CH_2Cl_2 at room temperature. The salts were precipitated by addition of ether to concentrated acetone solutions.

The low-temperature ^{31}P NMR spectra (in CD_3COCD_3 at $-95\,°C$) were explained by the presence of the two conformers IXa and IXb (δ in ppm upfield of $P(OCH_3)_3$, J in Hz):

conformer	IXa	IXb
No. 31	-5.95 (s)	-22.65, -14.80, $J(P,P)=95.7$
No. 32	-12.21 (s)	-24.12, -13.40, $J(P,P)=94.2$

The IXb conformers produce an AB pattern. The IXa:IXb ratio is 13:87 for No. 31 and 28:72 for No. 32. When the temperature is increased, all resonances broaden and eventually collapse to a single line because of ligand scrambling. Free energies of activation and rate constants at 220 K show that the scrambling is more rapid for No. 31 than for No. 32 [12].

References:

[1] R.E. Dessy, R.B. King, M. Waldrop (J. Am. Chem. Soc. **88** [1966] 5112/7). – [2] P.M. Treichel, R.L. Shubkin, K.W. Barnett, D. Reichard (Inorg. Chem. **5** [1966] 1177/81). – [3] R.B. King, K.H. Pannell, C.A. Eggers, L.W. Houk (Inorg. Chem. **7** [1968] 2353/6). – [4] M.L. Brown, T.J. Meyer, N. Winterton (Chem. Commun. **1971** 309). – [5] W.E. Silverthorn (Chem. Commun. **1971** 1310/1).

[6] M.L.H. Green, R.N. Whiteley (J. Chem. Soc. A **1971** 1943/6). – [7] R.B. King, R.N. Kapoor, M.S. Saran, P.N. Kapoor (Inorg. Chem. **10** [1971] 1851/60). – [8] M.L. Brown, J.L. Cramer, J.A. Ferguson, T.J. Meyer, N. Winterton (J. Am. Chem. Soc. **94** [1972] 8707/10). – [9] R.J. Haines, A.L. du Preez (Inorg. Chem. **11** [1972] 330/6). – [10] M.J. Mays, P.L. Sears (J. Chem. Soc. Dalton Trans. **1973** 1873/5).

[11] D. Sellmann, E. Kleinschmidt (Angew. Chem. **87** [1975] 595/6; Angew. Chem. Intern. Ed. Engl. **14** [1975] 571). – [12] T.H. Whitesides, R.A. Budnik (Inorg. Chem. **14** [1975] 664/76). – [13] R.B. King, J.C. Cloyd Jr. (Inorg. Chem. **14** [1975] 1550/4). – [14] E.E. Isaacs, W.A.G. Graham (J. Organometal. Chem. **120** [1976] 407/21). – [15] N.J. Grice (Diss. Univ. Texas 1977, pp. 1/171; Diss. Abstr. Intern. B **38** [1977] 2181/2).

[16] G. Balavoine, M.L.H. Green, J.P. Sauvage (J. Organometal. Chem. **128** [1977] 247/52). – [17] D.L. Reger, C. Coleman (J. Organometal. Chem. **131** [1977]

153/62). – [18] D.L. Reger, E.C. Culbertson (J. Organometal. Chem. **131** [1977] 297/300). – [19] D. Sellmann, E. Kleinschmidt (J. Organometal. Chem. **140** [1977] 211/9). – [20] H. Felkin, P.J. Knowles, B. Meunier (J. Organometal. Chem. **146** [1978] 151/67).

[21] P.E. Riley, C.E. Capshew, R. Pettit, R.E. Davis (Inorg. Chem. **17** [1978] 408/14). – [22] M.E. Grant, J.J. Alexander (J. Coord. Chem. **9** [1979] 205/10). – [23] D.L. Reger, C.J. Coleman, P.J. McElligott (J. Organometal. Chem. **171** [1979] 73/84). – [24] D.L. Reger, C.J. Coleman (Inorg. Chem. **18** [1979] 3155/60). – [25] N. Grice, S.C. Kao, R. Pettit (J. Am. Chem. Soc. **101** [1979] 1627/8).

[26] B.V. Johnson, P.J. Ouseph, J.S. Hsieh, A.L. Steinmetz, J.E. Shade (Inorg. Chem. **18** [1979] 1796/9). – [27] J. Wachter, F. Jeanneaux, J.G. Riess (Inorg. Chem. **19** [1980] 2169/72). – [28] P. Vierling, J.G. Riess, A. Grand (J. Am. Chem. Soc. **103** [1981] 2466/7). – [29] P.M. Treichel, D.A. Komar (J. Organometal. Chem. **206** [1981] 77/88). – [30] H. Brunner, A.F.M. Moklesur Rahman (J. Organometal. Chem. **214** [1981] 373/80).

[31] D.A. Komar (Diss. Univ. Wisconsin 1981, pp. 1/224; Diss. Abstr. Intern. B **42** [1982] 2828).

1.5.2.2.6 Compounds of the $^5LFe(CO)(^2D)X$ Type

Compounds of the $^5LFe(CO)(^2D)X$ type possess a chiral iron atom. A brief introduction and general references to the problem of chirality is given at the beginning of Chapter 1.5.2.2.12, p. 150, which describes the $^5LFe(CO)(^2D)^1L$ compounds.

1.5.2.2.6.1 Compounds with $^5L = C_5H_5$

1.5.2.2.6.1.1 Compounds with $^2D = PR_3$ (R = CH$_3$, C$_2$H$_5$, C$_4$H$_9$-n, C$_6$H$_{11}$-cyclo, CH$_2$C$_6$H$_5$, CH$_2$CH$_2$P(C$_6$H$_5$)$_2$, and N(CH$_3$)$_2$)

The compounds of this section are collected in Table 7. They are prepared by the following methods.

Method I: $C_5H_5Fe(CO)_2X$ is reacted with the appropriate PR_3.

      a. UV irradiation of $C_5H_5Fe(CO)_2X$ with the phosphine PR_3 in an inert solvent gives the product [3, 8, 10], which is purified by chromatography on Al_2O_3 (No. 3) [10] or recrystallized [8].

      b. The components are reacted without irradiation [4, 5]. Reaction of $C_5H_5Fe(CO)_2N_3$ with $P(C_4H_9-n)_3$ in CH_2Cl_2 at room temperature gives $C_5H_5Fe(CO)(P(C_4H_9-n)_3)NCO$ (No. 17). The reaction mixture is evaporated and the residue dissolved in C_6H_6. No. 10 is precipitated by adding pentane [15].

      c. Reaction of $C_5H_5Fe(CO)_2X$ at $-20\,°C$ in ether with PR_3 (R = CH$_3$, N(CH$_3$)$_2$) and reduction with $LiAlH_4$ followed by hydrolysis gives the product [7].

Method II: The X ligand in $C_5H_5Fe(CO)(PR_3)X$ is exchanged.

      a. Reaction of $C_5H_5Fe(CO)(PR_3)H$ (No. 1 with R = CH$_3$, No. 25 with R = N(CH$_3$)$_2$) with CCl_4 yields Nos. 2 and 26. Compound No. 2 is recrystallized from benzene or hexane, and No. 26 is chromatographed on Al_2O_3 [7].

b. For preparation of No. 6 an equimolar amount of $AgBF_4$ is added to a methanol solution of $C_5H_5Fe(CO)(P(CH_3)_3)I$ (No. 3). AgI precipitates immediately, and a twofold excess of NaCN is added to the solution. After four hours the reaction mixture is filtered and the solvent removed. The residue is dissolved in methylene chloride and chromatographed on Al_2O_3. A large green band is eluted slowly with petroleum ether/CH_2Cl_2 (2:1) followed closely by No. 6, which was eluted with CH_2Cl_2 as a smaller yellow band [10].

Method III: Compounds of the type $C_5H_5Fe(CO)(PR_3)(^1L,X)$ are the starting materials.

a. Compound No. 18 is prepared from a C_6H_6 solution of C_5H_5-$Fe(CO)(P(C_4H_9-n)_3)CH_2C_6H_5$, to which finely powdered $(NC)_2C=C(CN)_2$ is added at 5 to 10 °C. The reaction mixture is stirred for 15 min. Workup by chromatography [6, 12].

b. Addition of BH_3 in THF to a solution of $C_5H_5Fe(CO)(PR_3)COCH_3$ (R = C_6H_{11}-cyclo, $CH_2C_6H_5$) in C_6D_6 or THF-d_8 affords the compounds $C_5H_5Fe(CO)(PR_3)H$ and $C_5H_5(CO)(PR_3)C_2H_5$ [13].

c. $C_5H_5Fe(CO)(P(C_4H_9-n)_3)CH_3$ is dissolved in $CHCl_3$ and SO_2 is bubbled slowly through the solution at 27 °C for 10 min. The reaction with gaseous SO_2 for 2 d or condensed SO_2 for 1 h gives lower yields [2].

d. $C_5H_5Fe(CO)(P(CH_3)_3)As(CH_3)_2$ (No. 4) is reacted with C_5H_5-$(CO)_2MnAs(CH_3)_2Co(CO)_3$ (Formula I) in hexane (25 °C, 0.7 h). The mixture is concentrated, filtered, and the filtrate washed with a small amount of hexane. Recrystallization from C_6H_6/hexane [16].

$$C_5H_5(CO)_2Mn\diagdown{\overset{As(CH_3)_2}{\underset{Co(CO)_3}{|}}}$$

I

Method IV: Compound $(+)(R)-C_5H_5Fe(CO)(P(C_6H_5)_3)Si(CH_3)C_{10}H_7-1$ (Table 8, No. 34 in 1.5.2.2.6.1.2) and $P(C_6H_{11}-cyclo)_3$ in cyclohexane are exposed to UV radiation to give a mixture of two diastereomers, which is separated by fractional crystallization from CH_2Cl_2/hexane [14].

Method V: Reaction of compound No. 8 with HCl in benzene/ether at room temperature gives No. 9. With a mole ratio of 1:2 No. 10 and with excess HCl No. 11 are the products [8].

Method VI: Reaction of $(C_5H_5)_2Fe_2(CO)_3P(C_2H_5)_3$ with I_2 in CH_2Cl_2 yields C_5H_5-$Fe(CO)(P(C_2H_5)_3)I$, $C_5H_5Fe(CO)_2I$, and $[C_5H_5Fe(CO)_2P(C_2H_5)_3]I$ in approximately equal amounts [1].

A $C_5H_5Fe(CO)(P(CH_2CH_2P(C_6H_5)_2)_3)I$ compound is reported in [11]. However, this is an ionic product of the $[C_5H_5Fe(CO)^2D-^2D]X$ type as the ^{31}P NMR spectrum clearly shows two coordinated P atoms [5], see 1.5.2.2.5, Table 6, No. 28.

Explanations for Table 7: Not assigned IR bands in the 1900 to 2000 cm^{-1} region are v(CO) bands.

References on p. 91

7*

Table 7
Compounds of the $C_5H_5Fe(CO)(PR_3)X$ Type.
Further information on numbers preceded by an asterisk is given at the end of the table.
For abbreviations and dimensions, see p. 375.

No.	X ligand	method of preparation (conditions, yield), properties and remarks explanations on p. 87	Ref.

$C_5H_5Fe(CO)(P(CH_3)_3)X$ compounds

1	H	Ic (~50%), yellow crystals, subl. 60°/0.1 ^1H NMR (C_6H_6): −14.84 (H, J(P,H)=84), 0.92 (CH$_3$, J(P,H)=10), 4.32 (C$_5$H$_5$) IR (hexadecane): 1925 reaction with CCl$_4$ gives $C_5H_5Fe(CO)(P(CH_3)_3)Cl$ (cf. Method IIa)	[7]
2	Cl	IIa (~100%), green crystals ^1H NMR (C_6H_6): 1.00 (CH$_3$, J(P,H)=10), 4.16 (C$_5$H$_5$, J(P,H)=1.4) IR (hexadecane): 1975	[7]
3	I	Ia (C_6H_6, THF, or dioxane, 72%), dark green crystals, dec. 135 to 137° IR (CHCl$_3$): 1950 reaction with AgBF$_4$/NaCN in CH$_3$OH gives $C_5H_5Fe(CO)(P(CH_3)_3)CN$ (No. 6)	[10]
4	As(CH$_3$)$_2$	preparation and properties not reported; for a reaction, see Method IIId	[16]
5	Co(CO)$_3$ / (CH$_3$)$_2$As' As"(CH$_3$)$_2$ / C$_5$H$_5$Mn(CO)$_2$	IIId (93), black violet, m.p. 90 to 93° ^1H NMR (C_6H_6): 0.85 (PCH$_3$, J=9.5), 1.36, 1.42 (As'CH$_3$, for both J=0.5), 1.97 (As"CH$_3$), 4.03 (FeC$_5$H$_5$, J=1.8), 4.52 (MnC$_5$H$_5$) IR (C_6H_6): 1854, 1915 (CO at Mn), 1935 (CO at Fe), 1935, 1943, 2001 (CO at Co) soluble in benzene	[16]
6	CN	IIb (50%), yellow crystals, m.p. 122 to 125° IR: 1962 in CHCl$_3$, 1964 in C$_6$H$_{12}$	[10]

$C_5H_5Fe(CO)(P(C_2H_5)_3)X$ compounds

| 7 | I | Ib (30 min, C_6H_6 at 80°, 30%), VI
 ^1H NMR (CDCl$_3$): 4.61 (C$_5$H$_5$)
 IR (CH$_2$Cl$_2$): 1944 <br> air-stable crystals | [1, 4] |
| 8 | Sn(C$_6$H$_5$)$_3$ | Ia (5 h, C_6H_6 or CH$_3$COCH$_3$, 68%), orange crystals, m.p. 154 to 156°
 ^1H NMR (CDCl$_3$): 4.48 (C$_5$H$_5$)
 ^{119}Sn-γ (80 K): δ=1.51 (BaSnO$_3$), Δ=0.76±0.1
 IR (CHCl$_3$): 1905 | [8, 9] |

Table 7 [continued]

No.	X ligand	method of preparation (conditions, yield), properties and remarks explanations on p. 87	Ref.
		soluble in C_6H_{12}, C_6H_6, $CHCl_3$ reaction with HCl in ether/C_6H_6 gives $C_5H_5Fe(CO)(P(C_2H_5)_3)SnCl_n(C_6H_5)_{3-n}$ (Nos. 9 to 11), also see Method V	
9	$Sn(C_6H_5)_2Cl$	V (30%), orange crystals, m.p. 141 to 143° (dec.) ^{1}H NMR (CD_3COCD_3): 4.14 (C_5H_5) ^{119}Sn-γ (80 K): $\delta = 1.63$ ($BaSnO_3$), $\Delta = 2.59$ IR ($CHCl_3$): 1924 soluble in C_6H_6, $CHCl_3$, C_6H_{12}, recrystallization from C_6H_{12}	[8, 9]
10	$SnCl_2C_6H_5$	V (50%), red crystals, m.p. 110 to 112° ^{1}H NMR (CD_3COCD_3): 4.70 (C_5H_5) ^{119}Sn-γ (80 K): $\delta = 1.83$ ($BaSnO_3$), $\Delta = 3.03$ IR ($CHCl_3$): 1945 soluble in C_6H_6, $CHCl_3$, C_6H_{12}, recrystallization from C_6H_{12}/C_6H_6	[8, 9]
11	$SnCl_3$	V (90%), maroon crystals, m.p. 116 to 123° (dec.) ^{1}H NMR (CD_3COCD_3): 4.78 (C_5H_5) ^{57}Fe-γ (298 K): $\delta = 0.40$, $\Delta = 1.84$ ^{119}Sn-γ (80 K): $\delta = 1.95$ ($BaSnO_3$), $\Delta = 1.91$ IR ($CHCl_3$): 1968 soluble in C_6H_6, $CHCl_3$, C_6H_{12}, recrystallization from C_6H_6/pentane	[8, 9]

$C_5H_5Fe(CO)(P(C_4H_9-n)_3)X$ compounds

No.	X ligand	method of preparation (conditions, yield), properties and remarks	Ref.
12	Cl	I b (4 d, C_6H_6, small amount)	[4]
13	Br	I a (C_6H_6) IR: see [3] for 400 to 700 cm^{-1} range	[3]
14	I	I a (C_6H_6), I b (10 h, C_6H_6, 80°, 85%), crystalline ^{1}H NMR ($CDCl_3$): 4.58 (d, C_5H_5, J(P,H) = 1.4) IR (C_6H_{12}): 1953, see [3] for 400 to 700 cm^{-1} range air-stable crystals	[3, 4]
*15	SO_2CH_3	III c (80 to 90%), orange, m.p. 108° IR: 1946, ν(SO) 1036 and 1156 in Nujol, 1962 in $CHCl_3$	[2]
16	$Si(C_6H_5)_3$	I a (87%), yellow, m.p. 101 to 103° ^{1}H NMR ($CDCl_3$): 4.48 (C_5H_5, J(P,H) $\sim$1.5) IR (C_6H_{12}): 1910	[14]

References on p. 91

Table 7 [continued]

No.	X ligand	method of preparation (conditions, yield), properties and remarks explanations on p. 87	Ref.
17	NCO	Ib (4 h, CH_2Cl_2, 62%), red product, m.p. 84° ^1H NMR ($CDCl_3$): 0.93 to 1.57 (m, C_4H_9), 4.6 (C_5H_5) IR (CH_2Cl_2): 1943	[15]
*18	N=C=C(CN)-C(CN)$_2$CH$_2$-C$_6$H$_5$	IIIa (50%), red–brown oil at 25° ^1H NMR ($CDCl_3$): 0.85 to 1.95 (m, C_4H_9), 3.24 (d, CH_2), 4.65 (d, C_5H_5, J(P,H)=1.5), 7.4 (s, C_6H_5) IR (neat oil): 1973, ν(N=C=C) 1273, 2145, ν(CN) 2195	[6, 12]

$C_5H_5Fe(CO)(P(C_6H_{11}$-cyclo)$_3)X$ compounds

No.	X ligand		Ref.
19	H	IIIa (5 to 25%) ^1H NMR (C_6D_6): −13.7 (FeH, J(P,H)=70), 4.58 (C_5H_5, J(P,H)=0.2)	[13]
20	Si(C_6H_5)$_3$	Ia (85%), orange, m.p. 210 to 211° (dec.) ^1H NMR ($CDCl_3$): 4.47 (C_5H_5, J(P,H)~1.5) IR (C_6H_{12}): 1900	[14]
21	Si(C_6H_5)$_2$CH$_3$	Ia (84%), orange, m.p. 183 to 184° ^1H NMR ($CDCl_3$): 0.92 (CH$_3$), 4.47 (C_5H_5, J(P,H)~1.5) IR (C_6H_{12}): 1880	[14]
*22	−Si(CH$_3$)C$_6$H$_5$ 	IV (66%) (+)-diastereomer with $[\alpha]_D^{25} = +30$: orange, m.p. 184 to 185° (dec.) ^1H NMR ($CDCl_3$): 0.88 (CH$_3$), 4.55 (C_5H_5, J(P,H)~1.5) IR (C_6H_{12}): 1901 (+)-diastereomer with $[\alpha]_D^{25} = +122$: orange, m.p. 163° (dec.) ^1H NMR ($CDCl_3$): 0.08 (CH$_3$), 4.53 (C_6H_5, J(P,H)~1.5) IR (C_6H_{12}): 1901	[14]

$C_5H_5Fe(CO)(P(CH_2C_6H_5)_3)X$ compounds

No.	X ligand		Ref.
23	H	IIIb (5 to 25%) ^1H NMR (C_6D_6): −13.8 (FeH, J(P,H)=81), 3.98 (C_5H_5, J(P,H)=1.0)	[13]

Table 7 [continued]

No.	X ligand	method of preparation (conditions, yield), properties and remarks explanations on p. 87	Ref.

$C_5H_5Fe(CO)(P(N(CH_3)_2)_3)X$ compounds

| 24 | H | Ic ($\sim$50%), yellow liquid
^{1}H NMR (C_6H_6): $-$ 13.52 (FeH, J(P,H) = 8.4), 2.29 (CH_3, J(P,H) = 9.5), 4.35 (C_5H_5, J(P,H) = 0.9)
IR (hexadecane): 1928
reaction with CCl_4 gives $C_5H_5Fe(CO)-(P(N(CH_3)_2)_3)Cl$ (Method IIa) | [7] |
| 25 | Cl | IIa ($\sim$100%), green crystals
^{1}H NMR (C_6H_6): 2.00 (CH_3, J(P,H) = 9), 4.02 (C_5H_5, J(P,H) = 0.8)
IR (hexadecane): 1951 | [7] |

* Further information:

$C_5H_5Fe(CO)(P(C_4H_9-n)_3)SO_2CH_3$ (Table 7, No. **15**). Protonation with HCl in benzene yields $[C_5H_5Fe(CO)(P(C_4H_9-n)_3)SO(OH)HCH_3]^+$. Thermal or photolytic desulfurylation gives only the starting material [2].

$C_5H_5Fe(CO)(P(C_4H_9-n)_3)N=C=C(CN)C(CN)_2CH_2C_6H_5$ (Table 7, No. **18**). The molecular structure of this compound was determined by comparison of the infrared spectrum with that of $Ir(CO)(P(C_6H_5)_3)(C_6N_4)N=C=C(CN)C(CN)_2H$, whose structure was elucidated crystallographically. Compound No. 18 turns dark brown in a short time and decomposes in 1 h when exposed to air at 25 °C, it is also unstable in solution [6, 12].

$C_5H_5Fe(CO)(P(C_6H_{11}-cyclo)_3)Si(CH_3)(C_6H_5)C_{10}H_7-1$ (Table 7, No. **22**). The optical rotations for the two diastereomers (optical purity > 77%) are given:

(+) isomer	$[\alpha]_D^{25}$	$[\alpha]_{578}^{25}$	$[\alpha]_{546}^{25}$
1. isomer	+30°	+31°	+48°
2. isomer	+122°	+134°	+197°

The reaction of $(+)$-$C_5H_5Fe(CO)(P(C_6H_{11}-cyclo)_3)Si(CH_3)(C_6H_5)C_{10}H_7-1$ ($[\alpha_D^{25}] = +30°$) with Cl_2 or $Cl_2/AlCl_3$ gives $ClSi(CH_3)(C_6H_5)C_{10}H_7-1$ (57% yield, $[\alpha]_D^{25} = +2.8°$) with retention of the configuration (stereoselectivity 58%). Reaction of the isomer with $[\alpha]_D^{25} = +122°$ with Cl_2 in the presence of a 2.5-fold excess $P(C_6H_{11}-cyclo)_3$ gives $ClSi(CH_3)(C_6H_5)C_{10}H_7-1$ (78% yield, $[\alpha]_D^{25} = -2.9°$) with inversion (stereoselectivity 60%) [14].

References:

[1] R.J. Haines, A.L. du Preez (Inorg. Chem. **8** [1969] 1459/64). $-$ [2] M. Graziani, A. Wojcicki (Inorg. Chim. Acta **4** [1970] 347/50). $-$ [3] D.J. Parker (J. Chem. Soc. A **1970** 1382/6). $-$ [4] R.J. Haines, A.L. du Preez, I.L. Marais (J. Organometal. Chem. **28** [1971] 405/13). $-$ [5] R.B. King, R.N. Kapoor, M.S. Saran, P.N. Kapoor (Inorg. Chem. **10** [1971] 1851/60).

[6] S.R. Su (Diss. Ohio State Univ. 1971; Diss. Abstr. Intern. B **32** [1972] 6283). –
[7] P. Kalck, R. Poilblanc (Compt. Rend. C **274** [1972] 66/9). – [8] R.E.J. Bichler,
H.C. Clark, B.K. Hunter, A.T. Rake (J. Organometal. Chem. **69** [1974] 367/76). –
[9] G.M. Bancroft, A.T. Rake (Inorg. Chim. Acta **13** [1975] 175/9). – [10] J.W. Faller,
B.V. Johnson (J. Organometal. Chem. **96** [1975] 99/113).

[11] R.B. King, J.C. Cloyd (Inorg. Chem. **14** [1975] 1550/4). – [12] S.R. Su,
A. Wojcicki (Inorg. Chem. **14** [1975] 89/98). – [13] J.A. van Doorn, C. Masters,
H.C. Volger (J. Organometal. Chem. **105** [1976] 245/54). – [14] G. Cerveau, E.
Colomer, R. Corriu, W.E. Douglas (J. Organometal. Chem. **135** [1977] 373/86). –
[15] D.A. Brown, F.M. Hussein, C.L. Arora (Inorg. Chim. Acta **29** [1978] L215/L216).

[16] H.J. Langenbach, H. Vahrenkamp (Chem. Ber. **113** [1980] 2189/99,
2200/10).

1.5.2.2.6.1.2 Compounds with $^2D = P(C_6H_5)_3$

The compounds of the type $C_5H_5Fe(CO)(P(C_6H_5)_3)X$ are listed in Table 8. They
can be prepared in several ways. The following methods are of a general nature.
Reaction conditions (solvent, temperature, and time) are given in Table 8. Special
procedures are described in the further information on p. 103.

Method I: $C_5H_5Fe(CO)_2X$ reacts with $P(C_6H_5)_3$.

        a. The components are UV–irradiated [6, 9, 11, 12, 14, 23, 29, 31
to 33, 35, 36, 45, 49, 52]. For Nos. 5 and 6 quantum yields for
various wavelengths and solvents are given [52]. Compound No. 8
is crystallized from benzene/hexane [9].

        b. The components react without irradiation [3, 4, 8, 26, 44, 49, 52,
54, 55].

        c. $C_5H_5Fe(CO)_2N_3$ is treated with $P(C_6H_5)_3$. The product is crystallized
from CH_2Cl_2/petroleum ether [53].

Method II: $C_5H_5Fe(CO)(P(C_6H_5)_3)CH_2R$ is the starting compound.

        a. $C_5H_5Fe(CO)(P(C_6H_5)_3)CH_2R$ (R=H, CH_3 [46], $CH_2=CH_2$ [26])
reacts with CF_3COOH. The product is crystallized from benzene/
hexane (3:1) [26]. The treatment of $C_5H_5Fe(CO)(P(C_6H_5)_3)$-
$CH_2CH(CH_3)C_6H_5$ with HCl in CH_2Cl_2 for 20 min gives C_5H_5-
$Fe(CO)(P(C_6H_5)_3)Cl$ (No. 4) and $C_6H_5CH(CH_3)_2$ [30].

        b. $C_5H_5Fe(CO)(P(C_6H_5)_3)CH_2C_6H_5$ reacts with $Se(SeCN)_2$. The reac-
tion mixture is evaporated and the residue chromatographed (Al_2O_3
with CH_2Cl_2) to give three bands. Addition of hexane to the light
red band gives $C_5H_5Fe(CO)(P(C_6H_5)_3)NCSe$ (No. 24). Similar
treatment of the brown band gives $C_5H_5Fe(CO)(P(C_6H_5)_3)SeCN$
(No. 21), and elution of the last, yellow band with $CHCl_3$ gives
$C_5H_5Fe(CO)(P(C_6H_5)_3)CN$ (No. 29). $C_5H_5Fe(CO)(P(C_6H_5)_3)CH_3$
and $Se(SeCN)_2$ react exothermally to give $SeP(C_6H_5)_3$, No. 21,
No. 29, but no No. 24. The products are contaminated by CH_3SeCN
[8].

        c. $C_5H_5Fe(CO)(P(C_6H_5)_3)CH_2C_6H_5$ reacts with $(NC)_2C=C(CN)_2$
to give No. 28 and $C_5H_5Fe(CO)(P(C_6H_5)_3)CN$ (28% yield) [17, 39].

d. These compounds are prepared by thermal decomposition of $C_5H_5Fe(CO)(P(C_6H_5)_3)CH_2R$ $(R=CH_3$ [17, 18, 43, 50] or C_3H_7 [43, 50] for compound No. 1, CD_3 for No. 2 [17]).

e. $C_5H_5Fe(CO)(P(C_6H_5)_3)CH_3$ reacts with Br_2 or I_2 in $CHCl_3$ for 20 min. Crystallization from $CHCl_3$/heptane [3].

Method III: $C_5H_5Fe(CO)(P(C_6H_5)_3)H$ reacts with $CHCl_3$ or CCl_4 [17, 18, 20].

Method IV: $C_5H_5Fe(CO)(P(C_6H_5)_3)COR$ is the starting compound.

a. Reaction of $(+)-C_5H_5Fe(CO)(P(C_6H_5)_3)COCH_3$ with an excess of I_2 gives optically inactive $C_5H_5Fe(CO)(P(C_6H_5)_3)I$ (No. 6). Reaction of $(-)-C_5H_5Fe(CO)(P(C_6H_5)_3)COOC_{10}H_{19}$ $(C_{10}H_{19}=$ menthyl) with I_2 gives $[C_5H_5Fe(CO)_2P(C_6H_5)_3]I$ and traces of optically inactive $C_5H_5Fe(CO)(P(C_6H_5)_3)I$ [28].

b. Reduction of $C_5H_5Fe(CO)(P(C_6H_5)_3)COCH_3$ in C_6D_6 with BH_3 in THF gives $C_5H_5Fe(CO)(P(C_6H_5)_3)H$ (No. 1) and $C_5H_5Fe(CO)$-$(P(C_6H_5)_3)C_2H_5$. Reduction with BD_3 gives $C_5H_5Fe(CO)$-$(P(C_6H_5)_3)D$ (No. 2) and $C_5H_5Fe(CO)(P(C_6H_5)_3)CD_2CH_3$ [40].

c. $C_5H_5Fe(CO)(P(C_6H_5)_3)CONH_2$ reacts with I_2. Purification by chromatography on Al_2O_3 with C_6H_6/CH_2Cl_2 [19].

Method V: SO_2 insertion into $C_5H_5Fe(CO)(P(C_6H_5)_3)R$. This method allows a distinction between isomers. IR and NMR data from solutions obtained from $C_5H_5Fe(CO)(P(C_6H_5)_3)R$ $(R=CH_3, CH_2C_6H_5)$ in liquid SO_2 show an intermediate O-sulfinate $C_5H_5Fe(CO)(P(C_6H_5)_3)OS(O)R$ (see Table 8, Nos. 9, 10). These species convert to the S-sulfinates on storage [30].

a. The insertion is accomplished by refluxing $C_5H_5Fe(CO)(P(C_6H_5)_3)R$ in liquid SO_2. The reaction of $C_5H_5Fe(CO)(P(C_6H_5)_3)CH_2C_3H_5$-cyclo gives mainly $C_5H_5Fe(CO)(P(C_6H_5)_3)SO_2CH_2CH_2CH=CH_2$ (No. 16) [41].

b. SO_2 is slowly bubbled through a solution of C_5H_5Fe-$(CO)(P(C_6H_5)_3)R$ in $CHCl_3$ at 27 °C (No. 11) [11], CH_2Cl_2 at 0 °C (Nos. 11, 15, 17) [41], or dimethylformamide at 0 °C (No. 12) [41] or 25 °C (No. 11) [41]. Reaction of $C_5H_5Fe(CO)(P(C_6H_5)_3)$-$CH_2C_3H_5$-cyclo in CH_2Cl_2 at 0 °C gives mainly No. 16 [41].

c. $C_5H_5Fe(CO)(P(C_6H_5)_3)R$ is added to a solution of KI in liquid SO_2 and the mixture refluxed. Additional $C_5H_5Fe(CO)(P(C_6H_5)_3)I$ forms if $C_5H_5Fe(CO)(P(C_6H_5)_3)R$ with $R=CH_3$, C_2H_5, or CH_2COOR' $(R'=$menthyl) is the starting product [41].

d. Gaseous SO_2 is introduced into an evacuated flask containing finely ground starting material [41].

Method VI: $[C_5H_5Fe(CO)_2P(C_6H_5)_3]X$ is the starting compound.

a. $[C_5H_5Fe(CO)_2P(C_6H_5)_3]SeCN$ is refluxed in C_6H_6 for 0.5 h [8].

b. $[C_5H_5Fe(CO)_2P(C_6H_5)_3]I$ reacts with NaCN in C_2H_5OH for 24 h [37].

References on p. 108

    c. $[C_5H_5Fe(CO)_2P(C_6H_5)_3]BF_4$ reacts with NaCN in CH_3OH for 4 h [33].

    d. An aqueous solution of NaN_3 is added to a solution of $[C_5H_5Fe(CO)_2P(C_6H_5)_3]PF_6$ in acetone. The mixture is stirred for 3 h. Purification by chromatography on Al_2O_3 (CH_2Cl_2 eluent) and precipitation with hexane [16].

    e. Like Method VI d but with NH_2NH_2 in CH_2Cl_2 for 1 h. Precipitation with pentane [16].

    f. Method VI d but with NH_2NHCH_3 [16].

Method VII: The reaction mixture obtained from $C_5H_5Fe(CO)_2I$ with $P(C_6H_5)_3$ in ether is reduced with $LiAlH_4$ and hydrolyzed [20, 59]. A solution of $C_5H_5Fe(CO)(P(C_6H_5)_3)I$ in THF is cooled to 0 °C and a fourfold excess of $LiAlH_4$ is added. After 1 min of stirring, the reaction mixture is filtered and dried under reduced pressure. The resulting brown oil is dissolved in toluene and chromatographed. A rigorous reaction took place between the brown solution and alumina, but a yellow solution can be eluted with toluene. It is concentrated, chromatographed again, and the eluate dried [59].

Method VIII: The first step of this method is similar to Method II c: $C_5H_5Fe(CO)(P(C_6H_5)_3)R$ compounds are reacted with an equimolar amount or slight excess $(NC)_2C=C(CN)_2$ in benzene at 5 to 10 °C for up to 15 min. For $R=CH_3$, C_2H_5, and $n-C_3H_7$ and in contrast to $R=CH_2C_6H_5$, the reaction gives acyl complexes of the $C_5H_5Fe(P(C_6H_5)_3)(C_6N_4)COR$ type (C_6N_4 as ^2L ligand) in 70 to 78% yield. These solid compounds can be rearranged to some extent to give Nos. 25 to 27 along with $C_5H_5Fe(CO)(P(C_6H_5)_3)R$ and/or $C_5H_5Fe(CO)(P(C_6H_5)_3)CN$ [39], also see [17].

Method IX: $(C_5H_5Fe(CO)_2)_2$ is reacted with $P(C_6H_5)_3$ and C_6H_5I under UV irradiation [2, 5]. UV irradiation of a mixture of $(C_5H_5Fe(CO)_2)_2$ and $C_5H_5Fe(CO)_2C_6H_4CH_3-4$ with $P(C_6H_5)_3$ and $p-CH_3C_6H_4I$ gives $C_5H_5Fe(CO)(P(C_6H_5)_3)C_6H_4CH_3-4$ and $C_5H_5Fe(CO)(P(C_6H_5)_3)I$ (No. 6) [2].

Method X: $C_5H_5Fe(CO)(P(C_6H_5)_3)C_5H_4Mn(CO)_3$ is treated with I_2 in $CHCl_3$ for 7 h. Treatment of the reaction mixture with aqueous $Na_2S_2O_3$, separation of the organic layer, and chromatography on Al_2O_3 gives $C_5H_5Mn(CO)_3$, $IC_5H_4Mn(CO)_3$, and $C_5H_5Fe(CO)(P(C_6H_5)_3)I$ (No. 6) [51].

    The isomer shift values in the Mössbauer spectra for $C_5H_5Fe(CO)(P(C_6H_5)_3)X$ (X= Br (No. 5), I (No. 6), SCN (No. 20), NCS (No. 23), and CN (No. 29)) are significantly higher (0.05 to 0.10 mm·s^{-1}) than for $C_5H_5Fe(CO)_2X$ compounds. The isomer shift in the $C_5H_5Fe(CO)(P(C_6H_5)_3)X$ series measured at 78 K decrease in the order I > Br > SCN > NCS > CN [55]. The characterization of the chirality and the designations R and S are described in the introduction of 1.5.2.2.12 on p. 150.

Explanations for Table 8: Not assigned IR bands in the 1900 to 2000 cm^{-1} region are $\nu(CO)$ bands.

References on p. 108

Table 8

Compounds of the $C_5H_5Fe(CO)(P(C_6H_5)_3)X$ Type.

Further information on numbers preceded by an asterisk is given at the end of the table.

For abbreviations and dimensions, see p. 375.

No.	X ligand	method of preparation (conditions, yield), properties, and remarks explanations on p. 94	Ref.
*1	H	IId (refl. heptane, ~20%, refl. xylene, or heated in vacuo at 140°, 51%), IVb (~25%), VII (~50%), yellow crystals, m.p. 130° (dec.) ^1H NMR (C_6D_6 or C_6H_6): −12.8 (H, J(P,H) = 75), 4.27 (C_5H_5, J(P,H) = 1.2); 4.02 (C_5H_5, J(P,H) = 75), 6.78 to 7.02 (C_6H_5); −13.0 (H, J(P,H) = 75), 4.08 (C_5H_5, J(P,H) = 1) ^{13}C NMR: 80.48 (s, C_5H_5), 128.12 (d, C-3 in C_6H_5, J = 9.0), 129.58 (d, C-4 in C_6H_5, J = 2.4), 133.30 (d, C-2 in C_6H_5, J = 10.3), 138.67 (d, C-1 in C_6H_5, J = 42.6), 210.80 (d, CO, J = 27.9) ^{31}P NMR: 55.9 (d, J = 72.0) IR (C_6H_6): 1923 in C_6H_6, 1934 in $C_{16}H_{34}$	[17, 18, 20, 40, 43, 50, 59]
2	D	IId (20 min, refl. heptane), IVb, yellow crystals ^1H NMR (C_6D_6): 4.62 (s, C_5H_5), 6.95 to 7.30 (m, C_6H_5)	[17, 40]
3	F	Ib (3 h, CH_2Cl_2), green solid ^1H NMR ($CDCl_3$): 4.45 (d, C_5H_5), 7.40 (m, C_6H_5) IR ($CHCl_3$): 1956	[54]
*4	Cl	Ia (C_6H_6, CH_3CN, or CH_3NO_2), Ib (0.5 h, C_6H_6 at 80°), Ib (3 h, CH_2Cl_2), IIa, III (90 to 95%), green or olive–green crystals ^1H NMR: 4.43 (d, C_5H_5, J(P,H) = 1), 7.40 (m, C_6H_5) in $CDCl_3$, 4.94 (C_5H_5, J(P,H) = 1) in CH_2Cl_2 IR (Nujol, $C_{16}H_{34}$, $CHCl_3$): 1960 to 1963	[3, 17, 18, 20, 44, 52, 54]
*5	Br	Ia (C_6H_6, CH_3CN, or CH_3NO_2), Ib (0.5 to 18 h, C_6H_6 at 80°, 51%), Ib (24 h, pentane at 36°, 80%), Ib (3 h, CH_2Cl_2), IIe (22%), green crystals, m.p. 143 to 144, 160, 168 to 169° ^1H NMR ($CDCl_3$): 4.47 (d, C_5H_5), 7.40 (m, C_6H_5)	[3, 4, 22, 26, 44, 52, 54, 55]

Table 8 [continued]

No.	X ligand	method of preparation (conditions, yield), properties, and remarks explanations on p. 94	Ref.
*6	I	^{57}Fe-γ (78 K): $\delta = 0.290$, $\Delta = 1.89$ ^{57}Fe-γ (298 K): $\delta = 0.231$, $\Delta = 1.88$ IR (CHCl$_3$): 1965 UV: λ_{max} (ε) = 350 (1028), 386 (700) in C$_6$H$_6$, 346 (922), 385 (717) in CH$_3$CN, 444, 620 in C$_5$H$_5$N, 385 (674) in CH$_3$NO$_2$	[2 to 5, 19, 28, 33, 44, 47, 49, 52, 54, 55]
		Ia (C$_6$H$_6$, CH$_3$CN, or CH$_3$NO$_2$), Ia (18 h, C$_6$H$_6$, 60 to 75%), Ia (4 h, THF, 68%), Ib (0.5 to 22 h, C$_6$H$_6$ at 80°, 33.7 to 37%), Ib (3 h, CH$_2$Cl$_2$, 14%), IIe (36.8%), IVa (15 min, C$_6$H$_6$, 44%), Vc (35 to 40%), IX (17 h, THF, 51 to 87%), X (71%), green or dark green crystals, m.p. 140, 151, 178 to 179° (dec.), 186, 230 to 235° (dec.) ^{1}H NMR (CDCl$_3$): 4.43 (d, C$_5$H$_5$, J(P,H) = 2), 7.37 (m, C$_6$H$_5$) ^{31}P NMR (CH$_2$Cl$_2$): 66.9 ^{57}Fe-γ (78 K): $\delta = 0.308$, $\Delta = 1.87$ ^{57}Fe-γ (298 K): $\delta = 0.233$, $\Delta = 1.85$ IR: 1935 in Nujol or C$_6$H$_6$, 1951 in CH$_2$Cl$_2$, 1955, 1961, or 1962 in CHCl$_3$ UV: λ_{max} (ε) = 342 (2090) in C$_6$H$_6$, 323 (2300) in CH$_3$CN	
*7	I · (SCN)$_2$	reddish brown ^{31}P NMR (CH$_2$Cl$_2$): 65.5 IR: 1978 in Nujol, ν(CN) 2120, 2142 in CH$_2$Cl$_2$, 166, 192 (CS), 798 (SI) in Nujol	[47]
*8	OOCCF$_3$	Ia (1 h, C$_6$H$_6$/C$_6$H$_{14}$ (1:40), 47%), IIa (5 min, C$_6$H$_6$, 20°, 45%) IIa (0.5 h, C$_6$H$_6$ at 5°, 70 to 80%), green or dark green crystals, m.p. 115 to 117° 122 to 123° (dec.) ^{1}H NMR: 4.48 (C$_5$H$_5$) in CHCl$_3$, 4.64 (s, C$_5$H$_5$), 7.57 (m, C$_6$H$_5$) in CDCl$_3$, 4.60 (d, C$_5$H$_5$, J(P,H) = 1) in CH$_3$COCH$_3$ ^{19}F NMR (CH$_2$Cl$_2$): 75.6 (CF$_3$) IR: 1961 in CS$_2$, 1974 in CH$_2$Cl$_2$	[9, 26, 46]

No.	Ligand	Data	Ref.
*9	$OS(O)CH_3$	Va, not isolated reaction intermediate in preparation of No. 11, ^{1}H NMR (SO_2, $-37°$): 1.57, 1.78 (s's, intensity ratio 63:37), 4.69 (br s, C_5H_5), 7.41, 7.51 (complex m's, C_6H_5), IR (SO_2, $-30°$): 1957	[30]
*10	$OS(O)CH_2C_6H_5$	Va, not isolated reaction intermediate in preparation of No. 17, IR (SO_2, $-30°$): 1957	[30]
*11	SO_2CH_3	Va (10 min, 95%), Vb (10 min at 27°, 80 to 90%), Vb (at 0°, 43%), Vb (at 25°, 17%), Vc (10 min, 41%), Vd, orange crystals, m.p. 158° ^{1}H NMR: 2.56 (s, CH_3), 4.62 (d, C_5H_5, J=1), 7.42 (m, C_6H_5) in $CDCl_3$, 2.38 (s, CH_3), 4.69 (d, C_5H_5, J=1.2), 7.48 (m, C_6H_5) in SO_2, $-37°$ ^{13}C NMR ($CDCl_3$): 58.9 (s, CH_3) ^{57}Fe-γ: $\delta=0.32$, $\Delta=1.74$ IR: 1956 in $CHCl_3$, 1969 in SO_2 ($-30°$), $\nu(SO)$ 1030, 1150 in $CHCl_3$, $\nu(SO)$ 1025, 1152 in Nujol	[11, 25, 30, 34, 41]
*12	$SO_2CH_2CH_3$	Va (10 min, 70%), Va (1 h, 60 to 65%), Vb (at 0°, 56%), Vc (10 min, 41%), m.p. 169 to 169.5° ^{1}H NMR ($CDCl_3$): 1.1 (t, CH_3, J=7), 2.7 (m, CH_2), 4.6 (d, C_5H_5, $J(P,H)\sim1$), 7.5 (m, C_6H_5) IR: 1973 in CH_2Cl_2, 1945, $\nu(SO)$ 1040, 1170 in $CHCl_3$, $\nu(SO)$ 1026, 1169 in Nujol	[17, 41]
13	$SO_2CH_2CD_3$	Va (1 h) ^{1}H NMR ($CDCl_3$): 2.72 (s, CH_2), 4.62 (d, C_5H_5, J=1.5), 7.44 (m, C_6H_5)	[17]
14	$SO_2CH_2CH=CH_2$	Ia (2 h, C_6H_6, 25%), orange, m.p. 85° IR: 1968 in $CHCl_3$, $\nu(SO)$ 1022, 1157 in Nujol	[11]
*15	$SO_2CH_2CH(CH_3)_2$	Va (10 min, 65%), Vb (at 0°, 16%), bright red parallel epipeds, m.p. 171.5 to 172° ^{1}H NMR ($CDCl_3$): 0.8 (t, J=6), 0.9 (t, J=6), 1.8 to 2.7 (m), 4.6 (d, C_5H_5, $J(P,H)\sim1$), 7.6 (m, C_6H_5) IR ($CHCl_3$): 1950, $\nu(SO)$ 1040, 1175	[41, 56]

References on p. 108

Table 8 [continued]

No.	X ligand	method of preparation (conditions, yield), properties, and remarks explanations on p. 94	Ref.
*16	$SO_2CH_2CH_2CH=CH_2$	Va (30 min, 54%), Vb (at 0°, 44%), orange solid, m.p. 164 to 166° ^{1}H NMR (CDCl₃): 2.3 to 2.8 (m), 4.7 to 5.2 (m), 4.7 (d, C₅H₅, J(P,H) = 1), 7.6 (m, C₆H₅), IR (CHCl₃): 1950, ν(SO) 1040, 1160	[41]
*17	$SO_2CH_2C_6H_5$	Va (1 h, <90%), Vb (at 0°, 61%), m.p. 147° ^{1}H NMR (CDCl₃): 3.94, 4.26 (CH₂, J = 13.5), 4.42 (d, C₅H₅, J(P,H) = 1), 7.5 (m, C₆H₅) IR: 1957 in CHCl₃, 1965 in liquid SO₂, ν(SO) 1035, 1166 in Nujol	[11, 21, 30, 41]
*18	$SO_2CH_2CH(CH_3)C_6H_5$	Va (5 h, 70%) see further information	[30]
*19	SO_2CH_2COO 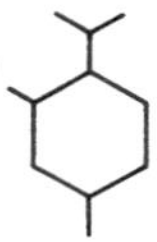 (cyclohexyl)	Va (3 h, 54%), Vc (3 h, 31%), yellow–orange glass ^{1}H NMR (CDCl₃): 1.6 to 2.4 (m, C₁₀H₁₈), 3.6 (s, CH₂), 4.6 (d, C₅H₅, J(P,H) ~ 1), 4.7 (m, CH₂O), 7.4 (m, C₆H₅) IR (film): 1970, ν(SO) 1040, 1175	[41]
*20	SCN	Ib (9 h, THF, ~20%, impure), olive brown ^{1}H NMR (CDCl₃): 4.60 (d, C₅H₅, J(P,H) = 2), 7.4 (m, C₆H₅) ^{57}Fe–γ (78 K): δ = 0.262, Δ = 1.789 ^{57}Fe–γ (293 K): δ = 0.184, Δ = 1.81 IR (CHCl₃): 1975, ν(CN) 2115	[52, 55]
*21	SeCN	IIb (17%), VIa (80%), dark red brown, m.p. 147 to 149° (dec.) ^{1}H NMR (CDCl₃): 4.60 (d, C₅H₅, J(P,H) = 1.3), 7.45 (m, C₆H₅) IR: 2108, ν(CN) 1974, 2007 in Nujol, 1993, ν(CN) 2120 in CHCl₃	[8]
*22	NCO	Ic (4 h, CH₂Cl₂, 75%), IVc (C₆H₆, 8%), IVd (66%), VIe (56%), VIf (37%), red or brownish red crystals, m.p. 129° (dec.), 131 to 132° (dec.) ^{1}H NMR (CDCl₃): 4.60 (C₅H₅), 7.48 (C₆H₅) IR: 1957, 1962, δ(NCO) 588, 600, νₛ(NCO) 1322, νₐₛ(NCO) 2245 in KBr, 1940 (CO), 1325, 2242 (NCO) in CHCl₃, 1953 (CO), 2231 (NCO) in CH₂Cl₂	[16, 19, 53]

References on p. 108

No.			Ref.
*23	NCS	Ib (2.5 h, C_6H_6 at 80°, 51%), for formation from No. 7, see p. 104, brown or dark brown, m.p. 141 to 143° ¹H NMR ($CDCl_3$): 4.52 (d, C_5H_5, $J(P,H) = 2$), 7.4 (m, C_6H_5) ⁵⁷Fe-γ (78 K): $\delta = 0.256$, $\Delta = 1.936$ ⁵⁷Fe-γ (293 K): $\delta = 0.1782$, $\Delta = 1.92$ IR: 1980, ν(CN) 2120 in $CHCl_3$, 1967 (CO), 817 (CS), 2112 (NC) in CH_2Cl_2 UV (THF): $\lambda_{max}(\varepsilon) = 308$ (2430), 435 (948), 550 (362) with $As(C_6H_5)_3$ UV irradiation gives $C_5H_5Fe(CO)(As(C_6H_5)_3)NCS$	[47, 52, 55]
*24	NCSe	IIb (20 min, C_6H_6 at 27°, 48%), light red brown, m.p. 139 to 141° (dec.) ¹H NMR ($CDCl_3$): 4.51 (d, C_5H_5, $J(P,H) = 1.3$), 7.4 (m, C_6H_5) IR: 1960, 1972 (CO), 663 (CSe), 2107 (CN) in Nujol, 1974 (CO), 2120 (CN) in $CHCl_3$	[8]
25	$N=C=C(CN)C(CN)_2CH_3$	VIII (65°, 2 h, trace), reddish brown oil IR (CH_2Cl_2): 1984, ν(N=C=C) 2134, ν(CN) 2212 solid and solutions very unstable in air	[17, 39]
26	$N=C=C(CN)C(CN)_2C_2H_5$	VIII (65°, 1 h, 33%), reddish brown oil ¹H NMR ($CDCl_3$): 5.3 (s, C_5H_5), 7.3 to 7.5 (m, C_6H_5) IR (CH_2Cl_2): 1990, ν(N=C=C) 1302, 2145, ν(CN) 2212 solid and solutions very unstable in air	[17, 39]
27	$N=C=C(CN)C(CN)_2C_3H_7\text{-}n$	VIII (27°, 12 h, 20%), brownish red oil IR (neat oil): 1979, ν(N=C=C) 1306, 2134, ν(CN) 2201 solid and solutions very unstable in air	[17, 39]
28	$N=C=C(CN)C(CN)_2CH_2C_6H_5$	IIc (2 to 3 h, C_6H_6, 5 to 10°, 45%), brownish red, m.p. 138 to 139° ¹H NMR ($CDCl_3$): 3.37 (d, CH_2), 4.87 (d, C_5H_5, $J(P,H) \sim 1$), 7.66 (m, $P(C_6H_5)_3$), 7.78 (s, CC_6H_5) IR (CH_2Cl_2): 1979, ν(N=C=C) 1298, 2145, ν(CN) 2201 solid appears to be stable indefinitely, no sign of dec. after 18 months, moderately stable in C_6H_6 solution	[13, 17, 39]

References on p. 108

Table 8 [continued]

No.	X ligand	method of preparation (conditions, yield), properties, and remarks explanations on p. 94	Ref.
*29	CN	Ib (C_6H_6 at 80°, 90%), IIb (7%), VIb (67.3%), VIc (60%), VIII, yellow powder or orange crystals, m.p. 225° (dec.), 228 to 230°, 230 to 235° (dec.) ^{1}H NMR ($CDCl_3$): 4.57 (d, C_5H_5, $J(P,H) = 1.2$), 7.40 (m, C_6H_5) ^{13}C NMR ($CDCl_3$): 84.17 (s, C_5H_5, $J(P,C) < 0.2$), 128.55 (d, C-3,5 in C_6H_5, $J(P,C) = 9.7$), 130.54 (s, C-4, $J(C,P) = 2.0$), 133.22 (d, C-2,6, $J(P,C) = 9.5$), 134.95 (d, C-1, $J(P,C) = 44.0$), 217.55 (d, CO, $J(P,C) = 27.2$) ^{13}C NMR ($CDCl_3$): 84.27 (s, C_5H_5), 128.44 (d, C-2,6 in C_6H_5, $J = 10.0$), 130.55 (d, C-4 in C_6H_5, $J = 2.0$), 133.34 (d, C-3,5 in C_6H_5, $J = 9.9$), 135.09 (d, C-1 in C_6H_5, $J = 46.3$), 218.21 (d, CO, $J = 28.3$), for the CN signals, see further information ^{57}Fe-γ (78 K): $\delta = 0.1347$, $\Delta = 1.93$ ^{57}Fe-γ (293 K): $\delta = 0.0576$, $\Delta = 1.89$ IR: 1945, 1972, ν(CN) 2094 in Nujol, 1969, 1971, 1975, ν(CN) 2090 in $CHCl_3$, 1959, ν(CN) 2099 in C_6H_6	[8, 17, 33, 37, 55]
*30	$Si(CH_3)_3$	Ia (48 h, hexane, 39%), m.p. 163 to 164° ^{1}H NMR ($CDCl_3$): 0.01 (CH_3), 4.17 (C_5H_5), 7.3, 7.5 (C_6H_5) IR (C_6H_{12}): 1916 mass spectrum: $[M]^+$, $[M-H]^+$, $[M-H_2]^+$, $[M-CO]^+$, $[M-CO-Si(CH_3)_3]^+$, $[M-CO-Si(CH_3)_3-H]^+$, $[M-CO-Si(CH_3)_3-H_2]^+$, $[M-P(C_6H_5)_3]^+$, $[M-CO-Si(CH_3)_3-C_6H_5-H]^+$, $[M-CO-Si(CH_3)_3-C_6H_5-H_2]^+$, $[M-C_5H_5-CO-Si(CH_3)_3]^+$	[6, 57]
31	$Si(CH_3)_2CH_2Si(CH_3)_3$	Ia ^{1}H NMR (CS_2): -0.22 (d, CH_2, $J = 4$), 0.02 ($Si(CH_3)_3$), 0.03, 0.16 ($Si(CH_3)_2$), 4.11 (d, C_5H_5, $J = 0.2$) IR (C_6H_{12}): 1901	[29]

References on p. 108

No.	Substituent	Data	Ref.
32	$Si(C_6H_5)_3$	Ia (C_6H_{14}, 90%), orange, m.p. 199° (dec.) ^1H NMR ($CDCl_3$): 4.03 (d, C_5H_5, J(P,H) ~1.5) IR (C_6H_{12}): 1916 no reaction with $[N(C_2H_5)_4]Cl$ or $[N(C_4H_9-n)_4]Br$ in CCl_4, only some decomposition (<10%)	[23, 45]
33	$Si(C_6H_5)_2CH_3$	Ia (C_6H_{14}, 80%), orange, m.p. 173 to 174° ^1H NMR ($CDCl_3$): −0.15 (CH_3), 4.16 (d, C_5H_5, J(P,H) ~1.5) IR (C_6H_{12}): 1895	[45]
*34		Ia (C_6H_{14}, 90%), orange, see further information	[32, 45]
*35	$Sn(CH_3)_3$	Ia (C_6H_6 or CH_3COCH_3), Ia (>5 h, C_6H_{14}, 40 to 61% depending on the excess of $P(C_6H_5)_3$), orange, m.p. 125 to 127°, 127 to 128° ^1H NMR: −0.07 (s, CH_3, $^2J(^{119}Sn, H) = 40.4$), 4.23 (d, C_5H_5, J(P,H) = 1.2), 7.40 (m, C_6H_5) in $CDCl_3$, 4.1 to 4.2 (C_5H_5), 7.2 to 7.4 (C_6H_5) in THF ^{57}Fe-γ (78 K): δ = 0.48, Δ = 1.87 ^{119}Sn-γ (78 K): δ = 1.41, Δ = 0.42 IR (C_6H_{12}): 1911, 1921	[6, 12, 14]
*36	$Sn(CH_3)_2C_6H_5$	Ia (like No. 35, 46%), red oil ^1H NMR (C_5D_5N): −0.09 (CH_3, $^2J(^{117}Sn, H) = 39.6$, $^2J(^{119}Sn, H) = 41.6$), −0.03 (CH_3, $^2J(^{117}Sn, H) = 38.8$, $^2J(^{119}Sn, H) = 40.8$), 4.18 (C_5H_5, J(P,H) ~2), 7.1 to 7.5 (C_6H_5) ^{13}C NMR (C_5D_5N): −5.08, −3.98 (CH_3), 81.63 (C_5H_5), 128 to 134 (C_6H_5), 224 (CO, J(P,C) = 30) IR (CS_2): 1908	[35]
*37	$Sn(C_6H_5)_3$	Ia (like No. 35, 46 to 69%), orange crystals, m.p. 177 to 180°, 192 to 194° ^1H NMR: 4.38 (C_5H_5), 6.93 to 7.43 (C_6H_5) in $CDCl_3$, 4.22 (d, C_5H_5, J(P,H) ~1.2) in CS_2 ^{57}Fe-γ (78 K): δ = 0.46, Δ = 1.84 ^{57}Fe-γ (298 K): δ = 0.39, Δ = 1.79 ^{119}Sn-γ (78 K): δ = 1.48, Δ = 0.69 ^{119}Sn-γ (80 K): δ = 1.53, Δ = 0.66 IR (C_6H_{12}, $CHCl_3$): 1911 to 1913, 1923 to 1924	[12, 14, 31]

Table 8 [continued]

No.	X ligand	method of preparation (conditions, yield), properties, and remarks explanations on p. 94	Ref.
38	$Sn(C_6H_5)_2Cl$	No. 37 and HCl (47%), orange, m.p. 172 to 175° ^{1}H NMR $(CDCl_3)$: 4.52 (d, C_5H_5, $J(P,H) = 1.5$) ^{57}Fe–γ (298 K): $\delta = 0.39$, $\Delta = 1.73$ ^{119}Sn–γ (80 K): $\delta = 1.63$, $\Delta = 2.74$ IR $(CHCl_3)$: 1932	[27, 31]
39	$SnCl_2C_6H_5$	No. 37 and HCl (77%), red, m.p. 164 to 168° ^{1}H NMR $(CDCl_3)$: 4.68 (d, C_5H_5) ^{57}Fe–γ (298 K): $\delta = 0.40$, $\Delta = 1.73$ ^{119}Sn–γ (80 K): $\delta = 1.80$, $\Delta = 3.00$ IR $(CHCl_3)$: 1949	[27, 31]
40	$SnCl_3$	Ia (10 to 20 min, C_6H_6, 16%), Ia (5 h, C_6H_6 or CH_3COCH_3, 21%), No. 37 and HCl (95%), red, maroon, or cherry-red needles, m.p. 175 to 185°, 183 to 185, 195° ^{1}H NMR: 4.79 (d, C_5H_5, $J(P,H) = -1.2$), 7.32 to 7.60 (C_6H_5) in $CDCl_3$, 4.67 (d, C_5H_5, $J(P,H) = \pm 0.5$), 7.40 (m, C_6H_5) in CS_2 ^{57}Fe–γ (78 K): $\delta = 0.47$, $\Delta = 1.85$ ^{57}Fe–γ ($\sim$80 K): $\delta = 0.248$ (Fe), $\Delta = 1.86$ ^{57}Fe–γ (298 K): $\delta = 0.41$, $\Delta = 1.80$ ^{119}Sn–γ ($\sim$80 K): $\delta = 1.86$ to 1.88, $\Delta = 1.88$ to 1.89 IR: 1965 in LiF, 1960, 1973 in KBr, 1969 in C_6H_{12} and CH_2Cl_2, 1973 in $CHCl_3$ solid stable in air, solutions decompose over a period of days	[12, 14, 27, 31, 36]
supplement			
41	$SO_2CH_2CH_2CH_3$	Va (10 min, 57%), m.p. 167.5 to 168° ^{1}H NMR $(CDCl_3)$: 0.8 (t, CH_3, $J = 7$), 1.7 (m, CH_2), 2.6 (m, CH_2), 4.6 (d, C_5H_5, $J(P,H) \sim 1$), 7.5 (m, C_6H_5) IR $(CDCl_3)$: 1950 (CO), 1045, 1170 (SO)	[41]

*Further information:

$C_5H_5Fe(CO)(P(C_6H_5)_3)H$ (Table **8**, No. **1**). A complete IR spectrum is given (Nujol, 510 to 1930 cm^{-1}) [17]. The pure compound is stable in air [17, 18]. Refluxing in heptane for 2 h or UV irradiation in petroleum ether causes extensive decomposition [17]. The reaction with $Mn_2(CO)_{10}$ gave the following products, separated on SiO_2 with hexane/toluene mixtures: $Mn_2(CO)_{10}$ eluted first, followed by $HMn(CO)_4P(C_6H_5)_3$, $Mn_2(CO)_8(P(C_6H_5)_3)_2$, and $(C_5H_5Fe(CO)_2)_2$ [59]. C_5H_5-$Fe(CO)(P(C_6H_5)_3)H$ reacts readily with $CHCl_3$ or CCl_4 to give $C_5H_5Fe(CO)$-$(P(C_6H_5)_3)Cl$ (No. **4**) [17, 18]. Reaction with $CF_2=CF_2$ in THF at 25 °C gives a yellow 1:1 adduct, which slowly loses HF and forms $C_5H_5Fe(CO)(P(C_6H_5)_3)CF=CF_2$ (Table **18**, No. **11**, in 1.5.2.2.12.3.4, p. 173) [17]. No reaction is observed with $CH_2=CH_2$ or $CH_2=CHCN$ in benzene [43]. IR monitoring showed that $C_5H_5Fe(CO)(P(C_6H_5)_3)H$ quantitatively reduces the $[C_5H_5Fe(CO)_2CH_2=CR'R'']PF_6$ salts in CH_3CN to C_5H_5-$Fe(CO)_2CR'R''CH_3$ and $[C_5H_5Fe(CO)(P(C_6H_5)_3)NCCH_3]^+$ with $R'=R''=H$ or CH_3, $R'=H$, $R''=CH_3$. The hydride reacts with vinyl ether complexes, $[C_5H_5Fe(CO)_2$-$CH_2=CHOR]PF_6$ by H$^-$ addition at the substituted vinyl carbon to give a 1:1 mixture of $C_5H_5Fe(CO)_2CH_2CH_2OR$ ($R=CH_3$, C_2H_5) and $C_5H_5Fe(CO)_2CH_2CH_3$. Alkoxy-carbene complexes of the $[C_5H_5Fe(CO)_2C(OCH_3)R]PF_6$ type quantitatively consume one equivalent of the hydride to give $C_5H_5Fe(CO)_2CH(OCH_3)R$ ($R=H$, CH_3, C_6H_5). There was no evidence of further reduction to $C_5H_5Fe(CO)_2CH_2R$ compounds [58].

$C_5H_5Fe(CO)(P(C_6H_5)_3)Cl$ (Table **8**, No. **4**) is also produced by decomposition of $C_5H_5Fe(CO)(P(C_6H_5)_3)C_6H_5$ (Table **19**, No. **1**, in 1.5.2.2.12.3.6, p.196) with the HCl generated during Friedel–Crafts acetylation [7]. It could not be purified by standard crystallization or chromatography techniques. These procedures always caused some decomposition to give a mixture of No. **4** and small amounts of $P(C_6H_5)_3O$ [3, 4].

$C_5H_5Fe(CO)(P(C_6H_5)_3)Br$ (Table **8**, No. **5**). Preparation from $C_5H_5Fe(CO)_2X$ ($X=Br$) with $P(C_6H_5)_3$ in refluxing benzene for 18 h forms only $C_5H_5Fe(CO)$-$(P(C_6H_5)_3)Br$. In the case of $X=Cl$ or I the ionic products $[C_5H_5Fe(CO)_2P(C_6H_5)_3]X$ form in approximately equimolar quantities with $C_5H_5Fe(CO)(P(C_6H_5)_3)X$ [3]. Solid $C_5H_5Fe(CO)(P(C_6H_5)_3)Br$ is stable in air, but its solutions are somewhat less stable [4]. Photolysis by visible, blue, or green light in pyridine or dimethyl sulfoxide yields $[C_5H_5Fe(CO)_2P(C_6H_5)_3]Br$ [22]. Reaction with a threefold excess of C_6H_5Li gives $C_5H_5Fe(CO)_2C_6H_5$ and $C_5H_5Fe(CO)(P(C_6H_5)_3)C_6H_5$. The latter is also formed with a twofold excess of C_6H_5MgBr in ether [26]. Heating No. **5** and KSeCN in refluxing acetone for 30 min forms $P(C_6H_5)_3Se$ [8].

$C_5H_5Fe(CO)(P(C_6H_5)_3)I$ (Table **8**, No. **6**) is reduced polarographically in di-methylformamide containing 0.1 N $[(C_4H_9)_4N]BF_4$: In a first step at $E_{1/2}=-1.26$ V to the radical $[C_5H_5Fe(CO)P(C_6H_5)_3]$, which shows no tendency to dimerize, and in a second step at $E_{1/2}=-1.59$ V to the anion $[C_5H_5Fe(CO)P(C_6H_5)_3]^-$ [15]. No. **6** is stable in air [4, 5], the solutions are less stable [4]. It is not decomposed in refluxing benzene over a period of 20 h although better crystals were produced [49]. It is sensi-tive to oxidation on heating in solution [5]. The reaction with excess KCN in refluxing ethanol yields $K[C_5H_5Fe(CN)_2P(C_6H_5)_3]$ [37]. Compound No. **6** reacts with $(SCN)_2$ to give $C_5H_5Fe(CO)(P(C_6H_5)_3)I\cdot(SCN)_2$ (No. **7**) and $[C_5H_5Fe(CO)_2P(C_6H_5)_3]$-$[I\cdot(SCN)_2]$ [47]. The reaction of No. **6** with LiR in THF is a direct route to C_5H_5-$Fe(CO)(P(C_6H_5)_3)R$ derivatives [42] giving reasonably high yields. In THF No. **6** reacts at -20 °C with $LiC_5H_4Mn(CO)_3$ to give $C_5H_5Fe(CO)(P(C_6H_5)_3)C_5H_4Mn(CO)_3$ (Table **20**, No. **3**, in 1.5.2.2.12.3.7, p. 200). The treatment of racemic $C_5H_5Fe(CO)$-

$(P(C_6H_5)_3)I$ with racemic $(CH_3)_3Si(C_6H_5)CHMgBr$ in benzene/ether (2:1) at 35 °C yields stable products having the formula $C_5H_5Fe(CO)(P(C_6H_5)_3)CH(C_6H_5)Si(CH_3)_3$ (Table 18, No. 40, in 1.5.2.2.12.3.4, p. 173) [38]. With C_6H_5NC in benzene at room temperature No. 6 gives a mixture of $C_5H_5Fe(CO)(CNC_6H_5)I$ and C_5H_5-$Fe(CNC_6H_5)_2I$ [1]. It reacts with $C_{10}H_8N_2$ (2,2′-bipyridine) or $C_{12}H_8N_2$ (1,10-phenanthroline) in boiling benzene to produce $[C_5H_5Fe(C_{10}H_8N_2)P(C_6H_5)_3]I$ and $[C_5H_5Fe(C_{12}H_8N_3)P(C_6H_5)_3]I$, respectively [44], see 1.5.1.2, p. 2.

$C_5H_5Fe(CO)(P(C_6H_5)_3)I \cdot (SCN)_2$ (Table 8, No. 7) is prepared from a solution of $C_5H_5Fe(CO)(P(C_6H_5)_3)I$ (No. 6) in ice-cooled CH_2Cl_2 to which an equimolar amount of $(SCN)_2$ in CH_2Cl_2 is added drop by drop. After 1 to 2 h the mixture is filtered and alcohol is added to the concentrated filtrate to precipitate No. 7. It is purified by recrystallization in CH_2Cl_2/C_2H_5OH (80% yield). The solid is stable in air at room temperature and is not particularly sensitive to hydrolysis. Heating several hours in CH_2Cl_2 decomposes it into I_2, $(SCN)_2$, and $C_5H_5Fe(CO)(P(C_6H_5)_3)NCS$ (No. 23) [47].

$C_5H_5Fe(CO)(P(C_6H_5)_3)OOCCF_3$ (Table 8, No. 8) is of 30 to 60% enantiomeric excess when prepared by Method IIa from the methyl and ethyl complexes. Excess starting material is significantly racemized. This was shown by NMR spectroscopy in the presence of optically active $Eu(opt)_3$ (Formula I): The CF_3COO group is basic enough to complex with this reagent and to form diastereomeric complexes which give rise to two diastereotopic C_5H_5 resonances separated by up to 0.2 ppm. No. 8 is racemized very slowly in the presence of CF_3COOH. Thus the observed racemization in the formation by cleavage Method IIa is not likely due to product instability to protic acids. No conclusion could be drawn from the circular dichroism spectrum whether No. 8 had retained or inverted stereochemistry at iron [46].

I   II

Compound No. 8 is quite stable in solution at room temperature, particularly when compared with $C_5H_5Fe(CO)_2CH_2=CHCH_3$ [26].

$C_5H_5Fe(CO)(P(C_6H_5)_3)OS(O)R$ (R = CH_3, $CH_2C_6H_5$, Table 8, Nos. 9 and 10). Dissolution of $C_5H_5Fe(CO)(P(C_6H_5)_3)R$ in liquid SO_2 gives a monocarbonyl whose IR $\nu(CO)$ band differs from that of the parent alkyl compound and from the final S-sulfinato compound. The ^1H NMR spectrum of this intermediate shows CH_3 signals at higher fields than for the S-sulfinato compound. In all cases investigated, Fe-$OS(O)CH_x$ protons have been found to absorb at higher fields than those of the isomeric Fe-$S(O)_2CH_x$. On storage Nos. 9 and 10 convert to the S-sulfinato compounds No. 11 and 17, respectively [30].

$C_5H_5Fe(CO)(P(C_6H_5)_3)SO_2CH_3$ (Table 8, No. 11). In the presence of about 70 wt % of the shift reagent $Eu(opt)_3$ (Formula I), the CH_3 and C_5H_5 resonances are doublets centered at $\delta = 3.78$ and 5.65 ppm, respectively. Peak integration thus provides a direct measure of optical purity [34]. Optical rotation (concentration

References on p. 108

~1 mg/mL in CH_2Cl_2): $[\alpha]_{578}^{25} = -138°$ (70% purity), $-177°$ (90% purity), $-179°$ (91% purity), and $-196°$ (pure) [41]. The combined $\nu(CO)$ frequencies and Mössbauer isomer shifts indicate that the σ-donor ability of RCO is greater than that of RSO_2 [25].

$C_5H_5Fe(CO)(P(C_6H_5)_3)SO_2C_2H_5$ (Table **8**, No. **12**). Optical rotation (concentration ~1 mg/mL in CH_2Cl_2): $[\alpha]_{578}^{25} = -255°$ (pure, as determined by NMR with Eu(opt)$_3$, Formula I) [41].

$C_5H_5Fe(CO)(P(C_6H_5)_3)SO_2CH_2CH(CH_3)_2$ (Table **8**, No. **15**). Optical rotation (concentration ~1 mg/mL in CH_2Cl_2): $[\alpha]_{578}^{25} = -229°$ (95% purity, as determined by NMR with Eu(opt)$_3$, Formula I) [41]. The absolute configuration was shown by X-ray anomalous dispersion techniques to be S (for sequencing rules, see Stanley, Baird [38]). The compound crystallizes in the orthorhombic space group $P2_12_12_1$-D_2^4 (No. 19) with a=13.800(4), b=13.523(5), c=13.649(5) Å, Z=4. D_c=1.39, D_m=1.38 g·cm^{-3} by flotation in $ZnCl_2$ solution. Refinement of the structure with two Friedel-related sets of data (3059 reflections) yielded R=6.4% for an S configuration and R=7.0% for an R configuration. This difference, according to Hamiltons' R-factor significance test, indicates that the probability of the S form being the correct one is well over 99.5%. The overall geometry and absolute stereochemistry of $(-)_{578}(S)$-$C_5H_5Fe(CO)(P(C_6H_5)_3)SO_2CH_2CH(CH_3)_2$ are shown in **Fig. 10**. This structure confirms that the stereochemistry of the SO_2 insertion involves retention of configuration at iron [56].

Fig. 10

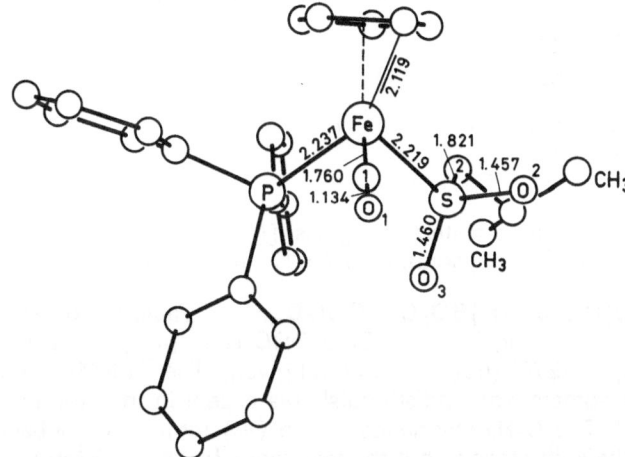

Molecular structure of $(-)_{578}$-(S)-$C_5H_5Fe(CO)(P(C_6H_5)_3)SO_2CH_2CH(CH_3)_2$ [56].

Bond angles (°):

C(1)–Fe–P	95.8(3)	Fe–C(1)–O	173.1(8)
C(1)–Fe–S	92.3(3)	O(2)–S–O(3)	114.4(4)
P–Fe–S	93.0(1)	O(2)–S–C(2)	104.6(4)
Fe–S–O(2)	113.0(3)	O(3)–S–C(2)	103.2(4)
Fe–S–O(3)	110.8(3)	C(2)–S–Fe	110.2(3)

References on p. 108

$C_5H_5Fe(CO)(P(C_6H_5)_3)SO_2CH_2CH_2CH=CH_2$ (Table **8**, No. **16**). Optical rotation (concentration ~ 1 mg/mL in CH_2Cl_2): $[\alpha]_{578}^{25} = -57°$ (40% purity) was determined by NMR with $Eu(opt)_3$ (Formula I on p. 104) [41].

$C_5H_5Fe(CO)(P(C_6H_5)_3)SO_2CH_2C_6H_5$ (Table **8**, No. **17**) is not decarbonylated by treatment with $Rh(P(C_6H_5)_3)_3Cl$ in benzene [21]. Optical rotation (concentration ~ 1 mg/mL in CH_2Cl_2): $[\alpha]_{578}^{25} = -275°$ (pure) was determined by NMR with $Eu(opt)_3$ (Formula I on p. 104) [41].

$C_5H_5Fe(CO)(P(C_6H_5)_3)SO_2CH_2CH(CH_3)C_6H_5$ (Table **8**, No. **18**) has two chiral centers: the Fe atom and the C atom of the CH group in the X ligand. These two centers permit four enantiomers but four pure forms have not been prepared and the steric arrangements are still uninvestigated. From the optically active alkyl compounds (Table 18, No. 26, in 1.5.2.2.12.3.4, p.173) the following red compounds were prepared:

| | m.p. in °C | ¹H NMR (CDCl₃) | | | | IR (ν in cm⁻¹) | |
		CH₃ (J in Hz)	CH or CH₂	C₅H₅ (J in Hz)	C₆H₅	ν(SO) (KBr)	ν(CO) (CH₂Cl₂)
$\dfrac{(RR)(SS)}{(RS)(SR)} = \dfrac{95}{5}$	66 to 83	1.37 d (6.5)	2.64 to 3.76 m	4.68 d (1.3)	7.34 m	1035 1160, 1180	1957
$\dfrac{(RR)(SS)}{(RS)(SR)} = \dfrac{16}{84}$	55 to 71	1.23 d (6.5)	2.67 to 3.63 m	4.68 d (1.3)	7.38 m	1035 1160, 1180	1956
$\dfrac{(RS)}{(SS)} = \dfrac{95}{5}$	149 to 155						
$\dfrac{(RS)}{(SS)} = \dfrac{26}{74}$	69 to 84						

The amorphous compound sinters to a glass. $[(RR)(SS)] : [(RS)(SR)] = 95:5$ is configurationally stable in $CDCl_3$ solution at 25 °C over a period of 10 d [30].

$C_5H_5Fe(CO)(P(C_6H_5)_3)SO_2CH_2COOC_{10}H_{19}$ (Table **8**, No. **19**). Optical rotations (concentration 2 mg in 1 mL $CH_3COOC_2H_5$): $[\alpha]_{578}^{25} = +206°$ (optical purity 78%) and $[\alpha]_{578}^{25} = -270°$ (optical purity not given). The ¹H NMR spectra of the (+)- and (−)-diastereomers were indistinguishable except in the presence of Eu shift reagents [24, 41]. The C_5H_5 resonance of the (+) form is not substantially affected by addition of the shift reagent, but the resonance for the (−) form is moved upfield. The pattern of the $Eu(fod)_3$-d_{27} (Formula II on p. 104) shifted NMR and the circular dichroism spectra are given [24].

$C_5H_5Fe(CO)(P(C_6H_5)_3)SCN$ (Table **8**, No. **20**). All attempts to remove traces of phosphine by chromatography and fractional crystallization proved unsuccessful. If the percentage of $P(C_6H_5)_3$ present in the sample is taken into account, the upper limit for the integrated absorption intensity of the CN stretching band is 6×10^4 $M^{-1} \cdot cm^{-2}$, indicative of an Fe-SCN linkage. Compare also the spectra and further information for No. 23 [52].

References on p. 108

C₅H₅Fe(CO)(P(C₆H₅)₃)SeCN (Table **8**, No. **21**). The ν(CO) and ν(CN) bands of the IR spectra in CH_3CN, C_6H_6, $CHCl_3$, CH_2Cl_2, and $(CH_3)_2NCHO$ are given in figures. The compound is an air-stable solid, soluble in chloroform, benzene, and acetone, sparingly soluble in methanol, ethanol, and ether, and insoluble in hexane and CCl_4. With respect to conversion to the No. 29 it is more stable than No. 24. After 3 h in boiling benzene only a trace of compound No. 29 and some unidentified, noncarbonyl decomposition material are detected. Addition of $P(C_6H_5)_3$ to a benzene solution at 27 °C gives no noticeable deselenation in 3 h; however, on refluxing ~50% conversion to compound No. 29 occurs in 3 h. No isomerization is detected in $CHCl_3$ or $(CH_3)_2NCHO$ at 27 °C after several hours or in Nujol suspension after 2 d [8].

C₅H₅Fe(CO)(P(C₆H₅)₃)NCO (Table **8**, No. **22**) is stable in air. It is soluble in benzene, ether, CH_2Cl_2, and $CHCl_3$ but only sparingly soluble in pentane and hexane. Its CH_2Cl_2 and $CHCl_3$ solutions decompose very slowly [16]. It reacts with NH_4I in benzene to give $C_5H_5Fe(CO)(P(C_6H_5)_3)I$ (No. 6) and NH_4NCO [19].

C₅H₅Fe(CO)(P(C₆H₅)₃)NCS (Table **8**, No. **23**). Photolysis of $C_5H_5Fe(CO)_2SCN$ and $P(C_6H_5)_3$ results in a mixture of $C_5H_5Fe(CO)_2SCN$, $C_5H_5Fe(CO)_2NCS$, and their substitution products $C_5H_5Fe(CO)(P(C_6H_5)_3)SCN$ and $C_5H_5Fe(CO)(P(C_6H_5)_3)NCS$. Neither $C_5H_5Fe(CO)(P(C_6H_5)_3)SCN$ nor $C_5H_5Fe(CO)(P(C_6H_5)_3)NCS$ isomerizes upon 366- or 436-nm irradiation in degassed THF for up to 4 h. The ^{13}CO incorporation and $P(C_6H_5)_3$ substitution, which parallel the behavior of halide complexes, strongly suggest dissociation of CO from the excited-state $C_5H_5Fe(CO)_2NCS^*$ and $C_5H_5Fe(CO)_2SCN^*$ as the primary photochemical process. Rearrangement of the thio-cyanato ligand and nucleophilic attack on the coordinatively unsaturated intermediates by CO or $P(C_5H_5)_3$ are likely secondary processes. The total inhibition of Fe–NCS → Fe–SCN but not Fe–SCN → Fe–NCS isomerization by CO or $P(C_6H_5)_3$ may be a consequence of the expected Fe–N–C and Fe–S–C bond angles of ~180° and ~105°, respectively [52].

C₅H₅Fe(CO)(P(C₆H₅)₃)NCSe (Table **8**, No. **24**). The figures of ν(CO) and ν(CN) of the IR spectra in CH_3CN, C_6H_6, $CHCl_3$, CH_2Cl_2, and $(CH_3)_2NCHO$ are given. The solid compound is stable in air. It is soluble in chloroform, benzene, and acetone, sparingly soluble in methanol, ethanol, and ether, and insoluble in hexane and CCl_4. No isomerization was detected in $CHCl_3$ or $(CH_3)_2NCHO$ at 27 °C after several hours or in Nujol suspension after two days. The deselenation, which forms C_5H_5Fe-$(CO)(P(C_6H_5)_3)CN$ (No. 29), occurs slowly in benzene at room temperature; however, on refluxing the process is complete in less than two hours. The same deselenation is found in CH_2Cl_2 in the presence of $P(C_6H_5)_3$ at −72 °C after ~85 h, at −23, and 0 °C [8].

C₅H₅Fe(CO)(P(C₆H₅)₃)CN (Table **8**, No. **29**) is formed from No. 24 by decomposition [8] (see further information for No. 24) and is also formed in refluxing ethanol from KCN and $[C_5H_5Fe(CO)(P(C_6H_5)_3)CNC_2H_5]PF_6$ [37]. The solid is air-stable but decomposes in $CHCl_3$ on standing. It is insoluble in C_6H_{12} [33]. In the ^{13}NMR spectrum, the CN signal could not be located. Apparently it lies under the phenyl resonances [37]. Relative shifts in the ^1H NMR spectra upon addition of $Eu(fod)_3$ (Formula II on p. 104) or $Yb(opt)_3$ (analogous to $Eu(opt)_3$, Formula I on p. 104) at room temperature and the relative shift in the ^{13}C NMR spectra upon addition of $Eu(fod)_3$ are given [33, 37]. Mass spectrum: 437 (molecular ion), 409, 383, and 262 ($P(C_6H_5)_3$) [37]. The compound can not be substituted further [37].

C₅H₅Fe(CO)(P(C₆H₅)₃)Si(CH₃)₃ (Table **8**, No. **30**). Appearence potential AP, ionization potential IP, and dissociation energie D in eV from mass spectrum [57]:

References on p. 108

$[M-CO]^+$:   AP=7.59,   D=0.73; $[M-P(C_6H_5)_3]^+$:   AP=7.43,   D=0.57;
$[M-Si(CH_3)_3]^+$: AP=9.48, IP=6.86, D=2.62.

$C_5H_5Fe(CO)(P(C_6H_5)_3)Si(CH_3)(C_6H_5)C_{10}H_7-1$ (Table 8, No. 34) has two chiral centers: the Fe atom and the Si atom. From these two centers there are four enantiomers (+)a, (−)a, (+)b, and (−)b. The steric arrangement in the molecule has not yet been investigated. Only the (+)a form has been obtained chemically pure, this by repeated recrystallization in a mixture of CH_2Cl_2/C_6H_{14} (30/70 volume ratio). The parent compound (=Pc) $C_5H_5Fe(CO)_2Si(CH_3)(C_6H_5)C_{10}H_7-1$ reacts with $P(C_6H_5)_3$ (Method Ia) to give the orange compound No. 34. (−)(S)-Pc affords a mixture of ~60% (+)a and ~40% (−)b, while (+)(R)-Pc affords a mixture of (−)a and (+)b. Each mixture is separated by fractional crystallization in CH_2Cl_2/C_6H_{14} (30:70 volume ratio). (−)a: m.p. 217 °C (dec.), (+)b: m.p. 173 °C (dec.).

Optical rotation: mixture ~60% (+)a and ~40% (−)b: $[\alpha]_D^{20}=-2.8°$, $[\alpha]_{578}^{25}=+0.7°$, $[\alpha]_{546}^{25}=+17.8°$ (c=0.45, cyclohexane); mixture (−)a and (+)b: $[\alpha]_D^{25}=+2.1°$; (+)a (pure): $[\alpha]_D^{25}=+139°$, $[\alpha]_{578}^{25}=+157.6°$, $[\alpha]_{546}^{25}=259.3°$ (c=0.22, cyclohexane); (−)a (with 9% (+)b): $[\alpha]_D^{25}=-112°$; (+)b (with 17% (−)a): $[\alpha]_D^{25}=+128°$; (−)b (with 8% (+)a): $[\alpha]_D^{25}=-163°$, $[\alpha]_{578}^{25}=-182°$, $[\alpha]_{546}^{25}=-262°$ (c=0.44, cyclohexane).

^{1}H NMR ($CDCl_3$): (+)a (pure): δ=4.14 ppm; (−)a (with 9% (+)b): δ=0.24 (CH_3), 4.14 (C_5H_5, J~1.5 Hz) ppm; (+)b (with 17% (−)a): δ=0.24 (CH_3), 4.28 (C_5H_5, J~1.5 Hz) ppm; (−)b (with 8% (+)a): δ=4.28 ppm. IR (C_6H_{12}): (−)a (with 9% (+)b): 1910 (CO) cm^{-1}; (+)b (with 17% (−)a): 1917 (CO) cm^{-1}.

The a forms are practically insoluble in hexane, while the b forms are quite soluble. The compound is quite stable towards HCl, $NaBH_4$, CH_3OH, CH_3ONa, KHF_2, Li_2O, MgO, Na, and Li. The Fe-Si σ bond is cleaved by H_2O or $LiAlH_4$ in ether. Also Br_2 or I_2 in CCl_4 cleave the bond. The stereoselectivity about the Si atom in this reaction is described in [32, 45].

$C_5H_5Fe(CO)(P(C_6H_5)_3)Sn(CH_3)_3$ (Table 8, No. 35). The reported m.p. of 95 to 103 °C (and 125 to 127 °C) are for a compound containing 0.5 mol of benzene per mole of complex [14]. A linear correlation exists between the geminal M proton coupling constants (M=^{117}Sn, ^{119}Sn) and the Taft σ* constants of the transition metal complex group [10].

$C_5H_5Fe(CO)(P(C_6H_5)_3)Sn(CH_3)_2C_6H_5$ (Table 8, No. 36). No coupling of the C atoms of the C_5H_5 and the CH_3 groups is observed with the P atom. The anisochronism of the CH_3 groups in the ^{1}H NMR spectra show that the compound is configurationally stable around both the iron atom and the tin atom [35].

$C_5H_5Fe(CO)(P(C_6H_5)_3)Sn(C_6H_5)_3$ (Table 8, No. 37) reacts in benzene with HCl in ether (mole ratio 1:1) to give $C_5H_5Fe(CO)(P(C_6H_5)_3)Sn(C_6H_5)_2Cl$ (No. 36). It reacts with HCl (mole ratio 1:2) to give $C_5H_5Fe(CO)(P(C_6H_5)_3)SnCl_2C_6H_5$ (No. 37) and with excess HCl to give $C_5H_5Fe(CO)(P(C_6H_5)_3)SnCl_3$ (No. 38) [27].

References:

[1] K.K. Joshi, P.L. Pauson, W.H. Stubbs (J. Organometal. Chem. **1** [1963] 51/7). − [2] I.V. Polovyanyuk, Yu.A. Chapovsky, L.G. Makarova (Izv. Akad. Nauk SSSR Ser. Khim. **1966** 387; Bull. Acad. Sci. USSR Div. Chem. Sci. **1966** 368). − [3] P.M. Treichel, R.L. Shubkin, K.W. Barnett, D. Reichard (Inorg. Chem. **5** [1966] 1177/81). − [4] K.W. Barnett (Diss. Univ. Wisconsin 1967; Diss. Abstr. B **28** [1968]

3203). − [5] A.N. Nesmeyanov, Yu.A. Chapovsky, I.V. Polovyanyuk, L.G. Makarova (J. Organometal. Chem. **7** [1967] 329/37).

[6] R.B. King, K.H. Pannell (Inorg. Chem. **7** [1968] 1510/3). − [7] E.S. Bolton, G.R. Knox, C.G. Robertson (Chem. Commun. **1969** 664). − [8] M.A. Jennings, A. Wojcicki (Inorg. Chim. Acta **3** [1969] 335/40). − [9] R.B. King, R.N. Kapoor (J. Inorg. Nucl. Chem. **31** [1969] 2169/77). − [10] G. Singh (Proc. 4th Intern. Conf. Organometal. Chem., Bristol 1969, Abstr. B 10).

[11] M. Graziani, A. Wojcicki (Inorg. Chim. Acta **4** [1970] 347/50). − [12] N.E. Kolobova, V.V. Skripkin, K.N. Anisimov (Izv. Akad. Nauk SSSR Ser. Khim. **1970** 2225/8; Bull. Acad. Sci. USSR Div. Chem. Sci. **1970** 2095/7). − [13] S.R. Su, J.A. Hanna, A. Wojcicki (J. Organometal. Chem. **21** [1970] P21/P22). − [14] W.R. Cullen, J.R. Sams, J.A.J. Thompson (Inorg. Chem. **10** [1971] 843/50). − [15] L.I. Denisovich, I.V. Polovyanyuk, B.V. Lokshin, S.P. Gubin (Izv. Akad. Nauk SSSR Ser. Khim. **1971** 1964/9; Bull. Acad. Sci. USSR Div. Chem. Sci. **1971** 1851/5).

[16] M. Graziani, L. Busetto, A. Palazzi (J. Organometal. Chem. **26** [1971] 261/5). − [17] S.R. Su (Diss. Ohio State Univ. 1971; Diss. Abstr. Intern. B **32** [1972] 6283). − [18] S.R. Su, A. Wojcicki (J. Organometal. Chem. **27** [1971] 231/40). − [19] J. Ellermann, H. Behrens, H. Krohberger (J. Organometal. Chem. **46** [1972] 119/38). − [20] P. Kalck, R. Poilblanc (Compt. Rend. C **274** [1972] 66/9).

[21] J.J. Alexander, A. Wojcicki (Inorg. Chem. **12** [1973] 74/6). − [22] D.M. Allen, A. Cox, T.J. Kemp, L.H. Ali (J. Chem. Soc. Dalton Trans. **1973** 1899/901). − [23] R.J.P. Corriu, W.E. Douglas (J. Organometal. Chem. **51** [1973] C3/C4). − [24] T.C. Flood, D.L. Miles (J. Am. Chem. Soc. **95** [1973] 6460/2). − [25] G. Ingletto, E. Tondello, L. DiSipio, G. Carturan, M. Graziani (J. Organometal. Chem. **56** [1973] 335/7).

[26] K.R. Aris, J.M. Brown, K.A. Taylor (J. Chem. Soc. Dalton Trans. **1974** 2222/8). − [27] R.E.J. Bichler, H.C. Clark, B.K. Hunter, A.T. Rake (J. Organometal. Chem. **69** [1974] 367/76). − [28] H. Brunner, J. Strutz (Z. Naturforsch. **29b** [1974] 446/7). − [29] K.H. Pannell, J.R. Rice (J. Organometal. Chem. **78** [1974] C35/C39). − [30] P. Reich-Rohrwig, A. Wojcicki (Inorg. Chem. **13** [1974] 2457/64).

[31] G.M. Bancroft, A.T. Rake (Inorg. Chim. Acta **13** [1975] 175/9). − [32] G. Cerveau, E. Colomer, R. Corriu, W.E. Douglas (J. Chem. Soc. Chem. Commun. **1975** 410/1). − [33] J.W. Faller, B.V. Johnson (J. Organometal. Chem. **96** [1975] 99/113). − [34] T.C. Flood, F.J. DiSanti, D.L. Miles (J. Chem. Soc. Chem. Commun. **1975** 336/7). − [35] M. Gielen, C. Hoogzand, I. Van Den Eynde (Bull. Soc. Chim. Belges **84** [1975] 939/45).

[36] T.N. Pecoraro (Diss. Rutgers State Univ. 1975; Diss. Abstr. Intern. B **36** [1975] 746). − [37] D.L. Reger (Inorg. Chem. **14** [1975] 660/4). − [38] K. Stanley, M.C. Baird (J. Am. Chem. Soc. **97** [1975] 6598/9). − [39] S.R. Su, A. Wojcicki (Inorg. Chem. **14** [1975] 89/98). − [40] J.A. van Doorn, C. Masters, H.C. Volger (J. Organometal. Chem. **105** [1976] 245/54).

[41] T.C. Flood, F.J. DiSanti, D.L. Miles (Inorg. Chem. **15** [1976] 1910/8). − [42] D.L. Reger, E.C. Culbertson (Syn. Reactiv. Inorg. Metal-Org. Chem. **6** [1976] 1/10). − [43] D.L. Reger, E.C. Culbertson (J. Am. Chem. Soc. **98** [1976] 2789/94). − [44] S.C. Tripathi, S.C. Srivastava, V.N. Pandey (Transition Metal Chem. [Weinheim] **1** [1976] 266/8). − [45] G. Cerveau, E. Colomer, R. Corriu, W.E. Douglas (J. Organometal. Chem. **135** [1977] 373/86).

[46] T.C. Flood, D.L. Miles (J. Organometal. Chem. **127** [1977] 33/44). –
[47] H.B. Kuhnen (Z. Naturforsch. **32b** [1977] 718/20). – [48] A.N. Nesmeyanov,
E.G. Perevalova, L.I. Leont'eva, E.V. Shumilina (Izv. Akad. Nauk SSSR Ser. Khim.
1977 1142/6; Bull. Acad. Sci. USSR Div. Chem. Sci. **1977** 1048/52). – [49] V.N.
Pandey (Inorg. Chim. Acta **22** [1977] L39/L41). – [50] D.L. Reger, E.C. Culbertson
(Inorg. Chem. **16** [1977] 3104/7).

[51] E.V. Shumilina (Vestn. Mosk. Univ. Khim. **32** [1977] 476/8; Moscow Univ.
Chem. Bull. **32** No. 4 [1977] 76/8). – [52] D.G. Alway, K.W. Barnett (Inorg. Chem.
17 [1978] 2826/31). – [53] D.A. Brown, F.M. Hussein, C.L. Arora (Inorg. Chim.
Acta **29** [1978] L215/L216). – [54] E.J. Kuhlman (Diss. Univ. Cincinnati 1978; Diss.
Abstr. Intern. B **39** [1978] 2296). – [55] G.J. Long, D.G. Alway, K.W. Barnett (Inorg.
Chem. **17** [1978] 486/9).

[56] S.L. Miles, D.L. Miles, R. Bau, T.C. Flood (J. Am. Chem. Soc. **100** [1978]
7278/82). – [57] T.R. Spalding (J. Organometal. Chem. **149** [1978] 371/5). –
[58] T. Bodnar, S.J. LaCroce, A.R. Cutler (J. Am. Chem. Soc. **102** [1980] 3292/4). –
[59] P.L. Bogdan, A. Wong, J.D. Atwood (J. Organometal. Chem. **229** [1982]
185/91).

1.5.2.2.6.1.3   Compounds with $^2D = PR_2R'$

The compounds listed in Table 9 can be prepared in several ways. The following
methods are of a general nature. Some reaction conditions (solvent, temperature, and
time) are given in Table 9. Other special procedures are described in the further information, p. 115.

Method I: The compound $C_5H_5Fe(CO)_2X$ reacts with the appropriate phosphine
PR_2R'.

    a. Mixtures of $C_5H_5Fe(CO)_2X$ and PR_2R' (mostly in excess) are UV irradi-
ated [4, 9, 15]. The preparation of No. 2 is an exception. In this case
$C_5H_5Fe(CO)_2H$ (obtained by reduction of $C_5H_5Fe(CO)_2I$ with $LiAlH_4$
in ether at $-20\,°C$ followed by hydrolysis) is reacted in statu nascendi
with $P(CH_3)_2C_6H_5$ [6]. No. 6 crystallizes from hexane at $-78\,°C$ [4].

    b. The components are stirred or refluxed in a solvent without irradiation
[2, 7, 13, 14].

Method II: Reaction of $[C_5H_5Fe(CO)PR_2R']^+$ (obtained by silver(I) assisted substitution of I in $C_5H_5Fe(CO)(PR_2R')I$ with $AgBF_4$) in situ with NaCN in CH_3OH
(Nos. 5 and 10) [9] or with KCN in C_2H_5OH (No. 18) [13] yields the
compound.

The compounds are generally chromatographed on Al_2O_3 and crystallized from
binary solvent mixtures.

Although all compounds described in Table 9 possess a chiral Fe atom, optically
active forms are only observed for Nos. 17, 18, and 22 to 25, which have another
chiral center in the R' group of PR_2R' and thus form diastereomers.

Table 9
Compounds of the $C_5H_5Fe(CO)(PR_2R')X$ Type.
Further information on numbers preceded by an asterisk is given at the end of the table.
For abbreviations and dimensions, see p. 375.

No.	PR_2R'	X	method of preparation (conditions, yield) and properties not assigned IR bands in the 1920 to 1970 cm^{-1} region are ν(CO)	Ref.
1	$PH_2C_6H_5$	I	Ib (22 h, C_6H_6, 80°), green ^{1}H NMR (CDCl$_3$): 4.55 (s, C_5H_5), 7.39 (m, C_6H_5) IR (C_6H_6): 1965 (CO), 2340 (PH) unstable	[7]
*2	$P(CH_3)_2C_6H_5$	H	Ia (ether, −20°), brown ^{1}H NMR (C_6D_6): −13.9 (FeH, J(P,H) = 79), 1.29 (d, CH$_3$, J(P,H) = 9), 4.20 (C_5H_5, J(P,H) = 1.2) IR ($C_{16}H_{34}$): 1927	[6, 12]
*3	$P(CH_3)_2C_6H_5$	Cl	formation from No. 2 and CCl$_4$, green ^{1}H NMR (C_6H_6): 1.13 (CH$_3$, J(P,H) = 10), 1.45 (CH$_3$, J(P,H) = 10), 3.87 (C_5H_5, J(P,H) = 1.5) IR ($C_{16}H_{34}$): 1956	[6]
*4	$P(CH_3)_2C_6H_5$	I	Ib (22 h, C_6H_6, 80°, 3% [2], 80% [9]), green or dark green, m.p. 98° ^{1}H NMR (CS$_2$): 1.73 (d, CH$_3$, J(P,H) = 9.3), 2.14 (d, CH$_3$, J(P,H) = 10), 4.26 (C_5H_5, J(P,H) = 1.5), 7.34 (C_6H_5) IR: 1946 in KBr, 1950 in CHCl$_3$	[1, 2, 9]
*5	$P(CH_3)_2C_6H_5$	CN	II (4 h, 20%), amber oil ^{1}H NMR (CDCl$_3$): 1.82 (d, CH$_3$, J(P,H) = 10.1), 1.98 (d, CH$_3$, J(P,H) = 10.8), 4.64 (d, C_5H_5, J(P,H) = 1.6), 7.8 (m, C_6H_5) ^{13}C NMR (CDCl$_3$): 18.03 (d, CH$_3$, J(P,C) = 31.8), 18.72 (d, CH$_3$, J(P,C) = 34.8), 82.95 (s, C_5H_5, J(P,C) <0.2), 128.79 (d, C-3,5 in C_6H_5, J(P,C) = 9.2), 130.10 (s, C-4 in C_6H_5, J(P,C) = 2.0), 138.31 (d, C-1 in C_6H_5, J(P,C) = 42.9), 217.31 (d, CO, J(P,C) = 29.4) IR (C_6H_{12}, CHCl$_3$): 1963	[9]

References on p. 118

Table 9 [continued]

No.	PR₂R'	X	method of preparation (conditions, yield) and properties not assigned IR bands in the 1920 to 1970 cm⁻¹ region are ν(CO)	Ref.
*6	$P(CH_3)_2C_6H_5$	$Sn(CH_3)_3$	Ia (5 h, C_6H_6 or CH_3COCH_3, ~20%), yellow, m.p. ~30°	[4]
*7	$P(CH_3)_2C_6H_5$	$Sn(C_6H_5)_3$	Ia (5 h, C_6H_6 or CH_3COCH_3) ^1H NMR (CS_2): 1.40, 1.66 (d's, CH_3, J=17.6), 4.33 (d, C_6H_5), 7.23 (m, C_6H_5). ^{57}Fe-γ (78 K): δ=0.43, Δ=1.71 ^{119}Sn-γ (78 K): δ=1.47, Δ=0.78 IR (C_6H_{12}):1917	[4]
*8	$P(C_6H_5)_2H$	Br	Ib (7 d, C_6H_6 in the dark, 1.7%), unstable green oil ^1H NMR ($CDCl_3$): 4.56 (s, C_5H_5), 5.93 (d, PH, J(P,H)~3.67), 7.44 (m, C_6H_5) IR (CH_2Cl_2):1961	[7]
9	$P(C_6H_5)_2CH_3$	I	Ia (61%), dark green, m.p. 145 to 148° (dec.) IR ($CHCl_3$):1952	[9]
*10	$P(C_6H_5)_2CH_3$	CN	II (4 h, 20%), yellow, m.p. 132 to 135° ^1H NMR ($CDCl_3$): 2.07 (d, CH_3, J(P,H) = 9.6), 4.52 (d, C_5H_5, J(P,H) = 1.8), 7.5 (m, C_6H_5) ^{13}C NMR ($CDCl_3$): 19.40 (d, CH_3, J(P,C) = 34.5), 83.61 (s, C_5H_5, J(P,C) < 0.2), 128.68 (d, C-3,5 in C_6H_5, J(P,C) = 7.4), 130.82 (s, C-4 in C_6H_5, J(P,C) = 2.0), 131.07 (d, C-2,6 in C_6H_5, J(P,C) = 9.9), 131.32 (s, C-4 in C_6H_5, J(P,C) = 2.0), 132.17 (d, C-2,6 in C_6H_5, J(P,C) = 9.5), 135.61 (d, C-1, J(P,C) = 44.0), 136.99 (d, C-1, J(P,C) = 41.9), 217.0 (d, CO, J(P,C) = 27.0) IR (C_6H_{12}, $CHCl_3$):1968	[9]
*11	$P(C_6H_5)_2CH_3$	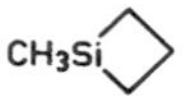 CH₃Si◇	Ia (8 h, C_6H_{14}, 13%), yellow, m.p. 106 to 107° ^1H NMR (C_6H_6): 0.70 (s, $SiCH_3$), 0.9 to 1.9 (m, CH_2), 1.69 (d, PCH_3, J(P,H)=8.0), 2.1 to 3.2 (m, CH_2), 4.20 (d, C_5H_5, J(P,H)=1.5) IR (C_6H_{12}):1916 mass spectrum: $[M]^+$, $[M-CO]^+$, $[M-CO-C_2H_4]^+$, $[M-SiC_3H_6]^+$, $[M-CO-SiC_4H_8]^+$, $[M-CO-P(C_6H_5)_2CH_2-SiC_4H_8]^+$	[8, 15]

References on p. 118

No.			Data	Ref.
*12	$P(C_6H_5)_2CH_3$	$Sn(CH_3)_3$	Ia (5 h, C_6H_6 or CH_3COCH_3), orange, m.p. 180 to 182°	[4]
*13	$P(C_6H_5)_2CH_3$	$Sn(C_6H_5)_3$	Ia (5 h, C_6H_6 or CH_3COCH_3) ^{1}H NMR (CS_2): 1.57 (d, CH_3), 4.43 (d, C_5H_5), 7.33 (m, C_6H_5) ^{57}Fe-γ (78 K): δ=0.47, Δ=1.76 ^{119}Sn-γ (78 K): δ=1.42, Δ=0.55 IR (C_6H_{12}): 1917	[4]
*14	$P(C_6H_5)_2CF_3$	$Sn(CH_3)_3$	Ia (5 h, C_6H_6 or CH_3COCH_3), orange, m.p. 108 to 110° ^{1}H NMR (CS_2): −0.12 (s, CH_3, J(^{119}Sn,H)=41.2), 4.57 (d, C_5H_5), 7.45 (m, C_6H_5) ^{19}F NMR (CS): 58.3 (d, J(P,F)=58) ^{57}Fe-γ (78 K): δ=0.44, Δ=1.91 ^{119}Sn-γ (78 K): δ=1.40 IR (C_6H_{12}): 1923, 1933	[4]
*15	$P(C_6H_5)_2CF_3$	$Sn(C_6H_5)_3$	Ia (5 h, C_6H_6 or CH_3COCH_3, 39%), orange, m.p. 142 to 144° ^{1}H NMR (CS_2): 4.53 (d, C_5H_5), 7.07 (m, C_6H_5) ^{19}F NMR (CS_2): 57.1 (d, J(P,F)=56) ^{57}Fe-γ (78 K): δ=0.46, Δ=1.83 ^{119}Sn-γ (78 K): δ=1.50, Δ=0.44 IR (C_6H_{12}): 1935	[4]
*16	$P(C_6H_5)_2CH_2C_6H_5$	H	^{1}H NMR (C_6D_6): −13.3 (FeH, J(P,H)=80), 4.14 (C_5H_5, J(P,H)=1.2)	[12]
*17	[menthyl-substituted $P(C_6H_5)_2$]	I	Ib (70 h, C_6H_6 at 80°, 44%), green, m.p. 128 to 130° ^{1}H NMR (CD_3COCD_3): 0.17 (d, CH_3, J=6.6), 0.79 (d, CH_3, J=6.5), 1.02 (d, CH_3, J=6.6), 1.69, 2.11 (br, C_6H_9, CH), 4.24 (C_5H_5, J=1.4), 7.63, 7.88 (br, C_6H_5) ^{31}P NMR (C_6H_6): 69.6 (s) IR (C_6H_6): 1935	[13]
*18	[menthyl-substituted $P(C_6H_5)_2$]	CN	II (1 h, 51%), yellow oil ^{1}H NMR ($CDCl_3$): 0.4 to 2.1 ($C_{10}H_{19}$), 4.22 (C_5H_5, J=1.2), 7.4 (d, C_6H_5) IR ($C_6H_5CH_3$): 1956, ν(CN) 2097	[13]

Table 9 [continued]

No.	PR₂R'	X	method of preparation (conditions, yield) and properties not assigned IR bands in the 1920 to 1970 cm⁻¹ region are $\nu(CO)$	Ref.
*19		$Sn(CH_3)_3$	Ia (5 h, C_6H_6 or CH_3COCH_3, 35%), brown red, m.p. 144 to 145° (dec.) ¹H NMR (CS_2): -0.08 (CH_3, $J(^{119}Sn,H)=41.2$), 4.28 (d, C_5H_5), 7.33 (m, C_6H_5) ⁵⁷Fe-γ (78 K): $\delta=0.46$, $\Delta=1.86$ ¹¹⁹Sn-γ (78 K): $\delta=1.46$ IR (C_6H_{12}): 1923	[4]
20		$Sn(C_6H_5)_3$	Ia (5 h, C_6H_6 or CH_3COCH_3, 21%), red, m.p. 170 to 175° (dec.) ¹H NMR (CS_2): 4.50 (d, C_5H_5), 7.19 (m, C_6H_5) ⁵⁷Fe-γ (78 K): $\delta=0.48$, $\Delta=1.78$ ¹¹⁹Sn-γ (78 K): $\delta=1.44$ IR (C_6H_{12}): 1931 solid stable to air or light, solutions slowly decompose, most stable in acetone or CS_2	[4]
21	$P(C_6H_5)_2C\equiv CC_6H_5$	Br	Ib (1 h, C_6H_6 at 80°, 31%), green, m.p. 127 to 128° ¹H NMR ($CHCl_3$): 4.68 (s, C_5H_5), 6.5 to 7.6 ($C_2C_6H_5$), 7.5 to 8.2 (PC_6H_5) IR (CH_2Cl_2): 1965	[3]
*22	$P(C_6H_5)_2N(CH_3)$-$CH(CH_3)C_6H_5$	Cl	see further information, dark green, m.p. 93 to 95.5°	[11]
*23	$P(C_6H_5)_2N(CH_3)$-$CH(CH_3)C_6H_5$	Br	see further information, dark green, m.p. 118 to 120°	[11]
*24	$P(C_6H_5)_2N(CH_3)$-$CH(CH_3)C_6H_5$	I	Ib (20 h, C_6H_6 at 80°, 57%), Ib (2 h, CH_3COCH_3 at 56°, 75%), dark green, m.p. 139 to 141° see further information	[14]
*25	$P(C_6H_5)_2N(C_2H_5)$-$CH(CH_3)C_6H_5$	I	see further information	[10]

*Further information:

$C_5H_5Fe(CO)(P(CH_3)_2C_6H_5)H$ (Table 9, No. 2) is also prepared by passing B_2H_6 through a solution of $C_5H_5Fe(CO)(P(CH_3)_2C_6H_5)COCH_3$ (see 1.5.2.2.12.4, Table 21, No. 3) in benzene or THF at $-80\,°C$ [12].

$C_5H_5Fe(CO)(P(CH_3)_2C_6H_5)Cl$ (Table 9, No. 3) arises from action of CCl_4 on $C_5H_5Fe(CO)(P(CH_3)_2C_6H_5)H$ in good yield. Separation by chromatography and crystallization in benzene/hexane [6].

$C_5H_5Fe(CO)(P(CH_3)_2C_6H_5)I$ (Table 9, No. 4). For the silver(I) assisted reaction, see Method II, p. 110 [9].

$C_5H_5Fe(CO)(P(CH_3)_2C_6H_5)CN$ (Table 9, No. 5). Addition of the shift reagent $Eu(fod)_3$ (Formula II on p. 104) produces downfield shifts and first–order coupling patterns in ^1H and ^{13}C NMR spectra and allows the determination of conformational effects and rotational barriers. The paramagnetic ^{13}C shifts do not fit the predictions of any of the individual conformations I to III. The similarity of carbonyl and cyanide groups implies that I and II have nearly equal energies. An average of these two conformations predicts shifts which are in reasonable agreement with the observed results and significantly different from those predicted for III. Maximum barrier to rotation about the Fe–P bond is estimated at 8.0 kcal·mol^{-1} [9].

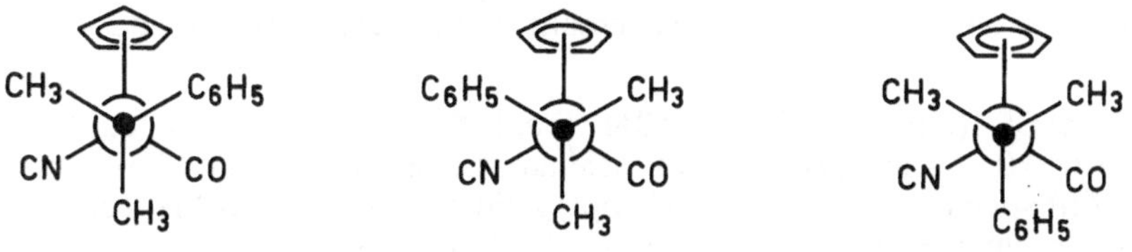

I   II   III

$C_5H_5Fe(CO)(P(CH_3)_2C_6H_5)SnR_3$ (Table 9, Nos. 6 and 7, $R=CH_3$ or C_6H_5) are indefinitely stable to air or light in the solid state. Both decompose slowly in solution, No. 6 being the less stable. Both are most stable in CH_3COCH_3 or CS_2 [4].

$C_5H_5Fe(CO)(P(C_6H_5)_2H)Br$ (Table 9, No. 8). Further bands of the IR spectrum in CH_2Cl_2 in the region 520 to 1700 cm^{-1} are given in [7]. LiC_4H_9-n yields a small amount of $(Z)-(C_5H_5Fe(CO)\mu-P(C_6H_5)_2)_2$ via the intermediate $C_5H_5Fe(CO)-(P(C_6H_5)_2Li)Br$ [7].

$C_5H_5Fe(CO)(P(C_6H_5)_2CH_3)CN$ (Table 9, No. 10). Addition of the shift reagent $Eu(fod)_3$ (Formula II on p. 104) produces downfield shifts and first–order coupling patterns in ^1H and ^{13}C NMR spectra and allows the determination of conformational effects and rotational barriers. For the preferred conformations IV to VI, the observed ^{13}C shifts are strikingly similar to those predicted for V and bear no resemblance

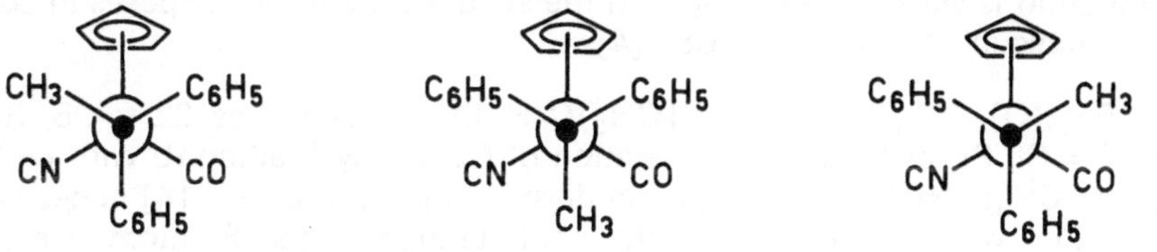

IV   V   VI

References on p. 118

to either IV or VI. Also the values of the phenyl and methyl shifts correlate with those expected for the conformation V. Maximum barrier to rotation about the Fe-P bond is estimated at 8.0 kcal$\cdot$mol^{-1}. For the silver(I) assisted reaction, see Method II, p. 110 [9].

$C_5H_5Fe(CO)(PCH_3(C_6H_5)_2)SiC_4H_9$ (Table 9, No. 11) is monomer in n-hexane. The compound is catalytically polymerized by $(C_2H_4PtCl_2)_2$ in benzene solution [15].

$C_5H_5Fe(CO)(P(C_6H_5)_2R')SnR_3$ (Table 9, Nos. 12 to 15, with R'$=CH_3$ or CF_3, $R=CH_3$ or C_6H_5) are stable to air or light in the solid state but decompose slowly in solution. The $Sn(CH_3)_3$ derivatives are less stable in solution and decomposed during NMR measurements. The CF_3 group enhances the stability in solution. All four compounds are most stable in CH_3COCH_3 or CS_2 solution [4].

$C_5H_5Fe(CO)(P(C_6H_5)_2CH_2C_6H_5)H$ (Table 9, No. 16) is prepared by passing B_2H_6 through a solution of $C_5H_5Fe(CO)(P(C_6H_5)_2CH_2C_6H_5)COCH_3$ (see 1.5.2.2.12.4, Table 21, No. 20) in benzene or THF at -80 °C [12].

$C_5H_5Fe(CO)(P(C_6H_5)_2C_{10}H_{19})I$ (Table 9, No. 17). The difficult separation from the starting material is only possible by careful chromatography. The circular dichroism spectrum ($c=0.028$ in C_6H_6) shows maxima and minima (wave length in nm) at 315 (max), 354 (min), 411 (max), 512 (min), and 634 (max). Careful examination of this spectrum and the ^1H, ^{31}P, and ^{13}C NMR spectra does not indicate any resonance doubling. The silver(I) assisted replacement of the iodide by cyanide yields No. 18 [13].

$C_5H_5Fe(CO)(P(C_6H_5)_2C_{10}H_{19})CN$ (Table 9, No. 18). The circular dichroism spectrum ($c=0.056$ in C_6H_6) shows maxima at 373 and 450 nm. This spectrum and the ^1H NMR spectrum (with added $Yb(opt)_3$, Formula I on p. 104) reveal the presence of equal amounts of two diastereomers. The C_5H_5 resonance splits into two peaks separated by 0.24 ppm the average shift being $\delta=3.37$ ppm. The two signals have the same intensity [13].

$C_5H_5Fe(CO)(P(C_6H_5)_2C_5F_6P(C_6H_5)_2)Sn(CH_3)_3$ (Table 9, No. 19) crystallizes in the triclinic space group $P1-C_1^1$ (No. 1) with four formula units in a A-centered cell of dimensions $a=10.831(1)$, $b=22.382(4)$, $c=15.459(2)$ Å, $\alpha=101.48(1)°$, $\beta=92.13(2)°$, and $\gamma=83.26(2)°$. $D_m=1.64(4)$ g$\cdot$cm^{-3}, $D_c=1.57$ g$\cdot$cm^{-3}. The structure was refined to $R=8.0\%$ for 1703 observed reflections. The molecule is shown in Fig. 11. The angles subtended at the iron atom by the covalently bonded atoms average $92.2(10)°$, which suggests a pseudo-octahedral geometry about the iron atom. If a vector is drawn from the Fe atom to the centroid of the C_5H_5 ring, an approximate tetrahedral arrangement results. However, the average angle subtended at the iron atom by the centroid position is $123.1(10)°$, indicating a considerable distortion. Conformational restriction of the rotation about the Fe-Sn bond may be due to steric factors rather than π-bond character since the degenerate d orbitals of a molecule of this type could not provide a π-bond with a significant barrier to rotation [5]. The compound is stable to air or light in the solid state and decomposes in solution. It is most stable in CH_3COCH_3 or CS_2 [4].

$C_5H_5Fe(CO)(P(C_6H_5)_2NR-CH(CH_3)C_6H_5)X$ (Table 9, Nos. 22 to 25, $R=CH_3$ or C_2H_5, $X=Cl$, Br, or I). In the preparation of No. 24 by Method Ib with $(+)-(S)-P(C_6H_5)_2N(CH_3)CH(CH_3)C_6H_5$ a pair of diastereomers forms [11, 14] in nearly equal amounts [11], which differ only in the configuration at the Fe atom (yield 95%) [11, 14]. Fractional crystallization from C_6H_6/C_6H_{14} (2:3) at 4 °C gives the sparingly

Fig. 11

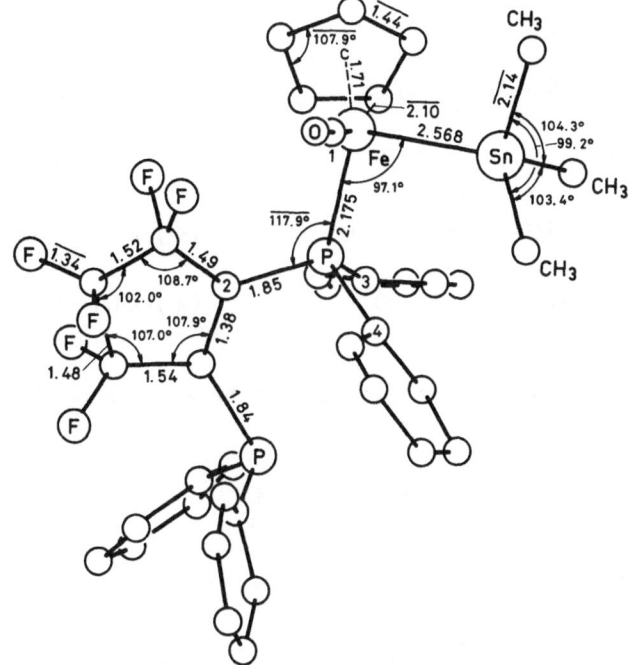

Molecular structure of $C_5H_5Fe(CO)(P(C_6H_5)_2C_5F_6P(C_6H_5)_2)Sn(CH_3)_3$ [5].

Other bond lengths (Å) and bond angles (°):

Fe–C(1)	1.63(3)	C(1)–O	1.29(3)
Sn–Fe–C(1)	83.5(9)	C(1)–Fe–C$_5$H$_5$(c)	127.9(10)
Sn–Fe–C$_5$H$_5$(c)	115.4(10)	C(2)–P–C(3)	102.6(10)
P–Fe–C(1)	96.1(10)	C(2)–P–C(4)	100.2(11)
P–Fe–C$_5$H$_5$(c)	126.0(10)	C(3)–P–C(4)	109.0(11)
Fe–C(1)–O	175.6(25)		

soluble, optically pure $(-)_{365}$-diastereomer No. 24 with $[\alpha]_{365}^{20} = -6350°$, $[\alpha]_{436}^{20} = +720°$ (c = 0.5 mg·mL^{-1} in $C_6H_5CH_3$). Elution of the mixture of isomers of No. 24 obtained by chromatographic workup with pentane gives the soluble $(+)_{365}$-diastereomer No. 24 with $[\alpha]_{365}^{20} = +2740°$ and $[\alpha]_{436}^{20} = -540°$ (0.5 mg·mL^{-1} in $C_6H_5CH_3$, 45% optically pure) [14]. Nos. 22 and 23 are prepared from No. 24 with the stoichiometric amount of Cl_2 (designated as reaction a) or Br_2 (reaction c) in CH_2Cl_2 at −78 °C (yield 90 and 94%, respectively). Similar reactions of C_5H_5-$Fe(CO)(P(C_6H_5)_2N(CH_3)CH(CH_3)C_6H_5)CH_3$ (designated by A in the following table) give compounds No. 22, 23 (reaction b), and with ice cooling No. 24 (reaction d). The reactions a to d occur predominantly with retention of configuration at the Fe atom. These reactions are summarized on p. 118.

The circular dichroism spectra of compounds $(-)$-22, $(-)$-23, and $(-)$-24 are reproduced in a figure in [11].

reaction	starting material $[\alpha]^{20}_{365}$ (c=1 mg·mL^{-1}, C$_6$H$_6$)	reaction product $[\alpha]^{20}_{365}$ (c=1 mg·mL^{-1}, C$_6$H$_6$)
a	(+)-24/(−)-24~5/95 −5600°	(+)-22/(−)-22=5/95 −2350°
b	(+)-A/(−)-A~15/85 −1020°	(+)-23/(−)-23=19/81 not given
c	(+)-24/(−)-24~10/90 −5330°	(+)-23/(−)-23=11/89 −4400°
d	(+)-A/(−)-A~10/90 −1260°	(+)-24/(−)-24=30/70 −2700°

The members of each pair of diastereomers have characteristic ^{1}H NMR shifts (δ in ppm, J in Hz) [11, 14]:

compound	C–CH$_3$ (d's)	N–CH$_3$ (d's)	C$_5$H$_5$ (d's)	CH (q's)	C$_6$H$_5$ (m's)
(+)-22	0.68 (J=7)	1.43 (J=8)	3.30 (J=1.8)	4.90 (J=7)	6.30
(−)-22	0.71 (J=7)	1.38 (J=8)	3.36 (J=1.8)	4.90 (J=7)	6.30
(+)-23	0.81 (J=7)	1.63 (J=8)	3.40 (J=1.8)	4.76 (J=7)	6.56
(−)-23	0.83 (J=7)	1.56 (J=8)	3.43 (J=1.8)	4.76 (J=7)	6.56
(+)-24	0.78 (J=7)	1.63 (J=8)	3.45 (J=1.8)	4.70 (J=7)	6.53
(−)-24	0.80 (J=7)	1.53 (J=7)	3.48 (J=1.8)	4.70 (J=7)	6.53

The IR spectra in CHCl$_3$ show the ν(CO) band at 1963 (No. 22), 1961 (No. 23), and 1956 (No. 24) cm^{-1}. The mass spectra of compounds No. 22 and 23 exhibit only the m/e values of the (C$_6$H$_5$)$_2$PN(CH$_3$)CH(CH$_3$)C$_6$H$_5$ ligand and its fragments [11]. The configuration of compound No. 24 is stable both at and somewhat above room temperature. In benzene, compounds No. (−)-22 and (−)-23 undergo a first-order epimerization by a change of configuration at the iron atom. The epimerization of No. (−)-22 is characterized by a final ratio (+)/(−)-22=62/38 and half-lifes $\tau_{1/2}$(20 °C)=181 min, $\tau_{1/2}$(30 °C)=29 min, and $\tau_{1/2}$(40 °C)=5.6 min. *For No. (−)-23: (+)/(−)-23=60/40, $\tau_{1/2}$(45 °C)~185 min [11]. In the presence of (+)(S)-(C$_6$H$_5$)$_2$PN(R)CH(CH$_3$)C$_6$H$_5$ (R=CH$_3$, C$_2$H$_5$) the following is found: (+)/(−)-24=50/50 and $\tau_{1/2}$(70 °C)=51 min for No. (−)-24, (+)/(−)-25=65/35 and $\tau_{1/2}$(70 °C)=11 min for No. (−)-25 [10]. The reaction of No. (−)-24 with LiCH$_3$ in ether at −78 °C gives a diastereomer mixture of C$_5$H$_5$Fe(CO)(P(C$_6$H$_5$)$_2$-N(CH$_3$)CH(CH$_3$)C$_6$H$_5$)CH$_3$ in a ratio (+)/(−)=32/68 [11].

References:

[1] H. Brunner, E. Schmidt (Angew. Chem. **81** [1969] 570/1; Angew. Chem. Intern. Ed. Engl. **8** [1969] 616/7). − [2] H. Brunner, H.-D. Schindler, E. Schmidt, M. Vogel (J. Organometal. Chem. **24** [1970] 515/26). − [3] R.B. King, A. Efraty (Inorg. Chim. Acta **4** [1970] 319/23). − [4] W.R. Cullen, J.R. Sams, J.A.J. Thompson (Inorg. Chem. **10** [1971] 843/50). − [5] F.W.B. Einstein, R. Restivo (Inorg. Chim. Acta **5** [1971] 501/10).

[6] P. Kalck, R. Poilblanc (Compt. Rend. C **274** [1972] 66/9). − [7] P.M. Treichel, W.K. Dean, W.M. Douglas (J. Organometal. Chem. **42** [1972] 145/58). − [8] C.S. Cundy, M.F. Lappert (J. Organometal. Chem. **57** [1973] C72/C74). − [9] J.W. Faller, B.V. Johnson (J. Organometal. Chem. **96** [1975] 99/113). − [10] H. Brunner, F. Rackl (J. Organometal. Chem. **118** [1976] C19/C22).

[11] H. Brunner, G. Wallner (Chem. Ber. **109** [1976] 1053/60). − [12] J.A. van Doorn, C. Masters, H.C. Volger (J. Organometal. Chem. **105** [1976] 245/54). − [13] D.L. Reger (J. Inorg. Nucl. Chem. **39** [1977] 1095/7). − [14] H. Brunner, M. Muschiol, W. Nowak (Z. Naturforsch. **33b** [1978] 407/11). − [15] C.S. Cundy, M.F. Lappert, C.-K. Yuen (J. Chem. Soc. Dalton Trans. **1978** 427/33).

1.5.2.2.6.1.4 Compounds with $^2D = P(OR)_3$

The compounds listed in Table 10 can be prepared in several ways. The following methods are of a general nature. Reaction conditions (solvent, temperature, and time) are given in Table 10. Other special procedures are described in the further information on p. 125.

Method I: $C_5H_5Fe(CO)_2X$ is reacted with $P(OR)_3$.

      a. The components are dissolved in an inert solvent and UV irradiated [2, 6, 16, 17, 19, 23, 24]. Such a reaction with $C_5H_5Fe(CO)_2H$ (obtained in statu nascendi by reduction of $C_5H_5Fe(CO)_2I$ with $LiAlH_4$ in ether at $-20\,°C$ followed by hydrolysis) gives $C_5H_5Fe(CO)$-$(P(OCH_3)_3)H$ (No. 1) [14].

      b. Mixtures of $C_5H_5Fe(CO)_2X$ and the appropriate $P(OR)_3$ (usually in excess) are stirred at room temperature or refluxed in an inert solvent without irradiation [3, 5, 8, 13, 15]. Treatment of $C_5H_5Fe(CO)_2Cl$ with a slight excess of $P(OR)_3$ yields $C_5H_5Fe(CO)(P(OR)_3)Cl$, $C_5H_5Fe(CO)_2PO(OR)_2$, and a small amount of $C_5H_5Fe(CO)(P(OR)_3)$-$PO(OR)_2$. $C_5H_5Fe(CO)(P(OC_6H_5)_3)Cl$ (No. 21) is the major product for $R = C_6H_5$. $C_5H_5Fe(CO)_2PO(OR)_2$ is the major product if R is CH_3, C_2H_5, $CH_2CH=CH_2$, or C_4H_9-n (Nos. 2 and 6, 8 and 10, 16 and 17, and No. 19). The compounds $C_5H_5Fe(CO)(P(OR)_3)Cl$ and C_5H_5-$Fe(CO)_2PO(OR)_3$ are separated by chromatography, but the derivatives with $R = CH_3$, C_2H_5 (Nos. 2 and 8) could not be separated in this way. The compounds $C_5H_5Fe(CO)(P(OC_4H_9$-n)$_3)X$ with $X = Cl$, I are not mentioned and therefore probably cannot be prepared by this method [9, 10].

Method II: The compounds $C_5H_5Fe(CO)(P(OR)_3)R'$ ($R = CH_3$, C_4H_9-n, C_6H_5, $R' = CH_3$, C_2H_5) are converted to the sulfinates (Nos. 5, 18, 25 to 27) by reaction with refluxing SO_2 for about 1 h. Purification is effected by chromatography on Al_2O_3 with $CHCl_3$ eluent [11, 12].

In contrast to $C_5H_5Fe(CO)(P(OC_6H_5)_3)OOCCF_3$ (No. 24) the compound C_5H_5Fe-$(CO)(P(OCH_3)_3)OOCCF_3$ could not be prepared by Method Ia [6]. The compounds $C_5H_5Fe(CO)(P(OC_4H_9$-n)$_3)X$ ($X = Cl$, I) [9], and $C_5H_5Fe(CO)(P(OC_6H_{11}$-cyclo)$_3)I$ [1] were said to have been prepared, but no details or properties were given [1, 9].

Although all of the compounds cited in Table 10 possess a chiral Fe atom, optical isomers were observed only in the case of Nos. 12, 23, and 27.

Explanations for Table 10: Not assigned IR bands in the 1900 to 2000 cm^{-1} region are $v(CO)$ vibrations.

Table 10
Compounds of the $C_5H_5Fe(CO)(P(OR)_3)X$ Type.
Further information on numbers preceded by an asterisk is given at the end of the table.
For abbreviations and dimensions, see p. 375.

No.	X	method of preparation (conditions, yield) and properties, explanations see above	Ref.
		$C_5H_5Fe(CO)(P(OCH_3)_3)X$ compounds	
*1	H	I a, amber liquid ^1H NMR (C_6H_6): -13.76 (FeH, J(P,H)$=93.6$), 3.38 $(CH_3, J(P,H)=12)$, 4.45 (C_5H_5) IR $(C_{16}H_{34})$: 1940, 1951 mass spectrum: $[M]^+$	[14]
*2	Cl	I b (16 h, C_6H_6, 30%), red crystals, poorly defined m.p. ^1H NMR: 4.76 (d, C_5H_5, J(P,H)$\sim$1) in $CDCl_3$, 3.44 $(CH_3, J(P,H)=11)$, 4.28 (C_5H_5) in C_6H_6 IR $(C_6H_{12}/C_{16}H_{34})$: 1970/1969, 1980/1982 mass spectrum: $[M]^+$ stable in air	[10, 14]
*3	Br	I b (24 h, C_6H_6, 80°), poorly defined m.p. at 84 to 85° (dec.) IR (CS_2): 1971, 1982 stable in air	[8]
*4	I	I b (24 h, C_6H_6, 80°, 70%), poorly defined m.p. at 82 to 84° ^1H NMR $(CDCl_3)$: 4.72 (d, C_5H_5, J(P,H)$=1.0$) IR: 1968, 1981 in C_6H_{12}, 1966, 1978 in CS_2 stable in air	[8, 10]
5	SO_2CH_3	II ($\sim$60%), yellow, m.p. 138° ^1H NMR $(CDCl_3)$: 3.0 (SO_2CH_3), 3.85 $(OCH_3,$ J$=11$), 4.90 (C_5H_5) IR: 1984 in CH_2Cl_2, 1030, 1165 (SO) in Nujol, for the 450 to 3100 cm^{-1} region (neat oil), see [11]	[11, 12]
6	$PO(OCH_3)_2$	I b (16 h, C_6H_6), oil IR (CH_2Cl_2): 1964	[10]
7	$SnCl_3$	I a (C_6H_6), red orange, not isolated IR (CH_2Cl_2): 1970 solid stable in air, solutions dec. over a period of days	[19]

Table 10 [continued]

No.	X	method of preparation (conditions, yield) and properties, explanations on p. 120	Ref.

$C_5H_5Fe(CO)(P(OC_2H_5)_3)X$ compounds

8	Cl	Ib (4 d, C_6H_6)	[10]
*9	I	Ib (16 h, C_6H_6, 80°, 70%), Ib (n-C_8H_{18}, 45 to 60°), Ib (O(C_4H_9-n)$_2$, 50 to 60°), poorly defined m.p. ^{1}H NMR (CDCl$_3$): 4.70 (C_5H_5) IR (C_6H_{12}): 1964, 1972 crystals stable in air	[1, 10, 13]
10	PO(OC$_2$H$_5$)$_2$	Ib (4 d, C_6H_6), oil IR (C_6H_{12}): 1966	[10]
11	Si(C$_6$H$_5$)$_3$	Ia (63%), yellow, m.p. 116 to 117° ^{1}H NMR (CDCl$_3$): 1.17 (tt, C_2H_5), 3.80 (qq, OCH$_2$), 4.42 (C_5H_5, J(P,H) ∼1.5) IR (C_6H_{12}): 1920 mass spectrum: [M]$^+$	[24]
*12	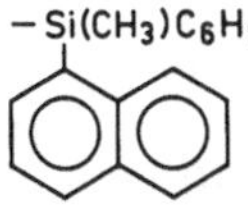 −Si(CH$_3$)C$_6$H$_5$ (naphthalene ring structure)	Ia (72%), yellow oil ^{1}H NMR (CDCl$_3$): 1.17 (tt, C_2H_5), 3.80 (qq, OCH$_2$), 4.34 (C_5H_5, J(P,H) ∼1.5) IR (C_6H_{12}): 1915	[24]
*13	Sn(C$_6$H$_5$)$_3$	Ia (5 h, C_6H_6)	[16]
*14	SnCl$_3$	Ia (C_6H_6), yellow crystals, m.p. 106 to 108° ^{1}H NMR (CDCl$_3$): 4.39 (d, C_5H_5) ^{57}Fe-γ (298 K): δ=0.35, Δ=1.79 ^{119}Sn-γ (80 K): δ=1.77, Δ=1.85 IR: 1977 to 1979 in CH$_2$Cl$_2$ or CS$_2$, 1985 in CHCl$_3$ solid stable in air, solutions dec. in several days	[16, 17, 19]

$C_5H_5Fe(CO)(P(OC_3H_7-i)_3)X$ compounds

*15	I	IR (CH$_2$Cl$_2$): 1944	[4]

$C_5H_5Fe(CO)(P(OCH_2CH=CH_2)_3)X$ compounds

16	Cl	Ib (1.5 h, C_6H_6, 20%), not isolated, identified by IR	[9, 10]
*17	PO(OCH$_2$-CH=CH$_2$)$_2$	Ib (1.5 h, C_6H_6, 20%), oil ^{1}H NMR (CDCl$_3$): 5.05 (t, C_5H_5, J(P,H) = 0.8) IR: 1970 in C_6H_{12}, 1159 (PO) in CS$_2$	[10]

$C_5H_5Fe(CO)(P(OC_4H_9-n)_3)X$ compounds

18	SO$_2$CH$_3$	II (60 to 65%), yellow oil ^{1}H NMR (CDCl$_3$): 1.0 to 1.6 (m, CH$_2$CH$_2$CH$_3$), 3.0 (s, SO$_2$CH$_3$), 4.15 (q, OCH$_2$), 4.85 (s, C_5H_5) IR: 1990 in CH$_2$Cl$_2$, 1030, 1176 (SO) in Nujol, for the 500 to 3100 cm^{-1} region (neat oil), see [11]	[11, 12]

References on p. 128

Table 10 [continued]

No.	X	method of preparation (conditions, yield) and properties, explanations on p. 120	Ref.
19	PO(OC$_4$H$_9$-n)$_2$	Ib (C$_6$H$_6$) IR (CH$_2$Cl$_2$): ~1961	[10]

C$_5$H$_5$Fe(CO)(P(OC$_6$H$_5$)$_3$)X compounds

No.	X	method of preparation (conditions, yield) and properties, explanations on p. 120	Ref.
*20	H	^{1}H NMR (C$_6$D$_6$): −12.9 (FeH, J(P,H) = 94), 4.07 (C$_5$H$_5$, J(P,H) = 0.9) for IR bands in the 400 to 3100 cm^{-1} region, see [11]	[22]
*21	Cl	neat or in solution unstable in air Ib (4 d, C$_6$H$_6$, 55%), poorly defined m.p. ^{1}H NMR (CDCl$_3$): 4.25 (s, C$_5$H$_5$) IR (CH$_2$Cl$_2$): 1986 crystals stable in air	[10]
*22	Br	Ib (24 h, C$_6$H$_6$, 80°) IR: 1990 in C$_7$H$_{16}$, 1978 in C$_5$H$_5$N	[3, 15]
*23	I	Ib (C$_8$H$_{18}$, 126°), Ib (24 h, C$_6$H$_6$, 80°), Ib (16 h, C$_6$H$_6$, 80°, >85%), Ib (40 h, C$_6$H$_6$, 80°, 78%) dark brown green, poorly defined m.p., 126 to 128°, 136 to 138° ^{1}H NMR (CDCl$_3$): 4.34 (s, C$_5$H$_5$), ~7.5 (m, C$_6$H$_5$) IR: 1979 in CH$_2$Cl$_2$, 1981 in Nujol, 1986 to 1988 in C$_8$H$_{18}$, C$_6$H$_{12}$, or CS$_2$, 1193 in CHCl$_3$ crystals stable in air	[1, 3 to 5, 8, 10, 13]
24	OOCCF$_3$	Ia (8 h, C$_6$H$_6$/C$_6$H$_{14}$ (1:40), 60%), red–brown crystals, m.p. 88 to 89° ^{1}H NMR (CH$_3$COCH$_3$): 4.39 (d, C$_5$H$_5$, J = 0.8) ^{19}F NMR (CH$_2$Cl$_2$): 74.9 (CF$_3$) IR (CH$_2$Cl$_2$): 1994, for the 700 to 3200 cm^{-1} region, see [6]	[6]
25	SO$_2$CH$_3$	II (60 to 65%), yellow IR: 1990 in CH$_2$Cl$_2$, 1038, 1190 (SO) in Nujol, for the 400 to 3100 cm^{-1} region, see [11]	[11, 12]
26	SO$_2$C$_2$H$_5$	II (60 to 65%), yellow ^{1}H NMR (CDCl$_3$): 1.35 (t, CH$_3$, J = 7), 3.20 (q, CH$_2$, J = 7), 4.46 (d, C$_5$H$_5$, J = 1), 7.44 (m, C$_6$H$_5$) IR (Nujol): 1984 (CO), 1050, 1184 (SO), for the 400 to 3100 cm^{-1} region, see [11]	[11]
*27	SO$_2$CH(C$_6$H$_5$)- Si(CH$_3$)$_3$	not isolated	[25]

References on p. 128

Table 10 [continued]

No.	X	method of preparation (conditions, yield) and properties, explanations on p. 120	Ref.
28	SeCN	Ib (3 min, C_6H_6, 80°, 70%), dark orange red, m.p. 119 to 121° (dec.) ^{1}H NMR ($CDCl_3$): 4.56 (d, C_5H_5, J(P,H) = 1), 7.4 (m, C_6H_5) IR: 1974, 2007 (CO), 2108 (CN) in mull, 1993 (CO), 2120 (CN) in $CHCl_3$	[5]
*29	NCO	Ib, red crystals, m.p. 123° (dec.) ^{1}H NMR ($CDCl_3$): 4.30 (C_5H_5), 7.40 (C_6H_5) IR (CH_2Cl_2): 1985, 2037 (CO), 2233 (NCO)	[26]
*30	N=C=C(CN)-C(CN)$_2$CH$_3$	red oil IR (neat oil): 2006 (CO), 1306, 2134, 2156 (N=C=C), 2201 (CN)	[11, 21]
31	CN	Ib (3 min, C_6H_6, 80°, traces), m.p. 185 to 189° (darkening above 160°) ^{1}H NMR ($CDCl_3$): 4.55 (d, C_5H_5, J(P,H) = 1.0), ~7.2 (m, C_6H_5) IR: 1988 (CO), 2105 (CN) in mull, 2002 (CO), 2102 (CN) in $CHCl_3$	[5]
*32	Si(CH$_3$)$_3$	Ia (96 h, C_6H_{14}, 12%), orange crystals, m.p. 95 to 96° ^{1}H NMR ($CDCl_3$): 0.41 (CH_3), 4.16 (d, C_5H_5), 7.24 (C_6H_5) IR (C_6H_{12}): 1953, for the 600 to 3100 cm^{-1} region (KBr), see [2]	[2]
*33	Sn(CH$_3$)$_3$	Ia (24 h, C_6H_{14}, 28%), orange crystals, m.p. 99 to 100° ^{1}H NMR ($CDCl_3$): 0.31 (CH_3, ^{2}J(^{119}Sn,H) = 44), 4.05 (C_5H_5), 7.22 (C_6H_5) IR (C_6H_{12}): 1949, for the 600 to 3100 cm^{-1} region, see [2]	[2]
*34	Sn(C$_6$H$_5$)$_3$	Ia (>5 h, C_6H_6 or ether, 65%), pale yellow crystals, m.p. 126 to 128° ^{1}H NMR ($CDCl_3$): 4.26 (C_5H_5) ^{119}Sn-γ (80 K): δ = 1.48, Δ = 0.53 IR ($CHCl_3$): 1940	[16, 17]
35	Sn(C$_6$H$_5$)$_2$Cl	see No. 34 for preparation, yellow crystals, m.p. 155 to 157° ^{1}H NMR ($CDCl_3$): 4.38 (d, C_5H_5) ^{119}Sn-γ (80 K): δ = 1.57, Δ = 2.69 IR ($CHCl_3$): 1957	[16, 17]

References on p. 128

Table 10 [continued]

No.	X	method of preparation (conditions, yield) and properties, explanations on p. 120	Ref.
36	$SnCl_2C_6H_5$	see No. 34 for preparation, yellow crystals, m.p. 168 to 171° ^1H NMR $(CDCl_3)$: 4.41 (d, C_5H_5) $^{119}Sn-\gamma$ (80 K): $\delta = 1.77$, $\Delta = 2.83$ IR $(CHCl_3)$: 1972	[16, 17]
37	$SnCl_3$	Ia (C_6H_6), see No. 34, yellow or yellow-orange crystals, m.p. 205 to 207° (dec.) ^1H NMR $(CDCl_3)$: 4.46 (d, C_5H_5) $^{57}Fe-\gamma$ (80 K): $\delta = 0.238$, $\Delta = 1.77$ $^{57}Fe-\gamma$ (298 K): $\delta = 0.36$, $\Delta = 1.77$ $^{119}Sn-\gamma$ (80 K): $\delta = 1.79$, $\Delta = 1.82$ IR: 1992 in CH_2Cl_2, 1996 in $CHCl_3$ solid stable in air, solutions dec. over a period of days	[16, 17, 19]
*38	$SnBr_3$	m.p. 204 to 206° IR (CS_2): 1988	[28]
*39	$SnBr_2I$	dec. 160° IR (CS_2): 1987	[28]

$C_5H_5Fe(CO)(P(OCH_2)_3CCH_3)X$ compounds

*40	Cl	Ia (1 h, C_6H_{12}), brown solid ^1H NMR $(CHCl_3)$: 0.87 (CH_3), 4.34 (d, OCH_2, $J(P,H) = 5.0$), 4.75 (d, C_5H_5, $J(P,H) = 0.8$) IR: 1994 in $CHCl_3$, 1981 to 1982 in CH_3COCH_3 or CH_3CN	[23]
*41	Br	Ia (1 h, C_6H_{12}), brown solid ^1H NMR $(CHCl_3)$: 0.85 (CH_3), 4.34 (d, OCH_2, $J(P,H) = 5.0$), 4.76 (d, C_5H_5, $J(P,H) = 1.0$) IR: 1989 in $CHCl_3$, 1980 to 1981 in CH_3COCH_3 or CH_3CN	[23]
*42	I	Ia (1 h, C_6H_{12}), dark brown solid ^1H NMR $(CHCl_3)$: 0.87 (CH_3), 4.36 (d, OCH_2, $J(P,H) = 5.0$), 4.78 (d, C_5H_5, $J(P,H) = 1.0$) IR: 1984 in $CHCl_3$, 1975 to 1976 in CH_3COCH_3 or CH_3CN	[23]

$C_5H_5Fe(CO)(P(OCH_2)_3CC_2H_5)X$ compounds

*43	Cl	Ia (1 h, C_6H_{12}), brown solid ^1H NMR $(CHCl_3)$: 0.90 (CH_3), 1.00 (CH_2), 4.36 (d, OCH_2, $J(P,H) = 5.0$), 4.74 (d, C_5H_5, $J(P,H) = 1.0$) IR: 1994 in $CHCl_3$, 1981 to 1982 in CH_3COCH_3 or CH_3CN	[23]

References on p. 128

Table 10 [continued]

No.	X	method of preparation (conditions, yield) and properties, explanations on p. 120	Ref.
*44	Br	Ia (1 h, C_6H_{12}), brown solid ^{1}H NMR (CHCl$_3$): 0.88 (CH$_3$), 1.00 (CH$_2$), 4.35 (d, OCH$_2$, J(P,H) = 5.0), 4.76 (d, C$_5$H$_5$, J(P,H) = 1.0) IR: 1989 in CHCl$_3$, 1980 to 1981 in CH$_3$COCH$_3$ or CH$_3$CN	[23]
*45	I	Ia (1 h, C_6H_{12}), Ib (20 h, C$_6$H$_6$, 80°), dark brown or green solid ^{1}H NMR (CHCl$_3$): 0.87 (CH$_3$), 0.96 (CH$_2$), 4.34 (d, OCH$_2$, J(P,H) = 5.0), 4.78 (d, C$_5$H$_5$, J(P,H) = 1.0) IR: 1982 in CHCl$_3$, 1975 to 1976 in CH$_3$COCH$_3$ or CH$_3$CN, 1960 (in KBr?)	[23, 27]

$C_5H_5Fe(CO)(P(OCH_2)_3CC^1H_2C^2H_2CH_3)X$ compounds

No.	X	method of preparation (conditions, yield) and properties	Ref.
*46	Cl	Ia (1 h, C_6H_{12}), brown solid ^{1}H NMR (CHCl$_3$): 0.94 (CH$_3$), 1.00 (H-2), 1.20 (H-1), 4.38 (d, OCH$_2$, J(P,H) = 5.1), 4.75 (d, C$_5$H$_5$, J(P,H) = 1.0) IR: 1992 in CHCl$_3$, 1981 to 1982 in CH$_3$COCH$_3$ or CH$_3$CN	[23]
*47	Br	Ia (1 h, C_6H_{12}), brown solid ^{1}H NMR (CHCl$_3$): 0.93 (CH$_3$), 1.01 (H-2), 1.20 (H-1), 4.37 (d, OCH$_2$, J(P,H) = 4.8), 4.76 (d, C$_5$H$_5$, J(P,H) = 0.9) IR: 1989 in CHCl$_3$, 1980 to 1981 in CH$_3$COCH$_3$ or CH$_3$CN	[23]
*48	I	Ia (1 h, C_6H_{12}), dark brown solid ^{1}H NMR (CHCl$_3$): 0.93 (CH$_3$), 1.03 (H-2), 1.18 (H-1), 4.34 (d, OCH$_2$, J(P,H) = 5.0), 4.78 (d, C$_5$H$_5$, J(P,H) = 1.0) IR: 1984 in CHCl$_3$, 1975 to 1976 in CH$_3$COCH$_3$ or CH$_3$CN	[23]

*Further information:

$C_5H_5Fe(CO)(P(OCH_3)_3)H$ (Table 10, No. 1). It is supposed that the second ν(CO) band originates from hindered rotation about the Fe–P bond or the P–O–C group [14].

$C_5H_5Fe(CO)(P(OCH_3)_3)Cl$ (Table 10, No. 2) is a monomer in benzene. It is supposed that the second ν(CO) band originates from hindered rotation about the Fe–P bond or the POCH$_3$ group [14]. Molar conductivity $\Lambda = 2\ \Omega^{-1} \cdot cm^2 \cdot mol^{-1}$ for 1 to 10×10^{-4} M in acetone solution [10].

References on p. 128

$C_5H_5Fe(CO)(P(OCH_3)_3)X$ (Table **10**, Nos. **3** and **4**, with X= Br, I). The two absorption bands in each IR spectra indicate a form of isomerism [8]. Conductivity of No. **4** $\Lambda = 1.5\ \Omega^{-1}\cdot cm^2\cdot mol^{-1}$ for 1 to 10×10^{-4} M in acetone solution [10].

$C_5H_5Fe(CO)(P(OC_2H_5)_3)I$ (Table **10**,   No. **9**).   Molar   conductivity $\Lambda = 1.9\ \Omega^{-1}\cdot cm^2\cdot mol^{-1}$ for 1 to 10×10^{-4} M in acetone solution [10].

$C_5H_5Fe(CO)(P(OC_2H_5)_3)Si(CH_3)(C_6H_5)C_{10}H_7$-1 (Table **10**, No. **12**). The oil consists of a mixture of two inseparable diastereomers. $[\alpha]_D^{25} = +13.6°$. The compound is quite stable towards HCl, $NaBH_4$, CH_3OH, CH_3ONa, KHF_2, Li_2O, MgO, Na, and Li. In an ether solution the Fe–Si bond is cleaved by H_2O or $LiAlH_4$ with retention of configuration at the Fe atom. In CCl_4 solution also Br_2 and I_2 cleave this bond. More about the stereoselectivity in respect to the configuration of the silane formed is given in [24].

$C_5H_5Fe(CO)(P(OC_2H_5)_3)Sn(C_6H_5)_3$ (Table **10**, No. **13**) is prepared with an excess of 50 to 100% of $P(OC_2H_5)_3$. The reaction with HCl gives No. 14 [16].

$C_5H_5Fe(CO)(P(OC_2H_5)_3)SnCl_3$ (Table **10**, No. **14**) is obtained from the 3-h reaction of $C_5H_5Fe(CO)(P(OC_2H_5)_3)Sn(C_6H_5)_3$ (No. 13) and HCl at room temperature in benzene or ether. Recrystallization in benzene or pentane ($\sim$85% yield) [16].

$C_5H_5Fe(CO)(P(OC_3H_7$-i$)_3)I$ (Table **10**, No. **15**) is prepared by reaction of $(C_5H_5)_2Fe_2(CO)_3P(OC_3H_7$-i$)_3$ with iodine in CH_2Cl_2. IR spectroscopy shows that the reaction affords $C_5H_5Fe(CO)_2I$, $C_5H_5Fe(CO)(P(OC_3H_7$-i$)_3)I$, and $[C_5H_5Fe(CO)_2$-$P(OC_3H_7$-i$)_3]^+$ in approximately equal amounts [4].

$C_5H_5Fe(CO)(P(OC_3H_5)_3)PO(OC_3H_5)_2$ (Table **10**, No. **17**) is monomeric in C_6H_6, and its molar conductivity is $\Lambda = 0.1\ \Omega^{-1}\cdot cm^2\cdot mol^{-1}$ for 1 to 10×10^{-4} M in acetone solution [10].

$C_5H_5Fe(CO)(P(OC_6H_5)_3)H$ (Table **10**, No. **20**). Addition of BH_3 in THF to a solution of $C_5H_5Fe(CO)(P(OC_6H_5)_3)COCH_3$ in C_6D_6 or THF-d_8 affords both C_5H_5-$Fe(CO)(P(OC_6H_5)_3)H$ (No. 20, 5% yield) and $C_5H_5Fe(CO)(P(OC_6H_5)_3)C_2H_5$ [22]. Attempts to prepare $C_5H_5Fe(CO)(P(OC_6H_5)_3)H$ by heating $C_5H_5Fe(CO)$-$(P(OC_6H_5)_3)R$ complexes in hydrocarbon solvent at 126 °C for 24 h or reduction of $[C_5H_5Fe(CO)(P(OC_6H_5)_3)^2L]^+$ complexes by $NaBH_3CN$ in THF failed [29].

$C_5H_5Fe(CO)(P(OC_6H_5)_3)Cl$ (Table **10**,   No. **21**).   Molar   conductivity $\Lambda = 3.3\ \Omega^{-1}\cdot cm^2\cdot mol^{-1}$ for 1 to 10×10^{-4} M in acetone solution. It crystallizes from benzene/petroleum ether [10].

$C_5H_5Fe(CO)(P(OC_6H_5)_3)Br$ (Table **10**, No. **22**). Photolysis with visible, blue, or green light in C_5H_5N or $(CH_3)_2SO$ yields $(C_5H_5Fe(CO)_2)_2$ and traces of C_5H_5-$Fe(CO)_2Br$ [15].

$C_5H_5Fe(CO)(P(OC_6H_5)_3)I$ (Table **10**, No. **23**) is prepared by the reaction of $(C_5H_5)_2Fe_2(CO)_3P(OC_6H_5)_3$ with I_2 in CH_2Cl_2. IR spectroscopy shows that the reaction produces $C_5H_5Fe(CO)_2I$, $C_5H_5Fe(CO)(P(OC_6H_5)_3)I$, and $[C_5H_5Fe(CO)_2$-$P(OC_6H_5)_3]^+$ in approximately equal amounts [4]. Resolved optically pure C_5H_5-$Fe(CO)(P(OC_6H_5)_3)I$ was prepared by I_2 cleavage of the resolved diastereomers of $C_5H_5Fe(CO)(P(OC_6H_5)_3)CH_2OC_{10}H_{19}$, prepared by methods previously described in [18]; for related compounds, see [25]. Molar conductivity $\Lambda = 2.6\ \Omega^{-1}\cdot cm^2\cdot mol^{-1}$ for 1 to 10×10^{-4} M in acetone solution. It crystallizes from benzene/petroleum ether [10]. $C_5H_5Fe(CO)(P(OC_6H_5)_3)I$ reacts with $(CH_3)_3Si(C_6H_5)CHMgBr$ in benzene/ether (2:1) to give $C_5H_5Fe(CO)(P(OC_6H_5)_3)CH(C_6H_5)Si(CH_3)_3$ (1.5.2.2.12.6,

Table 22, No. 15) [20, 25]. The green equimolar mixture of $C_5H_5Fe(CO)(P(OC_6H_5)_3)I$ and $AgBF_4$ in CH_2Cl_2 turns first brown and then dark green over a 30-min period. An intermediate has a $\nu(CO)$ band at 2000 cm^{-1} and is believed to be $[C_5H_5Fe(CO)-(P(OC_6H_5)_3)]^+$, see 1.5.2.2.4, p. 75.

$C_5H_5Fe(CO)(P(OC_6H_5)_3)SO_2CH(C_6H_5)Si(CH_3)_3$ (Table **10**, No. **27**). Initial attempts to insert SO_2 into $C_5H_5Fe(CO)(P(OC_6H_5)_3)CH(C_6H_5)Si(CH_3)_3$ (1.5.2.2.12.6, Table 22, No. 15) by Method II yielded no sulfinate complex but rather mixtures of the diastereomers from the starting products. The formation of an isomer (a) of No. 27 by reaction of $(RS)(SR)$ diastereomer of the starting product with neat SO_2 begins slowly at $-3\,°C$ and becomes rapid at $17\,°C$. With the $(RR)(SS)$ diastereomer the reaction, producing an isomer (b) of No. 27, runs similarly. Mixtures of both reaction products were isolated as orange-red solids if kept cold $(-10\,°C)$ while the SO_2 solvent was removed under reduced pressure. Both products, either as solids or dissolved in organic solvents, loose SO_2 at room temperature, and satisfactory analytical and infrared data could not be obtained. ^1H NMR for 27a: $\delta = 4.32$ (C_5H_5) and 4.89 (CH) in SO_2, $\delta = 0.15$ ($SiCH_3$), 4.43 (C_5H_5), 5.07 (CH), and ~ 7.2 (C_6H_5) in $CDCl_3/CD_2Cl_2$ at $0\,°C$; ^1H NMR for 29b: $\delta = 4.54$ (C_5H_5) and 5.27 (CH) in SO_2, $\delta = 0.31$ ($SiCH_3$), 4.65 (C_5H_5), 5.25 (CH), and ~ 7.2 (C_6H_5) in $CDCl_3/CD_2Cl_2$ at $0\,°C$. The isomerization 29a$\rightleftharpoons$29b reaches equilibrium after ~ 8 h at $20\,°C$ [25].

$C_5H_5Fe(CO)(P(OC_6H_5)_3)NCO$ (Table **10**, No. **29**) is prepared by stirring $C_5H_5Fe(CO)_2N_3$ in CH_2Cl_2 at room temperature with a slight excess of $P(OC_6H_5)_3$ for 4 h. After filtration and removal of the solvent, the product is washed with petroleum ether. Crystallization from CH_2Cl_2 was induced by adding petroleum ether and cooling in an ice-salt mixture (75% yield) [26].

$C_5H_5Fe(CO)(P(OC_6H_5)_3)N=C=C(CN)C(CN)_2CH_3$ (Table **10**, No. **30**). The green $C_5H_5Fe(P(OC_6H_5)_3)(C_6N_4)COCH_3$, which forms in the reaction of C_5H_5-$Fe(CO)(P(OC_6H_5)_3)CH_3$ (1.5.2.2.12.6, Table 22, No. 8) with $(CN)_2C=C(CN)_2$, turns reddish brown on storage in air to give No. 30. This is separated from $C_5H_5Fe(CO)-(P(OC_6H_5)_3)CH_3$ by chromatography. It is a reddish brown oil [11, 21].

$C_5H_5Fe(CO)(P(OC_6H_5)_3)M(CH_3)_3$ (Table **10**, No. **32** with M = Si and No. **33** with M = Sn). The stability of the Sn–Fe bond is greater than that of the Si–Fe bond. Silicon-iron cleavage occurs during UV irradiation (Method Ia) to give an orthometalated complex, see 1.5.2.1.1, p. 50, No. 32 [2]. The relation between the geminal M-proton coupling constants (M = ^{117}Sn, ^{119}Sn, ^{207}P) and the Taft σ^* constants of trimethyltin-transition metal complexes and trimethyllead complexes is linear [7].

$C_5H_5Fe(CO)(P(OC_6H_5)_3)Sn(C_6H_5)_3$ (Table **10**, No. **34**) reacts in benzene solution with HCl in ether (mole ratio 1:1) to give $C_5H_5Fe(CO)(P(OC_6H_5)_3)Sn(C_6H_5)_2Cl$ (No. **35**, yield 56%), with HCl (mole ratio 1:2) to give $C_5H_5Fe(CO)-(P(OC_6H_5)_3)SnCl_2C_6H_5$ (No. **36**, yield 58%), and with excess HCl to give $C_5H_5Fe(CO)(P(OC_6H_5)_3)SnCl_3$ (No. **37**, yield 88%) [16].

$C_5H_5Fe(CO)(P(OC_6H_5)_3)SnX_2Y$ (Table **10**, No. **38** with $X_2Y = Br_3$ and **39** with $X_2Y = Br_3$ and Br_2I) are prepared by the reaction of $C_5H_5Fe(CO)(P(OC_6H_5)_3)I$ and $SnBr_2$ in refluxing THF. Refluxing for 30 min affords No. 39, and refluxing for 90 min affords No. 38. The hot solutions are filtered, some solvent is removed at reduced pressure, the remaining solution is cooled to $-25\,°C$, and the compounds crystallize [28].

References on p. 128

$C_5H_5Fe(CO)(P(OCH_2)_3CR)X$ (Table **10**, Nos. **40** to **48**, with $R=CH_3$, C_2H_5, and n-C_3H_7, and $X=Cl$, Br, I). A single IR band in the 1980 to 1995 cm^{-1} range is strong evidence that the compounds are single conformers. More details on the IR spectra (in Nujol) between 200 and 800 cm^{-1}, including figures, are available in [23]. Molar conductivities in acetone at 25 °C, $\Lambda < 10$ $\Omega^{-1} \cdot$cm$^2 \cdot$mol^{-1} at $\sim 10^{-3}$ M, are typical for nonelectrolytes. The solids undergo slight decomposition upon standing for several weeks. At room temperature, all are insoluble in cyclohexane, n–pentane, and the other common hydrocarbons, but they are soluble in chloroform, acetone, and acetonitrile. The solubilities of the complexes depend on both X and R: $I > Br > Cl$ and $C_3H_7 > C_2H_5 > CH_3$ [23].

References:

[1] D.A. Brown, A.R. Manning, J.M. Rowley (New Aspects Chem. Metal Carbonyls Deriv., 1st Intern. Symp. Proc., Venice 1968, Ref. C5, pp. 1/10; C.A. **72** [1970] No. 11769). – [2] R.B. King, K.H. Pannell (Inorg. Chem. **7** [1968] 1510/3). – [3] D.A. Brown, H.J. Lyons, A.R. Manning, J.M. Rowley (Inorg. Chim. Acta **3** [1969] 346/50). – [4] R.J. Haines, A.L. du Preez (Inorg. Chem. **8** [1969] 1459/64). – [5] M.A. Jennings, A. Wojcicki (Inorg. Chim. Acta **3** [1969] 335/40).

[6] R.B. King, R.N. Kapoor (J. Inorg. Nucl. Chem. **31** [1969] 2169/77). – [7] G. Singh (Proc. 4th Intern. Conf. Organometal. Chem., Bristol, Engl., 1969, Ref. B10). – [8] D.A. Brown, H.J. Lyons, A.R. Manning (Inorg. Chim. Acta **4** [1970] 428/30). – [9] R.J. Haines, A.L. du Preez, L.L. Marais (J. Organometal. Chem. **24** [1970] C26/C28). – [10] R.J. Haines, A.L. du Preez, I.L. Marais (J. Organometal. Chem. **28** [1971] 405/13).

[11] S.R. Su (Diss. Ohio State Univ. 1971; Diss. Abstr. Intern. B **32** [1972] 6283). – [12] S.R. Su, A. Wojcicki (J. Organometal. Chem. **27** [1971] 231/40). – [13] D.J. Jones, R.J. Mawby (Inorg. Chim. Acta **6** [1972] 157/60). – [14] P. Kalck, R. Poilblanc (Compt. Rend. C **274** [1972] 66/9). – [15] D.M. Allen, A. Cox, T.J. Kemp, L.H. Ali (J. Chem. Soc. Dalton Trans. **1973** 1899/901).

[16] R.E.J. Bichler, H.C. Clark, B.K. Hunter, A.T. Rake (J. Organometal. Chem. **69** [1974] 367/76). – [17] G.M. Bancroft, A.T. Rake (Inorg. Chim. Acta **13** [1975] 175/9). – [18] T.C. Flood, F.J. DiSanti, D.L. Miles (J. Chem. Soc. Chem. Commun. **1975** 336/7). – [19] T.N. Pecoraro (Diss. Rutgers State Univ. 1975; Diss. Abstr. Intern. B **36** [1975] 746). – [20] K. Stanley, M.C. Baird (J. Am. Chem. Soc. **97** [1975] 6598/9).

[21] S.R. Su, A. Wojcicki (Inorg. Chem. **14** [1975] 89/98). – [22] J.A. van Doorn, C. Masters, H.C. Volger (J. Organometal. Chem. **105** [1976] 245/54). – [23] W.E. Stanclift, D.G. Hendricker (J. Organometal. Chem. **107** [1976] 341/9). – [24] G. Cerveau, E. Colomer, R. Corriu, W.E. Douglas (J. Organometal. Chem. **135** [1977] 373/86). – [25] K. Stanley, M.C. Baird (J. Am. Chem. Soc. **99** [1977] 1808/12).

[26] D.A. Brown, F.M. Hussein, C.L. Arora (Inorg. Chim. Acta **29** [1978] L215/L216). – [27] V.N. Pandey (Inorg. Chim. Acta **22** [1977] L39/L41). – [28] B. O'Dwyer, A.R. Manning (Inorg. Chim. Acta **38** [1980] 103/5). – [29] C.J. Coleman (Diss. Univ. South Carolina 1978; Diss. Abstr. Intern. B **39** [1979] 5909).

1.5.2.2.6.1.5   Compounds with $^2D = P(OR)_2R'$ or $PR_2'OR$

$C_5H_5Fe(CO)(P(OCH_2CH=CH_2)_2C_6H_5)X$ with $X=Cl$ or $PO(OCH_2CH=CH_2)$-C_6H_5 results together with $C_5H_5Fe(CO)_2PO(OCH_2CH=CH_2)C_6H_5$ from the reaction of $C_5H_5Fe(CO)_2Cl$ with excess $P(OCH_2CH=CH_2)_2C_6H_5$ in benzene at 25 °C for 1 h. $C_5H_5Fe(CO)(P(OCH_2CH=CH_2)_2C_6H_5)Cl$ (20% yield) was identified by means of IR (data not given). It was not crystallized.

$C_5H_5Fe(CO)(P(OCH_2CH=CH_2)_2C_6H_5)PO(OCH_2CH=CH_2)C_6H_5$ was isolated as an oil in a 25% yield. IR: $\nu(PO)$ 1140 cm^{-1} in CS_2, $\nu(CO)$ 1961 cm^{-1} in C_6H_{12}. ^{1}H NMR (CDCl$_3$): $\delta = 4.38$ ppm (t, C_5H_5, $J(P,H) = 1.2$ Hz). Conductivity $\Lambda = 0.1\ \Omega^{-1}\cdot cm^2\cdot mol^{-1}$ for 1 to 10×10^{-4} M in acetone solution. It is monomeric in benzene solution [1, 2].

$C_5H_5Fe(CO)(P(OC_6H_5)_2C_6H_5)H$ is obtained together with $C_5H_5Fe(CO)$-$(P(OC_6H_5)_2C_6H_5)C_2H_5$ by addition of BH$_3$ in THF to a solution of $C_5H_5Fe(CO)$-$(P(OC_6H_5)_2C_6H_5)COCH_3$ in C_6D_6 or THF-d$_8$. ^{1}H NMR (C$_6$D$_6$): $\delta = -13.2$ (FeH, $J(P,H) = 88$ Hz), 4.11 (C_5H_5, $J(P,H) = 1.0$ Hz) ppm [3].

$C_5H_5Fe(CO)(P(C_6H_5)_2OCH_2CH=CH_2)X$ with X = Cl or $PO(C_6H_5)_2$ results together with $C_5H_5Fe(CO)_2PO(C_6H_5)_2$ from the reaction of $C_5H_5Fe(CO)_2Cl$ with excess $P(C_6H_5)_2OCH_2CH=CH_2$ in refluxing benzene (3 h). $C_5H_5Fe(CO)(P(C_6H_5)_2OCH_2CH=CH_2)Cl$ (20% yield) was identified by its IR spectrum (data not given) and was not crystallized. $C_5H_5Fe(CO)(P(C_6H_5)_2OCH_2CH=CH_2)PO(C_6H_5)_2$ crystallizes from benzene/petroleum ether in a 15% yield. IR: $\nu(PO)$ 1116 cm^{-1} in CS_2, $\nu(CO)$ 1933 cm^{-1} in C_6H_{12}. ^{1}H NMR (CDCl$_3$): $\delta = 4.47$ ppm (s, C_5H_5). Conductivity $\Lambda = 2.6\ \Omega^{-1}\cdot cm^2\cdot mol^{-1}$ for 1 to 10×10^{-4} M in acetone solution. It is monomeric in benzene solution. The solid is stable in air but gives a poorly defined melting point [1, 2].

$C_5H_5Fe(CO)(P(C_6H_5)_2OC_6H_5)H$ is obtained together with $C_5H_5Fe(CO)$-$(P(C_6H_5)_2OC_6H_5)C_2H_5$ by addition of BH$_3$ in THF to a solution of $C_5H_5Fe(CO)$-$(P(C_6H_5)_2OC_6H_5)COCH_3$ in C_6D_6 or THF-d$_8$. ^{1}H NMR (C$_6$D$_6$): $\delta = -13.1$ (FeH, $J(P,H) = 82$ Hz), 4.12 (C_5H_5) ppm [3].

References:

[1] R.J. Haines, A.L. du Preez, L.L. Marais (J. Organometal. Chem. **24** [1970] C26/C28). − [2] R.J. Haines, A.L. du Preez, I.L. Marais (J. Organometal. Chem. **28** [1971] 405/13). − [3] J.A. Van Doorn, C. Masters, H.C. Volger (J. Organometal. Chem. **105** [1976] 245/54).

1.5.2.2.6.1.6 Compounds with ^{2}D = PF$_3$ or PF$_2$N(R,X)$_2$

$C_5H_5Fe(CO)(PF_3)I$ is prepared by refluxing a mixture of 3.29 mmol $C_5H_5Fe(CO)_2I$ and 8.8 mmol Ni(PF$_3$)$_4$ in toluene for 12 h. The solvent is removed from the filtered reaction mixture, and the residue dissolved in hexane. Cooling of the filtered hexane solution to -78 °C gives a dark crystalline precipitate, which is sublimed at 25 °C/ 0.1 Torr to give brown crystals, 77% yield, m.p. 102 to 103 °C [1, 3]. ^{1}H NMR (C$_6$D$_6$): $\delta = 4.12$ ppm (s, C_5H_5), ^{19}F NMR (C$_6$D$_6$): $\delta = 4.5$ ppm (d, $J(P,F) = 1325$ Hz). IR: $\nu(PF)$ 848, 868, and 888 cm^{-1} in Nujol (further bands are available from the original), $\nu(CO)$ 2021 cm^{-1} in C_6H_{14}. Heating to 220 °C results in decomposition to ferrocene [3].

$C_5H_5Fe(CO)(PF_2NHCH_3)Cl$ is prepared by boiling a mixture of 4.0 mmol $C_5H_5Fe(CO)_2Cl$ and 4.6 mmol $CH_3N(PF_2)_2$ in C_6H_6/CH_3OH (10:1) for 1 h. The solvents are removed, and the oily residue is chromatographed on SiO$_2$ with ether/hexane (3:1) [6].

Brown–black crystals, 38% yield, m.p. 112 to 114 °C. ^1H NMR (CDCl$_3$): $\delta = 2.76$ (ddt, CH$_3$, J = 1, 6, and 12 Hz), 4.91 (d, C$_5$H$_5$) ppm. ^{13}C NMR (CDCl$_3$): $\delta = 26.6$ (CH$_3$), 82.7 (C$_5$H$_5$), and 215.1 (d, CO, J = 44 Hz) ppm. ^{31}P NMR (CH$_2$Cl$_2$): $\delta = 195.1$ ppm (J(F,P) = 1138 Hz). These data show the P atom in monodentate PF$_2$NHCH$_3$ to be the donor atom. IR: ν(CO) 1980 cm^{-1} in C$_6$H$_{14}$, ν(NH) 3220 cm^{-1} in KBr [6].

C$_5$H$_5$Fe(CO)(PF$_2$N(C$_2$H$_5$)$_2$)I is formed together with C$_5$H$_5$Fe(PF$_2$N(C$_2$H$_5$)$_2$)$_2$I from C$_5$H$_5$Fe(CO)$_2$I and PF$_2$N(C$_2$H$_5$)$_2$ in C$_6$H$_{12}$/C$_6$H$_6$ by UV irradiation at 25 °C for 8 h. The crude product is purified by chromatography with petroleum ether and recrystallized in pentane by cooling to -78 °C [4].

Gray crystals, 62% yield, m.p. 90 to 91 °C. ^1H NMR (CDCl$_3$): $\delta = 1.20$ (t, C$_2$H$_5$, J = 7 Hz), $\sim$3.4 (C$_2$H$_5$), and 4.78 (s, C$_5$H$_5$) ppm. ^{19}F NMR (THF): $\delta = 25.7$ (J = 1102 Hz) and 29.8 (J = 1177 Hz) ppm. IR: ν(PF) 794 and 815 cm^{-1} in KBr, ν(CO) 1980 and 1999 cm^{-1} in C$_6$H$_{12}$ [4]. For a rapid analytic determination by X-rays and colorimetry, see [2].

C$_5$H$_5$Fe(CO)(PF$_2$N(CH$_3$)PF$_2$)Cl can be obtained by reacting C$_5$H$_5$Fe(CO)$_2$Cl and CH$_3$N(PF$_2$)$_2$ (mole ratio 1:1) in boiling hexane for 5 min or in boiling ether for 1 h, 45 to 55% yield [5]. Preparation in ether and purification by chromatography gives a 56% yield [6].

Red-purple crystals, m.p. 85 to 87 °C [5, 6]. ^1H NMR (CDCl$_3$): $\delta = 3.00$ (d, CH$_3$, J = 8 Hz) and 4.95 (s, C$_5$H$_5$) ppm. ^{13}C NMR (CDCl$_3$): $\delta = 25.7$ (CH$_3$), 83.5 (C$_5$H$_5$), and 213.9 (d, CO, J = 44 Hz) ppm. ^{31}P NMR (CH$_2$Cl$_2$): $\delta = 135.6$ (NPF, J(F,P) = 1258, J(P,P) = 165 Hz) and 194.4 (FePN, J(F,P) = 1168, J(P,P) = 165 Hz) ppm. The nonequivalent P atoms indicate monodentate CH$_3$N(PF$_2$)$_2$ with one P atom bonded to Fe [6]. IR (C$_6$H$_{14}$): ν(CO) 2000 to 2003 cm^{-1} [5, 6].

References:

[1] R.B. King, A. Efraty (J. Am. Chem. Soc. **93** [1971] 5260/1). – [2] J.M. McCall, D.E. Leyden, C.W. Blount (Anal. Chem. **43** [1971] 1324/5). – [3] R.B. King, A. Efraty (J. Am. Chem. Soc. **94** [1972] 3768/73). – [4] R.B. King, W.C. Zipperer, M. Ishaq (Inorg. Chem. **11** [1972] 1361/70). – [5] R.B. King, M.G. Newton, J. Gimeno, M. Chang (Inorg. Chim. Acta **23** [1977] L35/L36).

[6] R.B. King, J. Gimeno (Inorg. Chem. **17** [1978] 2396/400).

1.5.2.2.6.1.7   Compounds with $^2D = AsR_2R'$, As(C$_6$H$_5$)$_3$, or Sb(C$_6$H$_5$)$_3$

The compounds listed in Table 11 can be prepared in several ways. The following methods are of a general nature. Reaction conditions (solvent, temperature, and time) are given in the table. Other special procedures are described in the further information.

Method I: Mixtures of C$_5$H$_5$Fe(CO)$_2$X and an excess of the appropriate ^2D-ligand are UV irradiated for several hours [1, 3, 4, 5, 8].

Method II: Mixtures of C$_5$H$_5$Fe(CO)$_2$X and the appropriate ^2D-ligand are refluxed in a solvent without irradiation for 8 to 10 h [6, 7]. The reaction mixtures are freed from the precipitated [C$_5$H$_5$Fe(CO)$_2$^{2}D]I. On concentration of the filtrate, green solids are obtained [6].

Although all of the compounds cited in Table 11 possess a chiral Fe atom, no optical activities of these compounds have been reported.

Table 11
$C_5H_5Fe(CO)(^2D)X$ Compounds with $^2D = AsR_2R'$, $As(C_6H_5)_3$, and $Sb(C_6H_5)_3$.
Further information on numbers preceded by an asterisk is given at the end of the table.
For abbreviations and dimensions, see p. 375.

No. X	method of preparation (conditions, yield) and properties	Ref.

$C_5H_5Fe(CO)(As(C_6H_5)_2CH_3)X$ compounds

| 1 | I | I (20 h, C_6H_6, 80°), green
IR (C_6H_6): 1952 | [7] |

$C_5H_5Fe(CO)(As(C_6H_5)_2CF_3)X$ compounds

| 2 | $Sn(CH_3)_3$ | I (5 h, C_6H_6, 18%), orange crystals, m.p. 85 to 87°
^{1}H NMR (CS_2): −0.03 (s, CH_3, J(^{119}Sn,H) = 42.2),
    4.53 (s, C_5H_5), 7.49 (d, C_6H_5)
^{19}F NMR (CH_3COCH_3): 55.6
^{57}Fe-γ (78 K): δ = 0.53, Δ = 1.98
^{119}Sn-γ (78 K): δ = 1.41
IR (C_6H_{12}): 1921, 1927
solid stable in air and light, solutions dec. slowly,
    most stable in CH_3COCH_3 or CS_2 | [4] |

$C_5H_5Fe(CO)(As(C_6H_5)_3)X$ compounds

*3	I	I, II (C_6H_6, 80°, 32.5%), green crystals, m.p. 110° IR (Nujol): 1934	[3, 6]
*4	$SO_2CH_2C_6H_5$	m.p. 115° ^{1}H NMR ($CDCl_3$): 4.15, 4.44 (CH_2, J = 13.5), 4.48 (C_5H_5), 7.7 (C_6H_5) IR: 1962 (CO) in $CHCl_3$, 1035, 1166 (SO) in Nujol	[2]
*5	NCS	I (6 h, THF, 1%), brown crystals IR ($CHCl_3$): 1975 (CO), 2115 (CN)	[8]
6	$SnCl_3$	I (C_6H_6), red crystals ^{1}H NMR ($CDCl_3$): 4.77 to 4.78 (C_5H_5) ^{119}Sn-γ (∼80 K): δ = 1.899, Δ = 1.83 IR: 1964, 1975 in KBr, 1969 in CH_2Cl_2 solid stable in air, solutions dec. over a period of days	[5]
*7	$Sn(CH_3)_3$	I (5 h, C_6H_6, 50%), red–brown crystals, m.p. 120 to 122° ^{1}H NMR (CS_2): 0.00 (s, CH_3, J(^{119}Sn,H) = 40.2), 4.31 (s, C_5H_5), 7.40 (m, C_6H_5) ^{57}Fe-γ (78 K): δ = 0.53, Δ = 1.94 ^{119}Sn-γ (78 K): δ = 1.43 IR (C_6H_{12}): 1909, 1920	[4]
*8	$Sn(C_6H_5)_3$	I (5 h, C_6H_6, 28%), red crystals, m.p. 154 to 158° ^{1}H NMR (CS_2): δ = 4.28 (s, C_5H_5), 7.09 (m, C_6H_5) ^{57}Fe-γ (78 K): δ = 0.53, Δ = 1.90 ^{119}Sn-γ (78 K): δ = 1.46, Δ = 0.67 IR (C_6H_{12}): 1911	[4]

References on p. 133

Table 11 [continued]

No. X	method of preparation (conditions, yield) and properties	Ref.

$C_5H_5Fe(CO)(Sb(C_6H_5)_3)X$ compounds

*9	I	I (C_6H_6), II (C_6H_6, 80°, 30.8%), green crystals, m.p. 100, 128 to 129° IR: 1932 in Nujol, 1954 in CCl_4 solid stable in air, but less stable than No. 3	[1, 6]
10	$SnCl_3$	I (C_6H_6), red orange ^1H NMR ($CDCl_3$): 4.83 (C_5H_5) ^{57}Fe-γ (~80 K): $\delta = 0.331$, $\Delta = 1.81$ IR: 1955 in KBr, 1968 in CH_2Cl_2 solid stable in air, solutions dec. over a period of days	[5]
11	$Sn(CH_3)_3$	I (5 h, C_6H_6, 33%), orange-red crystals, m.p. 92 to 94° ^1H NMR (CS_2): 0.03 (s, CH_3, J(^{119}Sn,H) = 41.4), 4.42 (s, C_5H_5), 7.38 (m, C_6H_5) ^{57}Fe-γ (78 K): $\delta = 0.55$, $\Delta = 1.93$ ^{119}Sn-γ (78 K): $\delta = 1.44$ IR (C_6H_{12}): 1911, 1921 stability as No. 2	[4]
12	$Sn(C_6H_5)_3$	I (5 h, C_6H_6, 65%), red crystals, m.p. 127 to 129° ^1H NMR (CS_2): 4.40 (s, C_5H_5), 7.10 (m, C_6H_5) ^{57}Fe-γ (78 K): $\delta = 0.55$, $\Delta = 1.90$ ^{119}Sn-γ (78 K): $\delta = 1.42$, $\Delta = 0.56$ IR (C_6H_{12}): 1919 stability as No. 2	[4]

*Further information:

$C_5H_5Fe(CO)(As(C_6H_5)_3)I$ (Table 11, No. 3). Figures and intensities of the IR spectrum (Nujol) from 680 to 1486 cm^{-1} [6] and from 534 to 594 cm^{-1} [3] are available. Reaction with o-phenanthroline or 2,2'-bipyridine (^2D-^2D) in boiling C_6H_6 gives [$C_5H_5Fe(As(C_6H_5)_3)^2D$-$^2D$]I [6].

$C_5H_5Fe(CO)(As(C_6H_5)_3)SO_2CH_2C_6H_5$ (Table 11, No. 4). Freshly prepared $C_5H_5Fe(CO)(As(C_6H_5)_3)CH_2C_6H_5$ (see 1.5.2.2.12.8, Table 25, No. 3) is dissolved in $CHCl_3$, and SO_2 bubbled slowly through the solution at 27 °C. The reaction is monitored by IR. After ~10 min, the solvent is evaporated, and the residue chromatographed to isolate No. 4. Refluxing in THF or photolysis (UV, benzene, 30 °C, 72 h) to bring about desulfurylation gives only the starting materials and decomposition products [2].

$C_5H_5Fe(CO)(As(C_6H_5)_3)NCS$ (Table 11, No. 5). Attempts to prepare this compound by thermal substitution of $C_5H_5Fe(CO)_2NCS$ were unsuccesful. The integrated absorption intensity of the 2115 cm^{-1} IR band is 11.7×10^4 M^{-1}·cm^{-2}, typical for Fe-NCS bonding. The compound is soluble in C_6H_{14}, C_6H_6, and $CHCl_3$ [8].

$C_5H_5Fe(CO)(As(C_6H_5)_3)SnR_3$ (Table 11, Nos. 7 and 8, with R = CH_3, C_6H_5) in their solid states are indefinitely stable in air or light. They decompose in solution, which makes recrystallization difficult. The compounds are most stable in CH_3COCH_3 or CS_2 [4].

$C_5H_5Fe(CO)(Sb(C_6H_5)_3)I$ (Table 11, No. 9). Figures and intensities of the IR (Nujol) from 685 to 1580 cm^{-1} are available in [6]. The compound is soluble in C_6H_6, CH_2Cl_2, $CHCl_3$, and CCl_4 but is insoluble in aliphatic hydrocarbons and petroleum ether. Reaction with o-phenanthroline or 2,2'-bipyridine ($^2D-^2D$) in boiling C_6H_6 gives $[C_5H_5Fe(Sb(C_6H_5)_3)^2D-^2D]I$ [6].

References:

[1] D.A. Brown, H.J. Lyons, A.R. Manning (Inorg. Chim. Acta 4 [1970] 428/30). – [2] M. Graziani, A. Wojcicki (Inorg. Chim. Acta 4 [1970] 347/50). – [3] D.J. Parker (J. Chem. Soc. A 1970 1382/6). – [4] W.R. Cullen, J.R. Sams, J.A.J. Thompson (Inorg. Chem. 10 [1971] 843/50). – [5] T.N. Pecoraro (Diss. Rutgers State Univ. 1975; Diss. Abstr. Intern. B 36 [1975] 746).

[6] S.C. Tripathi, S.C. Srivastava, V.N. Pandey (Transition Metal Chem. [Weinheim] 1 [1976] 266/8; C. A. 86 [1977] No. 190170). – [7] V.N. Pandey (Inorg. Chim. Acta 22 [1977] L39/L41). – [8] D.G. Alway, K.W. Barnett (Inorg. Chem. 17 [1978] 2826/31).

1.5.2.2.6.2 $^5LFe(CO)(^2D)X$ Compounds with Substituted Cyclopentadienyl Ligands

1.5.2.2.6.2.1 Compounds with $^5L = C_5H_4CH_3$

The compounds are listed in Table 12. Except for No. 10, all compounds were prepared by refluxing $CH_3C_5H_4Fe(CO)_2X$ and a slight excess of PR_3 or $P(OR)_3$ in benzene for 24 h, >90% yield [1, 2]. No. 10 was obtained by UV irradiation of $CH_3C_5H_4Fe(CO)_2I$ and excess $As(C_6H_5)_3$ in solution [2].

Table 12
$CH_3C_5H_4Fe(CO)(^2D)X$ Compounds with $^2D = PR_3$, $P(OR)_3$, and $As(C_6H_5)_3$.
Further information for No. 4 is given at the end of the table.

No.	^2D	X	m.p. in °C	IR, ν(CO) band in cm^{-1} (in C_7H_{16})	Ref.
1	$P(C_4H_9-n)_3$	I	42 to 44	1950	[2]
2	$P(C_6H_5)_3$	I	114 to 115	1955, 1961	[2]
3	$P(OCH_3)_3$	Cl	unstable liquid	1965, 1977	[2]
*4	$P(OCH_3)_3$	I	44 to 46	1963, 1976	[2]
5	$P(OC_3H_7-i)_3$	I	36 to 38	1955, 1972	[2]
6	$P(OCH_2)_3CCH_3$	I	160 to 162 (dec.)	1976 (in $CHCl_3$)	[2]
7	$P(OC_6H_5)_3$	Cl	102 to 104	1987	[1, 2]
8	$P(OC_6H_5)_3$	Br	108 to 112 (dec.)	1986	[1, 2]
9	$P(OC_6H_5)_3$	I	102 to 103	1983	[1, 2]
10	$As(C_6H_5)_3$	I	94 (dec.)	1971 (in CCl_4)	[2]

*Further information:

$CH_3C_5H_4Fe(CO)(P(OCH_3)_3)I$ (Table 12, No. 4). The $\nu(CO)$ band of this compound dissolved in n-heptane is clearly resolved into two components. Variable-temperature studies of solutions in pentane at $-80\,°C$ and decaline at $\sim150\,°C$ show that the relative peak heights vary from $0.285:1$ to $0.657:1$, respectively. The two absorption bands in the spectrum are well resolved in heptane and cyclohexane but less well separated in xylene, carbon disulfide, or carbon tetrachloride. However, only one broad, but symmetrical, absorption band is observed when chloroform is the solvent [2]. The occurrence of two $\nu(CO)$ bands is probably due to restricted rotation about the P–C and P–O–C bonds [2]. References see under 1.5.2.2.6.2.2.

1.5.2.2.6.2.2 Compounds with $^5L = C_5H_4CH(C_6H_5)_2$

$(C_6H_5)_2CHC_5H_4Fe(CO)(P(C_6H_5)_3)Cl$. To a stirred solution of $(C_6H_5)_2CHC_5H_4$-$Fe(CO)_2Cl$ in benzene at 40 °C is slowly added a solution of $P(C_6H_5)_3$ in benzene. After 15 min the solution is filtered off from a yellow precipitate. The green filtrate is concentrated and a tenfold volume of petroleum ether is added to precipitate the compound, yield 18% [3].

References:

[1] D.A. Brown, H.J. Lyons, A.R. Manning, J.H. Rowley (Inorg. Chim. Acta **3** [1969] 346/50). – [2] D.A. Brown, H.J. Lyons, A.R. Manning (Inorg. Chim. Acta **4** [1970] 428/30). – [3] J. Ellermann, H. Behrens, H. Krohberger (J. Organometal. Chem. **46** [1972] 119/38).

1.5.2.2.6.2.3 Compounds with $^5L = 1\text{-}CH_3(3\text{-}C_6H_5)C_5H_3$

$1\text{-}CH_3(3\text{-}C_6H_5)C_5H_3Fe(CO)(P(C_6H_5)_3)I$. A solution of $1\text{-}CH_3(3\text{-}C_6H_5)$-$C_5H_3Fe(CO)(P(C_6H_5)_3)CH_3$ is treated dropwise while rapidly stirred with an approximately equimolar solution of I_2, with freshly prepared HI in CH_2Cl_2, or with an approximately equimolar solution of HgI_2 in THF. Solvent is removed, and the residue dissolved in benzene and chromatographed on Al_2O_3 with benzene eluent. The compound crystallized on addition of pentane. The three methods gave 60, 72, and 86% yield, respectively. The cleavage of the Fe–CH_3 bond proceeds with retention of configuration at iron. Another substitution method was carried out with $1\text{-}CH_3(3\text{-}C_6H_5)$-$C_5H_3Fe(CO)_2I$ and a slight excess of $P(C_6H_5)_3$ in THF which were irradiated for 4.5 h. Chromatographic workup as above gave a 50% yield [3].

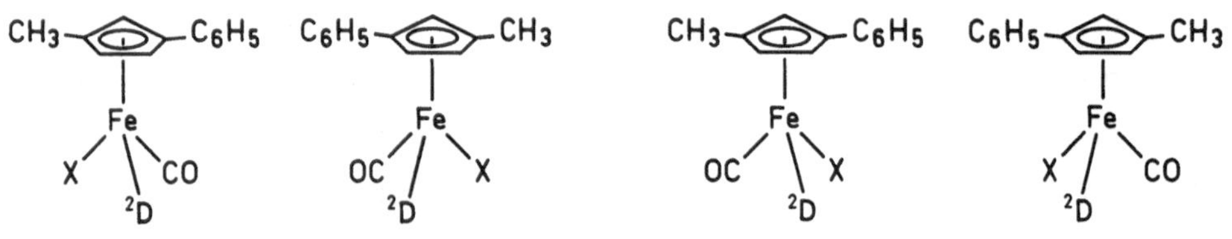

$^2D = P(C_6H_5)_3$, X = I

The black crystals, melting at 167 °C, are a 1:1 mixture of the diastereomeric pairs I and II. The diastereomers are partially separated by a combination of column chromatography (Al_2O_3/C_6H_6) and fractional crystallization in C_6H_6/C_6H_{14} (10:13). Three additional crystallizations afford the (RS)(SR) forms II pure enough for X-ray investigations [3].

The ^1H NMR spectra (in C_6D_6) of the racemates I and II show the following chemical shifts (δ in ppm):

	CH_3 (s's) 1.5 H each peak	C_5H_3 (m) 0.5 H each peak	C_6H_5 (m) 20 H total
I	2.18	3.59, 4.11, 4.65	7.0 to 7.6
II	1.95	3.18, 4.39, 4.83	

IR spectrum (racemate in $CHCl_3$): ν(CO) 1949 cm^{-1}. The (RS)(SR) form II crystallizes in the monoclinic system with a=9.773(7), b=15.635(8), c=17.235(8) Å, and β=92.43(4)°, space group $P2_1/n - C_{2h}^5$. Z=4 gives $D_c = D_m = 1.59$ g·cm^{-3}. The molecular structure is shown in **Fig. 12** [3].

Fig. 12

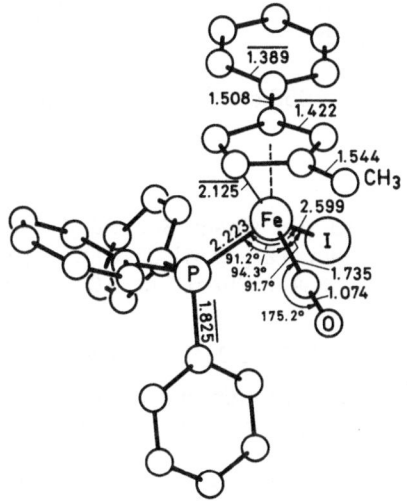

Molecular structure of $1-CH_3(3-C_6H_5)C_5H_3Fe(CO)(P(C_6H_5)_3)I$, SR isomer of the (RS)(SR) pair [3].

$1-CH_3(3-C_6H_5)C_5H_3Fe(CO)(P(C_6H_5)_3)SO_2CH_3$. A solution of 10 mmol SO_2 in CH_2Cl_2 at −75 °C is added dropwise to a stirred solution of 0.14 to 0.194 mmol of $1-CH_3(3-C_6H_5)Fe(CO)(P(C_6H_5)_3)CH_3$ in 25 mL of CH_2Cl_2 at −10 °C. The resulting solution is stirred for 1 h, the solvent removed, and the residue dissolved in a minimum of CH_2Cl_2. Chromatography on Al_2O_3 with $CHCl_3$ affords the compound in a 78 to 80% yield. Use of liquid sulfur dioxide with diastereomerically enriched $1-CH_3(3-C_6H_5)C_5H_3Fe(CO)(P(C_6H_5)_3)CH_3$ (A) at −78 °C, the solution maintained at about −60 °C for 2 h, gives a yield of ~55% for a stereochemically enriched product after workup as described above [3], also see [1, 2]. The insertion proceeds with essentially equal stereospecificity at Fe for the two diastereomers of the starting material. In CH_2Cl_2 it is almost completely stereoselective (≳95%) whereas in neat SO_2 it exhibits a lower stereospecificity (~79%). No separation of the diastereomers III and IV was attempted [3].

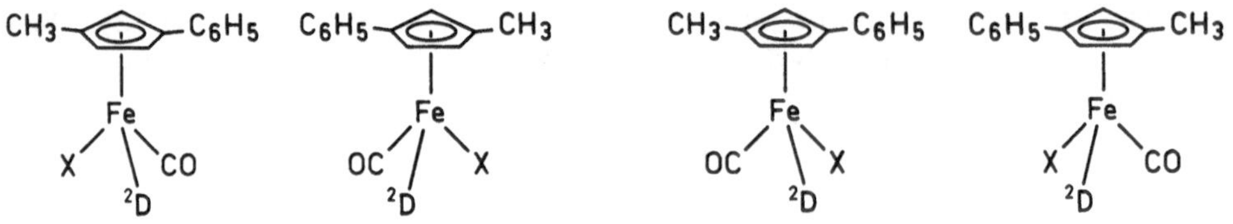

CH_3—⬡—C_6H_5 C_6H_5—⬡—CH_3 CH_3—⬡—C_6H_5 C_6H_5—⬡—CH_3

X—Fe—CO   OC—Fe—X   OC—Fe—X   X—Fe—CO
   ^2D ^2D ^2D ^2D

(RR)   (SS)   (RS)   (SR)

$^2D = P(C_6H_5)_3$, $X = SO_2CH_3$

III   IV

The ^1H NMR spectra of the racemates III and IV show the following chemical shifts (δ in ppm):

	CCH_3 (s's) 1.5 H each peak	SCH_3 (s's) 1.5 H each peak	C_5H_3 (m) 0.5 H each peak	C_6H_5 (m) 20 H total
III	1.80	2.39	4.50, 4.68, 4.85 ⎫	
IV	1.89	2.40	3.99, 4.34, 4.85 ⎭	7.0 to 8.6

IR spectrum (racemate): $v(CO)$ 1958 cm^{-1} in CHCl$_3$, $v(SO_2)$ 1031 and 1161 cm^{-1} in Nujol [3].

References:

[1] T.G. Attig, A. Wojcicki (J. Am. Chem. Soc. **96** [1974] 262/3). – [2] A. Wojcicki (Advan. Organometal. Chem. **12** [1974] 31/81). – [3] T.G. Attig, R.G. Teller, S.-M. Wu, R. Bau, A. Wojcicki (J. Am. Chem. Soc. **101** [1979] 619/28).

1.5.2.2.7 Compounds of the $^5LFe(CO)^2D$-X Type

The compounds in this section contain chelating 3-electron donor ligands. They are collected in Table 13 and are arranged by the size of the -Fe-^2D-X- ring system.

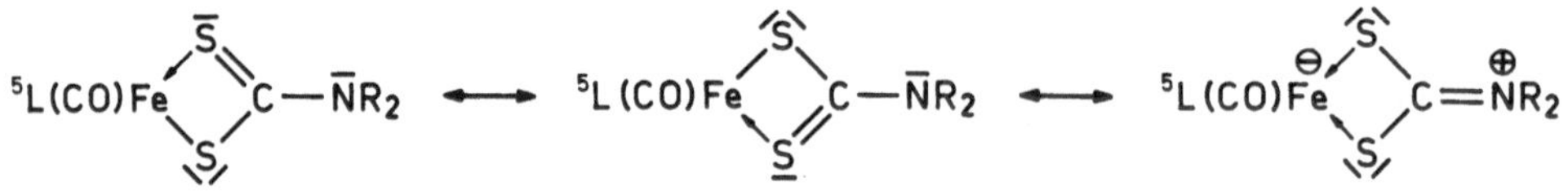

$^5L(CO)Fe\overset{S}{\underset{S}{\diagup\diagdown}}C—\bar{N}R_2$ ⟷ $^5L(CO)Fe\overset{S}{\underset{S}{\diagup\diagdown}}C—\bar{N}R_2$ ⟷ $^5L(CO)Fe\overset{\ominus S}{\underset{S}{\diagup\diagdown}}C=\overset{\oplus}{N}R_2$

a   b   c

I

The majority of these compounds are built up from 4-membered -Fe-^2D-X- rings (Nos. 1 to 29). They are arranged by the atomic sequences -Fe-N-N-N-, -Fe-N-C-S-, -Fe-P-C-S-, -Fe-S-C-S-, -Fe-Se-C-S-, and -Fe-Se-C-Se-. Except for the -Fe-P-C-S- rings, no distinction can be made between the ^2D and X sites of the ligand because of mesomerism examplified by Formulas Ia and Ib for the dithiocarbamate group [1, 4, 14]. In this case, due to the electron releasing effect of NR$_2$, a further mesomeric structure Ic must be considered consistent with $v(CN)$ frequencies of the dithiocarbamate ligand in the region of values between single- and double-bond frequencies [1, 4, 5, 10, 17]. The mesomeric electron-releasing ability of the N lone pair may be quantified by the pK$_a$ of the [R$_2$NH$_2$]$^+$ cation when R is alkyl. Nitrogen-phenyl conjugation, which substantially lowers the aqueous pK_a of phenyl amines, is insignificant in coordinated phenyl dithiocarbamates [10].

One cyclopentadienyl–substituted derivative of No. 2 was described:

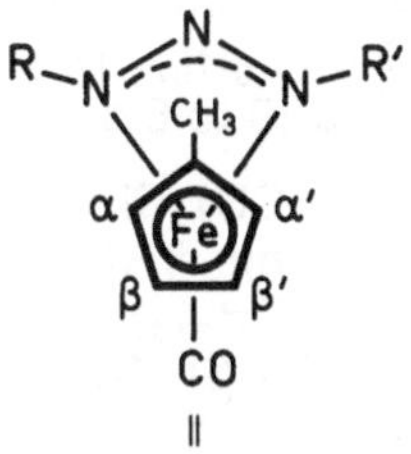

$CH_3C_5H_4Fe(CO)N_3(C_6H_4CH_3-4)C_6H_4Cl-4$, prepared by the method given for Nos. 1 to 3, is a brown substance. ^1H NMR spectrum ($CDCl_3$): $\delta = 2.28$ (CH_3) and 6.8 to 7.2 (m, C_6H_4) ppm. ^{13}C NMR spectrum ($CDCl_3$): $\delta = 13.2$ (CH_3), 89.1, 89.8, 96.6, 96.8, and 118.3 (C_5H_4) ppm. Two pairs of nonequivalent C atoms (α, α' and β, β' in Formula II) show that the triazenido ligand remains fixed with respect to the $CH_3C_5H_4$ ring. IR spectrum (in KBr): $v(CO)$ 1954 and $v(NNN)$ 1275 cm^{-1} [14].

The compounds described in Table 13 can be prepared by a number of methods.

Method I: Reactions of $C_5H_5Fe(CO)_2X$ compounds (X = Cl, Br, I) with salts of $[^2D-X]^-$ anions. Formally this involves a metathesis and internal replacement of one CO group.

$$C_5H_5Fe(CO)_2X + M[^2D-X] \xrightarrow[-MX]{} (C_5H_5Fe(CO)_2\text{-}X\text{-}^2D)$$

$$\xrightarrow{-CO} C_5H_5Fe(CO)^2D\text{-}X$$

   a. $C_5H_5Fe(CO)_2Cl$ is reacted with $Ag[RN_3R']$ in refluxing benzene for 4 h to give Nos. 1 to 3 [14].

   b. $C_5H_5Fe(CO)_2Cl$ is treated with a slight excess of M[SC(S)NHR] salts in refluxing acetone for 3 h. $[CH_3NH_3][SC(S)NHCH_3]$ was used for No. 10 [16], and $Na[SC(S)NHC_3H_7-i]$ for No. 11 [5]. The yields are low since $C_5H_5Fe(CO)_2SC(S)NHR$ compounds form as the main products [5, 16]. A better yield of No. 12 resulted from the reaction of $C_5H_5Fe(CO)_2Br$ and $[(CH_3)_2NH_2][SC(S)N(CH_3)_2]$ in refluxing benzene [17].

   c. $C_5H_5Fe(CO)_2Br$ or $C_5H_5Fe(CO)_2I$ are reacted with Na[SC(S)SR] salts in THF at room temperature for 6 to 14 h. This gives mainly the $C_5H_5Fe(CO)_2SC(S)SR$ type but also small amounts of Nos. 25 and 26 [2], also see Method IV.

   d. $C_5H_5Fe(CO)_2Br$ is treated in THF with $Na[SCH_2CH_2P(C_6H_5)_2]$, which can be obtained from the phosphinoethanethiol and NaH [18].

Method II: $C_5H_5Fe(CO)_2I$ reacts in the presence of $N(C_2H_5)_3$ with the amide proton of a thiocarbamoyl group. Complex No. 5 formed from $CH_3NH\text{-}C(S)P(C_6H_5)_2$ in THF at 40 °C. Nos. 33 and 34 were obtained in a long reaction from $RNH\text{-}C(S)P(C_6H_5)_2S$ (R = CH_3, C_6H_5) in benzene at room temperature (6 to 7 d) [19].

Method III: Attack of $C_5H_5Fe(CO)_2X$ (X=Cl, Br, I) at Si-N, Sn-S, or Sn-Se bonds of such compounds where N, S, and Se are part of a potentially chelating molecule. This gives silicon and tin halides and $C_5H_5Fe(CO)^2D-X$ complexes.

    a. Compounds No. 5 to 7 formed in high yields from $C_5H_5Fe(CO)_2I$ and $(CH_3)_3Si-NR-C(S)P(C_6H_5)_2$ (R=CH_3, C_2H_5, C_6H_5, mole ratio 1:1) in THF at 40 to 45 °C (5 to 6 d) [19].

    b. Reactions of $C_5H_5Fe(CO)_2Cl$ with $(CH_3)_3Sn-SC(S)N(CH_3)_2$ (mole ratio 1:1) in refluxing THF (13 h) [4], $C_5H_5Fe(CO)_2Br$ with $(CH_3)_2Sn(SeC(Se)N(CH_3)_2)_2$ (mole ratio 2:1) in refluxing benzene (2 h), and with $Cl(CH_3)_2Sn-SC(Se)N(CH_3)_3$ under similar conditions [17] gave Nos. 12, 29, and 28, respectively.

Method IV: $C_5H_5Fe(CO)_2SC(S)SR$ compounds (R=CH_3, C_2H_5, C_6H_5, see Method Ic) are irradiated in benzene at room temperature for 6 h to give Nos. 25 to 27 by intramolecular replacement of CO [2].

Method V: Oxidative cleavage of $(C_5H_5Fe(CO)_2)_2$ by the -S-S- group of disulfides which contain other sulfur atoms as potential donor sites. Reactions with di-thiocarbamoyl-disulfides, $R_2N(S)C-S-S-C(S)NR_2$, in refluxing cyclohexane (18 h) [1] or in refluxing benzene [10] gave Nos. 12 to 24. The disulfide III was used for Nos. 31 and 32. It reacted within 10 min both with irradiation or by heating [12].

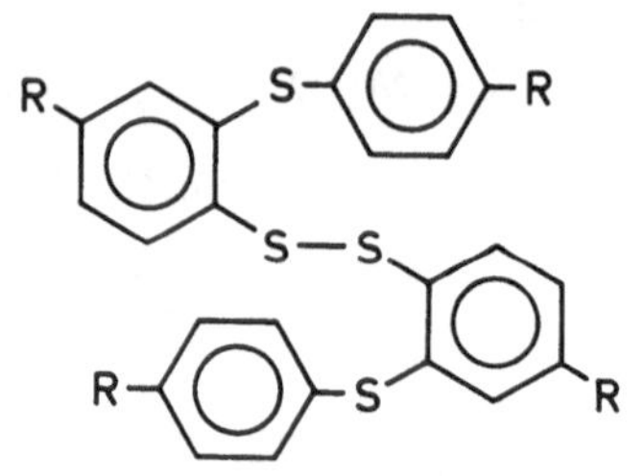

R=H, CH_3

III

Attempts to prepare $C_5H_5Fe(CO)C_6H_5N_3C_6H_5$ from $C_5H_5Fe(CO)_2I$ and $Na[C_6H_5N_3C_6H_5]$ in boiling 1,2-dimethoxyethane gave $(C_5H_5Fe(CO)_2)_2$ as the only identifiable iron carbonyl derivative [7]. Reactions of $C_5H_5Fe(CO)_2C(O)NHR$ (R=CH_3, C_2H_5) with COS gave $C_5H_5Fe(CO)_2SC(O)NHR$ complexes. There was no sign for the formation of a monocarbonyl complex with a chelating SC(O)NHR ligand [16], also see further information for No. 10.

Explanations for Table 13: Free valencies in the formulas show the bonding of the ligands to the Fe atom. For Nos. 1 to 4 and 8 to 29 only one mesomeric formula is given. The term special indicates other preparative methods described under further information.

[57]Fe Mössbauer spectra were measured at room temperature. Not assigned IR bands in the 1900 to 2000 cm^{-1} range belong to ν(CO) vibrations.

References on p. 146

Table 13
Compounds of the $C_5H_5Fe(CO)^2D-X$ Type.
Further information on numbers preceded by an asterisk is given at the end of the table.
For abbreviations and dimensions, see p. 375.

No.	^{2}D–X ligand method of preparation (yield in %)	properties and remarks explanations on p. 138	Ref.

four-membered Fe(^{2}D–X) rings

*1	C_6H_4Cl-4 la (50)	red brown ^{1}H NMR (CDCl$_3$): 4.72 (C$_5$H$_5$), 7.0 (m, C$_6$H$_4$) IR (KBr): 1968, ν(NNN) 1285	[14]
*2	$C_6H_4CH_3-4$ la (50)	red brown ^{1}H NMR (CDCl$_3$): 2.28 (CH$_3$), 4.71 (C$_5$H$_5$), 6.8 to 7.2 (m, C$_6$H$_4$) IR (KBr): 1960, ν(NNN) 1280	[14]
*3	$C_6H_4CH_3-4$ la (50)	red brown ^{1}H NMR (CDCl$_3$): 2.28 (CH$_3$), 4.70 (C$_5$H$_5$), 6.9 (m, C$_6$H$_4$) IR (KBr): 1954, ν(NNN) 1284	[14]
*4	special	green–brown needles, dec. 104° ^{1}H NMR (CDCl$_3$): 3.07 (s, NCH$_3$), 4.60 (s, C$_5$H$_5$), 7.26 (m, C$_6$H$_5$) IR (CH$_2$Cl$_2$): 1950	[11]
*5	$(C_6H_5)_2$ II, IIIa (71)	reddish brown, m.p. 168 to 169° ^{1}H NMR (CDCl$_3$): 3.25 (d, CH$_3$, J(P,H)=2.9), 4.485 (d, C$_5$H$_5$, J(P,H)=1.2) ^{31}P NMR (THF): 46.2 IR: 1945 in KBr, 1964 in CCl$_4$, ν(NCS) 1585 in KBr	[19]
*6	$(C_6H_5)_2$ IIIa (67)	reddish brown, m.p. 146 to 148° ^{1}H NMR (CDCl$_3$): 1.20 (t, CH$_3$, ^{3}J(H,H)=7.3), 3.55 (d,q, CH$_2$, J(P,H)=2.9), 4.475 (d, C$_5$H$_5$, J(P,H)=1.5) ^{31}P NMR (THF): 45.3 IR: 1940 in KBr, 1965 in CCl$_4$, ν(NCS) 1565 in KBR	[19]

References on p. 146

Table 13 [continued]

No.	^{2}D–X ligand method of preparation (yield in %)	properties and remarks explanations on p. 138	Ref.
*7	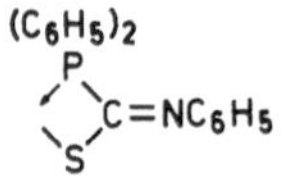 (C$_6$H$_5$)$_2$ P / C=NC$_6$H$_5$ S III a (85)	reddish brown, m.p. 158 to 159° ^{1}H NMR (CDCl$_3$): 4.49 (d, C$_5$H$_5$, J(P,H) = 1.2) ^{31}P NMR (THF): 47.0 IR: 1944 in KBr, 1966 in CCl$_4$, ν(NCS) 1549 in KBr	[19]
*8	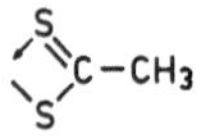 S C–CH$_3$ S special	red oil ^{1}H NMR (CDCl$_3$): 2.42 (CH$_3$), 4.72 (C$_5$H$_5$) IR: 1970 in C$_6$H$_{14}$, ν$_{as}$(CS$_2$) 1265 as film	[9]
*9	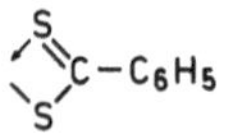 S C–C$_6$H$_5$ S special	green crystals, m.p. 79° ^{1}H NMR (CDCl$_3$): 4.92 (C$_5$H$_5$), 7.20 to 8.10 (C$_6$H$_5$) IR: 1972 in C$_6$H$_{14}$, ν$_{as}$(CS$_2$) 1270 in Nujol	[9]
*10	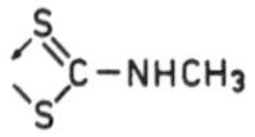 S C–NHCH$_3$ S Ib (5)	red, m.p. 108 to 109° ^{1}H NMR (CDCl$_3$): 2.82 (d, CH$_3$, ^{3}J(H,H) = 5.0), 4.62 (s, C$_5$H$_5$), 6.66 (NH) IR: 1960 in CHCl$_3$, ν(CN) 1520 in Nujol, ν(NH) 3395 in CCl$_4$	[16]
*11	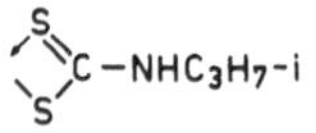 S C–NHC$_3$H$_7$-i S Ib (4)	m.p. 108° (dec.), sublimes at 110° in high vacuum ^{1}H NMR (C$_6$D$_5$CD$_3$, 38°): 0.81 (d, CH$_3$, ^{3}J(H,H) = 6), 4.07 (m, CH), 4.35 (s, C$_5$H$_5$), 7.20 (NH) IR (KBr): 1932, ν(CN) 1515, ν(NH) 3281	[5]
*12	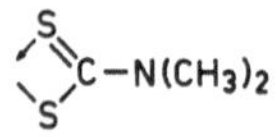 S C–N(CH$_3$)$_2$ S Ib (65), IIIb (43), V (57)	purple needles, m.p. between 141 and 153° (dec.), sublimes at 140°/0.1 Torr ^{1}H NMR (CCl$_4$): 3.10 (CH$_3$), 4.78 (C$_5$H$_5$) ^{57}Fe-γ: δ=0.51, Δ=1.81 IR (CH$_2$Cl$_2$): 1936, ν(CN) 1510	[1, 4, 10, 17]
13	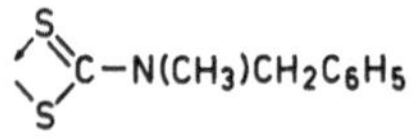 S C–N(CH$_3$)CH$_2$C$_6$H$_5$ S V	m.p. 82 to 86° ^{1}H NMR (CS$_2$, 35°): 4.417 (C$_5$H$_5$) ^{57}Fe-γ: δ=0.51, Δ=1.86 IR (CH$_2$Cl$_2$): 1937.0	[10]
14	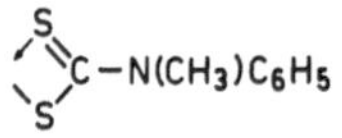 S C–N(CH$_3$)C$_6$H$_5$ S V	m.p. 126 to 128° ^{1}H NMR (CS$_2$, 35°): 4.487 (C$_5$H$_5$), 7.147 to 7.267 (m, C$_6$H$_5$) ^{57}Fe-γ: δ=0.51, Δ=1.81 IR (CH$_2$Cl$_2$): 1935.9	[10]

References on p. 146

Table 13 [continued]

No.	^{2}D-X ligand method of preparation (yield in %)	properties and remarks explanations on p. 138	Ref.
15	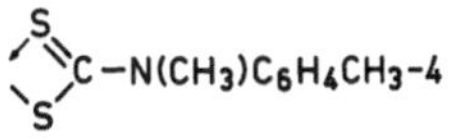S S C–N(CH3)C6H4CH3-4 V	m.p. 130 to 135° ^{1}H NMR (CS_2, 35°): 4.477 (C_5H_5) ^{57}Fe-γ: δ=0.51, Δ=1.84 IR (CH_2Cl_2): 1935.8	[10]
16	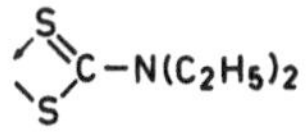S S C–N(C2H5)2 V	m.p. 96 to 98° ^{1}H NMR (CS_2, 35°): 4.380 (C_5H_5) ^{57}Fe-γ: δ=0.50, Δ=1.80 IR (CH_2Cl_2): 1934.2	[10]
17	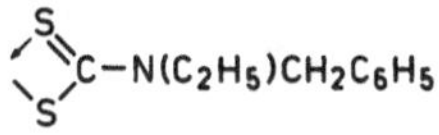S S C–N(C2H5)CH2C6H5 V	m.p. 82 to 85° ^{1}H NMR (CS_2, 35°): 4.413 (C_5H_5) ^{57}Fe-γ: δ=0.52, Δ=1.89 IR (CH_2Cl_2): 1935.9	[10]
18	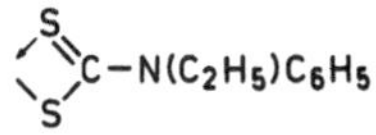S S C–N(C2H5)C6H5 V	m.p. 110 to 111° ^{1}H NMR (CS_2, 35°): 4.477 (C_5H_5), 6.990 to 7.360 (m, C_6H_5) ^{57}Fe-γ: δ=0.51, Δ=1.82 IR (CH_2Cl_2): 1936.2	[10]
19	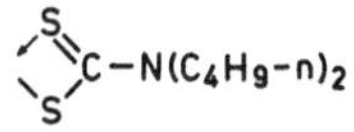S S C–N(C4H9-n)2 V	oil ^{1}H NMR (CS_2, 35°): 4.378 (C_5H_5) IR (CH_2Cl_2): 1933.5	[10]
20	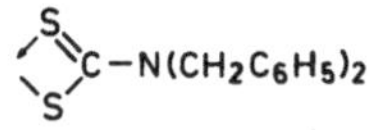S S C–N(CH2C6H5)2 V	m.p. 118 to 120° ^{1}H NMR (CS_2, 35°): 4.465 (C_5H_5) ^{57}Fe-γ: δ=0.52, Δ=1.89 IR (CH_2Cl_2): 1937.5	[10]
21	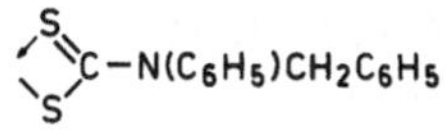S S C–N(C6H5)CH2C6H5 V	m.p. 100 to 105° ^{1}H NMR (CS_2, 35°): 4.412 (C_5H_5) ^{57}Fe-γ: δ=0.50, Δ=1.80 IR (CH_2Cl_2): 1937.2	[10]
22	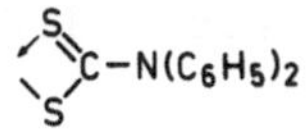S S C–N(C6H5)2 V	m.p. 194 to 196° ^{1}H NMR (CS_2, 35°): 4.393 (C_5H_5), 7.187 (C_6H_5) ^{57}Fe-γ: δ=0.50, Δ=1.80 IR (CH_2Cl_2): 1937.9	[10]
23	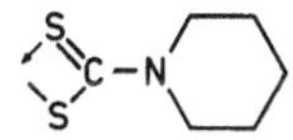S S C–N V	m.p. 149 to 150° ^{1}H NMR (CS_2, 35°): 4.378 (C_5H_5) ^{57}Fe-γ: δ=0.50, Δ=1.79 IR (CH_2Cl_2): 1933.8	[10]

References on p. 146

Table 13 [continued]

No.	²D–X ligand method of preparation (yield in %)	properties and remarks explanations on p. 138	Ref.
24	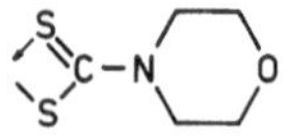 V	m.p. 132° ^1H NMR (CS_2, 35°): 4.412 (C_5H_5) $^{57}Fe-\gamma$: $\delta = 0.51$, $\Delta = 1.79$ IR (CH_2Cl_2): 1937.6	[10]
*25	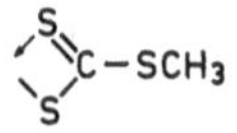 Ic (20), IV (78)	ruby-red prisms, m.p. 55°, subl. at 20°/0.001 Torr ^1H NMR (CS_2): 2.55 (CH_3), 4.60 (C_5H_5) IR: 1956 in CH_2Cl_2, 1964 in CCl_4	[2]
*26	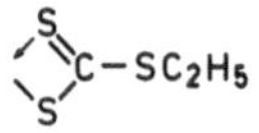 Ic (1), IV (64)	ruby-red, m.p. 60° ^1H NMR (CS_2): 1.37 (t, CH_3, J(H,H) = 7.5), 3.15 (q, CH_2), 4.60 (C_5H_5) IR (CCl_4): 1966	[2]
*27	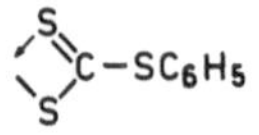 IV (62)	deep red, m.p. 92° ^1H NMR (CS_2): 4.57 (C_5H_5), 7.43 (C_6H_5) IR (CCl_4): 1964	[2]
*28	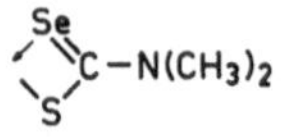 IIIb (60)	red, m.p. 140° (dec.) IR (CH_2Cl_2): 1931, ν(CN) 1515	[17]
*29	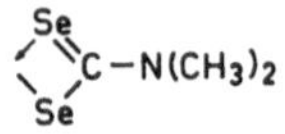 IIIb (55)	dark red, m.p. 145° (dec.) IR (CH_2Cl_2): 1926, ν(CN) 1520	[17]

five-membered Fe(²D–X) rings

No.	²D–X ligand method of preparation (yield in %)	properties and remarks explanations on p. 138	Ref.
*30	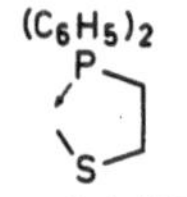 Id (34)	deep red, m.p. 100° ^1H NMR ($CDCl_3$): 2.63 (m, CH_2), 4.40 (d, C_5H_5, $^3J(P,H) = 1$), 7.37 to 7.53 (m, C_6H_5) ^{31}P NMR ($CDCl_3$): 67 IR (CH_2Cl_2): 1935	[18]
*31	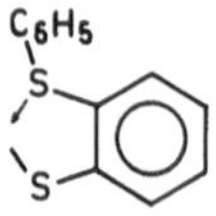 V (~100)	^1H NMR ($CDCl_3$): 4.62 (C_5H_5), 4.36 and 4.76 at −60°, intensity ratio 1 : 7.5	[12]
32	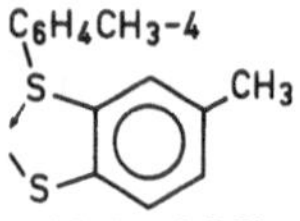 V (~100)	no properties reported	[12]

References on p. 146

Table 13 [continued]

No.	2D–X ligand method of preparation (yield in %)	properties and remarks explanations on p. 138	Ref.
*33	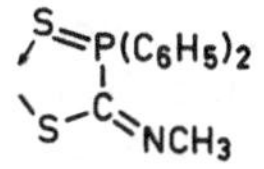 II (65)	brown, m.p. 112 to 115° (dec.) ^{1}H NMR (CDCl$_3$): 3.53 (d, CH$_3$, J(P,H) = 3.8), 4.33 (s, C$_5$H$_5$) ^{31}P NMR (THF): 81.3 IR (KBr): 1960, ν(PS) 625, ν(NCS) 1542, ν(CO) 1971 in CCl$_4$	[19]
*34	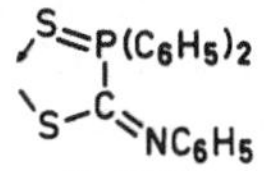 II (78)	brown, m.p. 118 to 121° (dec.) ^{1}H NMR (CDCl$_3$): 4.34 (s, C$_5$H$_5$) ^{31}P NMR (THF): 85.9 IR (KBr): 1945, ν(PS) 625, ν(NCS) 1512, ν(CO) 1973 in CCl$_4$	[19]

six-membered Fe(²D–X) rings

No.	2D–X ligand	properties and remarks	Ref.
*35	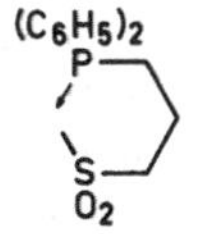 special	light brown, dec. p. >175° IR: 1967 in CH$_2$Cl$_2$, ν_s(SO$_2$) 1040, ν_{as}(SO$_2$) 1160 in KBr	[13]

* Further information:

C$_5$H$_5$Fe(CO)N$_3$(C$_6$H$_4$R-4)C$_6$H$_4$R′-4 (R = R′ = Cl or CH$_3$, R = Cl, R′ = CH$_3$, Table **13**, Nos. **1** to **3**) can also be prepared from C$_5$H$_5$Fe(CO)(P(C$_6$H$_5$)$_3$)X (X = Cl, I) by replacement of the X and phosphine ligand with Ag[RN$_3$R′] under conditions similar to Method Ia (70% yield for X = I). The reaction requires milder conditions when X = Cl than when X = I. This supports the assumption that a silver–iron bond is initially formed, followed by ring closure at the Fe atom and migration of chloride from iron to silver [14].

The compounds are stable under N$_2$ and dissolve in common organic solvents of low polarity. The aromatic groups are magnetically equivalent in the ^{1}H NMR spectra. The strong IR absorption near 1280 cm^{-1} is characteristic of a chelating tri-azenido ligand [14].

In cyclic voltammetry (CH$_2$Cl$_2$) irreversible oxidation occurred at E$_p$ = 1.065 for No. 1 and 0.940 V for No. 3 (vs. Ag/AgI). No cathodic wave was observed for No. 1, and the i_{anodic} to $i_{cathodic}$ ratio was 1.4 for No. 3. No oxidation occurred with AgPF$_6$. Reaction with NOPF$_6$ did not give the Fe^{III} species but rather the diamagnetic [C$_5$H$_5$Fe(CO)(NO)RN$_3$R]PF$_6$ compounds with a monodentate triazenido ligand [15], see 1.5.2.2.8, p. 146.

C$_5$H$_5$Fe(CO)N(CH$_3$)-C(C$_6$H$_5$)=S (Table **13**, No. **4**) formed in traces in a reaction of C$_5$H$_5$Fe(CO)$_2$Cl with C$_6$H$_5$CS-N(CH$_3$)Li in THF. But the main product of this reaction, C$_5$H$_5$Fe(CO)-CO-N(CH$_3$)-C(C$_6$H$_5$)=S (^{1}L-^{2}D ligand, see 1.5.2.2.14, p. 233), can be converted to No. 4 by irradiation in benzene at room temperature for 3 h,

References on p. 146

44% yield after chromatography on SiO_2 with benzene eluent. Recrystallization from ether/pentane.

The mass spectrum shows that the thioamide ligand is readily cleaved as a whole from the Fe atom. It contains the molecular ion $[M]^+$ and $[M-CO]^+$, $[(C_5H_5)_2Fe]^+$, and $[C_5H_5Fe]^+$ as the most abundant fragments [11].

$C_5H_5Fe(CO)S-C(=NR)-P(C_6H_5)_2$ (R = CH_3, C_2H_5, C_6H_5, Table **13**, Nos. **5** to **7**). By-products in Method IIIa are the $C_5H_5Fe(CO)(P(C_6H_5)_2C(S)NHR)I$ complexes. They can be separated by chromatography on SiO_2 with CH_2Cl_2 eluent or by crystallization from ether. Nos. **5** to **7** crystallize quantitatively at $-30\,°C$, while the iodo compounds remain in solution. Only small amounts of Nos. **5** and **7** were formed by using $C_5H_5Fe(CO)_2Br$ in Method IIIa; the method failed completely with $C_5H_5Fe(CO)_2Cl$. This can be explained by a primary attack of the thiocarbamoylphosphine through the P atom and the several ways of subsequent ligand elimination. The amine in Method II favors the formation of the chelate complex No. **5**. In the absence of amine, No. **5** formed as a by-product, with the iodo compound as the main product.

The mass spectra are completely reported. $[M]^+$ and $[M-CO]^+$ occur with low intensities. An important fragmentation is the elimination of $[RNCS]^+$ [19].

$C_5H_5Fe(CO)S_2CR$ (R = CH_3, C_6H_5, Table **13**, Nos. **8** and **9**) were prepared by stirring $C_5H_5Fe(CO)_2COR$ and P_4S_{10} (mole ratio 1:2) in ether for 4 d. The compounds (10% yield) were separated on Al_2O_3 with hexane eluent from the main products $C_5H_5Fe(CO)_2SC(S)R$. B_2S_3 can be used; it reacts generally faster and the yields are higher. Higher yields (30%) were also obtained with P_4S_{10} in refluxing hexane. Nos. **8** and **9** form on slow decomposition of the $C_5H_5Fe(CO)_2SC(S)R$ compounds in solution at room temperature. The observations suggest that the conversion sequence of the ligand is acyl, thioacyl, dithiocarboxyl, and dithiocarboxyl chelate [9], also see [8].

$C_5H_5Fe(CO)S_2CNHCH_3$ (Table **13**, No. **10**) also formed (<5% yield) along with $C_5H_5Fe(CO)_2SC(S)NHCH_3$ (44%) from $C_5H_5Fe(CO)_2CONHCH_3$ and CS_2 at room temperature (13 h). No. **10** is first eluted from Al_2O_3 with C_6H_6/CH_2Cl_2 and is recrystallized from benzene/hexane. Longer reaction time or boiling CS_2 increases the yield of No. **10** only slightly.

The mass spectrum shows $[M]^+$ (relative intensity 21), $[M-CO]^+$ (100), $[C_5H_5Fe]^+$ (40), and Fe^+ (31) [16].

$C_5H_5Fe(CO)S_2CNHC_3H_7-i$ (Table **13**, No. **11**) is eluted from SiO_2 with benzene before the dicarbonyl main product.

At lower temperature the ^1H NMR resonances of the CH_3 groups split into two doublets because of hindered rotation about the S_2C-N bond (Formula Ic on p. 136). The coalescence temperature, rate constant, and free enthalpy of activation for the racemization are $T_c = 314 \pm 2\,K$, $k = 19.0 \pm 1\,s^{-1}$, and $\Delta G^* = 16.6 \pm 0.3\,kcal \cdot mol^{-1}$.

The mass spectrum is similar to that of No. **10**. The solid is stable in air for a prolonged time, solutions slowly decompose in contact with air [5].

$C_5H_5Fe(CO)S_2CN(CH_3)_2$ (Table **13**, No. **12**). Other IR bands (CH_2Cl_2) from 811 to 1940 cm^{-1} are reported [1]. The stretching force constant, $k(CO) = 15.14\,mdyn \cdot Å^{-1}$ (Cotton-Kraihanzel method), is greater than for the thioselenocarbamato (No. **28**) and diselenocarbamato (No. **29**) complexes, suggesting an increasing electron-donor ability in this sequence of carbamato ligands [17].

References on p. 146

One-electron reduction at $E_{1/2} = -2.3$ V (vs. Ag/AgClO$_4$) was observed by polarography in CH$_3$OCH$_2$CH$_2$OCH$_3$ [3]. A quasireversible oxidation-reduction process with anodic $E_p = 0.710$ V (vs. SCE) was found by cyclic voltammetry in CH$_2$Cl$_2$. However, in CH$_3$CN the separation of the anodic and cathodic peak was $\Delta E_p = 0.80$ V and could thus not be associated with a reversible oxidation-reduction process. A mechanism involving the opening of the chelate ligand in the oxidized species, the addition of a CH$_3$CN molecule, and the reduction of this adduct was discussed [17].

C$_5$H$_5$Fe(CO)S$_2$CSR (R = CH$_3$, C$_2$H$_5$, C$_6$H$_5$, Table 13, Nos. 25 to 27) are produced by slow decomposition of C$_5$H$_5$Fe(CO)$_2$SC(S)SR compounds in solution at room temperature. Crude products obtained by Method IV were purified by chromatography on Al$_2$O$_3$ with petroleum ether/ether eluent and were recrystallized from the same solvent mixture or from ether (No. 26).

The IR spectra (CCl$_4$) show a characteristic strong band near 950 to 970 cm^{-1} (accompanied by weaker bands), which is probably associated with a ν(CS) of the chelating trithiocarbonate ligand. Two other characteristic bands of medium intensity are near 523 and 560 cm^{-1}.

The air-stable solids are sublimable at room temperature (10^{-3} Torr) but not without some decomposition, involving loss of CS$_2$. CS$_2$ is also eliminated in refluxing toluene to give (C$_5$H$_5$Fe(CO)μ-SR)$_2$ compounds, see C4, 2.5.2.3.1.2, p. 2 [2].

C$_5$H$_5$Fe(CO)SeSCN(CH$_3$)$_2$ and C$_5$H$_5$Fe(CO)Se$_2$CN(CH$_3$)$_2$ (Table 13, Nos. 28 and 29). The stretching force constants of the carbonyl group (Cotton-Kraihanzel method) are k(CO) = 15.06 (No. 28) and 14.98 (No. 29) mdyn·Å^{-1}, also see No. 12.

Quasireversible oxidation-reduction processes were observed by cyclic voltammetry in CH$_2$Cl$_2$: $E_p = 0.66$ V (No. 28) and 0.60 V (No. 29) vs. SCE. In CH$_3$CN, the voltammogramm shows a reversible one-electron oxidation-reduction only for No. 29 ($E_p = 0.15$ V vs. Ag/Ag$^+$). The oxidized species of No. 28 ($E_p = 0.20$ V vs. Ag/Ag$^+$) is unstable in CH$_3$CN. It appears to be converted into another product which is reduced around -0.50 V (vs. Ag/Ag$^+$). At a low sweep rate, the cathodic wave of the reversible part of the last process disappears completely [17].

C$_5$H$_5$Fe(CO)SCH$_2$CH$_2$P(C$_6$H$_5$)$_2$ (Table 13, No. 30). The very high ^{31}P resonance shift of the phosphinoethanethiol upon complexation ($\Delta\delta = 85$ ppm to low field) is noteworthy. The mass spectrum clearly established the monomeric chelated structure: [M]$^+$ (relative intensity 12), [M − CO]$^+$ (44), and [C$_5$H$_5$Fe]$^+$ (100) [18].

C$_5$H$_5$Fe(CO)SC$_6$H$_4$SC$_6$H$_5$-2 (Table 13, No. 31). The temperature dependence of the ^{1}H resonances of the C$_5$H$_5$ protons (coalescence at about -5 °C) is attributed to an inversion at the C$_6$H$_5$S → donor group. The free energy of activation of the inversion was calculated, $\Delta G^+ = 56.5 \pm 1.0$ kJ·mol^{-1} (13.5 kcal·mol^{-1}) [12].

The properties of No. 31 are identical to those of a nonidentified "C isomer" of (C$_5$H$_5$Fe(CO)μ-SC$_6$H$_5$)$_2$ (reported in [6], see C4, 2.5.2.3.1.2, No. 15, pp. 7 and 10). An X-ray study of an analogous Ru complex established the chelate structure [12].

C$_5$H$_5$Fe(CO)SC(=NR)P(C$_6$H$_5$)$_2$=S (R = CH$_3$, C$_6$H$_5$, Table 13, Nos. 33 and 34). The crude products of Method II were purified on SiO$_2$ with THF eluent and were crystallized from ether at -35 °C. The unusually large low-field shift of the ^{31}P resonance could not be explained [19].

C$_5$H$_5$Fe(CO)SO$_2$CH$_2$CH$_2$CH$_2$P(C$_6$H$_5$)$_2$ (Table 13, No. 35) is quantitatively formed from C$_5$H$_5$Fe(CO)CH$_2$CH$_2$CH$_2$P(C$_6$H$_5$)$_2$ (chelating ^{1}L-^{2}D ligand) and excess

References on p. 146

liquid SO_2 at $-60\,°C$ (1 h). Evaporation of SO_2 gives the pure compound, which dissolves only in polar solvents [13].

References:

[1] F.A. Cotton, J.A. McCleverty (Inorg. Chem. **3** [1964] 1398/402). – [2] R. Bruce, G.R. Knox (J. Organometal. Chem. **6** [1966] 67/75). – [3] R.E. Dessy, R.B. King, M. Waldrop (J. Am. Chem. Soc. **88** [1966] 5112/7). – [4] E.W. Abel, M.O. Dunster (J. Chem. Soc. Dalton Trans. **1973** 98/102). – [5] H. Brunner, T. Burgemeister, J. Wachter (Chem. Ber. **108** [1975] 3349/54).

[6] R.J. Haines, J.A. de Beer, R. Greatrex (J. Organometal. Chem. **85** [1975] 89/99). – [7] R.B. King, K.C. Nainan (Inorg. Chem. **14** [1975] 271/4). – [8] L. Busetto, A. Palazzi (Chim. Ind. [Milan] **58** [1976] 804). – [9] L. Busetto, A. Palazzi, E. Foresti Serantoni, L. Riva di Sanseverino (J. Organometal. Chem. **129** [1977] C55/ C58). – [10] J.B. Zimmerman, T.W. Starinshak, D.L. Uhrich, N.V. Duffy (Inorg. Chem. **16** [1977] 3107/11).

[11] H. Brunner, J. Wachter (J. Organometal. Chem. **142** [1977] 133/7). – [12] S.D. Killops, S.A.R. Knox, G.H. Riding, A.J. Welch (J. Chem. Soc. Chem. Commun. **1978** 486/8). – [13] E. Lindner, G. Funk, S. Hoehne (Angew. Chem. **91** [1979] 569/70; Angew. Chem. Intern. Ed. Engl. **18** [1979] 535). – [14] E. Pfeiffer, K. Vrieze (Transition Metal Chem. [Weinheim] **4** [1979] 385/8). – [15] J.G.M. van der Linden, A.H. Dix, E. Pfeiffer (Inorg. Chim. Acta **39** [1980] 271/4).

[16] L. Busetto, A. Palazzi, V. Foliadis (Inorg. Chim. Acta **40** [1980] 147/52). – [17] G. Nagao, K. Tanaka, T. Tanaka (Inorg. Chim. Acta **42** [1980] 43/8). – [18] M. Savignac, P. Cadiot, F. Mathey (Inorg. Chim. Acta **45** [1980] L43/L44). – [19] U. Kunze, A. Antoniadis (Z. Naturforsch. **36b** [1981] 1588/94).

1.5.2.2.8 Nitrosyl Compounds

$C_5H_5Fe(CO)(NO)I$ was prepared in a 60% yield by passing NO gas through a benzene solution of $C_5H_5Fe(CO)_2I$ at 25 °C (1 h).

The red-brown solid decomposes without melting. IR spectrum (KBr): $v(CO)$ 2025 and $v(NO)$ 1650 cm^{-1}, both strong. The position of the $v(NO)$ indicates that the bonding arises from transfer of the odd NO electron to Fe followed by lone-pair donation from NO^+.

The compound is soluble in water, alcohols, acetone, and THF, but insoluble in hydrocarbons [1].

$[C_5H_5Fe(CO)(NO)N_3(C_6H_4CH_3-4)_2]PF_6$ was obtained in an attempt to oxidize $C_5H_5Fe(CO)N_3(C_6H_4CH_3-4)_2$ (see 1.5.2.2.7, p. 136, No. 3) in CH_3CN with solid $NOPF_6$ at 0 °C. Addition of excess ether precipitated the salt. According to IR and ^1H NMR spectra, the diamagnetic cation contains a monodentate triazenido ligand. The spectra themselves are not reported [2].

$[C_5H_5Fe(CO)(NO)N_3(C_6H_4Cl-4)_2]PF_6$ was apparently also prepared like the previous complex [2].

References:

[1] V.N. Pandey (Transition Metal Chem. [Weinheim] **2** [1977] 48/50). – [2] J.G.M. van der Linden, A.H. Dix, E. Pfeiffer (Inorg. Chim. Acta **39** [1980] 271/4).

1.5.2.2.9 $C_5H_5Fe(CO)B_3H_8$

The compound was prepared in a 60% yield (86% given in [1]) by irradiating $C_5H_5Fe(CO)_2I$ and $[N(CH_3)_4][B_3H_8]$ (mole ratio 1:1.2) in CH_2Cl_2 at 15 °C for 14 h. The crude product from the rapidly filtered and evaporated solution gave the compound on sublimation at 10^{-4} Torr from room temperature to -30 °C [1, 2].

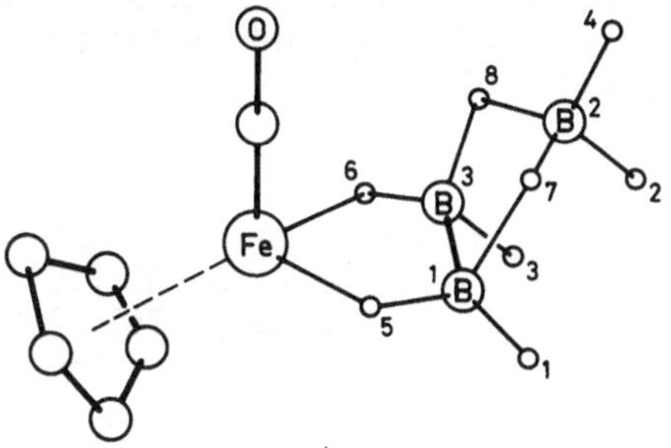

The brown-black solid melts at 53 to 54 °C. The structure proposed on the basis of the spectra is shown in Formula I (small circles are H atoms). The labelling is used for the assignment of the NMR spectra. The ^1H NMR data are listed below, δ in ppm (J(B,H) in Hz):

	H-5,6	H-7,8	H-1,3	H-2, H-4	C_5H_5
CD_2Cl_2, 20 °C	−15.76(60)	−1.90(54)	0.74(142)	1.67, 3.15(132)	4.76
CD_2Cl_2, −45 °C	−12.56(66)	−1.96(56)	0.65(146)	1.58, 3.06(148)	4.76
C_6D_6, 20 °C	−15.81(59)	−1.42(56)	1.19(143)	2.28, 3.91	3.83

The absolute assignments to H-2 and H-4 are uncertain. All resonances collapse to sharp singlets upon ^{11}B decoupling, indicating C_s symmetry. The normal ^{11}B NMR spectrum consists of a high-field doublet assigned to B-1,3 and a low-field triplet assigned to B-2 (J in Hz): −42.5(127) and +2.5(125) ppm in CD_2Cl_2 and −42.0(132) and +3.4(122) ppm in C_6D_6, both at ambient temperature. Line-narrowed spectra and selective coupling revealed a wealth of fine structure in the high-field "doublet". The IR spectrum (KBr) is completely given. Bands in the ν(CO) and ν(BH) region were observed (in CH_2Cl_2) at 1995, 2190, 2460, and 2525 cm^{-1}.

The mass spectrum exhibits the molecular ion. The compound slowly decomposes to give a whitish gray solid when stored in vacuum at room temperature. It can be handled in air for about 2 h. The deep purple-red solutions in hydrocarbons, ether, acetone, or CH_2Cl_2 are more sensitive to air. Irradiation in CH_2Cl_2 did not produce a tridentate B_3H_8 ligand [2].

References:

[1] D.F. Gaines, S.J. Hildebrandt (J. Am. Chem. Soc. **96** [1974] 5574/6). − [2] D.F. Gaines, S.J. Hildebrandt (Inorg. Chem. **17** [1978] 794/806).

1.5.2.2.10 Unstable $C_5H_5Fe(CO)^1L$ Species

$C_5H_5Fe(CO)COCH_3$ is produced by brief wavelength-selective photolysis of $C_5H_5Fe(CO)_2COCH_3$ in Ar, CH_4, N_2, or CO matrices at 12 K. This was concluded from the appearance of a weak $v(CO)$ band (doublet band in CO matrix) near 1950 cm^{-1}. The generally much weaker $v(C=O)$ of the acetyl group could not be observed. Continued irradiation (12 min) gave $C_5H_5Fe(CO)_2CH_3$ as the exclusive product. Similar photolysis of $C_5H_5Fe(CO)_2CH_3$ in Ar, CH_4, and N_2 gave no evidence for the formation of a monocarbonyl product [6].

Other $C_5H_5Fe(CO)R$ intermediates were assumed to form in thermal or photolytical decompositions of $C_5H_5Fe(CO)_2R$ compounds or the alkyl isomerization of C_5H_5-$Fe(CO)(P(C_6H_5)_3)C_4H_9$-s [1 to 5].

References:

[1] K. Nicholas, S. Raghu, M. Rosenblum (J. Organometal. Chem. **78** [1974] 133/7). − [2] J.J. Alexander (J. Am. Chem. Soc. **97** [1975] 1729/32). − [3] D.L. Reger, E.C. Culbertson (J. Am. Chem. Soc. **98** [1976] 2789/94). − [4] D.L. Reger, E.C. Culbertson (Inorg. Chem. **16** [1977] 3104/7). − [5] D.L. Reger, C.J. Coleman, P.J. McElligott (J. Organometal. Chem. **171** [1979] 73/84).

[6] D.J. Fettes, R. Narayanaswamy, A.J. Rest (J. Chem. Soc. Dalton Trans. **1981** 2311/6).

1.5.2.2.11 Compounds of $[C_5H_5Fe(CO)(CN)^1L]^-$ Anions and $C_5H_5Fe(CO)(CN)^1L$ Radicals

The ^1L ligands of the compounds listed in Table 14 are CO-R groups which form from Fe-bonded R groups by a cyanide-induced insertion of CO as described under the preparative Method I.

Method I: $C_5H_5Fe(CO)_2R$ compounds ($R=CH_3$, C_2H_5) and KCN (mole ratio about 1:1.3) are refluxed in CH_3OH/H_2O (3:1) for 6 h (No. 1) or 0.5 h (No. 4). The residues from the evaporation of solvents are dissolved in THF. Nos. 1 and 4 precipitate as THF solvates on addition of ether. Reprecipitation from methanol/ether gives the solvent-free potassium salts (Nos. 1 and 4) [1]. Details of the preparation of Nos. 2 and 6 (with NH_4CN?) were not reported [2].

Method II: Mixtures of potassium salts in CH_3OH and excess aqueous $[As(C_6H_5)_4]Cl$ are diluted with water until the solution becomes cloudy. Nos. 3 and 5 crystallize within several hours. They are recrystallized from THF by slow diffusion of ether/pentane (1:1) into the solution [1].

Method III: CH_3CN solutions of the salts No. 2 or 6 (containing 0.1 M $[N(C_2H_5)_4]ClO_4$) are electrolyzed at $+0.40$ V (vs. SCE) and 0 °C. The radicals can be separated from the supporting electrolyte by extracting the solution with n-hexane [2].

The K^+ salts are very hygroscopic and very soluble in water and polar solvents [1]. The $[N(C_2H_5)_4]^+$ salts dissolve in CH_3CN [2]. The $[As(C_6H_5)_4]^+$ salts dissolve in most organic solvents but are insoluble in water. They are much more stable in air than the K^+ salts.

Salts No. 1 and 4 react with excess CH_3I in CH_3CN at 40 to 60 °C to give C_5H_5-$Fe(CO)(CNCH_3)COR$ compounds. The corresponding aldehydes are formed in the decomposition with acids [1].

The $[N(C_2H_5)_4]^+$ salts No. 2 and 6 can be oxidized to give the radicals No. 7 and 8. Cyclic voltammograms of CH_3CN solutions at 25 °C show a reversible one-electron oxidation at the anodic potentials (vs. SCE, saturated NaCl) $E_p = 0.20$ V for No. 2 and 0.18 V for No. 6. Except for the cathodic wave on the reverse scan ($\Delta E_p \sim 0.1$ V), no other reductive process was evident up to -1.5 V [2].

Table 14
$[C_5H_5Fe(CO)(CN)COR]^-$ Salts and $C_5H_5Fe(CO)(CN)COR$ Radicals.
Further information for Nos. 7 and 8 is given at the end of the table.
For abbreviations and dimensions, see p. 375.

No.	^{1}L ligand method of preparation (yield in %)	anion	properties and remarks	Ref.
salts				
1	$COCH_3$ I (78)	K^+	^{1}H NMR (D_2O): 2.77 (s, CH_3), 4.85 (s, C_5H_5) IR (KBr): 1905 (CO), 1555, 1570 (C=O), 2080 (CN)	[1]
2	$COCH_3$ I	$[N(C_2H_5)_4]^+$	no properties reported, electrochemical oxidation gives No. 7	[2]
3	$COCH_3$ II	$[As(C_6H_5)_4]^+$	yellow needles ^{1}H NMR (CD_3COCD_3): 2.3 (s, CH_3), 4.3 (s, C_5H_5), 7.92 (s, C_6H_5) IR (KBr): 1880 (CO), 1560 (C=O), 2088 (CN)	[1]
4	COC_2H_5 I (92)	K^+	^{1}H NMR (D_2O): 1.0 (t, CH_3), 3.1 (q, CH_2), 4.8 (s, C_5H_5) IR (KBr): 1900 (CO), 1575 (C=O), 2085 (CN)	[1]
5	COC_2H_5 II	$[As(C_6H_5)_4]^+$	yellow needles ^{1}H NMR (CD_3COCD_3): 0.67 (t, CH_3), 2.77 (q, CH_2), 4.3 (s, C_5H_5), 7.92 (s, C_6H_5) IR (KBr): 1880 (CO), 1575 (C=O), 2085 (CN)	[1]
6	COC_6H_5 I	$[N(C_2H_5)_4]^+$	no properties reported, electrochemical oxidation gives No. 8	[2]
radicals				
7	$COCH_3$ III	–	dark brown in CH_3CN ESR (CH_3CN, 25°): g=2.059	[2]
8	COC_6H_5 III	–	ESR (CH_3CN, 25°): g=2.060	[2]

Further information:

C₅H₅Fe(CO)(CN)COR (R=CH₃, C₆H₅, Table **14**, Nos. **7** and **8**). The large isotropic g value indicates an odd-electron spin density, which is centered at the Fe atom. Thus no nitrogen or proton hyperfine splitting was observed.

A CH₃CN solution of the radical No. 7 is stable at −30 °C for several hours. At 20 °C, the decomposition follows first-order kinetics (half-life $\tau = 3.6$ min) and gives CH₃COCH₃, CH₃CHO (minor product), and small amounts of CH₄ and C₂H₆. Carbonyliron species could not be identified, even under 1 atm CO. High yields of acetone are only obtained at the beginning of the decomposition. This is explained by a cross-coupling with the very unstable C₅H₅Fe(CO)(CN)CH₃ radical, which can be efficiently trapped only at high concentrations of the acetyl-iron radical. There was no sign for a dimerization of the radical. Controlled-potential reduction at 0.0 V of the brown radical solution in CH₃CN regenerates the [C₅H₅Fe(CO)(CN)COCH₃]⁻ anion in a 90±5% yield [2].

References:

[1] F. Kruck, M. Höfler, L. Liebig (Chem. Ber. **105** [1972] 1174/83). − [2] R.J. Klingler, J.K. Kochi (J. Organometal. Chem. **202** [1980] 49/63).

1.5.2.2.12   Compounds of the ⁵LFe(CO)(²D)¹L Type

General References:

V. Prelog, G. Helmchen, Grundlagen des CIP-Systems und Vorschläge für eine Revision, Angew. Chem. **94** [1982] 614/31.

H. Brunner, Chiral Metal Atoms in Optically Active Organo-Transition-Metal Compounds, Advan. Organometal. Chem. **18** [1980] 151/206.

H. Brunner, in: E.A. Koerner von Gustorf, F.W. Grevels, I. Fischler, The Organic Chemistry of Iron, Vol. 1, New York − San Francisco − London 1978.

H. Brunner, Stereochemistry of the Reactions of Optically Active Organometallic Transition Metal Compounds, Top. Current Chem. **56** [1975] 67/90.

W.B. Jennings, Chemical Shift Nonequivalence in Prochiral Groups, Chem. Rev. **75** [1975] 307/22.

J.A. Schellman, Circular Dichroism and Optical Rotation, Chem. Rev. **75** [1975] 323/31.

A. Wojcicki, Insertion Reactions of Transition Metal-Carbon σ-Bonded Compounds II. Sulfur Dioxide and Other Molecules, Advan. Organometal. Chem. **12** [1974] 31/81.

E.L. Eliel, Grundlagen der Stereochemie, Basel 1972.

H. Brunner, Optische Aktivität an asymmetrischen Übergangsmetallatomen, Angew. Chem. **83** [1971] 274/85; Angew. Chem. Intern. Ed. Engl. **10** [1971] 249/60.

I. Ugi, D. Marquarding, H. Klusacek, G. Gokel, P. Gillespie, Chemistry and Logical Structures, Angew. Chem. **82** [1970] 741/71; Angew. Chem. Intern. Ed. Engl. **9** [1970] 703/30.

G. Snatzke, Circulardichroismus und optische Rotationsdispersion − Grundlagen und Anwendung auf die Untersuchung der Stereochemie von Naturstoffen, Angew. Chem. **80** [1968] 15/26; Angew. Chem. Intern. Ed. Engl. **7** [1968] 14/25.

K. Mislow, Einleitung in die Stereochemie, Weinheim/Bergstr. 1967.
R.S. Cahn, C. Ingold, V. Prelog, Spezifikation der molekularen Chiralität, Angew. Chem.
    78 [1966] 413/47; Angew. Chem. Intern. Ed. Engl. **5** [1966] 385/415.

Introduction

The stereochemistry of a reaction center has been a classic tool for studying reaction mechanisms. For the compounds $C_5H_5Fe(CO)(^2D)^1L$ (this section) and C_5H_5Fe-$(CO)(^2D)X$ (1.5.2.2.6, p. 86) four different ligands are bonded to the iron atom in a nonplanar "pseudotetrahedral" arrangement. The iron is a chiral center. Further these compounds are characterized by "18-electron" or "saturated" configurations, which are stable under ambient conditions. As a result the optical isomers are resolvable. These compounds thus offer models for the study of carbon–metal σ bonds, which appear to be of central importance in metal catalysis of reactions at a carbon atom [9].

A compound of the type $C_5H_5Fe(CO)(^2D)^1L$ has two stereoisomeric forms:

These are mirror images and thus enantiomers. A second chiral center can be introduced into ^1L or D, e.g., $C_5H_5Fe^*(CO)(P(C_6H_5)_3)CH_2C^*H(CH_3)C_6H_5$. Now there are a total of four stereoisomers, two diastereomeric pairs of enantiomers [5]:

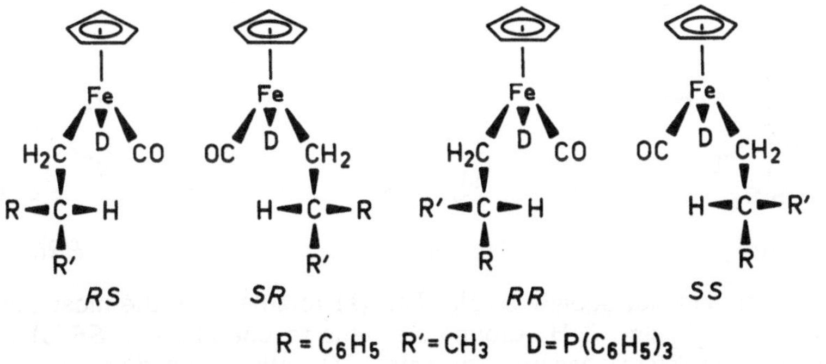

$$R = C_6H_5 \quad R' = CH_3 \quad D = P(C_6H_5)_3$$

The diastereomers RR and SR or SS and RS are not mirror images.

These compounds are named by an extended form of R and S nomenclature. It was suggested that the polyhapto C_5H_5 group be considered as a pseudopolyatom, i.e., a unit having a mass of (5×12) or 60, to determine priority. Thus it tends to have a high priority [7]. For the example given above the priority sequence around Fe is $C_5H_5 > P(C_6H_5)_3 > CO > CH_2CH(CH_3)C_6H_5$. The configuration symbol for the Fe atom is placed before that of the chiral center in ^1L or D.

Another way of introducing a second chiral center into these molecules is by putting two substituents on the cyclopentadienyl ring. Examples that have been studied are $1\text{-}CH_3(3\text{-}C_6H_5)C_5H_3Fe(CO)(P(C_6H_5)_3)^1L$ where ^1L is CH_3 or $COCH_3$ and $1\text{-}CH_3\text{-}(3\text{-}C_6H_5)C_5H_3Fe(CO)(P(C_6H_5)_3)I$. Then the priority about the Fe atom is $I > 1\text{-}CH_3\text{-}(3\text{-}C_6H_5)C_5H_3 > P(C_6H_5)_3 > CO > COCH_3 > CH_3$. The priority about the other center, a carbon atom of the ring, is determined in the usual way. The carbon atom used to determine configuration is C*, which has a higher priority than C** or any other C in the ring:

It was suggested that the R or S of the ring carbon be placed before that of the iron because the C_5H_5 group has higher priority [11]. Thus the sequence of configuration symbols is reverse to that of the first example.

In the case of $C_5H_5Fe(CO)(D)CH(C_6H_5)Si(CH_3)_3$ the analysis has been carried further. Again there are two diastereomeric pairs. But each of these can exist as three rotamers. This is illustrated here only for the SR and RR diastereomers:

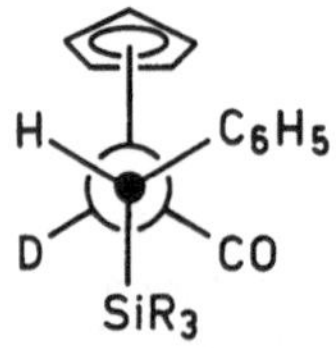

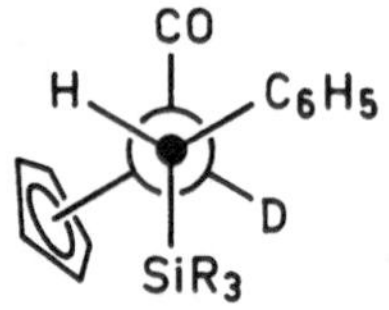

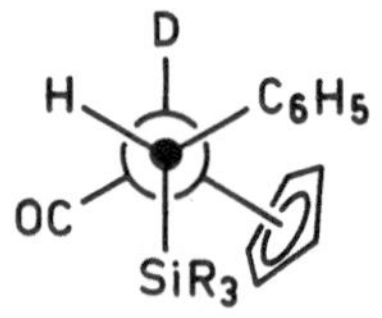

These rotamers are not equally stable. The (a) rotamers are the most stable because the bulky $Si(CH_3)_3$ and C_5H_5 groups are trans to one another. SR (a) is somewhat more stable than RR (a) because the small CO, rather than the L, is gauche to the bulky $Si(CH_3)_3$ and C_6H_5 groups. Thus the RS,SR pair is more stable than the RR,SS pair [7].

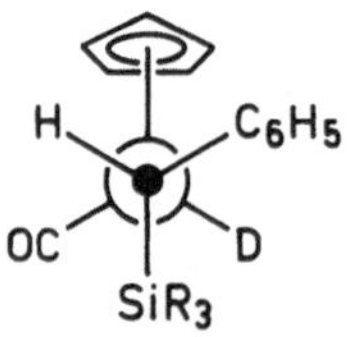

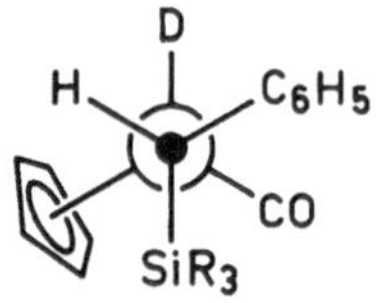

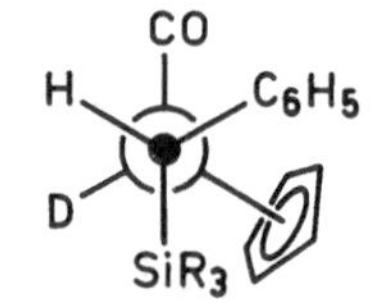

The absolute configuration of an isomer can be determined by single-crystal X-ray methods: The absolute configuration of $(+)\text{-}C_5H_5Fe(CO)(P(C_6H_5)_3)CH_2O\text{-}menthyl$ was determined to be S by anomalous X-ray scattering, and that of $(+)\text{-}C_5H_5Fe\text{-}(CO)[P(C_6H_5)_3]CH_2COO\text{-}menthyl$ was determined to be R by using $(-)\text{-}menthyl$ as reference [10]. Then the absolute configurations of compounds of the same type can be determined by comparing CD and ORD spectra [2, 9, 10]. NMR spectra can be used to distinguish between diastereoisomers because the H and C atoms have different environments in space [6].

A number of reactions of stereochemical interest have been carried out. Single enantiomers, both enantiomers of a diastereomeric pair, or all four stereoisomers have been used. CD and ORD are useful only in the first case, and NMR is useful in the last [9, 11]. Compounds with $^1L = COR$ can be decarbonylated. If this is carried out in boiling toluene or photochemically for an extended period, then the optical activity is lost. However, if the photochemical reaction is carried out quickly, some optical activity is retained. Comparison of the CD spectra of reactant and product suggest an inversion at the Fe center [4].

It is also possible to insert molecules into the metal-carbon σ bond. Generally these are unsaturated, electrophilic molecules, such as CO, SO_2, and TCNE (tetra-cyanoethylene). The first example is carbonylation. If SO_2, either as refluxing SO_2 or a solution in $CHCl_3$, is reacted with $C_5H_5Fe(CO)(^2D)^1L$, orange air-stable products, $C_5H_5Fe(CO)(^2D)SO_2{}^1L$, form. The S-sulfinato structure is revealed in the IR spectra and the ^1H NMR [1, 3]. These reactions demonstrate complete stereospecificity, with either inversion or retention. If reaction is carried out at $-60\,°C$ in neat SO_2, it is less stereospecific [5].

The powerful electrophile TCNE has also been inserted into metal-carbon σ bond. The rates of reaction differ somewhat from those of SO_2 insertion and thus suggest a somewhat different mechanism [8].

References:

[1] M. Graziani, A. Wojcicki (Inorg. Chim. Acta **4** [1970] 347/50). − [2] T.C. Flood, D.L. Miles (J. Am. Chem. Soc. **95** [1973] 6460/2). − [3] G. Ingletto, E. Tondello, L. Di Sipio, G. Carturan, M. Graziani (J. Organometal. Chem. **56** [1973] 335/7). − [4] H. Brunner, J. Strutz (Z. Naturforsch. **29b** [1974] 446/7). − [5] P. Reich-Rohrwig, A. Wojcicki (Inorg. Chem. **13** [1974] 2457/64).

[6] W.B. Jennings (Chem. Rev. **75** [1975] 307/22). − [7] K. Stanley, M.C. Baird (J. Am. Chem. Soc. **97** [1975] 6598/9). − [8] S.R. Su, A. Wojcicki (Inorg. Chem. **14** [1975] 89/98). − [9] T.C. Flood, F.J. DiSanti, D.L. Miles (Inorg. Chem. **15** [1976] 1910/8). − [10] C.-K. Chou, D.L. Miles, R. Bau, T.C. Flood (J. Am. Chem. Soc. **100** [1978] 7271/8).

[11] T.G. Attig, R.G. Teller, S.-M. Wu, R. Bau, A. Wojcicki (J. Am. Chem. Soc. **101** [1979] 619/28).

1.5.2.2.12.1 Compounds with Nitrogen Donors

A compound formulated in [1] as $C_5H_5Fe(CO)(NCCF_3)C(NH)CF_3$ does not exist. It was later found to be a compound of the $^5LFe(CO)^1L^{-2}D$ type with $^1L^{-2}D = -C(CF_3)=N(CF_3)=NH-$, which is described in 1.5.2.2.14, p. 233, Table 25, No. 15.

$C_5H_5Fe(CO)(NH_3)COCH_3$ is obtained from $C_5H_5Fe(CO)_2CH_3$ in liquid ammonia at -50 to $-60\,°C$ by the uptake of ammonia in about 2 to 5 d. The reddish brown solution is filtered off from the unreacted $C_5H_5Fe(CO)_2CH_3$, added to precooled ether, and the liquid ammonia is removed at $-60\,°C$. The compound is precipitated at $0\,°C$ from the concentrated ether solution by adding petroleum ether, 35% yield.

In its ^1H NMR spectrum in CD_3OD there are a broad signal at $\delta = 1.21$ (NH_3) ppm and two sharp singlets at $\delta = 2.56$ (CH_3) and 4.5 (C_5H_5) ppm. In the IR spectrum in KBr, the most important absorptions are $v(C=O)$ 1562 and 1578 cm^{-1} and $v(CO)$ 1895 cm^{-1}, while in CH$_3$OH these bands are at 1560 and 1928 cm^{-1}.

The mass spectrum shows the ions [C$_6$H$_8$FeCO]$^+$, [C$_6$H$_6$FeCO]$^+$, and [C$_6$H$_6$Fe]$^+$. The compound dissolves readily in liquid ammonia to give a reddish brown solution. Such solutions decompose at 70 to 80 °C. The formation reaction is reversed in THF or benzene as was confirmed by IR spectroscopy [2].

C$_5$H$_5$Fe(CO)(NC$_5$H$_5$)CH$_3$. When a solution of C$_5$H$_5$Fe(CO)$_2$CH$_3$ and NC$_5$H$_5$ in CD$_3$CN is UV-irradiated above 27 °C, C$_5$H$_5$Fe(CO)(C$_5$H$_5$N)CH$_3$ forms. The ^{1}H NMR spectrum of this solution has signals at $\delta = 0.27$ (CH$_3$) and 4.37 (C$_5$H$_5$) ppm [3].

References:

[1] R.B. King, K.H. Pannell (J. Am. Chem. Soc. **90** [1968] 3984/7). – [2] H. Behrens, A. Pfister, M. Moll, E. Sepp (Z. Anorg. Allgem. Chem. **428** [1977] 61/7). – [3] C.R. Folkes, A.J. Rest (J. Organometal. Chem. **136** [1977] 355/61).

1.5.2.2.12.2 Compounds with $^2D = PR_3$
   (R = CH$_3$, C$_2$H$_5$, n-C$_4$H$_9$, and cyclo-C$_6$H$_{11}$)

The compounds listed in Table 15 were prepared by the following methods.

Method I: C$_5$H$_5$Fe(CO)$_2$^{1}L and PR$_3$ are refluxed in THF for 2 d, and the products are separated chromatographically [8, 13]. Compound No. 8 is formed at 25 °C [2] and also in refluxing ether but with only 21% yield [8]. No. 9 is prepared in CH$_3$CN [7].

Method II: C$_5$H$_5$Fe(CO)$_2$^{1}L and PR$_3$ are UV-irradiated, and the products separated chromatographically. A variety of reaction conditions habe been reported, one hour in pentane for compound No. 1 [11], in benzene at 25 to 30 °C for No. 4 [4], 3 to 8 h in petroleum ether for No. 5 [3, 12], and 3 h at room temperature in petroleum ether for No. 7 [3].

Method III: A mixture of C$_5$H$_5$Fe(CO)$_2$C$_{10}$H$_{15}$ (C$_{10}$H$_{15}$ = adamant-1-yl) and benzene is condensed in P(CH$_3$)$_3$ and the mixture refluxed for 1 h. The solvent is evaporated off in a vacuum, and the product is sublimed at 100 °C/ 10^{-2} Torr [10].

Method IV: C$_5$H$_5$Fe(CO)$_2$COCH$_3$ is heated with P(C$_6$H$_{11}$)$_3$ to 100 °C for 3 h, and No. 11 is extracted from the reaction mixture with petroleum ether. Pure No. 11 can also be obtained from [C$_5$H$_5$Fe(CO)(P(C$_6$H$_{11}$)$_3$)- C(OH)CH$_3$]BF$_4$ by deprotonation with water [6].

Method V: C$_5$H$_5$Fe(CO)$_2$P(C$_2$H$_5$)$_3$ is allowed to react in liquid ammonia at −33 °C for 10 to 60 min. The excess ammonia is distilled off in a high vacuum at −50 °C, and No. 3 is extracted from the reaction mixture with benzene or CH$_2$Cl$_2$ [9].

Method VI: C$_5$H$_5$Fe(CO)(P(C$_6$H$_{11}$)$_3$)COCH$_3$ (No. 11) is reduced by B$_2$H$_6$ in C$_6$D$_6$ or THF. Alternatively, BH$_3$·THF in THF is added to No. 11. The completion of the reaction (2 min at 20 °C) is determined spectroscopically [15].

Table 15
Compounds of the $C_5H_5Fe(CO)(PR_3)^1L$ Type.
Further information for numbers preceded by an asterisk is given at the end of the table.
For abbreviations and dimensions, see p. 375.

No.	^1L ligand method of preparation (yield in %)	PR_3 ligand	properties and remarks	Ref.
*1	CH_3 II (49)	$P(CH_3)_3$	yellowish brown crystals, m.p. 45° ^1H NMR (CD_3COCD_3): −0.45 (d, $FeCH_3$, J(P,H) = 7.0), 1.35 (d, PCH_3, J(P,H) = 9.4), 4.45 (d, C_5H_5, J(P,H) = 1.2) IR (C_6H_{14}): 1910 (CO)	[11]
*2	III (96)	$P(CH_3)_3$	orange red, m.p. 200° (dec.) IR: 1909 (CO), 1591 (C=O)	[10]
*3	$CONH_2$ V	$P(C_2H_5)_3$	m.p. 36° (dec.)	[9]
*4	C_6H_4F-4 II (46)	$P(C_2H_5)_3$	orange oil IR: 1911 (CO)	[4]
*5	CH_3 II (65)	$P(n\text{-}C_4H_9)_3$	red solid IR: 1901 to 1905 (CO) in CH_2Cl_2 or $CHCl_3$	[3, 12]
6	$CO(CH_2)_2$-$CH=C=CHCH_3$ I	$P(n\text{-}C_4H_9)_3$	—	[13]
*7	$CH_2C_6H_5$ II	$P(n\text{-}C_4H_9)_3$	—	[3]
*8	$COCH_3$ I (96 to 99) I (21)	$P(n\text{-}C_4H_9)_3$	orange solid, m.p. 35 to 37° IR $(CHCl_3)$: 1916 (CO), 1590 (C=O), solid ^1H NMR $(CDCl_3)$: δ = 0.79 to 1.98 (m, C_4H_9), 2.75 (s, CH_3), 4.67 (s, C_5H_5) ^{13}C NMR $(CDCl_3)$: δ = 14.1, 26.4, 28.2, 29.1 (C_4H_9, J(P,C) = 17), 52.8 (CH_3), 83.9 (C_5H_5), 276.8 ($COCH_3$), 220.6 (CO) IR: 1911 (CO), 1601 (C=O) in THF, 1916 (CO), 1590 (C=O) in $CHCl_3$	[2, 8, 16, 17]
9	$COCH_2CH_3$ I (85)	$P(n\text{-}C_4H_9)_3$	low-melting red solid ^1H NMR: δ = 0.84 (t, CH_3 in ^1L, $^3J(H,H)$ = 7.5), 2.84 (m, CH_2 in ^1L, $^2J(H,H)$ = 15.8) IR (C_6H_{12}): 1913 (CO), 1614 (C=O)	[7]

References on p. 157

Table 15 [continued]

No.	¹L ligand method of preparation (yield in %)	PR₃ ligand	properties and remarks	Ref.
*10	CH₂CH₃ VI	P(C₆H₁₁)₃	¹H NMR (C₆D₆): δ=4.37 (C₅H₅, J<0.2)	[15]
*11	COCH₃ IV (95)	P(C₆H₁₁)₃	dark red crystals, m.p. 144° ¹H NMR: 1.28, 1.82 (m, C₆H₁₁), 2.54 (s, CH₃), 4.54 (d, C₅H₅) in CDCl₃, 2.18 (COCH₃, J<0.2), 4.39 (C₅H₅, J<0.2) in C₆D₆, 1.41, 1.94 (m, C₆H₁₁), 3.07 (s, CH₃), 5.05 (s, C₅H₅) in CF₃COOH IR: 1905 (CO), 1600 (C=O)	[6, 15]

*Further information:

$C_5H_5Fe(CO)(P(CH_3)_3)CH_3$ (Table **15**, No. **1**). The air-sensitive crystals can be recrystallized from pentane (-78 °C) [11].

$C_5H_5Fe(CO)(P(CH_3)_3)COC_{10}H_{15}$ (Table **15**, No. **2**) sublimes at 100 °C/ 10^{-2} Torr [10].

$C_5H_5Fe(CO)(P(C_2H_5)_3)CONH_2$ (Table **15**, No. **3**) is quite soluble in benzene or CH_2Cl_2 but insoluble in liquid ammonia. The following ions were observed in the mass spectrum:

$[C_5H_5Fe(CO)(P(C_2H_5)_3)C(O)NH_2]^+$, $[C_5H_5Fe(CO)(P(C_2H_5)_3)NH_2]^+$, $[C_5H_5Fe(P(C_2H_5)_3)NH_2]^+$, $[C_5H_5FeP(C_2H_5)_3]^+$, $[C_5H_5Fe]^+$, and $[P(C_2H_5)_3]^+$ [9].

$C_5H_5Fe(CO)(P(C_2H_5)_3)C_6H_4F\text{-}4$ (Table **15**, No. **4**). The half-wave potential $E_{1/2}$ of CH_3CN solutions was determined to be -2.4 V. The reduction reaction is $C_5H_5Fe(CO)(P(C_2H_5)_3)C_6H_4F\text{-}4 \rightarrow [C_5H_5Fe(CO)P(C_2H_5)_3]^- + [C_6H_4F\text{-}4]^\cdot$ [5].

$C_5H_5Fe(CO)(P(C_4H_9\text{-}n)_3)CH_3$ (Table **15**, No. **5**) is soluble in $CHCl_3$. Both the solid and the solution in $CHCl_3$ absorb gaseous SO_2, forming $C_5H_5Fe(CO)$- $(P(C_4H_9\text{-}n)_3)SO_2CH_3$ [3]. No. 5 reacts with $(CN)_2C=C(CN)_2$ in benzene at 5 to 10 °C to form $C_5H_5Fe(COCH_3)(P(C_4H_9\text{-}n)_3)C_6N_4$ [12, 14].

$C_5H_5Fe(CO)(P(C_4H_9\text{-}n)_3)CH_2C_6H_5$ (Table **15**, No. **7**) reacts with $(CN)_2C=C\text{-}$ $(CN)_2$ in C_6H_6 at 25 °C to form $C_5H_5Fe(COCH_2C_6H_5)(P(C_4H_9\text{-}n)_3)C_6N_4$ [12, 14].

$C_5H_5Fe(CO)(P(C_4H_9\text{-}n)_3)COCH_3$ (Table **15**, No. **8**) is stable in air. It is soluble in pentane, dichloromethane, chloroform, and benzene but insoluble in methanol or water. In a chloroform or benzene solution the compound is rapidly decomposed by air [1]. In the ¹H NMR spectrum of a 0.33 M solution in $C_6H_5CH_3$ containing Eu(fod)₃ (Formula II on p. 104) shifts the resonances downfield by 0.27 (CH₃), 0.50 (γ-CH₂), 2.60 (β-CH₂), 3.56 (COCH₃), 3.62 (C₅H₅), and 3.82 (α-CH₂) ppm [8].

$C_5H_5Fe(CO)(P(C_6H_{11})_3)C_2H_5$ (Table **15**, No. **10**) is not reduced at 25 °C by B_2H_6, either in C_6H_6 or in $C_6H_5CH_3$. At 100 °C it is reduced (e.g., in $C_6D_5CD_3$) in 45 min to give $C_5H_5Fe(CO)(P(C_6H_{11})_3)H$ and C_2H_6 [15].

$C_5H_5Fe(CO)(P(C_6H_{11})_3)COCH_3$ (Table **15**, No. **11**) reacts in dry CH_2Cl_2 with $[(C_2H_5)_3O]BF_4$ to give $[C_5H_5Fe(CO)(P(C_6H_{11})_3)C(OC_2H_5)CH_3]BF_4$. No. 11 reacts with HCl in petroleum ether to give a stable product. This could not be characterized on account of small quantities of HCl_2^-. But treatment with HBF_4 allowed isolation of $[C_5H_5Fe(CO)(P(C_6H_{11})_3)C(OH)CH_3]BF_4$ [6]. No. 11 is reduced to C_5H_5-$Fe(CO)(P(C_6H_{11})_3)C_2H_5$ (No. 10) by $BH_3 \cdot THF$ in THF or B_2H_6 in benzene or THF. Around 100 °C the reduction proceeds to $C_5H_5Fe(CO)(P(C_6H_{11})_3)H$ [15].

References:

[1] J.P. Bibler, A. Wojcicki (Inorg. Chem. **5** [1966] 889/92). − [2] I.S. Butler, F. Basolo, R.G. Pearson (Inorg. Chem. **6** [1967] 2074/9). − [3] M. Graziani, A. Woj-cicki (Inorg. Chim. Acta **4** [1970] 347/50). − [4] A.N. Nesmeyanov, L.G. Makarova, I.V. Polovyanyuk (J. Organometal. Chem. **22** [1970] 707/12). − [5] L.I. Denisovich, I.V. Polovyanyuk, B.V. Lokshin, S.P. Gubin (Izv. Akad. Nauk SSSR Ser. Khim. **1971** 1964/9; Bull. Acad. Sci. USSR Div. Chem. Sci. **1971** 1851/5).

[6] M.L.H. Green, L.C. Mitchard, M.G. Swanwick (J. Chem. Soc. A **1971** 794/7). − [7] M. Green, D.J. Westlake (J. Chem. Soc. A **1971** 367/71). − [8] T.J. Marks, J.S. Kristoff, A. Alich, D.F. Shriver (J. Organometal. Chem. **33** [1971] C35/C37). − [9] J. Ellermann, H. Behrens, H. Krohberger (J. Organometal. Chem. **46** [1972] 119/38). − [10] S. Moorhouse, G. Wilkinson (J. Organometal. Chem. **105** [1976] 349/55).

[11] H.G. Alt, M. Herberhold, M.D. Rausch, B.H. Edwards (Z. Naturforsch. **34b** [1979] 1070/7). − [12] S.R. Su (Diss. Ohio State Univ. 1971, pp. 1/287; Diss. Abstr. Intern. B **32** [1972] 6283). − [13] J. Benaim, J.Y. Merour, J.L. Roustan (Tetrahedron Letters **1971** 983/6). − [14] S.R. Su, A. Wojcicki (Inorg. Chem. **14** [1975] 89/98). − [15] J.A. van Doorn, C. Masters, H.C. Volger (J. Organometal. Chem. **105** [1976] 245/54).

[16] E.J. Kuhlmann (Diss. Univ. Cincinnati 1978, pp. 1/149; Diss. Abstr. Intern. B **39** [1978] 2296). − [17] E.J. Kuhlmann, J.J. Alexander (Inorg. Chim. Acta **34** [1979] L193/L195).

1.5.2.2.12.3 Compounds with $^2D = P(C_6H_5)_3$

1.5.2.2.12.3.1 Compounds with $^1L = COR$ and $CH=CHCOR$

$C_5H_5Fe(CO)(P(C_6H_5)_3)CHO$ has not been shown to exist, but it has been postulated as an intermediate [47] to explain the mechanism of the reaction

$$2[C_5H_5Fe(CO)(P(C_6H_5)_3)CHOR]^+ + I^-$$
$$\rightarrow [C_5H_5Fe(CO)_2P(C_6H_5)_3]^+ + C_5H_5Fe(CO)(P(C_6H_5)_3)CH_2OR + RI$$

The compounds in Table 16 have been prepared in a number of ways.

Method I: From $C_5H_5Fe(CO)_2{}^1L$ and $P(C_6H_5)_3$ or substances which generate $P(C_6H_5)_3$.

    a. $C_5H_5Fe(CO)_2CH_3$ is stirred together with $P(C_6H_5)_3$ at 120 °C for 1 h, washed with petroleum ether, and recrystallized [6].

    b. The starting material $C_5H_5Fe(CO)_2COR$ is refluxed for several hours with $P(C_6H_5)_3$ in a hydrocarbon [1, 5, 11, 12, 16 to 18, 20, 23 to 25, 35 to 38, 40, 43]. The reaction product is purified by chromatography [17, 23 to 25, 38] or by recrystallization in hexane (No. 4) [11, 12, 20]. For the separation of the No. 6 diastereomers, see further information [37].

c. The starting material $C_5H_5Fe(CO)_2COR$ is reacted with $P(C_6H_5)_3$ in benzene and irradiated for 5.5 to 8 h at 70 to 80 °C [7, 19]. The product is recrystallized [7].

d. $C_5H_5Fe(CO)_2CH_3$ and $P(C_6H_5)_3$ are simultaneously heated and UV–irradiated in petroleum ether for 20 h at 90 to 100 °C [3, 4].

e. $C_5H_5Fe(CO)_2{}^1L$ and $P(C_6H_5)_3$ are UV–irradiated for several hours in an inert solvent at room temperature [2, 10, 13, 41, 42, 45, 50, 51]. The reaction product is purified by chromatography [13, 41, 42, 51] or recrystallized from CH_2Cl_2/hexane (No. 2) [10].

f. $C_5H_5Fe(CO)_2{}^1L$ and $Rh(P(C_6H_5)_3)Cl$ are stirred for several hours at room temperature [28, 40, 45, 48, 50]. For No. 16 $Ir(P(C_6H_5)_3)Cl$ is used instead of a Rh compound [45, 50]. The product is purified by chromatography [28, 45, 48].

Method II: From $C_5H_5Fe(CO)(P(C_6H_5)_3)COOC_{10}H_{19}$ ($C_{10}H_{19}$ = menthyl).

a. Transesterification by stirring the starting material in absolute CH_3OH at room temperature for 20 min [30].

b. (+)- and (−)-$C_5H_5Fe(CO)(P(C_6H_5)_3)COOC_{10}H_{19}$ react with $LiCH_3$ to give (−)- and (+)-$C_5H_5Fe(CO)(P(C_6H_5)_3)COCH_3$. During this reaction, the configuration at the Fe atom is inverted [32]. To prepare pure (+)- or (−)-$C_5H_5Fe(CO)(P(C_6H_5)_3)COCH_3$, pure (−)- or (+)-$C_5H_5Fe(CO)(P(C_6H_5)_3)COOC_{10}H_{19}$ is used. These compounds are dissolved in THF, and an ether solution of $LiCH_3$ is added dropwise at −30 °C. The mixtures are stirred for 1 h at −30 °C and then at room temperature. The solvent is removed, and the brown residue chromatographed [26]. The (+)-isomer of No. 3 is obtained by a modification of this method [34].

Method III: $[C_5H_5Fe(CO)(P(C_6H_5)_3)C(OC_2H_5)CH_3]BF_4$ is dealkylated in THF by NaD [27] or in ether by $LiCH_3$ at −30 °C with slow warming to 0 °C [22]. Compound No. 1 forms.

Method IV: CO is passed for 18 h through a solution of $C_5H_5Fe(CO)(P(C_6H_5)_3)R$ in petroleum ether at 90 to 100 °C. The solvent is then removed, and the residue is chromatographed [3]. Such a reaction in CH_3CN gives No. 6 [37].

Method V: The compounds $C_5H_5Fe(CO)(P(C_6H_5)_3)COR$ with R = CH_3 (No. 1) or C_2H_5 (No. 3) form at room temperature from $[C_5H_5Fe(CO)(P(C_6H_5)_3)-C(OCH_3)R]PF_6$ in CH_2Cl_2 through the action of $[CH_3P(C_6H_5)_3]I$ [47].

Method VI: $C_5H_5Fe(CO)(P(C_6H_5)_3)COCH_2C_6H_5$ (No. 9) is formed from $C_5H_5-Fe(CO)(P(C_6H_5)_3)C{\equiv}CC_6H_5$ in THF through the action of dilute aqueous HBF_4 [44].

References on p. 168

Table 16

Compounds of the $C_5H_5Fe(CO)(P(C_6H_5)_3)$ ¹L Type with ¹L = COR or CH=CHCOR.
Further information on numbers preceded by an asterisk is given at the end of the table.
For abbreviations and dimensions, see p. 375.

No.	¹L ligand method of preparation (conditions, yield in %)	properties and remarks explanations on p. 164	Ref.
*1	$COCH_3$ Ia, Ib (60 to 72), Ib (48 h, C_6H_{14}, 96 to 99), Ib (ether, 50), Ib (60 h, C_7H_{16}), Ib (dioxane), Ib (48 h, THF), Id (59.7), If (3 h, C_6H_6 or CH_2Cl_2)	yellow to red brown m.p. 140 to 145°, 164° ¹H NMR: 2.57 (CH_3, J = 0.8), 4.25 (C_5H_5, J = 1.2) in C_6D_6, 2.52 to 2.57 (s, CH_3), 4.59 to 4.62 (s, C_5H_5), 7.59 (m, C_6H_5) in $CDCl_3$, 2.17 (CH_3), 4.24 (C_5H_5), 7.05 (C_6H_5) in CS_2, 2.59 (CH_3), 4.96 (C_5H_5), 7.56 (C_6H_5) in CF_3COOH ¹³C NMR: 51.6 (d, CH_3, J = 6.0), 85.2 (s, C_5H_5), 128.2 (d, C_6H_5, J = 9.3), 133.3 (s, C_6H_5), 133.51 (d, C_6H_5, J = 9.9), 136.9 (d, C_6H_5, J = 43.5), 222.1 (d, CO, J = 31.2), 274.1 (d, C=O, J = 21.4) in CH_2Cl_2, 52.3 (CH_3), 85.5 (C_5H_5), 220.3 (CO, ²J(P,C) = 27), 274.8 (C=O) 52.3 (CH_3), 85.5 (C_5H_5), 220.3 (CO, ²J(P,C) = 27), 274.8 (C=O) IR: 1920, 1595 in Nujol or $CHCl_3$, 1923, 1602 in CH_2Cl_2, 1917, 1598 in THF	[1, 3, 4 to 6, 22, 24, 27, 28, 37, 39, 43, 45, 47, 49, 50]
	(+) isomer IIb (11)	m.p. 142°, specific rotation $[\alpha]_{546}^{20} = +227°$ (10^{-3} M in C_6H_6) molar rotation $[M]_{436}^{22} = -1550°$ (C_6H_6)	
	(−) isomer IIb (30)	m.p. 140°, specific rotation $[\alpha]_{546}^{20} = -228°$ (10^{-3} M in C_6H_6)	
2	$COCF_3$ Ie (4 h, C_6H_{14}, 34)	yellow, m.p. 180 to 181°, dec. 181° ¹H NMR ($CDCl_3$): 4.50 (d, C_5H_5) ¹³C NMR ($CDCl_3$): 85.7 (C_5H_5), 219.1 (CO), 262.9 (C=O) ¹⁹F NMR (CH_2Cl_2): 79.3 (s, CF_3) IR: 1946, 1626 in Nujol, 1962, 1629 in C_6H_{12}	[10, 49]

Table 16 [continued]

No.	¹L ligand, method of preparation (conditions, yield in %)	properties and remarks, explanations on p. 164	Ref.
*3	$COCH_2CH_3$ Ib (12 h, C_7H_{16}, 62), Ib (CH_3CN, 85), V	orange, m.p. 159 to 160°, 166° ¹H NMR: 0.60 (t, CH_3, J=7), 2.70 (m, CH_2), 4.40 (s, C_5H_5), 7.3 (m, C_6H_5) in $CDCl_3$, 0.58 (t, CH_3, $^3J(H,H)=7.0$), 2.64 (m, CH_2, $^2J(H,H)=17.0$), 4.32 (m, C_5H_5, J=1.0) in CH_3CN IR: 1920, 1608 in C_6H_{12}, 1918, 1600 in CH_2Cl_2	[23 to 25, 47]
	(+) isomer IIb (26)	molar rotation $[M]_{436}^{27} = -800°$	[34]
*4	$COCH_2Si(CH_3)_3$ Ib (55 h, THF, 30), Ib (20 h, THF, 50)	red, m.p. 138 to 139° ¹H NMR: −0.01 (s, CH_3), 1.78, 2.18 (d's, CH_2), 4.42 (d, C_5H_5) in C_6H_{12}, 0.01 (s, CH_3), 2.15 (s, CH_2), 4.40 (s, C_5H_5) in $CDCl_3$, 1.7, 2.7 (d's, CH_2, J=12) in $CDCl_3$ IR: 1917, 1584, 1601 in C_6H_{12}, 1918, 1565, 1571 in KBr, 1555 in $CHCl_3$	[11, 14, 20, 38]
*5	$COCH_2Si(CH_3)_2Si(CH_3)_3$ Ib (THF)	¹H NMR (CS_2): 0.01 (CH_3), 1.84, 2.06, 2.68, 2.90 (CH_2, an AB quartet centered at 2.37, J=13), 4.38 (C_5H_5)	[36]
*6	$COCH_2CH(CH_3)C_6H_5$ Ib (15 h, CH_3CN, 86), Ib (48 h, THF, ~85), Ic (1 h, C_6H_6, 26), Ie (1 h, C_6H_6), IV (25)	see further information	[37]
*7	$COCHDCHDC(CH_3)_3$ Ib (THF, 62)	yellow, m.p. 143° (dec.) ¹H NMR: 2.72, 2.46 (J=4.4) IR: 1925, 1615	[16]
8	$COCH_2C_6H_{11}$–cyclo Ib (140 min, $OS(CD_3)_2$)	¹H NMR: 2.7 (CH_2), 4.33 (d, C_5H_5, J=1) in CD_3SOCD_3, 4.37 (C_5H_5) in $CDCl_3$ IR (CD_3SOCD_3): 1595	[35]

References on p. 168

No.	Compound / Reagents	Data	Ref.
9	COCH$_2$C$_6$H$_5$ Ib (42 h, dioxane, 9), VI	m.p. 156° ^{1}H NMR (CS$_2$): 3.49, 4.02 (CH$_2$, ^{2}J(H,H) = 13.5), 4.14 (C$_5$H$_5$, J(P,H) = 1.5), 7.01 (CC$_6$H$_5$), 7.31 (PC$_6$H$_5$) IR: 1912, 1603	[9, 17, 44]
*10	COCH(CH$_3$)$_2$ Ib (CH$_3$CN, 90)	orange, m.p. 167° ^{1}H NMR: 0.25 (d, CH$_3$, J(H,H) = 7), 0.99 (d, CH$_3$, J(H,H) = 7), 2.91 (sept, CH, J(H,H) = 7), 4.40 (d, C$_5$H$_5$, ^{3}J(P,H) = 2) in CDCl$_3$(?), 0.88 (d, CH$_3$, J = 7.0), 2.84 (h, CH, J(H,H) = 7.0), 4.34 (d, C$_5$H$_5$, J = 0.6) in CH$_3$CN IR (C$_6$H$_{12}$ or CH$_2$Cl$_2$): 1918, 1607	[23, 46]
11	COC(CH$_3$)$_3$ Ib (C$_6$H$_6$, 56)	—	[40]
*12	O=C–(furan-2-yl) Ie (1.5 h, ether, 42)	orange, m.p. 151 to 152° ^{1}H NMR (CDCl$_3$): 4.40 (C$_5$H$_5$), 6.38, 6.68 (C$_4$H$_3$O), 7.05 to 7.40 (C$_6$H$_5$) IR: 1944 in C$_6$H$_6$, 1565 in KBr	[41]
*13	O=C–(5-CH$_3$-furan-2-yl) Ie (1.5 h, ether, 30)	orange, m.p. 153 to 154° ^{1}H NMR (CDCl$_3$): 2.20 (CH$_3$), 4.38 (C$_5$H$_5$), 5.88, 6.62 (C$_4$H$_2$O), 7.10 to 7.50 (C$_6$H$_5$) IR: 1928 in C$_6$H$_6$, 1550 in KBr	[41]
*14	O=C–(C$_5$H$_4$)Mn(CO)$_3$ Ie (36 h, THF, 30)	yellow crystals, m.p. 167 to 168°, dec. p. 168° ^{1}H NMR (CS$_2$): 4.43 (d, C$_5$H$_5$, J(P,H) = 1), 4.47 (C$_5$H$_4$), 4.52 (C$_5$H$_4$), 7.30 (m, C$_6$H$_5$) IR (CCl$_4$): 1940, 1955, 2030, 1600	[42]
*15	COC$_6$H$_5$ Ib (22 h, C$_6$H$_6$, 29.2), Ic (8 h, C$_6$H$_6$), Ie (16 h, C$_6$H$_{12}$, 100), If (3 to 15 h, C$_6$H$_6$, 17), If (3 h, CH$_2$Cl$_2$, 44)	orange, m.p. 164° (dec.) ^{1}H NMR (CDCl$_3$): 4.51 (d, C$_5$H$_5$), 7.28 (m, C$_6$H$_5$) ^{13}C NMR (CDCl$_3$): 85.5 (C$_5$H$_5$), 127.2 (C$_6$H$_5$), 128.1 (C$_6$H$_5$), 128.2 (PC$_6$H$_5$), 130.0 (C$_6$H$_5$), 130.6 (PC$_6$H$_5$), 134.4 (PC$_6$H$_5$), 137.3 (PC$_6$H$_5$), 153.4 (C$_6$H$_5$), 222.1 (CO, ^{2}J(P,C) = 39) IR (solid): 1880, 1923, 1570, 1582, 1600, solid and in C$_6$H$_6$, 1920, 1595 in Nujol	[2, 7, 19, 28, 45, 50]

References on p. 168

Table 16 [continued]

No.	¹L ligand method of preparation (conditions, yield in %)	properties and remarks explanations on p. 164	Ref.
16	$COC_6H_4OCH_3$-4 If (24 h, C_6H_6, 39)	orange, m.p. 153 to 154° ¹H NMR ($CDCl_3$): 3.78 (s, CH_3), 4.52 (d, C_5H_5), 7.32 (m, C_6H_5) ¹³C NMR ($CDCl_3$): 55.2 (OCH_3), 85.7 (C_5H_5), 112.8 (C_6H_4), 128.7 (C_6H_5), 129.1 (C_6H_4), 130.2, 134.3, 137.0 (C_6H_5), 146.6, 160.5 (C_6H_4), 220.7 (CO), 270.2 (C=O) IR (Nujol): 1910, 1575	[45, 49, 50]
17	COC_6H_4OH-4 Ie (20 min, C_6H_6, 51), If (3 to 4 h, CH_2Cl_2, 51)	yellow–orange lump	[45, 50]
18	COC_6H_4F-4 Ic (6 h, C_6H_6, 7.84)	yellow lump, m.p. 154.5 to 156° IR (C_6H_6?): 1923, 1562	[19]
19	COC_6H_4Cl-4 If (3 to 4 h, C_6H_6 or CH_2Cl_2, 30)	—	[45, 50]
20	$COC_6H_4CH_3$-4 Ic (5.5 h, C_6H_6, 12.6)	yellow–orange lump, m.p. 161 to 163°, dec. 163° IR (C_6H_6?): 1922, 1573	[19]
21	COC_6H_{11}-cyclo Ib (140 min, CD_3SOCD_3)	¹H NMR: 2.7 (C_6H_{11}) in CD_3SOCD_3, 4.2 (C_5H_5) in $CDCl_3$	[35]
22	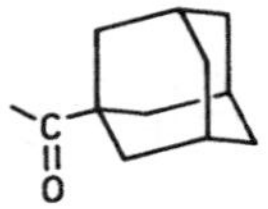 Ib (0.5 h, C_6H_6, 98), If (1 h, C_6H_6, 13)	red, m.p. 176 to 177° IR (C_6H_6): 1915, 1598	[40]

References on p. 168

No.	Structure / reagent	Data	Ref.
23		orange red, m.p. 176 to 177° IR (C$_6$H$_6$): 1917, 1603	[40]
*24	Ib (0.5 h, C$_6$H$_6$) —C=O Ie (1.5 h, ether, 32)	orange, m.p. 150 to 151° ^{1}H NMR (CDCl$_3$): 4.48 (C$_5$H$_5$), 6.56 (C$_8$H$_5$O), 7.05 to 7.25 (C$_6$H$_5$) IR: 1947 in C$_6$H$_6$, 1568 in KBr	[41]
*25	—C=O Ie (1.5 h, ether, 45)	orange, m.p. 155 to 156° ^{1}H NMR (CDCl$_3$): 4.49 (C$_5$H$_5$), 6.50 (C$_4$H$_3$S), 7.10 to 7.70 (C$_6$H$_5$) IR: 1925 in C$_6$H$_6$, 1560 in KBr	[41]
*26	CH=CHCOCH$_3$ Ie (6 to 6.5 h, THF, 62.5)	yellow, m.p. 154° (dec.) ^{1}H NMR (CDCl$_3$): 6.51 (CHCO), 9.68 (CH=, ^{3}J(H,H) = 16) IR: 1941 in C$_6$H$_{12}$, 1643 in KBr	[13]
*27	CH=CHCOC$_6$H$_5$ Ie (6 to 6.5 h, THF, 65)	yellow, m.p. 164° (dec.) ^{1}H NMR (CDCl$_3$): 10.61 (CH=, ^{3}J(H,H) = 16), signal of CHCO is overlapped IR: 1941 in C$_6$H$_{12}$, 1622 in KBr	[13]
28	COC$_3$H$_5$–cyclo Ie (57 h, C$_6$H$_{12}$, 30)	bright orange yellow, m.p. 161 to 163° IR (C$_6$H$_{12}$): 1922, 1602	[51]
*29	COCH(CH$_3$)CH$_2$CH$_3$ Ib	IR: 1920, 1610	[18]
30	COCH$_2$CH$_2$CH=CH$_2$ Ib	—	[53]

References on p. 168

Explanations for Table 16: The strong IR bands of the two kinds of carbonyl groups are not labelled in the table. They are arranged by $v(CO)$ in the 1900 to 1930 cm^{-1} range and $v(C=O)$ in the 1550 to 1630 cm^{-1} range.

*Further information:

$C_5H_5Fe(CO)(P(C_6H_5)_3)COCH_3$ (Table **16**, No. **1**) is also formed by reaction of $C_5H_5Fe(CO)_2{}^{13}COCH_3$ with $Rh(P(C_6H_5)_3)Cl$ in benzene at room temperature [28]. It is quantitatively formed in the reaction of $C_5H_5Fe(CO)(P(C_6H_5)_3)COCH_2Si(CH_3)_3$ (No. **4**) with HCl [20].

The circular dichroism and rotation-dispersion spectra are shown as graphs in [26]. The chemical shifts $\delta = 133.3$ and 133.51 ppm in the ^{13}C NMR spectra [43] were assigned to the p- and o-positions in the phenyl rings, which were labelled by deuteration. At $-80\,°C$, only one molecular species was observed in the ^1H NMR spectrum, but two ketone bands were found in the IR spectra [14]. The ^{57}Fe Mössbauer spectrum shows an isomer shift of $\delta = 0.31$ mm·s^{-1} and a quadrupole splitting of $\Delta = 1.83$ mm·s^{-1} [31].

The compound is soluble in pentane, benzene, and dichloromethane but is insoluble in methanol or water [1]. The compound decomposes in refluxing dioxane, and $C_5H_5Fe(CO)_2CH_3$ was found among the decomposition products [24]. When No. **1** is heated in benzene between 90 and 100 °C, an equilibrium is established: C_5H_5Fe-$(CO)(P(C_6H_5)_3)COCH_3 \rightleftharpoons C_5H_5Fe(CO)_2CH_3 + P(C_6H_5)_3$. The decarbonylation of No. **1**, often noted during heating, is attributed exclusively to a photochemical reaction accompanying the reaction in the equilibrium [52]. Decarbonylation gives C_5H_5Fe-$(CO)(P(C_6H_5)_3)CH_3$ in refluxing heptane, hexane, or cyclohexane by UV irradiation, in petroleum ether at 30 to 35 °C, or in benzene at 40 to 45 °C [24, 25], in refluxing petroleum ether (b.p. 90 to 100 °C) on simultaneous UV irradiation [3, 4], and on strong irradiation at room temperature in benzene/pentane [34]. Both decarbonylation of $(-)$-$C_5H_5Fe(CO)(P(C_6H_5)_3)COCH_3$ in boiling toluene (20 h) [32, 33] and photochemical decarbonylation of $(+)$-$C_5H_5Fe(CO)(P(C_6H_5)_3)COCH_3$ in n-hexane (5 to 30 min) at room temperature give inactive $C_5H_5Fe(CO)(P(C_6H_5)_3)CH_3$. Since a weakly optically active $(C_5H_5)Fe(CO)(P(C_6H_5)_3)CH_3$ ($[\alpha]_{579}^{25} = -12°$) is obtained for a 1-min irradiation and since the circular dichroism spectra of No. **1** and its decarbonylation product are mutually complementary, decarbonylation involves inversion at the Fe atom, I → II.

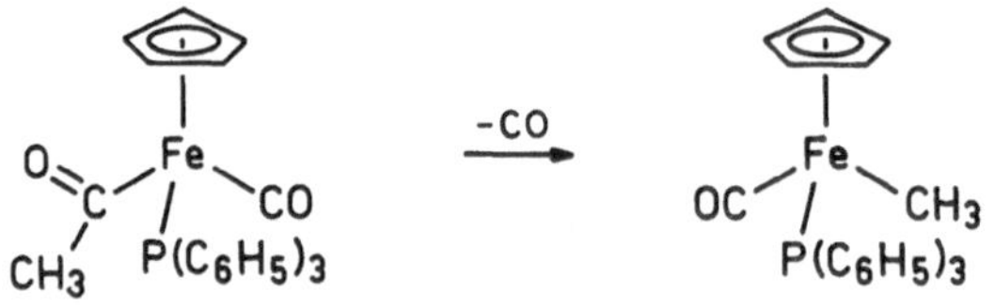

This inversion is in contrast to the retention of the configuration at the α-C atom during carbonylation or decarbonylation. The appearance of optical activity on shortening the reaction time during photochemical decarbonylation shows that the transformation of the configuration at the Fe atom occurs subsequently [33].

The solid No. **1** is stable in air. When dissolved in benzene or chloroform, it is quickly decomposed in air [1]. It reacts with excess I_2 in benzene (15 min) to form $C_5H_5Fe(CO)(P(C_6H_5)_3)I$. This reaction is not stereospecific [4, 33]. Protonation takes place in kerosene with HCl or HBr to give $[C_5H_5Fe(CO)(P(C_6H_5)_3)C(OH)CH_3]^+$ [6].

References on p. 168

Reduction to $C_5H_5Fe(CO)(P(C_6H_5)_3)CH_2CH_3$ is achieved through the action of $BH_3 \cdot THF$ in THF or of B_2H_6 in benzene or THF (compare 1.5.2.2.12.3.4, Table 18, No. 6). Around 100 °C, the reduction can proceed to $C_5H_5Fe(CO)(P(C_6H_5)_3)H$ [39]. No. 1 is converted to $[C_5H_5Fe(CO)(P(C_6H_5)_3)=C=CH_2]^+$ in CH_2Cl_2 by CF_3SO_3H through the $[C_5H_5Fe(CO)(P(C_6H_5)_3)C(CH_3)O_3SCF_3]^+$ [46]. $(+)-C_5H_5Fe(CO)-(P(C_6H_5)_3)COCH_3$ is converted by $NaBH_4$ in C_2H_5OH or also by $[(C_2H_5)_3O]BF_4$ through the intermediate $[(+)-C_5H_5Fe(CO)(P(C_6H_5)_3)C(OC_2H_5)CH_3]^+$ into $(+)-C_5H_5Fe(CO)(P(C_6H_5)_3)C_2H_5$. The Fe-center configuration is not changed [34].

$C_5H_5Fe(CO)(P(C_6H_5)_3)COC_2H_5$ (Table 16, No. 3). The solid is stable in air. In petroleum ether it is decarbonylized at 30 to 35 °C by UV irradiation to $C_5H_5Fe-(CO)(P(C_6H_5)_3)C_2H_5$ [24, 25]. Photolysis ($\lambda = 360$ nm) in THF causes the Fe atom in $(+)-C_5H_5Fe(CO)(P(C_6H_5)_3)COC_2H_5$ to undergo a configurational change as described for No. 1 [34]. The racemic mixture reacts easily with liquid SO_2 [25].

$C_5H_5Fe(CO)(P(C_6H_5)_3)COCH_2Si(CH_3)_3$ (Table 16, No. 4). In the ^1H NMR spectrum, the CH_2 protons are found to be nonequivalent. An attempt to resolve either doublet failed. The $\Delta\delta$ values vary from 80 Hz at -80 °C in $CDCl_3$ to 29 Hz at 110 °C in CH_3SOCH_3. The spectra of the individual rotamers could not be observed at low temperatures. Two acetyl bands in the IR spectra indicate a restricted rotation about the Fe-C bond [14].

In the mass spectra the following ions were observed: $[C_5H_5Fe(CO)(P(C_6H_5)_3)-COCH_2Si(CH_3)_3]^+$, $[C_5H_5Fe(CO)(P(C_6H_5)_3)CH_3Si(CH_3)_3]^+$, $[C_5H_5Fe(P(C_6H_5)_3)-CH_2Si(CH_3)_3]^+$, $[C_5H_5FeP(C_6H_5)_3]^+$, $[C_5H_5Fe(CO)(P(C_6H_5)_3)COCH_2Si(CH_3)_2]^+$, $[C_5H_5Fe(CO)(P(C_6H_5)_3)CH_2Si(CH_3)_2]^+$, $[C_5H_5Fe(P(C_6H_5)_3)CH_2Si(CH_3)_2]^+$, $[Fe(P(C_6H_5)_3)CH_2Si(CH_3)_3]^+$, and $[FeP(C_6H_5)_3]^+$ [38]. Treatment of No. 4 with HCl quantitatively gives $C_5H_5Fe(CO)(P(C_6H_5)_3)COCH_3$ (No. 1) [20].

$C_5H_5Fe(CO)(P(C_6H_5)_3)COCH_2Si(CH_3)_2Si(CH_3)_3$ (Table 16, No. 5). On thermal decomposition, No. 5 gives $C_5H_5Fe(CO)_2CH_2Si(CH_3)_2Si(CH_3)_3$ and $C_5H_5Fe-(CO)(P(C_6H_5)_3)Si(CH_3)_2CH_2Si(CH_3)_3$. Photochemical decarbonylation forms the latter exclusively [36].

$C_5H_5Fe(CO)(P(C_6H_5)_3)COCH_2CH(CH_3)C_6H_5$ (Table 16, No. 6). This compound possesses two chiral centers: the Fe atom and the C-3 (CH) atom of the ^1L ligand. Thus there are four possible forms, but these have not been obtained pure, and their steric configurations have not yet been clearly determined. The product obtained by Method I b, starting with optically inactive $C_5H_5Fe(CO)_2CH_2CH(CH_3)C_6H_5$, affords No. 6. The ^1H NMR spectrum of the product shows it to be approximately a 53:47 mixture of the $(RR)(SS)$ pair (a) and the $(RS)(SR)$ pair (b). $(-)(S)-C_5H_5Fe-(CO)_2CH_2CH(CH_3)C_6H_5$ gives the optically active $(-)-C_5H_5Fe(CO)(P(C_6H_5)_3)-COCH_2CH(CH_3)C_6H_5$, consisting of a 50:50 mixture of the diastereomers $(RR)-$ and $(SR)-C_5H_5Fe(CO)(P(C_6H_5)_3)COCH_2CH(CH_3)C_6H_5$ (A). The two pairs of enantiomers are generally obtained in a ratio of approximately 53:47 owing to incomplete elution of the more slowly moving pair during column chromatography. No. 6 enriched in the less soluble diastereomer b is obtained by fractional crystallization from C_6H_6/C_5H_{12} at 5 °C. The final batch of crystals was identified by ^1H NMR spectroscopy as a 9:91 mixture of a and b. No. 6 enriched in a is obtained by two successive chromatographic separations on Al_2O_3 with C_6H_6/C_5H_{12} (1:1); a 90:10 mixture of a and b is isolated. A similar ratio of the diastereomers (RR) and (SR) is obtained by chromatography of A. Repeated chromatography of A on Al_2O_3 with C_6H_6/C_5H_{12} finally gives a 94:6 mixture of $(RR)-$ and $(SR)-C_5H_5Fe(CO)(P(C_6H_5)_3)-$

COCH$_2$CH(CH$_3$)C$_6$H$_5$ (Aa) and an 8:92 mixture of (SR)- and (RR)-C$_5$H$_5$Fe-(CO)(P(C$_6$H$_5$)$_3$)COCH$_2$CH(CH$_3$)C$_6$H$_5$ (Ab). The physical properties of these isomers are shown in Table 17 [37].

Table 17
Properties of Various Forms of C$_5$H$_5$Fe(CO)(P(C$_6$H$_5$)$_3$)COCH$_2$CH(CH$_3$)C$_6$H$_5$.

isomer (see text)	m.p. in °C	specific rotation $[\alpha]_D^{23}$ (c, CHCl$_3$)	spectra
a (90:10, a:b)	149 to 151	—	^{1}H NMR (CDCl$_3$): 0.77 (d, CH$_3$, J= 6.5), 2.40 to 3.63 (CH, CH$_2$), 4.18 (d, C$_5$H$_5$, J=1.3), 7.33 (m, C$_6$H$_5$) IR (CH$_2$Cl$_2$): 1608, 1910
b (9:91, a:b)	159 to 164 (dec.)	—	^{1}H NMR (CDCl$_3$): 1.09 (d showing some additional lines, CH$_3$, J= 6.5), 2.45 to 3.40 (CH, CH$_2$), 4.34 (d, C$_5$H$_5$, J=1.3), 7.34 (m, C$_6$H$_5$) IR (CH$_2$Cl$_2$): 1608, 1910
A	155 to 157 (dec.)	−41.0° (0.48)	
Aa (94:6)	50 to 61	−127.0° (0.35)	
Ab (8:92)	54 to 66	+48.5° (0.35)	

These preparations are decarbonylized in benzene solution through the action of UV radiation to give (C$_5$H$_5$)Fe(CO)(P(C$_6$H$_5$)$_3$)CH$_2$CH(CH$_3$)C$_6$H$_5$. During this process the mixtures (RR)(SS) and (RS)(SR) undergo no epimerization. All these preparations were shown to be stable in CH$_3$CN up to 80 °C. They are stable in the dark in CDCl$_3$ for several days [37].

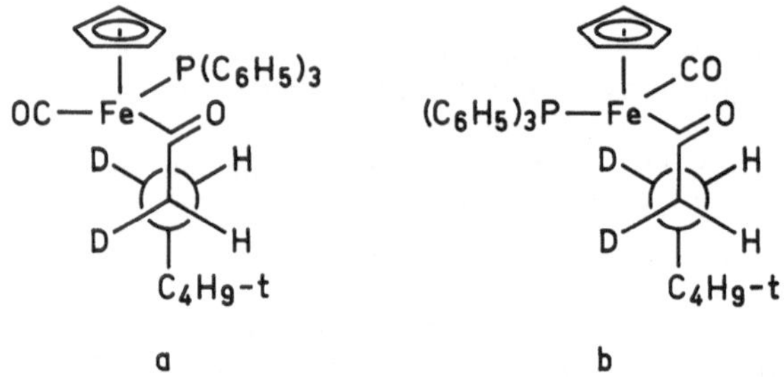

III

C$_5$H$_5$Fe(CO)(P(C$_6$H$_5$)$_3$)COCHDCHDC(CH$_3$)$_3$ (Table 16, No. 7). The ^{1}H NMR spectrum (see the graphical representation in the original work) consists of two equally strong AX patterns. They originate from the two pairs of diastereomers (see Formula III). The CHDCO resonance appears at δ=2.46 and 2.72 ppm, with the vicinal coupling constant J=4.4 Hz. A comparison of the coupling constant with those for the threo (J=4.2 Hz) and the erythro (J=12.5 Hz) diastereomers according to the

References on p. 168

analysis of the ABXY spectrum shows that the transformation of the alkyl–Fe bond in $C_5H_5Fe(CO)_2CHDCHDC(CH_3)_3$ to the corresponding C–C bond of No. 7 takes place with complete retention of configuration [16].

$C_5H_5Fe(CO)(P(C_6H_5)_3)COCH(CH_3)_2$ (Table 16, No. 10) reacts with CF_3SO_3H in CH_2Cl_2 to give $[C_5H_5Fe(CO)(P(C_6H_5)_3)C=C(CH_3)_2]^+$ [46] through the intermediate $[C_5H_5Fe(CO)(P(C_6H_5)_3)C(CH(CH_3)_2)O_3SCF_3]^+$ cation. No. 10 reacts with $[(C_6H_5)_3C]BF_4$ in CH_2Cl_2, ether precipitates $[C_5H_5Fe(CO)_2P(C_6H_5)_3]BF_4$ [23].

$C_5H_5Fe(CO)(P(C_6H_5)_3)COC_4H_2OR$ (Table 16, Nos. 12 and 13, with R=H and CH_3). Nos. 12 and 13 undergo partial decarbonylation through UV irradiation during preparation. They decompose quickly in solution [41].

$C_5H_5Fe(CO)(P(C_6H_5)_3)COC_5H_4Mn(CO)_3$ (Table 16, No. 14). On boiling in m–xylene for 2 h, No. 14 loses CO and forms $C_5H_5Fe(CO)(P(C_6H_5)_3)C_5H_4Mn(CO)_3$ and $C_5H_5Fe(CO)_2C_5H_4Mn(CO)_3$. When irradiated for 5 h in THF, CO is lost quantitatively, and $C_5H_5Fe(CO)(P(C_6H_5)_3)C_5H_4Mn(CO)_3$ forms. No. 14 is stable when boiled in CH_3CN or THF [42].

$(C_5H_5)Fe(CO)(P(C_6H_5)_3)COC_6H_5$ (Table 16, No. 15) is stable in air as a solid [2]. It crystallizes as bright red monoclinic prisms along the c axis, space group $P2_1/a-C_{2h}^5$: a=16.38, b=18.75, c=8.14 Å, β=94° 40'. Z=4 gives $D_c=1.36$, while $D_m=1.35$ g·cm^{-3}. The molecular structure is shown in Fig. 13. The distance from the Fe atom to the benzoyl C atom is conspicuously shorter than in the phenyl compound where the Fe–C distance is 2.11 Å (see compound No. 1, Table 19, in 1.5.2.2.12.3.6). This shortening is attributed to dπ–pπ effects. Bond lengths and angles are given in Fig. 13 [8, 15]. When dissolved in benzene and UV-irradiated at 25 °C, the compound is slowly decarbonylized to form $(C_5H_5)Fe(CO)-(P(C_6H_5)_3)C_6H_5$ [19].

Fig. 13

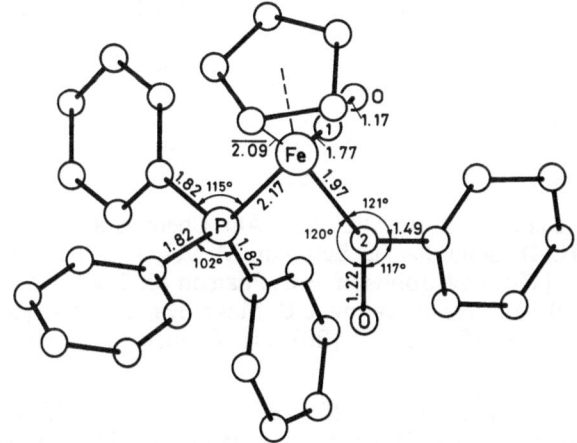

Molecular structure of $C_5H_5Fe(CO)(P(C_6H_5)_3)COC_6H_5$ [15].

Other bond angles (°):

C(1)–Fe–C(2)	93	C(2)–Fe–P	88
C(1)–Fe–P	86	Fe–C(1)–O	159

$C_5H_5Fe(CO)(P(C_6H_5)_3)COR$ (Table **16**, Nos. **24** and **25**, with $R=C_8H_5O$ or C_4H_3S). Solid No. 24 is stable in air, whereas No. 25 decomposes slowly. In solution, both decompose quickly. When UV-irradiated in ether, both are decarbonylized [41].

$C_5H_5Fe(CO)(P(C_6H_5)_3)CH{=}CHCOR$ (Table **16**, Nos. **26** and **27**, with $R=CH_3$ or C_6H_5). Nos. 26 and 27 are stable in air in the solid state but gradually decompose in solution. They are readily soluble in CH_2Cl_2, $CHCl_3$, and THF, less soluble in C_6H_6, and insoluble in petroleum ether or hexane. In benzene, both compounds are monomeric. No. 26 reacts slowly with $Fe_2(CO)_9$ in benzene at 4 to 50 °C to give $(CO)_4FeP(C_6H_5)_3$ [13].

$C_5H_5Fe(CO)(P(C_6H_5)_3)COCH(CH_3)CH_2CH_3$ (Table **16**, No. **29**). Cl_2 and H_2O split off the COC_4H_9 group at the Fe atom, C_4H_9COOH being formed [18].

References:

[1] J.P. Bibler, A. Wojcicki (Inorg. Chem. **5** [1966] 889/92). – [2] A.N. Nesmeyanov, Yu.A. Chapovsky, B.V. Lokshin, I.V. Polovyanyuk, L.G. Makarova (Dokl. Akad. Nauk SSSR **166** [1966] 1125/8; Dokl. Chem. Proc. Acad. Sci. USSR **166/171** [1966] 213/6). – [3] P.M. Treichel, R.L. Shubkin, K.W. Barnett, D. Reichard (Inorg. Chem. **5** [1966] 1177/81). – [4] K.W. Barnett (Diss. Univ. Wisconsin 1967; Diss. Abstr. B **28** [1968] 3203). – [5] I.S. Butler, F. Basolo, R.G. Pearson (Inorg. Chem. **6** [1967] 2074/9).

[6] M.L.H. Green, C.R. Hurley (J. Organometal. Chem. **10** [1967] 188/90). – [7] A.N. Nesmeyanov, Yu.A. Chapovsky, I.V. Polovyanyuk, L.G. Makarova (J. Organometal. Chem. **7** [1967] 329/37). – [8] Yu.A. Chapovsky, V.A. Semion, V.G. Andrianov, Yu.T. Struchkov (Zh. Strukt. Khim. **9** [1968] 1100/2; J. Struct. Chem. [USSR] **9** [1968] 990/2). – [9] H. Brunner, E. Schmidt (Angew. Chem. **81** [1969] 570/1). – [10] R.B. King, R.N. Kapoor, K.H. Pannell (J. Organometal. Chem. **20** [1969] 187/93).

[11] R.B. King, K.H. Pannell, C.R. Bennett, M. Ishaq (J. Organometal. Chem. **19** [1969] 327/37). – [12] R.B. King, K.H. Pannell, M. Ishaq, C.R. Bennett (Proc. 4th Intern. Conf. Organometal. Chem., Bristol, Engl., 1969, Ref. A3). – [13] A.N. Nesmeyanov, M.I. Rybinskaya, L.V. Rybin, V.S. Kaganovich, Yu.A. Ustynyuk, I.F. Leshcheva (Izv. Akad. Nauk SSSR Ser. Khim. **1969** 1100/3; Bull. Acad. Sci. USSR Div. Chem. Sci. **1969** 1004/6). – [14] K.H. Pannell (J. Chem. Soc. Chem. Commun. **1969** 1346/7). – [15] V.A. Semion, Yu.T. Struchkov (Zh. Strukt. Khim. **10** [1969] 664/71; J. Struct. Chem. [USSR] **10** [1969] 563/9).

[16] G.M. Whitesides, D.J. Boschetto (J. Am. Chem. Soc. **91** [1969] 4313/4). – [17] H. Brunner, H.-D. Schindler, E. Schmidt, M. Vogel (J. Organometal. Chem. **24** [1970] 515/26). – [18] R.W. Johnson, R.G. Pearson (J. Chem. Soc. Chem. Commun. **1970** 986/7). – [19] A.N. Nesmeyanov, L.G. Makarova, I.V. Polovyanyuk (J. Organometal. Chem. **22** [1970] 707/12). – [20] K.H. Pannell (J. Organometal. Chem. **21** [1970] P17/P18).

[21] H. Brunner (Chimia [Aarau] **25** [1971] 284/6). – [22] M.L.H. Green, L.C. Mitchard, M.G. Swanwick (J. Chem. Soc. A **1971** 794/7). – [23] M. Green, D.J. Westlake (J. Chem. Soc. A **1971** 367/71). – [24] S.R. Su (Diss. Ohio State Univ. 1971; Diss. Abstr. Intern. B **32** [1972] 6283). – [25] S.R. Su, A. Wojcicki (J. Organometal. Chem. **27** [1971] 231/40).

[26] H. Brunner, E. Schmidt (J. Organometal. Chem. **36** [1972] C18/C22). – [27] A. Davison, D.L. Reger (J. Am. Chem. Soc. **94** [1972] 9237/8). – [28] J.J.

Alexander, A. Wojcicki (Inorg. Chem. **12** [1973] 74/6). — [29] T.G. Attig, P. Reich-Rohrwig, A. Wojcicki (J. Organometal. Chem. **51** [1973] C21/C23). — [30] H. Brunner, E. Schmidt (J. Organometal. Chem. **50** [1973] 219/25).

[31] G. Ingletto, E. Tondello, L. DiSipio, G. Carturan, M. Graziani (J. Organometal. Chem. **56** [1973] 335/7). — [32] H. Brunner (Ann. N.Y. Acad. Sci. **239** [1974] 213/24). — [33] H. Brunner, J. Strutz (Z. Naturforsch. **29b** [1974] 446/7). — [34] A. Davison, N. Martinez (J. Organometal. Chem. **74** [1974] C17/C20). — [35] K. Nicholas, S. Raghu, M. Rosenblum (J. Organometal. Chem. **78** [1974] 133/7).

[36] K.H. Pannell, J.R. Rice (J. Organometal. Chem. **78** [1974] C35/C39). — [37] P. Reich-Rohrwig, A. Wojcicki (Inorg. Chem. **13** [1974] 2457/64). — [38] K.H. Pannell (Transition Metal Chem. [Weinheim] **1** [1976] 36/40). — [39] J.A. van Doorn, C. Masters, H.C. Volger (J. Organometal. Chem. **105** [1976] 245/54). — [40] S. Moorhouse, G. Wilkinson (J. Organometal. Chem. **105** [1976] 349/55).

[41] A.N. Nesmeyanov, N.E. Kolobova, L.V. Goncharenko, K.N. A'nisimov (Izv. Akad. Nauk SSSR Ser. Khim. **1976** 153/9; Bull. Acad. Sci. USSR Div. Chem. Sci. **1976** 142/6). — [42] A.N. Nesmeyanov, E.G. Perevalova, L.I. Leont'eva. E.V. Shumilina (Izv. Akad. Nauk SSSR Ser. Khim. **1977** 1142/6; Bull. Acad. Sci. USSR Div. Chem. Sci. **1977** 1048/52). — [43] D.L. Reger, D.J. Fauth, M.D. Dukes (Syn. Reactiv. Inorg. Metal-Org. Chem. **7** [1977] 151/5). — [44] A. Davison, J.P. Solar (J. Organometal. Chem. **155** [1978] C8/C12). — [45] E.J. Kuhlmann (Diss. Univ. Cincinnati 1978; Diss. Abstr. Intern. B **39** [1978] 2296).

[46] B.E. Boland, S.A. Fam, R.P. Hughes (J. Organometal. Chem. **172** [1979] C29/C32). — [47] A.R. Cutler (J. Am. Chem. Soc. **101** [1979] 604/6). — [48] E.J. Kuhlmann, J.J. Alexander (J. Organometal. Chem. **174** [1979] 81/7). — [49] E.J. Kuhlmann, J.J. Alexander (Inorg. Chim. Acta **34** [1979] L193/L195). — [50] E.J. Kuhlmann, J.J. Alexander (Inorg. Chim. Acta **34** [1979] 197/209).

[51] M.I. Bruce, M.Z. Iqbal, F.G.A. Stone (J. Organometal. Chem. **20** [1969] 161/8). — [52] H. Brunner, H. Vogt (Z. Naturforsch. **33b** [1978] 1231/4). — [53] J. Benaïm, J.Y. Merour, J.L. Roustan (Tetrahedron Letters **1971** 983/6).

1.5.2.2.12.3.2   Compounds with $^1L = COOR$ or CH_2COOR

$C_5H_5Fe(CO)(P(C_6H_5)_3)COOH$ is precipitated from an equimolar mixture of $[C_5H_5Fe(CO)_2(P(C_6H_5)_3)]Cl$ and KOH in benzene/water by the addition of pentane. The yellow crystalline powder melts at 130 °C with decomposition. IR spectrum (in cm^{-1}): 1565 (C=O) and 1930 (CO) in Nujol and 1935 (CO) in C_6H_6. In solvents having a low polarity the compound does not dissociate (according to the IR spectrum); by contrast, in formamide it ionizes to give $[C_5H_5Fe(CO)(P(C_6H_5)_3)CO]^+$ and OH^-. In solvents of intermediate polarity, say, acetone/formamide (15%), both the undissociated and ionic forms are present in comparable amounts.

The compound is clearly amphoteric: With excess KOH solution, the salt $[C_5H_5Fe(CO)(P(C_6H_5)_3)COO]K$ forms which gives back the starting material with dilute aqueous HCl (pH = 2.0). With enough HCl to lower the pH below 2.0 $[C_5H_5Fe(CO)(P(C_6H_5)_3)CO]Cl$ forms. Heating $C_5H_5Fe(CO)(P(C_6H_5)_3)COOH$ in benzene decarboxylates it to give $C_5H_5Fe(CO)(P(C_6H_5)_3)H$ [13].

$C_5H_5Fe(CO)(P(C_6H_5)_3)COOCH_3$ is obtained by the reaction of equimolecular amounts of $[C_5H_5Fe(CO)_2(P(C_6H_5)_3)]Cl$ and $NaOCH_3$ in methanol and evaporation of the solvent. It is a crystalline yellow powder [13]. It is formed by the reaction of $[C_5H_5Fe(CO)(P(C_6H_5)_3)CO]I$ with CH_3OD [5], and it can also be obtained by the reaction of $C_5H_5Fe(CO)(P(C_6H_5)_3)COOC_{10}H_{19}$ ($C_{10}H_{19}$=menthyl) with methanol at room temperature (60% yield). The compound decomposes above 90 °C. ^1H NMR in CS_2, δ in ppm: 2.84 (s, CH_3), 4.3 (d, C_5H_5, J(P,H)=1.5 Hz). IR (ν in cm^{-1}): 1630 (C=O) and 1942 (CO) in KBr [6] or 1605 (C=O) and 1935 (CO) in C_6H_6. The compound behaves in solvents of various polarities just as the previous compound [13]. It is monomeric [6].

$C_5H_5Fe(CO)(P(C_6H_5)_3)COOCH(CH_3)_2$ shows in solvents an amphoteric behavior like the methyl ester mentioned above. In acetone the CH_3 groups in the $CH(CH_3)_2$ unit are not equivalent for the undissociated molecule and show doublets in the ^1H NMR spectrum at δ=0.90 and 0.82 ppm (J=6.0 Hz). In CD_3COCD_3 with 20% formamide, the ^1H NMR spectrum shows the doublet δ=0.62 and 0.82 ppm (J= 6.0 Hz) for the CH_3 groups of the undissociated form and the singlet δ=1.05 ppm for the dissociated form. The C_5H_5 resonance occurs at δ=4.50 ppm for the undissociated form and 5.50 ppm for the dissociated form with equal intensities. The IR spectrum (in cm^{-1}) for the CD_3COCD_3 with 20% formamide solution shows 1605 (C=O) and 1935 (CO) for the undissociated and 2030 and 2080 for the dissociated forms [13].

$C_5H_5Fe(CO)(P(C_6H_5)_3)COOC_{10}H_{19}$ ($C_{10}H_{19}$=menthyl) is obtained as a mixture of stereoisomers by the reaction of $[C_5H_5Fe(CO)(P(C_6H_5)_3)CO]PF_6$ with sodium mentholate in ether in the absence of light. The ether is removed, and the glassy yellowish orange mass extracted into benzene. It is filtered off from the residues of $[(C_5H_5)Fe(CO)(P(C_6H_5)_3)CO]PF_6$ and $NaPF_6$, and the filtrate evaporated. The product left behind is crystallized by adding hexane. Through repeated elution of the mixture of diastereomers with pentane, $(+)-C_5H_5Fe(CO)(P(C_6H_5)_3)COOC_{10}H_{19}$ (up to 30% yield) is obtained. Specific rotation $[\alpha]_{546}^{20}=+72°$ (10^{-3} M, C_6H_6). $(-)-C_5H_5Fe$-$(CO)(P(C_6H_5)_3)COOC_{10}H_{19}$ is obtained from the residue by extraction with boiling pentane (50% yield). It has a specific rotation of $[\alpha]_{546}^{20}=-120°$ (10^{-3} M, C_6H_6). Both stereoisomers are yellowish orange [6]. Specific rotation in pentane solution as function of wavelength [2]:

in nm	436	546	579	589
$(+)-C_5H_5Fe(CO)(P(C_6H_5)_3)COOC_{10}H\mu=_{19}$	−1450°	+ 70°	+35°	+30°
$(-)-C_5H_5Fe(CO)(P(C_6H_5)_3)COOC_{10}H_{19}$	+1550°	−120°	−80°	−75°

The circular dichroism and rotation–dispersion spectra are presented graphically in [4]. Specific rotation $[\alpha]_{546}$ in benzene as function of the temperature:

t in °C	15	20	25	30	35	40
$(+)-C_5H_5Fe(CO)(P(C_6H_5)_3)-$ COOC$_{10}H_{19}$	+77°	+72°	+68°	+64°	+60°	+57°
$(-)-C_5H_5Fe(CO)(P(C_6H_5)_3)-$ COOC$_{10}H_{19}$	−127°	−120°	−116°	−112°	−108°	−104°

^1H NMR spectrum of both stereoisomers in CS_2: δ=4.35 (d, C_5H_5, J(P,H)=1 Hz) ppm. IR spectrum of both stereoisomers (KBr): 1615 cm^{-1} (C=O), 1938 (CO) cm^{-1} [6]. In comparison with $C_5H_5Fe(CO)(P(C_6H_5)_3)CH_2COOC_{10}H_{19}$ (the next compound) and $C_5H_5Fe(CO)(P(C_6H_5)_3)CH_2OC_{10}H_{19}$ (1.5.2.2.12.3.4, Table 18, No. 36), the (+) form has the (+)(R) configuration [12]. The (−) form crystals are rhombic,

References on p. 172

space group $P2_12_12_1 - D_2^4$, with lattice constants a=11.22 (4), b=14.817 (7), and c=18.958 (3) Å; Z=4 gives D_c=1.25, while D_m=1.25 (2) g·cm^{-3}. The configuration of the (−) form is shown in **Fig. 14**. The C_5H_5 ring is planar. The C_6H_5 rings are tilted towards one another. The short Fe-C(1) distance is unusual, and the Fe-COO group shows strong bond angle deviations; the Fe-C(2) bond (1.825 Å) is also short [14], preliminary communication [10].

Fig. 14

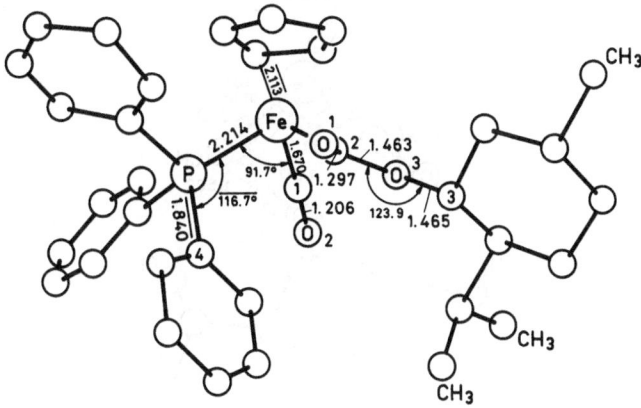

Molecular structure of $(-)-C_5H_5Fe(CO)(P(C_6H_5)_3)COOC_{10}H_9$ [14].

Other bond angles (°):

Fe-C(1)-O(2)	175.3(6)	C(1)-Fe-C(2)	90.9(3)
Fe-C(2)-O(1)	132.1(5)	P-Fe-C(2)	89.9(2)
Fe-C(2)-O(3)	117.7(4)	O(1)-C(2)-O(3)	108.8(5)

Both solid stereoisomers are stable in air [2] but decompose above 120 °C. Both are monomeric in benzene solution [6]. There is no configuration inversion in solution at room temperature, a contrast to the isoelectronic (+)- and (−)-$C_5H_5Mn(NO)(P(C_6H_5)_3)COOC_{10}H_{19}$ [1, 2]. The compound is cleaved in benzene at room temperature on treatment with iodine to give $[C_5H_5Fe(CO)_2P(C_6H_5)_3]I$ and traces of $C_5H_5Fe(CO)(P(C_6H_5)_3)I$ [9]. The (+) form reacts with $LiCH_3$ to give $(-)-C_5H_5Fe(CO)(P(C_6H_5)_3)COCH_3$, and the (−) form gives $(+)-C_5H_5Fe(CO)-(P(C_6H_5)_3)COCH_3$ (configuration inversion) [3, 4, 8, 9]. $C_5H_5Fe(CO)(P(C_6H_5)_3)-COOC_{10}H_{19}$ is converted in methanol to racemic $C_5H_5Fe(CO)(P(C_6H_5)_3)-COOCH_3$ [6].

$C_5H_5Fe(CO)(P(C_6H_5)_3)CH_2COOC_{10}H_{19}$ ($C_{10}H_{19}$=menthyl) is prepared by the addition of $Na[C_5H_5Fe(CO)_2]$ to (−)-menthol bromoacetate followed by reaction with $P(C_6H_5)_3$ through photolysis. Repeated recrystallization gives $(+)-C_5H_5Fe-(CO)(P(C_6H_5)_3)CH_2COOC_{10}H_{19}$, m.p. 129 to 130 °C, specific rotation $[\alpha]_{578}^{25}=+330°$ and $(-)-C_5H_5Fe(CO)(P(C_6H_5)_3)CH_2COOC_{10}H_{19}$, m.p. 115 to 116.5 °C, specific rotation $[\alpha]_{578}^{25}=-309°$. ^{1}H NMR spectrum (in CDCl$_3$): δ=4.26 (C_5H_5, J=1 Hz) ppm. The separation of the stereoisomers after addition of Eu(fod)$_3$ (Formula I, p. 104) was followed with ^{1}H NMR [7]. For plots of this spectrum and also for circular dichroism spectra, see [7].

Single crystal X-ray pattern (crystals repeated recrystallized) of the (+) form revealed the compound to be triclinic, space group P_1-C_1 with a=7.660(1),

References on p. 172

$b = 13.806(1)$, $c = 15.948(1)$ Å, $\alpha = 108.23(1)°$, $\beta = 88.62(1)°$, $\gamma = 95.06(1)°$; $Z = 2$, $D_c = 1.22$, $D_m = 1.20$ g·cm^{-3} (flotation). The molecular structure is given in **Fig. 15**. The structure and the circular dichroism spectra (plots in [12]), may be interpreted to imply an R configuration at the Fe atom [12].

Fig. 15

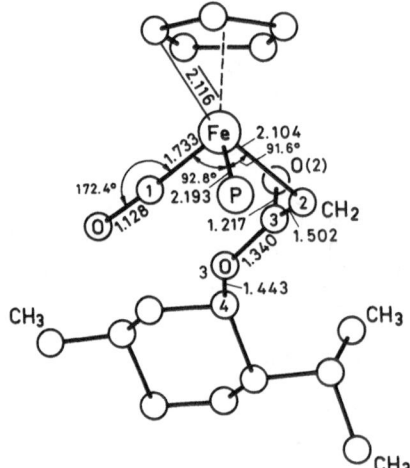

Molecular structure of $(+)$-$C_5H_5Fe(CO)(P(C_6H_5)_3)CH_2COOC_{10}H_{19}$ [12]. The phenyl groups are omitted.

Other bond angles (°):

C(1)–Fe–C(2)	100.0(6)	C(2)–C(3)–O(3)	113.7(10)
Fe–C(2)–C(3)	109.5(8)	C(3)–O(3)–C(4)	119.2(10)
C(2)–C(3)–O(2)	124.8(12)	O(2)–C(3)–O(3)	121.5(12)

The compound is decomposed by electrophilic reagents in benzene solution. With I_2, ICl, or HgI_2 (at 5 °C, C_6H_6), $C_5H_5Fe(CO)(P(C_6H_5)_3)I$ and $C_{10}H_{19}OOCCH_2X$ with X = I, Cl, or HgI form. In these reactions, the configuration at the Fe atom is not changed [11]. $C_5H_5Fe(CO)(P(C_6H_5)_3)CH_2COOC_{10}H_{19}$ reacts with refluxing SO_2 over several hours with insertion of SO_2, giving $C_5H_5Fe(CO)(P(C_6H_5)_3)SO_2CH_2$-$COOC_{10}H_{19}$ [7]. When $C_5H_5Fe(CO)(P(C_6H_5)_3)CH_2COOC_{10}H_{19}$ is partially oxidized by $FeCl_3·6H_2O$ in THF, the recovered compound showed no loss of optical purity. In methanol, the ester group is exchanged. Reaction with CF_3COOH gives C_5H_5Fe-$(CO)(P(C_6H_5)_3)OOCCF_3$ of ~50% enantiomeric excess and $CH_3COOC_{10}H_{19}$. The compound does not react with $FeCl_3·6H_2O$ in THF at 25 °C [11].

$C_5H_5Fe(CO)(P(C_6H_5)_3)CH_2COOCH_3$ is formed as the racemic modification from $C_5H_5Fe(CO)(P(C_6H_5)_3)CH_2COOC_{10}H_{19}$ by exchange in methanol to which a trace of H_2SO_4 has been added [11].

References:

[1] H. Brunner, H.-D. Schindler, E. Schmidt (Proc. 13th Intern. Conf. Coord. Chem., Cracow, Pol., 1970, Vol. 2, pp. 330/1). − [2] H. Brunner, E. Schmidt (J. Organometal. Chem. **21** [1970] P53/P54). − [3] H. Brunner (Chimia [Aarau] **25** [1971] 284/6). − [4] H. Brunner, E. Schmidt (J. Organometal. Chem. **36** [1972] C18/C22). − [5] J. Ellermann, H. Behrens, H. Krohberger (J. Organometal. Chem. **46** [1972] 119/38).

[6] H. Brunner, E. Schmidt (J. Organometal. Chem. **50** [1973] 219/25). − [7] T.C. Flood, D.L. Miles (J. Am. Chem. Soc. **95** [1973] 6460/2). − [8] H. Brunner (Ann. N.Y. Acad. Sci. **239** [1974] 213/24). − [9] H. Brunner, J. Strutz (Z. Naturforsch. **29b** [1974] 446/7). − [10] M.G. Reisner, I. Bernal, H. Brunner, M. Muschiol (Angew. Chem. **88** [1976] 847/8; Angew. Chem. Intern. Ed. Engl. **15** [1976] 776).

[11] T.C. Flood, D.L. Miles (J. Organometal. Chem. **127** [1977] 33/44). − [12] C.-K. Chou, D.L. Miles, R. Bau, T.C. Flood (J. Am. Chem. Soc. **100** [1978] 7271/8). − [13] N. Grice, S.C. Kao, R. Pettit (J. Am. Chem. Soc. **101** [1979] 1627/8). − [14] G.M. Reisner, I. Bernal, H. Brunner, M. Muschiol (Inorg. Chem. **17** [1978] 783).

1.5.2.2.12.3.3 Compounds with $^1L = CONH_nR_{2-n}$

$C_5H_5Fe(CO)(P(C_6H_5)_3)CONH_2$ is obtained when $[C_5H_5Fe(CO)_2P(C_6H_5)_3]I$ is reacted with liquid NH_3 at $-33\,°C$ for 10 min. Excess NH_3 is distilled off at $-50\,°C$ and the compound is extracted from the reaction mixture with C_6H_6 or CH_2Cl_2 (in which it is quite soluble).

IR spectrum: 1910 and 1920 (CO), $\rho(NH_2)$ 1235, $\delta(NH_2)$ 1556, $\nu(C=O$ and $C-N)$ 1561, 1600, and 1655 cm^{-1} in Nujol; 1919 (CO), $\delta(NH_2)$ 1555, and $\nu(C=O$ and $C-N)$ 1583 cm^{-1} in CH_2Cl_2.

The compound decomposes above 70 °C. The following fragments were observed in the mass spectrum (relative intensities): $[M]^+$ (3), $[M-CO]^+$ (9), $[M-2CO]^+$ (40), $[C_5H_5Fe(P(C_6H_5)_3)]^+$ (38), $[P(C_6H_5)_3]^+$ (100), and $[C_5H_5Fe]^+$ (17). The compound is insoluble in liquid NH_3. With CD_3COCD_3, it forms the next compound and small quantities of other CO-containing complexes. The reaction with CH_3OD gives $C_5H_5Fe(CO)(P(C_6H_5)_3)COOCH_3$. The compound decomposes in $CDCl_3$ with loss of the $CONH_2$ group [2].

$C_5H_5Fe(CO)(P(C_6H_5)_3)COND_2$ was obtained from the previous complex and CD_3COCD_3. IR spectrum: $\rho(ND_2)$ 1030 and $\delta(ND_2)$ 1185 cm^{-1} [2].

$C_5H_5Fe(CO)(P(C_6H_5)_3)CONHCH_3$ is formed by introducing CH_3NH_2 into a suspension of $[C_5H_5Fe(CO)_2P(C_6H_5)_3]PF_6$ in ether saturated with N_2. After 30 min the mixture is filtered and the filtrate evaporated. The crude product is recrystallized from ether/hexane (50% yield).

The compound melts at 114 °C. IR spectrum (CH_2Cl_2): 1915 (CO) and 1558 (C=O) cm^{-1}. The reaction with HCl in ether gives $[C_5H_5Fe(CO)_2P(C_6H_5)_3]Cl$ [1].

$C_5H_5Fe(CO)(P(C_6H_5)_3)CONHC_2H_5$ is prepared like the previous complex by use of $C_2H_5NH_2$. IR spectrum (CH_2Cl_2): 1911 (CO) and 1560 (C=O) cm^{-1} [1].

$C_5H_5Fe(CO)(P(C_6H_5)_3)CONHC_3H_7-i$ is prepared like the previous compounds by use of $i-C_3H_7NH_2$. IR spectrum (C_6H_{12}): 1924 (CO) and 1585 (C=O) cm^{-1} [1].

References:

[1] L. Busetto, R.J. Angelici (Inorg. Chim. Acta **2** [1968] 391/4). − [2] J. Ellerman, H. Behrens, H. Krohberger (J. Organometal. Chem. **46** [1972] 119/38).

1.5.2.2.12.3.4 Compounds with $^1L = $ Alkyl and Substituted Alkyl

The absolute configuration of some of these optically active compounds has been determined by comparison of their circular dichroism spectra with those of C_5H_5-

Fe(CO)(P(C₆H₅)₃)R, where R=COOC₁₀H₁₉ (previous section) and R=CH₂OC₁₀H₁₉ (this section, No. 36) whose molecular structures were known from X-ray diffraction. The following (+) forms ([α]₅₇₈) have an S configuration about Fe [42]: R=CH₃ (No. 1), R=n-C₃H₇ (No. 12), R=i-C₄H₉ (No. 20) [43], R=CH₂C₃H₅-cyclo (No. 22), R=CH₂OC₁₀H₁₉ (No. 36), and R=CH₂SO₂OC₁₀H₁₉ (No. 42). The (+) forms for R=CH₂Cl (No. 2), R=CH₂Br (No. 3), R=CH₂I (No. 4), and R=CH₂C₆H₅ (No. 25) have an R configuration [42].

The compounds in Table 18 have been prepared as follows.

Method I: C₅H₅Fe(CO)₂R is reacted with P(C₆H₅)₃.

a. C₅H₅Fe(CO)₂CH₃ and P(C₆H₅)₃ are refluxed in a solvent (temperature, reaction time, and yield): heptane (99 °C, 42 h, 45%), dioxane (102 °C, 41 h, 3%), or hexane (69 °C, 48 h, 2%). After solvent removal, No. 1 was purified chromatographically [12].

b. C₅H₅Fe(CO)₂R and P(C₆H₅)₃ are UV irradiated in a solvent at 20 to 30 °C, and after solvent removing the required compound is isolated from the residue by chromatography. Nos. 5, 6, 9, 15, 29, and 35 are purified by recrystallization. Petroleum ether is used for No. 1 (6 h [2], 3 h [12], or 20 h [1]), No. 6 (3 h) [12, 13], No. 12 (3 h) [12], No. 25 (3 h) [8, 12, 23, 29], No. 28 (3 h) [29], and No. 38 (3 h) [9, 29]. Hexane is used for No. 5 (12 h) [5], No. 9 (15 h) [5], No. 15 (13 h) [5], and Nos. 38 and 39 (both 7 h) [27]. Benzene is used for No. 26 (1 h) [22], No. 28 (2.5 h) [45], and No. 36 (2 h) [26, 34]. Compound No. 26, prepared from nearly pure optically active substances, is obtained as a pure optically active isomer [22]. Benzene/pentane (1:3) is used for Nos. 29 [20, 34] and 35 [44]. The first-order rate constant for the formation of No. 1 in C₆H₆ or CD₃CN is $k = (11.8$ to $12.6) \times 10^{-4}$ s^{-1} [38].

c. Photolysis of C₅H₅Fe(CO)₂COR and P(C₆H₅)₃ in C₆H₆ as described in Method I b gives C₅H₅Fe(CO)(P(C₆H₅)₃)R [22].

Method II: Decarbonylation of C₅H₅Fe(CO)(P(C₆H₅)₃)COR.

a. Decarbonylation of C₅H₅Fe(CO)(P(C₆H₅)₃)COCH₃ in petroleum ether (90 to 100 °C, 20 h) [2], toluene (reflux, 20 h) [19], or heptane (99 °C, 20 h) [12, 13].

b. Decarbonylation of compounds of the C₅H₅Fe(CO)(P(C₆H₅)₃)COR type (R=CH₃, C₂H₅, or CH₂CH(CH₃)C₆H₅) by irradiation, usually at room temperature. No. 1: 0.5 h in hexane [19], 0.5 h in benzene/pentane (1:4) [21], 1 h in petroleum ether at 30 to 35 °C, or at 40 to 45 °C in benzene [12, 13]. No. 6: 1 h in petroleum ether at 30 to 35 °C [12, 13]. No. 26: 2 h in benzene, to give a racemic mixture. The enantiomers are separated as described for No. 6, Table 16, in 1.5.2.2.12.3.1, p. 160 [22].

Method III: C₅H₅Fe(CO)(P(C₆H₅)₃)X (X=H, Br, or I) is reacted in a solvent with CF₂=CF₂ for X=H or with RLi, RMgCl, or RMgBr for X=Br or I. Nos. 1, 6, 7, 12, 16 [32, 36], 17, 18 [32, 39], 19, 20 [32, 36], and 21, 23, 24 [32, 39] are obtained from C₅H₅Fe(CO)(P(C₆H₅)₃)I and RLi. The reaction mixture containing a 10% excess of RLi in THF is

References on p. 195

allowed to warm from $-78\,°C$ to room temperature, the solvent removed, the mixture extracted with benzene/hexane (5:1), the residue from the evaporated filtrate dissolved in acetone/water (4:1), the solution filtered hot and then cooled to $0\,°C$ to obtain crystals. Nos. 23 and 24 are purified chromatographically. No. 8 is obtained from the compound with $X=I$ and $LiCH_2CN$ [28]. No. 10 is obtained from that with $X=H$ and CF_2CF_2 in tetrahydrofuran at $25\,°C$ [12]. No. 13 is formed from that with $X=Br$ and $RMgCl$ in ether at $0\,°C$ (30 min). It is precipitated with dioxane, filtered, washed with water, and dried [18]. No. 40 is formed from that with $X=I$ and $RMgBr$ at $35\,°C$ in benzene/ether. Only one diastereoisomer is formed, as was revealed by SO_2 insertion [30, 41].

Method IV: a. $C_5H_5Fe(CO)(P(C_6H_5)_3)CH_2Cl$ (No. 2) generally forms optically pure stereoisomers when the Cl atom is substituted by H or R groups: With LiH in THF or with $NaBH_4$ in $CH_3OCH_2CH_2OCH_3$ at $0\,°C$, the mixture is allowed to warm up to $25\,°C$, to give No. 1 [26, 34], with CH_3MgCl or CH_3Li in THF [26], or with CH_3Li in ether/THF to give No. 6 [34], with KCN and $[N(C_4H_9\text{-}n)_4]Br$ catalyst in benzene, to give No. 8 (chromatographic purification) [34], with a 90% excess of CH_3CH_2MgBr (or $(CH_3)_2CHMgBr$) ether/THF to give No. 12 (No. 20) [26, 34], with cyclo-C_3H_5Li in THF to give No. 22 [26, 34], with C_6H_5Li in ether/benzene/THF to give No. 25, or with an 80% excess of C_6H_5MgBr in THF to give a particularly high yield of No. 25, with CH_3OH/CH_3COOK to give No. 29 (chromatographic purification) [34], and with $NaOC_{10}H_{19}$ ($C_{10}H_{19}=$menthyl) in ether to give No. 36. In this last case the diastereomers are separated by fractional crystallization from pentane [20].

b. $C_5H_5Fe(CO)(P(C_6H_5)_3)CH_2Br$ also forms optically pure stereoisomers. No. 1 is obtained with $NaBH_4$ in methanol. The mixture is allowed to warm at room temperature, and the product is purified by chromatography [34].

Method V: $[C_5H_5Fe(CO)(P(C_6H_5)_3)C(OC_2H_5)CH_3]BF_4$ is deprotonated with $C_2H_5O^-$ in ethanol to give No. 30 [14].

Method VI: Reduction of $[C_5H_5Fe(CO)(P(C_6H_5)_3)^1L]BF_4$ by $NaBH_4$ or another such reducing agent in ethanol. Compounds No. 6 and 32 result in almost equal amounts from reduction of the starting material with $^1L=C(OC_2H_5)CH_3$ by $NaBH_4$ in ethanol [14], also see [11]. Nos. 7 and 33 are prepared from this same starting material by reduction with $NaBD_4$ [14]. Compound No. 32 is also obtained from $C_5H_5Fe(CO)(P(C_6H_5)_3)C(OC_2H_5)=CH_2$ by reduction with H_2/PtO or B_2H_6 in ethanol [14]. Compound No. 14 is obtained from the starting material with $R=CH_2=CH_2$ with KCN in ethanol [28]. No. 44 is formed by reduction of $[C_5H_5Fe(CO)(P(C_6H_5)_3)CH_2=C=CH_2]BF_4$ with $NaBH_4$ [50]. Nos. 45 and 48 are obtained by reduction of $[C_5H_5Fe(CO)(P(C_6H_5)_3)R]BF_4$ ($R=CH_2=C=CH_2$ or $C_2H_5C≡CC_2H_5$) with $Li[BH(C_4H_9\text{-}s)_3]$ in THF at 0 or $-78\,°C$. The suspensions are warmed to room temperature, the solvent evaporated, and the residue chromatographed with benzene/hexane [50]. Nos. 46 and 47 are obtained at $-78\,°C$ from $[C_5H_5Fe(CO)(P(C_6H_5)_3)CH_3C≡CCH_3]BF_4$ in THF with

NaH for No. 46, with $Li[Cu(CH_3)_2]$, with $Li[Cu(C_6H_5)_2]$, with $NaC\equiv$ CH, with $Na[CH(COOC_2H_5)]$, with $NaOC_2H_5$, with $NaSC_4H_9$, with $NaSC_6H_5$, and with KCN for all the derivatives collected under No. 47. The reaction mixture was allowed to warm to room temperature, the solvent evaporated, the residue extracted with CH_2Cl_2, and this solution chromatographed on Al_2O_3 with hexane/CH_2Cl_2 [51].

Method VII: $C_5H_5Fe(CO)(P(C_6H_5)_3)CH_2OR$, $R=CH_3$ or $C_{10}H_{19}$ (menthyl), reacts with HX (X=Cl, Br, or I) in ether at 0 °C to give C_5H_5Fe-$(CO)(P(C_6H_5)_3)CH_2X$ and ROH with inversion at the Fe atom: The $(+)(S)$ and $(-)(R)$ menthyl compounds give the $(+)(R)$ and $(-)(S)$ forms, respectively [20, 26, 34].

Method VIII: No. 11 is formed by splitting HF off from $C_5H_5Fe(CO)$-$(P(C_6H_5)_3)CF_2CHF_2$ [12].

Method IX: No. 27 is obtained rapidly from $C_5H_5Fe(CO)(P(C_6H_5)_3)C\equiv CC_6H_5$ by reaction with $(CN)_2C=C(CN)_2$ in CH_2Cl_2 [47].

Method X: Nos. 29 and 31 form along with $[C_5H_5Fe(CO)_2P(C_6H_5)_3]^+$ when equimolar amounts of iodine and $[C_5H_5Fe(CO)(P(C_6H_5)_3)CHOR]PF_6$, $R=CH_3$ or C_2H_5, are reacted in CH_2Cl_2 [46].

Method XI: $[C_5H_5Fe(CO)(P(C_6H_5)_3)CH_2=CH_2]BF_4\cdot C_6H_6$ and $Na[PO(OCH_3)_2]$ form No. 37 on stirring for 15 h in THF [40].

Method XII: $C_5H_5Fe(CO)(P(C_6H_5)_3)CH_2OR$ (No. 29, $R=CH_3$ and No. 36, $R=C_{10}H_{19}$) react with SO_2 in liquid SO_2 at about -10 °C for 2 h to form Nos. 41 and 42. The residue after evaporation of the SO_2 is taken up in $CHCl_3$ and chromatographed [34].

Method XIII: $C_5H_5Fe(CO)(P(C_6H_5)_3)_3COCH_3$ is reduced by passing B_2H_6 through its solution in C_6D_6 or THF, or it is treated with $BH_3\cdot THF$ in THF. The reaction is followed (20 °C, 2 min) by ^1H NMR [33].

Explanations for Table 18: For the assignment of the NMR resonances of the ^1L ligands, the labelling of the C atoms follows the longest carbon chain; carbon atom C-1 is bonded to Fe. Coupling constants J of the ^1H NMR spectra indicate P,H coupling if not otherwise indicated. Not designated IR bands in the 1870 to 1970 cm^{-1} range (usually close to 1900 cm^{-1}) are $\nu(CO)$. The compounds are crystalline if not stated otherwise.

References on p. 195

Table 18
$C_5H_5Fe(CO)(P(C_6H_5)_3)^1L$ Compounds with ^1L = Alkyl and Substituted Alkyl.
Further information on numbers preceded by an asterisk is given at the end of the table.
For abbreviations and dimensions, see p. 375.

No.	^1L ligand method of preparation (yield in %)	properties and remarks explanations on p. 176	Ref.
*1	CH_3 Ia (2), Ia (3), Ia (45), Ib (29.1), Ib (67), IIa (27), IIa (~40), IIb (51), IIb (60 to 65), III	red, m.p. 152°, 157 to 158° ^1H NMR: 0.35 (d, CH_3, J=6), 4.18 (C_5H_5) in C_6H_6, −0.02 to −0.17 (d, CH_3, J=6 to 6.5), 4.20 to 4.25 (C_5H_5), 7.33 (m, C_6H_5) in $CDCl_3$, CD_3CN, or C_6H_5CN $^{57}Fe-\gamma$: δ=0.37, Δ=1.83 IR ($CHCl_3$, CH_2Cl_2): 1905 to 1906	[1, 2, 12, 17, 19, 21, 22, 32, 36, 38]
	(+)(S) form IVa (71), IVb (33)	m.p. 173 to 174°, $[\alpha]_{578}^{25}$= +100° (c=0.002, C_6H_6) [26], corrected by [34]: +50° (c= 0.1, C_6H_6), reproducible value $[\alpha]_{578}^{25}$= +91° (c=0.1, CH_2Cl_2, pure) [34] ^1H NMR ($CDCl_3$): −0.18 (d, CH_3, J=6.5), 4.29 (d, C_5H_5, J~1), 7.43 (m, C_6H_5) ^{13}C NMR: −22.6 (d, CH_3, J(P,C)$\cong$20)	[26, 34]
	(−)(R) form V	$[\alpha]_{436}^{27}$= +140° (C_6H_6/C_5H_{12}?)	[21]
*2	CH_2Cl VII (70 to 75)	light orange, m.p. 123 to 124°	[26, 34]
	(+)(R) form VII	m.p. 115° (dec.), $[\alpha]_{578}^{25}$= +383° (c=0.1, C_6H_6)	[34]
	(−)(S) form VII (71)	m.p. 115° (dec.), $[\alpha]_{578}^{25}$= −380° (c=0.1, C_6H_6) ^1H NMR (C_6D_6): 4.33 (d, C_5H_5, J=1), 4.4 (m, CH_2), 6.9 (m, C_6H_5) IR (C_6H_6): 1930	[34]
*3	CH_2Br VII (70 to 80)	orange, m.p. 124 to 128° (dec.)	[34]
	(+)(R) form VII (54)	m.p. 129°, $[\alpha]_{578}^{25}$= +280° (c=0.1, $C_2H_5OOCCH_3$) ^1H NMR ($CDCl_3$): 4.22 (d, CH_2), 4.47 (s, C_5H_5), 7.4 (m, C_6H_5) IR (C_6H_6): 1930	[34]
*4	CH_2I (−)(S) form VII	red	[20]

Table 18 [continued]

No.	^{1}L ligand method of preparation (yield in %)	properties and remarks explanations on p. 176	Ref.
*5	CF$_3$ Ib (78)	yellow, m.p. 165 to 167° ^{1}H NMR (CDCl$_3$): 4.50 (d, C$_5$H$_5$, J=1.5), ~7.5 (m, C$_6$H$_5$) ^{19}F NMR: −13.7 (d, J=1.6) in CH$_2$Cl$_2$, −13.5 (d, J(P,F)=2.9) in CH$_2$Cl$_2$/CFCl$_3$ IR (C$_6$H$_{12}$): 1944 [25] or 1968 [5]	[5, 15, 25]
*6	C$_2$H$_5$ Ib (40), IIb, III (50 to 60), IV VI, XIII	orange red, m.p. 140 to 142°, 151.5 to 152°, 155°, dec. 140° ^{1}H NMR (C$_6$D$_6$): 1.08 (m, CH$_2$, J=12), 1.59 (m, CH$_3$), 1.88 (m, CH$_2$, J=2), 4.12 (C$_5$H$_5$, J=1.0) IR: 1912 in C$_6$H$_{14}$, 1901 in CH$_2$Cl$_2$	[12 to 14, 32 to 34]
	(+)(S) form IIb (43), V	$[\alpha]^{27}_{437}$ = −256°	[21]
	(−)(R) form IIb, IVa (67 to 78)	m.p. 140 to 142° (dec.), $[\alpha]^{27}_{437}$ = +110° (43% optically pure), $[\alpha]^{25}_{578}$ = +89°, +95° (~1 mg/mL, C$_6$H$_6$)	[21, 26, 34]
*7	CD$_2$CH$_3$ III (50 to 60), VI	−	[14, 32]
*8	CH$_2$CN III (10), IVa (33)	red, m.p. 164 to 165°, 170 to 172° ^{1}H NMR (CDCl$_3$): 0.61 (CH$_2$), 0.72 (CH$_2$), 4.40 (d, C$_5$H$_5$, J=1.2), 7.33 (C$_6$H$_5$), 7.42 (C$_6$H$_5$) ^{13}C NMR (CH$_2$Cl$_2$): 15.11 (d, CH$_2$, J=16.9), 85.11 (s, C$_5$H$_5$), 128.50 (d, C-2 in C$_6$H$_5$, J=9.6), 130.16 (d, C-4 in C$_6$H$_5$, J=2.1), 133.28 (d, C-3 in C$_6$H$_5$, J=9.8), 135.35 (d, C$_6$H$_5$, J(P,C)=41.3), 221.37 (d, CO, J=30.2) IR: 1928 (CO), 2190 (CN) in C$_6$H$_6$, 1930 (CO), 2188 (CN) in Kel-F	[28, 34]
*9	C$_2$F$_5$ Ib (44)	orange crystals, m.p. 143 to 144° ^{1}H NMR (CDCl$_3$): 4.49 (d, C$_5$H$_5$, J=1.4), ~7.5 (m) ^{19}F NMR (CH$_2$Cl$_2$): 59.9 (m, CF$_2$, J(P,F)= 1.4), 63.3 (m, CF$_2$, J(P,F)=43.6), 81.7 (t, CF$_3$, ^{3}J(F,F)=2.1) at room temp., 60.2 (J(P,F)=0.04), 65.3 (J(P,F)=41.8) at 193 K IR (C$_6$H$_{12}$): 1966	[5, 25]

References on p. 195

Table 18 [continued]

No.	^{1}L ligand method of preparation (yield in %)	properties and remarks explanations on p. 176	Ref.
*10	CF$_2$CHF$_2$ III	yellow solid	[12]
11	CF=CF$_2$ VIII	light yellow powder ^{1}H NMR (CDCl$_3$?): 4.50 (d, C$_5$H$_5$, J=1.5), 7.47 (m, C$_6$H$_5$) IR (CH$_2$Cl$_2$?): 1946	[12]
*12	CH$_2$CH$_2$CH$_3$ Ib (40), III (50 to 60)	red ^{1}H NMR (CDCl$_3$): 0.4 to 1.6 (m, C$_3$H$_7$), 4.1 (d, C$_5$H$_5$, J∼1), 7.4 (m, C$_6$H$_5$) IR (CH$_2$Cl$_2$): 1901	[12, 32]
	(+)(S) form IVa (56)	m.p. 117.5 to 118°, $[\alpha]_{578}^{25}$ = +93° (∼1 mg/mL, C$_6$H$_6$, optically pure)	[26, 34]
*13	CH$_2$CH=CH$_2$ III (90)	m.p. 104 to 105° ^{1}H NMR (C$_6$D$_6$): 1.8 (m, H-2), 2.3 (m, H-1), 4.04 (s, C$_5$H$_5$), 4.74 (d, H-3 cis), 4.89 (d, H-3 trans), 6.48 (q, H-2, J(H-2,3cis)=10, J(H-2,3trans)=17) IR (CS$_2$): 1917	[18]
*14	CH$_2$CH$_2$CN VIa (78)	red, m.p. 139 to 140° ^{1}H NMR (CDCl$_3$): 0.69, 1.67 (d's, H-1), 2.24 (m, H-2), 4.28 (d, C$_5$H$_5$, J=1.3), 7.32, 7.40 (C$_6$H$_5$) ^{13}C NMR (CH$_2$Cl$_2$): −2.67 (d, FeCH$_2$, J=20.1), 24.0 (d, CH$_2$CN, J=3.6), 84.80 (s, C$_5$H$_5$), 122.65 (d, CN, J=2.0), 128.38 (d, C-2 in C$_6$H$_5$, J=11.5), 129.87 (s, C-4 in C$_6$H$_5$), 133.19 (d, C-3 in C$_6$H$_5$, J=9.7), 136.3 (d, C-1 in C$_6$H$_5$, J(P,C)=40.7), 222.58 (d, CO, J=31.9) IR: 1897 in mull, 1905 in C$_6$H$_6$	[28]
*15	CF(CF$_3$)$_2$ Ib (44)	red, m.p. 152 to 153° ^{1}H NMR (CDCl$_3$, 300 K): 4.48 (d, C$_5$H$_5$, J=1.8), ∼7.5 (m) ^{19}F NMR (CH$_2$Cl$_2$): 66.3, 66.9 (m's, CF$_3$, ^{3}J(F,F)=11, ^{4}J(F,F)=9), 173.2 (m, CF, J(P,F)=37.9) IR (C$_6$H$_{12}$): 1956, 1972	[5, 24, 25]

References on p. 195

Table 18 [continued]

No.	^{1}L ligand method of preparation (yield in %)	properties and remarks explanations on p. 176	Ref.
*16	CH$_2$CH$_2$CH$_2$CH$_3$ III (50.1)	red needles, m.p. 116 to 117° ^{1}H NMR (C$_6$D$_6$): 0.98 (t, CH$_3$, J=7.0), 1.55, 1.80 (m's, CH$_2$), 4.15 (d, C$_5$H$_5$, J=1.2), 7.10, 7.55 (C$_6$H$_5$) ^{13}C NMR (CH$_2$Cl$_2$): 4.05 (d, C–1, J=18.6), 13.86 (s, C–4), 28.44 (d, C–3, J=1.9), 41.98 (d, C–2, J=3.2), 84.85 (s, C$_5$H$_5$), 128.07 (d, C–3 in C$_6$H$_5$, J=9.2), 129.45 (d, C–4 in C$_6$H$_5$, J=1.6), 133.33 (d, C–2 in C$_6$H$_5$, J=9.4), 137.28 (d, C–1 in C$_6$H$_5$, J=39.3), 223.46 (d, CO, J=32.8) IR: 1887 in mull, 1906 in C$_6$H$_{14}$	[32, 36]
*17	CD$_2$CH$_2$CH$_2$CH$_3$ III	–	[32, 35]
*18	CH$_2$CD$_2$CH$_2$CH$_3$ III	–	[32, 35]
*19	CH(CH$_3$)CH$_2$CH$_3$ III (31.7)	m.p. 109.5 to 111° ^{1}H NMR (C$_6$D$_6$): 0.98 (t, CH$_3$–4, J=7.0), 1.55 (m, CH–1, CH$_2$–3, CH$_3$–2), 4.18 (d, C$_5$H$_5$, J=1.4), 7.05, 7.54 (C$_6$H$_5$) ^{13}C NMR (CH$_2$Cl$_2$): 3.53 (d, C–1, J=18.0), 13.38 (s, C–3), 29.39 (d, CH$_3$ on C–1, J=2.5), 41.53 (d, C–3, J=4.0), 84.43 (s, C$_5$H$_5$), 127.7 (d, C–3 in C$_6$H$_5$, J=9.6), 129.11 (s, C–4 in C$_6$H$_5$), 132.89 (d, C–2 in C$_6$H$_5$, J=9.7), 136.82 (d, C–1 in C$_6$H$_5$, J=39.4), 223.60 (d, CO, J=32.1) IR: 1878 in mull, 1904 in C$_6$H$_{14}$	[32, 36]
*20	CH$_2$CH(CH$_3$)$_2$ III (56)	m.p. 120.5 to 121.5°, 131 to 132° ^{1}H NMR: 0.86 (m, CH), 1.12, 1.15 (d's, CH$_3$, J=5.85), 1.78 (m, CH$_2$), 4.16 (d, C$_5$H$_5$, J=1), 7.00, 7.48 (C$_6$H$_5$) in C$_6$D$_6$, 0.3 to 1.8 (m, 9 H, includes t centered at 0.8, J=7 and m centered at 1.5), 4.2 (d, C$_5$H$_5$, J(P,H) ~1), 7.4 (m, C$_6$H$_5$) in CDCl$_3$ ^{13}C NMR (CH$_2$Cl$_2$): 14.12 (d, C–1, J=18.5), 26.35, 27.14 (s's, C–3), 36.31 (d, C–2, J=1.8), 84.97 (s, C$_5$H$_5$), 128.16 (d, C–3 in C$_6$H$_5$, J=9.5), 129.1 (s, C–4 in C$_6$H$_5$), 133.44 (d, C–2 in C$_6$H$_5$, J=9.3), 137.19 (d, C–1 in C$_6$H$_5$, J=39.1), 224.09 (d, CO, J=35.0)	[32, 34, 36]

References on p. 195

Table 18 [continued]

No.	^{1}L ligand method of preparation (yield in %)	properties and remarks explanations on p. 176	Ref.
		IR: 1886 in mull, 1901 in C_6H_{12}, 1920 in $CHCl_3$	
	(+)(S) form IVa (74)	oil, does not crystallize, $[\alpha]_{578}^{25} = +71°$ (2×10^{-3} M, C_6H_6, purity >95%)	[26, 34]
*21	$CD_2CH(CH_3)_2$ III	–	[39]
*22	$CH_2C_3H_5$-cyclo IVa (40)	m.p. 126 to 126.5° ^{1}H NMR ($CDCl_3$): -0.2 to $+1.0$ (m, CH_2), 1.8 (m, CH), 4.4 (d, C_5H_5, J=1), 7.4 (m, C_6H_5) IR ($CHCl_3$): 1900	[34]
	(+)(S) form IVa (40)	oil, does not crystallize, $[\alpha]_{578}^{25} = +176°$ ($\sim$1 mg/mL, optically pure)	[34]
*23	$CH_2CH(CH_3)CH_2CH_3$ III (52)	red, m.p. 112 to 113° ^{1}H NMR ($C_6D_5CD_3$): 1.18 (m, C_5H_{11}), 4.18 (d, C_5H_5, J=1.2), 7.00, 7.48 (C_6H_5) ^{13}C NMR (CH_2Cl_2): 11.79, 12.15 (s's, C-1), 22.31, 23.17 (s's, CH_3 on C-2), 33.26, 34.08 (s's, C-3), 42.32, 42.90 (s, d, C-2, J=2.7), 84.88, 85.08 (s's, C_5H_5), 129.11 (d, C-3 in C_6H_5, J=8.9), 129.53 (s, C-4 in C_6H_5), 133.37 (d, C-2 in C_6H_5, J=10.0), 137.13 (d, C-4 in C_6H_5, J=39.6), 225.68, 226.06 (d's, CO, J=34.8) IR: 1890 in mull, 1895 in CH_2Cl_2	[32, 39]
*24	$CH_2CH_2CH(CH_3)_2$ III (50)	m.p. 111 to 112.5° ^{1}H NMR ($C_6D_5CD_3$): 1.20 (m, C_5H_{11}), 4.17 (d, C_5H_5, J=1.2), 7.08, 7.50 (C_6H_5) ^{13}C NMR (CH_2Cl_2): 1.79 (d, C-1, J=17.9), 22.37, 22.66 (s's, CH_3), 33.09 (s, C-3), 49.35 (d, C-2, J=2.9), 84.95 (s, C_5H_5), 128.17 (d, C-3 in C_6H_5, J=9.1), 129.55 (s, C-4 in C_6H_5), 133.43 (d, C-2 in C_6H_5, J=9.8), 137.33 (d, C-1 in C_6H_5, J=39.6), 223.5 (d, CO, J=33.3) IR: 1865 in mull, 1905 in C_6H_{14}	[32, 39]

Table 18 [continued]

No.	^{1}L ligand method of prepa- ration (yield in %)	properties and remarks explanations on p. 176	Ref.
*25	$CH_2C_6H_5$ Ib (65 to 70), IVa (13)	red solid, m.p. 130, 135 to 137° ^{1}H NMR $(CDCl_3)$: 4.07 $(C_5H_5, J=1)$, 7.02, 　7.35 (m's, C_6H_5) in $CDCl_3$, 　1.74 $(H_{trans}$ in $CH_2, J=10.7)$, 　2.56 $(H_{gauche}$ in $CH_2, J=3.9, J(H,H)=8.1)$, 　4.07 $(C_5H_5, J=1)$ in CH_2Cl_2 ^{57}Fe-γ: $\delta=0.34$, $\Delta=1.74$ IR: 1905 to 1907 in CH_2Cl_2 or $CHCl_3$, 1917, 　$\sim$1925 in petroleum ether	[8, 12, 17, 23, 29, 34]
	$(-)(S)$ form IVa (63)	m.p. 79 to 83°, $[\alpha]_{578}^{25}=-165°$ ($\sim$1 mg/mL, 　optically pure)	[26, 34]
*26	$CH_2CH(CH_3)C_6H_5$ Ib, Ic, IIb (53)	optical active forms, see further information	[16, 22]
*27	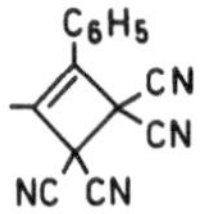 IX (64)	red solid ^{1}H NMR: see further information IR (mull): 1935, 1950	[47]
*28	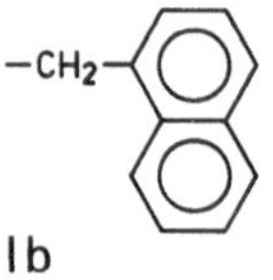 Ib	m.p. 128 to 132° ^{1}H NMR: 3.95 $(C_5H_5, J=1)$, 7.42 $(C_6H_5$ and 　$C_{10}H_7$) in $CDCl_3$, 2.12 $(H_{trans}$ in CH_2, 　$J=7.4)$, 2.99 $(H_{gauche}$ in $CH_2, J=7.2$, 　$J(H,H)=8.5)$ in CH_2Cl_2 IR (petroleum ether): 1916, 1923, 1930	[29]
*29	CH_2OCH_3 Ib (42), Ib (89), IVa (52), X (29)	orange needles, m.p. 150 to 152°, 156 to 　157.5° ^{1}H NMR: 3.10 (s, CH_3), 4.30 (m, CH_2), 4.35 　(d, C_5H_5), 4.65 (m, CH_2), 7.12 (m, C_6H_5), 　7.65 (m, C_6H_5) in C_6D_6, 3.02 (s, CH_3), 4.44 　(d, C_5H_5, $J\sim1$), 4.06 to 4.64 (m, CH_2), 7.4 　(m, C_6H_5) in $CDCl_3$ IR $(CHCl_3)$: 1895 [34] or 1900 [45]	[34, 45, 46]
30	$C(OCH_3)=CH_2$ V	—	[14]
*31	$CH_2OCH_2CH_3$ X (71)	—	[46]

References on p. 195

Table 18 [continued]

No.	¹L ligand method of preparation (yield in %)	properties and remarks explanations on p. 176	Ref.
*32	CH(OCH$_2$CH$_3$)CH$_3$ VI (80)	orange, m.p. 114 to 115°, 130.5 to 131° ¹H NMR (CD$_3$COCD$_3$): 0.96 (m, CH$_3$ on CH$_2$, J=6.8), 1.49 (d, CH$_3$-2, J=6.2), 3.24 (m, OCH$_2$), 3.92 (q, CH-1, J=7.5), 4.26 (d, C$_5$H$_5$, J=1.1), 7.22 (s, C$_6$H$_5$) IR (C$_6$H$_{14}$): 1912	[11, 14]
33	CD(OCH$_2$CH$_3$)CH$_3$ VI	–	[14]
*34	C(OC$_2$H$_5$)=CH$_2$ V	–	[14]
*35	CH(OCH$_3$)C$_6$H$_5$ Ib (78)	orange, m.p. 156 to 157° ¹H NMR (C$_6$D$_6$): 2.82 (s, CH$_3$), 4.22 (d, C$_5$H$_5$, J=1), 5.26 (d, CH, J=9), 6.96 to 7.88 (m, C$_6$H$_5$) ¹³C NMR (CDCl$_3$): 59.3 (OCH$_3$), 85.9 (C$_5$H$_5$), 122.2, 123.3, 127.4 (CC$_6$H$_5$), 127.8, 128.2, 129.4, 133.6, 134.0 (PC$_6$H$_5$), 135.8, 137.4 (PC$_6$H$_5$ or CC$_6$H$_5$), 155.8 (CH), 222.2 (d, CO, J(P,C)=30)	[44]
*36	−CH$_2$O- Ib, IVa	¹H NMR (CDCl$_3$): 0.62 to 1.92 (m, 18 H in C$_{10}$H$_{19}$), 2.60 (m, CH), 4.08 to 4.52 (m, OCH$_2$), 4.44 (d, C$_5$H$_5$, J~1), 7.16 to 7.95 (m, C$_6$H$_5$) IR (C$_6$H$_6$): 1900	[34]
	(+)(S) form Ib (22)	red orange, m.p. 146 to 146.5°, $[\alpha]_{578}^{25}$= +209° (c=0.1, C$_6$H$_6$, optically pure)	[26, 34]
	(−)(R) form Ib (mother liquor of the (+)(S) form, 22)	m.p. 119 to 120°, $[\alpha]_{578}^{25}$= −407° (c=0.1, C$_6$H$_6$, optically pure)	[26, 34]
37	CH$_2$CH$_2$PO(OCH$_3$)$_2$ XI (41)	red, m.p. 150 to 151.5° ¹H NMR (CD$_3$COCD$_3$): 2.10 (m, CH$_2$), 3.39, 3.48 (d, CH$_3$, J=10.8), 4.10 (d, C$_5$H$_5$, J=1.2), 7.05, 7.45 (m, C$_6$H$_5$) ¹³C NMR (CH$_2$Cl$_2$): −7.8 (m, C-1), 33.0 (d, C-2, J=12.0), 51.6 (d, CH$_3$, J=13.0), 85.2 (s, C$_5$H$_5$), 128.2 (d, C-3 in C$_6$H$_5$, J=8.0), 129.3 (s, C-4 in C$_6$H$_5$), 133.4 (d, C-2 in C$_6$H$_5$, J=6.0), 135.8 (d, C-1 in C$_6$H$_5$, J=36.0), 222.0 (d, CO, J=48.0) IR: 1898 in mull, 1911 in THF	[40]

Table 18 [continued]

No.	^{1}L ligand method of preparation (yield in %)	properties and remarks explanations on p. 176	Ref.
*38	CH$_2$Si(CH$_3$)$_3$ Ib (23)	red, m.p. 132 to 135°, 139 to 141° ^{1}H NMR: −1.25 (dd, CH$_2$, J=12, J=14), −0.25 (dd, CH$_2$, J=2, J=12), −0.4 (s, CH$_3$), 4.26 (d, C$_5$H$_5$), 7.2 to 7.8 (m, C$_6$H$_5$) in C$_6$H$_{12}$, −0.1 (s, CH$_3$), 4.17 (C$_5$H$_5$, J=1), 7.28 (m, C$_6$H$_5$) in CDCl$_3$, −1.2 (H$_{trans}$ in CH$_2$, J=13.7), −0.19 (H$_{gauche}$ in CH$_2$, J=2.0, J(H,H)=11.9) in CH$_2$Cl$_2$ IR (petroleum ether): 1915.5, 1922	[6, 23, 27, 29]
39	CH$_2$Si(CH$_3$)$_2$C$_6$H$_5$ Ib (~20)	^{1}H NMR (C$_6$H$_{12}$): −1.15 (dd, CH$_2$, J=12, J=13), 0.05 (dd, CH$_2$, J=2, J=12), 0.32 (d, CH$_3$, J=8), 4.25 (d, C$_5$H$_5$, J=1), 7.2 to 7.8 (m, C$_6$H$_5$) IR: 1915	[27]
*40	CH(C$_6$H$_5$)Si(CH$_3$)$_3$ III	^{1}H NMR (C$_6$D$_6$): 0.30 (s, CH$_3$), 1.40 (d, CH, J=7.7), 4.77 (d, C$_5$H$_5$, J=1.5), ~7.14 (m, C$_6$H$_5$) IR (C$_6$H$_{12}$): 1922 optically active forms, see further information	[30, 41]
41	CH$_2$SO$_2$OCH$_3$ XII (77)	m.p. 152 to 153° ^{1}H NMR (CDCl$_3$): 1.7 (d, CH$_2$, J(H,H)=11, J~1), 3.4 (s, CH$_3$), 4.5 (d, C$_5$H$_5$, J=1.4), 7.4 (m, C$_6$H$_5$) ^{13}C NMR: 20.2 (CH$_2$, J(P,C)≅20), 54.7 (CH$_3$) IR (CHCl$_3$): 1940	[34]
42	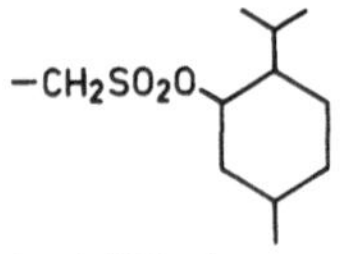 −CH$_2$SO$_2$O〈cyclohexyl menthyl group〉 (+)(S)-form XII (82)	m.p. 169 to 171°, [α]$^{25}_{578}$= +80° (c=0.1, CH$_2$Cl$_2$) ^{1}H NMR (CDCl$_3$): 0.5 to 2.4 (m, C$_{10}$H$_{19}$), 1.6 (dd, CH$_2$, J(H,H)=11, J=11), 2.7 (dd, CH$_2$, J(H,H)=11, J~1), 4.2 (m, CHO), 4.4 (d, C$_5$H$_5$, J=1), 7.4 (m, C$_6$H$_5$) IR (CHCl$_3$): 1940	[34]
*43	CH(CH$_3$)CN see further information	red solid ^{1}H NMR: 0.8 (complex m, H), 1.3 (d, CH$_3$, J=7.3) IR (C$_6$H$_5$CH$_3$): 2185 (CN)	[52]
44	CH(CH$_3$)$_2$	−	[50]

References on p. 195

Table 18 [continued]

No.	¹L ligand method of preparation (yield in %)	properties and remarks explanations on p. 176	Ref.
*45	$C(CH_3)=CH_2$ VI (50)	small orange crystals, m.p. 129 to 130° ^1H NMR $(CDCl_3)$: 2.22 (s, CH_3), 4.43 (s, C_5H_5), 4.65, 5.54 (s's, CH_2), 7.34 (m, C_6H_5) ^{13}C NMR (CH_2Cl_2): 40.01 (s, CH_3), 84.65 (s, C_5H_5), 124.10 (d, CH_2, J=3.1), 129.08 (d, C-3,5 in C_6H_5, J=9.7), 129.64 (s, C-4 in C_6H_5), 133.42 (d, C-2,6 in C_6H_5, J=9.6), 136.86 (d, C-1 in C_6H_5, J=39.9), 162.81 (d, Fe-C, J=25.5) IR (C_6H_{14}): 1918	[50]
*46	$(E)-C(CH_3)=CHCH_3$ VI	^1H NMR: 5.40 (CH, ^3J(H,H)=0.8)	[51]
*47	$(E)-C(CH_3)=C(CH_3)R$ VI	$R=CH_3$, C_6H_5, $C\equiv CH$, $CH(COOC_2H_5)_2$, OC_2H_5, SC_4H_9, SC_6H_5, CN see further information	[51]
*48	$(E)-C(C_2H_5)=$ $CHCH_2CH_3$ VI (52)	red, m.p. 128 to 129° ^1H NMR (CS_2): 0.81 (m, CH_3), 1.43 (m, CH_2), 4.32 (s, C_5H_5), 4.74 (t, CH, J=6.2), 7.23 (m, C_6H_5) ^{13}C NMR (C_6H_6): 15.32 (s, C-4), 23.41 (s, CH_3 in C_2H_5), 37.30 (s, C-3), 46.40 (s, CH_2 in C_2H_5), 84.80 (s, C_5H_5), 129.71 (s, C-4 in C_6H_5), 133.88 (d, C-2,6 in C_6H_5), 137.51 (d, C-1 in C_6H_5, J=40.8), 141.38 (d, C-2, J=7.3), 147.63 (d, C-1, J=28.0), the resonance of C-3 and C-5 in C_6H_5 are obscured by the benzene resonance IR (C_6H_{14}): 1915	[50]

*Further information:

$C_5H_5Fe(CO)(P(C_6H_5)_3)CH_3$ (Table 18, No. 1). The reduction $C_5H_5Fe(CO)-(P(C_6H_5)_3)CH_3 \rightarrow [C_5H_5Fe(CO)P(C_6H_5)_3]^- + CH_3^+$ in dimethylformamide or acetonitrile was observed polarographically, the half-wave potentials being $E_{1/2}= -2.10$ V and -2.16 V in these two solvents [10]. No. 1 is stable when dry at room temperature, but a solution decomposes slowly in air [32]. It decomposes in refluxing solvents: 30% in hexane in 27 h, 77% in heptane in 17 h, a little in dioxane in 5 h and 80% in 20 h [12, 13]. It is racemized on strong UV irradiation in C_6H_6/C_5H_{12} [21].

When CO is allowed into a solution of the compound in refluxing petroleum ether (95 to 100 °C), $C_5H_5Fe(CO)(P(C_6H_5)_3)COCH_3$ (25%), $C_5H_5Fe(CO)_2CH_3$, and $P(C_6H_5)_3$ are formed [1, 2]. However, in a later study no insertion of CO was observed [8].

References on p. 195

The compound reacts in CH_2Cl_2 at 5 to 10 °C with $(CN)_2C{=}C(CN)_2$ to give $C_5H_5Fe(COCH_3)(P(C_6H_5)_3)NCC(CN){=}C(CN)_2$ [7, 31]. The same product is also obtained in ether or THF. This reaction occurs in ice-cold C_6H_6 but stirring at 25 °C returns the starting materials. No. 1 does not react with $(C_6H_5)_2C{=}C(CN)_2$ in C_6H_6 or CH_2Cl_2. No reaction occurs in C_6H_6 or THF with CNCH=CHCN, either at 20 °C or on heating. In THF decomposition sets in at 66 °C [12].

The reaction with CF_3COOH (5 °C, 30 min) gives $C_5H_5Fe(CO)(P(C_6H_5)_3)OOCCF_3$ [37]. In $CHCl_3$ No. 1 reacts with chlorine at 0 °C [2] and with bromine or iodine at room temperature in 20 min to yield $C_5H_5Fe(CO)(P(C_6H_5)_3)X$ (X = Cl, Br, or I) [1, 2]. The iodine compound can be obtained also in C_6H_6 with I_2 or HgI_2 (5 °C, 30 min) or with ICl (5 °C, 5 min) [37]. On refluxing in THF for 12 h with $P(OCH_3)_3$, $P(OC_4H_9{-}n)_3$, or $P(OC_6H_5)_3$ the ligands on the P atom exchange [12, 13].

Compound No. 1 reacts with gaseous SO_2 or refluxing SO_2 or on bubbling SO_2 through a solution in $CHCl_3$ to give $C_5H_5Fe(CO)(P(C_6H_5)_3)SO_2CH_3$ [8], the SO_2 insertion occurring within 2 min. The dicarbonyl compound requires several hours [3]. The configuration at the Fe atom is not changed on insertion of SO_2 [22]. With liquid SO_2 in the presence of KI, $C_5H_5Fe(CO)(P(C_6H_5)_3)SO_2CH_3$ and $C_5H_5Fe(CO)$-$(P(C_6H_5)_3)I$ are formed in 10 min [34]. Statements about the thermal stability of No. 1 are conflicting [20, 34].

$C_5H_5Fe(CO)(P(C_6H_5)_3)CH_2Cl$ (Table 18, No. 2). If No. 2 is prepared by Method VII from a pure stereoisomer of $C_5H_5Fe(CO)(P(C_6H_5)_3)CH_2OC_{10}H_{19}$ (No. 36), it itself is a pure stereoisomer, hence an ideal starting material for other pure isomers [34]. It decomposes in solution to give $C_5H_5Fe(CO)(P(C_6H_5)_3)Cl$ [34]. It reacts with $NaOC_{10}H_{19}$ to form $C_5H_5Fe(CO)(P(C_6H_5)_3)CH_2OC_{10}H_9$ ($C_{10}H_9 =$ menthyl) [20]. The Cl atom of the chloromethyl group is very easily replaced by nucleophiles. For example, it reacts in benzene/water with KCN to form C_5H_5-$Fe(CO)(P(C_6H_5)_3)CH_2CN$ (No. 8). In methanol/ether it reacts with CH_3OH in the presence of CH_3COOK to give $C_5H_5Fe(CO)(P(C_6H_5)_3)CH_2OCH_3$ (No. 29). With Grignard or alkyl lithium reagents in THF it gives the compounds $C_5H_5Fe(CO)$-$(P(C_6H_5)_3)^1L$, with $^1L = C_2H_5$ (No. 6), n-C_3H_7 (No. 12), $CH_2CH(CH_3)_2$ (No. 20), $CH_2C_3H_5$-cyclo (No. 22), and $CH_2C_6H_5$ (No. 25). No. 1 is obtained with $NaBH_4$ [34].

$C_5H_5Fe(CO)(P(C_6H_5)_3)CH_2Br$ (Table 18, No. 3) is somewhat less stable than No. 2 on heating. In solution it decomposes to give $C_5H_5Fe(CO)(P(C_6H_5)_3)Br$. $NaBH_4$ reduces to $C_5H_5Fe(CO)(P(C_6H_5)_3)CH_3$ (No. 1). Methylene transfer was observed in a reaction with a large excess of β-methylstyrene at 25 °C, which yielded cyclo-$C_3H_4(C_6H_5{-}1)CH_3{-}2$ [34].

$C_5H_5Fe(CO)(P(C_6H_5)_3)CH_2I$ (Table 18, No. 4) is so unstable that it cannot be properly characterized [34]. According to [20], it is more stable than the chlorine compound. Its circular dichroism spectrum is available.

$C_5H_5Fe(CO)(P(C_6H_5)_3)CF_3$ (Table 18, No. 5). The ^{19}F NMR spectrum is reported for 163, 173, 243, 263, 273, and 303 K [15]. The coupling constant decreases with temperature; the coalescence temperature is approximately 250 K (in 1:1 CH_2Cl_2/$CFCl_3$). Extrapolation of $^3J(P,F)$ to infinite temperature yields limiting coupling constants of 6.4 Hz for 1:1 CH_2Cl_2/$CFCl_3$ and 10.5 Hz for 3:1 C_6H_6/$CFCl_3$. Thus the P,F coupling constants are solvent- and temperature-dependent, even for the individual conformations [15].

References on p. 195

C₅H₅Fe(CO)(P(C₆H₅)₃)C₂H₅ (Table **18**, No. **6**) is formed also from [C₅H₅Fe-(CO)(P(C₆H₅)₃)CH₂=CH₂]PF₆ by addition of hydride from C₅H₅Fe(CO)-(P(C₆H₅)₂CH₂CH₂P(C₆H₅)₂)H [49].

The circular dichroism spectra of the (+)(R) and (−)(S) forms are shown in figures in [21]. The compound is considerably less soluble in hexane than is C₅H₅Fe-(CO)(P(C₆H₅)₃)CH(OC₂H₅)CH₃ (No. 32) [14]. Molten No. 6 decomposes at 140 °C to yield C₅H₅Fe(CO)(P(C₆H₅)₃)H and CH₂=CH₂. The same decomposition products are obtained from a hexane solution under N₂ at atmospheric pressure and 61.2 °C (8 h) [32, 35]. In refluxing heptane a further product, (C₅H₅Fe(CO)₂)₂, is obtained [12, 13]. No. 6 reacts rapidly with I₂, ICl, or HgI₂ in C₆H₆ at 5 °C (5 min) to give C₅H₅Fe(CO)(P(C₆H₅)₃)I [37]. No reaction occurs with B₂H₆ at 25 °C in C₆H₆ or C₆H₅CH₃, but at 100 °C reduction does proceed (e.g., in 45 min in C₆H₅CH₃) to yield C₅H₅Fe(CO)(P(C₆H₅)₃)H [3]. No. 6 reacts with (CN)₂C=C(CN)₂ in C₆H₆ at 5 to 10 °C. When cooled with ice, the reaction mixture yields C₅H₅Fe-(COC₂H₅)(P(C₆H₅)₃)C₆N₄, which decomposes to the starting materials when the mixture is stirred at 35 to 40 °C [12, 31]. No. 6 reacts in CF₃COOH with the solvent (5 °C, 30 min) to give C₅H₅Fe(CO)(P(C₆H₅)₃)OOCCF₃ [37]. It absorbs SO₂ in CH₂Cl₂ to form C₅H₅Fe(CO)(P(C₆H₅)₃)SO₂C₂H₅ [26] and also reacts in liquid SO₂ in the presence of KI to give the same insertion product along with C₅H₅Fe(CO)-(P(C₆H₅)₃)I [34].

C₅H₅Fe(CO)(P(C₆H₅)₃)CD₂CH₃ (Table **18**, No. **7**) decomposes in the melt at 140 °C and in hexane at 61.2 °C into C₅H₅Fe(CO)(P(C₆H₅)₃)H and deuterated ethylene [32, 35].

C₅H₅Fe(CO)(P(C₆H₅)₃)CH₂CN (Table **18**, No. **8**). The NMR spectra were assigned with the aid of Yb(opt)₃ (Formula I on p. 104). Thereby the strong effect of the Fe atom on the CN group was observed. This is also apparent in the ν(CN) position in the IR spectrum. The solid is stable, but solutions decompose slowly in contact with air [28].

C₅H₅Fe(CO)(P(C₆H₅)₃)C₂F₅ (Table **18**, No. **9**). The two J(P,F) values are very different but depend little on the temperature. This is consistent with configuration I but does not exclude other rotamers [25].

C₅H₅Fe(CO)(P(C₆H₅)₃)CF₂CHF₂ (Table **18**, No. **10**) slowly loses HF and forms C₅H₅Fe(CO)(P(C₆H₅)₃)CF=CF₂ (No. 11) [12].

C₅H₅Fe(CO)(P(C₆H₅)₃)CH₂CH₂CH₃ (Table **18**, No. **12**) reacts with (CN)₂-C=C(CN)₂ in benzene at 5 to 10 °C to form C₅H₅Fe(COC₃H₇)(P(C₆H₅)₃)C₆N₄ [12, 31]. With liquid SO₂ at −10 °C insertion gives C₅H₅Fe(CO)(P(C₆H₅)₃)-SO₂CH₂CH₂CH₃ [34] (see 1.5.2.2.6.1.2, Table 8, No. 41).

C₅H₅Fe(CO)(P(C₆H₅)₃)CH₂CH=CH₂ (Table **18**, No. **13**). On thermolysis in cyclohexane or THF, P(C₆H₅)₃ is lost and compound II forms. Thermolysis follows first-order kinetics. The rate constants are in cyclo-C₆H₁₂ k = (13.9 ± 0.08) × 10⁻⁵ s⁻¹

at 60.9 °C and $(28.4\pm0.11)\times10^{-5}$ s^{-1} at 67.0 °C; activation parameters at 25 °C: $\Delta H^{\neq}=26.74\pm0.22$ kcal·mol^{-1} and $\Delta S^{\neq}=3.88\pm0.68$ cal·K^{-1}·mol^{-1}. In THF: k$=$ $(10.5\pm0.09)\times10^{-5}$ s^{-1} at 60.6 °C; $\Delta H^{\neq}=26.97\pm1.47$ kcal·mol^{-1} and $\Delta S^{\neq}=2.28\pm$ 4.42 cal·K^{-1}·mol^{-1} at 25 °C.

HCl or HBF$_4$ in ether react with No. 13 at room temperature to form the cation $[C_5H_5Fe(CO)(P(C_6H_5)_3)CH_2=CHCH_3]^+$ (Formulas III a and III b). With CF$_3$COOH the compound forms $C_5H_5Fe(CO)(P(C_6H_5)_3)OOCCF_3$ [18].

a   III   b

R = CH$_3$

$C_5H_5Fe(CO)(P(C_6H_5)_3)CH_2CH_2CN$ (Table **18**, No. **14**). The ^{1}H NMR spectrum is shifted only slightly with Yb(opt)$_3$ (Formula I on p. 104), and no resonance is split. The solid is stable but the solutions decompose gradually in contact with air [28]. Heating No. 14 at 95 °C for 3 h in C$_6$H$_5$CH$_3$ results in complete isomerization to $C_5H_5Fe(CO)(P(C_6H_5)_3)CH(CH_3)CN$ (No. 43) [52].

$C_5H_5Fe(CO)(P(C_6H_5)_3)CF(CF_3)_2$ (Table **18**, No. **15**). The slight temperature dependence of $^3J(P,F)$ in the ^{19}F NMR spectrum suggests that one rotamer predominates, namely, that shown in Formula IV, which is consistent with J(P,F)$=37.9$ [25]. The doubling of the v(CO) band in the IR spectrum cannot be assigned to the Fe–C bonding. It may be due to two possible rotations of the P–aryl bonds [24].

IV

$C_5H_5Fe(CO)(P(C_6H_5)_3)C_4H_9$-n (Table **18**, No. **16**) is formed on heating the i-C$_4$H$_9$ compound (No. 19) in xylene at 63 °C under N$_2$ for 4 h [39]. Molten, it decomposes at 140 °C, and in xylene it decomposes at 61.2 °C (8 h) to give C$_5$H$_5$Fe- (CO)(P(C$_6$H$_5$)$_3$)H and butene. On thermolysis in xylene at 61.2 °C (8 h) under CO butene and $(C_5H_5Fe(CO)_2)_2$ form in a first-order reaction. The rate constant k·10^3 (in min^{-1}) is 0.396$\pm$0.009 at 51.0 °C, 2.72$\pm$0.03 at 61.2 °C, 6.1$\pm$0.01 at 74.8 °C, and 9.81$\pm$0.02 at 91.0 °C [32, 35].

$C_5H_5Fe(CO)(P(C_6H_5)_3)CD_2CH_2CH_2CH_3$ (Table **18**, No. **17**) decomposes on heating in xylene (70 °C, 1 h) to yield C$_5$H$_5$Fe(CO)(P(C$_6$H$_5$)$_3$)H and deuterated butene, where the deuterium is distributed statistically, as revealed by NMR. The first-order rate constant is k$=(2.10\pm0.03)\times10^{-3}$ min^{-1} at 61.2 °C [32, 35]. Evaluation of the ^{2}H NMR spectrum shows that isomerization of the butyl group during the thermal decomposition also forms $C_5H_5Fe(CO)(P(C_6H_5)_3)C(CH_3)_2CHD_2$ [39]. The Raman spectrum is presented in a figure in [35].

References on p. 195

$C_5H_5Fe(CO)(P(C_6H_5)_3)CH_2CD_2CH_2CH_3$ (Table **18**, No. **18**) decomposes on heating much like No. 17, including the isomerization of the butene. The first-order rate constant is $k = (2.30 \pm 0.04) \times 10^{-3}$ min^{-1} at 61.2 °C [32, 35]. The ions having mass numbers 468 (parent ion), 412, 383, and 262 $(P(C_6H_5)_3)$ are observed in the mass spectrum [32, 36].

$C_5H_5Fe(CO)(P(C_6H_5)_3)CH(CH_3)CH_2CH_3$ (Table **18**, No. **19**) decomposes molten at 140 °C and in xylene at 62 °C (8 h) to form $C_5H_5Fe(CO)(P(C_6H_5)_3)H$ and butene. The first-order rate constant is $k = (2.48 \pm 0.05) \times 10^{-3}$ min^{-1} at 61.2 °C [32, 35]. Analysis of the decomposition products shows the presence of $CH_2=CHCH_2CH_3$ and (Z)- and (E)-$CH_3CH=CHCH_3$ in the gas phase, and the solid $C_5H_5Fe(CO)$-$(P(C_6H_5)_3)CH_2CH_2CH_2CH_3$ in addition to $C_5H_5Fe(CO)(P(C_6H_5)_3)H$ [39]. The mass spectrum shows ions having the mass numbers 468 (parent ion), 412, 383, and 262 $(P(C_6H_5)_3)$ [32, 36].

$C_5H_5Fe(CO)(P(C_6H_5)_3)CH_2CH(CH_3)_2$ (Table **18**, No. **20**) decomposes molten at 140 °C and in xylene at 62 °C (8 h) to form $C_5H_5Fe(CO)(P(C_6H_5)_3)H$ and 2-methyl-propene. The first-order rate constant is $k = (1.80 \pm 0.02) \times 10^{-3}$ min^{-1} at 61.2 °C [32, 35]. The mass spectrum shows ions with the mass numbers 468 (parent ion), 412, 383, and 262 $(P(C_6H_5)_3)$ [32, 36]. No. 20 reacts with liquid SO_2 at -10 °C to form $C_5H_5Fe(CO)(P(C_6H_5)_3)SO_2CH_2CH(CH_3)_2$ [34], see 1.5.2.2.6.1.2, Table 8, No. 15).

$C_5H_5Fe(CO)(P(C_6H_5)_3)CD_2CH(CH_3)_2$ (Table **18**, No. **21**). Isomerization of the isobutyl group and migration of D atoms (as for Nos. 17 and 18) are found to occur during decomposition in xylene at 65 °C (2.5 h). Raman and ^2H NMR spectroscopy serve to identify 2-methylpropene in the volatile decomposition products. The migration of the deuterium can be explained only by an intermediate, C_5H_5Fe-$(CO)(P(C_6H_5)_3)C(CH_3)_2CHD_2$ [39].

$$C_5H_5Fe(CO)(^2D)-CH_2-\triangleleft \quad \xrightarrow{SO_2}$$

$$\left[\begin{array}{c} C_5H_5\overset{+}{Fe}(CO)(^2D) \diagdown\diagup\diagdown SO_2^- \\ VI \\ \Updownarrow \\ C_5H_5Fe(CO)(^2D) \diagdown \underset{O_2}{\overset{H}{\diagdown S}} \\ VII \end{array}\right] \longrightarrow C_5H_5Fe(CO)(^2D)-\underset{O}{\overset{O}{S}}\diagup\diagdown\diagup$$

V

$C_5H_5Fe(CO)(P(C_6H_5)_3)CH_2C_3H_5$-cyclo (Table **18**, No. **22**). In liquid SO_2 or passing SO_2 through a solution of No. 22 in CH_2Cl_2, the product is mainly that having Formula V $(^2D = P(C_6H_5)_3)$ after long reaction times at room temperature or higher temperatures. After short periods or at low temperatures an unstable oily intermediate is obtained, which may have the structure VI or VII. The stereospecificity of the reaction is about 40% [34].

$C_5H_5Fe(CO)(P(C_6H_5)_3)^1L$ (Table **18**, Nos. **23** and **24**, having $^1L = CH_2CH(CH_3)CH_2CH_3$ and $CH_2CH_2CH(CH_3)_2$) yield $C_5H_5Fe(CO)(P(C_6H_5)_3)H$ and

References on p. 195

the isomers $CH_2=CHCH(CH_3)_2$ and $CH_2=C(CH_3)CH_2CH_3$ in the ratio 1.3:1 during thermal decomposition in xylene at 63 °C [39].

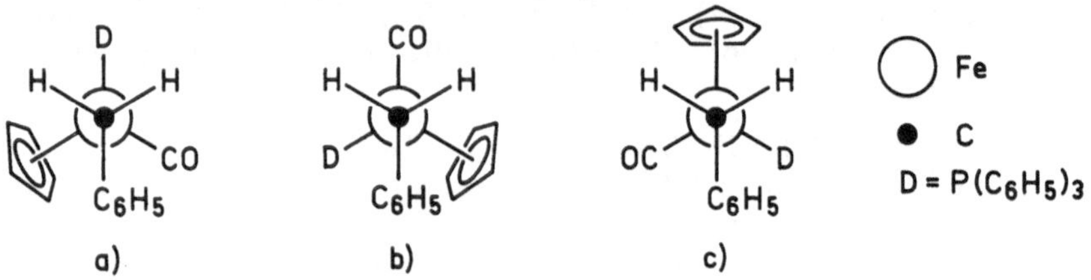

a)   b)   c)

VIII

$C_5H_5Fe(CO)(P(C_6H_5)_3)CH_2C_6H_5$ (Table **18**, No. **25**). The methylene protons are not equivalent in the ^1H NMR spectrum (see the figure in the paper) since the CH_2 group is attached to an asymmetric unit though rapidly rotating about the Fe–C bond [4]. The dependence of the coupling constants J(P,H) on the temperature shows that of the three rotamers VIII, form c is the most prevalent, particularly at low temperatures. At 298 K the distribution a:b:c is 23:18:59 [23, 29].

The uptake of SO_2 in $CHCl_3$ or CH_2Cl_2 requires 2 min, while the dicarbonyl compound requires several hours [3]. In these solvents or in liquid SO_2 compound No. 25 reacts somewhat slower than does No. 1, and a mixture of starting material and $C_5H_5Fe(CO)(P(C_6H_5)_3)SO_2CH_2C_6H_5$ always results [8, 34]. The configuration at the Fe atom is unchanged on insertion of SO_2 [22]. No. 25 reacts with $Se(SeCN)_2$ at room temperature to give the $C_5H_5Fe(CO)(P(C_6H_5)_3)X$ isomers with X = –SeCN and –NCSe. At higher temperature $C_5H_5Fe(CO)(P(C_6H_5)_3)CN$ is also found [48]. The reaction with $(CN)_2C=C(CN)_2$ in benzene at 5 to 10 °C (2 to 3 h) gives C_5H_5Fe-$(CO)(P(C_6H_5)_3)N=C=C(CN)C(CN)_2CH_2C_6H_5$, $C_5H_5Fe(CO)(P(C_6H_5)_3)CN$, and a compound that possibly has the composition $C_5H_5Fe(CO)(P(C_6H_5)_3)CH_2C_6H_5 \cdot C_6N_4$ [12, 31]. The same reaction in CH_2Cl_2 at 25 °C produces only the first two products [7].

$C_5H_5Fe(CO)(P(C_6H_5)_3)CH_2CH(CH_3)C_6H_5$ (Table **18**, No. **26**) has two chiral centers: the Fe atom and the C–2 atom in the ^1L ligand. Corresponding to the two centers, there are four different forms, which have not been obtained pure. Nor have their configurations yet been clarified.

Of the three photochemical reactions (Method Ib, Ic, and IIb) Method IIb is the recommended route because of the good yield and high purity of the product. The two pairs of enantiomers (RR) and (SS) 26 (a) and (RS) and (SR) 26 (b) are usually obtained in a ratio 54:46 because the last part of the chromatographic band is discarded upon purification (Al_2O_3, 3:1, pentane/benzene eluent). They can be separated by a combination of chromatography on Al_2O_3 and crystallization from benzene/pentane; in this fashion, a:b mixtures of 98:2 and 9:91 were isolated. Method IIb starting with $C_5H_5Fe(CO)(P(C_6H_5)_3)COCH_2CH(CH_3)C_6H_5$ enriched in the (RR)(SS) or (RS)(SR) forms (1.5.2.2.12.3.1, Table 16, No. 6), designated as c and d, yielded the stereochemical data shown below (15 mg of $(C_5H_5Fe(CO)_2)_2$ also isolated) [22]:

starting c:d	recovered c:d	isolated a:b (% stereospecificity)	rechromatographed a:b
90:10, 400 mg	91:9, 65 mg	82:18 (78), 195 mg	90:10, 160 mg
12:88, 500 mg	15:85, 160 mg	32:68 (51), 240 mg	33:67, 200 mg

References on p. 195

Optically active $C_5H_5Fe(CO)(P(C_6H_5)_3)CH_2CH(CH_3)C_6H_5$ (A) synthesized according to Method I c from (R)-$C_5H_5Fe(CO)_2COCH_2CH(CH_3)C_6H_5$ can be separated similarly into the (RS) and (SS) diastereomers, designated as A a and A b. Some properties of these two forms as well as a and b are given in the following table [22]:

compound	m.p. in °C	spectra and remarks
a a:b=98:2	123 to 125 (dec.)	¹H NMR (CDCl₃): 1.15 (d, CH₃, J=6.5), 0.63 to 2.09 (compl., CH₂), 2.64 (compl. m, CH), 4.03 (d, C₅H₅, J=1.3), 7.33 (compl. m, C₆H₅) IR (CH₂Cl₂): 1895 (CO)
b a:b=9:91	123 to 125 (dec.)	¹H NMR (CDCl₃): 1.25 (d, CH₃, J=6.5), 0.75 to 1.99 (compl., CH₂), 2.55 (compl. m, CH), 4.26 (d, C₅H₅, J=1.3), 7.30 (compl. m, C₆H₅) IR (CH₂Cl₂): 1898 (CO)
A a A a:A b=95:5 A b A a:A b=14:86	123 to 126 (dec.) 124 to 126 (dec.)	$[\alpha]_D^{23}$ could not be measured because of low stability

In benzene or THF epimerization occurs with time, accompanied by an uneven decomposition [16]. When the compounds a and b are kept in CDCl₃ solution (25 °C, 48 h) epimerization also occurs. Its extent is higher for b (36%) than for a (27%). Compound a epimerizes when irradiated in C_6H_6 solution (17% epimerization in 25 min). The epimerization is accompanied by considerable decomposition to an unidentified material. CHCl₃ solutions of optically active A a and A b darken too rapidly to permit measurement of their specific rotations. In CH₂Cl₂ solutions HCl splits No. 26 into $C_5H_5Fe(CO)(P(C_6H_5)_3)Cl$ and cumene [22]. In liquid SO₂ [22, 34] or in CH₂Cl₂ solution containing $\sim$10% SO₂, the SO₂ is inserted to form $C_5H_5Fe(CO)(P(C_6H_5)_3)$-$SO_2CH_2CH(CH_3)C_6H_5$ with a high stereospecificity. By contrast, the carbonylation (4 atm CO, 25 °C, 25 h) of No. 26 yields $C_5H_5Fe(CO)(P(C_6H_5)_3)COCH_2$-$CH(CH_3)C_6H_5$ in a 25% yield without any meaningful stereochemical information [22].

$C_5H_5Fe(CO)(P(C_6H_5)_3)C_4(CN)_4C_6H_5$ (Table **18**, No. **27**) occurs as a mixture of two isomers, Formulas IXa and IXb, p. 191. Two C_5H_5 resonances are seen in the NMR spectrum. Their intensities change with time until they reach the equilibrium ratio of 7:3. The compound forms solvates with CH_2Cl_2, $CHCl_2CHCl_2$, CH_3COCH_3, and $C_6H_5CH_3$. The form IXa does not react with dipolar compounds. The NMR spectrum depends on the solvent, which shows that at least one of the forms interacts with the solvent. In $CDCl_3$ $\delta = 4.44$ and 4.79 ppm (C_5H_5) and in acetone $\delta = 4.70$ and 4.75 ppm (C_5H_5). The intensities of two $\nu(CO)$ bands in the IR spectrum depend on the method of crystallization [47].

$C_5H_5Fe(CO)(P(C_6H_5)_3)CH_2C_{10}H_7$ (Table **18**, No. **28**, with $C_{10}H_7 = 1$-naphthyl). The temperature dependence of the $^3J(P,H)$ coupling constants was evaluated in terms of the relative abundancies of the rotamers (cf. Formula VIII on p. 190). The number of rotamers is doubled because of the asymmetry of the coordinated naphthyl methyl group [29].

$C_5H_5Fe(CO)(P(C_6H_5)_3)CH_2OCH_3$ (Table **18**, No. **29**) is split by anhydrous HCl or HBr ether at 0 °C into $C_5H_5Fe(CO)(P(C_6H_5)_3)CH_2X$ and CH_3OH. Surprisingly the insertion of SO_2 in liquid SO_2 leads to $C_5H_5Fe(CO)(P(C_6H_5)_3)CH_2SO_2OCH_3$ [34]. Abstraction of H^- with $[(C_6H_5)_3C]PF_6$ in CH_2Cl_2 gives $[C_5H_5Fe(CO)-(P(C_6H_5)_3)CHOCH_3]PF_6$ [46].

$C_5H_5Fe(CO)(P(C_6H_5)_3)CH_2OCH_2CH_3$ (Table **18**, No. **31**) forms with $[(C_6H_5)_3C]PF_6$ in CH_2Cl_2 the product $[C_5H_5Fe(CO)(P(C_6H_5)_3)CHOC_2H_5]PF_6$ [46].

$C_5H_5Fe(CO)(P(C_6H_5)_3)CH(OCH_2CH_3)CH_3$ (Table **18**, No. **32**) does not react with B_2H_6 or $B_2H_6/NaBH_4$ mixtures in ethanol [14].

$C_5H_5Fe(CO)(P(C_6H_5)_3)C(OC_2H_5)=CH_2$ (Table **18**, No. **34**) may be protonated by HBF_4 in propionic anhydride to form $[C_5H_5Fe(CO)(P(C_6H_5)_3)C(OC_2H_5)CH_3]BF_4$ [14].

$C_5H_5Fe(CO)(P(C_6H_5)_3)CH(OCH_3)C_6H_5$ (Table **18**, No. **35**). When the dark red solution of No. 35 in $CDCl_3$ is added to excess cooled CF_3SO_3H at -80 °C, then the dark brown $[C_5H_5Fe(CO)(P(C_6H_5)_3)CHC_6H_5]SO_3CF_3$ is formed on slow warming up to 0 °C [44].

$C_5H_5Fe(CO)(P(C_6H_5)_3)CH_2OC_{10}H_{19}$ (Table **18**, No. **36**, with $C_{10}H_{19} = $ menthyl). Circular dichroism spectra of the two forms in benzene are given in the paper [20]. The ^1H NMR spectra of the two forms cannot be distinguished. Addition of $Eu(fod)_3$-d_{27} (Formula II on p. 104) has no effect on the spectra [34]. According to an X-ray study of single crystals of the $(+)(S)$ form $(R = 6.2\%, R_w = 7.8\%)$, the crystals are monoclinic, space group $P2_1 - C_2^2$, with $a = 10.882$ (3), $b = 11.054$ (4), $c = 13.664$ (4) Å, and $\beta = 102.66°$. $Z = 2$ gives $D_c = 1.20$, while $D_m = 1.18$ g·cm^{-3}. The molecular structure is shown in **Fig. 16**. The (S) configuration is confirmed by the X-ray structure and the circular dichroism spectra [42].

The compound is soluble in hexane, the $(-)(R)$ form more so than the $(+)(S)$ form so that the two may be readily separated in a mixture of 30% ether/70% hexane [34].

Hydrogen halides split the compound easily in ether at 0 °C to form $C_5H_5Fe-(CO)(P(C_6H_5)_3)CH_2X$ (X = Cl, Br, or I) and $C_{10}H_{19}OH$ [26, 34]. In benzene I_2 and ICl react at 5 °C, HgI_2 at 25 °C, to form $C_5H_5Fe(CO)(P(C_6H_5)_3)I$ and $C_{10}H_{19}OCH_2X$ (X = I, Cl, or HgI, respectively). The configuration at the Fe atom is retained [20].

Fig. 16

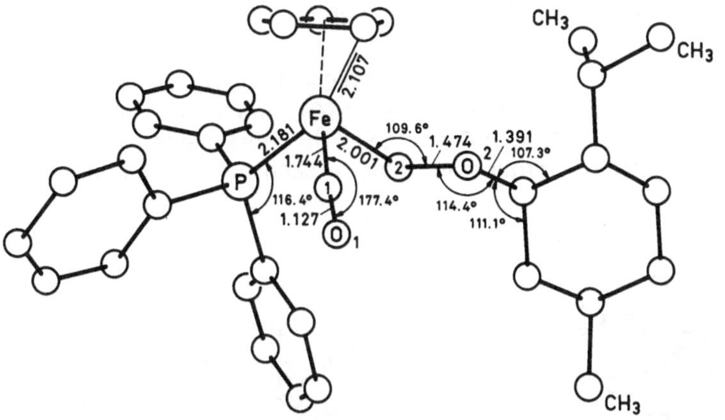

Molecular structure of $(+)(S)$-$C_5H_5Fe(CO)(P(C_6H_5)_3)CH_2OC_{10}H_{19}$ [42].

Other bond angles (°):

P–Fe–C(1)	91.7
P–Fe–C(2)	92.0
C(1)–Fe–C(2)	89.2

The insertion of SO_2 is fast in liquid SO_2 (-10 °C, 2 h) and yields C_5H_5Fe-$(CO)(P(C_6H_5)_3)CH_2$ $SO_2OC_{10}H_{19}$ [34]. When No. 36 is cleaved with HBF_4 in pure E-$C_6H_5CH=CHCH_3$, the $(+)(S)$ form favors the formation of $(1R,2R)$-1-methyl-2-phenylcyclopropane (Formula X), while the $(-)(R)$ form favors the formation of the $(1S,2S)$ isomer [20].

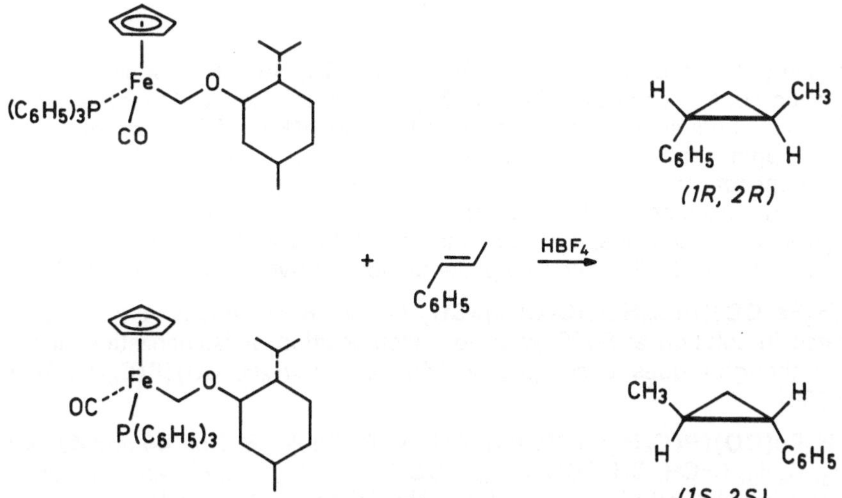

$C_5H_5Fe(CO)(P(C_6H_5)_3)CH_2Si(CH_3)_3$ (Table 18, No. 38). The temperature dependence of the ^1H NMR coupling constants J(P,H) shows that out of the three possible rotamers (cf. Formula VIII on p. 190) form c is the most abundant, particularly at low temperature. At 298 K the distribution of the forms a:b:c is 17:6:77 [23,

29], also see [24]. Similar conclusions, but without quantitative results, had already been made in [15].

$C_5H_5Fe(CO)(P(C_6H_5)_3)CH(C_6H_5)Si(CH_3)_3$ (Table **18**, No. **40**). This compound has the four diastereomers (RR), (SS), (SR), and (RS). Each of these has three rotamers (see Formula XI for SR and SS). The most stable rotamer is that in which the C_5H_5 and $Si(CH_3)_3$ groups are trans to each other. For Formula XI this is form a [30, 41].

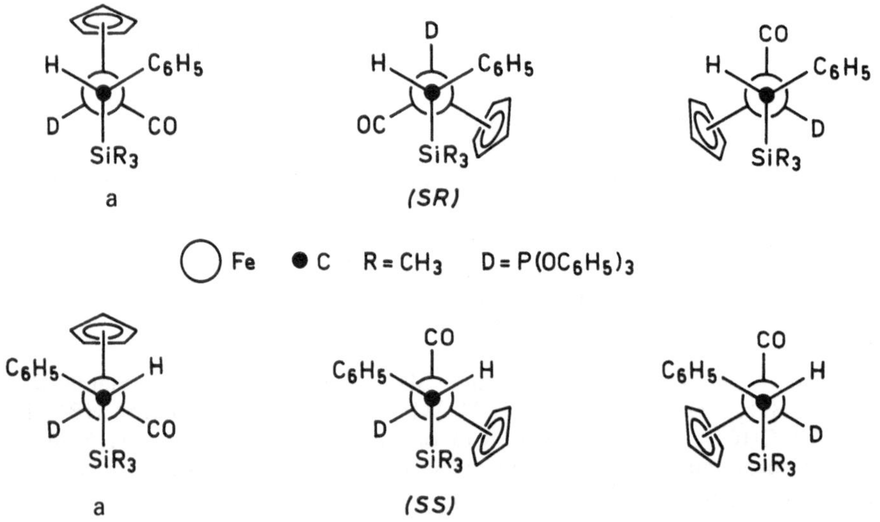

O Fe ● C R = CH₃ D = P(OC₆H₅)₃

a   (SR)

a   (SS)

XI

$C_5H_5Fe(CO)(P(C_6H_5)_3)CH(CH_3)CN$ (Table **18**, No. **43**). Forms quantitative by isomerization of No. 14 in $C_6H_5CH_3$ at 95 °C for 3 h. It is believed that the two diastereomers form in equal amounts because in the 400–MHz H NMR spectrum each half of the 1.3 ppm doublet consists of two peaks separated by 1.2 Hz. In the 60 and 200 MHz spectra, this resonance is a simple doublet. (A $J(P,H) = 1.0$ Hz is resolved on the C_5H_5 resonance.) Also the complex CH resonance in the 60 MHz spectrum is a quintet, i.e., two overlapping quartets, at 400 MHz and much different in appearance than at 60 MHz. Thus the two patterns indicate two diastereomers [52].

$C_5H_5Fe(CO)(P(C_6H_5)_3)C(CH_3)=CH_2$ (Table **18**, No. **45**) is very stable and can be heated in solution at 90 °C with no decomposition or isomerization of the vinyl group. Although it does decompose at 140 °C, no $C_5H_5Fe(CO)(P(C_6H_5)_3)H$ is produced [50].

$C_5H_5Fe(CO)(P(C_6H_5)_3)C(CH_3)=C(CH_3)R$ (Table **18**, Nos. **46** and **47**, with R = H, CH_3, C_6H_5, C≡CH, $CH(COOC_2H_5)_2$, OC_2H_5, SC_4H_9, SC_6H_5, and CN) are formed by preparative Method VI in the (E) form. The (Z) forms with R = H or C_6H_5 isomerize on heating toluene solutions at ~40 °C for 15 min to the presumably more stable E forms. The compounds are stable on heating and only slowly decompose in solution when exposed to air. Cleavage by I_2 or Br_2 affords $C_5H_5Fe(CO)(P(C_6H_5)_3)X$ (X = Br, I) and $R(CH_3)C=C(CH_3)X$ under retention of configuration at the Fe atom and the alkenyl group [51].

$C_5H_5Fe(CO)(P(C_6H_5)_3)C(CH_2CH_3)=CHC_2H_5$ (Table **18**, No. **48**) is very stable and can be heated in solution at 90 °C with no decomposition or isomerization of the vinyl group. Although it does decompose at 140 °C, no $C_5H_5Fe(CO)(P(C_6H_5)_3)H$ is produced [50].

References:

[1] P.M. Treichel, R.L. Shubkin, K.W. Barnett, D. Reichard (Inorg. Chem. **5** [1966] 1177/81). – [2] K.W. Barnett (Diss. Univ. Wisconsin 1967; C.A. **69** [1968] No. 27510). – [3] A. Wojcicki, J.J. Alexander, M. Graciani, J.E. Thomasson, F.A. Hartman (News Aspects Chem. Metal Carbonyls Deriv. Proc. 1st Intern. Symp., Venice 1968, Abstr. C6, pp. 1/6). – [4] J.W. Faller, A.S. Anderson (J. Am. Chem. Soc. **91** [1969] 1550/1). – [5] R.B. King, R.N. Kapoor, K.H. Pannell (J. Organometal. Chem. **20** [1969] 187/93).

[6] K.H. Pannell (J. Chem. Soc. Chem. Commun. **1969** 1346/7). – [7] A. Wojcicki, S.R. Su, J.A. Hanna (4th Intern. Conf. Organometal. Chem., Bristol 1969). – [8] M. Graziani, A. Wojcicki (Inorg. Chim. Acta **4** [1970] 347/50). – [9] K.H. Pannell (J. Organometal. Chem. **21** [1970] P17/P18). – [10] L.I. Denisovich, I.V. Polovyanyuk, B.V. Lokshin, S.P. Gubin (Izv. Akad. Nauk SSSR Ser. Khim. **1971** 1964/9; Bull. Acad. Sci. [USSR] Div. Chem. Sci. **1971** 1851/5).

[11] M.L.H. Green, L.C. Mitchard, M.G. Swanwick (J. Chem. Soc. A **1971** 794/7). – [12] S.R. Su (Diss. Ohio State Univ. 1971; Diss. Abstr. Intern. B **32** [1972] 6283). – [13] S.R. Su, A. Wojcicki (J. Organometal. Chem. **27** [1971] 231/40). – [14] A. Davison, D.L. Reger (J. Am. Chem. Soc. **94** [1972] 9237/8). – [15] J. Thomson, W. Keeney, M.C. Baird, W.F. Reynolds (J. Organometal. Chem. **40** [1972] 205/14).

[16] T.G. Attig, P. Reich-Rohrwig, A. Wojcicki (J. Organometal. Chem. **51** [1973] C21/C23). – [17] G. Ingletto, E. Tondello, L. Di Sipio, G. Carturan, M. Graziani (J. Organometal. Chem. **56** [1973] 335/7). – [18] K.R. Aris, J.M. Brown, K.A. Taylor (J. Chem. Soc. Dalton Trans. **1974** 2222/8). – [19] H. Brunner, J. Strutz (Z. Naturforsch. **29b** [1974] 446/7). – [20] A. Davison, W.C. Krusell, R.C. Michaelson (J. Organometal. Chem. **72** [1974] C7/C10).

[21] A. Davison, N. Martinez (J. Organometal. Chem. **74** [1974] C17/C20). – [22] P. Reich-Rohrwig, A. Wojcicki (Inorg. Chem. **13** [1974] 2457/64). – [23] K. Stanley, M.C. Baird (Inorg. Nucl. Chem. Letters **10** [1974] 1111/5). – [24] K. Stanley, R.A. Zelonka, J. Thomson, M.C. Baird (Proc. 16th Intern. Conf. Coord. Chem., Dublin 1974, pp. 4.19). – [25] K. Stanley, R.A. Zelonka, J. Thomson, P. Fiess, M.C. Baird (Can. J. Chem. **52** [1974] 1781/6).

[26] T.C. Flood, F.J. DiSanti, D.L. Miles (J. Chem. Soc. Chem. Commun. **1975** 336/7). – [27] K.H. Pannell (Transition Metal. Chem. [Weinheim] **1** [1975/76] 36/40). – [28] D.L. Reger (Inorg. Chem. **14** [1975] 660/4). – [29] K. Stanley, M.C. Baird (J. Am. Chem. Soc. **97** [1975] 4292/8). – [30] K. Stanley, M.C. Baird (J. Am. Chem. Soc. **97** [1975] 6598/9).

[31] S.R. Su, A. Wojcicki (Inorg. Chem. **14** [1975] 89/98). – [32] E.C. Culbertson (Diss. Univ. South Carolina 1976; Diss. Abstr. Intern. B **38** [1977] 181/2). – [33] J.A. van Doorn, C. Masters, H.C. Volger (J. Organometal. Chem. **105** [1976] 245/54). – [34] T.C. Flood, F.J. DiSanti, D.L. Miles (Inorg. Chem. **15** [1976] 1910/8). – [35] D.L. Reger, E.C. Culbertson (J. Am. Chem. Soc. **98** [1976] 2789/94).

[36] D.L. Reger, E.C. Culbertson (Syn. Reactiv. Inorg. Metal–Org. Chem. **6** [1976] 1/10). – [37] T.C. Flood, D.L. Miles (J. Organometal. Chem. **127** [1977] 33/44). – [33] C.R. Folkes, A.J. Rest (J. Organometal. Chem. **136** [1977] 355/61). – [39] D.L. Reger, E.C. Culbertson (Inorg. Chem. **16** [1977] 3104/7). – [10] D.L. Reger, E.C. Culbertson (J. Organometal. Chem. **131** [1977] 297/300).

[41] K. Stanley, M.C. Baird (J. Am. Chem. Soc. **99** [1977] 1808/12). – [42] C.-K. Chou, D.L. Miles, R. Bau, T.C. Flood (J. Am. Chem. Soc. **100** [1978] 7271/8). – [43] S.L. Miles, D.L. Miles, R. Bau, T.C. Flood (J. Am. Chem. Soc. **100** [1978] 7278/82). – [44] G.O. Nelson (Diss. Univ. North Carolina 1977; Diss. Abstr. Intern. B **39** [1978] 237/8). – [45] P.E. Riley, C.E. Capshew, R. Pettit, R.E. Davis (Inorg. Chem. **17** [1978] 408/14).

[46] A.R. Cutler (J. Am. Chem. Soc. **101** [1979] 604/6). – [47] A. Davison, J.P. Solar (J. Organometal. Chem. **166** [1979] C13/C17). – [48] M.A. Jennings, A. Wojcicki (Inorg. Chim. Acta **3** [1969] 335/40). – [49] T. Bodnar, S.J. Lacroce, A.R. Cutler (J. Am. Chem. Soc. **102** [1980] 3292/4). – [50] D.L. Reger, C.J. Coleman, P.J. McElligott (J. Organometal. Chem. **171** [1979] 73/84).

[51] D.L. Reger, P.J. McElligott (J. Am. Chem. Soc. **102** [1980] 5923/4). – [52] D.L. Reger, P.J. McElligott (J. Organometal. Chem. **216** [1981] C12/C14).

1.5.2.2.12.3.5 Compounds with $^1L = C \equiv CR$

$C_5H_5Fe(CO)(P(C_6H_5)_3)C \equiv CC_4H_9$ is formed by the reaction of $C_5H_5Fe(CO)_2$-$C \equiv CC_4H_9$ with $P(C_6H_5)_3$ at 160 °C as brown crystals, which are recrystallized from a mixture of benzene and petroleum ether, m.p. 129 to 132 °C. IR: $v(CO)$ 1945 cm^{-1} and $v(C \equiv C)$ 2095 cm^{-1} [1].

$C_5H_5Fe(CO)(P(C_6H_5)_3)C \equiv CC_6H_5$ is formed by the reaction of $C_5H_5Fe(CO)_2$-$C \equiv CC_6H_5$ with $P(C_6H_5)_3$ at 160 °C as brown crystals, which are recrystallized from a mixture of benzene and petroleum ether, m.p. 194 to 196 °C. ^1H NMR spectrum $(CDCl_3)$: $\delta = 4.57$ (d, C_5H_5, $J(P,H) = 1.1$ Hz) and 7 to 8 (m, C_6H_5) ppm. IR: $v(CO)$ 1950 and $v(C \equiv C)$ 2085 cm^{-1}. The mass spectrum shows ions with mass numbers 278 (parent ion), 157 ($[FeC \equiv CC_6H_5]^+$), and 124 [1].

The compound reacts in THF with dilute aqueous HBF_4 to yield C_5H_5Fe-$(CO)(P(C_6H_5)_3)COCH_2C_6H_5$. With $HBF_4 \cdot O(CH_3)_2$ in methanol it forms the yellow $[C_5H_5Fe(CO)(P(C_6H_5)_3)C(OCH_3)CH_2C_6H_5]BF_4$ [2]. It reacts with C_5H_5Co-$(P(C_6H_5)_3)C_6H_5C \equiv CCOOCH_3$ in benzene at room temperature to give $C_5H_5Fe(CO)$-$(P(C_6H_5)_3)C_4(C_6H_5)_2(COOCH_3)Co(C_5H_5)$ (1.5.2.2.12.3.7, Table 20, No. 2) [3].

References:

[1] M.L.H. Green, T. Mole (J. Organometal. Chem. **12** [1968] 404/6). – [2] A. Davison, J.P. Solar (J. Organometal. Chem. **155** [1978] C8/C12). – [3] K. Yasufuku, H. Yamazaki (J. Organometal. Chem. **121** [1976] 405/11).

1.5.2.2.12.3.6 Compounds with $^1L = C_6H_5$ and Substituted Phenyl

The compounds presented in Table 19 have been prepared by the following methods.

Method I: $C_5H_5Fe(CO)_2{}^1L$ and $P(C_6H_5)_3$ are UV irradiated in benzene, if not otherwise specified. Separation and purification are by chromatography [3, 4, 6, 8 to 10, 13, 17].

Method II: Decarbonylation of $C_5H_5Fe(CO)(P(C_6H_5)_3)COC_6H_5$ in benzene at 80 °C by UV irradiation for 30 min [13].

Method III: From $C_5H_5Fe(CO)(P(C_6H_5)_3)Br$ and LiC_6H_5 or C_6H_5MgBr in ether, chromatographic separation and purification [18].

Method IV: Reduction of $C_5H_5Fe(CO)(P(C_6H_5)_3)C_6H_4CHO-4$ with $NaBH_4$ or $LiAlH_4$ [11].

A linear relationship seems to exist between the Hammett substituent constants σ, the ^{1}H NMR chemical shifts of the C_5H_5 ring, and the values of the symmetric and antisymmetric $\nu(CO)$ for No. 1 and the para-substituted derivatives No. 3, 4, and 5 [11]. The variation of the chemical shifts of the C_5H_5 protons remains within experimental error for Nos. 1, 2, 5, 7, 8, and 9. No relationship was found between the shifts and the donor-acceptor properties of the substituents [16]. But a nearly linear relationship between the half-wave potentials $E_{1/2}$ and the Hammett σ values was found for Nos. 1, 2, 5, 7, 8, and 9: $E_{1/2} = 0.19 \sigma_p^\circ - 2.01$. The Fe-C bond is irreversibly broken with formation of the $[C_5H_5Fe(CO)P(C_6H_5)_3]^-$ anion and C_6H_4X radicals which react with the electrode mercury to give $Hg(C_6H_4R)_2$ compounds. The half-wave potentials (in V vs. SCE) for various ^1L ligands and two solvents are given below. In dimethylformamide a second half-wave is observed at potentials more negative than -2.12 V. This results from the reduction of the organomercury compounds.

compound No.	^1L ligand	in CH_3CN $-E_{1/2}$ in V	in $HCON(CH_3)_2$ $-E_{1/2}$ in V
9	$C_6H_4COOC_2H_5-4$	1.92	1.86
7	C_6H_4Cl-4	1.94	1.90
5	C_6H_4F-4	1.99	1.93
1	C_6H_5	2.01	1.96
8	$C_6H_4OCH_3-4$	2.03	1.98
2	$C_6H_4CH_3-4$	2.04	1.98

The effect of the para-substituent, seen in the table, agrees well with the hypothesis of a homolytic character of activation for the bond splitting [5, 7, 14].

Table 19
$C_5H_5Fe(CO)(P(C_6H_5)_3)^1L$ Compounds with $^1L = C_6H_5$ and Substituted Phenyl. Further information on numbers preceded by an asterisk is given at the end of the table. For abbreviations and dimensions, see p. 375.

No.	^1L ligand method of preparation (conditions, yield in %)	properties and remarks (IR bands in the $\nu(CO)$ region)	Ref.
*1	C_6H_5 I (16 h, C_6H_{12}, 30°, 46) I (2.5 h, 80°, 55) II (30.3) III (26 with LiC_6H_5, 68 with C_6H_5MgBr)	red or dark red, m.p. 166 to 166.5° or 168 to 169° ^{1}H NMR ($CHCl_3$): 4.394 (C_5H_5, $J(P,H) = 1.2$) ^{31}P NMR ($CHCl_3$): 70.3 IR: 1870, 1912 for the solid, 1928 in C_6H_6, 1931 in C_6H_{12}	[3, 8, 13, 16, 18]

References on p. 200

Table 19 [continued]

No.	¹L ligand method of preparation (conditions, yield in %)	properties and remarks (IR bands in the $\nu(CO)$ region)	Ref.
2	$C_6H_4CH_3$-4 I (2 h, 80°, 21), I (4 h, THF at 25°, 11)	red, m.p. 141.5 to 142.5° (dec.) ¹H NMR (CHCl₃): 4.393 (C_5H_5, J(P,H)=1)	[4,8, 10,16]
*3	C_6H_4CHO-4 I (compare No. 1)	—	[11]
4	$C_6H_4CH_2OH$-4 IV	—	[11]
*5	C_6H_4F-4 I (21 h, 26) I (2 h, 80°, 30.9)	red or orange red, m.p. 126.5 to 127° ¹H NMR (CHCl₃): 4.385 (C_5H_5, J(P,H)=0.9) ¹⁹F NMR (C_6H_6): 13.28 in C_6H_6, 13.84 in CHCl₃ IR: 1915 for the solid, 1921 in CHCl₃, 1925 in THF	[9,10, 16,17]
6	C_6H_4F-3 I (13 h, room temp., 32)	red orange, m.p. 70 to 71° ¹⁹F NMR (CHCl₃): 4.47 IR (THF): 1927	[16,17]
7	C_6H_4Cl-4 I (12 h, 20°, 87) I (1 h, 80°, 98.5)	red, m.p. 161 to 162° (dec.) ¹H NMR (CHCl₃): 4.387 (C_5H_5, J(P,H)=1.1) IR: 1903 (1972, 2022) for the solid, 1922 (1967, 2020) in CHCl₃	[8,9, 13,16]
8	$C_6H_4OCH_3$-4 I (21 h, room temp., 25)	red, m.p. 132 to 133° ¹H NMR (CHCl₃): 4.398 (C_5H_5, J(P,H)=0.9) IR: 1890 (1900, 1930) for the solid, 1919 in CHCl₃	[9,16]
9	$C_6H_4COOC_2H_5$-4 I (18 h, 20°, 16)	orange, m.p. 155° (dec.) ¹H NMR (CHCl₃): 4.40 (C_5H_5) IR: 1908 (1958) for the solid, 1925 in CHCl₃	[9,16]
10	C_6F_5 I (3 h, 80°, 60)	red, dec. 156 to 157°	[6]

*Further information:

$C_5H_5Fe(CO)(P(C_6H_5)_3)C_6H_5$ (Table **19**, No. 1) crystallizes as dark red monoclinic prisms elongated along the c axis [2, 12]. Space group $P2_1/n-C_{2h}^5$ (No. 11), a=11.32±0.03, b=14.44±0.03, c=15.24±0.03 Å, and β=91°±0.3°. Z=4 gives

$D_c = 1.31$ g·cm^{-3}, equal to D_m [12]. The molecule is shown in **Fig. 17**. Earlier data can be found in [1, 2].

Fig. 17

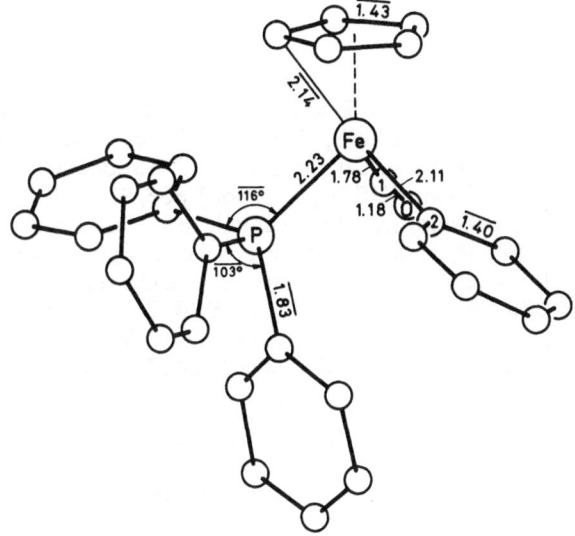

Molecular structure of $C_5H_5Fe(CO)(P(C_6H_5)_3)C_6H_5$ [12].

Other bond angles (°):

Fe–C(1)–O	174	P–Fe–C(1)	87
C(1)–Fe–C(2)	96	P–Fe–C(2)	88

The exchange of the H atoms of the C_5H_5 ring was studied in a mixture of $C_2H_5OD/C_2H_5ONa/C_6H_6$ (14:1:8) at 100 °C. The first–order rate constant for the exchange is $k = 2.6 \times 10^{-5}$ s^{-1} [15]. The solid is stable in air [3, 8]; solutions are stable for a prolonged period [8]. When CO is passed for 7 h through a solution in refluxing benzene being irradiated No. 1 is partly converted to $C_5H_5Fe(CO)_2C_6H_5$, which decomposes rapidly [13]. In a chlorobenzene solution in the presence of freshly sublimed AlBr$_3$, compound No. 1 disproportionates at room temperature to give C_5H_5Fe-$(CO)_2Br$, $[C_5H_5Fe(CO)_2P(C_6H_5)_3]^+$, and $[(C_5H_5Fe(CO)_2)_2Br]^+$ [18]. The Fe–C bond is split by HCl to give $C_5H_5Fe(CO)(P(C_6H_5)_3)Cl$ and C_6H_6. The phenyl group is much more reactive towards electrophilic reagents than is benzene and thus is suitable for synthesis. The intermediate $C_5H_5Fe(CO)(P(C_6H_5)_3)C_6H_4COCH_3$-4 is obtained with $CH_3COCl/AlCl_3$ at 0 °C; the intermediate $C_5H_5Fe(CO)(P(C_6H_5)_3)C_6H_4CHO$-4 (No. 3) is obtained with $C_6H_5N(CH_3)CHO/POCl_3$ or $Cl_2CHOC_2H_5/AlCl_3$, but these are immediately decomposed by the HCl released during their formation to yield $C_5H_5Fe(CO)(P(C_6H_5)_3)Cl$ and C_6H_5R (R = COCH$_3$ or CHO). Substitution does not occur in the C_5H_5 ring, and it occurs in the C_6H_5 ring only in the 2- and 4-positions [11]. Compound No. 1 reacts with acetic anhydride in the presence of AlCl$_3$ to yield $C_5H_5Fe(CO)(P(C_6H_5)_3)Cl$ and $C_6H_5COCH_3$ [15]. Refluxing in C_6H_6 with $C_6H_5OP(OCH_2-)_2$ converts No. 1 to $C_5H_5Fe(CO)(P(OCH_2-)_2OC_6H_5)C_6H_5$ [19].

$C_5H_5Fe(CO)(P(C_6H_5)_3)C_6H_4CHO$-4 (Table **19**, No. 3) is reduced by NaBH$_4$ to $C_5H_5Fe(CO)(P(C_6H_5)_3)C_6H_4CH_2OH$-4 (No. 4) [11].

References on p. 200   14*

$C_5H_5Fe(CO)(P(C_6H_5)_3)C_6H_4F-4$ (Table 19, No. 5). The solid is stable in air, but solutions are less stable [10]. When CO is passed for 11 h through a refluxing solution of No. 5 in benzene being UV irradiated, 66% is converted into $C_5H_5Fe(CO)_2C_6H_4F-4$ [13].

References:

[1] N.G. Bokii, R.L. Avoyan, G.N. Zakharova, M.Kh. Minasyan, Z.A. Akopyan, Yu.T. Struchkov (Zh. Strukt. Khim. **6** [1965] 795/6; J. Struct. Chem. [USSR] **6** [1965] 762/3). — [2] R.L. Avoyan, Yu.A. Chapovsky, Yu.T. Struchkov (Zh. Strukt. Khim. **7** [1966] 900/2; J. Struct. Chem. [USSR] **7** [1966] 838/9). — [3] A.N. Nesmeyanov, Yu.A. Chapovsky, B.V. Lokshin, I.V. Polovyanyuk, G.L. Makarova (Dokl. Akad. Nauk SSSR **166** [1966] 1125/8; Dokl. Chem. Proc. Acad. Sci. USSR **166/171** [1966] 213/6). — [4] I.V. Polovyanyuk, Yu.A. Chapovsky, L.G. Makarova (Izv. Akad. Nauk SSSR Ser. Khim. **1966** 387; Bull. Acad. Sci. USSR Div. Chem. Sci. **1966** 368). — [5] L.I. Denisovich, S.P. Gubin, Yu.A. Chapovsky (Izv. Akad. Nauk SSSR Ser. Khim. **1967** 2378/84; Bull. Acad. Sci. USSR Div. Chem. Sci. **1967** 2271/5).

[6] A.N. Nesmeyanov, Yu.A. Chapovsky (Izv. Akad. Nauk SSSR Ser. Khim. **1967** 2075/7; Bull. Acad. Sci. USSR Div. Chem. Sci. **1967** 1988/90). — [7] A.N. Nesmeyanov, Yu.A. Chapovsky, L.I. Denisovich, B.V. Lokshin, I.V. Polovyanyuk (Dokl. Akad. Nauk SSSR **174** [1967] 1342/4; Dokl. Chem. Proc. Acad. Sci. USSR **172/177** [1967] 576/8). — [8] A.N. Nesmeyanov, Yu.A. Chapovsky, I.V. Polovyanyuk, L.G. Makarova (J. Organometal. Chem. **7** [1967] 329/37). — [9] A.N. Nesmeyanov, I.V. Polovyanyuk, B.V. Lokshin, Yu.A. Chapovsky, L.G. Makarova (Zh. Obshch. Khim. **37** [1967] 2015/7; J. Gen. Chem. [USSR] **37** [1967] 1911/3). — [10] A.N. Nesmeyanov, Yu.A. Chapovsky, I.V. Polovyanyuk, L.G. Makarova (Izv. Akad. Nauk SSSR Ser. Khim. **1968** 1628/9; Bull. Acad. Sci. USSR Div. Chem. Sci. **1968** 1536/8).

[11] E.S. Bolton, G.R. Knox, C.G. Robertson (Chem. Commun. **1969** 664). — [12] V.A. Semion, Yu.T. Struchkov (Zh. Strukt. Khim. **10** [1969] 88/94; J. Struct. Chem. [USSR] **10** [1969] 80/5). — [13] A.N. Nesmeyanov, L.G. Makarova, I.V. Polovyanyuk (J. Organometal. Chem. **22** [1970] 707/12). — [14] L.I. Denisovich, I.V. Polovyanyuk, B.V. Lokshin, S.P. Gubin (Izv. Akad. Nauk SSSR Ser. Khim. **1971** 1964/9; Bull. Acad. Sci. USSR Div. Chem. Sci. **1971** 1851/5). — [15] T.Yu. Orlova, V.N. Setkina, L.G. Makarova, I.V. Polovyanyuk, D.N. Kursanov (Dokl. Akad. Nauk SSSR **201** [1971] 622/3; Dokl. Chem. Proc. Acad. Sci. USSR **196/201** [1971] 966/7).

[16] A.N. Nesmeyanov, I.F. Leshcheva, I.V. Polovyanyuk, Yu.A. Ustynyuk, L.G. Makarova (J. Organometal. Chem. **37** [1972] 159/65). — [17] A.N. Nesmeyanov, L.G. Makarova, I.V. Polovyanyuk (Izv. Akad. Nauk SSSR Ser. Khim. **1972** 607/9; Bull. Acad. Sci. USSR Div. Chem. Sci. **1972** 567/9). — [18] K.R. Aris, J.M. Brown, K.A. Taylor (J. Chem. Soc. Dalton Trans. **1974** 2222/8). — [19] R.P. Stewart, L.R. Isbrandt, J.J. Benedict, J.G. Palmer (J. Am. Chem. Soc. **98** [1976] 3215/9).

1.5.2.2.12.3.7 Compounds with Other Cyclic ^1L Ligands

The compounds listed in Table 20 have been prepared as follows.

Method I: $C_5H_5Fe(CO)_2{}^1L$ is reacted with $P(C_6H_5)_3$.

a. The components are heated together in a solvent. Compound No. 3 is isolated chromatographically [5].

b. The components are UV irradiated in a solvent (also with heating as the case may be) and the product is purified by chromatography [2, 5]. The compounds Nos. 7 to 10 are recrystallized from pentane/ether (5:1) [3].

c. $C_5H_5Fe(CO)_2COR$ and $P(C_6H_5)_3$ are irradiated. The product is purified by chromatography [5].

Method II: $C_5H_5Fe(CO)(P(C_6H_5)_3)I$ is stirred for 1.5 h in THF at 0 °C with $LiC_5H_4Mn(CO)_3$, water is added, and the compound extracted with ether. The ether is evaporated and the residue chromatographed [5].

Method III: $C_5H_5Co(P(C_6H_5)_3)C_6H_5C\equiv CCOOCH_3$ and $C_5H_5Fe(CO)(P(C_6H_5)_3)$-$C\equiv CC_6H_5$ are stirred together in benzene at room temperature for 2 d. The benzene is evaporated and the residue chromatographed [4].

Table 20
$C_5H_5Fe(CO)(P(C_6H_5)_3)^1L$ Compounds with Other Cyclic ^1L Ligands.
Further information on numbers preceded by an asterisk is given at the end of the table.
For abbreviations and dimensions, see p. 375.

No.	^1L ligand method of preparation (conditions, yield in %)	properties and remarks (IR bands in the $v(CO)$ region)	Ref.
*1	Ib (12 h, C_6H_6, 80°, 45)	orange, m.p. 198 to 199° (dec.) ^1H NMR (CDCl$_3$): 4.61 (s, C_5H_5), 7.2 to 7.4 (m, C_6H_5) ^{19}F NMR (CH$_2$Cl$_2$): 107.1 (s, CF$_2$), 116.2 (d, CF$_2$, J=15) IR: 1945 to 1971 in KBr, 1961, 1965 in C_6H_{12}	[1]
2	III	m.p. 185 to 187° ^1H NMR (CDCl$_3$): 3.70 (s), 3.79 (s), 4.64 (s, CoC$_5H_5$), 4.82 (d, FeC$_5H_5$, J(P,H)=1.2) IR (KBr): 1952	[4]
*3	Ia (13 h, CH$_3$CN, 80°, 100) Ib (25 h, THF, 96) Ic (36 h, THF, 40) II (88)	red, m.p. 177 to 179° (dec.) IR (CS$_2$): 1917, 1944, 2004	[5]
*4	Ib (4 h, THF, 22)	bright red, m.p. 161 to 162° ^1H NMR (CDCl$_3$): 2.0 to 2.9 (m, H-2), 4.33 (s, C$_5H_5$), 4.4 (m, H-4) IR (C_6H_{14}): 1930	[2]

References on p. 204

Table 20 [continued]

No.	¹L ligand method of preparation (conditions, yield in %)	properties and remarks (IR bands in the ν(CO) region)	Ref.
*5	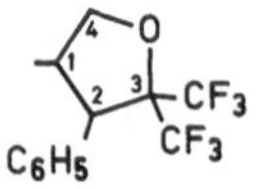 Ib (2 h, THF, 40)	orange, m.p. 145 to 146° ¹H NMR (CDCl₃): 2.0 to 2.9 (m, H-1), 3.2 to 3.6 (m, H-2), 4.27 (t, C₅H₅, J(P,H)=1.2), 4.5 to 4.8 (m, H-4), 7.1 (m, C₆H₅) ¹⁹F NMR (CFCl₃): 74.7 (q), 74.9 (q), 76.5 (q, J=9.7) IR (C₆H₁₄): 1930	[2]
*6	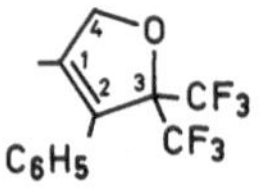 Ib (2 h, THF, 47)	red, m.p. 193° (dec.) ¹H NMR (CDCl₃): 4.17 (d, C₅H₅, J(P,H)= 1.3), 4.36 (s, H-4), 7.03 (m, C₆H₅) ¹⁹F NMR (CFCl₃): 72.9 (q), 74.0 (q, J=9.0) IR (C₆H₁₄): 1948	[2]
*7	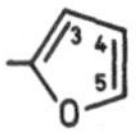 Ib (30 min, pentane/ ether (1:1), 20°, 42)	red, m.p. 148 to 149° ¹H NMR (CDCl₃): 4.30 (C₅H₅), 5.85 (H-3), 6.08 (H-4), 7.0 to 7.20 (C₆H₅) IR (C₆H₆): 1948	[3]
*8	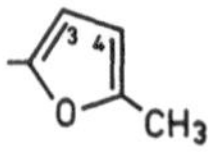 Ib (30 min, pentane/ ether (1:1), 20°, 52)	red, m.p. 141 to 142° ¹H NMR (CDCl₃): 4.40 (C₅H₅), 5.86 (H-3), 6.90 (H-4), 7.0 to 7.25 (C₆H₅) IR (C₆H₆): 1950	[3]
*9	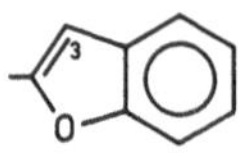 Ib (30 min, pentane/ ether (1:1), 20°, 35)	red, m.p. 174 to 175° ¹H NMR (CDCl₃): 4.30 (C₅H₅), 6.50 (H-3), 7.0 to 7.30 (C₆H₅) IR (C₆H₆): 1950	[3]
*10	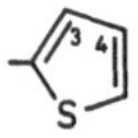 Ib (30 min, pentane/ ether (1:1), 20°, 45)	red, m.p. 161 to 163° ¹H NMR (CDCl₃): 4.40 (C₅H₅), 6.56 (H-3), 7.02 to 7.30 (C₆H₅) IR (C₆H₆): 1946	[3]

*Further information:

$C_5H_5Fe(CO)(P(C_6H_5)_3)C_4F_4Cl$ (Table **20**, No. **1**). The appearance of two ¹⁹F NMR resonances indicates that the cyclobutene ring has a plane of symmetry, permitting free rotation of the C_4F_4Cl group around the Fe–C σ bond. Further IR bands are given in the paper. The following ions are observed in the mass spectrum (210 °C sample temperature): $[C_5H_5Fe(CO)(P(C_6H_5)_3)C_4F_4Cl]^+$, $[C_5H_5Fe(P(C_6H_5)_3)C_4F_4Cl]^+$, $[C_5H_5Fe(P(C_6H_5)_3)C_4F_3Cl]^+$, $[C_5H_5Fe(P(C_6H_5)_3)C_4F_4]^+$, $[Fe(P(C_6H_5)_3)C_4F_4Cl]^+$, $[C_5H_5Fe(P(C_6H_5)_2)C_4F_3Cl]^+$, $[C_4F_3C_5H_5P(C_6H_5)_3]^+$, $[C_5H_5FeP(C_6H_5)_3]^+$, $[C_4F_3P(C_6H_5)_3]^+$, $[Fe(P(C_6H_5)_3)Cl]^+$, $[Fe(P(C_6H_5)_3)F]^+$,

$[FePC(C_6H_5)_3]^+$, $[ClP(C_6H_5)_3]^+$, $[P(C_6H_5)_3]^+$, $[FePC_{12}H_9]^+$, $[P(C_6H_5)_2]^+$, $[PC_{12}H_8]^+$, $[C_5H_5C_4F_3]^+$, $[C_5H_4C_4F_3]^+$, $[C_{12}H_8]^+$ and/or $[C_6H_5FeF]^+$, $[C_5H_5FeF]^+$, $[C_6H_5Fe]^+$, $[C_5H_5Fe]^+$, $[C_6H_5P]^+$, $[C_6H_4P]^+$, $[C_6H_5]^+$, $[C_6H_3]^+$, $[C_5H_6]^+$, and $[C_5H_5]^+$. Some metastable ions are also found: m/e = 516, 344, 181, 129, and 44.5. The responsible species have been assigned. The appearance of a neutral metal halide fragment, FeFCl, is rare [1].

$C_5H_5Fe(CO)(P(C_6H_5)_3)C_5H_4Mn(CO)_3$ (Table **20**, No. 3). The $\nu(CO)$ bands at 1917 and 2004 cm^{-1} are assigned to the antisymmetric and symmetric modes of the CO groups at the Mn atom [5]. The cleavage of No. 3 by I_2 in CHCl$_3$ at room temperature (7 h) gave $C_5H_5Fe(CO)(P(C_6H_5)_3)I$, $C_5H_5Mn(CO)_3$, and $C_5H_4IMn(CO)_3$ [6].

$C_5H_5Fe(CO)(P(C_6H_5)_3)C_4H_5O(CF_3)_2$ (Table **20**, No. 4) is not appreciably soluble in pentane but is soluble in benzene, CH_2Cl_2, and CHCl$_3$ [2].

$C_5H_5Fe(CO)(P(C_6H_5)_3)C_4H_4O(CF_3)_2C_6H_5$ (Table **20**, No. 5) is insoluble in pentane but is soluble in benzene and CHCl$_3$ [2].

$C_5H_5Fe(CO)(P(C_6H_5)_3)C_4H_2O(CF_3)_2C_6H_5$ (Table **20**, No. 6) does not react with liquid SO$_2$ in CHCl$_3$ when refluxed for 6 h. Nor does it react with HCl in acetone within 24 h [2].

$C_5H_5Fe(CO)(P(C_6H_5)_3)^1L$ (Table **20**, Nos. 7 to 10, with $^1L = C_4H_3O$, $C_4H_2OCH_3$, C_8H_5O, and C_4H_3S). The compounds are more stable towards heat and oxygen than are the corresponding $C_5H_5Fe(CO)_2{}^1L$ compounds, but they decompose rapidly in solution. In hexane/CH$_2$Cl$_2$ they are cleaved by dry HCl at the Fe-C σ bond, C_5H_5Fe-$(CO)(P(C_6H_5)_3)Cl$ forming. However, the σ bond is stable towards attack by HgCl$_2$ [3]. No. 10 crystallizes in needle-like monoclinic crystals along the c axis: space group $Pc-C_s^2$, a = 8.26, b = 16.39, c = 8.96 Å, and β = 107.2°. Z = 2 gives D_c = 1.43, while D_m = 1.42 g·cm^{-3}. The molecular structure is shown in **Fig. 18**. The bond between the Fe atom and the thiophene ring is strikingly short [7].

Fig. 18

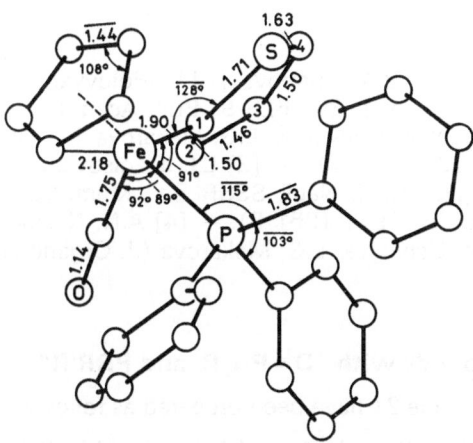

Molecular structure of $C_5H_5Fe(CO)(P(C_6H_5)_3)C_4H_3S$ [7].

Other bond angles (°):

C(1)–C(2)–C(3)	118	C(4)–S–C(1)	101
C(2)–C(3)–C(4)	105	S–C(1)–C(2)	103
C(3)–C(4)–S	112		

References on p. 204

References:

[1] R.B. King, A. Efraty (J. Fluorine Chem. **1** [1972] 283/94). – [2] D.W. Lichtenberg, A. Wojcicki (Inorg. Chem. **14** [1975] 1295/1301). – [3] A.N. Nesmeyanov, N.E. Kolobova, L.V. Goncharenko, K.N. Anisimov (Izv. Akad. Nauk SSSR Ser. Khim. **1976** 153/9; Bull. Acad. Sci. USSR Div. Chem. Sci. **1976** 142/6). – [4] K. Yasufuku, H. Yamazaki (J. Organometal. Chem. **121** [1976] 405/11). – [5] A.N. Nesmeyanov, E.G. Perevalova, L.I. Leont'eva, E.V. Shumilina (Izv. Akad. Nauk SSSR Ser. Khim. **1977** 1142/6; Bull. Acad. Sci. USSR Div. Chem. Sci. **1977** 1048/52).

[6] E.V. Shumilina (Vestn. Mosk. Univ. Khim. **32** [1977] 476/8; Moscow Univ. Chem. Bull. **32** No. 4 [1977] 76/8). – [7] V.G. Andrianov, G.N. Sergeeva, Yu.T. Struchkov, K.N. Anisimov, N.E. Kolobova, A.S. Beshastnov (Zh. Strukt. Khim. **11** [1970] 168/9; J. Struct. Chem. [USSR] **11** [1970] 163/4).

1.5.2.2.12.4 Compounds with $^2D = P(C_6H_4CH_3\text{-}4)_3$

$C_5H_5Fe(CO)(P(C_6H_4CH_3\text{-}4)_3)C_6H_5$ is formed when $C_5H_5Fe(CO)_2C_6H_5$ and $P(C_6H_5CH_3\text{-}4)_3$ in benzene are UV irradiated for 2 h at 80 °C, 96.5% yield. The red crystals melt at 171 to 173 °C with decomposition [2]. ^1H NMR $(CHCl_3)$: $\delta = 4.387$ (C_5H_5) ppm. ^{31}P NMR $(CHCl_3)$: $\delta = 70.1$ ppm [4]. IR (C_6H_6): $\nu(CO)$ 1923 cm^{-1} [2].

$C_5H_5Fe(CO)(P(C_6H_4CH_3\text{-}4)_3)C_6H_4F\text{-}4$ is formed when $C_5H_5Fe(CO)_2C_6H_4F\text{-}4$ and $P(C_6H_4CH_3\text{-}4)_3$ in benzene are UV irradiated for 5 h at 25 to 30 °C, 38.3% yield. The compound melts at 163 to 165 °C [1]. ^{19}F NMR (C_6H_6): $\delta = 13.88$ ppm (vs. C_6H_5F) [4]. The irreversible polarographic one-electron reduction at $E_{1/2} = -2.06$ V (vs. SCE) in CH_3CN or -2.01 V in $HCON(CH_3)_2$ results in cleavage of the Fe-C bond to give the $[C_5H_5Fe(CO)P(C_6H_4CH_3\text{-}4)_3]^-$ anion [3], for more comments, see 1.5.2.2.12.3.6, p. 196, introductory remarks.

References:

[1] A.N. Nesmeyanov, Yu.A. Chapovsky, I.V. Polovyanyuk, L.G. Makarova (Izv. Akad. Nauk SSSR Ser. Khim. **1968** 1628/9; Bull. Acad. Sci. USSR Div. Chem. Sci. **1968** 1536/8). – [2] A.N. Nesmeyanov, L.G. Makarova, I.V. Polovyanyuk (J. Organometal. Chem. **22** [1970] 707/12). – [3] L.I. Denisovich, I.V. Polovyanyuk, B.V. Lokshin, S.P. Gubin (Izv. Akad. Nauk SSSR Ser. Khim. **1971** 1964/9; Bull. Acad. Sci. USSR Div. Chem. Sci. **1971** 1851/5). – [4] A.N. Nesmeyanov, I.F. Leshcheva, I.V. Polovyanyuk, Yu.A. Ustynyuk, L.G. Makarova (J. Organometal. Chem. **37** [1972] 159/65).

1.5.2.2.12.5 Compounds with $^2D = PR_2R'$ and PRR'R″

The compounds in Table 21 have been prepared as follows.

Method I: $C_5H_5Fe(CO)_2{}^1L$ and ^2D are UV irradiated in C_6H_6 at room temperature unless otherwise stated. The products are separated and purified chromatographically [1, 5, 6, 12, 14 to 16].

Method II: $C_5H_5Fe(CO)_2{}^1L$ and ^2D are refluxed together in THF [2, 4, 6, 10, 12, 15, 17]. Nos. 3 to 5 and 9 form three times as fast in CH_3CN as in THF [10].

Method III: Reduction of $C_5H_5Fe(CO)(^2D)COCH_3$ to $C_5H_5Fe(CO)(^2D)C_2H_5$ where $^2D = P(CH_3)_2C_6H_5$, $P(C_6H_5)_2CH_2C_6H_5$, or $P(CH_2C_6H_5)_3$. To the starting material either $BH_3 \cdot THF$ in THF may be added or B_2H_6 is passed through the solution in C_6D_6 or THF-d_8. The progress of the reaction can be followed by ^1H NMR [17].

Method IV: No. 21 is prepared by UV irradiating a solution of No. 22 in benzene/pentane (4:1) for 3 h. After removal of the solvent, the residue is chromatographed with a mixture of the same solvents but in the proportion 1:4 [12].

Method V: No. 6 is prepared by adding an excess of C_5H_5Na to $[C_5H_5Fe(CO)-P(CH_3)_2C_6H_5]I$ in THF at $-78\,°C$ and allowing the constantly stirred mixture to warm to room temperature over 3 h. After removal of the solvent, the compound is purified by chromatography on SiO_2 (decomposition on Al_2O_3) [7].

The NMR spectra of compounds No. 3, 4, 5, and 6 show that the two CH_3 groups attached to the P atom are not equivalent [10, 16]. This is caused by the diastereotopic shielding by the asymmetric Fe atom. This effect is not observed for those compounds where the P atom is bound to three identical groups [10].

Explanations for Table 21: ^1H NMR coupling constants are J(P,H) values if not otherwise indicated. The unassigned IR bands belong to ν(CO) vibrations.

Table 21
Compounds of the $C_5H_5Fe(CO)(^2D)^1L$ Type with PR_2R' and $PRR'R''$ Donors.
Further information on numbers preceded by an asterisk is given at the end of the table.
For abbreviations and dimensions, see p. 375.

No.	^1L ligand method of preparation (conditions, yield in %)	properties and remarks explanations see above	Ref.
\multicolumn{4}{l}{$C_5H_5Fe(CO)(P(CH_3)_2C_6H_5)^1L$ compounds}			
*1	C_2H_5 III	^1H NMR (C_6D_6): 1.17 (d, PCH_3, $J=9$), ~ 1.5 (m, C_2H_5), 4.05 (C_5H_5, $J=1.0$)	[17]
*2	$CH_2C_6H_5$ I (36 h, petroleum ether)	oil ^1H NMR: 1.57 (d, CH_3, $J=8.5$), 1.85 (d, CH_3, $J=8.5$), 3.98 (C_5H_5, $J=1.5$), 7.09 (m, CC_6H_5), 7.42 (m, PC_6H_5) in $CDCl_3$, 1.69 (H_{trans} in CH_2, $J=8.0$), 2.37 (H_{vic} in CH_2, $J=5.0$, $J(H,H)=8.5$) in CH_2Cl_2 IR (petroleum ether): 1916, ~ 1925	[16]

References on p. 210

Table 21 [continued]

No.	^{1}L ligand method of preparation (conditions, yield in %)	properties and remarks explanations on p. 205	Ref.
*3	COCH$_3$ II (48 h, 76 to 85)	yellow or orange, m.p. 68, 77° ^{1}H NMR: 1.21 (CH$_3$, J=10), 1.50 (CH$_3$, J=10), 2.66 (COCH$_3$, J<0.2), 4.08 (C$_5$H$_5$, J=1.2) in C$_6$D$_6$, 1.40 (CH$_3$, J=9.3), 1.76 (CH$_3$, J=10), 2.37 (COCH$_3$), 4.22 (C$_5$H$_5$, J=1.5), 7.32 (C$_6$H$_5$, J=3) in CS$_2$ IR: 1894 in KBr, 1917 in C$_6$H$_{12}$	[2, 4, 10, 17]
4	COCH$_2$CH$_3$ II (48 h, 95)	orange, m.p. 65° ^{1}H NMR: 0.84 (t, CH$_3$, J=7.3), 1.44 (d, PCH$_3$, J=9.0), 1.76 (d, PCH$_3$, J=9.5), 1.78 (m, CH$_2$, ^{2}J(H,H)=16.1, ^{3}J(H,H)=7.3), 4.29 (d, C$_5$H$_5$, J=1.4) IR (C$_6$H$_{12}$): 1916	[10]
*5	COCH(CH$_3$)$_2$ II (48 h, 90)	red lump, low m.p. ^{1}H NMR: 0.8 (d, CH$_3$, J(H,H)=6.5), 1.00 (d, CH$_3$, J(H,H)=6.5), 1.44 (d, PCH$_3$, J=9.0), 1.76 (d, PCH$_3$, J=9.5), 3.12 (sextet, COCH, J(H,H)=6.5), 4.30 (d, C$_5$H$_5$, J=1.4) IR (C$_6$H$_{12}$): 1913	[10]
6	C$_5$H$_5$-σ V	red ^{1}H NMR (CDCl$_3$): 1.63 (J=9.0), 1.74 (J=9.8), 1.82 (J=9.2), 2.12 (J=10.0), 2.26 (J=10.9), 3.59 (J=1.7), 4.35 (J=1.7), 5.67 (J=1.7), 7.55 IR (C$_6$H$_{12}$): 1925	[7]

C$_5$H$_5$Fe(CO)(P(C$_6$H$_5$)$_2$CH$_3$)^{1}L compounds

No.	^{1}L ligand	properties and remarks	Ref.
*7	CH$_2$C$_6$H$_5$ I (6 h, petroleum ether)	m.p. 34 to 37° ^{1}H NMR (CDCl$_3$): 4.05 (C$_5$H$_5$, J=1), 7.01 (m, CC$_6$H$_5$), 7.35 (m, PC$_6$H$_5$) in CDCl$_3$, 1.62 (H$_{trans}$ in CH$_2$, J=9.9), 2.35 (H$_{vic}$ in CH$_2$, J=5.1, J(H,H)=8.2) in CH$_2$Cl$_2$ IR (petroleum ether): 1925	[13, 16]
*8	CF(CF$_3$)$_2$ I	orange brown, m.p. 100 to 108° ^{1}H NMR (CDCl$_3$): 1.93 (dd, CH$_3$, J(F,H)=2.3, J=9.2), 4.50 (d, C$_5$H$_5$, J=1.6), ~7.4 (m) ^{19}F NMR (CH$_2$Cl$_2$): 66.1, 67.4 (m, CF$_3$, ^{3}J(F,F)=10.7, ^{2}J(F,F)=10), 175.6 (m, CF, J(P,F)=26.4) IR (C$_6$H$_{12}$): 1952, 1965	[14]

References on p. 210

Table 21 [continued]

No.	^{1}L ligand method of preparation (conditions, yield in %)	properties and remarks explanations on p. 205	Ref.
*9	COCH$_2$CH$_3$ II (48 h, 95)	orange, m.p. 109° ^{1}H NMR: 0.68 (t, CH$_3$, J = 7.3), 2.02 (d, PCH$_3$, J = 9.3), 2.38 (m, CH$_2$, ^{2}J(H,H) = 16.8, ^{3}J(H,H) = 7.3), 4.30 (d, C$_5$H$_5$, J = 1.2) IR (C$_6$H$_{12}$): 1920	[10]
*10	CH$_2$Si(CH$_3$)$_3$ I (7 h, C$_6$H$_{14}$, ~20)	^{1}H NMR (C$_6$H$_{12}$): −1.00 (dd, CH$_2$, J = 12, J = 13), −0.75 (dd, CH$_2$, J = 12.2), −0.05 (s, SiCH$_3$), 1.80 (d, PCH$_3$, J = 8), 4.30 (d, C$_5$H$_5$, J = 1.5), 7.3 to 7.8 (m, C$_6$H$_5$) IR (CH$_2$Cl$_2$ or C$_6$H$_{12}$): 1910	[6, 15]
*11	COCH$_2$Si(CH$_3$)$_3$ II (20 h, ~50)	^{1}H NMR (C$_6$H$_{12}$): 0.02 (s, SiCH$_3$), 2.04 (d, CH$_2$, J = 11), 2.06 (d, PCH$_3$, J = 9.5), 2.64 (d, CH$_2$), 4.47 (d, C$_5$H$_5$, J = 1.5), 7.2 to 7.8 (m, C$_6$H$_5$) IR (CH$_2$Cl$_2$ or C$_6$H$_{12}$): 1917	[15]

C$_5$H$_5$Fe(CO)(P(C$_2$H$_5$)$_2$C$_6$H$_5$)^{1}L compounds

12	C$_6$H$_5$ I (11 h, 80°, 61)	orange, m.p. 140.5 to 141.5° ^{1}H NMR (CHCl$_3$): 4.337 (C$_5$H$_5$) ^{31}P NMR (CHCl$_3$): 63.0 IR: 1912	[5, 11]
*13	C$_6$H$_4$F-4 I (10 h, 25 to 30°, 53), I (3 h, 80°, 35.2)	orange red, m.p. 122 to 124° ^{19}F NMR (C$_6$H$_6$): 13.43	[1, 11]
14	COC$_6$H$_5$ I (11 h, 80°, 32.3)	orange oil ^{1}H NMR (CHCl$_3$): 4.393 (C$_5$H$_5$) IR: 1918	[5, 11]

C$_5$H$_5$Fe(CO)(P(C$_6$H$_5$)$_2$C$_2$H$_5$)^{1}L compounds

15	C$_6$H$_5$ I (3 h, 80°, 87), I (14 h, 80°, 52)	red yellow, m.p. 119 to 121° ^{1}H NMR (CHCl$_3$): 4.313 (C$_5$H$_5$) ^{31}P NMR (CHCl$_3$): 67.0 IR: 1916	[5, 11]
*16	C$_6$H$_4$F-4 I (6 h, 25 to 30°, 47.5), I (3 h, 80°, 69.5)	orange red, m.p. 118 to 118.5° ^{19}F NMR (C$_6$H$_6$): 13.03	[1, 11]

References on p. 210

Table 21 [continued]

No.	^{1}L ligand method of preparation (conditions, yield in %)	properties and remarks explanations on p. 205	Ref.
*17	CH$_2$Si(CH$_3$)$_3$ I (7 h, C$_6$H$_{14}$, ~20)	^{1}H NMR (C$_6$H$_{12}$): −0.8 (dd, SiCH$_2$, J= 12), −0.2 (dd, SiCH$_2$, J=12.3), 0.06 (s, SiCH$_3$), 1.0, 2.15 (m's, PC$_2$H$_5$), 4.25 (d, C$_5$H$_5$, J=1.5), 7.2 to 7.7 (m, C$_6$H$_5$) IR (C$_6$H$_{12}$ or CH$_2$Cl$_2$): 1911	[15]
*18	COCH$_2$Si(CH$_3$)$_3$ II (20 h, 52)	m.p. 109 to 111° ^{1}H NMR (C$_6$H$_{12}$): 0.05 (s, SiCH$_3$), 0.9, 2.3 (m's, PC$_2$H$_5$), 3.95 (d, SiCH$_2$, J=12), 4.39 (d, C$_5$H$_5$, J=1.5), 7.2 to 7.7 (m,C$_6$H$_5$) IR (C$_6$H$_{12}$ or CH$_2$Cl$_2$): 1918	[15]

C$_5$H$_5$Fe(CO)(P(C$_6$H$_5$)$_2$CH$_2$C$_6$H$_5$)^{1}L compounds

No.	^{1}L	properties	Ref.
19	C$_2$H$_5$ III	^{1}H NMR (C$_6$D$_6$): ~1.6 (m, C$_2$H$_5$), 4.01 (C$_5$H$_5$, J=1.2)	[17]
20	COCH$_3$ II (48 h)	^{1}H NMR (C$_6$D$_6$): 2.57 (CH$_3$, J=0.7), 4.05 (C$_5$H$_5$, J=1.2)	[17]

C$_5$H$_5$Fe(CO)[P(C$_6$H$_5$)$_2$^{1}CH$_2$^{2}CH(C$_6$H$_5$)^{3}CH$_2$^{4}CH$_3$-(S)(+)]^{1}L compounds

No.	^{1}L	properties	Ref.
*21	CH$_3$ IV (55)	orange oil ^{1}H NMR (CDCl$_3$): −0.18 (d, CH$_3$, J=6.2), −0.17 (d, CH$_3$, J=6.2), 0.42 (t, H-4, J(H,H)=7), 2.2 (m, H-3), 2.5 (m, H-1,2), 4.08 (s, C$_5$H$_5$) IR (CHCl$_3$): 1905	[12]
22	COCH$_3$ II (72 h, 73)	orange oil ^{1}H NMR (CDCl$_3$): 0.48 (t, H-4, J(H,H)= 7), 1.32 (m, H-3), 2.47, 2.58 (s's, CH$_3$), 2.7 (m, H-1), 4.23, 4.30 (s's, C$_5$H$_5$) IR (CHCl$_3$): 1917	[12]

C$_5$H$_5$Fe(CO)(P(CH$_2$C$_6$H$_5$)$_3$)^{1}L compounds

No.	^{1}L	properties	Ref.
*23	CH$_2$CH$_3$ III	^{1}H NMR (C$_6$D$_6$): 1.08 (FeCH$_2$, J=12), 1.59 (CH$_3$, J=2, J(H,H)=9.5, J(H,H)= 9), 1.88 (FeCH$_2$, J=2), 3.85 (C$_5$H$_5$, J=1.0)	[17]
*24	COCH$_3$ II (48 h)	^{1}H NMR (C$_6$D$_6$): 2.75 (CH$_3$, J=0.7), 3.14 (d, CH$_2$, J=9), 3.90 (C$_5$H$_5$, J=1.0)	[17]

C$_5$H$_5$Fe(CO)(P(CH$_3$)(C$_6$H$_5$)C$_{10}$H$_7$-2)^{1}L compounds (C$_{10}$H$_7$-2=naphth-2-yl)

No.	^{1}L	properties	Ref.
*25	CH$_3$ I (2 h, kerosene, 54)	red oil ^{1}H NMR (CDCl$_3$): −0.23 (d, CH$_3$, J=6.5), 1.72 (PCH$_3$, J=8.5), 4.12 (s, C$_5$H$_5$) IR (CH$_2$Cl$_2$): 1903	[12]

References on p. 210

*Further information:

$C_5H_5Fe(CO)(P(CH_3)_2C_6H_5)C_2H_5$ (Table 21, No. 1). In the preparation by Method III the compound reacts further with $BH_3 \cdot THF$ to give $(C_5H_5)Fe(CO)\text{-}(P(CH_3)_2C_6H_5)H$ [17].

$C_5H_5Fe(CO)(P(CH_3)_2C_6H_5)CH_2C_6H_5$ (Table 21, No. 2). The temperature dependence of the P,H coupling constants in the ^1H NMR spectrum shows that of the three possible rotamers (see Formula VIII, p. 190) form c is probably the most abundant, particularly at low temperatures. At 298 K the distribution is a:b:c = 34:24:42. The estimated potential energies above the level of rotamer c are $\Delta E = 0.18$ kcal·mol^{-1} for rotamer a and 0.33 kcal·mol^{-1} for rotamer b [16].

$C_5H_5Fe(CO)(P(CH_3)_2C_6H_5)COCH_3$ (Table 21, No. 3). The diastereotope CH_3 groups at the P atom produce two doublets in the ^1H NMR spectrum. The nonequivalence of the CH_3 groups remains and no signal broadening occurs when the compound is heated over 100 °C. This gives $k = 70.3$ s^{-1} as the upper limit of the rate of isomerization [4]. The detection of distinct signals of the three conformers requires temperatures below -100 °C. At this temperature, only broadening of the CH_3 signals was observed [2]. No. 3 is reduced by $BH_3 \cdot THF$ to No. 1 (preparative Method III) [17]. It reacts with trifluoromethyl sulfonic anhydride in CH_2Cl_2 or $CDCl_3$ to form $[C_5H_5Fe(CO)\text{-}(P(CH_3)_2C_6H_5)C=CH_2]CF_3SO_3$ [18].

$C_5H_5Fe(CO)(P(CH_3)_2C_6H_5)COCH(CH_3)_2$ (Table 21, No. 5) forms $[C_5H_5\text{-}Fe(CO)_2P(CH_3)_2C_6H_5]BF_4$ when treated for 5 min with $[(C_6H_5)_3C]BF_4$ in CH_2Cl_2 at room temperature. Ether induces crystallization [10].

$C_5H_5Fe(CO)(P(C_6H_5)_2CH_3)CH_2C_6H_5$ (Table 21, No. 7). The temperature dependence of the P,H coupling constants in the ^1H NMR spectrum shows that of the three possible rotamers (Formula VIII, p. 190) form c is the most abundant. At 298 K the distribution is a:b:c = 21:25:54. The estimated energy of a and b over that of c is $\Delta E = 0.48$ kcal·mol^{-1} [16].

$C_5H_5Fe(CO)(P(C_6H_5)_2CH_3)CF(CF_3)_2$ (Table 21, No. 8). The strong temperature dependence of the coupling constants J(P,F) in the ^{19}F NMR spectra makes plausible that the compound is present as a mixture of rotamers. This is supported by the appearance of two v(CO) bands [16].

$C_5H_5Fe(CO)(P(C_6H_5)_2CH_3)COC_2H_5$ (Table 21, No. 9) forms $[C_5H_5Fe(CO)_2\text{-}P(C_6H_5)_2CH_3]BF_4$ when treated with $[(C_6H_5)_3C]BF_4$ in CH_2Cl_2 at room temperature for 5 min. Ether induces crystallization [10].

$C_5H_5Fe(CO)(P(C_6H_5)_2CH_3)^1L$ (Table 21, Nos. 10 and 11, with $^1L = CH_2Si(CH_3)_3$ or $COCH_2Si(CH_3)_3$). The ^1H NMR spectrum shows nonequivalent CH_2 protons. In the mass spectrum the molecular ion is only found for No. 11. For both compounds the following fragments were identified: $[C_5H_5Fe(P(C_6H_5)_2CH_3)CH_2\text{-}Si(CH_3)_3]^+$, $[C_5H_5Fe(P(C_6H_5)_2CH_3)CH_2Si(CH_3)_2]^+$, $[C_5H_5Fe(P(C_6H_5)_2CH_3)]^+$, $[Fe(P(C_6H_5)_2CH_3)CH_2Si(CH_3)_2]^+$, and $[Fe(P(C_6H_5)_2CH_3)]^+$ [15].

$C_5H_5Fe(CO)(^2D)C_6H_4F\text{-}4$ (Table 21, Nos. 13 and 16, with $^2D = P(C_2H_5)_2C_6H_5$ or $P(C_6H_5)_2C_2H_5$). The dry crystals are stable in air, but solutions are less stable [1]. Polarographic one-electron reduction in CH_3CN (or in $HCON(CH_3)_2$) was observed for No. 13 at $E_{1/2} = -2.20$ (-2.14) V (vs. SCE) and for No. 16 at $E_{1/2} = -2.11$ (-2.05) V. The Fe-aryl bond is irreversibly cleaved [9]. For more information, see 1.5.2.2.12.3.6, p. 196, introductory remarks.

References on p. 210

$C_5H_5Fe(CO)(P(C_6H_5)_2C_2H_5)^1L$ (Table 21, Nos. 17 and 18, where $^1L=$ $CH_2Si(CH_3)_3$ or $COCH_2Si(CH_3)_3$). The CH_2 protons are nonequivalent [3, 15]. In the mass spectrum the parent ion was found for compound No. 18 and for both compounds the following ions were found: $[C_5H_5Fe(CO)(P(C_6H_5)_2C_2H_5)CH_2$-$Si(CH_3)_3]^+$, $[C_5H_5Fe(P(C_6H_5)_2C_2H_5)CH_2Si(CH_3)_3]^+$, $[C_5H_5FeP(C_6H_5)_2C_2H_5]^+$, $[C_5H_5Fe(P(C_6H_5)_2C_2H_5)CH_2Si(CH_3)_2]^+$, $[Fe(P(C_6H_5)_2C_2H_5)CH_2Si(CH_3)_2]^+$, and $[FeP(C_6H_5)_2C_2H_5]^+$ [15].

$C_5H_5Fe(CO)[P(C_6H_5)_2CH_2CH(C_6H_5)CH_2CH_3$-$(S)(+)]CH_3$ (Table 21, No. 21). The two diastereomers have nearly identical ^1H NMR spectra [12].

$C_5H_5Fe(CO)(P(CH_2C_6H_5)_3)C_2H_5$ (Table 21, No. 23). The CH_3 protons of the ethyl group yield an unresolved multiplet at about 1.5 ppm, while for the $P(C_6H_5)_3$ compound (1.5.2.2.12.3.4, Table 18, No. 6) three completely resolved multiplets are ascribed to the CH_3 group. The compound reacts with $BH_3 \cdot THF$ to give No. 24 and $C_5H_5Fe(CO)(P(CH_2C_6H_5)_3)H$ [17].

$C_5H_5Fe(CO)(P(CH_2C_6H_5)_3)COCH_3$ (Table 21, No. 24). $BH_3 \cdot THF$ reduces the acetyl group to give No. 23 [17].

$C_5H_5Fe(CO)(P(CH_3)(C_6H_5)C_{10}H_7$-2)$CH_3$ (Table 21, No. 25, $C_{10}H_7$-2= naphth-2-yl). The two diastereomers have identical ^1H NMR spectra [12].

References:

[1] A.N. Nesmeyanov, Yu.A. Chapovsky, I.V. Polovyanyuk, L.G. Makarova (Izv. Akad. Nauk SSSR Ser. Khim. **1968** 1628/9; Bull. Acad. Sci. USSR Div. Chem. Sci. **1968** 1536/8). − [2] H. Brunner, E. Schmidt (Angew. Chem. **81** [1969] 570/1; Angew. Chem. Intern. Ed. Engl. **8** [1969] 616). − [3] K.H. Pannell (J. Chem. Soc. Chem. Commun. **1969** 1346/7). − [4] H. Brunner, H.-D. Schindler, E. Schmidt, M. Vogel (J. Organometal. Chem. **24** [1970] 515/26). − [5] A.N. Nesmeyanov, L.G. Makarova, I.V. Polovyanyuk (J. Organometal. Chem. **22** [1970] 707/12).

[6] K.H. Pannell (J. Organometal. Chem. **21** [1970] P17/P18). − [7] A.S. Anderson (Diss. Yale Univ. 1971; Diss. Abstr. Intern. B **32** [1972] 2581). − [8] H. Brunner (Chimia [Aarau] **25** [1971] 284/6). − [9] L.I. Denisovich, I.V. Polovyanyuk, B.V. Lokshin, S.P. Gubin (Izv. Akad. Nauk SSSR Ser. Khim. **1971** 1964/9; Bull. Acad. Sci. USSR Div. Chem. Sci. **1971** 1851/5). − [10] M. Green, D.J. Westlake (J. Chem. Soc. A **1971** 367/71).

[11] A.N. Nesmeyanov, I.F. Leshcheva, I.V. Polovyanyuk, Yu.A. Ustynyuk, L.G. Makarova (J. Organometal. Chem. **37** [1972] 159/65). − [12] T.G. Attig, A. Wojcicki (J. Organometal. Chem. **82** [1974] 397/415). − [13] K. Stanley, M.C. Baird (Inorg. Nucl. Chem. Letters **10** [1974] 1111/5). − [14] K. Stanley, R.A. Zelonka, J. Thomson, P. Fiess, M.C. Baird (Can. J. Chem. **52** [1974] 1781/6). − [15] K.H. Pannell (Transition Metal. Chem. [Weinheim] **1** [1975/76] 36/40).

[16] K. Stanley, M.C. Baird (J. Am. Chem. Soc. **97** [1975] 4292/8). − [17] J.A. van Doorn, C. Masters, H.C. Volger (J. Organometal. Chem. **105** [1976] 245/54). − [18] B.E. Boland, S.A. Fam, R.P. Hughes (J. Organometal. Chem. **172** [1979] C29/ C32).

1.5.2.2.12.6 Compounds with $^2D = P(OR)_3$

The compounds in Table 22 have been prepared as follows.

Method I: $C_5H_5Fe(CO)_2{}^1L$ is UV irradiated in a solvent together with the phosphorus compound. The product is separated or purified by chromatography if not otherwise stated [2, 7 to 9, 11, 14, 16, 17, 27].

Method II: $C_5H_5Fe(CO)_2R$ is refluxed, if not otherwise stated, with the phosphorus compound in a solvent without irradiation. The compounds are separated or purified by chromatography [1, 4, 10, 22, 24, 28]. The reaction (Nos. 4 and 14) in CH_3CN is three times as fast as in tetrahydrofuran [10]. $(C_5H_5Fe(CO)_2)_2$ is formed as a by-product in the preparation of $C_5H_5Fe(CO)(P(OC_6H_5)_3)C_5H_5$ (No. 16) [24]. This preparation must be carried out in the dark because in light other products are formed [28].

Method III: Reduction of $[C_5H_5Fe(CO)(P(OC_6H_5)_3)^2L]^+$ (^2L = alkene, $CH_2=CH_2$ for No. 9, $CH_2=CHCH_2CH_3$ for No. 10, and $CH_3CH=CHCH_3$ for No. 11) with $Na[BH_3CN]$ (when not otherwise stated) in THF at room temperature [30]. For No. 12, ^2L is $CH_3CH_2CH=CHCH_2CH_3$, and the reduction is carried out with $Li[B(C_4H_9-i)_3H]$ at $-78\,^\circ C$ [30].

Method IV: The compounds No. 1, 6, 8, and 17 are obtained by the reaction $C_5H_5Fe(CO)(P(C_6H_5)_3)^1L + P(OR)_3 \rightarrow C_5H_5Fe(CO)(P(OR)_3)^1L + P(C_6H_5)_3$, where $R = CH_3$, C_4H_9-n, or C_6H_5. After completion of the exchange the solvent is removed at reduced pressure, and the required compound separated from the residue by chromatography [11, 12, 20].

Method V: $C_5H_5Fe(CO)(P(OC_6H_5)_3)I$ is reacted with RLi or RMgBr in a suitable solvent. For No. 10, a suspension of the starting material in THF is treated at $-78\,^\circ C$ with n-C_4H_9Li in hexane. The stirred mixture is warmed to room temperature over 30 min, the solvent removed, and the compound recovered from the resulting oil by benzene extraction and chromatography [30]. No. 15: The method for No. 10 is used with $(CH_3)_3SiCH(C_6H_5)MgBr$ in ether and extraction with CH_2Cl_2/petroleum ether (1:15). When prepared at $35\,^\circ C$, the stereomer A is obtained. For $0\,^\circ C$ and 1-h reaction time, a mixture of the more stable A and the less stable B, in the ratio 3:1, is obtained. Stereoisomer B decomposes when subjected to chromatography [25].

Method VI: $C_5H_5Fe(CO)(P(OC_6H_5)_3)COCH_3$ in C_6D_6 or in THF is treated with B_2H_6 or with $BH_3\cdot THF$ in THF. The progress of the reaction is followed by ^1H NMR [22].

Method VII: $C_5H_5Fe(CO)(P(OC_6H_5)_3)COCH_3$ is decarbonylized by refluxing in heptane for 60 h [12].

Method VIII: $C_5H_5Fe(CO)_2COCH_3$ is treated in benzene at $80\,^\circ C$ with $P(OC_6H_5)_3$ and UV irradiated. Small amounts of No. 8 and large amounts of the orthometalated $C_5H_5Fe(C_6H_4OP(OC_6H_5)_2)P(OC_6H_5)_3$ (Formula IV on p. 27) are obtained [6].

The n- and iso-C_3H_7 isomers of $C_5H_5Fe(CO)(P(OC_6H_5)_3)C_3H_7$ form in a 4:1 ratio by reduction of $[C_5H_5Fe(CO)(P(OC_6H_5)_3)CH_2=CHCH_3]BF_4$ with $Na[BH_3CN]$

in THF. The proportion of the iso compound increases when $NaBH_4$ is used and becomes predominant with $Li[B(C_4H_9-s)_3H]$ as the reducing agent. The two substances were not further characterized [30].

A comparison of the $v(CO)$ values of the compounds having $^2D = P(C_6H_5)_3$ or $P(OC_6H_5)_3$ (Nos. 18 and 19 in Table 22 and Nos. 5 and 6 in Table 19, p. 197) shows that as the π acceptor strength of the ^2D ligand increases the Fe–CO bonding is weakened [14].

Explanations for Table 22: ^1H NMR coupling constants are J(P,H) values if not otherwise stated. The unassigned IR bands belong to $v(CO)$ vibrations.

Table 22
Compounds of the $C_5H_5Fe(CO)(P(OR)_3)^1L$ Type.
Further information on numbers preceded by an asterisk is given at the end of the table.
For abbreviations and dimensions, see p. 375.

No.	^1L ligand method of preparation (conditions, yield in %)	properties and remarks explanations see above	Ref.
	$C_5H_5Fe(CO)(P(OCH_3)_3)^1L$ type		
*1	CH_3 I (C_5H_{12}, 38), IV (THF, 66°)	yellow oil, yellow solid, m.p. 35° ^1H NMR: -0.13 (CH_3, J=5), 3.57 (OCH_3, J=11), 4.52 (C_5H_5) in $CDCl_3$, -0.23 (d, CH_3, J=4.8), 3.52 (d, OCH_3, J=11.2), 4.47 (d, C_5H_5, J=0.8) in CD_3COCD_3 IR: 1930 in CH_2Cl_2, 1926, 1938 in C_6H_{14}	[11, 12, 27]
*2	$CH_2C_6H_5$ I (24 h, petroleum ether)	oil ^1H NMR: 3.58 (d, OCH_3, J=11), 4.27 (C_5H_5, J=1), 7.05 (m, C_6H_5) in $CDCl_3$, 1.98 (H_{trans} in CH_2, J=9.6, J(H,H) = 8.4), 2.42 (H_{vic}, J=5.0) in CH_2Cl_2 IR (petroleum ether): 1928, 1937	[15, 17]
*3	$-CH_2-$ (naphthyl) I (24 h, petroleum ether)	oil ^1H NMR: 3.66 (d, OCH_3, J=11.5), 4.17 (C_5H_5, J=1.5), 7.37 (m, $C_{10}H_7$) in $CDCl_3$, 2.40 (H_{trans} in CH_2, J=9.0, J(H,H)=8.6), 3.02 (H_{vic} in CH_2, J= 5.5) in CH_2Cl_2 IR (petroleum ether): 1928, 1936	[17]
4	$COCH_2CH_3$ II (CH_3CN)	yellow solid, m.p. 63° ^1H NMR: 0.86 (t, CH_3, J=7.0), 2.82 (q, CH_3, J=7.0), 3.56 (d, OCH_3, J=11.5), 4.56 (d, C_5H_5, J=0.9) IR (C_6H_{12}): 1936	[10]

References on p. 219

Table 22 [continued]

No.	^{1}L ligand method of preparation (conditions, yield in %)	properties and remarks explanations on p. 212	Ref.
*5	$CH_2Si(CH_3)_3$ I (24 h, petroleum ether)	oil ^{1}H NMR: −0.1 (s, $SiCH_3$), 3.44 (d, OCH_3, J=11), 4.37 (C_5H_5, J<1) in $CDCl_3$, −1.22 (H_{trans} in CH_2, J=11.4, J(H,H)=11.9), −0.50 (H_{vic}, J=3.6) in CH_2Cl_2 IR (petroleum ether): 1928, 1938	[15, 17]

$C_5H_5Fe(CO)(P(OC_4H_9-n)_3)$ ^{1}L type

6	CH_3 IV (THF, 66°)	yellow oil ^{1}H NMR ($CDCl_3$): −0.17 (CH_3, J=5), 0.89 to 1.65 ($CH_2CH_2CH_3$), 3.89 (OCH_2), 4.54 (C_5H_5) IR (CH_2Cl_2): 1923	[11, 12]
*7	$COCH_3$ II (30 to 48 h, THF, 96 to 99), II (48 h, ether, 21)	yellow–orange oil ^{1}H NMR ($CDCl_3$): 0.83 to 1.92 (m, $CH_2CH_2CH_3$), 2.57 (s, OCH_3), 3.97 (q, OCH_2), 4.67 (s, C_5H_5) ^{13}C NMR ($CDCl_3$): 14.6, 19.8, 32.9 (C_4H_9), 51.4 (CH_3), 65.1 (C_4H_9), 84.7 (C_5H_5), 218.1 (CO, J=46), 268.4 ($COCH_3$, J=32) IR: 1959 in $CHCl_3$, 1937 in THF	[1, 4, 26, 29]

$C_5H_5Fe(CO)(P(OC_6H_5)_3)$ ^{1}L type

*8	CH_3 I (petroleum ether, 20°, 100°), IV (THF, 66°), VI, VII, VIII	yellow, m.p. 106°, 107.5 to 109° ^{1}H NMR ($CDCl_3$): 0.10 (d, CH_3, J=6), 4.02 (s, C_5H_5), 7.30 (m, C_6H_5) IR (CH_2Cl_2): 1940	[6, 8, 11, 12]
*9	C_2H_5 I, III (59), VI	oil or solid, m.p. 86 to 88° ^{1}H NMR: 1.35 (t, CH_3, J=7), 1.75 (m, CH_2), 4.04 (d, C_5H_5), 7.30 (m, C_6H_5) in $CDCl_3$, ~1.5 (unresolved m), 3.93 (C_5H_5, J=0.6) in C_6D_6 ^{13}C NMR (CH_2Cl_2): −4.74 (d, CH_2, J=29.2), 22.8 (CH_3), 83.35 (C_5H_5), 121.65 (d, C–2 in C_6H_5, J=4.0), 124.7 (C–4 in C_6H_5), 129.71 (C–3 in C_6H_5), 152.13 (d, OC, J=7.3) IR (oil): 1940	[11, 22, 30]

Table 22 [continued]

No.	^{1}L ligand method of preparation (conditions, yield in %)	properties and remarks explanations on p. 212	Ref.
*10	C_4H_9-n III (64), V (44)	yellow, m.p. 95 to 96°, dec. 170° ^{1}H NMR ($CDCl_3$): 0.92 (t, CH_3, J=5.9), 1.42 (m, C_4H_9), 4.01 (d, C_5H_5, J=1.1), 7.25 (m, C_6H_5) ^{13}C NMR (CH_2Cl_2): 1.76 (d, C_4H_9, J=28.8), 14.02 (C_4H_9), 28.53 (C_4H_9), 41.51 (C_4H_9), 83.24 (C_5H_5), 121.62 (d, C-2 in C_6H_5, J=4.5), 124.65 (C-4 in C_6H_5), 129.65 (C-3 in C_6H_5), 152.06 (d, OC, J=7.4) IR (C_6H_6): 1925	[30]
*11	$CH(CH_3)CH_2CH_3$ III (56)	yellow oil, diastereomers A and B ^{1}H NMR (CCl_4): 0.98 (t, C_4H_9, J=6.2), 1.50 (m, C_4H_9), 4.13 (s, C_5H_5), 7.13 (m, C_6H_5) ^{13}C NMR (CH_2Cl_2): 16.15 (B)+16.26 (A) (C_4H_9), 21.13 (B)+21.35 (A) (d, C_4H_9, J=25.6, J=26.0), 28.42 (B)+28.57 (A) (d, C_4H_9), 28.42 (B)+28.57 (A) (C_4H_9), 39.30 (B)+40.28 (A) (s, d, C_4H_9, J=2.4), 84.01 (C_5H_5), 121.87 (C-2 in C_6H_5, J=4.1), 124.82 (C-4 in C_6H_5), 129.81 (C-3 in C_6H_5), 152.32 (d, OC, J=10.2), 220.73 (A)+220.74 (B) (pair of d, CO, J=48.1, J=48.1) IR (C_6H_{14}): 1943	[30]
*12	$C(C_2H_5)=CHC_2H_5$ III (52)	yellow oil ^{1}H NMR (CS_2): 0.73 to 1.82 (m, C_6H_{11}), 5.22 (t, C_6H_{11}, J=8.1), 7.15 (m, C_6H_5) ^{13}C NMR (CH_2Cl_2): 14.44 (C_6H_{11}), 22.71 (C_6H_{11}), 36.30 (d, C_6H_{11}, J=2.6), 45.10 (d, C_6H_{11}, J=2.8), 84.07 (C_5H_5), 121.90 (d, C-2 in C_6H_5, J=3.9), 124.71 (C-4 in C_6H_5), 129.65 (C-3 in C_6H_5), 132.78 (d, C_6H_{11}, J=6.5), 141.39 (d, C_6H_{11}, J=7.5), 152.30 (d, OC, J=9.0), 220.05 (d, CO, J=49.5) IR (C_6H_{12}): 1943	[30]
*13	$COCH_3$ II (48 h, THF, 96 to 99), II (THF, 50.7°)	yellow orange, m.p. 65° ^{1}H NMR: 2.68 (CH_3), 4.32 (s, C_5H_5), 7.46 (m, C_6H_5) in $CDCl_3$, 2.76 (CH_3, J=0.9), 4.03 (C_5H_5, J=0.9) in C_6D_6	[1, 4, 22, 26, 29]

References on p. 219

Table 22 [continued]

No.	¹L ligand method of preparation (conditions, yield in %)	properties and remarks explanations on p. 212	Ref.
		¹³C NMR (CDCl₃): 52.2 (CH₃), 84.6 (C₅H₅), 122.7, 125.2, 129.8, 151.9 (C₆H₅), 217.5 (CO, J=44), 264.5 (C=O, J=32) IR (CHCl₃): 1950	
*14	COCH₂CH₃ II (CH₃CN, 90)	yellow, m.p. 104° ¹H NMR: 0.88 (t, CH₃, J=7.3), 2.82 (m, CH₂, ²J(H,H)=9.5, ³J(H,H)=7.3), 5.02 (d, C₅H₅, J=0.6) IR (petroleum ether): 1949	[10]
*15	CH(C₆H₅)Si(CH₃)₃ isomer A (see text) V (45)	yellow ¹H NMR (C₆D₆): 0.33 (s, CH₃), 2.24 (d, CH, J=10.2), 4.05 (d, C₅H₅, J=1.0), ∼7.14 (m, C₆H₅) ¹³C NMR (C₆H₆): 2.27 (s, CH₃), 2.9 (d, CH, J∼18), 82.8 (d, C₅H₅, J=1.5), 121 to 157 (C₆H₅), 219.5 (d, CO, J=50) IR (C₆H₁₂): 1950	[25]
*16	C₅H₅-σ II (C₆H₆, room temp.)	¹H NMR (CD₃COCD₃): 3.49 (d, C₅H₅, J=0.8), 5.76 (d, σ-C₅H₅, J=1.8) IR (Nujol): 1935	[24]
*17	C₆H₅ I (2 h, C₆H₁₄, 30°, 73), I (C₆H₁₄, 70°, <73), I (2.5 h, C₆H₆, 80°, 46), IV (C₆H₆, 44)	yellow, m.p. 144.5 to 145.5°, 147 to 148° ¹H NMR: 4.28 (d, C₅H₅, J=0.8), 7.08 (m, C₆H₅) in CDCl₃, 4.238 (C₅H₅) in CHCl₃ ¹³C NMR (CDCl₃): 84.2 (C₅H₅), 121.7 (C-4 in C₆H₅), 121.7 (C-2 in OC₆H₅, J=5), 124.6 (C-4 in OC₆H₅), 126.4 (C-3 in C₆H₅), 129.4 (s, C-3 in OC₆H₅), 146.5 (C-2 in C₆H₅), 150.0 (C-1 in C₆H₅), 151.7 (d, OC, J=10), 219.3 (d, CO, J=46) IR: 1935 for the solid, 1944 in C₆H₆, 1950 in C₆H₁₂	[2, 7, 9, 13, 20, 23]
*18	C₆H₄F-4 I (5.5 h, C₆H₆, 25 to 30°, 42.8), I (C₆H₆, 80°, <42)	yellow, m.p. 130 to 132° ¹⁹F NMR: 13.10 in CHCl₃, 12.37 in C₆H₆ IR (C₆H₆, THF): 1949	[9, 13, 14]
*19	C₆H₄F-3 I (15.5 h, C₆H₆, room temp., 33)	green yellow, m.p. 148 to 150° ¹⁹F NMR (CHCl₃): 4.29 IR (THF): 1957	[13, 14]

References on p. 219   15*

Table 22 [continued]

No.	^{1}L ligand method of preparation (conditions, yield in %)	properties and remarks explanations on p. 212	Ref.
*20	C_6H_4Cl-4 I (15 h, C_6H_6, 80°, 14)	yellow, m.p. 90.5 to 91.5°	[7]
*21	C_6F_5 I (3 h, C_6H_6, 80°, ~95)	orange, m.p. 128.5 to 130° IR (solid): 1910, 1940	[7]
*22	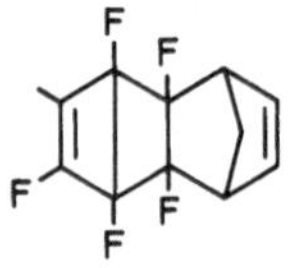 I (2 h, C_6H_{14}, room temp., ~1.5)	yellow IR: 1986, 1993	[16]

*Further information:

$C_5H_5Fe(CO)(P(OCH_3)_3)CH_3$ (Table **22**, No. 1). The solid is stable in air. It is soluble in pentane, hexane, acetone, CH_2Cl_2 [27], and $CHCl_3$ [12]. It reacts readily with refluxing SO_2 to yield $C_5H_5Fe(CO)(P(OCH_3)_3)SO_2CH_3$-(S) [12].

$C_5H_5Fe(CO)(P(OCH_3)_3)R$ (Table **22**, Nos. 2, 3, and 5, where $R=CH_2C_6H_5$, $CH_2C_{10}H_7$, and $CH_2Si(CH_3)_3$). Evaluation of the temperature dependence of the J(P,H) coupling constants of Nos. 2 and 5 shows that out of the three possible rotamers each (cf. Formula VIII, 1.5.2.2.12.3.4, p. 190) form c is the most abundant, in particularly at low temperatures. The distribution among the rotamers at 298 K is a:b:c = 24:24:52 for No. 2 and 21:16:63 for No. 5 [17], also see [15]. No such statement can be made about No. 3 since six rotamers are possibly because of the nonsymmetrical $CH_2C_{10}H_7$ ligand [17].

$C_5H_5Fe(CO)(P(OC_4H_9-n)_3)COCH_3$ (Table **22**, No. 7) is sensitive to air both as a solid and in solution. It is soluble in pentane, benzene, and CH_2Cl_2 but insoluble in methanol and water [1].

$C_5H_5Fe(CO)(P(OC_6H_5)_3)CH_3$ (Table **22**, No. 8) is stable in air as a solid. It reacts readily with liquid SO_2 under reflux to form $C_5H_5Fe(CO)(P(OC_6H_5)_3)SO_2CH_3$-(S) [11, 12]. In CH_2Cl_2 it reacts with $(CN)_2C=C(CN)_2$ at room temperature in 15 to 20 min to give the green $C_5H_5Fe(COCH_3)(P(OC_6H_5)_3)C_6N_4$ which gradually turns reddish brown at room temperature and forms $C_5H_5Fe(CO)(P(OC_6H_5)_3)N=CC(CN)_2=C(CN)_2CH_3$ (1.5.2.2.6.1.4, Table 10, No. 30) [11, 21].

$C_5H_5Fe(CO)(P(OC_6H_5)_3)C_2H_5$ (Table **22**, No. 9) is stable on heating. It is soluble in hexane, benzene, and CH_2Cl_2 [30]. With refluxing SO_2 it forms $C_5H_5Fe(CO)$-

$(P(OC_6H_5)_3)SO_2C_2H_5$ [11]. It reacts with $BH_3 \cdot THF$ to yield $C_5H_5Fe(CO)$-$(P(OC_6H_5)_3)H$ [22].

$C_5H_5Fe(CO)(P(OC_6H_5)_3)C_4H_9$-n (Table 22, No. 10) is quite stable towards thermal decomposition. No appreciable decomposition is observed in a hydrocarbon boiling at 126 °C for 24 h; the melt is stable at 135 °C for 30 min. No. 10 is soluble in hexane, benzene, and CH_2Cl_2 [30].

$C_5H_5Fe(CO)(P(OC_6H_5)_3)CH(CH_3)CH_2CH_3$ (Table 22, No. 11). The ratio of the two diastereomers A and B is 3:2 according to the ^{13}C NMR spectrum. The compound does not isomerize to the n-butyl compound when heated in hydrocarbons at 65 °C for 10 h, but the amount of less stable diastereomer diminishes appreciably. No. 11 is soluble in hexane, heptane, benzene, CCl_4, and CH_2Cl_2 [30].

$C_5H_5Fe(CO)(P(OC_6H_5)_3)C(C_2H_5)=CHC_2H_5$ (Table 22, No. 12) is soluble in hexane, benzene, CH_2Cl_2, THF, and CS_2 [30].

$C_5H_5Fe(CO)(P(OC_6H_5)_3)COCH_3$ (Table 22, No. 13) is soluble in pentane, benzene, CH_2Cl_2, and $CHCl_3$ but insoluble in CH_3OH and water [1]. When refluxed in heptane for 24 h, the starting material and traces of $C_5H_5Fe(CO)_2CH_3$ are found. After 60 h No. 8 is obtained. When a solution of No. 13 in petroleum ether is UV irradiated at 30 to 35 °C, there is no decomposition for 3 h, but after 7 h carbonyl groups cannot be observed in the IR spectrum [11, 12]. The solid is stable in air, but the $CHCl_3$ and C_6H_6 solutions rapidly decompose in air [1]. Reduction with $BH_3 \cdot$ THF gives No. 8 [22].

$C_5H_5Fe(CO)(P(OC_6H_5)_3)COC_2H_5$ (Table 22, No. 14) reacts with $[(C_6H_5)_3C]$-BF_4 in CH_2Cl_2 at room temperature to give $[C_5H_5Fe(CO)_2P(OC_6H_5)_3]BF_4$, which may be precipitated as colorless crystals by ether [10].

$C_5H_5Fe(CO)(P(OC_6H_5)_3)CH(C_6H_5)Si(CH_3)_3$ (Table 22, No. 15) is produced by Method V at 0 °C as two diastereomers, the stable A and the unstable B.

The NMR spectra of B were measured in ether/benzene (1:1). ^1H NMR: $\delta = 0.35$ (s, $SiCH_3$), 2.13 (d, CH, $^3J(P,H) = 10.5$ Hz), and 4.30 (d, C_5H_5, $^3J(P,H) = 1.0$ Hz) ppm. ^{13}C NMR: $\delta = 1.6$ (d, CH, J(P,C) = 18 Hz), 2.20 (s, $SiCH_3$), 87.70 (d, C_5H_5, J(P,C) = 1.4 Hz), 121 to 157 (C_6H_5), and 219.5 (d, CO, J(P,C) = 50 Hz) ppm. Isomer B is so unstable that it decomposes in solution at room temperature [20]. The identification of A as (SR)(RS) and B as (SS)(RR) diastereomers (Formulas I and II for one enantiomer each) was made by the temperature dependence of the vicinal coupling constants $^3J(P,H)$ [25].

Form A isomerizes in refluxing SO_2 or when SO_2 is passed through a solution of A to give a mixture of the two diastereomers. The quantity of B that is formed increases at lower temperatures, while A is favored at room temperature. At 22 °C, A is more stable than B by about 1.1 kcal $\cdot$ mol^{-1}. Petroleum ether is the only solvent found so far in which epimerization does not occur [25] (preliminary communication in [18]). Since A and B do not equilibrate in the absence of SO_2, it is assumed that the epimerization occurs between intermediate SO_2 insertion products.

The reaction with SO_2 was studied in closed tubes by NMR. Stereoisomer A does not react with SO_2 at -23 °C, reacts slowly at -3 °C, and rapidly at 17 °C to give $C_5H_5Fe(CO)(P(OC_6H_5)_3)SO_2CH(C_6H_5)Si(CH_3)_3$-(S) (denoted C). B behaves similarly, forming the SO_2 insertion product D. C and D release SO_2 slowly at low temperature to give back A and B. The equilibrium constants are rather dependent on the

References on p. 219

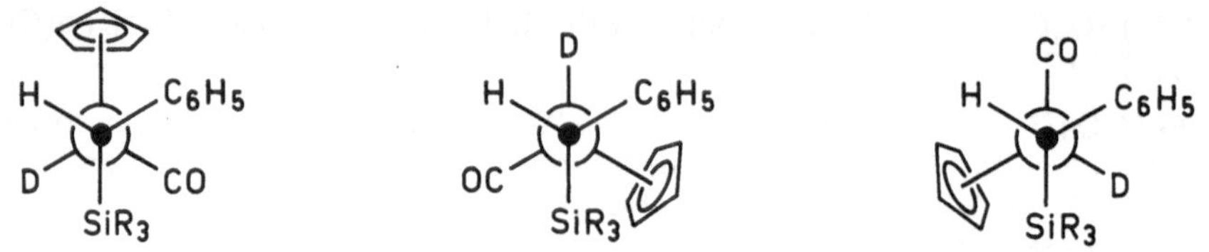

a   (SR)   I

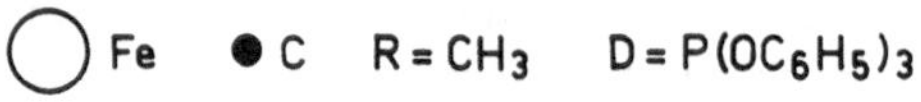

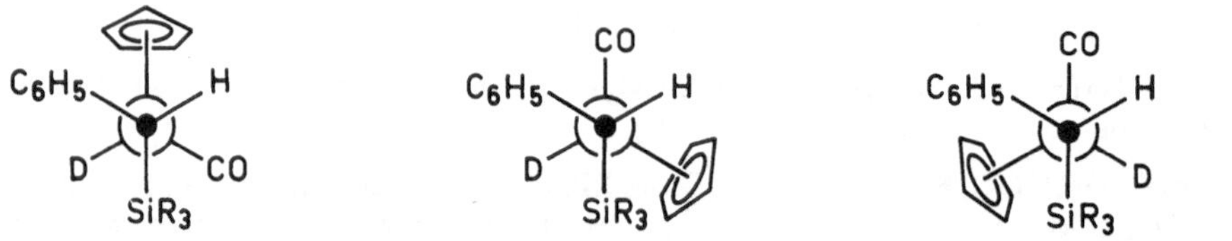

Fe   ● C   R = CH₃   D = P(OC₆H₅)₃

$$\text{Fe} \quad \bullet\, C \quad R = CH_3 \quad D = P(OC_6H_5)_3$$

a   (SS)   II

XI

SO₂ concentration. The second-order rate constants for the insertion in CDCl₃ at 22 °C are k(M⁻¹·s⁻¹)=2.1×10⁻⁴ for A and 35×10⁻⁴ for B. The activation parameters for the reaction in liquid SO₂ were determined:

SO_2 concentration. The second-order rate constants for the insertion in $CDCl_3$ at 22 °C are $k(M^{-1}\cdot s^{-1})=2.1\times10^{-4}$ for A and 35×10^{-4} for B. The activation parameters for the reaction in liquid SO_2 were determined:

	A → C	B → D
$\Delta E^{\ast}$ in kcal·mol⁻¹	20.1	19.9
$\Delta H^{\ast}$ in kcal·mol⁻¹	22.2	20.4
$\Delta S^{\ast}$ in cal·K⁻¹·mol⁻¹	7±3	2±4

These values are considerably more positive than those reported for other iron compounds [25], also see [19].

The crystals are stable in air for several hours, but the solutions decompose in air in a few minutes [25].

C₅H₅Fe(CO)(P(OC₆H₅)₃)C₅H₅ (Table **22**, No. **16**). The δ=5.76 ppm signal of the ¹H NMR spectrum shifts and splits on cooling the sample to −50 °C to give two equally intense, broad signals at δ=6.16 and 6.34 ppm [24].

$C_5H_5Fe(CO)(P(OC_6H_5)_3)C_5H_5$ (Table **22**, No. **16**). The $\delta=5.76$ ppm signal of the ^1H NMR spectrum shifts and splits on cooling the sample to -50 °C to give two equally intense, broad signals at $\delta=6.16$ and 6.34 ppm [24].

C₅H₅Fe(CO)(P(OC₆H₅)₃)C₆H₅ (Table **22**, No. **17**). The solid is stable in air [2]. It forms with P(OC₆H₅)₃ under UV irradiation in benzene at 80 °C compound IV and biphenyl [3, 7], see 1.5.2.1.1, p. 50, No. 32. When UV irradiated in cyclohexane at 27 °C, No. 17 is converted to complex III [20], see 1.5.2.2.14, p. 233, No. 20.

$C_5H_5Fe(CO)(P(OC_6H_5)_3)C_6H_5$ (Table **22**, No. **17**). The solid is stable in air [2]. It forms with $P(OC_6H_5)_3$ under UV irradiation in benzene at 80 °C compound IV and biphenyl [3, 7], see 1.5.2.1.1, p. 50, No. 32. When UV irradiated in cyclohexane at 27 °C, No. 17 is converted to complex III [20], see 1.5.2.2.14, p. 233, No. 20.

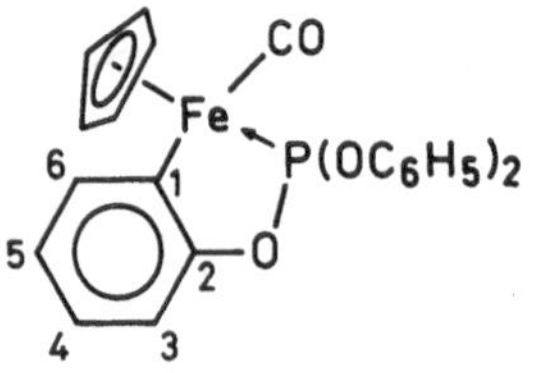

III 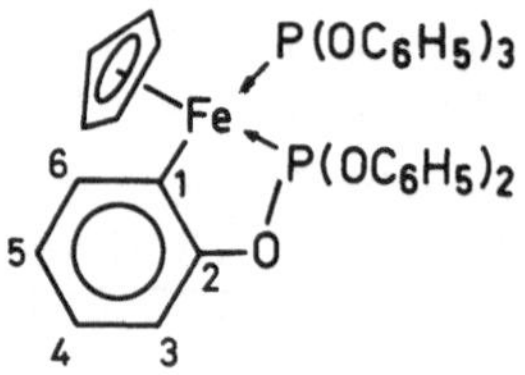   IV

$C_5H_5Fe(CO)(P(OC_6H_5)_3)C_6H_4F$ (Table 22, Nos. 18 and 19). The isomers are soluble in benzene, heptane, $CHCl_3$, and CH_2Cl_2 [14].

$C_5H_5Fe(CO)(P(OC_6H_5)_3)C_6H_4Cl$-4 (Table 22, No. 20) is soluble in petroleum ether, heptane, and benzene [7]. Irreversible polarographic one-electron reduction with cleavage of the Fe-aryl bond occurs at $E_{1/2} = -1.97$ V (vs. SCE) in CH_3CN and -1.92 V in $HCON(CH_3)_2$ [5], also see 1.5.2.2.12.3.6, p. 196.

$C_5H_5Fe(CO)(P(OC_6H_5)_3)C_6F_5$ (Table 22, No. 21) is soluble in petroleum ether, heptane, and benzene [7].

$C_5H_5Fe(CO)(P(OC_6H_5)_3)C_{11}H_6F_5$ (Table 22, No. 22) has a strong IR band at 1592 cm^{-1}. On the basis of the IR spectrum and the method of preparation, the ligand structure shown in Table 22 has been chosen. Two rotamers are expected in view of the bands in the $v(CO)$ region [16].

References:

[1] J.P. Bibler, A. Wojcicki (Inorg. Chem. **5** [1966] 889/92). – [2] A.N. Nesmeyanov, Yu.A. Chapovsky, B.V. Lokshin, I.V. Polovyanyuk, G.L. Makarova (Dokl. Akad. Nauk SSSR **166** [1966] 1125/8; Dokl. Chem. Proc. Acad. Sci. USSR **166** [1966] 213/6). – [3] A.N. Nesmeyanov, Yu.A. Chapovsky, Yu.A. Ustynyuk (Izv. Akad. Nauk SSSR Ser. Khim. **1966** 1870/1; Bull. Acad. Sci. USSR Div. Chem. Sci. **1966** 1814). – [4] I.S. Butler, F. Basolo, R.G. Pearson (Inorg. Chem. **6** [1967] 2074/9). – [5] L.I. Denisovich, S.P. Gubin, Yu.A. Chapovsky (Izv. Akad. Nauk SSSR Ser. Khim. **1967** 2378/84; Bull. Acad. Sci. USSR Div. Chem. Sci. **1967** 2271/5).

[6] A.N. Nesmeyanov, Yu.A. Chapovsky (Izv. Akad. Nauk SSSR Ser. Khim. **1967** 2075/7; Bull. Acad. Sci. USSR Div. Chem. Sci. **1967** 1988/90). – [7] A.N. Nesmeyanov, Yu.A. Chapovsky, Yu.A. Ustynyuk (J. Organometal. Chem. **9** [1967] 345/53). – [8] D.A. Brown, H. Lyons, J. Rowley, R. Sane (4th Intern. Conf. Organometal. Chem., Bristol 1969, p. J3). – [9] A.N. Nesmeyanov, L.G. Makarova, I.V. Polovyanyuk (J. Organometal. Chem. **22** [1970] 707/12). – [10] M. Green, D.J. Westlake (J. Chem. Soc. A **1971** 367/71).

[11] S.R. Su (Diss. Ohio State Univ. 1971; Diss. Abstr. Intern. B **32** [1972] 6283). – [12] S.R. Su, A. Wojcicki (J. Organometal. Chem. **27** [1971] 231/40). – [13] A.N. Nesmeyanov, I.F. Leshcheva, I.V. Polovyanyuk, Yu.A. Ustynyuk, L.G. Makarova (J. Organometal. Chem. **37** [1972] 159/65). – [14] A.N. Nesmeyanov, L.G. Makarova, I.V. Polovyanyuk (Izv. Akad. Nauk SSSR Ser. Khim. **1972** 607/9; Bull. Acad. Sci. USSR Div. Chem. Sci. **1972** 567/9). – [15] K. Stanley, M.C. Baird (Inorg. Nucl. Chem. Letters **10** [1974] 1111/5).

[16] B.L. Booth, R.N. Haszeldine, N.I. Tucker (J. Chem. Soc. Dalton Trans. **1975** 1446/8). – [17] K. Stanley, M.C. Baird (J. Am. Chem. Soc. **97** [1975] 4292/8). – [18] K. Stanley, M.C. Baird (J. Am. Chem. Soc. **97** [1975] 6598/9). – [19] K. Stanley, D. Groves, M.C. Baird (J. Am. Chem. Soc. **97** [1975] 6599/600). – [20] R.P. Stewart, J.J. Benedict, L. Isbrandt, R.S. Ampulski (Inorg. Chem. **14** [1975] 2933/6).

[21] S.R. Su, A. Wojcicki (Inorg. Chem. **14** [1975] 89/98). – [22] J.A. van Doorn, C. Masters, H.C. Volger (J. Organometal. Chem. **105** [1976] 245/54). – [23] R.P. Stewart, L.R. Isbrandt, J.J. Benedict, J.G. Palmer (J. Am. Chem. Soc. **98** [1976] 3215/9). – [24] J.A. Labinger (J. Organometal. Chem. **136** [1977] C31/C36). – [25] K. Stanley, M.C. Baird (J. Am. Chem. Soc. **99** [1977] 1808/12).

[26] E.J. Kuhlmann (Diss. Univ. Cincinnati 1978; Diss. Abstr. Intern. B **39** [1978] 2296). — [27] H.G. Alt, M. Herberhold, M.D. Rausch, B.H. Edwards (Z. Naturforsch. **34b** [1979] 1070/7). — [28] B.D. Fabian, J.A. Labinger (J. Am. Chem. Soc. **101** [1979] 2239/40). — [29] E.J. Kuhlmann, J.J. Alexander (Inorg. Chim. Acta **34** [1979] L193/L195). — [30] D.L. Reger, C.J. Coleman (Inorg. Chem. **18** [1979] 3155/60).

1.5.2.2.12.7 Compounds with Various Other P Donors

The compounds in Table 23 have been prepared as follows.

Method I: $C_5H_5Fe(CO)_2{}^1L$ compounds are reacted with the ^2D ligand (in equimolar amounts for Nos. 7 to 10, 12 to 14, 18, 21, and 22) in a refluxing solvent or at lower temperature [2 to 6, 9, 13]. Absolute darkness is necessary for Nos. 7 to 10 and 12 to 14 [13]. The reaction mixture is filtered (Nos. 6 [6], 18 [2], 22 [4]), the solvent removed (Nos. 6 [6], 15 to 18, 21 [2, 5]), and the residue chromatographed (No. 22 in CH_2Cl_2 on Al_2O_3 [4]) [4, 5] or recrystallized [2, 3, 6]. No. 6 is dissolved in C_6H_6, precipitated with hexane, and recrystallized [6], No. 21 from CH_2Cl_2/hexane [3]. The isomers of No. 12 can be separated by chromatography on SiO_2 with pentane/ether (5:1) under 63.5 atm. The two products consist of the (+) and (−) forms in the 91:9 and 13:87 ratio [12]. Nos. 7 to 9 and 12 to 14 are also separated on SiO_2 in two successive columns with pentane/benzene/ether (5:1:1) under 1.5 to 2.5 bar to give the pure diastereomers [13].

Method II: $C_5H_5Fe(CO)_2{}^1L$ is irradiated in a solution containing the phosphine at room temperature [1, 11, 12]. Compound No. 11 is purified chromatographically (SiO_2, benzene). Its diastereomers are separated by repeated fractional crystallization in ether/pentane (1:1) at −20 °C. The (−) form crystallizes out, the (+) form remains in solution [11, 12]. Nos. 13 and 14 are recrystallized from CH_2Cl_2/hexane [1].

Method III: $(−)-C_5H_5Fe(CO)(P(C_6H_5)_2N(CH_3)CH(CH_3)C_6H_5)I$ is treated in ether at −78 °C with an ether solution of $LiCH_3$. After the reaction mixture has warmed to room temperature, it is decanted from the insoluble LiI and chromatographed on SiO_2 with benzene. The orange-red zone is eluted and the solvent removed. The residual oil precipitates as a powder when stirred into pentane. This powder contains a mixture of the diastereomers (No. 11) in the ratio 32:68 of the (+) and (−) forms. The powder is dissolved in as little ether as possible, one third the volume of pentane is added, and the material allowed to crystallize overnight at −20 °C. Three crystallizations give the (−) form pure, in the ^1H NMR sense, while the (+) form remains in solution [8].

Method IV: For the preparation of No. 5, $C_5H_5Fe(CO)(P(C_6H_5)_3)C_6H_5$ and $C_6H_5OP(OCH_2)_2$ are heated together in refluxing benzene for 10 h, the solvent evaporated, and the residue chromatographed with ether/CH_2Cl_2. The pure compound is obtained from the third band by recrystallization from ether/heptane [10].

Method V: $C_5H_5Fe(CO)(^2D)COCH_3$ compounds $(^2D = P(C_6H_5)_2OC_6H_5$ or $P(OC_6H_5)_2C_6H_5$, Nos. 2 and 4) are reduced in C_6D_6 or THF with B_2H_6 or $BH_3 \cdot THF$, and the progress of the reaction (20 °C, 2 min) followed by ^1H NMR [9].

Explanations for Table 23: The coupling constants J in the ^{1}H NMR spectra give the P,H coupling. The IR bands in the 1900 to 1970 cm^{-1} range belong to ν(CO) vibrations.

Table 23
Compounds of the $C_5H_5Fe(CO)(^2D)^1L$ Type with Various Other P Donors.
Further information on numbers preceded by an asterisk is given at the end of the table.
For abbreviations and dimensions, see p. 375.

No. P donor method of preparation (conditions, yield in %)	^{1}L ligand	properties and remarks explanations see above	Ref.
*1 $P(C_6H_5)_2OC_6H_5$ V	C_2H_5	^{1}H NMR (C_6D_6): ~1.5 (unresolved m, C_2H_5), 4.07 (C_5H_5, J=1,0)	[9]
*2 $P(C_6H_5)_2OC_6H_5$ I (48 h, THF)	$COCH_3$	^{1}H NMR (C_6D_6): 2.30 (CH_3, J=0.8), 4.13 (C_5H_5, J=1.2)	[9]
*3 $P(OC_6H_5)_2C_6H_5$ V	C_2H_5	^{1}H NMR (C_6D_6): ca. 1.6 (unresolved m, C_2H_5), 4.05 (C_5H_5, J=1.0)	[9]
*4 $P(OC_6H_5)_2C_6H_5$ I (48 h, THF)	$COCH_3$	^{1}H NMR (C_6D_6): 2.68 (CH_3, J=1.0), 4.15 (C_5H_5, J=1.2)	[9]
*5 (structure: C_6H_5–P in dioxaphospholane ring) IV (66)	C_6H_5	yellow, m.p. 74 to 75° ^{1}H NMR $(CDCl_3)$: 3.71 (m, CH_2), 4.68 (d, C_5H_5, J=0.9), 7.17 (compl. m, C_6H_5) ^{31}P NMR $(CHCl_3)$: 194.5 IR (C_6H_{12}): 1953	[10]
*6 (structure: dioxaphospholane–O–menthyl) I (48 h, THF, 65)	$COCH_3$	light yellow crystals, dec. >130° IR (KBr): 1930	[6]
*7 $P(C_6H_5)_2NHCH-$ $(CH_3)C_6H_5$ diastereomers I (60 h, THF or C_6H_6, 65°, 40 (+)-7, 50 (−)-7)	$COCH_3$	yellow needles, m.p. 140 to 141° IR: 1895, 1908 in KBr, 1924 in $CHCl_3$, 1927, 1946 in $C_6H_5CH_3$	[13]
*8 $P(C_6H_5)_2NHCH-$ $(CH_3)C_6H_5$ diastereomers I (60 h, CH_3CN, 50°, 70)	COC_2H_5	yellow powder, m.p. 126° IR: 1916 in KBr, 1922 in $CHCl_3$, 1924, 1943 in $C_6H_5CH_3$	[13]

Table 23 [continued]

No. P donor method of preparation (conditions, yield in %)	^{1}L ligand	properties and remarks explanations on p. 221	Ref.
*9 $P(C_6H_5)_2NHCH$-$(CH_3)C_6H_5$ diastereomers I (80 h, CH_3CN, 40°, 50)	COC_3H_7-i	yellow solid, m.p. 40 to 42° IR: 1918 in KBr, 1918 in $CHCl_3$, 1923 in $C_6H_5CH_3$	[13]
*10 $P(C_6H_5)_2NHCH$-$(CH_3)C_6H_5$ I (60 h, CH_3CN, 50°, <1)	$COCH_2C_6H_5$	yellow solid IR ($CHCl_3$): 1922 characterized only by IR	[13]
*11 $P(C_6H_5)_2N(CH_3)$-$CH(CH_3)C_6H_5$ diastereomers II (3 h, C_6H_6, 4), III (70)	CH_3	yellow– or red orange, m.p. 131 to 133°	[8, 11]
*12 $P(C_6H_5)_2N(CH_3)$-$CH(CH_3)C_6H_5$ diastereomers I (40 h, THF, 65°, 65)	$COCH_3$	orange red, m.p. 86° IR: 1912 to 1916 in KBr, $CHCl_3$, or $C_6H_5CH_3$	[11 to 13]
*13 $P(C_6H_5)_2N(C_2H_5)$-$CH(CH_3)C_6H_5$ diastereomers I (48 h, THF or CH_3CN, 50°), 35 (+)-13, 45 (−)-13	$COCH_3$	orange red, m.p. 65 to 67° IR: 1910 to 1914 in KBr, $CHCl_3$, or $C_6H_5CH_3$	[13]
*14 $P(C_6H_5)_2N(CH_2$-$C_6H_5)CH$-$(CH_3)C_6H_5$ diastereomers I (60 h, THF or CH_3CN, 50°), 10 (+)-14, 20 (−)-14	$COCH_3$	orange red, m.p. 48 to 50° IR: 1915 to 1923 in KBr, $CHCl_3$, or $C_6H_5CH_3$	[13]
*15 $PF_2N(CH_3)_2$ I (23 h, $C_6H_{11}CH_3$, 23)	$COCH_3$	yellow–brown liquid ^{19}F NMR (CH_2Cl_2): 33.2 (J(P,F) = 1132, J(F,F) ~10), 34.4 (J(P,F) = 1152) IR (C_6H_{12}): 1949, 1965	[5]

References on p. 227

Table 23 [continued]

No. P donor method of preparation (conditions, yield in %)	[1]L ligand	properties and remarks explanations on p. 221	Ref.
*16 $PF_2N(C_2H_5)_2$ I (23 h, $C_6H_{11}CH_3$, 46)	$COCH_3$	yellow wax ^{1}H NMR $(CDCl_3)$: 1.15 (t, C_2H_5, J=7), 2.55 (s, CH_3), 3.3 (br, C_2H_5), 4.71 (d, C_2H_5, J=1) ^{19}F NMR (THF): 30.0 (J(P,F)=1128, J(F,F)~20), 31.8 (J(P,F)=1146) IR (C_6H_{12}): 1948, 1962	[5]
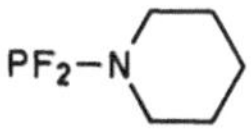 *17 $PF_2-N\langle\text{C}_5\text{H}_{10}\rangle$ I (23 h, $C_6H_{11}CH_3$, 56)	$COCH_3$	yellow solid, m.p. 41 to 42° ^{1}H NMR $(CDCl_3)$: 1.58 (s, NC_5H_{10}), 2.52 (s, CH_3), ~3.3 (br, NC_5H_{10}), 4.68 (s, C_5H_5) ^{19}F NMR (CH_2Cl_2): 32.6 (J(P,F)= 1136, J(F,F)~30), 34.0 (J(P,F)=1156) IR (C_6H_{12}): 1939, 1955	[5]
*18 $P(CH_2CH_2P\text{-}(C_6H_5)_2)_2C_6H_5$ I (76 h, CH_3CN, 93)	$COCH_3$	yellow brown, m.p. 62 to 64° ^{1}H NMR $(CDCl_3)$: 2.25 (CH_3), 2.41 (d, CH_3, J=6), 4.17 (d, C_5H_5, J=1) for ^{31}P NMR, see further information IR (CH_2Cl_2): 1913	[2]
*19 $P(C_6H_5)_2CH_2\text{-}CH_2P(C_6H_5)_2$ II (22 h, C_6H_{14}, 70)	C_2F_5	orange, dec. 165° ^{1}H NMR (CD_3COCD_3): 4.40 (s, C_5H_5) ^{19}F NMR (CH_2Cl_2): 65.8 (CF_2), 81.5 (CF_3) IR (CH_2Cl_2): 1950	[1]
*20 $P(C_6H_5)_2CH_2\text{-}CH_2P(C_6H_5)_2$ II (22 h, C_6H_{14}, 52)	$CF(CF_3)_2$	orange, dec. 192° ^{19}F NMR (CH_2Cl_2): 65.0 (CF_3) IR (CH_2Cl_2): 1955	[1]
*21 $(P(C_6H_5)_2CH_2CH_2\text{-}P(C_6H_5)CH_2)_2$ I (20 h, CH_3CN, 63)	$COCH_3$	orange, m.p. 57 to 60° IR (CH_2Cl_2): 1913	[3]
*22 $P(CH_2CH_2As\text{-}(C_6H_5)_2)_2C_6H_5$ I (10 h, CH_3CN, 57)	$COCH_3$	orange, m.p. 134 to 136° ^{1}H NMR $(CDCl_3)$: 1.8 to 2.0 (CH_2), 2.42 (CH_3), 4.18 (C_5H_5), 7.15 (C_6H_5) IR (CH_2Cl_2): 1909	[4]

References on p. 227

Table 23 [continued]

No. P donor method of preparation (conditions, yield in %)	[1]L ligand	properties and remarks explanations on p. 221	Ref.
23 Pfc$_3$ (fc = ferrocene)	C$_6$H$_4$F-4	see "Eisen-Organische Verbindungen" A6, 1977, p. 257, Table 35, No. 13	

supplement

*24 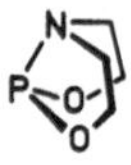	C$_6$H$_5$	amber ^{31}P NMR: 218 IR (THF): 1920	[14]

*Further information:

C$_5$H$_5$Fe(CO)(^{2}D)^{1}L (Table **23**, Nos. **1** to **4**, with ^{2}D = P(C$_6$H$_5$)$_2$OC$_6$H$_5$ or P(OC$_6$H$_5$)$_2$C$_6$H$_5$ and ^{1}L = C$_2$H$_5$ or COCH$_3$) are soluble in benzene and THF. Nos. 1 and 3 are reduced with BH$_3$·THF in THF to C$_5$H$_5$Fe(CO)(^{2}D)H. Nos. 2 and 4 can be converted by Method V to Nos. 1 and 3, respectively [9].

C$_5$H$_5$Fe(CO)(P(OCH$_2$)$_2$C$_6$H$_5$)C$_6$H$_5$ (Table **23**, No. **5**) is soluble in petroleum ether, heptane, cyclohexane, benzene, ether, CH$_2$Cl$_2$, and chloroform [10].

C$_5$H$_5$Fe(CO)(P(OCH$_2$)$_2$OC$_{10}$H$_{19}$)COCH$_3$ (Table **23**, No. **6**) could not be separated into its two diastereomers. The specific rotation [α]$_{546}^{20}$ is between −60° and −70°. The compound is monomeric in benzene. It is readily soluble in benzene and THF, less so in hexane [6].

C$_5$H$_5$Fe(CO)(P(C$_6$H$_5$)$_2$NRCH(CH$_3$)C$_6$H$_5$)^{1}L (Table **23**, Nos. **7** to **14**, with R = H, CH$_3$, C$_2$H$_5$, CH$_2$C$_6$H$_5$ and ^{1}L = CH$_3$, COCH$_3$, COC$_2$H$_5$, COC$_3$H$_7$-i, COCH$_2$C$_6$H$_5$). No. 11 can be obtained in higher yield by photochemical decarbonylation of the COCH$_3$ group of No. 12 [11]. The compounds appear as two diastereomers which can be separated by preparative Methods I and III. The two forms are designated as (+) and (−) in accord with their specific rotations [α]$_{436}^{20}$ (λ = 436 nm and t = 20 °C); the concentration was ∼3.5 × 10^{-3} M in toluene solution [13]:

No.	7	8	9	12	13	14
(+) forms	+2430	+2140	+2275	+1345	+1220	+1040
(−) forms	−2300	−2190	−2295	−1320	−1425	−1000

For diastereomers (−)-11 and (−)-12 the values [α]^{20} are:

λ in nm	365	436	546	578	579	589
(−)-11 in C$_6$H$_6$ (1 mg/mL)	−1340	−2030	+720	—	+410	+340
(−)-12 in C$_6$H$_5$CH$_3$ (4 × 10^{-3} M)	−4725	−1200	—	−1250	—	—

The circular dichroism spectra are presented in [8, 12]. The chemical shifts δ(in ppm) of the ^{1}H NMR spectra are listed in Table 24; in parentheses the multiplicity and/or

Table 24
¹H NMR spectra of $C_5H_5Fe(CO)(P(C_6H_5)_2NRCH(CH_3)C_6H_5)^1L$ Compounds.
Explanations are given in the text under Nos. 7 to 14.

diastereomer (solvent)	phosphine ligand ¹CH	²CH (t's)	³CH (m's)	⁴CH (d's)	¹L ligand ⁵CH	⁶CH	C₅H₅ (d's)	C₆H₅ (m's)	Ref.
(+)-7 (B)	—	—	5.43	1.28(*7)	2.71(d)	—	4.08(1)	7.13	[13]
(−)-7 (A)	—	—	5.43	0.98(*7)	2.87(d)	—	4.07(1)	7.03	[13]
(+)-8 (B)	—	—	5.21	1.04(*6.9)	3.26(m)	1.23(t, 6.5)	4.12(1.3)	7.15	[13]
(+)-8 (A)	—	—	5.49	1.33(*7.3)	2.95(m)	0.87(t, 7)	4.38(1.4)	7.34	[13]
(−)-8 (B)	—	—	5.21	1.34(*6.9)	3.14(m)	1.19(t, 6.5)	4.12(1.3)	7.04	[13]
(−)-8 (A)	—	—	5.49	1.03(*6.7)	2.95(m)	0.84(t, 7)	4.36(1.4)	7.22	[13]
(+)-9 (A)	—	—	4.07	1.32(*7)	3.30(m)	0.82(d, 7) 1.04	4.38(1.4)	7.23	[13]
(−)-9 (A)	—	—	4.07	1.15(*7)	3.28(m)	0.82(d, 7) 1.04	4.38(1.4)	7.40	[13]
(+)-11 (B)	1.52(d, 7) 1.52(d, 7)	—	4.03 4.03(q, *7)	0.72(*7) 0.72(*7)	−0.36(d, 2) −0.36(d, 6)	—	3.41(1) 3.41(1.8)	6.50 6.50	[11] [8]
(−)-11 (B)	1.43(d, 7)	—	4.03(*7)	0.90(*7)	−0.36(d, 6)	—	3.45(1)	6.50	[8, 11]
(+)-12 (B)	2.03(d, 7) 1.50(d, 7)	—	5.47 4.90	1.58(*7) 0.76(*7)	2.57(d) 2.06(s)	—	4.17(1) 3.62(1)	7.15 6.64	[13] [8]
(−)-12 (A)	2.07(d, 7) 1.53(d, 7)	—	5.43 4.90	1.32(*7) 1.03(*7)	2.58(d) 2.06(s)	—	4.12(1) 3.64(1)	7.13 6.64	[13] [8]
(+)-13 (B)	2.9(m)	0.39(t, *7)	5.58	1.72(*7)	2.63(d, 1)	—	4.13(1)	7.2	[13]
(−)-13 (A)	2.9(m)	0.01(t, *7)	5.58	1.40(*7)	2.67(d, 1)	—	4.18(1)	7.2	[13]
(+)-14 (B)	4.20(m)	—	5.13	1.43(*7)	2.76(d)	—	4.10(1)	6.97	[13]
(−)-14 (B)	4.20(m)	—	5.13	1.10(*7)	2.60(d)	—	3.97(1)	6.97	[13]

the coupling constants (in Hz), J(P,H) or, with an asterisk, $^3J(H,H)$ of the $CHCH_3$ or CH_2CH_3 group. The atoms are labelled $N(^1C-^2C)-^3C(C_6H_5)-^4C$ in the P ligand and $^5CH_3$ or $^5CH_2-^6CH_3$ in the ^1L ligand; A and B denote the solvents CD_3COCD_3 and C_6D_6, respectively.

The pure optical isomers are configurationally stable at room temperature. At higher temperatures they equilibrate with $C_5H_5Fe(CO)_2{}^1L$ and epimerize with respect to the configuration about iron [13]. No change is noted when No. 12 is heated in C_6D_6 to 77 °C, but in a thermostatic bath, a cleavage into $C_5H_5Fe(CO)_2CH_3$ and $P(C_6H_5)_2N(CH_3)CH(CH_3)C_6H_5$ begins at 80 °C, and a more rapid equilibrium is established at 90 to 100 °C between No. 8 and the phosphine employed for its preparation. The ratio of the diastereomers changes during the equilibration, and after 10 h at 97 °C attains a value of 42:58 for the $(+)$ and $(-)$ forms, which does not vary further [12]. Compound No. 11 is soluble in pentane, benzene, and CH_2Cl_2, while Nos. 7 to 10 and 12 to 14 are soluble in pentane, benzene, toluene, and ether [11, 13]. The mass spectra of Nos. 7 to 9 and 12 to 14 show the molecular ion $[M]^+$, $[M-R]^+, [M-CO]^+, [M-2CO]^+, [M-COR]^+, [M-2CO-R]^+, [M-COR-C_5H_5]^+,$ $[M-2CO-R-C_5H_5]^+, [C_5H_5FeP(C_6H_5)_2NR']^+, [P(C_6H_5)_2N(R')CH(CH_3)C_6H_5]^+,$ $[C_5H_5FeP(C_6H_5)_2]^+, [C_5H_5FeCH(CH_3)CH_2(C_6H_5)]^+, [P(C_6H_5)_3]^+, [C_5H_5FeNCH-$ $(C_6H_5)CH_3]^+, [P(C_6H_5)_2NR']^+,$ and $[C_5H_5Fe(CO)_2R]^+$ [13]. The mass spectrum of No. 11 contains $[M]^+, [M-CH_3]^+, [M-CO]^+, [M-CO-CH_3]^+, [M-CH_3-C_5H_5]^+,$ $[C_5H_5Fe(CO)CH_3]^+, [M-CH_3-C_5H_5-CO]^+, [P(C_6H_5)_2C_9H_{12}N]^+, [P(C_6H_5)_2-$ $NCH_3]^+,$ and $[Fe]^+$. No. 11 reacts with iodine in ice-cooled CH_2Cl_2 to form C_5H_5- $Fe(CO)(P(C_6H_5)_2N(CH_3)CH(CH_3)C_6H_5)I$. It reacts with chlorine and bromine to form the corresponding halogen compounds, which epimerize [8].

$C_5H_5Fe(CO)(^2D)COCH_3$ (Table 23, Nos. 15 to 17, with $^2D=PF_2N(CH_3)_2$, $PF_2N(C_2H_5)_2$, and $PF_2NC_5H_{10}$). The $v(CO)$ doublets indicate the presence of two stereoisomers each. The compounds are soluble in petroleum ether, pentane, cyclohexane, and CH_2Cl_2; THF is mentioned as solvent for No. 16. The mass spectrum of No. 17 shows the following ions: $[C_5H_5Fe(CO)_n(PF_2NC_5H_{10})COCH_3]^+$ $(n=0, 1), [C_5H_5Fe(CO)_nPF_2NC_5H_{10}]^+ (n=0$ to 2), $[C_5H_5Fe(PF_2NC_5H_{10})CH_3]^+,$ $[C_5H_5FePF_2NC_5H_9]^+, [C_5H_5Fe(CO)_nCH_3]^+ (n=1, 2), [PF_2C_5H_n]^+ (n=9, 10),$ $[C_6H_nFe]^+ (n=6, 8), [PFNC_5H_{10}]^+,$ and/or $[C_6H_6Fe]^+, [C_5H_5Fe]^+, [C_6H_6]^+$ and/or $[CH_2NPF]^+,$ and $[PF_2]^+$ [5].

$C_5H_5Fe(CO)(P(CH_2CH_2P(C_6H_5)_2)_2C_6H_5)COCH_3$ (Table 23, No. 18) is present as a mixture of isomers. Their structures were assigned to Formulas I and II on the basis of the ^{31}P NMR spectra (in CH_2Cl_2, P-1 is coordinated): $\delta=-12.7$ (P-2) and $+69.5$ (P-1) ppm for I ($R=C_6H_5$) and $\delta=-15.5$ (P-2), -12.7 (P-2), and $+74.8$ (P-1) ppm for II [7]. The reverse assignment in [7] must be an error. The compound is soluble in petroleum ether, hexane, benzene, CH_3CN, CH_2Cl_2, and $CHCl_3$ [2].

I

II

$C_5H_5Fe(CO)(P(C_6H_5)_2CH_2CH_2P(C_6H_5)_2)^1L$ (Table 23, Nos. 19 and 20, with $^1L=C_2F_5$ and $CF(CF_3)_2$) are readily soluble in CH_2Cl_2 and soluble in acetone and hexane. When irradiated in C_6H_6, No. 19 and 20 give the $C_5H_5Fe(P(C_6H_5)_2CH_2-)_2{}^1L$ compounds with chelating diphosphine [1], see 1.5.2.1.1, p. 50.

$C_5H_5Fe(CO)[(P(C_6H_5)_2CH_2CH_2P(C_6H_5)CH_2-)_2]COCH_3$ (Table 23, No. 21) is very readily soluble in CH_2Cl_2 and is soluble in hexane and acetonitrile [3].

$C_5H_5Fe(CO)(P(CH_2CH_2As(C_6H_5)_2)_2C_6H_5)COCH_3$ (Table 23, No. 22) has a very broad CH_2 resonance in the ^1H NMR spectrum, but it is attributed to paramagnetic impurities. The compound is soluble in hexane, benzene, $CHCl_3$, CH_2Cl_2, CH_3CN, and ethanol [4].

$C_5H_5Fe(CO)(P(OCH_2CH_2)_2N)C_6H_5$ (Table 23, No. 24) was recently obtained in an attempt to deprotonate the quaternary nitrogen atom in $[C_5H_5Fe(CO)-NH(C_2H_4O)_2PC_6H_5]PF_6$; details are given under compound No. 14 in Table 6, pp. 80 and 83. The structure has been confirmed by an X-ray study. The monoclinic crystals have the parameters $a = 7.776(1)$, $b = 14.477(3)$, $c = 14.083(1)$ Å, and $\beta = 102.04(1)°$, space group $P2_1/n - C_{2h}^5$. $Z = 4$ gives $D_c = 1.539$, while $D_m = 1.52$ g·cm^{-3}. The molecular structure is illustrated. Rather short bonds were noted for Fe–$C(C_6H_5)$ 2.04 Å, Fe–P 2.105 Å, and P–N 1.692 Å. HCl gas in THF converts the compound to the cationic starting material [14].

References:

[1] R.B. King, R.N. Kapoor, K.H. Pannell (J. Organometal. Chem. **20** [1969] 187/93). – [2] R.B. King, P.N. Kapoor, R.N. Kapoor (Inorg. Chem. **10** [1971] 1841/50). – [3] R.B. King, R.N. Kapoor, M.S. Saran, P.N. Kapoor (Inorg. Chem. **10** [1971] 1851/60). – [4] R.B. King, P.N. Kapoor (Inorg. Chim. Acta **6** [1972] 391/4). – [5] R.B. King, W.C. Zipperer, M. Ishaq (Inorg. Chem. **11** [1972] 1361/70).

[6] H. Brunner, E. Schmidt (J. Organometal. Chem. **50** [1973] 219/25). – [7] R.B. King, J.C. Cloyd (Inorg. Chem. **14** [1975] 1550/4). – [8] H. Brunner, G. Wallner (Chem. Ber. **109** [1976] 1053/60). – [9] J.A. van Doorn, C. Masters, H.C. Volger (J. Organometal. Chem. **105** [1976] 245/54). – [10] R.P. Stewart, L.R. Isbrandt, J.J. Benedict, J.G. Palmer (J. Am. Chem. Soc. **98** [1976] 3215/9).

[11] H. Brunner, M. Muschiol, W. Nowak (Z. Naturforsch. **33b** [1978] 407/11). – [12] H. Brunner, H. Vogt (Z. Naturforsch. **33b** [1978] 1231/4). – [13] H. Brunner, H. Vogt (J. Organometal. Chem. **191** [1980] 181/92). – [14] P. Vierling, J.G. Riess, A. Grand (J. Am. Chem. Soc. **103** [1981] 2466/7).

1.5.2.2.12.8 Compounds with As, Sb, O, and S Donors

The compounds listed in Table 25 have been prepared as follows.

Method I: A compound of the type $(C_5H_5)Fe(CO)_2{}^1L$ is reacted in solution between room temperature and 30 °C with the ligand ^2D under UV irradiation [1, 3, 4]. The crystals of Nos. 2 and 6 that precipitate are washed repeatedly with cold CH_3OH and are recrystallized from hexane/methanol [3].

Method II: $(C_5H_5)Fe(CO)_2{}^1L$, where $^1L = C_6H_{11}$ or $CH_2C_6H_{11}$, is dissolved in $(CH_3)_2SO$ or in 30 mol % $(CH_3)_2SO$ in $CDCl_3$ and stirred at 37 °C. Equilibration between the starting compound and Nos. 8 or 9 is established in 4 h, the solutions having become red [2].

Table 25
Compounds of the $C_5H_5Fe(CO)(^2D)^1L$ Type with As, Sb, O, and S Donors.
Further information on numbers preceded by an asterisk is given at the end of the table.
For abbreviations and dimensions, see p. 375.

No. ^2D method of preparation (conditions, yield in %)	^1L ligand	properties and remarks IR bands are $\nu(CO)$	Ref.
with As and Sb donors			
1 $As(C_6H_5)_3$ I (15 min, C_6H_6, CD_3CN, or C_6H_5CN)	CH_3	^1H NMR: 0.41 (CH_3), 4.23 (C_5H_5) in C_6H_6, 0.15 (CH_3), 4.36 (C_5H_5) in CD_3CN, 0.08 (CH_3), 4.34 (C_5H_5) in C_6H_5CN	[4]
*2 $As(C_6H_5)_3$ I (30 min, CH_3CN, 42)	$COCH_3$	red brown, m.p. 127 to 128° ^1H NMR (CS_2): 1.96 (CH_3), 4.66 (C_5H_5), 7.20 to 7.5 (C_6H_5) IR (CH_2Cl_2): 1917	[3]
*3 $As(C_6H_5)_3$ I (6 h, petroleum ether)	$CH_2C_6H_5$	red oil IR ($CHCl_3$): 1910	[1]
*4 $1,2-(As(CH_3)_2)_2C_6H_4$ I (15 min, C_6H_6, CD_3CN, or C_6H_5CN)	CH_3	no properties reported	[4]
5 $Sb(C_6H_5)_3$ I (15 min, C_6H_6, CD_3CN, or C_6H_5CN)	CH_3	^1H NMR: 0.38 (CH_3), 4.33 (C_5H_5) in C_6H_6, -0.19 (CH_3), 4.52 (C_5H_5) in CD_3CN, 0.06 (CH_3), 4.49 (C_5H_5) in C_6H_5CN	[4]
*6 $Sb(C_6H_5)_3$ I (30 min, CH_3CN)	$COCH_3$	dark brown ^1H NMR (CS_2): 1.89 (CH_3), 4.72 (C_5H_5), 7.25 to 7.45 (C_6H_5) IR (CH_2Cl_2): 1916	[3]
with O and S donors			
*7 $S(C_2H_5)_2$ I	CH_3	–	[4]
*8 $OS(CH_3)_2$ II	$COC_6H_{11}-$ cyclo	^1H NMR: 2.92 (m, H–1 in C_6H_{11}), 4.67 (s, C_5H_5) in CH_3SOCH_3, 2.88, 3.22 (s's, CH_3) in $CDCl_3/CH_3SOCH_3$ IR (CH_3SOCH_3): 2010	[2]
*9 $OS(CH_3)_2$ II	$COCH_2-$ $C_6H_{11}-$ cyclo	^1H NMR: 2.75 (d, CH_2, J=6), 4.67 (s, C_5H_5) in CH_3SOCH_3, 2.88, 3.22 (s's, CH_3) IR (CH_3SOCH_3): 2010	[2]

*Further information:

$C_5H_5Fe(CO)(As(C_6H_5)_3)COCH_3$ (Table **25**, No. **2**) readily decomposes in the common solvents, e.g., $CHCl_3$, ether, or C_6H_6, a fact which hinders purification [3].

$C_5H_5Fe(CO)(As(C_6H_5)_3)CH_2C_6H_5$ (Table **25**, No. **3**) reacts in $CHCl_3$ at 27 °C with SO_2 passed through the solution to form $C_5H_5Fe(CO)(As(C_6H_5)_3)SO_2CH_2C_6H_5$ [1] (1.5.2.2.6.1.7, p. 130, Table **11**, No. **4**).

$C_5H_5Fe(CO)(1,2-(As(CH_3)_2)_2C_6H_4)CH_3$ (Table **25**, No. **4**) is mentioned in [4], but its existence was demonstrated only qualitatively by NMR.

$C_5H_5Fe(CO)(Sb(C_6H_5)_3)COCH_3$ (Table **25**, No. **6**) readily decomposes in the common solvents, e.g., $CHCl_3$, ether, or C_6H_6, a fact which hinders purification [3].

$C_5H_5Fe(CO)(S(C_2H_5)_2)CH_3$ (Table **25**, No. **7**) is mentioned in [4], but its existence was demonstrated only qualitatively by NMR.

$C_5H_5Fe(CO)(OS(CH_3)_2)COR$ (Table **25**, Nos. **8** and **9**, with $R = C_6H_{11}$-cyclo and $CH_2C_6H_{11}$-cyclo) are found only in equilibrium with the starting materials (preparative Method II). They could not be isolated. They react in solution in $(CH_3)_2SO$ or in 30 mol % $(CH_3)_2SO$ in $CDCl_3$ with $P(C_6H_5)_3$ to form the stable $C_5H_5Fe(CO)$-$(P(C_6H_5)_3)COR$, with a rate constant of 4.3×10^{-4} $L \cdot mol^{-1} \cdot s^{-1}$ for $^1L = COCH_2C_6H_{11}$ [2].

References:

[1] M. Graziani, A. Wojcicki (Inorg. Chim. Acta **4** [1970] 347/50). – [2] K. Nicholas, S. Raghu, M. Rosenblum (J. Organometal. Chem. **78** [1974] 133/7). – [3] A.C. Giugell, A.J. Rest (J. Organometal. Chem. **99** [1975] C27/C28). – [4] C.R. Folkes, A.J. Rest (J. Organometal Chem. **136** [1977] 355/61).

1.5.2.2.12.9 Compounds of the $C_5H_{5-n}R_nFe(CO)(^2D)^1L$ Type

$C_5H_4CH(C_6H_5)_2Fe(CO)(P(C_6H_5)_3)CONH_2$ is prepared by treating $[C_5H_4CH(C_6H_5)_2Fe(CO)_2P(C_6H_5)_3]Cl$ for 20 min at -33 °C with liquid NH_3. The NH_3 is distilled off in a vacuum at -50 °C. The compound is extracted from the reaction mixture with C_6H_6 or CH_2Cl_2. IR spectrum: $\nu(CO)$ 1919 cm^{-1} in both Nujol and CH_2Cl_2. (See paper for details of the IR spectrum and its analysis.) The compound is very stable, insoluble in liquid NH_3, but quite soluble in C_6H_6 and CH_2Cl_2 [1].

$C_5H_3(CH_3-1)(C_6H_5-3)Fe(CO)(P(C_6H_5)_3)CH_3$ is obtained when a solution of $C_5H_3(CH_3-1)(C_6H_5-3)Fe(CO)(P(C_6H_5)_3)COCH_3$ (following compound) in benzene is UV irradiated for 3 h at room temperature. It is separated from the reaction mixture by chromatography on Al_2O_3 with pentane/benzene (4:1), 75% yield.

The separation of the diastereomeric forms a and b did not succeed completely since the compounds are not stable in the presence of Al_2O_3. They are best prepared from optically pure starting materials, see the acetyl compound on p. 230. An 85:15 mixture of a to b converts to a 97:3 mixture of a to b after standing for 24 h at 5 °C under N_2. The pure a form may be obtained from this by fractional crystallization.

The melting point of a 97% a + 3% b mixture is 137 to 138 °C (dec.) and that of a 8% a + 92% b mixture is 139 to 141 °C (dec.). ^1H NMR spectra (in C_6D_6, δ in ppm): 0.09 (d, CH_3, J(P,H) = 6.4 Hz), 1.78 (s, CH_3C_5), 3.81, 4.10, and 4.27 (m's, C_5H_3) for form a, 0.18, 1.47, 3.21, 4.30, and 4.4 (same multiplicities and assignments) for form b [5], also see [4] and figures in [2]. IR spectrum (in CH_2Cl_2): $v(CO)$ 1904 cm^{-1} [5].

A solution of a and b in a 70:30 ratio in C_6H_6 is epimerized by UV irradiation, in 5 min to 64:36, in 10 min to 60:40, in 15 min to 55:45, and in 20 min to 54:46 [5] (preliminary communication [2]). The compound retains its configuration in C_6H_6, THF, or CH_2Cl_2 under N_2 at 25 °C, either in the dark or in laboratory light, but epimerizes and partly decomposes when held for 3 h in $CDCl_3$ [5]. No epimerization is observed in the presence of $C_5H_3(CH_3-1)(C_6H_5-3)Fe(CO)(P(C_6H_5)_3)I$ or CH_3HgI in CH_2Cl_2 solution. In the presence of I_2, HI, or HgI_2, the Fe–CH_3 bond is cleaved to give $C_5H_3(CH_3-1)(C_6H_5-3)Fe(CO)(P(C_6H_5)_3)I$. The course of this reaction is stereospecifically different for the a and b forms. The a to b ratios obtained from a-rich (or b-rich) mixtures are given below:

reaction with	I_2	HI	HgI_2
initial mixture	100:0 (8:92)	90:10 (14:86)	100:0 (8:92)
starting material in end product	90:10 (20:80)	73:27 (27:73)	85:15 (17:83)
iodine compound in end product	69:31 (22:78)	53:47 (33:67)	50:50 (34:66)

A complete conversion of the starting material is obtained on reaction with SO_2 to form $C_5H_3(CH_3-1)(C_6H_5-3)Fe(CO)(P(C_6H_5)_3)SO_2CH_3$. The course of the reaction is stereospecific although the solvent also has an influence. The iodine compound $C_5H_3(CH_3-1)(C_6H_5-3)Fe(CO)(P(C_6H_5)_3)I$ is formed in liquid SO_2 in the presence of iodine in the ratio 37:63 from the 90:10 starting material. In the absence of I_2, the SO_2 insertion compound is formed. The a to b ratios from a-rich (or b-rich) mixtures are:

solvent	CH_2Cl_2	SO_2, −60 °C	SO_2, −10 °C
initial mixture	97:3 (8:92)	90:10 (8:92)	77:23
SO_2 insertion product	95:5 (10:90)	83:17 (18:82)	71:29

Somewhat different mechanisms are assumed for the SO_2 insertion and the iodine cleavage. The contact ion pair $[C_5H_3(CH_3-1)(C_6H_5-3)Fe(CO)P(C_6H_5)_3]^+$-$[O_2SCH_3]^-$ is first formed by dissociation on reaction with SO_2, and then it reacts to complete the SO_2 insertion. When iodide is present in the SO_2 solution, it combines with the cation, preventing the SO_2 insertion. Epimerization in this case occurs via the cation. The configuration at the Fe atom is retained to a large degree, both in SO_2 insertion and iodine cleavage [6], also see [3, 4].

$C_5H_3(CH_3-1)(C_6H_5-3)Fe(CO)(P(C_6H_5)_3)COCH_3$ is prepared from C_5H_3-$(CH_3-1)(C_6H_5-3)Fe(CO)_2CH_3$ and $P(C_6H_5)_3$ by refluxing in CH_3CN for 16.5 h. The solvent is removed, the residue taken up in a minimal quantity of CH_2Cl_2, transferred to an Al_2O_3 column, and the red band eluted with CH_2Cl_2/C_6H_6 (3:1). The yield of the orange crystals is 53% [5]. The compound consists of the diastereoisomers a and b, the a form being the racemate (RS)(SR) and the b form being the racemate (RR)(SS) as determined in an X-ray study [6]. The stereoisomers may be separated by chromatography on Al_2O_3 of a solution of the crystals in a minimal quantity of benzene. The band is first developed with pentane and then eluted with pentane/benzene (2:1), to yield a slowly moving broad orange band, the first half of which is collected, the second half of which is eluted with CH_2Cl_2. The solvents are removed,

References on p. 232

and the fractions repeatedly chromatographed in a similar manner. Recovered are a diastereomeric mixture of a and b in a 95:5 ratio, m.p. 136 to 138 °C (dec.) and in a 2:98 ratio, m.p. 135 to 137 °C (dec.). The 50:50 mixture melts at 129 to 134 °C (dec.) [2, 3, 5].

^{1}H NMR (in CDCl$_3$, δ in ppm): 1.68 (s, CH$_3$C$_5$), 2.28 (s, CH$_3$CO), 3.72, 4.75, and 4.80 (m's, C$_5$H$_3$) for form a, 1.56, 2.04, 4.17, 4.45, and 4.65 for form b. Figures of the spectra are given. IR spectrum (in CHCl$_3$): ν(CO) 1914, ν(C=O) 1595 cm^{-1} [5].

The b form as 1-C$_6$H$_6$ solvate crystallizes in the triclinic system with a = 15.359(4), b = 9.072(3), c = 14.112(4) Å, α = 86.22(3)°, β = 119.64(2)°, and γ = 109.38(2)°, space group P$\bar{1}$ – C$_i^1$. Z = 2 gives D$_c$ = 1.29 g·cm^{-3}. No reproducible densities could be measured due to loss of the bonded benzene. The molecular structure of the (SS) form is shown in **Fig. 19**.

Fig. 19

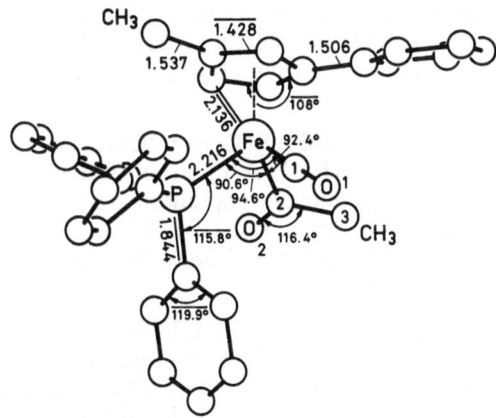

Molecular Structure of C$_5$H$_3$(CH$_3$-1)(C$_6$H$_5$-3)Fe(CO)(P(C$_6$H$_5$)$_3$)COCH$_3$(SS). The same number of the (RR) enantiomer exists in a crystal of the racemic mixture [6].

Other bond angles (°):

P-Fe-C$_5$H$_5$(c)	122.0(5)	Fe-C(1)-O(1)	178.1(10)
C(1)-Fe-C$_5$H$_5$(c)	122.0(3)	Fe-C(2)-O(2)	123.7(8)
C(2)-Fe-C$_5$H$_5$(c)	115.0(4)	Fe-C(2)-C(3)	119.7(8)

The compound is decarbonylized on UV irradiation to give C$_5$H$_3$(CH$_3$-1)(C$_6$H$_5$-3)- Fe(CO)(P(C$_6$H$_5$)$_3$)CH$_3$. The a to b ratios of the starting material (initial and final COCH$_3$) and product (CH$_3$) for various solvents are as follows:

solvent (time in min)	THF(5)	C$_6$H$_6$(5)	C$_6$H$_6$(5)	C$_6$H$_6$(15)	C$_6$H$_6$(40)
initial COCH$_3$	0:100	0:100	95:5	95:5	95:5
final COCH$_3$	0:100	0:100	95:5	95:5	95:5
CH$_3$ product	8:92	8:92	88:12	84:16	59:41

The part of the compound that had not reacted did not change its configuration during the irradiation. The rate of the decarbonylation of the a and b forms is the same, and the reaction is predominantly stereospecific [2, 3, 5].

References on p. 232   16•

$C_9H_6(CH_3-1)Fe(CO)(P(C_6H_5)_3)CH_2C_6H_5$ ($C_9H_6(CH_3-1) = 1$-methylindene, Formula I) is prepared by UV irradiation of a solution of $C_9H_6(CH_3-1)Fe(CO)$-$(P(C_6H_5)_3)COCH_2C_6H_5$ in benzene/pentane (4:1) for 1.5 h and is separated chromatographically in a 48% yield. ^1H NMR (in $CDCl_3$, δ in ppm): 1.8, 2.3 (m's, CH_2), 2.85 (s, CH_3), 4.05, 4.80 (m's, C_9H_6, $J(H,H) = 16$ Hz) ppm. IR (in $CHCl_3$): $\nu(CO)$ 1907 cm^{-1} [5].

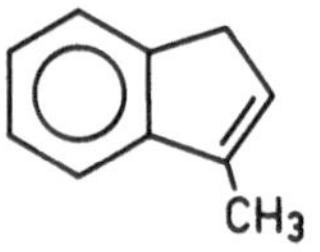

I

$C_9H_6(CH_3-1)Fe(CO)(P(C_6H_5)_3)COCH_2C_6H_5$ ($C_9H_6(CH_3-1)$, Formula I) is prepared by refluxing $C_9H_6(CH_3-1)Fe(CO)_2CH_2C_6H_5$ and $P(C_6H_5)_3$ in CH_3CN for 70 h. The residue left after the solvent was removed is chromatographed to give orange crystals in a yield of 52%. ^1H NMR (in $CDCl_3$, δ in ppm): 2.18 (s, CH_3), 3.05, 3.49 (CH_2, $J(H,H) = 16$ Hz), 4.37, 4.68 (m's, C_9H_6). IR (in CH_2Cl_2): $\nu(CO)$ 1912, $\nu(C=O)$ 1612 cm^{-1} [5].

$C_5H(CH_3-1)(C_6H_5-2,3,4)_3Fe(CO)(P(C_6H_5)_3)CH_3$ is obtained by UV irradiation for 3.5 h in a benzene/pentane (5:1) solution of $C_5H(CH_3-1)(C_6H_5-2,3,4)_3Fe$-$(CO)(P(C_6H_5)_3)COCH_3$, with purification by chromatography and recrystallization, 51% yield. ^1H NMR spectrum (δ in ppm): 1.29, 1.38 (s's, CH_3C_5) in $CDCl_3$ and -0.22, -0.28 (d's, CH_3, $J(P,H) = 6.5$ Hz), 4.03, 4.20 (s's, C_5H) in CH_2Cl_2. IR spectrum (in CH_2Cl_2): $\nu(CO)$ 1901 cm^{-1} [5].

$C_5H(CH_3-1)(C_6H_5-2,3,4)_3Fe(CO)(P(C_6H_5)_3)COCH_3$ is obtained by 24-h refluxing of $C_5H(CH_3-1)(C_6H_5-2,3,4)_3Fe(CO)_2CH_3$ and $P(C_6H_5)_3$ in CH_3CN. The residue left after removal of solvent is chromatographed to give a 58% yield of orange crystals. ^1H NMR spectrum (δ in ppm): 1.78 (s, CH_3C_5) in $CDCl_3$ and 2.65 (s, CH_3CO) and 4.78 (s, C_5H) in C_6D_6. IR spectrum (in $CHCl_3$): $\nu(CO)$ 1914, $\nu(C=O)$ 1590 cm^{-1} [5].

References:

[1] J. Ellermann, H. Behrens, H. Krohberger (J. Organometal. Chem. **46** [1972] 119/38). − [2] T.G. Attig, P. Reich-Rohrwig, A. Wojcicki (J. Organometal. Chem. **51** [1973] C21/C23). − [3] A. Wojcicki, T.G. Attig (6th Intern. Conf. Organometal. Chem., Amherst, Mass., 1973, Abstr. No. 123). − [4] T.G. Attig, A. Wojcicki (J. Am. Chem. Soc. **96** [1974] 262/3). − [5] T.G. Attig, A. Wojcicki (J. Organometal. Chem. **82** [1974] 397/415).

[6] T.G. Attig, R.G. Teller, S.-M. Wu, R. Bau, A. Wojcicki (J. Am. Chem. Soc. **101** [1979] 619/28).

1.5.2.2.13 Compounds of $[C_5H_5Fe(CO)(P(C_6H_5)_3)^1L]^+$ Cations

The positive charge of the cations in this section is due to the presence of a quaternary P atom in the ^1L ligand originating in a phosphorous ylide. Another type of cation, $[C_5H_5Fe(CO)(P(C_6H_5)_3)CO]^+$, has been mentioned as forming from $C_5H_5Fe(CO)(P(C_6H_5)_3)COOR$ compounds (R = H or CH_3) and strong bases, see 1.5.2.2.12.3.2, p. 169.

[C₅H₅Fe(CO)(P(C₆H₅)₃)CH₂P(C₆H₅)₃]BF₄ was prepared by removal of iodide from C₅H₅Fe(CO)(P(C₆H₅)₃)I with AgBF₄ in THF (15 min stirring) followed by dropwise addition of a benzene solution of CH₂P(C₆H₅)₃ (prepared from [CH₃P(C₆H₅)₃]I and C₄H₉Li) and further stirring for 15 h. The filtrate from the reaction mixture is concentrated to half its volume, and the compound is precipitated with ether, 41% yield. For another preparation, see the next compound.

The yellow powder decomposes at 68 to 70 °C. ^{1}H NMR spectrum (CDCl₃): δ (in ppm) = 3.18 (d, CH₂, J = 14.4), 5.65 (d, C₅H₅, J = 1.5), 7.65 and 7.85 (m's, C₆H₅). ^{13}C NMR spectrum (CH₂Cl₂, P' = ylide P): δ (in ppm) = 9.6 (dd, CH₂, J(P,C) = 43.0 Hz, J(P',C) = 102.0 Hz), 89.35 (s, C₅H₅), 118.48 (d, C-P', J = 88.4 Hz), 129.08 to 134.93 (m, C₆H₅), and 208.99 (d, CO, J = 24.1 Hz). IR spectrum (mull): ν(CO) 1951 cm⁻¹.

[C₅H₅Fe(CO)(P(C₆H₅)₃)CH₂CH₂CH₂P(C₆H₅)₃]BF₄ was obtained from a solution of [C₅H₅Fe(CO)(P(C₆H₅)₃)CH₂=CH₂]BF₄ in THF which was added to CH₂P(C₆H₅)₃ in benzene. The mixture was stirred for 15 h. Hexane first precipitates the previous complex (~35% yield), and after filtration more hexane precipitates the product, 40% yield.

The orange powder decomposes at 81 to 83 °C. ^{1}H NMR spectrum (CDCl₃): δ (in ppm) = 1.30, 3.40 (m's, CH₂), 4.26 (s, C₅H₅), 7.35, and 7.72 (m's, C₆H₅). In the following assignment C-1 is C-Fe and P' denotes the P atom of the ^{1}L ligand. ^{13}C NMR spectrum (in CH₂Cl₂): δ (in ppm) = 4.17 (dd, C-1, ^{1}J(P,C) = 19.2 Hz, ^{3}J(P',C) = 8.2), 26.78 (d, C-3, J = 42.6 Hz), 30.71 (C-2, ^{2}J(P,C) = ^{2}J(P',C) = 3.0 Hz), 84.24 (s, C₅H₅), 118.40 (d, C-1 in C₆H₅P', J = 85.4 Hz), 127.46 to 136.86 (m, C₆H₅), 222.33 (d, CO, J = 32.1 Hz). IR spectrum (mull): ν(CO) 1900 cm⁻¹.

Reference:

D.L. Reger, E.C. Culbertson (J. Organometal. Chem. **131** [1977] 297/300).

1.5.2.2.14 Compounds of the ^{5}LFe(CO)^{1}L-^{2}D Type

The compounds in this section contain ligands bonded to iron both through an sp³ carbon atom and a donor site consisting of groups with N, P, O, or S atoms (^{2}D in Formula I). The compounds are listed in Table 26, p. 237 and are arranged by the size of the ring system. Most of the compounds have four- and five-membered rings (Nos. 3 to 11 and Nos. 12 to 26).

I

C₅H₅Fe(CO)C(CH₃)OAlBr₃ represents a particular case presumed to contain also a five-membered ferra ring system. This product forms when a toluene solution of AlBr₃ is slowly added to a stirred solution of C₅H₅Fe(CO)₂CH₃ cooled to 0 °C. But it could not be isolated since removal of solvent under reduced pressure left a gummy, dark brown solid which did not completely redissolve.

The brown toluene solutions show ^{1}H NMR shifts $\delta = 2.23$ (s, CH_3) and 3.84 (s, C_5H_5) ppm and IR bands at 1986 (CO) and 1368 (C–O) cm^{-1}. The assumed ligand arrangement was confirmed by an X-ray diffraction study of the analogous Mn compound, Formula II with $M = Mn(CO)_4$. Thus $AlBr_3$ induces an alkyl migration to produce an acyl group, coordinating to the acyl oxygen, and one Br atom from $AlBr_3$ fills a coordination site of the metal. For the Mn compound, the expected decrease in CO bond order (IR, X-ray) was observed, but not a corresponding increase of the metal–carbon bond order.

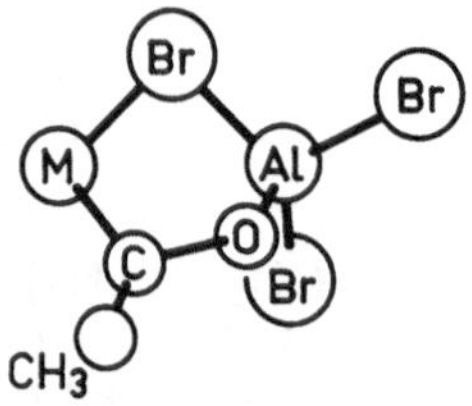

II

Toluene solutions of the iron compound slowly absorb CO to give $C_5H_5Fe(CO)_2$-$C(CH_3)OAlBr_3$. The initial reaction rate at 20 °C and $p(CO) = 383$ Torr is $k = 2.5 \times 10^{-6}$ M·s^{-1} for 1 M solutions [28].

Several other attempts to synthesize Fe-^{1}L-^{2}D ring systems are briefly summarized and illustrated by the formulas below.

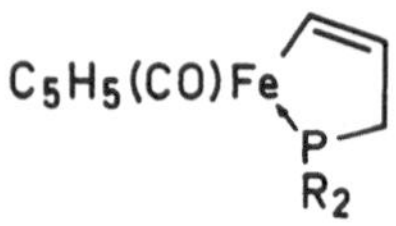

So far most attempts to obtain compounds III to VI [30] and VII have failed [2]. Irradiation of a mixture of $C_5H_5Fe(CO)_2Si(C_6H_5)_2CH_3$ with $P(CH_2CH=CH_2)_2C_6H_5$, $P(CH_2C_6H_5)_3$, or $P(CH_2CH_2C_6H_5)_2C_6H_5$ in hexane probably cleaves the Fe–Si bond

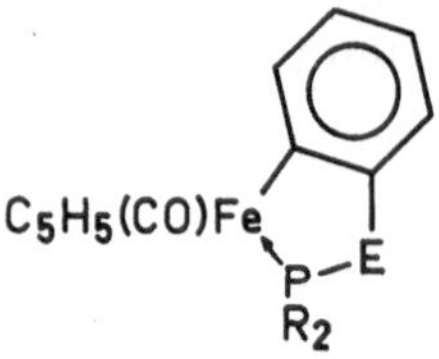

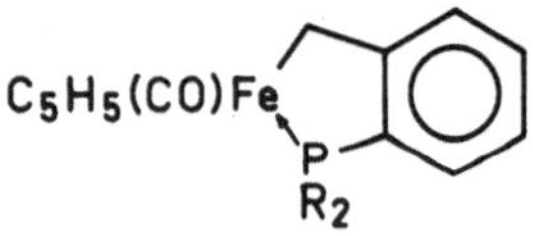

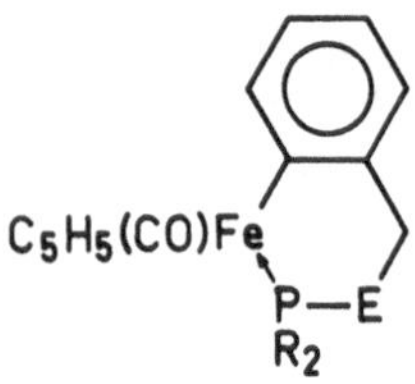

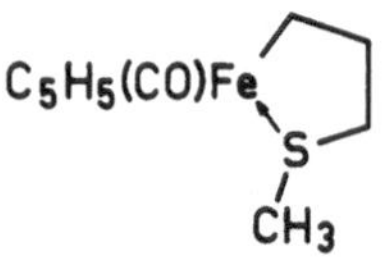

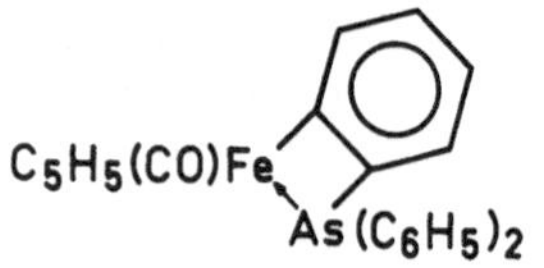

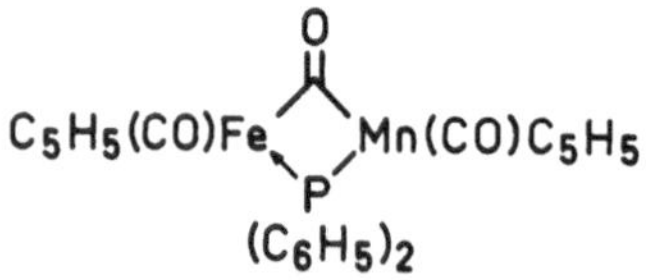

References on p. 249

and gives compounds III ($R=C_6H_5$ and $CH_2CH=CH_2$, ^{31}P NMR: $\delta=58$ ppm), IV (E= CH_2, $R=CH_2C_6H_5$, brown solution, ^{31}P NMR: $\delta=96$ ppm), and VI (E=CH_2, $R=C_6H_5$ and $CH_2CH_2C_6H_5$, ^{31}P NMR: $\delta=71$ ppm), respectively. However, attempts to isolate these products have been unsuccessful. Compound VIII was probably formed by irradiation of $C_5H_5Fe(CO)_2Si(C_6H_5)_2CH_3$ and $As(C_6H_5)_3$ in the presence of $(CH_3)_3NO$, but it is too unstable to be isolated [30]. Irradiation of $C_5H_5Fe(CO)_2CH_2CH_2CH_2SCH_3$ did not yield VII but only $C_5H_5Fe(CO)_2SCH_3$ [2]. Compound IX is mentioned in [31] and was reported to be prepared by splitting off one CO ligand from C_5H_5Fe-$(CO)_2P(C_6H_5)_2Mn(CO)_2C_5H_5$, but the structure is uncertain and no physical properties are reported.

C6H5
C5H5(CO)Fe   N–CH(CH3)C6H5 ⟶ C5H5(CO)Fe   C(CH3)C6H5
         P   P—N
      (C6H5)2   C6H5 C6H5
                                   CH2C6H5

X   XI

C5H5Fe(CO)2–CH2–[pyridine ring, N]   C5H5(CO)Fe [fused ring system with N and O]

XII   XIII

Compound X is supposed to be an unstable intermediate in the preparation of the isomer XI from $C_5H_5Fe(CO)_2CH(OCH_3)C_6H_5$ and $N(CH(CH_3)C_6H_5)P(C_6H_5)_2$. The structure of X is similar to that of the stable Nos. 6 and 7 (Table 26). Probably because of steric factors a 1,3-hydrogen shift takes place to give the noncyclic product XI [27]. In the preparation of XII from $Na[C_5H_5Fe(CO)_2]$ and 2-chloromethylpyridine some evidence for the formation of a trace of XIII was observed during the chromatography of the products (orange band, strongly adsorbed on Al_2O_3) [3].

C6H11N   C6H11N
              NC6H11   NHC6H11
C5H5(CO)Fe   C5H5(CO)Fe
          N   CH2R   R
        C6H11   C6H11HN

XIV   XV

Compounds XIV with $R=C_6H_5$ or C_6H_4Cl-4 were reported [9]. Later they were reformulated as structure XV where the ring is rearranged to a cyclic ^2L ligand [10, 16].

The most important preparative methods for the compounds in Table 26 are given below. Special procedures are described under further information.

References on p. 249

Method I: Compounds of the $C_5H_5Fe(CO)_2{}^1L$ type which possess a potential donor site in the ^1L ligand (in β or γ position relative to the Fe atom) are converted to $C_5H_5Fe(CO)^1L{-}^2D$ complexes by UV irradiation or heating. One of the two carbonyl groups of the starting material is (a) replaced by the donor atom or (b) enclosed in the $Fe{-}^1L{-}^2D$ ring.

    a. Irradiation of $C_5H_5Fe(CO)_2{}^1L$ complexes with $^1L=CH_2N(CH_3)_2$, $C(C_6H_5)=NC_6H_5$, $C(CF_3)=CHSCH_3$, $C(CF_3)=C(CF_3)SCF_3$, and $C(CF_3)=C(CF_3)SC_6F_5$ leads to Nos. 1 [25], 2 [18], 8 [26], 10 and 11 [14, 15], respectively. The reactions are carried out in ether [25], THF [26], or pentane [14, 15] for 3.5 to 40 h.

    b. Rearrangement of the starting materials by heating or irradiation gives Nos. 13, 14, and 23. No. 13 is prepared by heating complex XVI in THF for 19 h. No. 14 is obtained from complex XVII (1:1 mixture of diastereomers) in CH_3CN with a little $P(C_4H_9{-}n)_3$ at 60 to 65 °C for 7.5 h [29]. No. 23 forms from $C_5H_5Fe(CO)_2CH_2CH_2SCH_3$ under irradiation in benzene for 24 h or more [1, 2].

$C_5H_5Fe(CO)_2CH_2$—[ring with N=C(CH$_3$)] $C_5H_5Fe(CO)_2CH_2$—[ring with N(H)–CH$_3$]

           XVI   XVII

Method II: $C_5H_5Fe(CO)_2{}^1L$ complexes are reacted with (a) N or P donors and (b) $C_6H_5C{\equiv}CC_6H_5$. Method (a) leads to the replacement of one CO group and in some cases to the loss of the initial ^1L ligand. In Method (b) the alkyne is inserted into one of the Fe–CO bonds without loss of CO.

    a. The compounds $C_5H_5Fe(CO)_2{}^1L$ with $^1L=CH_3$, $CH(OCH_3)C_6H_5$, or C_6H_5 are treated with various N donors [6, 8, 10, 11] or P donor molecules [12, 17, 27]. The conditions vary. They include irradiation [10 to 12, 17, 27], heating in solution [8], or heating neat [6].

    b. The insertion of tolane is achieved by irradiation in pentane for 100 min [20].

Method III: Treatment of $C_5H_5Fe(CO)_2SiR_3$ compounds with 1-naphthylphosphines or $P(OC_6H_5)_3$ results in substitution of one CO ligand and metalation of one naphthyl ring in peri position or one C_6H_5 ring in ortho position. R_3SiH is formed. The reactions were carried out with $C_5H_5Fe(CO)_2{-}Si(C_6H_5)_2CH_3$ and $P(CH_3)(C_6H_5)C_{10}H_7{-}1$ or $P(C_{10}H_7{-}1)_2C_6H_5$ in benzene/hexane by UV irradiating the solutions for 8 h. The reactions are accelerated by $(CH_3)_3NO$. Irradiation of $C_5H_5Fe(CO)_2Si(CH_3)_3$ or $C_5H_5Fe(CO)_2Si(CH_3)_2C_6H_5$ in the presence of $P(OC_6H_5)_3$ in hexane for 12 to 15 h gave a mixture of No. 20 and $C_5H_5Fe(C_6H_4OP{-}(OC_6H_5)_2)P(OC_6H_5)_3$ (compound No. 32 in 1.5.2.1.1, p. 50) [30].

Method IV: Reactions of $C_5H_5Fe(CO)_2SCH_3$ with alkynes which have an activated triple bond [23, 24, 26]. No. 24 is prepared by heating a mixture of $C_5H_5Fe(CO)_2SCH_3$ and $CF_3C{\equiv}CH$ in THF for 24 h or by UV irradiation of this solution for 48 h. No. 27 forms when irradiation is carried out in hexane. Heating $C_5H_5Fe(CO)_2SCH_3$ and $CF_3C{\equiv}CCF_3$ in THF to 60 °C for 48 h or irradiation of this solution gives No. 25 [26].

References on p. 249

Method V: Isonitrile complexes of the $C_5H_5Fe(CO)(CNR)^1L$ type are irradiated in benzene for 8 to 12 h in the presence of RNC or R'NC. The chelating ligand forms by insertion of a coordinated isonitrile into the $Fe^{-1}L$ σ bond. The following starting materials are used:

$C_5H_5Fe(CO)(CNC_4H_9-t)CH_3$ and $t-C_4H_9NC$ for No. 3,
$C_5H_5Fe(CO)(CNC_6H_{11})COCH_3$ and $C_6H_{11}NC$ for No. 5, and
$C_5H_5Fe(CO)(CNC_4H_9-t)CH_3$ and $C_6H_{11}NC$ for No. 4.

Nos. 3 and 5 also formed, although in lower yields, when the corresponding $C_5H_5Fe(CO)(CNR)COCH_3$ complexes were irradiated in benzene in the absence of free RNC [10, 11].

Method VI: This is similar to Method IV but $(Z)-(C_5H_5Fe(CO)\mu-SCH_3)_2$ and $CF_3-C\equiv CCF_3$ are used in a 1:2 mole ratio and are irradiated in THF for 120 h to give Nos. 9 and 25 [26].

Method VII: $C_5H_5Fe(CO)^1L-^2D$ compounds obtained by the preceding methods can be used as starting materials. Decarbonylation of the $^1L-^2D$ ligand of No. 25 by irradiation in THF for 300 h gives No. 9. No. 27 is obtained when No. 24 in THF is irradiated for 120 h in the presence of excess $CF_3C\equiv CH$. No. 28 forms under similar conditions from Nos. 9 or 25 and $CF_3C\equiv CH$ [26].

Other compounds with Fe-CO-CR=CR-SR' rings analogous to No. 25 were also prepared by Method IV for the substituents $R = CF_3$ or $COOCH_3$ and $R' = CH_3$, C_2H_5, and $i-C_3H_7$. But properties are not reported [23].

Explanations for Table 26: For the assignment of the NMR resonances the atoms of the $^1L-^2D$ ligand are labelled beginning with the Fe-C atom as C-1. Not designated IR bands in the 1900 to 2000 cm^{-1} range are $\nu(CO)$ vibrations.

Table 26
Compounds of the $C_5H_5Fe(CO)^1L-^2D$ Type.
Further information for numbers preceded by an asterisk is given at the end of the table.
For abbreviations and dimensions, see p. 375.

No.	$^1L-^2D$ ligand method of preparation (yield in %)	properties and remarks explanations see above	Ref.
three-membered Fe($^1L-^2D$) rings			
*1	$CH_2-N(CH_3)_2$ Ia (63)	dark red needles ^1H NMR: 1.79, 1.84 (d's, CH_3), 1.98, 2.61 (d's, CH_2), 4.26 (s, C_5H_5) in C_6D_6, 2.17, 2.37 (d's, CH_3), 1.97, 2.85 (d's, CH_2), 4.39 (s, C_5H_5) in CD_3CN IR (C_6H_{12}): 1903	[25]
2	$C(C_6H_5)=NC_6H_5$ Ia	no properties reported	[18]

Table 26 [continued]

No.	^{1}L-^{2}D ligand method of preparation (yield in %)	properties and remarks explanations on p. 237	Ref.

four-membered Fe(^{1}L-^{2}D) rings

*3	C(=NC$_4$H$_9$-t)-C(CH$_3$)=N-C$_4$H$_9$-t V (10)	brown, m.p. 115 to 120° (dec.) ^{1}H NMR (CDCl$_3$): 1.14, 1.30 (s's, t-C$_4$H$_9$), 1.83 (s, CH$_3$), 4.41 (s, C$_5$H$_5$) IR (C$_6$H$_6$): 1918, ν(C=N) 1587, 2632	[11]
*4	C(=NC$_6$H$_{11}$)-C(CH$_3$)=N-C$_4$H$_9$-t V (9)	m.p. 97 to 100° (dec.) ^{1}H NMR (CDCl$_3$): 0.8 to 2.2 (m, C$_6$H$_{11}$, 10 H), 1.13 (s, t-C$_4$H$_9$), 1.76 (s, CH$_3$) 2.70 to 3.10 (br, C$_6$H$_{11}$, 1 H), 4.44 (s, C$_5$H$_5$) IR (C$_6$H$_6$): 1915, ν(C=N) 1580, 1633	[11]
*5	C(=NC$_6$H$_{11}$)-C(CH$_3$)=N-C$_6$H$_{11}$ IIa (62), V (35)	m.p. 122 to 125° (dec.) ^{1}H NMR (CDCl$_3$): 0.1 to 2.1 (m, C$_6$H$_{11}$, 10 H), 1.74 (s, CH$_3$), 2.64 to 3.05 (br, C$_6$H$_{11}$, 1 H), 4.46 (s, C$_5$H$_5$) IR (C$_6$H$_6$): 1919, ν(C=N) 1592, 1632	[10, 11]
*6	CH(C$_6$H$_5$)-N(C$_2$H$_5$)-P(C$_6$H$_5$)$_2$ IIa (43)	red orange, m.p. 134 to 136° ^{1}H NMR (C$_6$D$_6$): 0.77 (t, CH$_3$), 2.77 (dq, CH$_2$, ^{3}J(H,H) = 7, ^{3}J(P,H) = 10.2), 4.06 (d, C$_5$H$_5$, J = 1.4), 5.11 (d, CH, J(P,H) = 5.6), 6.98 to 7.83 (m, C$_6$H$_5$) ^{13}C NMR (CDCl$_3$): 40.85 (d, CH, J = 39.81), 46.27 (t, CH$_2$, J = 6.63), 81.04 (C$_5$H$_5$), 126.37 to 132.86 and 136.08 to 154.34 (C$_6$H$_5$), 221.09 (CO, J = 33.17) ^{31}P NMR: 85.02 IR (C$_6$H$_6$): 1923	[27]
*7	CH(C$_6$H$_5$)-N(CH$_2$C$_6$H$_5$)-P(C$_6$H$_5$)$_2$ IIa (40)	orange, m.p. 62 to 64° ^{1}H NMR (C$_6$D$_6$): 3.95 to 3.98 (ABX pattern, CH$_2$, ^{2}J(H,H) = 13.5), 4.07 (d, C$_5$H$_5$, J = 1.5), 5.22 (d, CH, J = 4.5), 6.76, to 7.72 (m, C$_6$H$_5$) ^{31}P NMR: 89.56 IR (C$_6$H$_6$): 1939	[27]
*8	C(CF$_3$)=CH-SCH$_3$ Ia (28)	red, m.p. 147° ^{1}H NMR: 2.35 (s, CH$_3$), 4.53 (s, C$_5$H$_5$), 6.20 (q, CH) ^{13}C NMR (CDCl$_3$): 29.0 (CH$_3$), 82.0 (C$_5$H$_5$), 128.0 (q, CF$_3$, ^{1}J(F,C) = 274.9), 137.4 (q, C-2, ^{3}J(F,C) = 11.7), 163.4 (q, C-1, ^{2}J(F,C) = 29.2) 220.6 (CO) ^{19}F NMR: 58.8 (d, J(H,F) = 1.8) IR: 1960 in CCl$_4$, 1940, 1952, ν(C=C) 1540 in Nujol	[26]

References on p. 249

Table 26 [continued]

No.	¹L–²D ligand method of preparation (yield in %)	properties and remarks explanations on p. 237	Ref.
*9	C̦(CF₃)=C(CF₃)-ȘCH₃ VI (15), VII (50, 65)	red brown, m.p. 40° ¹H NMR: 2.18 (s, CH₃), 4.65 (s, C₅H₅) ¹³C NMR (CDCl₃): 24.7 (CH₃), 79.8 (C₅H₅), 114.7 (qq, CF₃, ¹J(F,C)=275.0, ⁴J(F,C)=4.8), 123.6 (qq, C-2, ²J(F,C)=39.0, ³J(F,C)=8.0), 123.9 (qq, CF₃, ¹J(F,C)=273.5, ⁴J(F,C)=4.8), 176.1 (q, C-1, ²J(F,C)=37.1), 218.9 (CO) ¹⁹F NMR: 58.8, 61.2 (q's, CF₃, J(F,F)=8.5) IR (Nujol): 1970, ν(C=C) 1605, 1618	[26]
*10	C̦(CF₃)=C(CF₃)-ȘCF₃ Ia (72)	brown black ¹⁹F NMR: ⁵J(F,F)=8.5 for CCF₃ IR (C₆H₁₂): 1986, 1995, ν(C=C) 1595, ν(CF) 1097, 1135, 1148, 1156, 1187, 1194, 1238, 1256	[14, 15]
*11	C̦(CF₃)=C(CF₃)-ȘC₆F₅ Ia (73)	brown oil ¹⁹F NMR: ⁵J(F,F)=7.9 IR (C₆H₁₂): 1981, ν(C=C) 1598, ν(CF) 1090, 1096, 1135, 1145, 1158, 1241, 1256, 1296	[14, 15]

five-membered Fe(¹L-²D) rings

No.	¹L–²D ligand	properties and remarks	Ref.
*12	C̦O-CH₂CH₂-N̦(CH₃)₂ see further information	dark red, m.p. 108 to 111° (dec.), subl. 80°/0.1 Torr ¹H NMR (CHCl₃): 2.41 (CH₃ and CH₂, 7 H), 2.84 (CH₃, 3 H), 4.46 (C₅H₅) IR (KBr): 1890, ν(C=O) 1590	[3]
*13	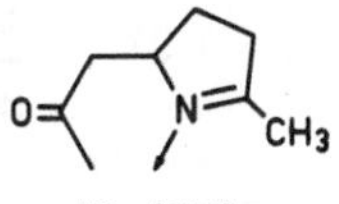 Ib (85)	thick red–orange oil, 55:45 mixture of diastereomers ¹H NMR (CDCl₃): 1.5 to 3.0 (m's, CH₂), 2.15, 2.27 (d's, CH₃, J=1), 3.90 (m, CH), 4.45, 4.51 (s's, C₅H₅) IR (CH₂Cl₂): 1916, ν(C=O) 1611	[29]
*14	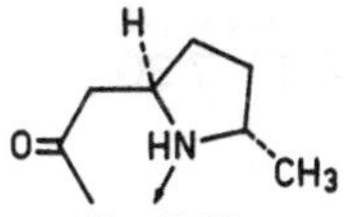 Ib (30)	red ¹H NMR (CDCl₃): 4.48 (s, C₅H₅) IR (CH₂Cl₂): 1907, ν(C=O) 1610	[29]

References on p. 249

Table 26 [continued]

No.	^{1}L-^{2}D ligand method of preparation (yield in %)	properties and remarks explanations on p. 237	Ref.

*15	$\underset{\uparrow}{C}(CF_3)=NC(CF_3)=\underset{\downarrow}{N}H$
	
	IIa (30)
		black needles, m.p. 149 to 151°,
		  subl. 70°/0.1 Torr
		^{1}H NMR (CH$_3$COCH$_3$): 5.00 (s, C$_5$H$_5$),
		  11.3 (br, NH)
		^{19}F NMR (CH$_2$Cl$_2$): 67.5, 70.5 (s's)
		IR (KBr): 1995, ν(CN) 1552, ν(NH) 3340,
		  ν(CF) 1105, 1120, 1145, 1159, 1184,
		  1252, 1283
			[6,
			13]

*16	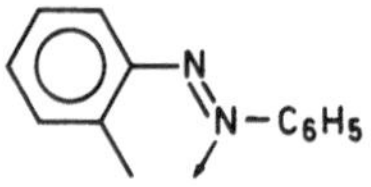
	IIa (10)
		green oil
		^{1}H NMR (CS$_2$): 4.40 (s, C$_5$H$_5$), 6.94 (H-3),
		  7.28 (br m, H-4), 7.28, 7.76 (C$_6$H$_5$), 7.76
		  (br m, H-5), 8.08 (dd, H-6)
		IR (C$_6$H$_{12}$): 1916, 1961
			[8]

*17	$\underset{\uparrow}{C}H_2CH_2CH_2\underset{\downarrow}{P}(C_6H_5)_2$ see further information	red brown IR (C$_6$H$_{14}$): 1925	[22]

*18	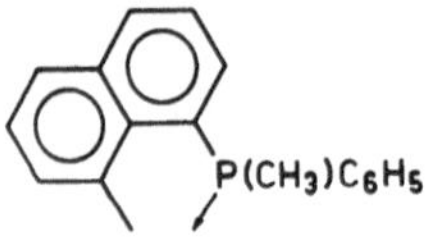
	III (85)
		orange oil, mixture of diastereomers
		^{1}H NMR (CDCl$_3$): 2.07, 2.11 (d's, CH$_3$,
		  J(P,H)=5), 4.41, 4.74 (d's, C$_5$H$_5$,
		  J(P,H)=1.5), 7.41 to 7.87 (m, C$_6$H$_5$,
		  C$_{10}$H$_6$)
		^{31}P NMR (CDCl$_3$): 79.7 and 81.3
			[30]

*19	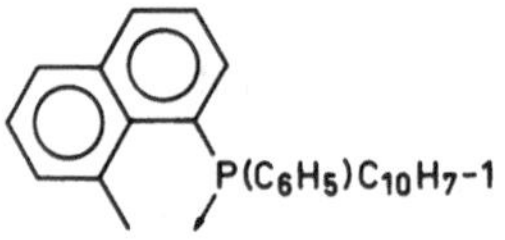
	III (23)
		orange, m.p. 218°
		^{1}H NMR (CDCl$_3$): 4.67 (d, C$_5$H$_5$, J=1.5),
		  7.40, 7.87 (m's, 17 H), 8.10 (m, H-8 of
		  C$_{10}$H$_7$)
		^{31}P NMR: 90
		IR (CH$_2$Cl$_2$): 1920
			[30]

*20	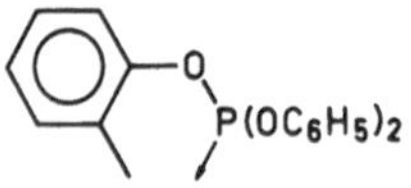
	IIa (53),
	III (33 to 38)
		golden, m.p. 118 to 119°
		^{1}H NMR (CDCl$_3$): 4.62 (d, C$_5$H$_5$, J=1.2),
		  6.88 (m, 3 H), 7.22 (m, 11 H)
		^{31}P NMR (CHCl$_3$): 203.4 (s)
		IR: 1973 in C$_6$H$_{12}$, C$_6$H$_4$ bands at 796 and
		  1098 in Nujol
			[12,
			21,
			30]

*21	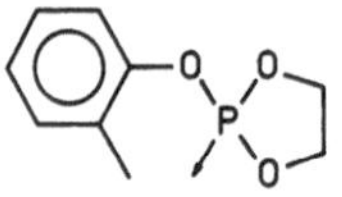
	IIa (54)
		golden, m.p. 116 to 117°
		^{1}H NMR (CDCl$_3$): 4.35 (m, CH$_2$), 4.72 (d,
		  C$_5$H$_5$, J=1.2), 6.75 (m, H-3, 4, 5), 7.30
		  (m, H-6)
		^{31}P NMR (CHCl$_3$): 223.4
		IR (C$_6$H$_{12}$): 1972
			[17]

References on p. 249

Table 26 [continued]

No.	¹L–²D ligand method of preparation (yield in %)	properties and remarks explanations on p. 237	Ref.
*22	$\underset{|}{C}(C_6H_5)=C(C_6H_5)-$ $C(CH_3)=\underset{\downarrow}{O}$ IIb (73)	m.p. 104° (dec.) ¹H NMR (CD₃COCD₃): 2.07 (s, CH₃), 4.53 (s, C₅H₅), 7.18 (m, C₆H₅) ¹³C NMR (CD₃COCD₃): 25.83 (CH₃), 84.11 (C₅H₅), 154.04 (C-2), 211.15 (C-3), 218.31 (CO), 257.97 (C-1) IR: 1939 in C₆H₁₄, ν(C=O) 1525 in KBr	[20]
*23	$\underset{|}{C}O-CH_2CH_2\underset{\downarrow}{S}CH_3$ Ib	red orange, m.p. 71 to 73° ¹H NMR (CS₂): 2.2 to 2.3 (br, CH₂), 2.22 (CH₃), 4.40 (C₅H₅) IR (halocarbon): 1935, ν(C=O) 1618	[1, 2]
*24	$\underset{|}{C}O-C(CF_3)=CH\underset{\downarrow}{S}CH_3$ IV	red, m.p. 65° ¹H NMR: 2.52 (s, CH₃), 4.65 (s, C₅H₅), 7.80 (q, CH) ¹³C NMR (CDCl₃): 27.7 (q, SCH₃, ¹J(H,C) = 138.0), 83.8 (m, C₅H₅, ¹J(H,C) = 130.7), 121.0 (q, CF₃, ¹J(F,C) = 278.3), 147.4 (q, C-2, ²J(F,C) = 26.0), 152.6 (dq, C-3, ¹J(H,C) = 186.7), 215.7 (s, CO), 257.4 (d, C-1, ³J(H,C) = 3.7) ¹⁹F NMR: 62.0 (d, J(H,F) = 0.9) IR: 1940, 1960 sh, ν(C=C) 1570 in Nujol, 1960 in CHCl₃, ν(C=O) 1630 in CCl₄	[23, 24, 26]
*25	$\underset{|}{C}O-C(CF_3)=C(CF_3)\underset{\downarrow}{S}CH_3$ IV, VI (17)	brown, m.p. 100° ¹H NMR: 2.80 (s, CH₃), 4.80 (s, C₅H₅) ¹³C NMR (CDCl₃): 29.8 (SCH₃), 84.2 (C₅H₅), 119.6, 121.1 (q's, CF₃, ¹J(F,C) = 278.5 and 276.0), 149.2, 152.6 (q's, C-2,3, ²J(F,C) = 26.0 and 39.0), 214.6 (CO), 254.0 (C-1) ¹⁹F NMR: 55.2, 59.8 (q's, CF₃, J(F,F) = 11.2) IR: 1955, 1965, ν(C=C) 1565 in Nujol, 1975 in CHCl₃, ν(C=O) 1625 in CCl₄	[23, 24, 26]
*26	$\underset{|}{C}O-N(CH_3)-C(C_6H_5)=\underset{\downarrow}{S}$ see further information	dark brown needles, dec. 126° ¹H NMR (CDCl₃): 3.13 (s, CH₃), 4.70 (s, C₅H₅), 7.36 (m, C₆H₅) IR (KBr): 1942, ν(C=O) 1643	[19]

References on p. 249

Table 26 [continued]

No.	^{1}L-^{2}D ligand method of preparation (yield in %)	properties and remarks explanations on p. 237	Ref.

six-membered Fe(^{1}L-^{2}D) rings

*27 $\underset{\downarrow}{C(CF_3)}=CHC(CF_3)=$
$\quad CH-SCH_3$

IV (12), VII (18)

brown, m.p. 76°
^{1}H NMR: 2.22 (s, CH_3), 4.59 (s, C_5H_5),
6.12 (s, CH), 6.98 (q, CH)
^{19}F NMR: 58.0 (s), 68.0 (d, J(H,F) = 2.0)
IR: 1960 sh, 1970, ν(C=C) 1550 in Nujol,
1965 in CCl_4

[26]

*28 $\underset{\downarrow}{C(CF_3)}=CHC(CF_3)=$
$\quad C(CF_3)SCH_3$
or

$\underset{\downarrow}{C(CF_3)}=C(CF_3)CH=$
$\quad C(CF_3)SCH_3$

VII

chestnut brown, m.p. 67°
^{1}H NMR: 2.01 (s, CH_3), 4.59 (s, C_5H_5),
7.50 (CH)
^{19}F NMR: 50.5 (q, J(F,F) = 17.0, J(H,F) =
17), 57.5 (q, J(F,F) = 17.0), 66.5 (s)
IR: 1960 sh, 1980, ν(C=C) 1510, 1640 in
Nujol, 1970 in CCl_4

[26]

*Further information:

$C_5H_5Fe(CO)CH_2N(CH_3)_2$ (Table **26**, No. **1**) is isolated from the crude solid by extraction with medium petroleum ether. The low-melting oily solid obtained is sublimed in a vacuum at 55 °C. The ^{1}H NMR spectrum in toluene remains unchanged up to ~100 °C, where the compound begins to decompose. The chelate ring is opened by phosphines: Reactions with a slight excess $P(CH_3)_2C_6H_5$ or $P(C_6H_5)_2CH_3$ (^{2}D) in THF at room temperature give the $C_5H_5Fe(CO)(^2D)CH_2N(CH_3)_2$ compounds [25].

$C_5H_5Fe(CO)C(=NC_4H_9\text{-}t)C(CH_3)=NC_4H_9\text{-}t$ (Table **26**, No. **3**). Method Va gives No. 3 along with $C_5H_5Fe(CO)(CNC_4H_9\text{-}t)CH_3$ (main product), $C_5H_5Fe\text{-}(CO)_2CH_3$, and $(C_5H_5Fe(CO)_2)_2$ (trace), which are separated by chromatography on Al_2O_3. The last eluate (C_6H_6/CH_2Cl_2 2:1) contains No. 3.

The mass spectrum shows the molecular ion [M]$^+$ (relative intensity 12) and the fragments [M − CO]$^+$, $[C_5H_5Fe(CNC_4H_9)CH_3]^+$ (100), $[C_5H_5Fe(CNH)CH_3]^+$ (99), $[C_5H_5FeCNCH_3]^+$ (90), and $[C_5H_5Fe]^+$ (57); possible fragmentation pathways are described [11].

$C_5H_5Fe(CO)C(=NC_6H_{11})C(CH_3)=NC_4H_9\text{-}t$ (Table **26**, No. **4**). The mass spectrum shows the molecular ion [M]$^+$ (relative intensity 9) and the fragments [M − CO]$^+$ (16), $[C_5H_5Fe(CNC_4H_9)CH_3]^+$ (52), and $[C_5H_5Fe]^+$ (38). The [M − CO]$^+$ fragment loses the $C_6H_{11}NC$ group, but no loss of t-C_4H_9NC from [M − CO]$^+$ is observed [11].

$C_5H_5Fe(CO)C(=NC_6H_{11})C(CH_3)=NC_6H_{11}$ (Table **26**, No. **5**). Method IIa with $C_5H_5Fe(CO)_2CH_3$ and $C_6H_{11}NC$ (1:4 mole ratio) irradiated in benzene at 25 °C for 13 h. Method Va gives No. 5 along with $C_5H_5Fe(CO)_2CH_3$ (main product) and $(C_5H_5Fe(CO)_2)_2$. Chromatography on activated Al_2O_3 gives No. 5 (with C_6H_6/CH_2Cl_2 8:1) in the last, reddish brown band [11].

References on p. 249

The reaction with t-C_4H_9NC in C_6H_6 at 40 °C gives compound XVIII in a 47% yield [10, 11]. However, compound XIX, having the same empirical formula, is obtained when the reaction is carried out at 70 °C. The reaction between No. 5 and $C_6H_{11}NC$ (temperature not given) probably forms a complex analogous to XIX because no signal assignable to a methyl group was observed in the ^1H NMR spectrum [16].

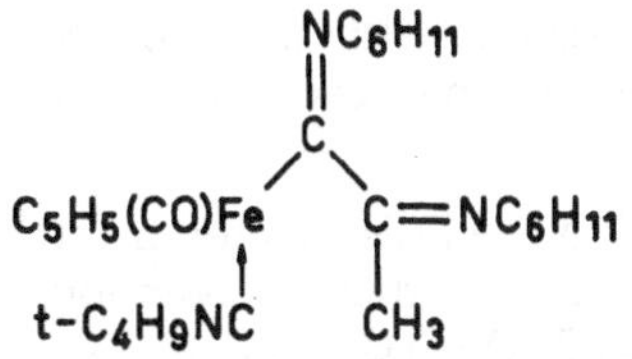

XVIII 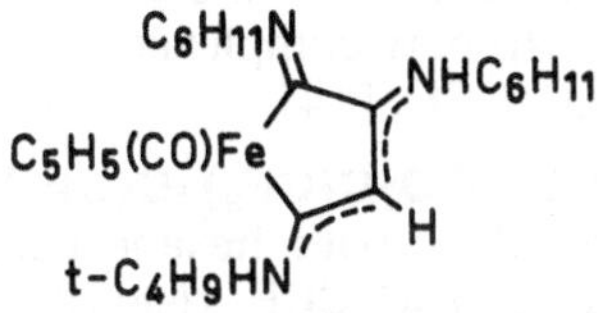   XIX

$C_5H_5Fe(CO)CH(C_6H_5)N(CH_2R)P(C_6H_5)_2$ (R=CH_3 or C_6H_5 (Table **26**, Nos. **6** and **7**). Method IIa with $C_5H_5Fe(CO)_2CH(OCH_3)C_6H_5$ and $N(CH_2R)(P(C_6H_5)_2)H$ irradiated in C_6H_6/pentane at 25 °C for 1 h. The compounds were formed along with $(C_5H_5Fe(CO)_2)_2$ and isolated on SiO_2 with benzene eluent. The compounds are stable; refluxing in C_6H_6 for 1 d caused no detectable rearrangement and very little decomposition [27].

Even though the methylene protons in the ethyl group of No. 6 are diastereotopic, they degenerate in the ^1H NMR spectrum and appear as a doubled quartet due to coupling to the P atom and the CH_3 group. This behavior is unlike that of the methylene protons of No. 7, which show an eight line ABX pattern [27].

In the ^{13}C NMR spectrum of No. 6 four signals ranging from 136.08 to 139.07 ppm are assigned to the two phenyl carbons adjacent to the P atom (J(P,C)=11.94 Hz). The signal at 154.34 ppm is assigned to the phenyl carbon adjacent to the CH group [27]. The value given for the CH_3 resonance must be an error.

$C_5H_5Fe(CO)C(CF_3)=CHSCH_3$ (Table **26**, No. **8**). When the starting material $C_5H_5Fe(CO)_2C(CF_3)=CHSCH_3$ is irradiated in the presence of $CF_3C\equiv CH$ (5:1 ratio), the yield can be increased to 50%. The compound is isolated by chromatography [26].

$C_5H_5Fe(CO)C(CF_3)=C(CF_3)SCH_3$ (Table **26**, No. **9**) is isolated by chromatography and crystallized from hexane. It does not react when it is irradiated in the presence of $CF_3C\equiv CCF_3$. However, irradiation for several days in the presence of excess CF_3-$C\equiv CH$ in THF gives No. 28 (17% yield) and a further product (13% yield), which probably has structure XX [26].

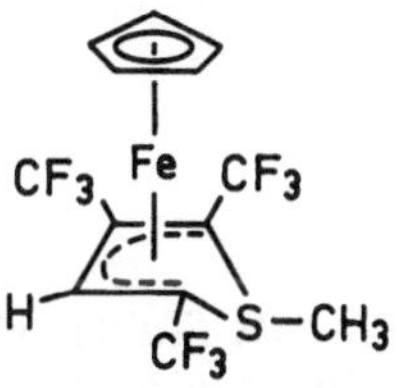

XX

References on p. 249

C₅H₅Fe(CO)C(CF₃)=C(CF₃)SCF₃ (Table **26**, No. **10**) is obtained along with $C_5H_5Fe(CO)_2SCF_3$ (5% yield) and is isolated by chromatography on SiO_2 with elution by hexane/benzene (5:1) [14, 15]. Similar products were also formed by thermal decomposition of $C_5H_5Fe(CO)_2C(CF_3)=C(CF_3)SCF_3$ in hexane at 60 °C although this reaction is more complicated and other fluorocarbon derivatives were detected, but not isolated [14].

The IR spectrum shows two CO modes, consistent with the presence of two isomers. No new complexes were isolated from the thermal reaction between No. 10 and $CF_3C\equiv CCF_3$ [14].

C₅H₅Fe(CO)C(CF₃)=C(CF₃)SC₆F₅ (Table **26**, No. **11**) is isolated by chromatography on SiO_2 with hexane/benzene (5:1) eluent. The thermal reaction between No. 11 and $CF_3C\equiv CCF_3$ yields only $C_5H_5Fe(CO)_2C(CF_3)=C(CF_3)SC_6F_5$ as an identifiable product. A product analogous to XX forms on irradiation of the same reactants [14].

C₅H₅Fe(CO)COCH₂CH₂N(CH₃)₂ (Table **26**, No. **12**) is prepared by stirring a mixture of $Na[C_5H_5Fe(CO)_2]$ and $N(CH_3)_2CH_2CH_2Cl$ in THF at room temperature for 16 h, removal of the solvent, extraction of the residue with CH_2Cl_2, evaporation of the CH_2Cl_2, and storage of the orange-brown partially solid residue under N_2 for 3 d. Formation of No. 12 appears to occur during storage. The residue is washed with pentane, triturated with benzene, and filtered. The filtrate is evaporated. Washing the residue with ether and subliming it for 16 h gives a 2.5% yield [3].

The compound is rather air sensitive. It is sparingly soluble in pentane and ether but soluble in benzene and CH_2Cl_2.

The complete IR spectrum is given in [3]. The mass spectrum exhibits intense ions arising from ferrocene and its fragmentation products but only rather weak ions arising directly from No. 12. The iron-containing ions are of types commonly found in the mass spectra of $C_5H_5Fe(CO)_2$ derivatives. The complete mass spectrum is reported [7].

C₅H₅Fe(CO)C₇H₁₀NO (Table **26**, No. **13**) is isolated by chromatography on neutral Al_2O_3 with ether/CH_2Cl_2 (1:1) eluent. Preliminary experiments designed to close the ligand of No. 13 to the carbopenem system XXI through oxidation in basic media have not been successful. No. 13 also resists reduction with $NaBH_4$ [29].

C₅H₅Fe(CO)C₇H₁₂NO (Table **26**, No. **14**) can only be obtained from the (E) isomer of XVII, while the (Z) isomer of XVII cannot be rearranged under the conditions of Method I b. No. 14 is isolated by chromatography on neutral Al_2O_3 with ether eluent. Oxidation of No. 14 with Ag_2O (1:2.5 mole ratio) in THF in the dark gives 2-methyl-carbopenam (Formula XXII) as the single stereoisomer in a 72% yield [29].

XXI   XXII

C₅H₅Fe(CO)C(CF₃)=NC(CF₃)=NH (Table **26**, No. **15**) was originally formulated as a $C_5H_5Fe(CO)(^2D)^1L$ complex [6], but an X-ray study revealed its true structure [13]. It is prepared by the reaction of $C_5H_5Fe(CO)_2CH_3$ with CF_3CN (1:6 mole ratio) at 90 °C for 16 h in 30% yield, sometimes along with $C_5H_5Fe(CO)_2COCH_3$ [6]. Addition of the free-radical inhibitor galvinoxal to this reaction improves the yields [13]. The contents of the reaction vessel are rinsed out with several small portions of CH_2Cl_2. The CH_2Cl_2 is evaporated. Chromatography of the residue on Al_2O_3 with ether eluent and sublimation gives the pure compound [6]. A speculative reaction scheme is described in [13]. It was impossible to prepare No. 15 from the starting materials by UV irradiation or by thermal reactions between $C_5H_5Fe(CO)_2R$ ($R=C_2H_5$ or $CH_2C_6H_5$) and CF_3CN [13].

The complete IR spectrum is given in [6]. The monoclinic crystals have the parameters $a=8.975(7)$, $b=14.696(13)$, $c=9.785(4)$ Å, and $\beta=104.84(5)°$; space group $P2_1/n-C_{2h}^5$. $Z=4$ gives $D_c=1.810$ g·cm^{-3}, while $D_m=1.825$ g·cm^{-3}. The arrangement of the molecules in the unit cell is shown in a figure in [13]. The molecular structure is shown in **Fig. 20**.

Fig. 20

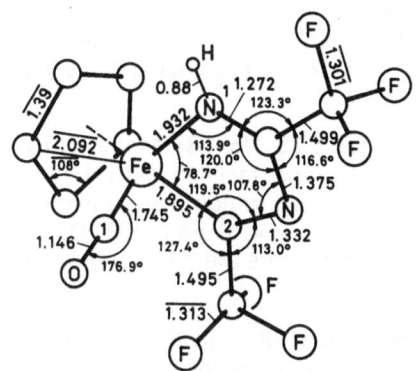

Molecular structure of $C_5H_5Fe(CO)C(CF_3)=NC(CF_3)=NH$ [13].

Other bond angles (°):

C(1)–Fe–C(2)   89.5(3)   C(1)–Fe–N(1)   97.1(3)

The central part of the molecule with the Fe–C(2)–N–C–N(1) ring, C(CF$_3$) atoms, and the H atom is substantially planar. The planarity of the ring leads to extensive delocalization. Therefore the bond angles at the ring C atoms are close to 120°, and all ring bonds except Fe–N(1) are shorter than expected for the formal structure. The N(1)–C bond is especially short. The mean dimensions of the $C_5H_5Fe(CO)$ moiety are those found in other compounds containing this grouping. But it is noteworthy that differences among the individual Fe–C$_5$ bond lengths appear to be significant, the longest of these bonds lie trans to the very short Fe–C(2) bond [13]. The complete mass spectrum is tabulated and discussed [6]. The most interesting fragmentation step is the loss of a CF$_2$ fragment from $[C_5H_5Fe(C(=NH)CF_3)]^+$ to form the ion $[C_5H_5Fe(CNH)F]^+$.

Solid No. 15 is stable in air for at least a few days. It is soluble in polar solvents such as CH_2Cl_2, ether, or THF. The black-green solutions are oxidized in air over a period of hours. Treatment of No. 15 with excess Br$_2$ cleaves the C_5H_5 ring to give a 35% yield of a single isomer of 1,2,3,4,5-pentabromocyclopentane [6]. The carbonyl group can be replaced by $P(CH_3)_2C_6H_5$ by reacting No. 15 in hexane with a large excess of the phosphine at room temperature for 10 d [13].

$C_5H_5Fe(CO)C_{12}H_9N_2$ (Table **26**, No. **16**). Method IIa with a 1:1 mixture of $C_5H_5Fe(CO)_2CH_3$ and azobenzene in ligroin (b.p. 100 to 120 °C) refluxed for 24 h. No. 16 can also be prepared by refluxing a mixture of $(C_5H_5Fe(CO)_2)_2$ and azobenzene in ligroin for 94 h. It is isolated by chromatography with petroleum ether/ether (19:1) eluent and vacuum distillation (bath temperature 110 to 120 °C/0.01 Torr) [8].

Compound No. 16 is slightly light and air sensitive. It is potentially resolvable into optical isomers although attempts to do so were unsuccessful. The complete IR spectrum is given in [8]. Replacement of CO by $P(C_6H_5)_3$ was achieved by UV irradiation in cyclohexane for 43 h [8].

$C_5H_5Fe(CO)C_3H_6P(C_6H_5)_2$ (Table **26**, No. **17**) is prepared from $C_5H_5Fe(CO)_2Br$ and $(C_6H_5)_2PC_3H_5MgCl$; the conditions and the yield are not given. The compound is surprisingly stable to heating and is sublimable in a high vacuum. It dissolves in all common organic solvents. The Fe-C σ bond is very reactive. Even at −60 °C SO_2 can be inserted quantitatively to form $C_5H_5Fe(CO)SO_2C_3H_6P(C_6H_5)_2$ (compound No. 35 in 1.5.2.2.7, p. 136) [22].

$C_5H_5Fe(CO)C_{17}H_{14}P$ (Table **26**, No. **18**). Method III with $C_5H_5Fe(CO)_2Si(CH_3)_3$ or $C_5H_5Fe(CO)_2Si(C_6H_5)_2CH_3$ and racemic $P(C_6H_5)(1-C_{10}H_7)CH_3$. A mixture of diastereomers was obtained. Crystallization from toluene/hexane isolated one of the diastereomers. ^1H NMR: $\delta = 2.07$ and 4.41 ppm. ^{31}P NMR: $\delta = 81.3$ ppm. IR (CH_2Cl_2): $\nu(CO)$ 1910 cm^{-1}. Attempts to prepare the other pure diastereomer were unsuccessful [30].

$C_5H_5Fe(CO)C_{26}H_{18}P$ (Table **26**, No. **19**). Method III with $C_5H_5Fe(CO)_2$-$Si(C_6H_5)_2CH_3$. The compound is purified by recrystallization from CH_2Cl_2/hexane [30].

$C_5H_5Fe(CO)C_{18}H_{14}O_3P$ (Table **26**, No. **20**). Method IIa with a mixture of $C_5H_5Fe(CO)_2C_6H_5$ and $P(OC_6H_5)_3$ in benzene which is irradiated for 3.5 h at 32 °C. The reaction time is increased by a factor of 8 when cyclohexane is the solvent [12]. Method III with $C_5H_5Fe(CO)_2SiR_3$ compounds (with $Si(CH_3)_3$, $Si(C_6H_5)_2CH_3$, or $Si(C_6H_5)(1-C_{10}H_7)CH_3$) [30]. No. 20 can also be prepared by irradiation of C_5H_5Fe-$(CO)(P(OC_6H_5)_3)C_6H_5$, an intermediate in Method IIa, in cyclohexane for 12 min [12] or by deprotonation of $[C_5H_5Fe(CO)P(OC_6H_5)_3]BF_4$ (1.5.2.2.4, p. 75) with $(C_2H_5)_2NH$ in CH_2Cl_2 for 10 min (57% yield) [21]. The complex is purified by chromatography on Al_2O_3 with petroleum ether/CH_2Cl_2 (9:1) [12] or benzene [21]; recrystallization from CH_2Cl_2/heptane at −10 °C [12]. Chromatography of a CH_2Cl_2 solution with hexane eluent did not separate No. 20 from its co-product $C_5H_5Fe(C_6H_4OP$-$(OC_6H_5)_2)P(OC_6H_5)_3$ (No. 32 in 1.5.2.1.1, p. 50) [30].

Earlier experiments using preparative Method IIa [4] or III [5] gave $C_5H_5Fe(CO)$-$(P(OC_6H_5)_3)C_6H_5$ [4], $C_5H_5Fe(CO)(P(OC_6H_5)_3)Si(CH_3)_3$, and $C_5H_5Fe(C_6H_4OP$-$(OC_6H_5)_2)P(OC_6H_5)_3$ [5] instead of No. 20, probably because of the reaction conditions.

^1H decoupled ^{13}C NMR spectrum $(CHCl_3)$, δ in ppm (assignment, multiplicity, and J(P,C) in Hz) [17]:

References on p. 249

$P(OC_6H_5)_2$	POC_6H_4Fe
120.7 (o-C, d, 4)	110.9 (C-3, d, 15)
121.3 (o-C, d, 5)	122.1, 123.9 (C-4,5, s's)
125.3 (p-C, s)	141.6 (C-1, d, 28)
129.3 (m-C, s)	144.2 (C-6, s)
129.5 (m-C, s)	161.7 (C-2, d, 22)
151.4 (CO, d, 7)	
151.8 (CO, d, 10)	

The resonances are $\delta = 83.0$ (s) ppm for C_5H_5 and 216.2 (d, $J = 38$ Hz) ppm for CO [17]. IR bands (Nujol) at 796 and 1098 cm^{-1} are evidence for the presence of the ortho-metalated $C_6H_4OP(OC_6H_5)_2$ moiety [12].

The mass spectrum shows the molecular ion $[M]^+$ (relative intensity 18) and the fragments $[M-CO]^+$ (100), $[C_5H_5Fe(CO)P(OC_6H_4)_2]^+$ (16), $[C_{11}H_9Fe]^+$ (17), and $[C_{10}H_{11}Fe]^+$ (12). Further fragments are given in [12].

$C_5H_5Fe(CO)C_8H_8O_3P$ (Table **26**, No. **21**). Method II a with a mixture of C_5H_5Fe-$(CO)_2C_6H_5$ and $(CH_2O)_2POC_6H_5$ in benzene irradiated for 4 h at 28 °C. No. 21 is isolated by chromatography on SiO_2 with petroleum ether/benzene (3:1) eluent; recrystallization from CH_2Cl_2/heptane at -10 °C [17].

^{13}C NMR spectrum ($CHCl_3$): δ (in ppm) $= 110.3$ (d, C-3, $J = 13$ Hz), 121.5 and 123.6 (s's, C-4,5), 141.9 (d, C-1, $J = 27$ Hz), 143.8 (s, C-6), and 161.5 (d, C-2, $J = 21$ Hz) for the C_6H_4 ring; 82.9 (s) for C_5H_5; and 216.4 (d, $J = 43$ Hz) for CO [17].

$C_5H_5Fe(CO)C(C_6H_5)=C(C_6H_5)C(CH_3)=O$ (Table **26**, No. **22**) is obtained along with compound No. XXIII (17% yield) and isolated by chromatography with benzene eluent (dark red-brown band); crystallization from pentane at -78 °C. The IR spectrum (hexane) shows no band in the region 1600 to 1700 cm^{-1}, where bands of ketonic carbonyl groups would be expected [20].

$C_5H_5Fe(CO)COC_2H_4SCH_3$ (Table **26**, No. **23**). Chromatography of a C_6H_6 solution on Al_2O_3 and successive elution with C_6H_6, CH_2Cl_2, and acetone gives No. 23 in the last orange band. Removal of the solvent and washing with pentane leaves oily crystals in a 1.2 to 4.4% yield (based on the $Na[C_5H_5Fe(CO)_2]$, used to prepare the rather unstable starting compound $C_5H_5Fe(CO)_2C_2H_4SCH_3$). Recrystallization from ether at -78 °C gives pure well-formed crystals [1, 2].

The solid is air sensitive. Further IR bands (KBr) are given in [2].

$C_5H_5Fe(CO)COC(CF_3)=CHSCH_3$ (Table **26**, No. **24**). The yields and the nature of the by-products depend on the reaction conditions. Reactions at 20 and 60 °C give yields of 25 and 30%, respectively, while at higher temperature (80 °C) the yield diminishes. By-products are $(Z)-(C_5H_5Fe(CO)\mu-SCH_3)_2$, $(C_5H_5Fe(CO)_2)_2$, and $C_5H_5Fe(CO)_2C\equiv CCF_3$. Photolysis gives an 11(?)% yield along with XXIV and $C_5H_5Fe(CO)_2C\equiv CCF_3$. No. 24 is isolated by elution from SiO_2 with hexane/CH_2Cl_2 and then pure CH_2Cl_2; recrystallization from hexane/CH_2Cl_2 [26].

The structure is based on an X-ray analysis [23, 24]; the J(H,F) coupling constant in the ^{19}F NMR spectrum confirms a (Z) configuration at the C=C double bond [26].

The monoclinic crystals have the parameters $a = 8.573(2)$, $b = 14.263(2)$, $c = 10.355(2)$ Å, and $\beta = 105.77(2)°$; space group $P2_1/n - C_{2h}^5$ and $Z = 4$ [24].

The solid is stable in air [26]. Solutions are thermally stable [23]. The solutions are stable in air for several hours [26].

The mass spectrum shows the molecular ion $[M]^+$ and the fragments $[M - CO]^+$ and $[M - 2CO]^+$ [24, 26]. On irradiation in THF the compound undergoes rearrangement to $C_5H_5Fe(CO)_2C(CF_3)=CHSCH_3$ (60% yield) [24, 26]. Photolysis in the presence of excess $CF_3C \equiv CH$ gives No. 27. There is no reaction when the compound is irradiated in the presence of $CF_3C \equiv CCF_3$ [26].

XXIII   XXIV

$C_5H_5Fe(CO)COC(CF_3)=C(CF_3)SCH_3$ (Table 26, No. 25) is prepared by Method IV in a 78% (thermal reaction) or 20% (photolysis) yield along with a F-containing whitish polymer. The crude product is dissolved in CH_2Cl_2 and separated from the polymer by filtration. Crystallization from CH_2Cl_2/hexane by cooling [26].

The mass spectrum shows the molecular ion $[M]^+$ and the fragments $[M - CO]^+$ and $[M - 2CO]^+$ [24]. Compound No. 25 is stable on heating in solution [23] and does not decompose when refluxed in THF under N_2 for several hours. The solid is stable in air, and solutions are air stable for several hours. The compound does not react with $CF_3C \equiv CCF_3$ on irradiation or heating and no reaction occurs with $P(C_6H_5)_3$ [26]. It is used as starting material for Nos. 9 and 28, see Method VII.

$C_5H_5Fe(CO)CON(CH_3)C(C_6H_5)=S$ (Table 26, No. 26) is prepared by dropwise addition of $C_6H_5CS-N(Li)CH_3$ to an equimolar amount of $C_5H_5Fe(CO)_2Cl$, both in THF at $-78°C$. The solution is stirred at room temperature for 1 h. Chromatography on SiO_2 with benzene/ether (1:1) (dark red band) and crystallization from ether/pentane gives a 29% yield [19].

The mass spectrum shows the molecular ion $[M]^+$ (relative intensity 14) and the fragments $[M - CO]^+$ (15), $[M - 2CO]^+$ (99), $[C_5H_5FeC(S)C_6H_5]^+$ (4), $[(C_5H_5)_2Fe]^+$ (44), $[C_5H_5Fe]^+$ (100), and $[FeC(S)C_6H_5]^+$ (13). It is noteworthy that dissociation of the thioamide ligand and the C_5H_5 group compete. Irradiation in benzene gives XXV (No. 4 in 1.5.2.2.7, p. 136) in a 44% yield [19].

XXV

$C_5H_5Fe(CO)C_7H_5F_6S$ (Table **26**, No. **27**) can be prepared in a 50% yield by heating $C_5H_5Fe(CO)_2C(CF_3)=CHSCH_3$ in the presence of $CF_3C\equiv CH$. It is isolated by chromatography on SiO_2 with CH_2Cl_2/hexane (1:2) and CH_2Cl_2; crystallization from hexane [26].

Four different isomers with the empirical formula $C_5H_5Fe(CO)C_7H_5F_6S$ and a cyclic structure are imaginable. The structure given in Table **26** is the preferred one on the basis of the NMR spectra [26].

$C_5H_5Fe(CO)C_8H_4F_9S$ (Table **26**, No. **28**) is obtained in 17% yield from compound No. 9 and in 15% yield from compound No. 25. It is isolated by chromatography on SiO_2. The available physical properties are not sufficient to decide between the two structures given in Table **26**. Other structures can be excluded because of the NMR spectra [26].

References:

[1] R.B. King, M.B. Bisnette (J. Am. Chem. Soc. **86** [1964] 1267/8). – [2] R.B. King, M.B. Bisnette (Inorg. Chem. **4** [1965] 486/93). – [3] R.B. King, M.B. Bisnette (Inorg. Chem. **5** [1966] 293/300). – [4] A.N. Nesmeyanov, Yu.A. Chapovsky, Yu.A. Ustynyuk (J. Organometal. Chem. **9** [1967] 345/53). – [5] R.B. King, K.H. Pannell (Inorg. Chem. **7** [1968] 1510/3).

[6] R.B. King, K.H. Pannell (J. Am. Chem. Soc. **90** [1968] 3984/7). – [7] R.B. King (Org. Mass Spectrom. **2** [1969] 387/99). – [8] M.I. Bruce, M.Z. Iqbal, F.G.A. Stone (J. Chem. Soc. A **1970** 3204/9). – [9] Y. Yamamoto, H. Yamazaki (Inorg. Chem. **11** [1972] 211/4). – [10] Y. Yamamoto, K. Aoki, H. Yamazaki (J. Am. Chem. Soc. **96** [1974] 2647/8).

[11] Y. Yamamoto, H. Yamazaki (J. Organometal. Chem. **90** [1975] 329/34). – [12] R.P. Stewart Jr., J.J. Benedict, L. Isbrandt, R.S. Ampulski (Inorg. Chem. **14** [1975] 2933/6). – [13] M. Bottrill, R. Goddard, M. Green, R.P. Hughes, M.K. Lloyd, S.H. Taylor, P. Woodward (J. Chem. Soc. Dalton Trans. **1975** 1150/5). – [14] J.L. Davidson, D.W.A. Sharp (J. Chem. Soc. Dalton Trans. **1975** 2283/7). – [15] J.L. Davidson, D.W.A. Sharp (B.L.L.D. [Brit. Libr. Lend. Div.] Suppl. Publ. J. Chem. Soc. Dalton Trans. **1975** No. SUP-21 457, pp. 1/3).

[16] K. Aoki, Y. Yamamoto (Inorg. Chem. **15** [1976] 48/52). – [17] R.P. Stewart Jr., L.R. Isbrandt, J.J. Benedict, J.G. Palmer (J. Am. Chem. Soc. **98** [1976] 3215/9). – [18] R.D. Adams, D.F. Chodosh, N.M. Golembeski (J. Organometal. Chem. **139** [1977] C39/C43). – [19] H. Brunner, J. Wachter (J. Organometal. Chem. **142** [1977] 133/7). – [20] H.G. Alt, M. Herberhold, M.D. Rausch, B.H. Edwards (Z. Naturforsch. **34b** [1979] 1070/7).

[21] D.L. Reger, C.J. Coleman (Inorg. Chem. **18** [1979] 3155/60). – [22] E. Lindner, G. Funk, S. Hoehne (Angew. Chem. **91** [1979] 569/70; Angew. Chem. Intern. Ed. Engl. **18** [1979] 535). – [23] J.L. Davidson, M. Shiralian, L. Manojlovic-Muir, K.W. Muir (J. Chem. Soc. Chem. Commun. **1979** 30/2). – [24] J.E. Guerchais, F. Le Floch-Perennou, F.Y. Petillon, A.N. Keith, L. Manojlovic-Muir, K.W. Muir, D.W.A. Sharp (J. Chem. Soc. Chem. Commun. **1979** 410/1). – [25] E.K. Barefield, D.J. Sepelak (J. Am. Chem. Soc. **101** [1979] 6542/9).

[26] F.Y. Petillon, F. Le Floch-Perennou, J.E. Guerchais, D.W.A. Sharp (J. Organometal. Chem. **173** [1979] 89/106). – [27] H. Brunner, G.O. Nelson (J. Organometal.

Chem. **173** [1979] 389/95). − [28] S.B. Butts, S.H. Strauss, E.M. Holt, R.E. Stimson, N.W. Alcock, D.F. Shriver (J. Am. Chem. Soc. **102** [1980] 5093/100). − [29] S.R. Berryhill, M. Rosenblum (J. Org. Chem. **45** [1980] 1984/6). − [30] G. Cerveau, G. Chauviere, E. Colomer, R.J.P. Corriu (J. Organometal. Chem. **210** [1981] 343/51).

[31] A.J. Carty (private communication from P.E. Garrou, Chem. Rev. **81** [1981] 229/66, 243).

1.5.2.2.15 Compounds with ^{1}L Ligands Containing Other Transition Metals Bonded to Iron

This section deals with a neutral compound I and the complex cations II and III. Because of the bridging C-CH$_3$ and CO groups, the right sides of the heteronuclear molecules are regarded as ^{1}L ligands.

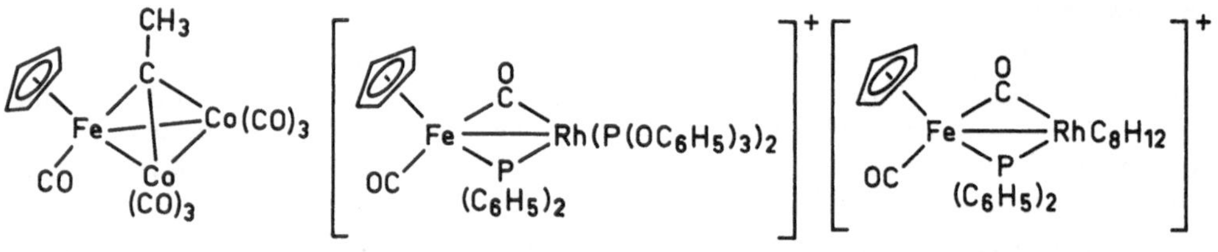

$C_5H_5Fe(CO)C(CH_3)Co_2(CO)_6$ (Formula I) was obtained in a very small yield from the reaction of $(C_5H_5Fe(CO)_2)_2$ with $CH_3CCo_3(CO)_9$ in C_6H_6 under irradiation for 3 d. Only an impure sample could be isolated by chromatography on SiO_2 with C_6H_{14}/C_6H_6 (9:1). An improvement of the yield was not possible since the substance itself decomposes in UV light.

^{1}H NMR spectrum (C_6D_6): $\delta = 3.74$ (CH_3) and 3.98 (C_5H_5) ppm. IR spectrum (C_6H_{12}): $\nu(CO)$ bands at 1920, 1988, 1996, 2020, 2030, and 2073 cm^{-1} [2, 3].

$[C_5H_5Fe(CO)(\mu\text{-}CO)(\mu\text{-}P(C_6H_5)_2)RhL_2]SbF_6$ with $L = P(OC_6H_5)_3$ (Formula II) and $L_2 =$ cyclooctadiene (Formula III) formed from $[RhL_2(solvent)_n]SbF_6$ compounds and equimolar amounts of $C_5H_5Fe(CO)_2P(C_6H_5)_2$. The IR spectra (CH_2Cl_2) showed the presence of a bridging and a terminal CO ligand: 1850 and 2028 cm^{-1} for II and 1834 and 2013 cm^{-1} for III [1].

References:

[1] R.J. Haines, J.C. Burckett-St. Laurent, C.R. Nolte (J. Organometal. Chem. **104** [1976] C27/C30). − [2] H. Beurich, H. Vahrenkamp (Angew. Chem. **93** [1981] 128/9; Angew. Chem. Intern. Ed. Engl. **20** [1981] 98). − [3] H. Beurich, R. Blumhofer, H. Vahrenkamp (Chem. Ber. **115** [1982] 2409/22).

1.5.2.2.16 Compounds of $[C_5H_5Fe(CO)(^1L)_2]^-$ Anions

The compounds in this section listed in Table 27 contain various acyl groups as ^{1}L ligands bonded to Fe through the carbon atom, Formula I. Compounds with R = alkyl can generally be protonated to afford formal enol tautomers of a 1,3-diketo-2-metallo system, Formula II with M = H. However, anions with formyl groups (one R = H, Nos. 1 to 3 in Table 27) cannot be protonated because the formyl group acts as an H⁻ donor towards protons to give H_2 and the corresponding $C_5H_5Fe(CO)_2COR$ complexes [2]. Even the Li compounds of these formyl anions are very unstable and could not be isolated.

The compounds of Table 27 are prepared as follows.

Method I: This is the addition of H⁻ to one carbonyl group of $C_5H_5Fe(CO)_2COR$ compounds. A slight excess of $Li[B(C_2H_5)_3H]$ is added to THF solutions of the acyl complexes kept at − 50 °C. Repeated isolation attempts failed. Yields were determined by ¹H NMR relative to a tetrachlorobenzene internal standard [2].

Method II: Analogous to Method I, CH_3^- is added to one carbonyl group of C_5H_5Fe-$(CO)_2COR$ compounds, with R = CH_3, i-C_3H_7, and $CH(C_6H_5)_2$. The reactions are carried out in ether with CH_3Li at −55 to −40 °C for 10 to 45 min. The unisolated $Li[C_5H_5Fe(CO)(COR)COCH_3]$ intermediates are converted to the enolates No. 4, 6, and 8 by treatment with HCl in ether between − 78 and 0 °C. After evaporation of ether, the products are extracted from the residues with pentane at 0 °C (for Nos. 4 and 6) or with benzene (for No. 8) [1].

Method III: Ether solutions of the Li intermediates of Method II are treated with $AlCl_3$ in ether at 0 °C with stirring for 45 min. The products are isolated by evaporation of solvent, extraction of the residue with pentane (for No. 7), benzene (for No. 5), or CH_2Cl_2 (for No. 9), filtration, and removal of the solvent [1].

Explanations for Table 27: In the ¹H NMR spectra, the cationic hydrogen of the enol tautomers is designated as H ⋯ O.

Table 27
Compounds of the $[C_5H_5Fe(CO)(^1L)_2]^-$ Anions.
Further information for all compounds is given at the end of the table.
For abbreviations and dimensions, see p. 375.

No.	¹L ligands method of preparation (yield in %)	cation	properties and remarks explanations see above	Ref.
1	CHO COCH₃ I (96.8)	Li	¹H NMR (THF, − 50°): 12.83 (CHO)	[2]

Table 27 [continued]

No.	¹L ligands method of prepa- ration (yield in %)	cation	properties and remarks explanations on p. 251	Ref.
2	CHO COC$_6$H$_5$ I (76.9)	Li	¹H NMR (THF, −50°): 12.91 (CHO) IR (THF): 1932 (CO), 1555 (C=O of CHO)	[2]
3	CHO COC$_6$H$_4$OCH$_3$-4	Li	¹H NMR (THF, −50°): 13.10 (CHO)	[2, 3]
	I (96.0)			
4	COCH$_3$ COCH$_3$ II (22)	H	golden, m.p. 87 to 89° (dec.) ¹H NMR (CS$_2$): 2.60 (s, CH$_3$), 4.81 (s, C$_5$H$_5$), 19.45 (br s, H···O) IR (C$_5$H$_{12}$): 1968 (CO)	[1]
5	COCH$_3$ COCH$_3$ III (20)	Al	golden, m.p. >250° ¹H NMR (CS$_2$): 2.68 (br s, CH$_3$), 4.56, 4.59, 4.62 (s's of different intensities, C$_5$H$_5$) IR (C$_5$H$_{12}$): 1939 (CO)	[1]
6	COCH$_3$ COC$_3$H$_7$-i II (13)	H	golden–brown liquid, m.p. >−78° ¹H NMR (CS$_2$): 1.00 (d, 6 H, CH$_3$), 2.69 (br s, 3 H, CH$_3$), 3.38 (br m, CH), 4.73 (s, C$_5$H$_5$), 20.08 (br s, H···O) IR (C$_5$H$_{12}$): 1930 (CO)	[1]
7	COCH$_3$ COC$_3$H$_7$-i III (44)	Al	golden, m.p. 53 to 54° (dec.) ¹H NMR (CS$_2$): 0.34 to 0.98 (m, 6 H, CH$_3$), 2.49 (4 s's, 3 H, CH$_3$), 3.02 to 3.65 (br m, CH), 4.39 to 4.59 (4 s's, C$_5$H$_5$) IR (C$_5$H$_{12}$): 1930 (CO), 1506 (C=O)	[1]
8	COCH$_3$ COCH(C$_6$H$_5$)$_2$ II (3)	H	golden, m.p. 27 to 28° ¹H NMR (CS$_2$): 2.58 (s, CH$_3$), 4.77 (s, C$_5$H$_5$), 5.90 (s, CH), 7.12 (m, C$_6$H$_5$), 19.11 (br, H···O) IR (C$_6$H$_6$): 1940 (CO)	[1]
9	COCH$_3$ COCH(C$_6$H$_5$)$_2$ III (59)	Al	golden brown, m.p. 120 to 190° (dec.) ¹H NMR (CS$_2$): 2.54, 2.62 (s's, relative intensity ∼3:1, CH$_3$), 4.06 to 4.46 (4 s's, C$_5$H$_5$), 6.20 (s, CH), 6.90 to 7.18 (m, C$_6$H$_5$) IR (C$_6$H$_{12}$): 1945 (CO), 1499 (C=O)	[1]

Further information:

Li[C$_5$H$_5$Fe(CO)(CHO)COR] (Table **27**, Nos. **1** to **3** with R=CH$_3$, C$_6$H$_5$, or C$_6$H$_4$OCH$_3$-4). The very low-field ^{1}H NMR resonances are characteristic of anionic metal formyl complexes, and they are extremely temperature dependent. A ^{13}C NMR spectrum of No. 2 (in THF/C$_6$D$_6$ at −40 °C) showed broadened C$_5$H$_5$ and C$_6$H$_5$ resonances, but the Fe-bonded C atoms gave only broad resonances barely above baseline ($\delta \sim 221$, 293, and 297 ppm, tentative).

The rates of thermal decomposition below room temperature were found to be first order and reproducible. The following rate constants were determined by ^{1}H NMR monitoring (k·10^4 in s^{-1}):

t in °C	−18	−16	+7	+14	+20
No. 1	–	9.94±0.52	–	–	–
No. 2	–	–	6.29±0.18	–	–
No. 3	0.353±0.02	–	5.67±0.15	10.66±0.14	24.87±0.93

A rather negative entropy of activation ($\Delta S^{\neq} = -17.6 \pm 2$ cal·K^{-1}·mol^{-1} for No. 3) eliminates CO dissociation as an initial rate-determining step. Hydride transfer from the formyl to the acyl ligand may be the first step. Decomposition of No. 2 at room temperature for 2 h and addition of (C$_6$H$_5$)$_3$SnCl gave C$_5$H$_5$Fe(CO)$_2$COC$_6$H$_5$, C$_5$H$_5$Fe(CO)$_2$Sn(C$_6$H$_5$)$_3$, (C$_5$H$_5$Fe(CO)$_2$)$_2$, and C$_6$H$_5$CH$_2$OH. B(C$_2$H$_5$)$_3$ had no effect on the decomposition rate [2], also see [3]. Protonation of No. 2 with CF$_3$SO$_3$H (mole ratio 1:1) at −50 °C results in immediate evolution of H$_2$ and formation of C$_5$H$_5$Fe(CO)$_2$COC$_6$H$_5$ in high yield [2].

H[C$_5$H$_5$Fe(CO)(COCH$_3$)COR] (Table **27**, Nos. **4**, **6**, and **8** with R=CH$_3$, i-C$_3$H$_7$, and CH(C$_6$H$_5$)$_2$). The ^{1}H NMR resonances of the enol protons at very low field do not reflect a high acidity of this proton. This deshielding is very dependent of the kind of metallo group. The broadening of this resonance (half-heigth width close to 50 Hz) probably indicates intra- and intermolecular exchange processes. The IR ν(C=O) vibration probably occurs below 1500 cm^{-1} and is not observed due to solvent interference.

The compounds undergo significant decomposition over a 2-day period even when stored at −20 °C under Ar. They decompose rapidly in air [1].

Al[C$_5$H$_5$Fe(CO)(COCH$_3$)COR]$_3$ (Table **27**, Nos. **5**, **7**, and **9** with R=CH$_3$, i-C$_3$H$_7$, or CH(C$_6$H$_5$)$_2$). The ^{1}H NMR spectra are very complex since both the C$_5$H$_5$ ligand and, for Nos. 7 and 9, the different chelate ring substituents may define (Z) and (E) isomers. The fine structure of the resonances was discussed and some tentative assignment was made.

The compounds are very sensitive to air oxidation, particularly in solution, although they have slightly better thermal and air stability than the hydrogen enolates [1].

References:

[1] C.M. Lukehart, J.V. Zeile (J. Am. Chem. Soc. **99** [1977] 4368/72). − [2] J.A. Gladysz, J.C. Selover (Tetrahedron Letters **1978** 319/22). − [3] J.C. Selover (Diss. Univ. California 1979 from Diss. Abstr. Intern. B **40** [1979] 258).

1.5.2.2.17   A C$_5$H$_5$Fe(CO) Compound with Unknown Ligand Structure

C$_5$H$_5$Fe(CO)C$_7$H$_5$F$_6$S was obtained on UV irradiation of a mixture of CF$_3$C≡CH and excess C$_5$H$_5$Fe(CO)$_2$SCH$_3$ in THF. Other products were complex I (No. 24, Table 26 in 1.5.2.2.14, p. 233) and a binuclear compound described in C4, 2.5.3.1.6, p. 183. C$_5$H$_5$Fe(CO)C$_7$H$_5$F$_6$S is not identical with complex II, which has the same empirical formula (No. 27, Table 26 in 1.5.2.2.14, p. 233).

The yellow crystals melt at 84 °C. ^{1}H NMR spectrum (solvent not given): δ (in ppm) = 2.30 (s, CH$_3$), 4.13 (q, CH, J(F,H) = 8.2 Hz), 4.80 (s, C$_5$H$_5$), and 6.60 (q, CH, J(F,H) = 6.7 Hz). ^{19}F NMR spectrum: δ (in ppm) = 56.6 (d, J(H,F) = 8.2 Hz) and 59.3 (d, J(H,F) = 6.7 Hz).

IR spectrum: ν(CO) 2005 and 2025 cm^{-1} in Nujol, 2020 cm^{-1} in CHCl$_3$. The mass spectrum shows the molecular ion. The spectra would be consistent with structure III, but there is not any evident reaction mechanism which could explain the formation of this ligand system.

Reference:

F.Y. Petillon, F. Le Floch–Perennou, J.E. Guerchais, D.W.A. Sharp (J. Organometal. Chem. **173** [1979] 89/106).

1.5.2.3 Compounds with Two CO Ligands

This large section on cyclopentadienyliron carbonyls, the dicarbonyls, begins with $C_5H_5Fe(CO)_2H$. The $C_5H_5Fe(CO)_2$ moiety occurs in numerous complexes containing one further ligand of the X or ^1L type (both neutral and cationic compounds) or of the ^2D or ^2L type (only cationic compounds).

In the text and the tables of all subsections of 1.5.2.3, the $C_5H_5Fe(CO)_2$ unit will be abbreviated Fp. [Fp]$^-$ and Fp$_2$ stand for the $[C_5H_5Fe(CO)_2]^-$ anion and the $C_5H_5Fe(CO)(\mu-CO)_2(CO)FeC_5H_5$ dimer, respectively.

The present volume deals only with that part of the $C_5H_5Fe(CO)_2X$ compounds where X represents hydrogen, halogens, and groups bonded to iron through elements of Main Groups V and VI. The remaining $C_5H_5Fe(CO)_2X$ compounds and those with ^5L ligands other than cyclopentadienyl will be in the next volume, B12. The $[^5LFe(CO)_2]^-$ anions, including [Fp]$^-$, will also be described in a subsequent volume although systematically [Fp]$^-$ should stand here at the beginning of 1.5.2.3.

1.5.2.3.1 $C_5H_5Fe(CO)_2H$ and $C_5H_5Fe(CO)_2D$

$C_5H_5Fe(CO)_2H$ has been prepared in high yields from Na[Fp] and $(CH_3)_3CCl$ in THF in a 16-h reaction. After the solvent is removed in vacuo, the residue is chromatographed on Al_2O_3 and distilled [8], also see [12, 25]. The compound is also obtained by treatment of FpCl with $NaBH_4$ in THF for 15 min followed by pouring the mixture into water and extracting the hydride into petroleum ether. Removal of the solvent at 0 °C/0.1 mm leaves a pale yellow liquid, which is distilled in a vacuum [1, 6], also see [7]. The action of $NaBH_4$ on a suspension of $[FpCO][B(C_6H_5)_4]$ in THF at -20 °C and workup as just described gives some volatile FpH, which condenses in the cold trap on removal of solvent. The residue, the main product consists of Fp$_2$ [4]. It is suggested that FpH is mainly an intermediate in this reaction [13]. Another method involves the electrochemical reduction of FpX (X = halogen) or Fp$_2$ in acidified THF, where the [Fp]$^-$ initially formed is quickly protonated. In a typical experiment, a 3×10^{-3} M solution of Fp$_2$ in THF/0.1 M $[N(C_4H_9)_4]ClO_4/3 \times 10^{-3}$ M HCl is electrolyzed for 30 to 40 min with a Pt anode and a Hg pool cathode. The potential of the Hg is kept constant at -1.7 V vs. Ag/AgClO$_4$ in 0.1 M $[N(C_4H_9)_4]ClO_4$ in THF [15]. Cleavage of the Fe–E bond of $FpE(CH_3)_3$ (E = Si or Ge) by HCl in CH_2Cl_2 at 60 °C (1 h) gives FpH and Fp$_2$ [20, 27]; solid $FpSi(CH_3)_3$ and anhydrous HCl undergo an extremely rapid reaction while warming from -196 °C to room temperature to produce FpH [21, 22]. In a similar reaction in C_6H_6 at 10 °C $FpSi(CH_3)C_3H_6$-cyclo gives FpH (20% yield) along with $FpSi(CH_3)(C_3H_7-n)Cl$ (70% yield) [26, 30]. Treatment of $FpSi(CH_3)_3$ with HF in ether also yields FpH. However, it decomposes almost totally to Fp$_2$ during the workup [28]. FpH and Fp$_2$ are also products of the slow reaction of $FpSi(CH_3)_3$ with $(CH_3)_2NH$ at room temperature [21, 22]. FpH is formed in small amounts in the reactions of FpBr with SiH_3K in monoglyme at -40 °C/1 h and 24 °C/2 h (the main product is FpSiH$_3$) [18], of $[FpP(C_6H_5)_3]Cl \cdot 3H_2O$ with $NaBH_4$ in THF/ether at -10 °C/0.5 h (the main product is cyclo-C_5H_6-$Fe(CO)_2P(C_6H_5)_3$) [4], of Na[Fp] with $[(CH_3)_2CHCH=N(CH_2CH_2)_2O]Cl$ in THF at -78 °C/0.5 h [32] and presumably of Na[Fp] with $(C_6F_5)_2CHBr$ in THF [14]. In all of these cases FpH could be identified by IR or ^1H NMR spectroscopy. In the formation of Fp$_2$ from $Fe(CO)_5$ and cyclopentadiene, FpH is assumed to be an intermediate, which transfers H_2 to cyclopentadiene giving cyclopentene and cyclopentane [3]. In the Fp$_2$-catalyzed formation of aldehydes from olefins, CO and H_2, a mechanism involving an FpH intermediate was proposed [23, 31]. FpH may also be involved in the formation of 1-Fp-benzocyclobutene and 1-Fp-naphthocyclobutene from Na[Fp] and the corresponding 1,2-dihalogenocyclobutenes [33].

Spectroscopic and polarographic studies of the equilibrium $Fp^- + HX \rightleftharpoons FpH + X^-$ (X = Cl, HCOO, C_6H_5COO, CH_3COO, $(CH_3)_3CCOO$) in THF have been carried out. FpH forms with every acid stronger than about benzoic acid even at stoichiometric concentration, and no oxidation wave of $[Fp]^-$ can be observed [24].

FpH is a yellow, petroleum-soluble oil which melts at ca. $-5\,°C$ [1]. It decomposes rapidly at room temperature giving hydrogen and Fp_2 [1, 8], at 0 °C the decomposition is complete within 13 h [18].

The ^1H NMR spectrum shows resonances (δ in ppm) at -11.91 in C_6H_{12} or -11.61 in C_6H_6 (assigned to FeH) and 4.74 (C_5H_5) in C_6H_{12} [1, 6, 30]. The IR spectrum (in CS_2) exhibits bands at 1900 w, 1930 sh, 1960 vs, and 2014 vs cm^{-1} in the $\nu(CO)$ region. A medium intensity band at 1835 cm^{-1} is assigned to $\nu(FeH)$. The spectrum is completely given [6].

Reactions are mostly carried out with FpH prepared in situ, either from FpCl and $NaBH_4$ in THF [7] or from Na[Fp] and $(CH_3)_3CCl$ in THF [12, 25]. Different types of addition products are obtained with unsaturated hydro- or fluorocarbons. Thus treatment with $CH_2=CHCH=CH_2$ produces a mixture of $FpCH_2CH=CHCH_3$ isomers [7], while with $CH_2=C=C=CH_2$, the compounds $FpC(=CH_2)CH=CH_2$ and $FpCH_2-C\equiv CCH_3$ are produced [29]. Reaction with $CF_2=CF_2$ at atmospheric pressure yields $FpCF_2CF_2H$ [9, 11], and reaction with $CF_3C\equiv CH$ in benzene in a sealed vessel at room temperature (4 d) yields a mixture of cis-$FpCH=CHCF_3$ and $FpC(CF_3)=CH_2$, in which the latter predominates [16, 17]. $FpCH(CN)CH_3$ and Fp_2 are the products of the reaction with $CH_2=CHCN$ in petroleum ether/THF after 2 h. In this case the course of the addition of the Fe-H system across a double bond adjacent to a group with a strong I effect is thought to be an indication that the Fe hydrogen is essentially hydridic in nature [5]. The relative acidities of the metal carbonyl hydrides are $(CO)_4CoH \gg (CO)_5MnH \gg FpH$ [19].

Substitution at the C_5H_5 ring takes place on treatment with $C_6H_5CH=CH_2$ in the presence of $AlCl_3$ giving $C_6H_5CH_2CH_2C_5H_4Fe(CO)_2H$ [2]. Addition of $(CH_3)_3CNCO$ to FpH in THF at $-78\,°C$, followed by stirring at room temperature for 12 h results in formation of $C_5H_5(CNC(CH_3)_3)Fe(\mu-CO)_2Fe(CO)C_5H_5$ as the main product along with $C_5H_5Fe(CO)(CNC(CH_3)_3)(CONHC(CH_3)_3)$ and the minor product $FpCONHC(CH_3)_3$ [25]. Treatment with CH_3SSCH_3 under similar conditions affords $FpSCH_3$ and Fp_2 [10, 12]. FpH reacts with N-methyl-N-nitroso-p-toluenesulfonamide to give Fp_2 without loss of CO. No nitrosyl ligand is incorporated [12].

$C_5H_5Fe(CO)_2D$ has been prepared by reduction of FpI in ether by $LiAlD_4$. After removal of solvent at 0 °C/0.1 Torr the liquid is fractionally distilled but complete separation from ether has been difficult [6].

The IR spectrum of the CS_2 solution shows bands at 1900 w, 1932 sh, 1960 vs, and 2014 vs cm^{-1} in the $\nu(CO)$ region. A medium intensity band at 1330 cm^{-1} is probably associated with the $\nu(FeD)$. The spectrum is completely reported [6].

References:

[1] M.L.H. Green, C.N. Street, G. Wilkinson (Z. Naturforsch. **14b** [1959] 738). – [2] J. Kozikowski, Ethyl Corp. (U.S. 2916503 [1959]). – [3] H.W. Sternberg, I. Wender (Chem. Soc. [London] Spec. Publ. No. 13 [1959] 35/55). – [4] A. Davison,

M.L.H. Green, G. Wilkinson (J. Chem. Soc. **1961** 3172/7). − [5] J.K.P. Ariyaratne, M.L.H. Green (J. Chem. Soc. **1963** 2976/83).

[6] A. Davison, J.A. McCleverty, G. Wilkinson (J. Chem. Soc. **1963** 1133/8). − [7] M.L.H. Green, P.L.I. Nagy (J. Chem. Soc. **1963** 189/97). − [8] M.L.H. Green, P.L.I. Nagy (J. Organometal. Chem. **1** [1963] 58/69). − [9] M.L.H. Green (personal communication from P.M. Treichel, F.G.A. Stone, Advan. Organometal. Chem. **1** [1964] 143/220, 181). − [10] R.B. King, M.B. Bisnette (J. Am. Chem. Soc. **86** [1964] 1267/8).

[11] M.L.H. Green, A.N. Stear (Z. Naturforsch. **20b** [1965] 812). − [12] R.B. King, M.B. Bisnette (Inorg. Chem. **4** [1965] 482/5). − [13] R.K. Kochhar, R. Pettit (J. Organometal. Chem. **6** [1966] 272/8). − [14] M.I. Bruce (J. Organometal. Chem. **10** [1967] 495/504). − [15] D. Grešová, A.A. Vlček (Inorg. Chim. Acta **1** [1967] 482).

[16] D.A. Harbourne, F.G.A. Stone (J. Chem. Soc. A. **1968** 1765/71). − [17] D.A. Harbourne, F.G.A. Stone (unpublished results from M.I. Bruce, F.G.A. Stone, Preparative Inorg. Reactions **4** [1968] 177/235, 191). − [18] E. Amberger, E. Mühlhofer, H. Stern (J. Organometal. Chem. **17** [1969] P5/P6). − [19] J.F. Bald, A.D. Berry, R.E. Highsmith, A.G. MacDiarmid, M.A. Nasta (Proc. 4th Intern. Conf. Organometal. Chem., Bristol 1969, Abstr. A5). − [20] R.E.J. Bichler, H.C. Clark (Proc. 4th Intern. Conf. Organometal. Chem., Bristol 1969, Abstr. O4).

[21] A.D. Berry, J.R. Bergerund, R.E. Highsmith, A.G. MacDiarmid, M.A. Nasta (Proc. 4th Intern. Conf. Organometal. Chem., Bristol 1969, Abstr. A4). − [22] M.A. Nasta, A.G. MacDiarmid (J. Organometal. Chem. **18** [1969] P11/P13). − [23] J. Tsuji, Y. Mori (Bull. Chem. Soc. Japan **42** [1969] 527/9). − [24] D. Miholova, A.A. Vlček (Proc. 3rd Conf. Coord. Chem., Smolenice-Bratislava 1971, pp. 221/6). − [25] W. Jetz, R.J. Angelici (J. Organometal. Chem. **35** [1972] C37/C39).

[26] C.S. Cundy, M.F. Lappert (J. Organometal. Chem. **57** [1973] C72/C74). − [27] R.E.J. Bichler, H.C. Clark, B.K. Hunter, A.T. Rake (J. Organometal. Chem. **69** [1974] 367/76). − [28] W. Malisch, P. Panster (Chem. Ber. **108** [1975] 2554/73). − [29] T.E. Bauch, W.P. Giering (J. Organometal. Chem. **114** [1976] 165/74). − [30] C.S. Cundy, M.F. Lappert, C.K. Yuen (J. Chem. Soc. Dalton Trans. **1978** 427/33).

[31] E. Cesarotti, A. Fusi, R. Ugo, G.M. Zanderighi (J. Mol. Catal. **4** [1978] 205/15). − [32] E.K. Barefield, D.J. Sepelak (J. Am. Chem. Soc. **101** [1979] 6542/9). − [33] T. Bauch, A. Sanders, C.V. Magatti, P. Waterman, D. Judelson, W.P. Giering (J. Organometal. Chem. **99** [1975] 269/79).

1.5.2.3.2 Compounds of the $C_5H_5Fe(CO)_2X$ Type with X = F, Cl, Br, I

In view of the close similarity of some methods of preparation, the physical properties, and the chemical behavior, the halides FpX are collectively treated. This implies a single large collection of references on pp. 294/303, which avoids too frequent recitations of one and the same reference.

1.5.2.3.2.1 Preparation and Formation

Generally Fp_2 is used as starting material for the preparation of the halides FpX (X = Cl, Br, I), whereas FpF is prepared by cleavage of $FpSi(CH_3)_2CH=CH_2$ with HF (see p. 260).

From $(C_5H_5Fe(CO)_2)_2$

By Oxidation with Oxygen in the Presence of X^-. Air is bubbled through a solution of Fp$_2$ in C_2H_5OH, $CHCl_3$, and concentrated HCl. After evaporation of solvent in vacuo and extraction of the residue with H_2O, the filtered red solution is extracted with $CHCl_3$. FpCl is crystallized from $CHCl_3$/petroleum ether, 75% yield [1, 3]. FpBr is prepared similarly by employing 34% aqueous HBr. Addition of ligroin to the $CHCl_3$ extract affords FpBr in a 65% yield [4, 6].

Replacing HX by concentrated H_2SO_4 and then adding H_2O, $Ba(OH)_2$, and $BaCl_2$ slowly forms FpCl [6].

Oxidation of Fp$_2$ by O_2 in acetone in the presence of 48 to 50% aqueous HBF$_4$ followed by addition of an aqueous solution of NaX (X = Cl, Br) to the intermediate $[Fp(H_2O)]^+$ leads also to formation of the FpX complexes, 50% yield of FpCl, 39% yield of FpBr. Attempts to prepare FpF by this method have been unsuccessful [200].

FpBr is also obtained by air oxidation of an absolute C_2H_5OH solution of Fp$_2$, KBr, and 65% HPF$_6$ at 25 °C for 3 h, 52% yield [92].

By Oxidation with Halogens. A solution of Cl_2 in CCl_4 is dropped to a solution of Fp$_2$ in CH_2Cl_2 at 20 to 25 °C giving a 42% yield of FpCl [11]. In the same way FpI is obtained by treatment of Fp$_2$ with I_2 [11, 170]. A solution of Br_2 in $CHCl_3$ is added dropwise to a solution of Fp$_2$ in $CHCl_3$ cooled in ice–salt mixture. Addition of ligroin gives FpBr in an 82% yield [6]. For preparation of FpI, Fp$_2$ and I_2 are reacted at room temperature (72% yield) [7] or refluxed in $CHCl_3$ (65% yield) [33]. Iodination reactions in $CHCl_3$, CH_2Cl_2, or C_6H_6 may proceed via the ionic intermediate $[Fp(\mu-I)Fp]X$ (X = I_3 or I) [112, 113, 132], while the corresponding bromination and chlorination reactions proceed by asymmetric cleavage of Fp$_2$ giving $[C_5H_5Fe(CO)_3]X$ in addition to the final product FpX [113, 132]. With n–hexane solvent there is no evidence for the formation of ionic intermediates [112]. The reaction of Fp$_2$ with Cl_2 in liquid HCl at −196 to −84 °C affords a mixture of FpCl, $[C_5H_5Fe(CO)_3]Cl$, and other decomposition products [149].

By Oxidation with Fe^{III} Salts. Fp$_2$ is oxidized in acetone by $[Fe(H_2O)_6]Cl_3$ (1:2 mole ratio) to give the solvent-bound complex $[FpOC(CH_3)_2]^+$, in which the coordinated acetone is easily displaced by Cl^- and Br^- to produce a 50% yield of FpCl or a 48.5% yield of FpBr [150].

For preparation of FpI, Fp$_2$ is oxidized with $Fe(ClO_4)_3$ in nitrogen–saturated acetone and then treated with $[N(C_4H_9-n)_4]I$ in acetone, 51.2% yield [150].

By Oxidation with AgX (X = ClO_4, PF_6, SbF_6). Fp$_2$ and AgX in acetone give the cation $[FpOC(CH_3)_2]^+$ and, on addition of aqueous KI, a 53% yield of FpI [201].

By Reactions with Halides. Reactions of Fp$_2$ with SnX$_2$ in refluxing C_6H_6, CH_3OH, or THF as well as with SnX$_4$ (X = Cl, Br, I) in refluxing xylene or butanol lead to formation of FpX in addition to complexes with Fe–Sn bonds [57, 171, 172]. When Fp$_2$ is treated with HgX$_2$ (X = F, Cl, I) in absolute C_2H_5OH at room temperature up to 45 °C the FpX compounds are produced together with FpHgX [11]. The formation of FpX by reaction of Fp$_2$ with As, Sb, and Bi halides is mentioned in [151, 152]. Reaction of Fp$_2$ with RSO_2Cl (R = alkyl or aryl) in THF yields FpCl and $FpSO_2R$ [58, 155]. FpCl was isolated from the reaction of Fp$_2$ with NOCl in liquid HCl at −196 to −84 °C [149] or in CH_2Cl_2 at room temperature (51% yield) [253]. Refluxing a mixture of Fp$_2$ and $RCOCH=CHI$ (R = C_6H_5, $4-CH_3C_6H_4$, $4-BrC_6H_4$) [343, 344]

References on p. 294

or ROCOCH=CHI (R=H, CH_3, C_2H_5) [345] in C_6H_6 gives the products FpI, FpCH=CHCOR, and FpCH=CHCOOR. Irradiation (350 to 600 nm) of Fp_2 in CHX_3 (X=Cl, Br), CCl_4, or CH_2Cl_2 forms FpX [275].

By Electrochemical Oxidation. FpCl is isolated in 35% yield when $\dot{F}p_2$, dissolved in a 0.1 M solution of $[N(C_2H_5)_4]Cl$ in CH_3CN, is oxidized at a carbon electrode at +0.75 V vs. the saturated NaCl calomel electrode.

Carrying out the oxidation in 0.25 M $[N(C_4H_9\text{-}n)_4]Cl$ in acetone at +0.50 V or in 0.2 M $[N(C_4H_9\text{-}n)_4]PF_6$ in CH_2Cl_2, followed by addition of excess $[N(C_4H_9\text{-}n)_4]Cl$ produces 43 and 86% yield, respectively, according to spectroscopic analysis [153].

From $C_5H_5Fe(CO)_2R$ by Cleavage of Fe-C Bonds

Many reports on formation of the complexes involve cleavage of an Fe-C bond by halogens or halogen compounds: Reported is the cleavage of $FpCH_2C_6H_5$ with stoichiometric amounts of Br_2 or I_2 in CCl_4 with approximately quantitative yields of FpX [173], FpC_6H_5 with I_2 in $CHCl_3$ [75], $FpCH(CH_3)C_2H_5$ with Br_2 in pyridine [133], FpCONHR (R=CH_3, cyclo-C_6H_{11}) with Br_2 in pentane or with I_2 in hexane [174], $FpCONH_2$ with I_2 in benzene [175], FpR (R=$CH_2CH_2C_6H_5$, $CH_2COC_5H_4$-FeC_5H_5, $CH_2CH_2C_5H_4FeC_5H_5$) with I_2 in CH_2Cl_2 at -65 °C to room temperature [225, 281], $[FpCH_2C(OC_2H_5)C_5H_4FeC_5H_5]BF_4$ with NaI in acetone at 20 °C [310], FpR (R=$C_5H_4Mn(CO)_3$, $C_5H_4Mn(CO)_2P(C_6H_5)_3$) with I_2 in $CHCl_3$ [282, 283], FpR (R=thien-2-yl, fur-2-yl, 5-methylfur-2-yl, benzofur-2-yl) with Cl_2, Br_2, or HCl in hexane [260], $FpC_5H_4FeC_5H_5$ with conc. HCl in dioxane or with Br_2 in C_6H_6 at 5 °C [176], $FpCH_2C\equiv CH$ with anhydrous HCl in petroleum ether [32], FpC_5H_5 with boiling HCl in ligroin [6], $Fp(4\text{-}CH_3C_6H_{10}\text{-}cyclo)$ with DCl in CH_2Cl_2 [334], FpR (R=alkyl, aryl, $CH_2Si(CH_3)_3$) with KX (X=I, Br) in liquid SO_2 at -10 °C, and $FpCH_2Si(CH_3)_3$ with SO_2 in CH_2Cl_2 or C_6H_6 in the presence of I^- [202].

Cleavage of the R ligand with $HgCl_2$ of FpC_6H_5 in C_6H_6 at 50 °C [75], FpR (R=CH_3, C_2H_5, $CH_2Si(CH_3)_3$, $CH_2CH_2C(CH_3)_3$, C_6H_5, 4-$C_6H_4OCH_3$) in THF [242], FpR (R=$CH_2CH_2C_6H_5$, $CH_2CH_2C_5H_4FeC_5H_5$) in acetone at 35 to 40 °C [281], and $FpCH_2C\equiv CC_6H_5$ in THF [284]. Cleavage with $HgBr_2$ of $FpCH_2C_6H_5$ in acetone at ~50 °C [173] and FpR (R=$CH_2COC_5H_4FeC_5H_5$, $C_5H_4Mn(CO)_3$) in refluxing acetone [261, 282].

Further cleavage reactions of such FpR where R=$CHDCHDC_6H_5$, $CD_2CH_2C_6H_5$, and $CHDCHDC(CH_3)_3$ with Cl_2, Br_2, I_2, ICl, and HBr were studied in various solvents and at various temperatures [154, 226, 243, 262]. For reactions of the same compounds with $HgCl_2$, affording FpCl, see [154, 227, 228, 243, 329]. For reactions of FpR (R=primary alkyl or aryl group) with $HgCl_2$ or of $FpCH_3$ with HgX_2 (X=Cl, Br, I), affording FpX, see [244].

Formation of FpX from cleavage of FpR (R=1,2,3-triphenylcyclopropenyl) carried out with X_2 (X=Br, I), $HgCl_2$, HCl, $LiBr/HBF_4\cdot O(C_2H_5)_2$, and $[N(C_4H_9)_4]I/HBF_4\cdot O(C_2H_5)_2$ is reported [331]. Treatment of $FpCH_2CH=CH_2$ with SnI_2 in C_6H_6 [342] or with KI in liquid SO_2 [311] forms FpI. Oxidation of $FpCH_2C_6H_5$ with excess $CuCl_2\cdot 2H_2O$ in CH_2Cl_2 gives FpCl [312], whereas the reactions with $CuCl_2\cdot 2H_2O$ or $CuBr_2$ in CH_2I_2 yields up to 65% FpI [330]. While $FpCH_3$ behaves similarly, good yields of FpX (X=Cl, Br) have been obtained from FpC_2H_5 and FpC_4H_9-n only with anhydrous $CuCl_2$ or $CuBr_2$ in CH_2Cl_2 [330].

FpI is produced by UV irradiation of FpC_6H_5 and n-C_3H_7I in C_6H_6 at 40 °C [60].

References on p. 294

Electrochemical (including cyclic voltammetry and coulometry) one-electron oxidation reactions of FpR ($R = CH_3$, $CH_2C_6H_5$, $CH_2CH_2C_6H_5$) in CH_2Cl_2 in the presence of added chloride ion yield FpCl in the range 72 to 82% [312].

From $C_5H_5Fe(CO)_2X$ or $C_5H_5Fe(CO)_2(\mu\text{-}X)Fe(CO)_2C_5H_5$ by Cleavage of Fe-X Bonds

By halogen exchange FpI may be obtained from FpCl and NaI or KI in refluxing acetone for 24 h in 65% yield [4, 6].

Cleavage reactions involving one Fe-S bond of the SO_2-bridged complexes Fp_2SO_2 and $(FpSO_2)_2$ by neat CH_3I or CH_3I in refluxing THF, yielding FpI and $FpSO_2CH_3$, are reported in [178, 245, 263, 313].

Treatment of $FpP(CF_3)_2$ with excess X_2 (X = Cl, Br) or ICl in $CFCl_3$ gives the cationic complexes $[Fp(P(CF_3)_2)X]X_3$ (X = Cl, Br), which decompose in CH_3CN solution to FpX [229].

Cleavage of $FpSi(CH_3)_2CH=CH_2$ with an excess of HX at $-20\,°C$ (X = F), $-78\,°C$ (X = Cl, Br), or $-40\,°C$ (X = I) for 1.5 to 2 d is the only method found in literature by which all four halides can be prepared. Yields: 82.1% FpF, 81.3% FpCl, 82.5% FpBr, 73.0% FpI. The reaction with HF/BF_3 in ether at $-30\,°C$ or with HI in SO_2 at $-40\,°C$ produces lower yields: 48.1% FpF, 62.3% FpI [246], also see [230]. Similar cleavage reactions have been carried out with $FpE(CH_3)_3$ (E = Si, Ge) and Cl_2 in $CHCl_3$, $FpSn(CH_3)_3$ and Cl_2 in CCl_4 or I_2 in C_5H_{12}, $FpE(CH_3)_3$ (E = Si, Ge, Sn) and ICl in C_6H_{12} or CF_3I in CF_3I [233], $FpSiC_3H_6CH_3$ ($SiC_3H_6CH_3 = 1$-methylsiletane-1-yl) and Cl_2 [203, 314] or Br_2 [314] in CCl_4, $FpSi(CH_3)(C_6H_5)C_{10}H_7$-1 and Cl_2 in CCl_4 at $-20\,°C$ [285], and $FpER_3$ (E = Si, Ge, Sn, R = alkyl, aryl) and I_2 in CCl_4 [247, 264]. Also $Fp_2Sn(CH_3)_2$ is cleaved by I_2 in $(CD_3)_2SO$ to give FpI [204]. The reaction of $FpSn(CH_3)_3$ with $(CH_3)_3SiCl$ in acetone results in formation of FpCl [204].

FpHgBr is cleaved on treatment with Br_2 in C_6H_6 to give FpBr, and FpHgI is cleaved with I_2 in $CHCl_3$ at $50\,°C$ to give FpI [179]. Action of concentrated HCl in dioxane at 50 to $75\,°C$ or gaseous HCl in dioxane at room temperature on FpHgI affords both FpI and FpCl, while with gaseous HCl in C_6H_6 only traces of FpCl are obtained [180]. FpCl and FpHgCl are the products of cleavage of one Fe-Hg bond in Fp_2Hg by concentrated HCl in dioxane at $40\,°C$ [180]. FpI has been isolated from the reaction of $FpHgCo(CO)_4$ with an aqueous solution of KI and I_2 [45]; FpBr was isolated from the reaction of FpHgCl with $[(C_5H_5)_2Co]Br_3$ in CH_2Cl_2 [11].

Photolysis of $FpM(CO)_3C_5H_5$ (M = Mo, W) in CCl_4 or of $FpMo(CO)_3C_5H_5$ in C_6H_6 in the presence of $n\text{-}C_4H_9CH_2I$ produces FpCl and FpI, respectively, in addition to other complexes [315].

By Further Reactions

FpX is formed from Na[Fp] by the reactions with RSO_2Cl (R = alkyl, aryl) in THF [44, 58, 155], with CH_3I and SO_2 in THF [313], and with SPF_2Br yielding FpBr [177].

A nonoxidative preparation of the halogen compounds starts with $[C_5H_5Fe(CO)_3]^+$-BF_4^-, which is stirred in acetone with potassium halides at 20 to $25\,°C$. Yields: 50% for FpI and 80% for each FpCl and FpBr [59].

Quantitative conversion to FpI is observed when $[Fp^2L]^+$ cations (^2L = olefin) are reacted with NaI in acetone/C_6H_6 or acetone [205, 252, 269, 286].

References on p. 294

FpI is formed in low yield ($\sim$10%) from $Fe(CO)_4I_2$ by treating $Fe(CO)_5$ with I_2 in ether and subsequently with NaC_5H_5 in THF [8]. $C_5H_5{}^{57}Fe(CO)_2I$ is obtained by a similar procedure from $^{57}Fe(CO)_4I_2$ [316]. Also the reaction of $Fe(CO)_4I_2$ with $C_5H_5Sn(CH_3)_3$ in ether affords FpI [206].

The reaction of $(C_5H_5)_2Fe_2(CO)_3{}^2D$ (^2D = phosphine or phosphite) with I_2 in CH_2Cl_2 is shown by IR to yield FpI, $C_5H_5Fe(CO)(^2D)I$, and $[Fp^2D]^+$, in approximately equal amounts, while in C_6H_6 the last compound is the main product in addition to trace quantities of the other two [95, 114].

1.5.2.3.2.2   Physical Properties and Structure

FpF, a light red powder, melts with decomposition above 98 °C [230, 246].

FpCl crystallizes from $CHCl_3$ on addition of petroleum ether in well-defined red crystals, which decompose without melting above 87 °C [1, 3]. The rather different m.p. of 94 to 95 °C is given in [171], see also [189].

FpBr is precipitated by adding ligroin to a $CHCl_3$ solution as red-brown crystals, which melt in air with decomposition at 98 to 102 °C and in vacuo at 105 to 107 °C [6, 65]. According to [171] the compound melts at 82 to 83 °C.

FpI is obtained from $CHCl_3$/petroleum ether as black crystals, which melt with decomposition at 118 to 120 °C [7, 59, 171] and sublime at about 90 °C/0.1 Torr [33].

The **^{1}H NMR spectra** of the complexes show only one singlet, corresponding to the chemical shift δ (in ppm) of the C_5H_5 protons. The shifts referred to $(CH_3)_3SiO$-$Si(CH_3)_3$ are indicated by an asterisk:

solvent	FpF	FpCl	FpBr	FpI	Ref.
$(CD_3)_2SO$ ·	5.75	–	–	–	[230, 246]
CH_2Cl_2	–	4.98	5.05	5.07	[190]
CH_3CN or CD_3CN	–	5.01	5.03	–	[138, 174]
$CDCl_3$	–	5.02 to 5.09	5.07	5.01 to 5.07	[78, 92, 200, 270]
CCl_4	–	–	–	5.05	[59]
$CHCl_3{}^*$	–	4.979	4.972	4.971	[191]

A systematic correlational analysis of the C_5H_5 ring proton chemical shifts for FpX (X = Cl, Br, I, CN, CH_3, C_6H_5, $COCH_3$) in terms of the Hammett-Taft σ constants for substituents X has revealed a linear relationship between $\delta(C_5H_5)$ and the inductive substituent constants [191]. In addition, a nearly linear correlation was found between the chemical shifts of the C_5H_5 protons and the Taft's polar constants σ^* or the electronegativities of the ligands X in FpX compounds (X = halogen, alkyl), while no such relation was obtained using Hammett's σ_H or resonance σ_R constants [78].

The **^{13}C NMR spectra** exhibit one signal for the two identical CO groups, and, under conditions of proton decoupling, one C_5H_5 resonance. The carbonyl chemical shifts are observed to be linearly dependent on the Taft σ_I values of X and on the carbonyl stretching frequencies. Comparison with other FpX complexes, $FpCH_3$, FpC_6H_5, or FpCN, reveals that the presence of strongly electron-withdrawing ligands results in carbonyl resonances being detected at higher magnetic fields, an indication of enhanced paramagnetic screening [190]. The following shifts are reported:

FpX	solvent	$\delta(C_5H_5)$	$\delta(CO)$	Ref.
FpCl	CH_2Cl_2	85.9	213.3	[190]
	$CHCl_3$	85.6 ± 0.3	212.9 ± 0.3	[139]
FpBr	CH_2Cl_2	85.9	213.5	[190]
	$CHCl_3$	85.4 ± 0.3	213.2 ± 0.3	[139]
FpI	CH_2Cl_2	84.7	213.6	[190]
	$CHCl_3$	84.8 ± 0.3	213.8 ± 0.3	[139]

Resonances of the ^{17}O **NMR spectra** (in CH_2Cl_2, downfield from $H_2^{17}O$) are at $\delta = 380.5$ and 379.2 ppm for FpCl and FpI, respectively [348].

The ^{19}F **NMR spectrum** of FpF in acetone shows a signal at 148.24 ppm [230]. For $(CD_3)_2SO$ solutions the signal is at 148.3 ppm [246].

Data of the **NQR spectra** of the halides recorded at the temperature of liquid nitrogen are reported as follows [207]:

FpX	nucleus	transition frequency (in MHz, ± 0.05)		asymmetry parameter η (in %, ± 0.5)	e^2Qq_{zz} (in MHz, ± 0.8)
		$^1/_2$ to $^3/_2$	$^3/_2$ to $^5/_2$		
FpCl	^{35}Cl	18.39	–	–	–
FpBr	^{81}Br	123.80	–	–	–
	^{79}Br	148.15	–	–	–
FpI	^{127}I	159.30	317.92	4.0	1060.1
		160.85	321.12	3.7	1070.7

The NQR frequencies of the halogens indicate some double bonding in the Fe–X bond. Introduction of the electron-acceptor o-carborane group into the C_5H_5 ring decreases the Br frequency in $(C_5H_4C_2H_2B_{10}H_9)Fe(CO)_2Br$. This is explained by a decrease in the population of the p_π orbitals of the bromine due to electron density transfer to the vacant metal orbitals. Since the asymmetry parameter on the iodine atom in FpI is close to zero, it is assumed that both halogen p_π orbitals participate equally in the additional π bonding with the central atom [207]. Comparison of the NQR data with those of $RC_3H_4Fe(CO)_3X$ ($R = H$, CH_3, C_6H_5, Br; $X =$ halogen) reveals that the Fp group is a stronger electron acceptor than the π-allyl-$Fe(CO)_3$ moiety [287].

The ^{57}Fe **Mössbauer spectra** consist of two lines arising from the quadrupole moment of the excited ^{57}Fe nucleus. Values for the isomer shift δ and the quadrupole splitting Δ (in mm/s) are summarized in the table on p. 263, the reference substance sodium nitroprusside being abbreviated as NPNa.

Data somewhat different to that given in the table (liquid nitrogen temperature, relative to natural iron) are reported in [254].

The differences in the isomer shifts of a series of FpX compounds are used for calculation of the partial isomer shift parameters of the ligands, e.g., Br^- or C_5H_5 [140]. For a discussion of the isomer shift data of various FpX complexes in terms of σ- and π-bonding contributions of the ligand X, see [140, 254, 317].

References on p. 294

temper- ature in K	δ CI	Br	I	Δ CI	Br	I	reference substance	Ref.
78	0.2277(9)	0.227(1)	0.215(1)	1.863(9)	1.87(1)	1.84(1)	natural α–Fe	[317]
	0.39	0.40	0.38	1.88	1.87	1.83	Co⁵⁷/Cr	[43]
	0.55	0.56	0.54	1.88	1.87	1.83	NPNa	[234]
85	–	0.26	–	–	2.00	–	metallic Fe	[140]
~293	0.1415(6)	0.1496(8)	0.1329(8)	1.83(1)	1.84(2)	1.82(2)	natural α–Fe	[317]
300	0.51	0.52	0.49	1.86	1.85	1.83	NPNa	[234]

It is difficult to find a direct relationship between isomer shifts and carbonyl stretching force constants [317].

Extensive data of the **IR spectra** are available in the $\nu(CO)$ region, where two intense bands of comparable intensities are observed, corresponding to symmetric (at ~ 2050 cm^{-1}) and asymmetric (at ~ 2000 cm^{-1}) CO vibrations. Most precise CO frequencies in cm^{-1} (intensities in M$^{-1} \cdot$cm$^{-2} \cdot 10^{-4}$):

solvent	FpF	FpCl	FpBr	FpI	Ref.
THF	–	1998(9) 2045(7)	1996(9) 2041(7)	1989(9) 2034(8)	[79]
CS₂	–	2007.8 2051.4	2004.0 2046.5	1997.7 2038.8	[115]
C₆H₁₂	–	2010.5 2053.0	2007.5 2048.5	2001.5 2041.0	[96, 271]
CHCl₃	–	2012.0 2057.6	2007.3 2052.9	2000.0 2043.8	[115]
CHCl₃	–	2012.1 (9.69±0.15) 2057.6 (7.30±0.02)	– –	2000.2(9.63±0.17) 2043.8(7.15±0.06)	[189]
CCl₄	–	2012.1±0.2 2055.3±0.2	– –	2001.4±0.2 2042.6±0.2	[127]
acetone	2000 2047	–	–	–	[230, 246]

Further $\nu(CO)$ data are reported for solutions in C₆H₁₄ [138], C₆H₁₂ [128, 318], C₇H₁₄ [116, 288], (CH₃)₂CO [270], CH₃CN [270], petroleum ether [255], C₆H₆ [315], CH₂Cl₂ [132, 190, 341], CH₂X₂ (X=Cl, Br, I) [192, 271], CCl₄ [64, 315], CS₂ [97, 141, 171], and for KBr [335]. The most striking differences in reported frequencies result from measurements in CHCl₃, e.g., for FpCl 2002, 2050 [317] and 2019, 2061 [92], see also [7, 64, 200, 270]. The spectrum of FpI (CHCl₃, $\nu(CO)$ region) is reproduced in [208]. The spectra (C₆H₁₂) of C₅H₅Fe(^{12}CO)(^{13}CO)X show the following bands (in cm^{-1}±2): 1978 and 2035 (X=Cl), 1975 and 2030 (X=Br), and 1968 and 2025 (X=I) [318].

The relationship between the frequencies of the symmetric and asymmetric CO vibrations of FpX (X=halogen, CN, CH$_3$, C$_6$H$_5$, C$_6$F$_5$) and the inductive substituent constants, σ_I [79], is linear, as is that between the average CO stretching frequencies or CO band intensities and the Taft polarity constants, σ^*, of the halogen in FpX [115]. No such correlation has been obtained using Hammett's σ_H or resonance constants σ_R, while a linear relation was also found between the A'' type of the CO stretching frequencies and the electronegativities of the ligands X (halogen, alkyl, aryl) in FpX compounds [78]. From the correlations it is concluded that π bonding is less important than σ bonding in the Fe-X bonds. The integrated intensities of the ν(CO) vibrations reveal that the extent of Fe-CO π bonding, and therefore the net electronic effect of X in FpX complexes where X=Cl, I, CN, and SnCl$_3$ are approximately the same [189].

In CHCl$_3$ the carbonyl absorption frequencies of FpX (X=Cl, Br, I) are higher and the splitting between A' and A'' modes is greater than in CS$_2$ [115].

The ν(CO) frequencies of FpI are 2029 (A') and 1964 (A'') at $\sim$0 kbar and 2037 (A') and 1969 (A'') at 30 kbar, all measured in CsI. The observed shift, small in comparison to that of cyano complexes, is explained by the stabilizing effect of back-bonding from metal to CO ligands which maintains the symmetry of metal-carbonyl complexes even at high pressures [272].

The carbonyl stretching frequencies of FpI are compared with those of Fe(CO)$_5$, Fe$_2$(CO)$_9$, Fp$_2$, and free CO and discussed with respect to the nature of the Fe-CO bonding system [208].

Three bands observed in the near-IR region are assigned to the binary stretching modes 2 A', A'+A'', and 2 A''. They are lower than expected. Reasons for the discrepancy are discussed in detail. A tentative interpretation of the relative intensities of these bands is also presented [115].

A graphical representation of the spectrum of FpI in the range 3800 to 4100 cm^{-1}, which shows the overtones of ring modes, is given in [80].

Selected IR frequencies (in cm^{-1}) of FpX in the 200 to 700 cm^{-1} region in Nujol [127, 129] and in C$_6$H$_6$ or CS$_2$ solutions [141] are given on p. 265.

The two ν(CO) absorptions of FpI are of similar intensities. However, those due to ν(Fe-CO) vibrations are not, probably resulting from a mixing of ν(Fe-CO) and δ(Fe-CO) vibrations [97]. The shoulder at 302 cm^{-1} in the spectrum of FpCl is attributed to ν(Fe-^{37}Cl), while the shoulder at 474 cm^{-1} has no obvious explanation [141].

The Nujol spectra show bands at 162 cm^{-1} for FpCl and 153 cm^{-1} for FpI, which are attributed to ring tilt. The FpI 137-cm^{-1} band is assigned to ν(FeI) [98].

The spectra of FpX (X=Cl, Br, I) show that the in-plane CH bending mode (a$_2$), which is inactive under the selection rules of local C$_{5v}$ symmetry, is observed in the mull spectrum only for FpI (at 1261 cm^{-1}) and with an intensity decreasing in the order I>Br>Cl in the spectra of the CCl$_4$, C$_2$Cl$_4$, and CS$_2$ solution (at 1255 to 1259 cm^{-1}). This behavior is interpreted as a consequence of a lower local symmetry of the C$_5$H$_5$Fe moiety of FpI and generally of a symmetry lower in solution than in the solid state. That the strong band at $\sim$845 cm^{-1} in the spectra for mulls is

References on p. 294

| FpCl | | FpBr | | FpI | | assignment |
Nujol	C_6H_6	Nujol	C_6H_6	Nujol	C_6H_6	
298	302 sh, 307	229	234	202		$\nu(FeX)$
348	349	350	351	358	356	$\nu(FeC_5H_5)$
370	373		375		377 ⎱	
414	413	408 sh	407	408 sh	406 ⎰	ring tilt
			CS_2		CS_2	
436	440	434	439	434	440 ⎱	$\nu(FeCO)$
479	474 sh, 481	479	479	488	484 ⎰	
	CS_2					
532	536	535	540	543	545 ⎫	
567	565	564	565	566	565 ⎪	
	595 sh		596 sh		598 sh ⎬	$\delta(FeCO)$
603	604	603	606	605	610 ⎭	

observed at ca. 835 cm^{-1} for solutions perhaps indicates less interaction of CH out-of-plane bending modes with the metal [130].

Further measurements have been carried out for FpX (X=Cl, Br, and I, 200 to 700 cm^{-1}) in C_6H_6, CCl_4 [127, 129], and CH_2Cl_2 [141]; for FpX (X=Cl and Br, 400 to 700 cm^{-1}) in CH_2X_2 [271]; for FpX (X=Cl and I, 100 to 700 cm^{-1}) in Nujol [98]; for FpCl (800 to 3200 cm^{-1}, ν(CH) 3110 cm^{-1}, solvent not unequivocally given) [3]; for FpI in KBr (400 to 3100 cm^{-1}) [7, 96], CS_2 (400 to 700 cm^{-1}) [97], and polyethylene (40 to 300 cm^{-1}) [271].

The solution spectra in the **UV and visible region** show the following absorptions (in nm, ε in M$^{-1} \cdot$cm^{-1}):

FpX	solvent	absorptions λ_{max} (ε)				Ref.
FpCl	CH_2Cl_2	475(155) sh		335(1070)	280(2055)	[234]
	C_6H_6		388(565) sh	336(935)		[318]
FpBr	CH_2Cl_2	473(330) sh		347(1330)	292(4690)	[234]
	C_6H_6		386(700) sh	350(1028)		[318]
	CH_3NO_2		385(674) sh			[318]
	CH_3CN		385(717) sh	346(922)		[318]
	cyclo-C_6H_{12}	∼490(140) sh	∼400(600) sh	354(990)		[142]
FpI	CH_2Cl_2	498(570) sh		343(2330)	260(4740)	[234]
	C_6H_6			342(2090)		[318]
	CH_3CN			323(2300)		[318]

The markedly distorted band at ∼350 nm is thought to consist of more than one transition [234]. For data on FpX (X=Cl and Br) in aqueous $HClO_4$, see [142].

Molecular Properties. The following stretching (k) and interaction (k$_i$) force constants (in mdyn·Å^{-1}) for the CO groups have been calculated by the Cotton–Kraihanzel method. Only extreme values of k are cited. Other k values are reported for FpCl [99, 115, 317, 341], FpBr [317], and FpI [115, 189].

FpX	k	k_i	Ref.
FpCl	16.58	0.393	[190]
	16.70	0.37	[189]
FpBr	16.51	0.376	[190]
	16.57	0.35	[115]
FpI	16.45	0.326	[190, 317]
	16.82	0.35	[99]

The remarkably high value of 16.82 mdyn·$Å^{-1}$ for FpI results from the fact that the authors [99] based their calculations on the highest $\nu(CO)$ frequencies found in literature (2020 and 2062 cm^{-1} for $CHCl_3$ solutions) [64].

From analysis of bonding in cyclopentadienyl metal–carbonyl derivatives, also including the compounds described in this chapter, it is suggested that increased C_5H_5 ^1H NMR chemical shifts and decreased CO stretching force constants indicate increased negative charge on the metal atom. In most cases the relationship between $\delta(C_5H_5)$ and k_{CO} could be approximated by a straight line [99]. The low CO force constants in FpX, in comparison to $Mn(CO)_5X$, is seen in the special bonding character of complexes with π-bonded ring ligands [273].

Integrated IR intensities of the $\nu(CO)$ stretching vibrations in $CHCl_3$ have been determined and used to calculate FeCO group dipole moment derivatives. The dipole moment derivatives for the symmetric stretching motion $\mu'_{FeCO}(1)$ are smaller than those for the corresponding asymmetric stretching vibration $\mu'_{FeCO}(2)$. This shows that the C_5H_5 ligand primarily acts as a donor in these complexes. Variations in the derivatives as a function of the angle between MCO groupings are considered in detail [189].

The He(I) photoelectron spectra of FpX show the following vertical ionization potentials (in eV) [256, 273]:

FpCl		FpBr		FpI	
8.00	8.00	7.95	7.93	7.81	7.77
8.27	8.38	8.27	8.30	8.18	8.17
8.95	8.99	8.91	8.89	8.8	8.73
					9.18
			9.57	9.35	9.37
	9.90	9.8	9.78		
10.15	10.17			10.12	10.03
10.65	10.5 to 11.1	10.45	10.4 to 10.8	10.48	10.40, 10.76
12.7		12.79		12.89	
13.88		13.95		13.69	
16.72		16.95		16.81	
[256]	[273]	[256]	[273]	[256]	[273]

The ionization-band envelopes are well represented by the asymmetric Gaussian functional forms. Approximate molecular orbital calculations of the complexes and an ab-initio calculation on the cyclopentadienide ion were used to interprete the spectra. The assignment of the molecular orbitals, represented in a diagram for FpCl and the ionization potentials were made by comparison with those of $Mn(CO)_5X$ complexes. The first two ionization peaks are attributed to ionizations from the metal-halogen antibonding π levels. The third peak may either result from ionization of the predominantly metal b_2-type orbital or from ionization of the metal-halogen σ bond, the a_1-type orbital. The next band, a shoulder in the spectra of the chloride and the bromide complexes and an unresolved and broad band in the spectrum of FpI, represents ionizations from the metal-halogen π-bonding orbitals. The ionizations between 10 and 11 eV are associated primarily with the C_5H_5 ring $\pi e_1''$ levels. In the spectrum of FpI, two separate peaks are observed in this range. That near 10 eV is tentatively attributed to ionization from the metal-halogen σ bond; that at 10.4 eV, to the ring e_1'' orbitals. The broad intense band from 13 to 16 eV has been assigned to ionization of the σ and π levels of the carbonyl groups and the C_5H_5 ring [273], also see [256].

Ionization potentials taken from mass spectrometric studies are 7.8_5 eV for FpCl and 7.6_0 eV for FpI [274].

The fine structure of the X-ray K-absorption spectrum of FpBr exhibits one maximum at 17 eV, which is attributed to a 5 p transition, and another smaller absorption at 3 eV, which is tentatively ascribed to the CO groups [12].

Crystal structures. The chloride and bromide are isomorphous, crystallizing from ligroin in the orthorhombic system, space group $P2_12_12_1 - D_2^4$ (No. 19) with 4 molecules in the unit cell [4].

1.5.2.3.2.3 Chemical Behavior

Solubility and Properties of Solutions. FpCl is very soluble in $CHCl_3$, alcohol, and ether, moderately soluble in C_6H_6, CS_2, CCl_4, and H_2O, sparingly soluble in petroleum ether [3, 11]. The red aqueous solutions do not give immediate precipitates with $AgNO_3$, silicotungstic acid, or Reinecke's salt. In acidic solutions $AgNO_3$ forms a precipitate within a minute, thus the compound is essentially undissociated in neutral aqueous solution [3].

FpBr is soluble in organic solvents and slightly soluble in water [11].

FpI is very slightly soluble in water, somewhat soluble in petroleum ether, and readily soluble in $CHCl_3$ and acetone to give brown solutions. The aqueous acetone solution precipitates AgI only slowly on addition of $AgNO_3$. In acidic solutions precipitation is rapid, and the resulting orange solution of the $[Fp]^+$ cation gives precipitates with silicotungstic acid and Reinecke's salt [7].

The molar conductivities of FpCl and FpI in 10^{-3} M CH_2Cl_2 solutions at 20 °C are 1.1 and 0.2 $\Omega^{-1} \cdot cm^2 \cdot mol^{-1}$, respectively [271]. The molar conductivities of FpCl in acetone are 1.2 $\Omega^{-1} \cdot cm^2 \cdot mol^{-1}$ for 1.17×10^{-3} M and 0.6 $\Omega^{-1} \cdot cm^2 \cdot mol^{-1}$ for 3.91×10^{-3} M [138].

Thermal Decomposition and Stability towards Air. Refluxing FpCl in THF for 15 h under nitrogen affords ferrocene and $FeCl_2$ [17]. Ferrocene is also obtained when FpCl or FpBr are heated in a glass tube over a free flame [6].

FpI is stable with respect to oxidation in the solid state, but decomposes slowly in $CHCl_3$ solution [208].

Mass Spectrum. Electron impact reveals the stepwise loss of the CO groups to be the most significant reaction. Elimination of the halogen atom seems less favored [65, 100]. Metastable peaks are observed in the FpBr spectrum for the following two transitions involving fragmentation: $[C_5H_5FeBr]^+ \rightarrow [C_3H_3FeBr]^+ + C_2H_2$ and $[C_5H_5Fe]^+ \rightarrow [C_3H_3Fe]^+ + C_2H_2$ [65]. The ionization-yield curves for FpBr [65] permit only a raw estimate of the appearance potentials. The relative intensities of the ions in the mass spectra are given in the following table.

ions	relative intensities			
	FpF	FpCl	FpBr	FpI
$[C_5H_5Fe(CO)_2X]^+$	3	12	47	81
$[C_5H_5Fe(CO)X]^+$	2	24	41.5	90
$[C_5H_5FeX]^+$	3	100	100	100
$[C_5H_5Fe(CO)_2]^+$	1	7.5	18.4	9
$[C_5H_5Fe(CO)]^+$	2	5.5	4.5	–
$[C_5H_5Fe]^+$	2	70	92.1	>100
$[(C_5H_5)_2Fe]^+$	–	71.2	4.6	27
$[C_3H_3FeX]^+$	–	–	6.3	–
$[C_3H_3Fe]^+$	1	9.5	5.9	34
$[C_2H_2Fe]^+$	2	–	–	–
$[C_2HFe]^+$	–	5.5	4.9	28
$[FeC]^+$	–	–	–	–
$[FeX]^+$	–	50	45.7	100
Fe^+	–	48	8.9	>100
X^+	–	1	–	9
$[C_5H_5]^+$	4	22	–	22
$[C_3H_3]^+$	–	24.5	–	>100
$[CO]^+$	100	–	–	–
energy in eV	70	70	20	70
Ref.	[246]	[100]	[65]	[100]

Photolysis and Radiolysis. The main product resulting from irradiation with wavelengths greater than 400 nm of FpCl in $(CH_3)_2SO$ or pyridine is Fp_2. By-products are free cyclopentadiene and chloride ions. Wavelengths greater than 280 nm also produce a dimeric substance, which is assumed to consist of Fp_2 in which one CO molecule is displaced by solvent. Isolation was not possible [209]. While Giannotti, Merle [275] report that irradiation (300 to 700 nm) of FpX (X=Cl, Br) in C_6H_6 or $CHCl_3$ affords ferrocene and FeX_2, Alway, Barnett [318] observed no changes in the IR- or UV-visible spectra of FpX (X=Cl, Br, I) when these complexes are irradiated at 366 or 436 nm in C_6H_6, CH_3CN, CH_3NO_2, or THF solutions. Photolysis of FpX (X=Cl, Br, I) in ^{13}CO-saturated C_6H_6 solutions forms $C_5H_5Fe(CO)(^{13}CO)X$ [318]. Photolysis of FpX (X=Cl, Br, I; Hg high pressure arc) in a 2-methyltetrahydrofuran matrix at about 77 K results in expulsion of CO, which becomes matrix-trapped, and formation of an unsaturated derivative characterized by IR. Warming the photolyzed matrix to room temperature restores the spectrum of starting material. At 30 K, FpCl undergoes no perceptible change during a 28-min irradiation, a contrast to its Mo and W analogues [289].

Neutron irradiation of FpI at $-78\,^\circ$C causes formation of $C_5H_5{}^{59}Fe(CO)_2^{\cdot}$, $(C_5H_5Fe(CO)_2)_2{}^{59}Fe$, $^{59}Fe(CO)_5$, and $^{59}Fe(C_5H_5)_2$. Subsequent treatment at 75 °C

References on p. 294

increases the yields of $^{59}Fe(CO)_5$ and the radioactive target compound but does not affect the yields of the other two products. The observed radioactivity of the Fpl carrier is supposed to result from the exchange reaction

$$C_5H_5Fe(CO)_2I + C_5H_5{}^{59}Fe(CO)_2 \rightleftharpoons C_5H_5{}^{59}Fe(CO)_2I + C_5H_5Fe(CO)_2{}^{\cdot}$$

where the Fpl molecule acts as a scavenging carrier for the Fp radical [235].

Studies of $^{59}Fe^+$ bombardment of Fpl target afforded modest yield of the labelled target and small yields of labelled $Fe(C_5H_5)_2$ and $Fe(CO)_5$ [276]. The results are discussed and compared with those of neutron bombardment of Fpl [276].

The **polarographic reduction** of FpX (X=Cl, Br, I) in $CH_3CN/[N(C_2H_5)_4]ClO_4$, takes place irreversibly in two steps: (1) Cleavage of the Fe-X bond and formation of the $C_5H_5Fe(CO)_2$ radical, and (2) reaction of the radical with the electrode material, forming Fp_2Hg, which is subsequently reduced to the $[C_5H_5Fe(CO)_2]^-$ anion. The half-wave potentials for the first wave are $E_{1/2} = -0.77$ V (X=Cl), -0.66 V (X=Br), and -0.54 V (X=I) vs. SCE, showing that the ease of reduction increases from Cl to I [131, 137]. For similar reactions in dimethylformamide/$[N(C_2H_5)_4]X'$ (X'=Cl, I, ClO$_4$), see [137]; in $CH_3OCH_2CH_2OCH_3/[N(C_4H_9-n)_4]ClO_4$, see [66]. The electrolytical reductions at an Hg pool cathode of FpX (X=halogen) in THF to form $[Fp]^-$ and in THF in the presence of HCl to give FpH are reported in [81].

Reduction of FpCl with $NaBH_4$ in THF affords FpH [14, 35]. The deuteride, FpD, is obtained by treatment of Fpl with $LiAlD_4$ in ether [35]. In contrast, reduction of FpCl with C_6H_5MgBr leads to Fp_2 [2].

Kinetic studies on the **^{14}CO exchange** with FpX in toluene at 31.8 °C only yielded approximate apparent rate constants because of some decomposition. The relative rates of exchange (in parentheses) decrease in the order: FpCl (500) > FpBr (25) > Fpl (1) [18].

Chloride exchange in FpCl with ^{36}Cl proceeds very slowly in C_6H_6, but in 5% C_2H_5OH in H_2O it is found to be of second order in the complex and independent of the chloride ion concentration. The rate constant at 25.1 °C is $k = 1.26 \times 10^3$ $M^{-1} \cdot s^{-1}$ [19].

Reaction of FpX with halogens of equal or lower electronegativity, X_2', yields adducts of the type $FpX \cdot X_2'$ (see 1.5.2.3.2.6, p. 293). These are thought to be charge transfer complexes, while with halogens of higher electronegativity, X_2'', exchange takes place, affording FpX'' [96].

Reactions with Brönsted and Lewis acids. Adduct formation of FpX with Lewis acids MX_n giving $FpXMX_n$ compounds is described in detail in 1.5.2.3.2.6, p. 293.

Reaction of FpX (X=Cl, Br, I) with concentrated H_2SO_4 liberates HX with formation of red solutions, from which the salts $[Fp(\mu\text{-}X)Fp]PF_6$ have been isolated by subsequent addition of 65% HPF_6 [234]. Action of concentrated H_2SO_4 on FpCl yields HCl and $FpOSO_3H$. The expected protonation of the Fe atom takes place neither with H_2SO_4 nor with 100% CF_3COOH [25].

Other reactions forming dinuclear cations $[Fp(\mu\text{-}X)Fp]^+$ have been reported: Fpl with $AlCl_3$ in a melt at 60 °C [48], FpBr with $AlBr_3$ in liquid SO_2 [51], FpX with $BF_3 \cdot O(C_2H_5)_2$ at 40 to 50 °C (X=Cl) and at 60 to 70 °C (X=Br, I) for ~2 h [52, 53], Fpl with $AgBF_4$ in C_6H_6 [210] or toluene [294], and FpX (X=Cl, Br, I) with $AgPF_6$ in toluene [234].

References on p. 294

The halides FpX (X = Cl, Br, I) dissolve in liquid HCl to give orange-red solutions, in which no halide exchange with the solvent has been observed for the bromide or iodide. Treatment of the FpCl-containing solution with BCl_3 at -196 to $-84\,°C$ forms $[Fp(\mu\text{-}Cl)Fp]BCl_4$, while PF_5 gives no reaction [149].

Many **reactions with phosphines, amines, arsines, and stibines** (^2D) have been reported leading to the products $[Fp^2D]X$ (described in the next volume B12) and $C_5H_5Fe(CO)(^2D)X$ described in 1.5.2.2.6, pp. 86/136. Therefore, Table 28 cites only extensive studies of typical reaction behavior of FpX towards ^2D molecules, i.e., either displacement of the halogen or replacement of a carbonyl group, giving the ionic and covalent compounds. There is no rule for the type of product or, if both types are formed, their proportions. Pandey [295], who investigated the reaction of FpI with $P(C_6H_5)_3$, $As(C_6H_5)_3$, and $Sb(C_6H_5)_3$ by changing the reaction conditions, showed that the proportion of the products depends on the reaction time. Thus only extensive refluxing (20 to 22 h) or UV irradiation (~ 60 h) affords $[Fp^2D]I$ in noticeable yield, whereas $C_5H_5Fe(CO)(^2D)I$ is the main product of short reaction times (8 to 18 h). The ionic derivative forms in a larger amount when excess ^2D is used [259]. The yield increases up to 80% when the reaction of FpI with $P(C_6H_5)_3$ in THF at room temperature/30 min is carried out in the presence of a molar equivalent of $C_2H_5Ni(P(C_6H_5)_3)(CH_3COCHCOCH_3)$, while only a 3.2% yield of $[Fp(P(C_6H_5)_3)]I$ is obtained in the absence of the Ni complex and after 25 d [265]. Haines et al. [123, 143] deduced that halogen displacement is favored over carbonyl replacement in the series $I < Br < Cl$ and as the basicity of the ligand PR_3 increases.

Generally the reactions with polydentate phosphines proceed in a similar manner but lead to other types of products. Displacement of the chloro group by $^2D\text{-}^2D$ gives the dinuclear ligand-bridged complex $[Fp(\mu\text{-}^2D\text{-}^2D)Fp]Cl_2$, while carbonyl replacement gives either the cationic derivative $[C_5H_5Fe(CO)^4D]^+$ with a bidentate ligand or the neutral phosphine-bridged product $C_5H_5Fe(CO)I(\mu\text{-}^2D\text{-}^2D)I(CO)\text{-}FeC_5H_5$. Under special conditions both CO groups are replaced to yield derivatives of the types $C_5H_5Fe(^4D)X$ with a covalent Fe-X bond and $[C_5H_5Fe^6D]^+$. Detailed reaction conditions are given in Table 28.

Both ionic and covalent derivatives are obtained from the reactions of FpX (X = Cl, I) with various phosphinohydrazines. In all cases the mode of coordination of the ligand to the metal has been found to be through phosphorus [266], also see [267].

The formation of $[Fp^2D]X$ compounds (isolated as PF_6^- salts) from FpX and ammonia or amines was also reported in [349], e.g., FpCl and NH_3 in refluxing toluene, FpCl and refluxing NH_2R (R = CH_3, C_2H_5, $t\text{-}C_4H_9$), or FpBr and refluxing $NH(CH_3)_2$ or $N(CH_3)_3$.

In Table 28, 20 °C is used for room temperature.

References on p. 294

Table 28
Reactions of $C_5H_5Fe(CO)_2X$ with Phosphines, Amines, Arsines, and Stibines.
Further information on reactions preceded by an asterisk is given at the end of the table.
For abbreviations and dimensions, see p. 375.

No.	X	reactant D and conditions	products, yields, and remarks for the use of D symbols, see text above the table		Ref.

monophosphines

No.	X	reactant D and conditions	$[Fp^2D]X$	$C_5H_5Fe(CO)(^2D)X$	Ref.
1	I	$H_2P(C_6H_5)$ in refluxing C_6H_6/22 h	13%, isolated as PF_6^- salt	no yield given, identified by IR	[198]
2	Br	$HP(C_6H_5)_2$ in C_6H_6/7 d	85%, isolated as PF_6^- salt	1.7%, identified by IR and NMR	[198]
3	Cl	$P(C_2H_5)_3$ in C_6H_6 or THF at 20° immediately	ca. 50%, isolated as PF_6^- salt	–	[61]
4	Cl	in C_6H_6 at 20°/15 h	>90%, isolated as $[B(C_6H_5)_4]^-$ salt	–	[143]
5	I	$P(C_2H_5)_3$ in C_6H_6 at 20°	>80%		[143]
6	I	in refluxing C_6H_6/30 min	55%	30%	[143]
7	Cl	$P(C_4H_9-n)_3$ in C_6H_6 at 20°/4 d	65%, isolated as $[B(C_6H_5)_4]^-$ salt	small amount, identified by IR	[143]
8	I	$P(C_4H_9-n)_3$ in refluxing C_6H_6/10 h	–	>85%	[143]
9	Cl	$P(C_6H_5)_3$ in refluxing C_6H_6/0.5 h	85%	small amount, identified by IR	[56]
10	Br	$P(C_6H_5)_3$ in refluxing C_6H_6/18 h	occasionally –	51.2% high yield	[56] [318]
11	I	$P(C_6H_5)_3$ in refluxing C_6H_6/18 to 20 h	50.2% 25%	33.7% 60%	[56] [295]
12	Cl	$P(C_6H_5)_3$ in refluxing THF/10 min	30%, obtained as trihydrate	–	[17]
*13	Cl	$P(C_6H_5)_3$ heated in a sealed tube	yield not given, isolated as Reinecke salt	–	[17]

References on p. 294

Table 28 [continued]

No.	X	reactant D and conditions	products, yields, and remarks for the use of D symbols, see text above the table		Ref.
*14	Br, I	$P(C_6H_5)_3$ in C_6H_6, CH_3CN, or CH_3NO_2, photolysis at 366 or 436 nm	–	high yield	[318]
15	I	$P(C_6H_5)_3$ in C_6H_6, UV irradiation/18 h	by-product	main product 81%, when X = Br	[295] [346]
16	I	in THF, UV irradiation at 20°/4 h	–	68%	[62, 75]
17	Br	$P(CH_3)_2C_6H_5$ (1:1 mole) in C_6H_6, UV irradiation/3.5 h	–	52%	[346]
18	Br	(1:2 mole) in C_6H_6, UV irradiation/17 h	$C_5H_5Fe(^2D)_2X$, 34%	–	[346]
19	Br	(1:1 mole) in refluxing C_6H_6/2 h	56%, isolated as PF_6^- salt	small amount	[346]
20	Cl	$P(C_6H_5)_2CH_2CH=CH_2$ in THF at 20°/2 d	ca. 40%, isolated as BF_4^- or $[B(C_6H_5)_4]^-$ salt	– –	[240]
21	Cl	$FpP(C_6H_5)_2$ in C_6H_6/5 min	~75% $[Fp(\mu-P(C_6H_5)_2)Fp]^+$, isolated as $[B(C_6H_5)_4]^-$ salt		[212]

diphosphane and bisphosphines

No.	X	reactant D and conditions	products, yields, and remarks		Ref.
22	Br	$P_2(CH_3)_4$ in refluxing toluene/16 h	16% $[Fp(\mu-P(CH_3)_2)Fp]^+$, isolated as $[B(C_6H_5)_4]^-$ salt		[46]
23	Cl	$(C_6H_5)_2PCH_2P(C_6H_5)_2$ in refluxing C_6H_6/8 h	~60% $C_5H_5Fe(CO)(^2D)X$		[199]
24	Cl, Br	$(CH_3)_2PCH_2CH_2P(CH_3)_2$ in C_6H_6, UV irradiation/3 to 6 h	$C_5H_5Fe(^4D)X$; but according to [346] FpBr does not react in this way		[93, 117, 346]
25	I	$(CH_3)_2PCH_2CH_2P(CH_3)_2$ with or without UV irradiation in C_6H_6 at 20°/24 h	6.4% $C_5H_5Fe(CO)X(\mu-^2D-^2D)-X(CO)FeC_5H_5$		[117]

References on p. 294

Table 28 [continued]

No.	X	reactant D and conditions	products, yields, and remarks for the use of D symbols, see text above the table	Ref.
26	Cl	$(C_6H_5)_2PCH_2CH_2P$-$(C_6H_5)_2$, components mixed with some drops THF at 20°/2 to 3 s	approximately quantitative $[C_5H_5Fe(CO)^4D]^+$, isolated as PF_6^- salt	[259, 304, 305]
27	Cl	in C_6H_6 at 20°/15 h	~75% $[Fp(\mu\text{-}^2D\text{-}^2D)Fp]X_2$	[199]
28	Cl	in refluxing C_6H_6/3 h in the presence of $AlCl_3$	35% $[Fp(\mu\text{-}^2D\text{-}^2D)Fp]^{2+}$, isolated as PF_6^- salt	[56]
29	Cl	in THF, UV irradiation/5 h	~75% $[C_5H_5Fe(CO)^4D]^+$, isolated as BF_4^- or $[B(C_6H_5)_4]^-$ salts	[199]
30	Cl, Br	$(C_6H_5)_2PCH_2CH_2P$-$(C_6H_5)_2$ in C_6H_6, UV irradiation/3 to 6 h	$C_5H_5Fe(^4D)X$	[93, 117]
31	Br	$(C_6H_5)_2PCH_2CH_2P$-$(C_6H_5)_2$ in CH_3CN, UV irradiation/30 to 60 min	87% $[C_5H_5Fe(^4D)NCCH_3]X$	[340]
32	I	$(C_6H_5)_2PCH_2CH_2P$-$(C_6H_5)_2$ in refluxing C_6H_6/8 h	~65% $C_5H_5Fe(CO)X(\mu\text{-}^2D\text{-}^2D)$-$X(CO)FeC_5H_5$	[199]
33	Cl,Br	cis-$(C_6H_5)_2PCH=CHP$-$(C_6H_5)_2$ in C_6H_6, UV irradiation	$C_5H_5Fe(^4D)X$	[93, 117]
34	Cl, Br, I	$(C_6H_5)_2PC\equiv CP(C_6H_5)_2$ in refluxing C_6H_6/2.5 h	$C_5H_5Fe(CO)X(\mu\text{-}^2D\text{-}^2D)X(CO)$-$FeC_5H_5$	[134]

polydentate phosphines

No.	X	reactant D and conditions	products, yields, and remarks	Ref.
35	Cl, Br	$CH_3C(CH_2P(C_6H_5)_2)_3$ in C_6H_6, UV irradiation/12 to 20 h	$[C_5H_5Fe^6D]X$	[117]
36	Br	$C_6H_5P(CH_2CH_2P$-$(C_6H_5)_2)_2$ in C_6H_6, UV irradiation/15 h	52% $[C_5H_5Fe^6D]X$	[166]
37	I	$[(C_6H_5)_2PCH_2CH_2P$-$(C_6H_5)CH_2]_2$ in C_6H_6, UV irradiation/29 h	77% $[C_5H_5Fe^6D]X$	[167]

References on p. 294

Table 28 [continued]

No.	X	reactant D and conditions	products, yields, and remarks for the use of D symbols, see text above the table	Ref.
38	I	$P(CH_2CH_2P(C_6H_5)_2)_3$ in refluxing C_6H_6/48 h	40% $[C_5H_5Fe(CO)^4D]X$	[167]
39	I	in C_6H_6, UV irradiation/45 min	$[C_5H_5Fe(CO)X(\mu-^2D-^6D)-FeC_5H_5]X$	[338]
40	I	$\{[(C_6H_5)_2PCH_2CH_2]_2-PCH_2\}_2$ in refluxing xylene/48 h	$(C_5H_5Fe(CO)X)_{1.5}D$, probably approximately equimolar mixture of $[C_5H_5Fe(CO)^4D]X$ and $[C_5H_5Fe(CO)X(\mu-^4D-^4D)-X(CO)FeC_5H_5]X_2$	[168]

phosphites, phosphonite, and phosphinite

No.	X	reactant D and conditions	products, yields, and remarks for the use of D symbols, see text above the table	Ref.
*41	Cl	$P(OR)_3$ (R = CH_3, C_2H_5, C_4H_9-n, $CH_2CH=CH_2$) in C_6H_6 at 20°	~20% $C_5H_5Fe(CO)(^2D)X$, main product $FpP(O)(OR)_2$, and $C_5H_5Fe(CO)(P(OR)_3)-P(O)(OR)_2$ in small amount	[123, 143]
*42	Br, I	$P(OR)_3$ (R = CH_3, C_2H_5, n-C_6H_{13}, C_6H_5), various solvents and temperatures	$C_5H_5Fe(CO)(^2D)X$ in high yield	[94, 116, 118, 123, 135, 143, 170]
43	Cl	$P(OC_6H_5)_3$ in C_6H_6 at 20°/4 d	7% $[Fp^2D]X$, isolated as $[B(C_6H_5)_4]^-$ salt, and 55% $C_5H_5Fe(CO)(^2D)X$	[143]
44	I	$P(OC_6H_5)_3$ in boiling C_6H_6, UV irradiation/10 h	$C_5H_5Fe(^2D)_2X$	[77]
45	Cl	$P(OCH_2CH=CH_2)_2C_6H_5$ or $P(OCH_2CH=CH_2)-(C_6H_5)_2$ in C_6H_6 at 25°/1 h and at 80°/3 h, respectively	similar type products as in reaction No. 41	[123, 143]
46	I	$P[(OCH_2)_3CC_2H_5]$ in refluxing C_6H_6/~20 h	$[Fp^2D]X$ and $C_5H_5Fe(CO)(^2D)X$	[295]
47	Cl, Br, I	$P[(OCH_2)_3CR]$ (R = CH_3, C_2H_5, n-C_3H_7) in C_6H_{12}, UV irradiation/1.5 h	$C_5H_5Fe(CO)(^2D)X$	[270]

References on p. 294

Table 28 [continued]

No.	X	reactant D and conditions	products, yields, and remarks for the use of D symbols, see text above the table	Ref.
48	I	$(C_2H_5O)_2POP(OC_2H_5)_2$ no further detail given	$[Fp(\mu\text{-}^2D\text{-}^2D)Fp]^{2+}$ and $C_5H_5Fe(CO)X(\mu\text{-}^2D\text{-}^2D)\text{-}X(CO)FeC_5H_5$	[306]

ammonia, amines, and hydrazine

No.	X	reactant D and conditions	products, yields, and remarks	Ref.
49	Cl	liquid NH_3/overnight	$\sim$40% $[Fp^2D]X$, isolated as $[B(C_6H_5)_4]^-$ salt	[61]
50	I	cyclo-$C_6H_{11}NH_2$, piperidine, morpholine, 1,10-phenanthroline, 2,2'-bipyridyl in refluxing C_6H_6/2 to 4 h	30 to 60% $[Fp^2D]X$	[279]
51	I	n-$C_4H_9NH_2$, cyclo-$C_6H_{11}NH_2$, $C_6H_5CH_2NH_2$, piperidine, morpholine in refluxing C_6H_6, 1:1 mole, 2 h	$[Fp^2D]X$	[307]
52		excess ^2D, 4 h	60 to 68% $[C_5H_5Fe^2D_2]X$	[307]
53	Cl	96% N_2H_4 in C_6H_6 at $\sim$7°/30 min	$\sim$40% $[Fp^2D]X$, isolated as $[B(C_6H_5)_4]^-$ salt	[61]

arsines and stibines

No.	X	reactant D and conditions	products, yields, and remarks	Ref.
54	Cl	$E(C_6H_5)_3$ (E=As, Sb) heated in a sealed tube	$[Fp^2D]X$, isolated as $[PtCl_6]^-$ salt	[17]
55	I	$E(C_6H_5)_3$ (E=As, Sb) in refluxing C_6H_6/22 h	$[Fp^2D]X$ and $C_5H_5Fe(CO)(^2D)X$ 38.4 and 32.5% for As 48.6 and 30.8% for Sb	[280, 295, 308]
56	Cl	$As(C_6H_5)_2CH_2$=$CHCH_2$ in THF at 20°/2 d	$\sim$40% $[Fp^2D]X$, isolated as BF_4^- or $[B(C_6H_5)_4]^-$ salt	[240]
57	Br	$As_2(CH_3)_4$ in refluxing toluene/16 h	31% $[Fp(\mu\text{-}As(CH_3)_2)Fp]^+$, isolated as ClO_4^- salt	[63]
58	J	$(C_6H_5)_2AsCH_2As\text{-}(C_6H_5)_2$ in refluxing C_6H_6/$\sim$20 h	$[Fp^2D]X$ and $C_5H_5Fe(CO)(^2D)X$	[295]

References on p. 294

Table 28 [continued]

No.	X	reactant D and conditions	products, yields, and remarks for the use of D symbols, see text above the table	Ref.
59	Cl	$Sb(C_4H_9-n)_3$ in toluene at 50°/5 min	$[Fp^2D]^+$, isolated as PF_6^- salt	[241]
60	Cl	$SbRR'CH_2CH=CH_2$ ($R=R'=CH_3$, $CH_2CH=CH_2$, C_6H_5; $R=CH_2CH=CH_2$, $R'=CH_3$, C_6H_5) in THF at 20°/2 d	~70% $[Fp(\mu-SbRR')Fp]^+$, isolated as $[B(C_6H_5)_4]^-$ salt	[240]

*Further information:

Reaction No. 13. $Fe(CO)_3(P(C_6H_5)_3)_2$ has also been isolated from this reaction in low yield [17]. It is mentioned as sole product in [6], where it was obtained from equimolar amounts of FpCl and $P(C_6H_5)_3$ in an autoclave at 120 °C/3 h.

Reaction No. 14. The quantum yields are essentially independent of irradiation wavelength but they increase with the $P(C_6H_6)_3$ concentration and decrease on addition of CO. The primary photochemical process is dissociation of one CO [318]. The results correlate with photoelectron spectral data and molecular orbital calculations reported by [273].

Reaction No. 41. The yield of $C_5H_5Fe(CO)(P(OR)_3)P(O)(OR)_2$ increases with the $P(OR)_3$ to FpX ratio and prolonged refluxing in C_6H_6. However, $FpP(O)(OR)_2$ remains the major product. Similar compounds were obtained from FpCl and $NaOP(OC_2H_5)_2$ in THF. By IR monitoring the reactions of FpCl with $P(OR)_3$ in C_6H_6 at room temperature, $[FpP(OR)_3]^+$ was identified as intermediate in the formation of $FpP(O)(OR)_2$. It can be isolated when the reactions are carried out in the presence of $Na[B(C_6H_5)_4]$ in C_6H_6 containing a little CH_3OH [123, 143].

Reaction No. 42. The kinetics of the substitution reactions $C_5H_5Fe(CO)_2X + {}^2D \rightarrow C_5H_5Fe(CO)({}^2D)X + CO$ (X = Br, I; $^2D = P(OR)_3$ with $R = C_2H_5$, $n-C_6H_{13}$, and C_6H_5) were studied in a variety of solvents [94, 116, 118, 135]. In the case of FpCl, no satisfactory kinetic results could be obtained. The formation of a mixture of covalent and ionic products was assumed [94]. Similar reactions of FpCl in C_6H_6 at room temperature gave $FpP(O)(OR)_2$ as the main product and only ~20% of $C_5H_5Fe(CO)-(P(OR)_3)Cl$, see reaction 41 [123, 143]. The kinetic parameters for the reactions of FpBr and FpI with $P(OC_6H_5)_3$ in various solvents at 70 °C are summarized below (k in s^{-1}, $\Delta H^{\neq}$ in $kcal \cdot mol^{-1}$, and $\Delta S^{\neq}$ in $cal \cdot K^{-1} \cdot mol^{-1}$). For comparison, data for the $FpI/P(OC_2H_5)_3$ pair are included in parentheses in the last row.

FpX	solvent	$k \times 10^5$	$\Delta H^{\neq}$	$\Delta S^{\neq}$	Ref.
FpBr	$n-C_8H_{18}$	27.6	31.2	14.1	[116, 135]
	$C_6H_4(CH_3)_2$	10.8	34.6	21.5	[135]
	$O(C_4H_9-n)_2$	23.4	26.4	− 0.5	[116, 135]
	$C_6H_5NO_2$	4.4	33.7	17.5	[135]

References on p. 294

FpX	solvent	$k \times 10^5$	ΔH^+	ΔS^+	Ref.
FpI	$n\text{-}C_8H_{18}$	1.04	26.8	-4.2	[116, 135]
	$C_6H_4(CH_3)_2$	0.44	33.2	11.6	[135]
	$O(C_4H_9\text{-}n)_2$	0.9	23.8	-12.5	[116, 135]
	$C_6H_5NO_2$	0.17	33.5	10.5	[135]
	$(n\text{-}C_8H_{18})$	—	(30.4 ± 1.0)	(7.9 ± 2.5)	[170]

There is a relation between k or ΔH^+ and the polarity of the solvent. This is consistent with a first-order dissociative mechanism involving a transition state less polar than the ground state. The anomalous behavior in $O(C_4H_9\text{-}n)_2$ is attributed to its coordinating ability [135]. Further rate constants for the reactions of FpBr with $P(OC_6H_5)_3$ and FpI with $P(OR)_3$ ($R = C_2H_5$, $n\text{-}C_6H_{13}$, C_6H_5) in the above listed solvents at 70 to 117 °C are also reported [94, 116, 118, 135, 170]. The results of the kinetic measurements are compared with those for $C_5H_5Ru(CO)_2X$ [135]. The variation of rate with the size of the ^5L ligand is in the sequence: cyclopentadienyl < cycloheptadienyl < cyclohexadienyl [288].

Other reactions are listed in Table 29. The reactants are primarily arranged by the elements which react with the Fp group. In Table 29 20° C is used for room temperature.

Further information for Table 29:

Reaction of FpI with $SnCl_2$: Isolation of $FpSnCl_2 \cdot CH_3OH$ has been cited as evidence of an insertion mechanism where the iodine atom remains associated with the molecule [68]. However, Mays, Pearson [110] do not exclude a substitution mechanism for two reasons: (1) $FpSnI_3$ exchanges halogen almost instantaneously with an excess of $SnCl_2$ in CH_3OH at room temperature, and (2) the only product they were able to isolate from the reaction of FpI with $SnCl_2$, either anhydrous or hydrated, was $FpSnCl_3$.

Reaction of FpI with HgR_2: The behavior depends on the stability of the Hg-C bond. FpR compounds are formed with $R = C_6H_4Cl\text{-}4$ and $1\text{-}C_4H_3S$. Insertion of Hg into the Fe-I bond to give FpHgI occurs for $R = CH_2CHO$, $CH_2CH=CH_2$, and $C\equiv CC_6H_5$ [144].

Reaction of FpX with NaR: Most reactions yield at 25 °C for less than 1 h the symmetrical 2-N substituted compound, whereas higher temperatures and longer times give the unsymmetrical 1-N substituted derivatives [301].

1.5.2.3.2.4 Catalysis and Use

FpCl combined with $Al(C_2H_5)_3$ or C_6H_5MgBr has been used as a soluble catalyst for hydrogenation of butadiene (40 to 45 °C/60 atm/6 h in C_6H_6). The main products are cis- and trans-2-butene. In addition, 1 to 2% of 1-butene were obtained, but no butane. The conversion is ca. 99% with $Al(C_2H_5)_3$ and ca. 85% with C_6H_5MgBr [88]. FpX (X = Br, I) also catalyze the reaction of alkylmagnesium halides with alkyl halides in ether (Kharasch reaction), e.g., $n\text{-}C_3H_7MgBr + n\text{-}C_3H_7Br \rightarrow C_3H_8 + C_3H_6 + MgBr_2$. The reactions of C_2H_5MgBr with C_2H_5Br and of $n\text{-}C_4H_9MgBr$ with $n\text{-}C_4H_9Br$ take place similarly [76]. Use of the halides as fuel additives and catalysts in CO reactions is mentioned in [11]. As antiknock agents in combustion engines the halides improve the burning characteristics of fuels without producing injurious emissions

References on p. 294   [continued on p. 293]

Table 29
Further Reactions of FpX Compounds with X = Cl, Br, and I.
For abbreviations and dimensions, see p. 375.

X	reactant and conditions	products and remarks	Ref.
with metals			
Cl	excess Mg in THF	highly colored, air–sensitive solution containing FpMgCl	[103]
Cl, Br, I	Hg in C_6H_6, UV irradiation at 35 to 38°, 20 to 32 h	FpHgX and ferrocene	[102, 144]
I	Hg in C_6H_6, 35 h	FpHgI and ferrocene	[144]
with CO			
Cl	CO stream, $AlCl_3$, in C_6H_6 at 40 to 80°, 3 h	$[C_5H_5Fe(CO)_3]^+$	[124]
Br	CO at 240 atm, $AlBr_3$, in C_6H_6 at 60°, 16 h	$[C_5H_5Fe(CO)_3]^+$	[20, 21, 26]
Cl	CO at 90 atm, $Na[B(C_6H_5)_4]$, in acetone at 25 to 30°, 48 h	$[C_5H_5Fe(CO)_3][B(C_6H_5)_4]$	[17]
Cl	CO at 120 atm, NH_4PF_6, in THF/H_2O (6:1) at 60°, 4 h	$[C_5H_5Fe(CO)_3]PF_6$	[72]
I	CO at 364 atm, $HPF_6/(C_2H_5CO)_2O$, at 150°, 2 h	$[C_5H_5Fe(CO)_3]PF_6$	[27]
with O compounds			
Br	$NaOCH_3$ in CH_3OH at 0 to 20°, 1 h	Fp_2 and FpC_5H_5 (25%)	[73]
Cl	$NaOCH(CF_3)_2$ in THF, 1 h	Fp_2 and an unstable yellow solid, possibly FpC_5H_5	[125]

References on p. 294

—	AgOOCR (R=CF$_3$, C$_3$F$_7$) in CH$_2$Cl$_2$ or C$_6$H$_6$, 4 h	FpOOCR, no products in THF	[90]
Cl	CF$_3$COCF$_3$, irradiation in THF, 18 h	yellow, H$_2$O-soluble crystals giving [C$_5$H$_5$Fe(CO)$_3$]PF$_6$ with NH$_4$PF$_6$	[125]
—	C$_4$H$_8$O (THF) and AgBF$_4$, 3 h	[Fp(C$_4$H$_8$O)]BF$_4$	[296]
with S and Se compounds			
Cl	LiSC$_6$H$_4$SCH$_3$-2 in THF, 30 min	FpSC$_6$H$_4$SCH$_3$-2	[326]
Cl	Li$_2$S$_2$C$_6$H$_4$-1,2 in THF at −78 to +20°, 90 min	Fp(μ-SC$_6$H$_4$S)Fp	[326]
Br	NaSR (R=CH$_3$, C$_2$H$_5$, C$_6$H$_5$) in ether or THF, ~12 h	FpSR	[54]
Cl	NaSC$_4$H$_9$-t	FpSC$_4$H$_9$-t	[193]
Br	NaSC$_6$H$_4$CH$_3$-4, no details	FpSC$_6$H$_4$CH$_3$-4	[104]
—	NaSC$_6$H$_4$F-4, no details	FpSC$_6$H$_4$F-4	[104]
—	AgSCF$_3$ in acetone, 24 h	FpSCF$_3$	[105]
with halides and pseudohalides			
Cl	NaI or KI in refluxing acetone, 24 h	65% FpI	[4, 6]
Cl	NaCN in refluxing CH$_3$OH, 24 h	FpCN	[1, 3]
Br	KCN in C$_2$H$_5$OH	FpCN	[36]
—	KCN in refluxing C$_2$H$_5$OH, 1 h	FpCN	[248]
Br, I	KCN in refluxing C$_2$H$_5$OH/H$_2$O, 0.5 to 1 h	K[C$_5$H$_5$Fe(CO)(CN)$_2$]	[36, 248]
—	AgCN in refluxing CHCl$_3$, 93 h	FpCN	[335]

Table 29 [continued]

X	reactant and conditions	products and remarks	Ref.
I	KSCN in refluxing CH_3OH (?), 18 h	FpSCN, small amounts (also see [92])	[101]
Cl	KSeCN in refluxing acetone, 1 h	82% FpSeCN and 2.1% FpCN	[89]
Cl	NaN_3 in refluxing 10% aqueous acetone, 3 h	FpN_3	[231]
Br, I	$Na[SC(S)SR]$ (R=CH_3, C_2H_5) in THF at 0 to 20°, 6 to 14 h	FpSC(S)SR	[64]
Br	$Na[SC(S)SC_6H_5]$ in CS_2/THF, 14 h	56% $FpSC_6H_5$ and 35% $FpSC(S)SC_6H_5$	[64]
Cl	$Na[SC(S)N(CH_3)_2]$ in acetone, 24 h	$FpSC(S)N(CH_3)_2$	[106]
Cl	$Na[SC(S)NHC_3H_7-i]$ in refluxing acetone, 3 h	34% $FpSC(S)NHCH(CH_3)_2$ and 4% $C_5H_5Fe(CO)SC(S)NHCH(CH_3)_2$	[249]
Cl	$Na_2[SC(CN)=(CN)CS]$ in H_2O, 12 h	FpSC(CN)=(CN)CSFp	[55]
Cl	$Na[SC(O)NC_4H_8]$ in refluxing benzene, 1 h	$FpSC(O)NC_4H_8$	[297]
Cl	$[S_2CR]^-$ (R=CN, $NHNH_2$, piperidino, 1,2,3,6-tetrahydropyrid-1-yl, morpholino)	FpSC(S)R	[211]
Cl	$NaSO_2R$ (R=C_6H_5, $C_6H_4CH_3$-4) in CH_3OH at 27°, 12 h	$FpSO_2R$	[58]
Cl	HPS_2F_2 (1:2 mole) at 20°, 48 h and 50°, 18 h	$FpSP(S)F_2$	[128]
Cl	$CsPS_2F_2$ (1:1 mole) in acetone, 48 h	$FpSP(S)F_2$	[107, 128]
Cl	$(CH_3)_3SnSCH_3$ in C_6H_6, 4 d	$FpSCH_3$	[194]

References on p. 294

Cl	$(CH_3)_3SnSSn(CH_3)_3$ in $(CH_3OCH_2-)_2$, C_6H_6, or CCl_4	$(CH_3)_3SnCl$ only isolable product	[126]
Cl	$(CH_3)_3SnS(R)M(CO)_5$ (M=Cr, W; R=CH$_3$, C$_2$H$_5$) in C_6H_6, 1 to 2 d	$FpS(R)M(CO)_5$	[136, 145, 194]
Cl	$(CH_3)_3SnS(CH_3)Mn(CO)_2C_5H_5$ in C_6H_6, 2 h	$FpS(CH_3)Mn(CO)_2C_5H_5$	[145, 194]
Cl	$[(CH_3)_3SnSMn(CO)_4]_2$ in C_6H_6, 2 d	$(FpSMn(CO)_4)_2$	[126]
Cl	$LiS(CH_3)Cr(CO)_5$ in ether/C_6H_6	$FpS(CH_3)Cr(CO)_5$	[136]
Cl	$BrMgS(CH_3)Cr(CO)_5$ in C_6H_6, 2 h	$FpS(CH_3)Cr(CO)_5$	[136]
Cl	$NaSeO_2CH_3$ in CH_2Cl_2 or CH_3OH	$FpSeO_2CH_3$	[327]
–	NO in C_6H_6 at 25°, 1 h	$C_5H_5Fe(CO)(NO)I$	[298]
Cl	2,2'–bipyridyl or 1,10–phenanthroline (^{4}D) in refluxing THF, 15 h	Fe complexes with ^{4}D ligands	[17]
Cl	2,2'–bipyridyl (^{4}D) and AlCl$_3$ in refluxing C_6H_6, 3 h	$[C_5H_5Fe(CO)^4D]^+$ identified by IR and NMR	[56]
Cl	CH$_3$CN and AlCl$_3$ in refluxing C_6H_6, 1 h	$[FpNCCH_3]^+$	[56]
–	CH$_3$CN and NOPF$_6$, 5 min	$[FpNCCH_3]PF_6$	[195]
–	N–methylpyrrole at 110°, 24 h	$[C_5H_5FeC_4H_4NCH_3]I$	[91]
Cl, Br, I	diazocyclopentadiene (excess) in refluxing C_6H_6, 2 h	$C_5H_5FeC_5H_4Br$ and ferrocene	[328]

References on p. 294

Table 29 [continued]

X	reactant and conditions	products and remarks	Ref.
Cl	LiN(CH₃)C(S)C₆H₅ in THF at −78 to +20°, 1 h	(29%) and (traces)	[299]
Cl, Br, I	LiN=C(C₄H₉-t)₂ in ether added to frozen FpX in ether, then 20°, 30 min	55% $C_5H_5Fe(CO)N=C(C_4H_9-t)_2$ and Fp_2	[232]
Cl	LiN=C(C₆H₅)₂ in hexane/ether added to frozen FpCl in ether (−196°), then 20°, 1 h	Fp_2 and a monocarbonyl species revealed by IR	[232]
Cl	$(CH_3)_3SiN=CR_2$ (R = t-C₄H₉, C₆H₅) in refluxing (CH₃OCH₂−)₂, 48 h	Fp_2 only isolable product	[232]
Cl, Br, I	Na compounds (NaR) of imidazole, benzimidazole, 5,6-dimethyl-benzimidazole, 1,2,3-triazole, 4-phenyl-1,2,3-triazole, and benzotriazole in CH₃CN at 45 to 55°, 2 h	FpR, R = 1-N-substituted heterocycle	[300]
Cl, Br, I	Na compounds (NaR) of 1,2,3-triazole, 4-phenyl-1,2,3-triazole, and benzotriazole in THF or CH₃CN at 20 to 45°, 1 to 2.5 h	FpR, R = 1- and 2-N-substituted heterocycle, further information on p. 277	[301]
I	benzotriazole (= RH) and Na in THF at 5 to 60°	FpR, only 2-N-substituted below 20°, 1- and 2-N-substituted at 60°	[146, 147, 250]
I	K compounds (KR) of pyrrole, substituted pyrrole, indole, 1,2,3,4-tetrahydrocarbazole, and carbazole in C₆H₆ at 50 to 80°, 3 to 5 h	FpR, R = N-substituted heterocycle (below 60°), some isomerization to azaferrocenes at reflux temperature	[47, 74]

Cl, Br, I	K compounds (KR) of imidazole, pyrazole, and triazole in liquid NH_3, 3 to 4 h	FpR, R = N-substituted heterocycle	[91]
—	$AgNO_3$ in CH_2Cl_2, 6 h	$FpNO_3$	[296]
—	$AgN(SO_2F)_2$ in CH_2Cl_2, 48 h	$FpN(SO_2F)_2$	[268]

with P and As compounds

Cl	$LiP(C_6H_5)_2$, no details	$FpP(C_6H_5)_2$	[193]
—	$NaP(C_6H_5)_2$ in C_6H_6	$FpP(C_6H_5)_2$	[212]
Cl	$LiP(Si(CH_3)_3)_2$ (1.8-THF solvate) in cyclopentane	$FpP(Si(CH_3)_3)_2$	[302]
—	$K[PC_4(C_6H_5)_4]$ (tetraphenyl-phospholide) in THF, overnight	$FpPC_4(C_6H_5)_4$	[148]
—	$Ni(PF_3)_4$ in refluxing $C_6H_5CH_3$, 12 h	$C_5H_5Fe(CO)(PF_3)I$	[196]
Cl	$(PF_2)_2NCH_3$ (1:1) in boiling ether, 1 h	$C_5H_5Fe(CO)(PF_2N(PF_2)CH_3)Cl$	[251, 303, 309]
Cl	$(PF_2)_2NCH_3$ (1:1) in refluxing C_6H_6	ferrocene only isolable product	[251]
Cl	$(PF_2)_2NCH_3$ (slight excess) in refluxing C_6H_6/CH_3OH, 1 h	$C_5H_5Fe(CO)(PF_2NHCH_3)Cl$	[251]
Cl	$(PF_2)_2NCH_3$ (excess) in refluxing hexane, 90 min	$C_5H_5Fe(PF_2N(PF_2)CH_3)_2Cl$	[251, 303, 309]
—	$(P(C_6H_5)_2CH_2CH_2)_3N$ (2:1) in C_6H_6, irradiation, 45 min	R = C_6H_5	[338]

Table 29 [continued]

X	reactant and conditions	products and remarks	Ref.
Cl, I	P(N(CH₃)₂)₃ in refluxing C₆H₆, 6 h	[FpP(N(CH₃)₂)₃]X	[34]
Cl	P(Si(CH₃)₃)₃ in THF		[302]
I	PF₂N(C₂H₅)₂ (20% excess) in C₆H₁₂/C₆H₆, irradiation, 8 h	62% C₅H₅Fe(CO)(PF₂N(C₂H₅)₂)I and 30% C₅H₅Fe(PF₂N(C₂H₅)₂)₂I	[197, 213]
Br, I	PF₂NC₅H₁₀–cyclo (excess) in C₆H₆, irradiation for several hours	C₅H₅Fe(PF₂NC₅H₁₀–cyclo)₂X	[197, 213]
	PF₂NC₅H₁₀–cyclo (excess) after treatment with AgBF₄, in THF, 15 h	[FpPF₂NC₅H₁₀–cyclo]⁺	[197]
Cl	P(CF₃)₂SH in CFCl₃ at 50°, 4 d	FpP(S)(CF₃)₂	[214]
	(R = CH₃) in (CH₃OCH₂–)₂, 30 min	FpAsC₄H₂R₂	[337]

with alkenes, alkynes, and aromatic compounds

X	reactant and conditions	products and remarks	Ref.
Cl, Br	mono– or diolefins (²L) and Lewis acids	[Fp²L]⁺ compounds, diolefins coordinating with only one double bond	[20 to 23, 26, 30, 31]
I	(1:1) in THF at 0 to 20°, 15 h, then in CH₂Cl₂ with aqueous KClO₄ R = i-C₃H₇	ClO₄⁻ salt	[319]

I	alkenes or alkynes (²L ligands) and AgBF₄ in CH₂Cl₂	[Fp²L]BF₄	[336]
Cl	fluoroolefins or CF₃C≡CCF₃ in THF, irradiation, 14 h	golden–yellow solids, soluble in H₂O giving [C₅H₅Fe(CO)₃]⁺	[125]
Cl	C₆H₅C≡CH or C₆H₅C≡CC₆H₅ in refluxing THF, 15 h	only ferrocene and FeCl₂ by thermal decomposition	[17]
Cl	C₆H₆ or C₆H₃(CH₃)₃–1,3,5 (⁶L ligands) and AlCl₃ at reflux temperature, 3 h	[⁶LFeC₅H₅]⁺	[10, 13, 15, 24]
with isocyanides			
Cl	C₆H₅NC (1:3) in refluxing C₆H₆, 1.5 h	68% [C₅H₅Fe(CNC₆H₅)₃]Cl	[37]
Cl	4–CH₃OC₆H₄NC in refluxing C₆H₆, 1.5 h	[C₅H₅Fe(CNC₆H₄OCH₃–4)₃]⁺	[119]
Br	C₆H₅NC (1:2) in refluxing C₆H₆, 3 h	37% C₅H₅Fe(CNC₆H₅)₂Br	[37]
–	C₆H₅NC (1:14.5) in refluxing C₆H₆, 6 h	35% C₅H₅Fe(CO)(CNC₆H₅)I and 8% C₅H₅Fe(CNC₆H₅)₂I	[37]
–	C₆H₅NC in THF, no details	C₅H₅Fe(CO)(CNC₆H₅)I	[29]
–	cyclo–C₆H₁₁NC (1:1.5) in refluxing C₆H₆, 5 h	85% C₅H₅Fe(CO)(CNC₆H₁₁)I	[181]
–	cyclo–C₆H₁₁NC (1:10)	85% C₅H₅Fe(CNC₆H₁₁)₂I	[181]
–	(+)–CH₃CH(C₆H₅)NC (1:1.5) in refluxing C₆H₆, 6 h	35% (+)– and (−)–C₅H₅Fe(CO)(CNCH(C₆H₅)CH₃)I and C₅H₅Fe(CNCH(C₆H₅)CH₃)₂I	[181, 236]
with Si, Ge, and Sn compounds			
Br	KSiH₃ in monoglyme at −40°, 1 h, and at 24°, 2 h, exothermic reaction	FpSiH₃	[109]

References on p. 294

Table 29 [continued]

X	reactant and conditions	products and remarks	Ref.
Cl	HGeCl₃ in C₆H₆/THF/ether at 35°, 1 h	FpGeCl₃	[67, 108]
Cl, I	excess [(CH₃)₃NH]GeCl₃ in refluxing THF, 18 h	FpGeCl₃	[120]
Br	KGeH₃ in (CH₃OCH₂−), no details	FpGeH₃	[257]
Cl, Br	Hg(Ge(C₂H₅)₃)₂ in C₆H₆ at 100°, 6 h in evacuated tube	FpGe(C₂H₅)₃	[82]
Cl	Hg(Ge(C₆H₅)₃)₂ in toluene at 100°, 18 h	FpGe(C₆H₅)₃	[121]
Cl	GeCl₂ heated in dioxane/ether, 40 min	FpGeCl₃	[108]
I	GeCl₂ in dioxane at 40 to 110°	FpGeCl₂I and disproportionation products	[108]
Cl	SnCl₂·2H₂O in refluxing CH₃OH, 5 h	FpSnCl₃	[38]
I	SnCl₂ (excess) in refluxing CH₃OH	FpSnCl₂I·CH₃OH or FpSnCl₃(?) further information on p. 277	[68, 110]
Cl	Sn(NCS)₂ and NH₄SCN in refluxing THF, ~2 h	FpSn(NCS)₃	[156]
Cl	Sn(OOCH)₂ in HCOOH, heated for ~4 h	FpSn(OOCH)₃	[156]
Cl	Sn(OOCCH₃)₂, heated in acetic anhydride	FpSn(OOCCH₃)₃	[156]
Cl	SnR₂ (R=CH(Si(CH₃)₃)₂) in C₆H₆	FpSnR₂Cl	[277]
Cl	(CH₃)₃SnCH₂CH=CH₂ in ether at 25°, 300 h, or in dioxane at 100°, 100 h	Fe(C₅H₅)₂, FpSn(CH₃)₃, and Fp₂, all below 10%	[206]

References on p. 294

with B, In, and Tl compounds

Cl	Na[B(C$_6$H$_5$)$_4$] in acetone at 25°, 6 d	[C$_5$H$_5$Fe(CO)$_3$][B(C$_6$H$_5$)$_4$], intermolecular transfer of CO is assumed	[17]
I	Na[B(C$_6$H$_5$)$_4$] in refluxing (CH$_3$OCH$_2$CH$_2$–)$_2$O	C$_5$H$_5$FeB(C$_6$H$_5$)$_4$ with B(C$_6$H$_5$)$_4$ η^6-π-complexed to Fe	[325]
Cl	[(C$_6$H$_5$)$_2$PCH$_2$CH$_2$P(C$_6$H$_5$)$_2$]$_2$Co(B(C$_6$H$_5$)$_2$)$_2$ in THF, 4 h	FpB(C$_6$H$_5$)$_2$	[69]
I	Na[C$_2$B$_4$H$_7$] in THF, 4 h	(μ-Fp)C$_2$B$_4$H$_7$	[182, 215]
I	K[B$_5$H$_8$] in (CH$_3$)$_2$O at −40°, 4 h	2-FpB$_5$H$_8$	[320]
I	K[2-FpB$_5$H$_7$] in monoglyme at −196 to +20°, ~8 h	2,4-Fp$_2$B$_5$H$_7$	[320]
I	1-LiCB$_8$H$_8$CCH$_3$-10 in refluxing ether, 30 min	1-FpCB$_8$H$_8$CCH$_3$-10	[122, 157]
I	1-LiCB$_8$H$_8$CLi-10 in ether, 4 h	1-FpCB$_8$H$_8$CFp-10 and 1-FpCB$_8$H$_8$CH-10	[111, 122, 157]
I	1-LiCB$_{10}$H$_{10}$CR-2 (R=CH$_3$, C$_6$H$_5$) in refluxing (CH$_3$OCH$_2$–)$_2$, 1 h	1-Fp-2-R-1,2-C$_2$B$_{10}$H$_{10}$	[111, 122, 157]
Br	1-CuCB$_{10}$H$_{10}$CH-n (n=2, 7) in ether/benzene/THF at 0 to 60°, 5 h	1-FpCB$_{10}$H$_{10}$CH-n	[291]
Br	1,2- or 1,7-Cu$_2$C$_2$B$_{10}$H$_{10}$ in THF/C$_6$H$_6$ at 0 to 60°, 5 h	1,2- or 1,7-Fp$_2$C$_2$B$_{10}$H$_{10}$	[291]
I	[N(CH$_3$)$_4$][B$_3$H$_8$], irradiation in CH$_2$Cl$_2$ at 15 to 20°, 3 to 14 h	C$_5$H$_5$Fe(CO)B$_3$H$_8$ with bidentate B$_3$H$_8$	[237, 321]

Table 29 [continued]

X	reactant and conditions	products and remarks	Ref.
Cl	$Cl_2BCo(^8D)P(C_6H_5)_3$ in ether, 2 h $^8D = \left((C_6H_5)_2B{\overset{ON=C(CH_3)-}{\underset{ON=C(CH_3)-}{\diagdown\diagup}}} \right)_2$	$FpBCl_2$ and $ClCo(^8D)P(C_6H_5)_3$	[83]
Cl, Br	InX (X=Cl, Br) in THF, 30 min	$FpInX_2$	[158, 183]
Cl	$Tl[HBR_3]$ (R=pyrazol-1-yl)	$Fe(HBR_3)_2$	[216]
—	$TlC_5H_4CR=C(CN)_2$ (R=CN, $C_5H_4FeC_5H_5$) in refluxing THF	$C_5H_5FeC_5H_4CR=C(CN)_2$, in the presence of TlC_5H_5 also $(C_5H_5FeC_5H_4)_2C=C(CN)_2$	[290]
with organic Mg compounds			
Br, I	$RMgX$ in ether, 2 to 3 h	50% FpR for R=CH$_3$; 3% FpR and 60% Fp$_2$ for R=C$_6$H$_5$	[8, 9]
Cl	$CH_2=CHMgCl$ in THF, no details	$FpCH=CH_2$	[50]
Cl	$CH_3C{\equiv}CMgBr$ in THF at -78 to $+20°$	$FpC{\equiv}CCH_3$	[238]
Cl	$RC{\equiv}CMgBr$ (R=n-C$_4$H$_9$, C$_6$H$_5$) in THF at -60 to $+20°$	$FpC{\equiv}CR$	[84]
I	C_6F_5MgBr in ether	FpC_6F_5 (~7%) and Fp$_2$	[39]
Br	$2\text{-}HCB_{10}H_{10}CCH_2MgBr\text{-}1/CuCl$ in ether/THF at 0 to 20°, 5 h	$1\text{-}FpCH_2CB_{10}H_{10}CH\text{-}2$	[291]

References on p. 294

with organic Li and Na compounds

I	$LiC{\equiv}CCF_3$ in THF at -78 to $+20°$, 5 h	$FpC{\equiv}CCF_3$	[85]
Cl	$LiC{\equiv}CC_6F_5$ in THF at $-78°$, 15 h	$FpC{\equiv}CC_6F_5$	[85]
Cl	$LiC{\equiv}CC_6H_5$ in hexane/ether at -78 to $+20°$, 15 h	$FpC{\equiv}CC_6H_5$	[85]
Br	LiC_3H_5-cyclo	FpC_3H_5-cyclo	[184]
Cl	$[LiC_3R_2$-cyclo$]ClO_4$ (R$=$N(C$_3$H$_7$-i)$_2$) in THF at $-78°$	$[FpC_3R_2$-cyclo$]ClO_4$	[322]
I	LiC_6F_4Br-3 in ether/hexane at $-78°$	FpC_6F_4Br-3	[159, 160, 217]
I	$1,4$-$Li_2C_6H_4$	$1,4$-$Fp_2C_6F_4$	[159]
I	1-, 2-, and 3-LiC_7H_7-cyclo mixture in THF at $-78°$	mixture of FpC_7H_7-cyclo isomers	[323]
Cl	LiC_8H_7-cyclo in ether	FpC_8H_7-cyclo	[258]
I	3,4,5-trichlorothien-2-yl lithium in ether at -60 to $+20°$, overnight	2-FpC_4Cl_3S	[86]
Br	$LiC_5H_4FeC_5H_5$ in ether/THF at -70 to $+20°$, 2 h	$FpC_5H_4FeC_5H_5$	[218]
I	$LiC_5H_3(Cl$-2$)FeC_5H_4Cl$ in THF at -78 to $+20°$, 18 h	$FpC_5H_3(Cl$-2$)FeC_5H_4Cl$	[339]
I	$LiC_5H_4Mn(CO)_3$ in THF at -50 to $-40°$, 45 min	$FpC_5H_4Mn(CO)_3$	[225]

References on p. 294

Table 29 [continued]

X	reactant and conditions	products and remarks	Ref.
Br	NaC$_5$H$_5$ in THF	ferrocene, FpC$_5$H$_5$, and Fp$_2$	[5]
Br, I	NaC$_5$H$_5$ in THF, 2 to 3 h	FpC$_5$H$_5$ and Fp$_2$	[8]
with transition metal compounds			
Br	CuR (R=C$_6$H$_5$, C$_6$H$_4$F-3, C$_6$H$_4$F-4 in ether/THF at 0 to 20°, 5 h	FpR	[291]
Cl, Br	CuC≡CR (R=C$_6$H$_5$, C$_6$H$_4$CH$_3$–4, C$_6$H$_4$F-4, C$_6$F$_5$) in refluxing acetone or THF, several hours	(Fp(C≡CR)CuX)$_2$ (see C4, pp. 216/9)	[185, 238], also see [161, 239]
—	AgC(CF$_3$)=C(CF$_3$)F in CH$_2$Cl$_2$, 5 h	FpC(CF$_3$)=C(CF$_3$)F	[186]
—	excess Ag[C(CN)$_3$] in CH$_3$OH at 20° 2 h	FpN=C=C(CN)$_2$(?), see 1.5.2.3.8.1, p. 342	[40, 70]
—	Hg(CH$_2$CHO)$_2$ in C$_6$H$_6$ at 70 to 80°, 6 h	19% FpHgI, further information on p. 277	[144]
—	Hg(CH$_2$R)$_2$ (R=CHO, CH=CH$_2$) in C$_6$H$_6$, irradiation 13 to 24 h	10 to 13% FpHgI and ferrocene, further information on p. 277	[144]
—	Hg(CH$_2$CH=CH$_2$)$_2$ in C$_6$H$_6$, 14 d	22% FpHgI and ferrocene, further information on p. 277	[144]
—	Hg(C≡CC$_6$H$_5$)$_2$ in C$_6$H$_6$, 19 h	8% FpHgI, further information on p. 277	[144]
—	Hg(C$_5$H$_4$FeC$_5$H$_5$)$_2$ in C$_6$H$_6$, irradiation, 9 h, and without irradiation in C$_6$H$_6$ at 70 to 80°, 10 h, or in boiling xylene, 8 h	FpC$_5$H$_4$FeC$_5$H$_5$	[162, 176]
—	in THF, irradiation, 16 h	ferrocene	[176]

References on p. 294

I	$Hg(2-C_4H_3S)_2$ in THF, UV irradiation, 22 h	17% $2-FpC_4H_3S$ and ferrocene, further information on p. 277	[144]
I	$Hg(C_6H_5)_2$ in C_6H_6, irradiation, 23 h	FpC_6H_5	[71]
I	$Hg(C_6H_4Cl-4)_2$ in THF, irradiation, 40 h	17% FpC_6H_4Cl-4 and ferrocene, further information on p. 277	[144]
I	$Hg(C_6H_4COOC_2H_5-4)_2$ in C_6H_6, irradiation, 44 h	ferrocene and presumably $FpC_6H_4COOC_2H_5-4$	[71]
I	$Na[C_5H_5Cr(CO)_3]$	no evidence for $FpCr(CO)_3C_5H_5$	[292]
Cl	$(C_5H_5Cr(NO)_2)_2$ in refluxing THF, 4.5 h	$C_5H_5Cr(NO)_2Cl$ and Fp_2	[293, 324, 332]
-	$Na[Mo(CO, D)_5]$ and $Na[W(CO,D)_5]$ (D = amine)	no $FpMo(CO,D)_5$ or $FpW(CO,D)_5$	[292]
-	$Na[C_5H_5Mo(CO)_3]$ in petroleum ether/C_2H_5OH, 30 min	$FpMo(CO)_3C_5H_5$	[16]
-	$Na[C_5H_5W(CO)_3]$ in refluxing THF, 1 h	$FpW(CO)_3C_5H_5$	[292]
-	$Na[Mn(CO)_5]$ in pentane/THF, 2 d	$FpMn(CO)_5$	[16]
Cl	$Na[Re(CO)_5]$. 2.5 h	$FpRe(CO)_5$	[41]
Cl	$Na_2[Fe(CO)_4]$ and $(C_6H_5)_2AsCl$ in THF	$Fp(\mu-As(C_6H_5)_2)Fe(CO)_4$	[219]
Cl	$(CO)_4FeP(C_6H_5)_2H$ and $(C_2H_5)_2NH$ (excess) in C_6H_6/ether, 2 d	$Fp(\mu-P(C_6H_5)_2)Fe(CO)_4$	[163]
Cl	$(CO)_4FeP(CH_3)_2Si(CH_3)_3$ in refluxing C_6H_6, 10 min	Fp_2 and $(CO)_3Fe(\mu-P(CH_3)_2)_2Fe(CO)_3$	[220]
Br	$RC_5H_4Fe(CO)_2Br$ (R = $1,2-C_2B_{10}H_{11}-3$) in decalin at 180°, 3 h	$Fe(C_5H_4R)_2$, ferrocene, and $C_5H_5FeC_5H_4R$	[187, 188]

References on p. 294

Table 29 [continued]

X	reactant and conditions	products and remarks	Ref.
I	Na[Co(CO)$_4$] in THF	C$_5$H$_5$Fe(CO)(μ-CO)$_2$Co(CO)$_3$	[28]
Cl, I	Na[Co(CO)$_3$^{2}D] (^{2}D = tertiary phosphine or arsine) in THF	C$_5$H$_5$FeCo(CO)$_5$^{2}D with bridging and nonbridging CO's proposed	[333]
Br	cyclo-C$_4$R$_4$Co(CO)$_2$Br (R = C$_6$H$_5$, C$_6$H$_4$CH$_3$-4) in refluxing C$_6$H$_6$, 2.5 h	C$_5$H$_5$CoC$_4$R$_4$-cyclo	[42, 49], also see [221, 347]
Cl	(CO)$_3$NiP(C$_6$H$_5$)$_2$H and (C$_2$H$_5$)$_2$NH (excess) in C$_6$H$_6$/ether, 2 d	Fp(μ-P(C$_6$H$_5$)$_2$)Ni(CO)$_3$	[163]
Br	cyclo-C$_4$R$_4$MBr$_2$ (M = Ni, Pd; R = C$_6$H$_5$, C$_6$H$_4$CH$_3$-4) in refluxing C$_6$H$_6$, 2 to 3 h	[C$_5$H$_5$MC$_4$R$_4$-cyclo][FeBr$_4$]	[42, 49], also see [221, 347]
Br	^{4}LPdBr$_2$ (^{4}L = cycloocta-1,5-diene) in refluxing C$_6$H$_6$, 2 h	[C$_5$H$_5$Pd^{4}L][FeBr$_4$]	[42, 49], also see [221, 347]
Cl	[C$_5$H$_5$Ru(CO)$_2$]$^-$ in THF	Fp$_2$ and (C$_5$H$_5$Ru(CO)$_2$)$_2$	[87]

References on p. 294

[278]. In contrast to Fp_2, FpI is not suitable as initiator for photopolymerization of olefins [169].

1.5.2.3.2.5 Analysis

Determination of iron in FpI [164] and FpX (X=Cl, I) [222] can be carried out by X-ray fluorescence. The results have been compared with values obtained by a standard calorimetric method [164] or by chemical analysis [222].

1.5.2.3.2.6 Adducts with Halogens and Lewis Acids

Adducts of FpX compounds are listed in Table 30. Halogen adducts form in cyclohexane with the same halogen molecules X_2 or other halogens X_2' when their electronegativity is lower than that of X. Lewis acid adducts were observed in solutions of CH_2Cl_2.

Only $FpI \cdot I_2$ (Table 30, No. 6) could be isolated from the reaction of FpI with an excess of I_2 in a closed system (25°C under 0.3 atm N_2 for 50 d) [96], whereas the other products have been identified in solution by their IR spectra. The adducts

Table 30
Adducts of FpX (X=Cl, Br, I) with Halogens and Lewis Acids.

No.	adduct	IR spectra (wave numbers in cm^{-1}) bands above 2000 cm^{-1} are $\nu(CO)$	Ref.
halogen adducts (IR in cyclohexane)			
1	$FpCl_3$	2022, 2063	[96, 271]
2	$FpClBr_2$	2017, 2057	[96]
3	$FpClI_2$	2018.5, 2058.5	[96]
4	$FpBr_3$	2015, 2054	[96]
5	$FpBr_5$	2029, 2069	[96]
6	FpI_3	2010, 2048; other bands: $\delta(I-I-I)$ 93, $\nu_{as}(I-I)$ 169; additional bands below 200 cm^{-1} are given [271]	[96, 271]
Lewis acid adducts (IR in CH_2X_2)			
7	$FpClSbCl_5$	2037, 2076	[192, 271]
8	$FpClSnCl_4$	2027, 2069	[271]
9	$FpClAlCl_3$	2035, 2077	[192, 271]
10	$FpClGaCl_3$	2034, 2076	[271]
11	$FpClInCl_3$	2024, 2070	[271]
12	$FpClTiCl_4$	2025, 2069	[271]
13	$FpClNbCl_5$	2029, 2071	[271]
14	$FpClMoCl_5$	2034, 2076	[271]

References on p. 294

Table 30 [continued]

No.	adduct	IR spectra (wave numbers in cm^{-1}) bands above 2000 cm^{-1} are $\nu(CO)$	Ref.
15	FpClFeCl$_3$	2035, 2076; other bands: $\nu(FeC)$ 465, $\delta(FeCO)$ 523, 555, 585	[192, 271]
16	FpBrAlBr$_3$	2028, 2070; other bands: $\nu(FeC)$ 470, $\delta(FeCO)$ 523, 554	[192, 271]
17	FpBrFeBr$_3$	2027, 2068; other bands: $\nu(FeC)$ 429, 471, $\delta(FeCO)$ 530, 557, 595	[271]
18	FpIAlI$_3$	2017, 2056	[192, 271]
19	FpIGaI$_3$	2016, 2056	[271]
20	FpIFeCl$_3$	2020, 2062	[192]

show higher $\nu(CO)$ frequencies (B$_1$ and A$_1$) but lower $\nu(FeC)$ and $\delta(FeCO)$ frequencies than the original FpX compounds [96, 192, 271]. The increase of $\nu(CO)$ frequencies with the electron–acceptor properties of the Lewis acid is thought to be evidence of charge transfer from FpX to the Lewis acid, the adduct being formulated as a halogen donor–acceptor complex [192]. These formulations are supported by a strong UV band of FpI$_3$ (in CHCl$_3$) at 367 nm [96] and by the low molar conductivities in 10^{-3} M CH$_2$Cl$_2$ solutions, $\Lambda = 0.53 \ \Omega^{-1} \cdot cm^2 \cdot mol^{-1}$ for FpI$_3$ (No. 6) and $2.3 \ \Omega^{-1} \cdot cm^2 \cdot mol^{-1}$ for FpClFeCl$_3$ (No. 15) at 20 °C [271]. Dilution of the solutions gives the starting compounds [192, 271]. An apparent [(FpCl)$_2$SbCl$_3$]$_2$ adduct turned out to be an ionic compound without Fe–Sb bonds.

Compounds of lanthanides also can act as Lewis acids towards FpX (X = Cl, I), but no defined adducts are formulated. Adduct formation between FpI and Nd(C$_5$H$_4$CH$_3$)$_3$ in benzene was detected by an upfield shift of the cyclopentadienyl ^{1}H resonance of 1.09 ppm, whereas the IR spectrum did not exhibit changes in the $\nu(CO)$ frequencies. Therefore, it is believed that Nd is attached at the halide site while it is attached at the CO ligands in related Mo and W adducts [224]. Treatment of FpCl with Eu(fod)$_3$ (Formula II on p. 104) in C$_6$D$_5$CD$_3$ or CDCl$_3$ causes an observable downfield shift of the cyclopentadienyl ^{1}H resonance, namely, 3.12 ppm, indicating an interaction. The spectra of FpBr and FpI are not changed [165].

References:

[1] J.E. Brown, H. Shapiro, E.G. de Witt, Ethyl Corp. (Ger. Offen. 1217952 [1953/66]). – [2] T.S. Piper, G. Wilkinson (Chem. Ind. [London] 1955 1296). – [3] T.S. Piper, F.A. Cotton, G. Wilkinson (J. Inorg. Nucl. Chem. 1 [1955] 165/74). – [4] B.F. Hallam, O.S. Mills, P.L. Pauson (J. Inorg. Nucl. Chem. 1 [1955] 313/6). – [5] B.F. Hallam, P.L. Pauson (Chem. Ind. [London] 1955 653).

[6] B.F. Hallam, P.L. Pauson (J. Chem. Soc. 1956 3030/7). – [7] T.S. Piper, G. Wilkinson (J. Inorg. Nucl. Chem. 2 [1956] 38/45). – [8] T.S. Piper, G. Wilkinson (J. Inorg. Nucl. Chem. 3 [1956] 104/24). – [9] T.S. Piper, G. Wilkinson (Natur-

wissenschaften **43** [1956] 15/6). – [10] T.H. Coffield, V. Sandel, R.D. Closson (J. Am. Chem. Soc. **79** [1957] 5826).

[11] J.C. Thomas (U.S. 2849471 [1958]). – [12] E.O. Fischer, G. Joos, E. Vogg (Z. Physik. Chem. [Frankfurt] **18** [1958] 80/9). – [13] Ethyl Corp. (Brit. 896391 [1958/62]). – [14] M.L.H. Green, C.N. Street, G. Wilkinson (Z. Naturforsch. **14b** [1959] 738). – [15] M.L.H. Green, L. Pratt, G. Wilkinson (J. Chem. Soc. **1960** 989/97).

[16] R.B. King, P.M. Treichel, F.G.A. Stone (Chem. Ind. [London] **1961** 747/8). – [17] A. Davison, M.L.H. Green, G. Wilkinson (J. Chem. Soc. **1961** 3172/7). – [18] A. Wojcicki, F. Basolo (J. Am. Chem. Soc. **83** [1961] 525/8). – [19] A. Wojcicki, F. Basolo (J. Inorg. Nucl. Chem. **17** [1961] 77/83). – [20] K. Fichtel (Diss. Univ. München 1961).

[21] E.O. Fischer, K. Fichtel (Chem. Ber. **94** [1961] 1200/4). – [22] E.O. Fischer, K. Fichtel, Badische Anilin- und Soda-Fabrik (Belg. 613119 [1961/2]). – [23] E.O. Fischer, K. Fichtel, Badische Anilin- und Soda-Fabrik (Ger. 1159444 [1961/3]). – [24] T.H. Coffield, R.D. Closson, Ethyl Corp. (U.S. 3130214 [1961/4]). – [25] A. Davison, W.M. McFarlane, L. Pratt, G. Wilkinson (J. Chem. Soc. **1962** 3653/66).

[26] E.O. Fischer, K. Fichtel, K. Öfele (Chem. Ber. **95** [1962] 249/52). – [27] R.B. King (Inorg. Chem. **1** [1962] 964/5). – [28] K.K. Joshi, P.L. Pauson (Z. Naturforsch. **17b** [1962] 565). – [29] P.L. Pauson, W.H. Stubbs (Angew. Chem. **74** [1962] 466). – [30] E.O. Fischer, K. Fichtel (Chem. Ber. **95** [1962] 2063/9).

[31] E.O. Fischer, K. Fichtel, Union Carbide Corp. (U.S. 3268565 [1962/6]). – [32] J.K.P. Ariyaratne, M.L.H. Green (J. Organometal. Chem. **1** [1963] 90/3). – [33] R.B. King, F.G.A. Stone (Inorg. Syn. **7** [1963] 99/115). – [34] R.B. King (Inorg. Chem. **2** [1963] 936/44). – [35] A. Davison, J.A. McCleverty, G. Wilkinson (J. Chem. Soc. **1963** 1133/8).

[36] C.E. Coffey (J. Inorg. Nucl. Chem. **25** [1963] 179/85). – [37] K.K. Joshi, P.L. Pauson, W.H. Stubbs (J. Organometal. Chem. **1** [1963] 51/7). – [38] F. Bonati, G. Wilkinson (J. Chem. Soc. **1964** 179/81). – [39] M.D. Rausch (Inorg. Chem. **3** [1964] 300/1). – [40] W. Beck, R.E. Nitzschmann, G. Neumair (Angew. Chem. **76** [1964] 346).

[41] A.N. Nesmeyanov, K.N. Anisimov, N.E. Kolobova, V.N. Khandozhko (Dokl. Akad. Nauk SSSR **156** [1964] 383/5; Dokl. Chem. Proc. Acad. Sci. USSR **154/159** [1964] 502/4). – [42] P.M. Maitlis, A. Efraty, M.L. Games (J. Organometal. Chem. **2** [1964] 284/6). – [43] R.H. Herber, R.B. King, G.K. Wertheim (Inorg. Chem. **3** [1964] 101/7). – [44] J.P. Bibler, A. Wojcicki (J. Am. Chem. Soc. **86** [1964] 5051/3). – [45] S.V. Dighe, M. Orchin (J. Am. Chem. Soc. **86** [1964] 3895/6).

[46] R.G. Hayter, L.F. Williams (Inorg. Chem. **3** [1964] 613). – [47] K.K. Joshi, P.L. Pauson, A.R. Qazi, W.H. Stubbs (J. Organometal. Chem. **1** [1964] 471/5). – [48] E.O. Fischer, E. Moser (Z. Naturforsch. **20b** [1965] 184/5). – [49] P.M. Maitlis, A. Efraty, M.L. Games (J. Am. Chem. Soc. **87** [1965] 719/24). – [50] M.L.H. Green, M. Ishaq, T. Mole (Z. Naturforsch. **20b** [1965] 598).

[51] E.O. Fischer, E. Moser (J. Organometal. Chem. **3** [1965] 16/24). – [52] E.O. Fischer, E. Moser, Badische Anilin- und Soda-Fabrik (Ger. 1248655 [1965/7]). – [53] E.O. Fischer, E. Moser (Z. Anorg. Allgem. Chem. **342** [1966] 156/64). – [54] M. Ahmad, R. Bruce, G.R. Knox (J. Organometal. Chem. **6** [1966] 1/10). – [55] J. Locke, J.A. McCleverty (Inorg. Chem. **5** [1966] 1157/61).

[56] P.M. Treichel, R.L. Shubkin, K.W. Barnett, D. Reichard (Inorg. Chem. **5** [1966] 1177/81). – [57] R.C. Edmondson, M.J. Newlands (Chem. Ind. [London] **1966** 1888/9). – [58] J.P. Bibler, A. Wojcicki (J. Am. Chem. Soc. **88** [1966] 4862/70). – [59] R.K. Kochhar, R. Pettit (J. Organometal. Chem. **6** [1966] 272/8). – [60] A.N. Nesmeyanov, Yu.A. Chapovskii, B.V. Lokshin, I.V. Polovyanyuk, L.G. Makarova (Dokl. Akad. Nauk SSSR **166** [1966] 1125/8; Dokl. Chem. Proc. Acad. Sci. USSR **166/171** [1966] 213/6).

[61] E.O. Fischer, E. Moser (J. Organometal. Chem. **5** [1966] 63/72). – [62] I.V. Polovyanyuk, Yu.A. Chapovskii, L.G. Makarova (Izv. Akad. Nauk SSSR Ser. Khim. **1966** 387; Bull. Acad. Sci. USSR Div. Chem. Sci. **1966** 368). – [63] P.W. Jolly, R. Pettit (J. Am. Chem. Soc. **88** [1966] 5044/5). – [64] R. Bruce, G.R. Knox (J. Organometal. Chem. **6** [1966] 67/75). – [65] E. Schumacher, R. Taubenest (Helv. Chim. Acta **49** [1966] 1447/55).

[66] R.E. Dessy, F.E. Stary, R.B. King, M. Waldrop (J. Am. Chem. Soc. **88** [1966] 471/6). – [67] A.N. Nesmeyanov, K.N. Anisimov, N.E. Kolobova, F.S. Denisov (Izv. Akad. Nauk SSSR Ser. Khim. **1966** 2246; Bull. Acad. Sci. USSR Div. Chem. Sci. **1966** 2185). – [68] A.R. Manning (Chem. Commun. **1966** 906). – [69] G. Schmid, H. Nöth (Chem. Ber. **100** [1967] 2899/907). – [70] W. Beck, R.E. Nitzschmann, H.S. Smedal (J. Organometal. Chem. **8** [1967] 547/50).

[71] A.N. Nesmeyanov, I.V. Polovyanyuk, B.V. Lokshin, Yu.A. Chapovskii, L.G. Makarova (Zh. Obshch. Khim. **37** [1967] 2015/7; J. Gen. Chem. [USSR] **37** [1967] 1911/3). – [72] M.L.H. Green, M. Ishaq, R.N. Whiteley (J. Chem. Soc. A **1967** 1508/15). – [73] A.N. Nesmeyanov, Yu.A. Chapovskii (Izv. Akad. Nauk SSSR Ser. Khim. **1967** 2075/7; Bull. Acad. Sci. USSR Div. Chem. Sci. **1967** 1988/90). – [74] P.L. Pauson, A.R. Qazi (J. Organometal. Chem. **7** [1967] 321/4). – [75] A.N. Nesmeyanov, Yu.A. Chapovskii, I.V. Polovyanyuk, L.G. Makarova (J. Organometal. Chem. **7** [1967] 329/37).

[76] L.I. Zakharkin (Izv. Akad. Nauk SSSR Ser. Khim. **1967** 956; Bull. Acad. Sci. USSR Div. Chem. Sci. **1967** 932). – [77] A.N. Nesmeyanov, Yu.A. Chapovskii, Yu.A. Ustynyuk (J. Organometal. Chem. **9** [1967] 345/53). – [78] R. Ugo, S. Čenini, F. Bonati (Inorg. Chim. Acta **1** [1967] 451/61). – [79] A.N. Nesmeyanov, Yu.A. Chapovskii, L.I. Denisovich, B.V. Lokshin, I.V. Polovyanyuk (Dokl. Akad. Nauk SSSR **174** [1967] 1342/4; Dokl. Chem. Proc. Acad. Sci. USSR **172/177** [1967] 576/8). – [80] F.A. Cotton, G. Yagupsky (Inorg. Chem. **6** [1967] 15/20).

[81] D. Grešová, A.A. Vlček (Inorg. Chim. Acta **1** [1967] 482). – [82] E.N. Gladyshev, V.I. Ermolaev, Yu.A. Sorokin, O.A. Kruglaya, N.S. Vyazankin, G.A. Razuvaev (Dokl. Akad. Nauk SSSR **179** [1968] 1333/5; Dokl. Chem. Proc. Acad. Sci. USSR **178/183** [1968] 350/2). – [83] G. Schmid, P. Powell, H. Nöth (Chem. Ber. **101** [1968] 1205/14). – [84] M.L.H. Green, T. Mole (J. Organometal. Chem. **12** [1968] 404/6). – [85] M.I. Bruce, D.A. Harbourne, F. Waugh, F.G.A. Stone (J. Chem. Soc. A **1968** 356/9).

[86] M.D. Rausch, T.R. Criswell, A.K. Ignatowicz (J. Organometal. Chem. **13** [1968] 419/30). – [87] T. Blackmore, J.D. Cotton, M.I. Bruce, F.G.A. Stone (J. Chem. Soc. A **1968** 2931/6). – [88] Y. Tajima, E. Kunioka (J. Org. Chem. **33** [1968] 1689/90). – [89] M.A. Jennings, A. Wojcicki (J. Organometal. Chem. **14** [1968] 231/4). – [90] R.B. King, R.N. Kapoor (J. Organometal. Chem. **15** [1968] 457/69).

[91] F. Seel, V. Sperber (J. Organometal. Chem. **14** [1968] 405/10). – [92] T.E. Sloan, A. Wojcicki (Inorg. Chem. **7** [1968] 1268/73). – [93] R.B. King, K.H. Pannell,

L.W. Houk, R.N. Kapoor (New Aspects Chem. Metal Carbonyls Deriv. 1st Intern. Symp. Proc., Venice 1968, Abstr. E8, pp. 1/6). − [94] D.A. Brown, A.R. Manning, J.M. Rowly (New Aspects Chem. Metal Carbonyls Deriv. 1st Intern. Symp. Proc., Venice 1968, Abstr. C5, pp. 1/10). − [95] R.J. Haines, A.L. du Preez (Chem. Commun. 1968 1513/4).

[96] M. Pankowski, M. Bigorgne (Compt. Rend. C **267** [1968] 1809/12). − [97] A.R. Manning (J. Chem. Soc. A **1968** 1670/3). − [98] D.M. Adams, J.N. Crosby, R.D.W. Kemmitt (J. Chem. Soc. A **1968** 3056/8). − [99] R.B. King (Inorg. Chim. Acta **2** [1968] 454). − [100] M.I. Bruce (Intern. J. Mass Spectrom. Ion Phys. **1** [1968] 141/55).

[101] P.M. Treichel, W.M. Douglas (J. Organometal. Chem. **19** [1969] 221/4). − [102] A.N. Nesmeyanov, L.G. Makarova, V.N. Vinogradova (Izv. Akad. Nauk SSSR Ser. Khim. **1969** 1398; Bull. Acad. Sci. USSR Div. Chem. Sci. **1969** 1303). − [103] J.M. Burlitch, S.W. Ulmer (J. Organometal. Chem. **19** [1969] P21/P23). − [104] M. Dekker, G.R. Knox, C.G. Robertson (J. Organometal. Chem. **18** [1969] 161/7). − [105] R.B. King, N. Welcman (Inorg. Chem. **8** [1969] 2540/3).

[106] C. O'Connor, J.D. Gilbert, G. Wilkinson (J. Chem. Soc. A **1969** 84/7). − [107] M. Lustig, L.W. Houk (Inorg. Nucl. Chem. Letters **5** [1969] 851/3). − [108] A.N. Nesmeyanov, K.N. Anisimov, N.E. Kolobova, F.S. Denisov (Izv. Akad. Nauk SSSR Ser. Khim. **1969** 1520/4; Bull. Acad. Sci. USSR Div. Chem. Sci. **1969** 1409/12). − [109] E. Amberger, E. Mühlhofer, H. Stern (J. Organometal. Chem. **17** [1969] P5/P6). − [110] M.J. Mays, S.M. Pearson (J. Chem. Soc. A **1969** 136/8).

[111] J.C. Smart, P.M. Garrett, M.F. Hawthorne (J. Am. Chem. Soc. **91** [1969] 1031/2). − [112] D.A. Brown, A.R. Manning, D.J. Thornhill (Chem. Commun. **1969** 338). − [113] R.J. Haines, A.L. du Preez (J. Am. Chem. Soc. **91** [1969] 769/70). − [114] R.J. Haines, A.L. du Preez (Inorg. Chem. **8** [1969] 1459/64). − [115] J. Dalton, I. Paul, F.G.A. Stone (J. Chem. Soc. A **1969** 2744/9).

[116] D.A. Brown, H.J. Lyons, A.R. Manning, J.M. Rowley (Inorg. Chim. Acta **3** [1969] 346/50). − [117] R.B. King, L.W. Houk, K.H. Pannell (Inorg. Chem. **8** [1969] 1042/8). − [118] D.A. Brown, H. Lyons, J. Rowley, F. Sane (Proc. 4th Intern. Conf. Organometal. Chem., Bristol 1969, Abstr. J3). − [119] R.J. Angelici, L.M. Charley (J. Organometal. Chem. **24** [1970] 205/9). − [120] J.D. Cotton, R.M. Peachey (Inorg. Nucl. Chem. Letters **6** [1970] 727/31).

[121] N.S. Vyazankin, G.S. Kalinina, V.I. Ermolaev, Yu.A. Sorokin (Zh. Obshch. Khim. **40** [1970] 1757/60; J. Gen. Chem. [USSR] **40** [1970] 1742/4). − [122] D.A. Owen, J.C. Smart, P.M. Garrett, M.F. Hawthorne (AD-710759 [1970]). − [123] R.J. Haines, A.L. du Preez, L.L. Marais (J. Organometal. Chem. **24** [1970] C26/C28). − [124] A.E. Kruse, R.J. Angelici (J. Organometal. Chem. **24** [1970] 231/9). − [125] T. Blackmore, M.I. Bruce, P.J. Davidson, M.Z. Iqbal, F.G.A. Stone (J. Chem. Soc. A **1970** 3153/8).

[126] H. Vahrenkamp (Chem. Ber. **103** [1970] 3580/90). − [127] D.J. Parker, M.H.B. Stiddard (J. Chem. Soc. A **1970** 1040/9). − [128] L.W. Houk, M. Lustig (Inorg. Chem. **9** [1970] 2462/5). − [129] D.J. Parker (J. Chem. Soc. A **1970** 1382/6). − [130] D.J. Parker, M.H.B. Stiddard (J. Chem. Soc. A **1970** 480/90).

[131] S.P. Gubin (Pure Appl. Chem. **23** [1970] 463/87). − [132] R.J. Haines, A.L. du Preez (J. Chem. Soc. A **1970** 2341/6). − [133] R.W. Johnson, R.G. Pearson (Chem. Commun. **1970** 986/7). − [134] A.J. Carty, A. Efraty, T.W. Ng, T. Birchall

(Inorg. Chem. **9** [1970] 1263/8). − [135] D.A. Brown, H.J. Lyons, R.T. Sane (Inorg. Chim. Acta **4** [1970] 621/5).

[136] W. Ehrl, H. Vahrenkamp (Chem. Ber. **103** [1970] 3563/79). − [137] S.P. Gubin, L.I. Denisovich (Elektrokhim. Protsessy Uchastiem Org. Veshchestv **1970** 61/8; C.A. **74** [1971] No. 18800). − [138] W.A.G. Graham, W. Jetz (Inorg. Chem. **10** [1971] 1159/65). − [139] L.F. Farnell, E.W. Randall, E. Rosenberg (Chem. Commun. **1971** 1078/9). − [140] K. Burger, L. Korecz, P. Mag, U. Belluco, L. Busetto (Inorg. Chim. Acta **5** [1971] 362/4).

[141] A.R. Manning (J. Chem. Soc. A **1971** 106/10). − [142] M.D. Johnson, D. Dodd (J. Chem. Soc. B **1971** 662/7). − [143] R.J. Haines, A.L. du Preez, I.L. Marais (J. Organometal. Chem. **28** [1971] 405/13). − [144] A.N. Nesmeyanov, L.G. Makarova, V.N. Vinogradova (Izv. Akad. Nauk SSSR Ser. Khim. **1971** 1984/7; Bull. Acad. Sci. USSR Div. Chem. Sci. **1971** 1869/72). − [145] H. Vahrenkamp, W. Ehrl (5th Intern. Conf. Organometal. Chem., Moscow 1971, Vol. 1, Abstr. No. 31).

[146] V.N. Babin, L.A. Fedorov, N.S. Kochetkova, Yu.A. Belousov (5th Intern. Conf. Organometal. Chem., Moscow 1971, Vol. 2, Abstr. No. 390). − [147] A.N. Nesmeyanov, V.N. Babin, N.S. Kochetkova, E.I. Mysov, Yu.A. Belousov, L.A. Fedorov (Dokl. Akad. Nauk SSSR **200** [1971] 1112/5; Dokl. Chem. Proc. Acad. Sci. USSR **196/201** [1971] 838/41). − [148] E.H. Braye, K.K. Joshi (Bull. Soc. Chim. Belges **80** [1971] 651/3). − [149] D.A. Symon, T.C. Waddington (J. Chem. Soc. A **1971** 953/7). − [150] E.C. Johnson, T.J. Meyer, N. Winterton (Inorg. Chem. **10** [1971] 1673/5).

[151] E. Eisner, M.J. Newlands, L.K. Thompson (5th Intern. Conf. Organometal. Chem., Moscow 1971, Vol. 2, Abstr. No. 303). − [152] W.R. Cullen, D.J. Patmore, J.R. Sams, M.J. Newlands, L.K. Thompson (Chem. Commun. **1971** 952/3). − [153] J.A. Ferguson, T.J. Meyer (Inorg. Chem. **10** [1971] 1025/8). − [154] G.M. Whitesides, D.J. Boschetto (J. Am. Chem. Soc. **93** [1971] 1529/31). − [155] M.I. Bruce, A.D. Redhouse (J. Organometal. Chem. **30** [1971] C78/C80).

[156] S.R.A. Bird, J.D. Donaldson, A.F. Le C. Holding, B.J. Senior, M.J. Tricker (J. Chem. Soc. A **1971** 1616/21). − [157] M.F. Hawthorne, D.A. Owen, J.C. Smart, P.M. Garrett (J. Am. Chem. Soc. **93** [1971] 1362/8). − [158] A.T.T. Hsieh, M.J. Mays (Inorg. Nucl. Chem. Letters **7** [1971] 223/5). − [159] S.C. Cohen (5th Intern. Conf. Organometal. Chem., Moscow 1971, Vol. 2, Abstr. No. 397). − [160] S.C. Cohen (J. Organometal. Chem. **30** [1971] C15/C16).

[161] M.I. Bruce, R.F. May (5th Intern. Conf. Organometal. Chem., Moscow 1971, Vol. 1, Abstr. No. 115). − [162] A.N. Nesmeyanov, L.G. Makarova, V.N. Vinogradova (Izv. Akad. Nauk SSSR Ser. Khim. **1971** 892; Bull. Acad. Sci. USSR Div. Chem. Sci. **1971** 818). − [163] K. Yasufuku, H. Yamazaki (J. Organometal. Chem. **28** [1971] 415/21). − [164] J.M. McCall, D.E. Leyden, C.W. Blount (Anal. Chem. **43** [1971] 1324/5). − [165] T.J. Marks, J.S. Kristoff, A. Alich, D.F. Shriver (J. Organometal. Chem. **33** [1971] C35/C37).

[166] R.B. King, P.N. Kapoor, R.N. Kapoor (Inorg. Chem. **10** [1971] 1841/50). − [167] R.B. King, R.N. Kapoor, M.S. Saran, P.N. Kapoor (Inorg. Chem. **10** [1971] 1851/60). − [168] R.B. King, M.S. Saran (Inorg. Chem. **10** [1971] 1861/7). − [169] H. Barzynski, F.J. Müller, M.J. Jun, M. Velic, Badische Anilin- und Soda-Fabrik (Ger. Offen. 2142105 [1971/73]). − [170] D.J. Jones, R.J. Mawby (Inorg. Chim. Acta **6** [1972] 157/60).

[171] P. Hackett, A.R. Manning (J. Chem. Soc. Dalton Trans. **1972** 1487/91). –
[172] P. Hackett, A.R. Manning (J. Organometal. Chem. **34** [1972] C15/C17). –
[173] A.N. Nesmeyanov, S.S. Churanov, I.Sh. Guzman, E.G. Perevalova (Izv. Akad.
Nauk SSSR Ser. Khim. **1972** 570/1; Bull. Acad. Sci. USSR Div. Chem. Sci. **1972**
526/7). – [174] W. Jetz, R.J. Angelici (Inorg. Chem. **11** [1972] 1960/2). – [175] J.
Ellermann, H. Behrens, H. Krohberger (J. Organometal. Chem. **46** [1972] 119/38).

[176] A.N. Nesmeyanov, L.G. Makarova, V.N. Vinogradova (Izv. Akad. Nauk SSSR
Ser. Khim. **1972** 1600/4; Bull. Acad. Sci. USSR Div. Chem. Sci. **1972** 1541/4). –
[177] C.B. Colburn, W.E. Hill, D.W.A. Sharp (Inorg. Nucl. Chem. Letters **8** [1972]
625/7). – [178] M.R. Churchill, B.G. DeBoer, K.L. Karla, P. Reich-Rohrwig, A. Woj-
cicki (J. Chem. Soc. Chem. Commun. **1972** 981/2). – [179] A.N. Nesmeyanov, L.G.
Makarova, V.N. Vinogradova (Izv. Akad. Nauk SSSR Ser. Khim. **1972** 2796/7; Bull.
Acad. Sci. USSR Div. Chem. Sci. **1972** 2723/4). – [180] A.N. Nesmeyanov, L.G.
Makarova, V.N. Vinogradova (Izv. Akad. Nauk SSSR Ser. Khim. **1972** 2798/9; Bull.
Acad. Sci. USSR Div. Chem. Sci. **1972** 2725/7).

[181] H. Brunner, M. Vogel (J. Organometal. Chem. **35** [1972] 169/77). –
[182] L.G. Sneddon, R.N. Grimes (J. Am. Chem. Soc. **94** [1972] 7161/2). –
[183] A.T.T. Hsieh, M.J. Mays (J. Organometal. Chem. **37** [1972] 9/14). – [184] A.
Cutler, R.W. Fish, W.P. Giering, M. Rosenblum (J. Am. Chem. Soc. **94** [1972]
4354/5). – [185] M.I. Bruce, R. Clark, J. Howard, P. Woodward (J. Organometal.
Chem. **42** [1972] C107/C109).

[186] R.B. King, W.C. Zipperer (Inorg. Chem. **11** [1972] 2119/25). – [187] L.I.
Zakharkin, L.V. Orlova (Izv. Akad. Nauk SSSR Ser. Khim. **1972** 209; Bull. Acad. Sci.
USSR Div. Chem. Sci. **1972** 207). – [188] L.I. Zakharkin, L.V. Orlova, B.V. Lokshin,
L.A. Fedorov (J. Organometal. Chem. **40** [1972] 15/28). – [189] D.J. Darensbourg
(Inorg. Chem. **11** [1972] 1606/9). – [190] O.A. Gansov, D.A. Schexnayder, B.Y.
Kimura (J. Am. Chem. Soc. **94** [1972] 3406/8).

[191] A.N. Nesmeyanov, I.F. Leshcheva, I.V. Polovyanyuk, Yu.A. Ustynyuk, L.G.
Makarova (J. Organometal. Chem. **37** [1972] 159/65). – [192] M. Pankowski, B.
Demerseman, G. Bouquet, M. Bigorgne (J. Organometal. Chem. **35** [1972] 155/9). –
[193] R.J. Haines, C.R. Nolte (J. Organometal. Chem. **36** [1972] 163/75). –
[194] W. Ehrl, H. Vahrenkamp (Chem. Ber. **105** [1972] 1471/85). – [195] N.G. Con-
nelly, J.D. Davies (J. Organometal. Chem. **38** [1972] 385/90).

[196] R.B. King, A. Efraty (J. Am. Chem. Soc. **94** [1972] 3768/73). – [197] R.B.
King, W.C. Zipperer, M. Ishaq (Inorg. Chem. **11** [1972] 1361/70). – [198] P.M. Treichel, W.K. Dean, W.M. Douglas (J. Organometal. Chem. **42** [1972] 145/58). –
[199] R.J. Haines, A.L. du Preez (Inorg. Chem. **11** [1972] 330/6). – [200] B.D.
Dombek, R.J. Angelici (Inorg. Chim. Acta **7** [1973] 345/7).

[201] W.E. Williams, F.J. Lalor (J. Chem. Soc. Dalton Trans. **1973** 1329/32). –
[202] S.E. Jacobson, P. Reich-Rohrwig, A. Wojcicki (Inorg. Chem. **12** [1973]
717/23). – [203] C.S. Cundy, M.F. Lappert (J. Organometal. Chem. **57** [1973] C72/
C74). – [204] R.M.G. Roberts (J. Organometal. Chem. **47** [1973] 359/66). –
[205] M. Rosenblum, J.M. Tancrede (Org. Syn. **53** [1973] 1876).

[206] E.W. Abel, S. Moorhouse (J. Chem. Soc. Dalton Trans. **1973** 1706/11). –
[207] E.V. Bryukhova, I.M. Alymov, G.K. Semin (Izv. Akad. Nauk SSSR Ser. Khim.
1973 1898; Bull. Acad. Sci. USSR Div. Chem. Sci. **1973** 1849). – [208] P.W.
Wiggans (Educ. Chem. **10** [1973] 52/3, 65). – [209] L.H. Ali, A. Cox, T.J. Kemp

(J. Chem. Soc. Dalton Trans. **1973** 1475/8). − [210] F.A. Cotton, B.A. Frenz, A.J. White (J. Organometal. Chem. **60** [1973] 147/52).

[211] L.W. Houk, L.J. Epley (Abstr. Papers 165th Natl. Meeting Am. Chem. Soc., Dallas, Texas, 1973, INOR 80). − [212] R.J. Haines, A.L. du Preez, C.R. Nolte (J. Organometal. Chem. **55** [1973] 199/203). − [213] W.C. Zipperer (Diss. Univ. Georgia 1972; Diss. Abstr. Intern. B **33** [1973] 2990). − [214] R.C. Dobbie, P.R. Mason (J. Chem. Soc. Dalton Trans. **1973** 1124/8). − [215] L.G. Sneddon, D.C. Beer, R.N. Grimes (J. Am. Chem. Soc. **95** [1973] 6623/9).

[216] D.J. O'Sullivan, F.J. Lalor (J. Organometal. Chem. **57** [1973] C58/C60). − [217] S.C. Cohen (J. Chem. Soc. Dalton Trans. **1973** 553/5). − [218] A.N. Nes-meyanov, L.G. Makarova, V.N. Vinogradova (Izv. Akad. Nauk SSSR Ser. Khim. **1973** 2796/8; Bull. Acad. Sci. USSR Div. Chem. Sci. **1973** 2731/3). − [219] J.P. Collman, R.G. Komoto, W.O. Siegl (J. Am. Chem. Soc. **95** [1973] 2389/90). − [220] W. Ehrl, H. Vahrenkamp (J. Organometal. Chem. **63** [1973] 389/98).

[221] A. Efraty (J. Organometal. Chem. **57** [1973] 1/28). − [222] N.E. Gel'man, O.L. Lependina, E.A. Bozhevol'nov, K.I. Nikolaeva (Zh. Analit. Khim. **28** [1973] 1231/3; J. Anal. Chem. [USSR] **28** [1973] 1099/101). − [223] W.R. Cullen, D.J. Patmore, J.R. Sams (Inorg. Chem. **12** [1973] 867/72). − [224] A.E. Crease, P. Legz-dins (J. Chem. Soc. Dalton Trans. **1973** 1501/7). − [225] A.N. Nesmeyanov, E.G. Perevalova, L.I. Leont'eva, S.A. Eremin, O.V. Grigor'eva (Izv. Akad. Nauk SSSR Ser. Khim. **1974** 2645/7; Bull. Acad. Sci. USSR Div. Chem. Sci. **1974** 2558/60).

[226] P.L. Bock, D.J. Boschetto, J.R. Rasmussen, J.P. Demers, G.M. Whitesides (J. Am. Chem. Soc. **96** [1974] 2814/25). − [227] D. Slack, M.C. Baird (J. Chem. Soc. Chem. Commun. **1974** 701/2). − [228] P.L. Bock, G.M. Whitesides (J. Am. Chem. Soc. **96** [1974] 2826/9). − [229] R.C. Dobbie, P.R. Mason (J. Chem. Soc. Dalton Trans. **1974** 2439/42). − [230] W. Malisch, P. Panster (J. Organometal. Chem. **64** [1974] C5/C9).

[231] A. Rosan, M. Rosenblum (J. Organometal. Chem. **80** [1974] 103/7). − [232] M. Kilner, C. Midcalf (J. Chem. Soc. Dalton Trans. **1974** 1620/4). − [233] R.E.J. Bichler, H.C. Clark, B.K. Hunter, A.T. Rake (J. Organometal. Chem. **69** [1974] 367/76). − [234] D.A. Symon, T.C. Waddington (J. Chem. Soc. Dalton Trans. **1974** 78/81). − [235] W. Kanellakopulos-Drossopulos, D.R. Wiles (Can. J. Chem. **52** [1974] 894/9).

[236] H. Brunner (Ann. N.Y. Acad. Sci. **239** [1974] 213/24). − [237] D.F. Gaines, S.J. Hildebrandt (J. Am. Chem. Soc. **96** [1974] 5574/6). − [238] O.M. Abu Salah, M.I. Bruce (J. Chem. Soc. Dalton Trans. **1974** 2302/4). − [239] R. Clark, J. Howard, P. Woodward (J. Chem. Soc. Dalton Trans. **1974** 2027/9). − [240] Y. Matsumara, M. Harakawa, R. Okawara (J. Organometal. Chem. **71** [1974] 403/6).

[241] W.R. Cullen, D.J. Patmore, J.R. Sams, J.C. Scott (Inorg. Chem. **13** [1974] 649/55). − [242] L.J. Dizikes, A. Wojcicki (J. Am. Chem. Soc. **97** [1975] 2540/2). − [243] T.C. Flood, F.J. DiSanti (J. Chem. Soc. Chem. Commun. **1975** 18/9). − [244] A. Wojcicki, L.J. Dizikes (7th Intern. Conf. Organometal. Chem., Venice, Italy, 1975, Abstr. S 1 C). − [245] N.H. Tennent, S.R. Su, C.A. Poffenberger, A. Wojcicki (J. Organometal. Chem. **102** [1975] C46/C48).

[246] W. Malisch, P. Panster (Chem. Ber. **108** [1975] 2554/73). − [247] J.R. Chipperfield, J. Ford, D.E. Webster (J. Chem. Soc. Dalton Trans. **1975** 2042/7). − [248] D.L. Reger (Inorg. Chem. **14** [1975] 660/4). − [249] H. Brunner, T. Burgemei-

ster, J. Wachter (Chem. Ber. **108** [1975] 3349/54). − [250] A.N. Nesmeyanov, Yu.A. Belousov, V.N. Babin, E.B. Zavelovich, N.S. Kochetkova (Tezisy Dokl. 12th Vses. Chugaevskoe Soveshch. Khim. Kompleksn. Soedin., Novosibirsk 1975, Vol. 3, pp. 476/7; C.A. **86** [1977] No. 5590).

[251] R.B. King, J. Gimeno (Inorg. Chem. **17** [1978] 2396/400). − [252] K.M. Nicholas, A.M. Rosan (J. Organometal. Chem. **84** [1975] 351/6). − [253] P. Legzdins, J.T. Malito (Inorg. Chem. **14** [1975] 1875/8). − [254] J.R. Dickinson, R.V. Parish, P.J. Rowbotham, A.R. Manning, P. Hackett (J. Chem. Soc. Dalton Trans. **1975** 424/8). − [255] K. Stanley, M.C. Bird (J. Am. Chem. Soc. **97** [1975] 4292/8).

[256] D.A. Symon, T.C. Waddington (J. Chem. Soc. Dalton Trans. **1975** 2140/3). − [257] E. Amberger (unpublished results from N.S. Vyazankin, G.A. Razuvaev, O.A. Kruglaya, Organometal. React. **5** [1975] 279). − [258] M. Cooke, C.R. Russ, F.G.A. Stone (J. Chem. Soc. Dalton Trans. **1975** 256/9). − [259] D. Sellmann, E. Kleinschmidt (Angew. Chem. **87** [1975] 595/6). − [260] A.N. Nesmeyanov, N.E. Kolobova, L.V. Goncharenko, K.N. Anisimov (Izv. Akad. Nauk SSSR Ser. Khim. **1976** 153/9; Bull. Acad. Sci. USSR Div. Chem. Sci. **1976** 142/6).

[261] A.N. Nesmeyanov, E.G. Perevalova, L.I. Leont'eva, O.V. Grigor'eva (Izv. Akad. Nauk SSSR Ser. Khim. **1976** 1171/3; Bull. Acad. Sci. USSR Div. Chem. Sci. **1976** 1140/2). − [262] D.A. Slack, M.C. Baird (J. Am. Chem. Soc. **98** [1976] 5539/46). − [263] N.H. Tennent, S.R. Su, C.A. Poffenberger, A. Wojcicki (Proc. 17th Intern. Conf. Coord. Chem., Hamburg, 1976, p. 130). − [264] J.R. Chipperfield, J. Ford, A.C. Hayter, D.E. Webster (J. Chem. Soc. Dalton Trans. **1976** 360/6). − [265] A.N. Nesmeyanov, L.S. Isaeva, L.N. Lorens (Dokl. Akad. Nauk SSSR **229** [1976] 634/6; Dokl. Chem. Proc. Acad. Sci. USSR **226/231** [1976] 498/500).

[266] G.E. Graves (Diss. Memphis State Univ. 1975; Diss. Abstr. Intern. B **36** [1976] 3946). − [267] D.W. McKennon (Diss. Memphis State Univ. 1975; Diss. Abstr. Intern. B **36** [1976] 3908). − [268] R. Froböse, R. Mews, O. Glemser (Z. Naturforsch. **31 b** [1976] 1497/1500). − [269] A. Cutler, D. Ehntholt, W.P. Giering, P. Lennon, S. Raghu, A. Rosan, M. Rosenblum, J. Tancrede, D. Wells (J. Am. Chem. Soc. **98** [1976] 3495/507). − [270] W.E. Stanclift, D.G. Hendricker (J. Organometal. Chem. **107** [1976] 341/9).

[271] M. Pankowski, M. Bigorgne, Y. Chauvin (J. Organometal. Chem. **110** [1976] 331/8). − [272] A. Klopsch, E. Hellner, K. Dehnicke (Ber. Bunsenges. Physik. Chem. **80** [1976] 500/3). − [273] D.L. Lichtenberger, R.F. Fenske (J. Am. Chem. Soc. **98** [1976] 50/63). − [274] G. Innorta, A. Foffani, S. Torroni (Inorg. Chim. Acta **19** [1976] 263/6). − [275] C. Giannotti, G. Merle (J. Organometal. Chem. **105** [1976] 97/100).

[276] W. Kanellakopulos-Drossopulos, D.R. Wiles (J. Inorg. Nucl. Chem. **38** [1976] 947/50). − [277] J.D. Cotton, P.J. Davidson, M.F. Lappert (J. Chem. Soc. Dalton Trans. **1976** 2275/86). − [278] F.J. Hamm, F. Rackl (Ger. Offen. 2627157 [1976/78]). − [279] S.C. Tripathi, S.C. Srivastava, V.N. Pandey (Transition Metal Chem. [Weinheim] **1** [1976] 58/60). − [280] S.C. Tripathi, S.C. Srivastava, V.N. Pandey (Transition Metal Chem. [Weinheim] **1** [1976] 266/8).

[281] A.N. Nesmeyanov, E.G. Perevalova, L.I. Leont'eva, S.A. Eremin, E.A. Zhdanova (Izv. Akad. Nauk SSSR Ser. Khim. **1977** 2557/61; Bull. Acad. Sci. USSR Div. Chem. Sci. **1977** 2368/72). − [282] A.N. Nesmeyanov, E.G. Perevalova, L.I. Leont'eva, E.V. Shumilina (Izv. Akad. Nauk SSSR Ser. Khim. **1977** 1142/6; Bull. Acad.

Sci. USSR Div. Chem. Sci. **1977** 1048/52). – [283] E.V. Shumilina (Vestn. Mosk. Univ. Ser. II Khim. **18** [1977] 476/8; Moscow Univ. Chem. Bull. **32** [1977] 76/7). – [284] L.J. Dizikes, A. Wojcicki (J. Organometal. Chem. **137** [1977] 79/80). – [285] G. Cerveau, E. Colomer, R. Corriu, W.E. Douglas (J. Organometal. Chem. **135** [1977] 373/86).

[286] A.N. Nesmeyanov, E.G. Perevalova, L.I. Leont'eva (Izv. Akad. Nauk SSSR Ser. Khim. **1977** 2582/4; Bull. Acad. Sci. USSR Div. Chem. Sci. **1977** 2391/3). – [287] S.I. Kuznetsov, E.V. Bryukhova, N.P. Avakyan, I.I. Kritskaya (Teor. Eksperim. Khim. **13** [1977] 83/5; Theor. Exptl. Chem. [USSR] **13** [1977] 61/2). – [288] D.A. Brown, S.K. Chawla (Inorg. Chim. Acta **24** [1977] 71/6). – [289] D.M. Allen, A. Cox, T.J. Kemp, Q. Sultana (Inorg. Chim. Acta **21** [1977] 191/4). – [290] M.B. Freeman, L.G. Sneddon, J.C. Huffman (J. Am. Chem. Soc. **99** [1977] 5194/6).

[291] L.I. Zakharkin, A.I. Kovredov, M.G. Meiramov, A.V. Kazantsev (Izv. Akad. Nauk SSSR Ser. Khim. **1977** 1673/5; Bull. Acad. Sci. USSR Div. Chem. Sci. **1977** 1544/5). – [292] V.N. Pandey (Inorg. Chim. Acta **23** [1977] L 26). – [293] B.W.S. Kolthammer, P. Legzdins (8th Intern. Conf. Organometal. Chem., Kyoto, Japan, 1977, Abstr. 1 A 33). – [294] A.N. Nesmeyanov, L.G. Makarova, V.N. Vinogradova, N.A. Ustynyuk (Koord. Khim. **3** [1977] 62/6; Soviet J. Coord. Chem. **3** [1977] 47/50). – [295] V.N. Pandey (Inorg. Chim. Acta **22** [1977] L39/L41).

[296] D.L. Reger, C. Coleman (J. Organometal. Chem. **131** [1977] 153/62). – [297] K.R.M. Springstein, D.L. Greene, B.J. McCormick (Inorg. Chim. Acta **23** [1977] 13/8). – [298] V.N. Pandey (Transition Metal Chem. [Weinheim] **2** [1977] 48/50). – [299] H. Brunner, J. Wachter (J. Organometal. Chem. **142** [1977] 133/7). – [300] A.N. Nesmeyanov, Yu.A. Belousov, V.N. Babin, G.G. Aleksandrov, Yu.T. Struchkov, N.S. Kochetkova (Inorg. Chim. Acta **23** [1977] 155/62).

[301] A.N. Nesmeyanov, Yu.A. Belousov, V.N. Babin, N.S. Kochetkova, S.Yu. Sil'-vestrova, E.I. Mysov (Inorg. Chim. Acta **23** [1977] 173/9). – [302] H. Schäfer (8th Intern. Conf. Organometal. Chem., Kyoto, Japan, 1977, Abstr. 2A 14). – [303] R.B. King, M.G. Newton, J. Gimeno, M. Chang, K.N. Chen (8th Intern. Conf. Organometal. Chem., Kyoto, Japan, 1977, Abstr. 1 A 35). – [304] D. Sellmann, E. Kleinschmidt (J. Organometal. Chem. **140** [1977] 211/9). – [305] D. Sellmann, K. Jödden, E. Kleinschmidt (8th Intern. Conf. Organometal. Chem., Kyoto, Japan, 1977, Abstr. 3A 24).

[306] A.L. du Preez, I.L. Marais, R.J. Haines, A. Pidcock, M. Safari (J. Organometal. Chem. **141** [1977] C10/C12). – [307] V.N. Pandey (Inorg. Chim. Acta **25** [1977] L37/L38). – [308] V.N. Pandey (8th Intern. Conf. Organometal. Chem., Kyoto, Japan, 1977, Abstr. 2A 08). – [309] R.B. King, M.G. Newton, J. Gimeno, M. Chang (Inorg. Chim. Acta **23** [1977] L35/L36). – [310] A.N. Nesmeyanov, E.G. Perevalova, L.I. Leont'eva, S.A. Eremin (Izv. Akad. Nauk SSSR Ser. Khim. **1978** 2121/4; Bull. Acad. Sci. USSR Div. Chem. Sci. **1978** 1873/6).

[311] L.S. Chen, S.R. Su, A. Wojcicki (Inorg. Chim. Acta **27** [1978] 79/89). – [312] W. Rogers, J.A. Page, M.C. Baird (J. Organometal. Chem. **156** [1978] C37/C42). – [313] P. Reich-Rohrwig, A.C. Clark, R.L. Downs, A. Wojcicki (J. Organometal. Chem. **145** [1978] 57/68). – [314] C.S. Cundy, M.F. Lappert, C.K. Yuen (J. Chem. Soc. Dalton Trans. **1978** 427/33). – [315] H.B. Abrahamson, M.S. Wrighton (Inorg. Chem. **17** [1978] 1003/8).

[316] H. Brunner, M. Muschiol, W. Nowak (Z. Naturforsch. **33b** [1978] 407/11). – [317] G.J. Long, D.G. Alway, K.W. Barnett (Inorg. Chem. **17** [1978]

486/9). − [318] D.G. Alway, K.W. Barnett (Inorg. Chem. **17** [1978] 2826/31). − [319] Z. Yoshida, Mitsubishi Chemical Industries Co., Ltd. (Japan. Kokai 78-25543 [1976/78]; C.A. **89** [1978] No. 43790). − [320] N.N. Greenwood, J.D. Kennedy, C.G. Savory, J. Staves, K.R. Trigwell (J. Chem. Soc. Dalton Trans. **1978** 237/44).

[321] D.F. Gaines, S.J. Hildebrandt (Inorg. Chem. **17** [1978] 794/809). − [322] R. Gompper, E. Bartmann (Angew. Chem. **90** [1978] 490/1). − [323] N.T. Allison, Y. Kawada, W.M. Jones (J. Am. Chem. Soc. **100** [1978] 5224/6). − [324] B.W.S. Kolthammer, P. Legzdins (J. Chem. Soc. Dalton Trans. **1978** 31/5). − [325] M.B. Moronski (Diss. Southern Illinois Univ. 1977; Diss. Abstr. Intern. B **39** [1978] 754/5).

[326] D. Sellmann, E. Unger (Z. Naturforsch. **33b** [1978] 1438/42). − [327] I.P. Lorenz (Angew. Chem. **90** [1978] 60/1). − [328] W.A. Herrmann, M. Huber (Chem. Ber. **111** [1978] 3124/35). − [329] D. Dong, D.A. Slack, M.C. Baird (Inorg. Chem. **18** [1979] 188/91). − [330] W.N. Rogers, J.A. Page, M.C. Baird (Inorg. Chim. Acta **37** [1979] L539/L540).

[331] R. Gompper, E. Bartmann, H. Nöth (Chem. Ber. **112** [1979] 218/33). − [332] B.W.S. Kolthammer (Diss. Univ. Brit. Columbia 1979; Diss. Abstr. Intern. B **40** [1979] 1709). − [333] D.J. Thornhill, A.R. Manning (Inorg. Chim. Acta **33** [1979] 45/9). − [334] W.N. Rogers, M.C. Baird (J. Organometal. Chem. **182** [1979] C65/C68). − [335] G.E. Ryschkewitsch, M.A. Mathur (J. Inorg. Nucl. Chem. **41** [1979] 1563/4).

[336] D.L. Reger, C.J. Coleman, P.J. McElligott (J. Organometal. Chem. **171** [1979] 73/84). − [337] G. Thiollet, F. Mathey, R. Poilblanc (Inorg. Chim. Acta **32** [1979] L67/L68). − [338] J.P. Barbier, P. Dapporto, L. Sacconi, P. Stoppioni (J. Organometal. Chem. **171** [1979] 185/93). − [339] A.G. Osborne, R.H. Whiteley (J. Organometal. Chem. **181** [1979] 425/37). − [340] P.M. Treichel, D.C. Molzahn (Syn. Reactiv. Inorg. Metal-Org. Chem. **9** [1979] 21/9).

[341] R.J. Angelici, P.A. Christian, B.D. Dombek, G.A. Pfeffer (J. Organometal. Chem. **67** [1974] 287/94). − [342] C.V. Magatti, W.P. Giering (J. Organometal. Chem. **73** [1974] 85/92). − [343] A.N. Nesmeyanov, M.I. Rybinskaya, V.S. Kaganovich, T.V. Popova, E.A. Petrovskaya (Izv. Akad. Nauk SSSR Ser. Khim. **1973** 2087/9; Bull. Acad. Sci. USSR Div. Chem. Sci. **1973** 2031/3). − [344] A.N. Nesmeyanov, M.I. Rybinskaya, E.A. Petrovskaya, T.V. Popova (Izv. Akad. Nauk SSSR Ser. Khim. **1972** 1646/7; Bull. Acad. Sci. USSR Div. Chem. Sci. **1972** 1588/9). − [345] A.N. Nesmeyanov, M.I. Rybinskaya, L.V. Rybin, E.A. Petrovskaya, V.A. Svoren (Izv. Akad. Nauk SSSR Ser. Khim. **1976** 1592/6; Bull. Acad. Sci. USSR Div. Chem. Sci. **1976** 1511/4).

[346] B. Meunier (Diss. Univ. Paris-Sud 1977, No. 1881). − [347] P.M. Maitlis, M.L. Games, A. Efraty (Proc. 8th Intern. Conf. Coord. Chem., Vienna 1964, pp. 218/21). − [348] J.P. Hickey, J.R. Wilkinson, L.J. Todd (J. Organometal. Chem. **179** [1979] 159/68). − [349] J.P. Stenson (Diss. Univ. Wisconsin 1970).

1.5.2.3.3 $C_5H_5Fe(CO)_2X$ Compounds with X = CN, NCO, NCS, SCN, SeCN, and N_3

The compounds listed in Table 31 are prepared by the following methods:

Method I: Oxidation of Fp_2 with air in absolute C_2H_5OH in the presence of 65% HPF_6 and excess KSCN at 25 °C for 1 h gives FpNCS and FpSCN, which are separated by chromatography [8].

Method II: Oxidation of Fp_2 with oxygen in acetone in the presence of $\sim 50\%$ aqueous HBF_4, followed by reaction with aqueous solutions of NaCN, KNCO, or NaN_3 [25].

Method III: Formation of $[FpOC(CH_3)_2]^+$ from Fp_2 and anhydrous $Fe(ClO_4)_3$ in acetone and further reaction of the cation with an aqueous solution of NH_4SCN [18].

Method IV: Reaction of Fp_2 in acetone with AgNCO at room temperature or with AgSCN under reflux for 24 h. As in Method III, the cation $[FpOC(CH_3)_2]^+$ is formed as an intermediate [30].

Method V: From FpCl by treatment with NaCN in refluxing CH_3OH for 24 h [1, 4, 5] or at room temperature for 1.5 h [26], with KSeCN in refluxing acetone for 1 h [7], or with NaN_3 in 10% aqueous acetone at 56 °C for 3 h [33].

Method VI: Exchange of ethylene in $[FpCH_2=CH_2]BF_4$ by treatment with a tenfold excess of KCN or KNCO in acetone at room temperature for 10 h [12].

             In a similar manner FpN_3 was prepared from various other olefin complexes, $[Fp^2L]^+$, in acetone and aqueous solutions of NaN_3 at 0 °C for 1 h, best yield with $^2L=$ cycloheptene [33].

Method VII: A mixture of $[FpCO]PF_6$ and NaCN, KNCS, or KNCO in acetone stirred at room temperature for 2 to 3 h yields FpCN, FpNCS, and FpNCO, respectively [9].

Method VIII: An acetone solution of $[FpCS]PF_6$ and KNCO or KNCS is stirred at room temperature for 1 h. FpCN is obtained according to $[FpCS]^+ + NCO^-$ (or NCS^-) $\rightarrow$ FpCN + COS (or CS_2) [14, 15].

Method IX: Reaction of $[FpCO]PF_6$ or $[FpCS]PF_6$ with NH_2NH_2 in CH_2Cl_2 at room temperature for 1.5 h leads to FpNCO [9] or FpNCS [14, 15]. Substituted hydrazines give lower yields or no reaction [9].

Method X: An aqueous solution of NaN_3 is added to $[FpCO]PF_6$ or $[FpCS]PF_6$ in acetone and stirred at room temperature for 15 min. The acetone is removed, H_2O added, and the products FpNCO [9] or FpNCS [14, 15] extracted with $CHCl_3$.

Method XI: Formation of adducts $FpX(MX'_m)_n$ from FpX and MX'_m [26, 27, 29, 40, 42].

Mössbauer spectra of the complexes show well-resolved quadrupole doublets. The isomer shifts, which decrease in the order $NCO > NCS > SCN \sim SeCN > CN$, are discussed in terms of σ- and π-bonding contributions of the ligand X. They are used for calculation of the partial isomer shift parameters of ligands like CN^-, NCO^-, NCS^-, and C_5H_5 [16]. The difficulty in quantifying the relationship between the isomer shifts and the carbonyl stretching force constants is discussed [37], also see [16]. The IR spectra in the $\nu(CO)$ region are characterized by two intense absorptions resulting from symmetric and asymmetric stretching modes, the former lying at higher frequencies.

The FpCN adducts, Nos. 7 to 16, exhibit higher $\nu(CN)$ and $\nu(CO)$ frequencies than the parent compound FpCN. For reasons, which are discussed in detail, $\nu_{as}(CO)$ is the most suitable measure of the acceptor strengths of the Lewis acids [26, 32].

References on p. 311

Explanations for Table 31: Mössbauer isomer shifts δ are referred to Fe at room temperature. For the adducts No. 7 to 19, the solvents and temperatures in Method XI are given in parentheses. K is the stepwise dissociation constant of the FpCN adducts. For the fod ligand in Nos. 17 to 19, see Formula II on p. 104.

Table 31
Pseudo Halides of the $C_5H_5Fe(CO)_2X$ Type and FpCN Adducts.
Further information on numbers preceded by an asterisk is given at the end of the table.
For abbreviations and dimensions, see p. 375.

No.	compound method of preparation (yield in %)	properties and remarks explanations see above	Ref.
*1	FpCN II (53) V (50) VI (65) VII (70) VIII (20)	yellow acicular crystals, dec. at ~120° ^1H NMR: 5.20 in $CDCl_3$, 5.11 in CH_2Cl_2 ^{13}C NMR $(CHCl_3)$: 85.9 (C_5H_5), 154.7 (CN), 211.1 (CO) ^{57}Fe-γ: δ=0.0696(3), Δ=1.899(3) at 78 K, δ=−0.0097(4), Δ=1.89(2) at ~293 K IR: 2018.6, 2062.2 (CO), and 2120.8 (CN) in $CHCl_3$, 2006.9, 2054.7, 2121.0 in $C_6H_5NO_2$, 2000, 2058, 2117.4 in Nujol	[1, 5, 6, 9, 12, 14 to 16, 20, 21, 23 to 26, 28, 31, 37, 40]
*2	FpNCO II (32) IV (70) VI (72) VII (80) IX (55) X (75)	red, m.p. 98 to 99° (dec.) ^1H NMR $(CDCl_3)$: 5.09 ^{57}Fe-γ: δ=0.2113(9), Δ=1.880(7) at 78 K, δ=0.1258(6), Δ=1.84(1) at ~293 K IR $(CHCl_3)$: 2013, 2060 (CO), v_s(NCO) 1330, v_{as}(NCO) 2235, δ(NCO) 635, δ(FeCO) 525, 570, 585, 597, v(FeC) 475	[9, 12, 16, 22, 25, 30, 37]
*3	FpNCS I (18.5) III VII (45) IX (35) X (55)	golden yellow, m.p. 119 to 121° (dec.) ^1H NMR $(CDCl_3)$: 5.14 ^{57}Fe-γ: δ=0.2015(8), Δ=1.878(7) at 78 K, δ=0.1227(6), Δ=1.87(1) at ~293 K IR: 2025, 2070 (CO), 2120 (CN) in $CHCl_3$, 2008, 2066 (CO), 2123 (CN), 826 (CS) in KBr UV (THF): λ_{max}(ε)=270 (4650, sh), 340 (1590), 418 (795)	[8, 9, 14 to 16, 18, 36 to 38]
*4	FpSCN I (39) III IV (50)	dark red, m.p. 36 to 40° ^1H NMR $(CDCl_3)$: 5.06 ^{57}Fe-γ: δ=0.1858(7), Δ=1.811(8) at 78 K, δ=0.1138(6), Δ=1.80(1) at ~293 K IR: 2010, 2050 (CO), 2115 (CN) in $CHCl_3$, 698 (CS) in Nujol UV (THF): λ_{max}(ε)=278 (7420), 345 (1604), 525 (1094)	[8, 18, 30, 36 to 38]

References on p. 311

Table 31 [continued]

No.	compound method of preparation (yield in %)	properties and remarks explanations on p. 305	Ref.
*5	FpSeCN V (82)	dark brown ^{57}Fe-γ: $\delta = 0.1867(9)$, $\Delta = 1.75(1)$ at 78 K, $\delta = 0.1057(5)$, $\Delta = 1.72(1)$ at ~ 293 K IR (CHCl$_3$): 2004, 2056 (CO), 2126 (CN)	[7, 37]
*6	FpN$_3$ II (43) V (18) VI (up to 82)	dark purple, m.p. 72.5 to 73° (dec.) ^{1}H NMR: 5.10 (CHCl$_3$), 5.30 (CD$_3$COCD$_3$) IR: 2005, 2060 (CO), ν_{as}(N$_3$) 2010 in CHCl$_3$, 1995, 2050 (CO), ν_{as}(N$_3$) 2010 in CS$_2$, 1985, 2055 (CO), ν_s(N$_3$) 1276, ν_{as}(N$_3$) 2000 in KBr	[25, 33]

adducts of FpCN

No.	compound method of preparation (yield in %)	properties and remarks explanations on p. 305	Ref.
7	FpCNBH$_3$ XI (with B$_2$H$_6$ in C$_6$H$_5$Cl or CH$_2$Cl$_2$)	pale yellow to brown, m.p. 95° (broad range), dec. p. 130° IR: 2016.9, 2070 (CO), 2194.2 (CN) in Nujol, 2010, 2065 (CO), 2195 (CN), 2280 to 2380 (BH) in KBr more stable towards hydrolysis and air than Nos. 8 to 10	[26, 40]
8	FpCNBF$_3$ XI (in CH$_2$Cl$_2$, 0°)	pale yellow IR: ~ 2025, 2076.0 (CO), 2195.9 (CN) in Nujol, 2030.8, 2073.0 (CO), 2201.0 (CN) in C$_6$H$_5$NO$_2$ sensitive to hydrolysis and air	[26]
9	FpCNBCl$_3$ XI (in C$_6$H$_5$Cl, 0°)	off white IR: 2023.7, 2071.2 (CO), 2205.4 (CN) in Nujol, 2037.4, 2076.5 (CO), 2191.4 (CN) in C$_6$H$_5$NO$_2$ more stable towards hydrolysis and air than No. 8	[26]
10	FpCNBBr$_3$ XI (in C$_6$H$_6$, 0°)	pale yellow IR: 2026.6, 2071.0 (CO), 2190.5 (CN) in Nujol, 2038.9, 2077.0 (CO), 2182.5 (CN) in C$_6$H$_5$NO$_2$ briefly stable in air	[26]
11	FpCNB(CH$_3$)$_3$ XI (in xylene, 0 to 20°)	IR (C$_6$H$_5$CH$_3$): 2023.1, 2066.1 (CO), 2194.5 (CN) dissociates at room temperature under vacuum	[26]
*12	FpCNB(C$_6$H$_5$)$_3$ see further information	yellow ^{1}H NMR (CD$_3$COCD$_3$): 5.59 (C$_5$H$_5$) IR (CH$_2$Cl$_2$): 2039, 2079 (CO), 2203 (CN)	[42]

References on p. 311

Table 31 [continued]

No.	compound method of prepa- ration (yield in %)	properties and remarks explanations on p. 305	Ref.
13	FpCNAlCl$_3$ XI (in C$_6$H$_5$NO$_2$)	IR: 2014.5, 2063.7 (CO), 2172.1 (CN) in Nujol, 2025.9, 2068.3 (CO), 2168.1 (CN) in C$_6$H$_5$NO$_2$ not isolated, existence indicated by ^{1}H NMR spectrum	[26]
14	FpCNAl(CH$_3$)$_3$ XI (in xylene, 0°)	yellow brown IR: 2032.3, 2068.9 (CO), 2164.9 (CN) in Nujol, 2027.8, 2069.1 (CO), 2166.2 (CN) in C$_6$H$_5$CH$_3$	[26]
15	FpCNGaCl$_3$ XI (in ether, 0°)	light yellow IR: 2034.0, 2072.4 (CO), 2151.7 (CN) in Nujol, 2035.1, 2073.2 (CO), 2152.1 (CN) in C$_6$H$_5$NO$_2$	[26]
16	FpCNGa(CH$_3$)$_3$ XI (in C$_6$H$_6$, 10°)	light yellow IR: 2012.2, ~2058 (CO), 2155.8 (CN) in Nujol, 2022.0, 2064.7 (CO), 2157.3 (CN) in C$_6$H$_5$CH$_3$	[26]
17	FpCNLn(fod)$_3$ (Ln = Pr, Eu, Ho, Yb) XI (in C$_6$H$_6$)	not isolated, K > 10^3 mol^{-1} in C$_6$H$_6$ at 37°	[27, 29]
18	FpCN(Ln(fod)$_3$)$_2$ (Ln = Pr, Eu, Ho, Yb) XI (in C$_6$H$_6$)	not isolated, K > 10^3 (Pr), K = 290 ± 145 (Eu), 80.7 ± 40 (Ho), 39.2 ± 20 (Yb) mol^{-1} in C$_6$H$_6$ at 37°	[27, 29]
19	FpCNPr(fod)$_3$ XI (in C$_6$H$_6$)	not isolated, K = 25.7 ± 13 mol^{-1} in C$_6$H$_6$ at 37°	[27, 29]

*Further information:

C$_5$H$_5$Fe(CO)$_2$CN (Table **31**, No. 1) is also obtained by stirring a CH$_2$Cl$_2$ solution of Fp$_2$ and (CN)$_2$C=C(CN)$_2$ at 25 °C for 3 h (70% yield). When this reaction is carried out in C$_6$H$_6$ at 25 °C for 48 h, there is considerable decomposition and only a trace of FpCN is obtained [35]. FpCN is formed in 7% yield by UV irradiation of Fp$_2$ and tetracyano-ethylene oxide in THF for 2 h and subsequent refluxing for 10 h [28]. Reactions of FpR (R = alkyl) with (CN)$_2$C=C(CN)$_2$ in CH$_2$Cl$_2$ at room temperature yield in addition to FpC(CN)$_2$C(CN)$_2$R and FpN=C=C(CN)C(CN)$_2$R up to 15% FpCN, and 5 to 10% have been isolated upon workup of solutions of the main products stored for 1 to 2 d [35]. Reactions of FpCl with KSeCN in refluxing acetone or of FpCH$_2$C$_6$H$_5$ with Se(SeCN)$_2$ in C$_6$H$_6$ at 75 °C give FpCN in 2.1 and 1.8% yield, respectively [7].

References on p. 311

FpCN was isolated as dark brown crystalline solid, m.p. 115 to 118 °C, from the reaction of FpI with AgCN (1:1.25 mole) in refluxing $CHCl_3$ for 93 h (65% yield) [40]. In [15] FpCN is described as red. Rather different melting points are mentioned: 113 °C in [28], 127 to 130 °C in [21].

The integrated intensities of the $\nu(CO)$ vibrations show that the extent of Fe-CO π-bonding is approximately the same in FpCN as in FpCl, FpI, and $FpSnCl_3$ [21]. Dipole moment derivatives for the carbonyl stretching modes A' and A" were calculated in [21].

Stretching and interaction force constants (in $mdyn \cdot Å^{-1}$): $k(CO) = 16.77$, $k(CO,CO) = 0.37$ [23, 31, 37] according to the Cotton-Kraihanzel method, $k(CO)$ 16.88, $k(CO,CO) = 0.44$, $k(CN) = 17.01$, and $k(CN,CO) = 0.24$ from $\nu(^{13}CO)$ 1986.9 cm^{-1} [21] by the iterative method of [13].

The compound is soluble in $CHCl_3$, C_6H_6, and H_2O and is sparingly soluble in petroleum ether [1, 4, 5]. The polarographic reduction at 25 °C proceeds irreversibly in two steps, at $E_{1/2} = -1.45$ and -2.03 V in $CH_3CN/[N(C_2H_5)_4]ClO_4$ and at $E_{1/2} = -1.45$ and -2.07 V in dimethylformamide/$[N(C_2H_5)_4]ClO_4$ (vs. SCE) [6].

The ^{14}CO exchange is slower than in FpX (X = Cl, Br, I), in toluene at 31.8 °C. Because of some decomposition, only approximate apparent rate constants were obtained [3]. The compound reacts with freshly prepared $(CH_3)_3NBH_2I$ in benzene at 60 °C to give $[FpCNBH_2N(CH_3)_3]I$ [40]. A refluxed mixture of FpCN and KCN in ethanol/water yields $K[C_5H_5Fe(CO)(CN)_2]$ [34]. Treatment with excess $P(C_6H_5)_3$ in refluxing C_6H_6 affords $C_5H_5Fe(CO)(P(C_6H_5)_3)CN$ [10]. Reaction with $n\text{-}C_3H_7OH$ between 0 °C and room temperature using HF as a carrier and catalyst leads to $n\text{-}C_3H_7C_5H_4Fe(CO)_2CN$ [2].

Interaction with group III Lewis acids yields adducts of the type $FpCNMX_3$ (M = B, Al, Ga; X = H, halogen, CH_3) [26, 32]; see Nos. 7 to 16 in Table 31. The stoichiometry of interactions between the NMR shift reagents $Ln(fod)_3$ (Ln = Pr, Eu, Ho, and Yb; fod = 1,1,1,2,2,3,3-heptafluoro-7,7-dimethyl-4,6-octanedionate, Formula II on p. 104) and FpCN was studied by ^1H NMR shift and vapor-pressure osmometry. While ^1H NMR shifts might be interpreted in terms of simple 1:1 interaction, the osmometric measurements clearly reveal the formation of 1:2 and, in some cases, higher adducts, see Nos. 17 to 19 in Table 31. The discrepancy between these two techniques is discussed [27, 29]; see also [19]. The identity of the donor site, CN, has been verified by the characteristic increase in $\nu(CN)$ [19].

FpCN can be employed as a fuel additive. It improves the drying properties of paints, varnish, printing inks, and synthetic resins; it reduces the soot formation of candles. It may be used as antioxidant and catalyst [4].

$C_5H_5Fe(CO)_2NCO$ (Table 31, No. 2) is formed in a 37% yield by oxidation of $FpCONH_2$ with the stoichiometric amount of I_2 in C_6H_6: $2C_5H_5Fe(CO)_2CONH_2 + I_2 \rightarrow C_5H_5Fe(CO)_2NCO + [C_5H_5Fe(CO)_3]I + NH_4I$ [22].

Stretching and interaction force constants were calculated by the Cotton-Kraihanzel method: $k(CO) = 16.82$ and $k(CO,CO) = 0.40$ $mdyn \cdot Å^{-1}$ [37].

The compound is an air-stable solid, which is soluble in C_6H_6, ethyl ether, CH_2Cl_2, and $CHCl_3$ and only sparingly soluble in pentane and hexane. Its CH_2Cl_2 and $CHCl_3$ solutions decompose very slowly to give a green material. Fp_2 is found to be the main product of decomposition in H_2O, CH_3OH, and aqueous 7 M KOH solutions [9].

References on p. 311

The major fragments (relative intensities) observed in the mass spectrum (16 eV) are $[C_5H_5Fe(CO)_2NCO]^+$ (59), $[C_5H_5Fe(CO)NCO]^+$ (70), $[(C_5H_5)_2Fe]^+$ (25), and $[C_5H_5FeNCO]^+$ (100). The lack of ions containing CO but not containing NCO reveals that the NCO group is more strongly bonded to Fe than the CO ligands [9].

A 14% isocyanate exchange takes place when FpNCO is treated with $KN^{14}CO$ in THF for 3 h [17].

$C_5H_5Fe(CO)_2NCS$ and $C_5H_5Fe(CO)_2SCN$ (Table 31, Nos. 3 and 4). According to [11] a small amount of FpSCN may be obtained from the reaction of FpI with KSCN on prolonged refluxing (18 h, solvent not stated), but it is also reported [8] that reactions of FpX $(X=Cl, I)$ and SCN^- are either unsuccessful or impractical because of the difficulties encountered in a complete removal of the unreacted carbonyl from the product.

Comparison of the $v(CO)$ frequencies with those of FpX $(X=halogen)$ reveals an increase in the order $I < Br \sim SCN < Cl < NCS$. This sequence parallels a decrease in polarizability of the anions, which diminishes the electron density at the metal and thus reducing the extent of Fe carbonyl π bonding [8]. FpNCS and FpSCN may also be distinguished by their $v(CS)$ absorptions (Nujol), 830 and 698 cm^{-1}, respectively [8]. Approximate CO stretching and interaction force constants were calculated by the Cotton–Kraihanzel method: $k(CO)=16.90$ and $k(CO,CO)=0.37$ for FpNCS, $k(CO)=16.61$ and $k(CO,CO)=0.38$ mdyn·$Å^{-1}$ for FpSCN [37].

In [15] FpNCS is described as red although melting point and the $v(CS)$ absorption at 829 cm^{-1} are consistent with FpSCN. A melting point of 105 to 107 °C is reported for FpNCS in [9]. Both FpNCS and FpSCN are soluble in acetone, acetonitrile, chloroform, and methanol; moderately soluble in benzene and ether, and sparingly soluble in saturated hydrocarbons [8]. FpSCN does not undergo linkage isomerization in $CHCl_3$ at 36 °C, in $(CH_3OCH_2CH_2)_2O$ at 50 °C, or in refluxing CH_3CN. Only some decomposition, in the last case to ferrocene, has been noted. But when the solid is heated at ~40 °C it liquifies to an oil, with concomitant conversion to FpNCS. Mechanisms were proposed [8]. Both isomers interconvert upon 366- or 436-nm irradiation in THF. The photostationary state being reached after 2 h is approximately a 1:1 mixture [36, 38]. Photoisomerization proceeds via the isomerization $C_5H_5Fe(CO)SCN \rightleftharpoons C_5H_5Fe(CO)NCS$ and is completely or partly inhibited by donors ^2D like CO, $P(C_6H_5)_3$, and $As(C_6H_5)_3$ that react with these decarbonylation products to give $C_5H_5Fe(CO)(^2D)SCN$ or $C_5H_5Fe(CO)(^2D)NCS$, respectively. Attempted thermal substitution of FpNCS by $As(C_6H_5)_3$ has been unsuccessful [38].

FpNCS does not give an observable ^1H NMR shift of the C_5H_5 signal by reaction with $Eu(fod)_3$ (see No. 1), thus indicating a basicity lower than FpCN [19].

$C_5H_5Fe(CO)_2SeCN$ (Table 31, No. 5) was also prepared by stirring a benzene slurry of $FpCH_2C_6H_5$ and $Se(SeCN)_2$ at 75 °C for 30 min (67% yield) [7]. On the basis of its IR spectrum the compound has been assigned an Se-bonded structure [7]. Force constants (Cotton–Kraihanzel model) were calculated: $k(CO)=16.52$ and $k(CO,CO)=0.37$ mdyn·$Å^{-1}$ [37]. FpSeCN resists isomerization in the molten state around 80 °C and in a number of organic solvents [7]. It reacts with the ^2D ligand $P(C_6H_5)_3$ $(\sim1:1$ mole ratio) in benzene at 27 °C for 15 min to give $[C_5H_5Fe(CO)_2(^2D)]SeCN$ (82%) and $C_5H_5Fe(CO)(^2D)SeCN$ (3%), the relative amounts not being affected when a threefold excess of $P(C_6H_5)_3$ is employed. Under similar conditions there is no reaction with $P(OC_6H_5)_3$, and with $P(C_6H_{11}$-cyclo$)_3$ only the ionic product $[C_5H_5Fe(CO)_2(^2D)]SeCN$ could be isolated. Refluxing FpSeCN and

References on p. 311

P(C₆H₅)₃ (1:1 mole ratio) in benzene for 30 min forms C₅H₅Fe(CO)(²D)SeCN, FpCN, and C₅H₅Fe(CO)(²D)CN, while a twofold excess of P(C₆H₅)₃ yields only SeP(C₆H₅)₃ and C₅H₅Fe(CO)(²D)CN. Similar treatment with P(OC₆H₅)₃ affords C₅H₅Fe-(CO)(²D)SeCN and a trace of C₅H₅Fe(CO)(²D)CN. The yield of the latter increases when excess P(OC₆H₅)₃ is used and the refluxing prolonged [10].

C₅H₅Fe(CO)₂N₃ (Table **31**, No. **6**) is very resistent to both thermal and photo-chemical decomposition. It does not react in refluxing acetone in the presence of ethylene for 1 h. Heating at 100 °C for 4 h in toluene saturated with CO gives only Fp₂. Photolysis in benzene with λ=253.7 nm in the presence of CO for 5 h leads to FpNCO, FpC₅H₅-σ, and ferrocene whereas with λ=350 nm no FpNCO is formed. The mass spectrum provides some evidence for both the formation of a nitrene and competing initial loss of N₃. FpN₃ reacts with CH₃OOCC≡CCOOCH₃ in refluxing ben-zene to give the triazole complex I [33].

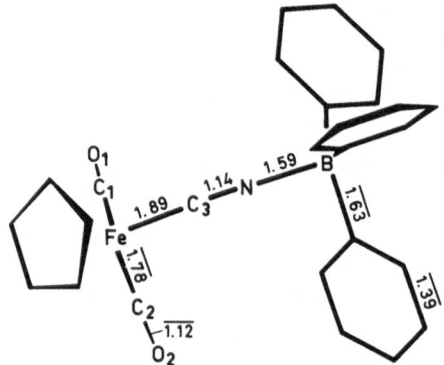

I

Reactions with the ²D ligands P(C₆H₅)₃, P(C₄H₉-n)₃, or P(OC₆H₅)₃ in CH₂Cl₂ at room temperature afford C₅H₅Fe(CO)(²D)NCO. A proposed mechanism involves attack of the displaced azide on a carbonyl C atom of the intermediate [C₅H₅Fe(CO)₂(²D)]⁺ cation and a subsequent Curtius rearrangement to form an isocyanate group [39].

C₅H₅Fe(CO)₂CNB(C₆H₅)₃ (Table **31**, No. **12**) was prepared from [FpOC(CH₃)₂]⁺ (obtained from Fp₂ and AgSbF₆ in acetone) and Na[B(C₆H₅)₃CN], 60% yield [42]. The compound crystallizes in the triclinic system with a=11.58, b=10.20, c=10.25 (±0.01) Å, α=111.7°, β=79.9°, and γ=97.4° (±0.1°); space group

Fig. 21

Molecular structure of C₅H₅Fe(CO)₂CNB(C₆H₅)₃ [41].

Bond angles (°):

Fe-C(3)-N	175±1	C(1)-Fe-C(3)	91
C(3)-N-B	169±1	C(2)-Fe-C(3)	92
C(1)-Fe-C(2)	95		

$P\bar{1} - C_i^1$. $Z=2$ gives $D_c=1.34$ and $D_m=1.34$ $g \cdot cm^{-3}$. A projection of the molecule down the z axis is shown in **Fig. 21**. The structure determination reveals that the bonding arrangement is Fe-C-N-B. The five C atoms of the C_5H_5 ring are coplanar within 0.005 Å, with the C-C bond lengths ranging from 1.39 to 1.41 Å. The Fe-C(C_5H_5) distances range from 2.07 to 2.11 Å. The CN group and the two CO groups are mutually perpendicular [41].

The electrical conductivity in 10^{-3} M acetone solution is 0.7 $\Omega^{-1} \cdot cm^2 \cdot mol^{-1}$ [42]. Unlike its Ru analogue the compound does not decompose to FpCN on heating in acetone [42].

References:

[1] T.S. Piper, F.A. Cotton, G. Wilkinson (J. Inorg. Nucl. Chem. **1** [1955] 165/74). – [2] J. Kozikowski, Ethyl Corp. (U.S. 2916503 [1959]). – [3] A. Wojcicki, F. Basolo (J. Am. Chem. Soc. **83** [1961] 525/8). – [4] J.E. Brown, E.G. De Witt, H. Shapiro (U.S. 3185718 [1957/65]). – [5] J.E. Brown, H. Shapiro, E.G. De Witt (Ger. Offen. 1217952 [1953/66]).

[6] A.N. Nesmeyanov, Yu.A. Chapovskii, L.I. Denisovich, B.V. Lokshin, I.V. Polovyanyuk (Dokl. Akad. Nauk SSSR **174** [1967] 1342/4; Dokl. Chem. Proc. Acad. Sci. USSR **172/177** [1967] 576/8). – [7] M.A. Jennings, A. Wojcicki (J. Organometal. Chem. **14** [1968] 231/4). – [8] T.E. Sloan, A. Wojcicki (Inorg. Chem. **7** [1968] 1268/73). – [9] R.J. Angelici, L. Busetto (J. Am. Chem. Soc. **91** [1969] 3197/200). – [10] M.A. Jennings, A. Wojcicki (Inorg. Chim. Acta **3** [1969] 335/40).

[11] P.M. Treichel, W.M. Douglas (J. Organometal. Chem. **19** [1969] 221/4). – [12] L. Busetto, A. Palazzi, R. Ros, U. Belluco (J. Organometal. Chem. **25** [1970] 207/11). – [13] D.J. Darensbourg (Inorg. Chim. Acta **4** [1970] 597/601). – [14] L. Busetto, M. Graziani, U. Belluco (Ger. Offen. 2116226 [1970/71]). – [15] L. Busetto, M. Graziani, U. Belluco (Inorg. Chem. **10** [1971] 78/80).

[16] K. Burger, L. Korecz, P. Mag, U. Belluco, L. Busetto (Inorg. Chim. Acta **5** [1971] 362/4). – [17] L.M. Charley, R.J. Angelici (Inorg. Chem. **10** [1971] 868/70). – [18] E.C. Johnson, T.J. Meyer, N. Winterton (Inorg. Chem. **10** [1971] 1673/5). – [19] T.J. Marks, J.S. Kristoff, A. Alich, D.F. Shriver (J. Organometal. Chem. **33** [1971] C35/C37). – [20] L.F. Farnell, E.W. Randall, E. Rosenberg (Chem. Commun. **1971** 1078/9).

[21] D.J. Darensbourg (Inorg. Chem. **11** [1972] 1606/9). – [22] J. Ellermann, H. Behrens, H. Krohberger (J. Organometal. Chem. **46** [1972] 119/38). – [23] O.A. Gansow, D.A. Schexnayder, B.Y. Kimura (J. Am. Chem. Soc. **94** [1972] 3406/8). – [24] A.N. Nesmeyanov, I.F. Leshcheva, I.V. Polovyanyuk, Yu.A. Ustynyuk, L.G. Makarova (J. Organometal. Chem. **37** [1972] 159/65). – [25] B.D. Dombek, R.J. Angelici (Inorg. Chim. Acta **7** [1973] 345/7).

[26] J.S. Kristoff, D.F. Shriver (Inorg. Chem. **12** [1973] 1788/93). – [27] T.J. Marks, R. Porter, D.F. Shriver (Proc. 10th Rare Earth Res. Conf., Carefree, Ariz., 1973, Vol. 1, pp. 372/9; CONF-730402-P1 [1973]). – [28] R.B. King, M.S. Saran (J. Am. Chem. Soc. **95** [1973] 1811/7). – [29] R. Porter, T.J. Marks, D.F. Shriver (J. Am. Chem. Soc. **95** [1973] 3548/52). – [30] W.E. Williams, F.J. Lalor (J. Chem. Soc. Dalton Trans. **1973** 1329/32).

[31] R.J. Angelici, P.A. Christian, B.D. Dombek, G.A. Pfeffer (J. Organometal. Chem. **67** [1974] 287/94). – [32] J.S. Kristoff (Diss. Northwestern Univ. 1973; Diss.

Abstr. Intern. B **34** [1974] 4272). − [33] A. Rosan, M. Rosenblum (J. Organometal. Chem. **80** [1974] 103/7). − [34] D.L. Reger (Inorg. Chem. **14** [1975] 660/4). − [35] S.R. Su, A. Wojcicki (Inorg. Chem. **14** [1975] 89/98).

[36] D.G. Alway, K.W. Barnett (J. Organometal. Chem. **99** [1975] C52/C54). − [37] G.J. Long, D.G. Alway, K.W. Barnett (Inorg. Chem. **17** [1978] 486/9). − [38] D.G. Alway, K.W. Barnett (Inorg. Chem. **17** [1978] 2826/31). − [39] D.A. Brown, F.M. Hussein, C.L. Arora (Inorg. Chim. Acta **29** [1978] L215/L216). − [40] G.E. Ryschkewitsch, M.A. Mathur (J. Inorg. Nucl. Chem. **41** [1979] 1563/4).

[41] M. Laing, G. Kruger, A.L. du Preez (J. Organometal. Chem. **82** [1974] C40/C42). − [42] R.J. Haines, A.L. du Preez (J. Organometal. Chem. **84** [1975] 357/67).

1.5.2.3.4 $C_5H_5Fe(CO)_2X$ Compounds with $X = NO_3$ and NO_2

The literature offers nothing about how NO_3 and NO_2 ligands are bound to the iron atom.

$C_5H_5Fe(CO)_2NO_3$ is prepared by oxidation of Fp_2 in acetone, with oxygen in the presence of $\sim 50\%$ aqueous HBF_4 (49%) [3], with $Fe(ClO_4)_3$ (22%) [1, 2], or with $AgNO_3$ (60%) [4]. Further treatment with an aqueous solution of KNO_3 or $NaNO_3$ is required in the first two cases. A method where FpI is reacted with $AgNO_3$ (1:1 mole ratio) in CH_2Cl_2 for 6 h to give a 64% yield is reported in [5].

The red-orange needles melt at 97 to 98 °C [5]. The ^1H NMR spectrum (CDCl$_3$) shows a C_5H_5 singlet at $\delta = 5.14$ ppm [3]. IR spectrum: the $\nu(CO)$ bands are at 2025, 2069 cm^{-1} in CHCl$_3$ [3] but at 2007, 2060 cm^{-1} in KBr [2].

$C_5H_5Fe(CO)_2NO_2$. $[Fp(OH_2)]^+$, which was prepared by oxidation of Fp_2 with oxygen in the presence of $\sim 50\%$ aqueous HBF_4 in acetone, is reacted with $NaNO_2$ to yield golden crystals of a product believed to be $FpNO_2$. It decomposes within a few hours at room temperature, even under vacuum. The ^1H NMR spectrum (CDCl$_3$?) shows a single resonance at $\delta = 5.17$ ppm. The IR spectrum (CHCl$_3$?) shows $\nu(CO)$ bands at 2023 and 2067 cm^{-1} [3].

References:

[1] E.C. Johnson, T.J. Meyer, N. Winterton (Chem. Commun. **1970** 934/5). − [2] E.C. Johnson, T.J. Meyer, N. Winterton (Inorg. Chem. **10** [1971] 1673/5). − [3] B.D. Dombek, R.J. Angelici (Inorg. Chim. Acta **7** [1973] 345/7). − [4] W.E. Williams, F.J. Lalor (J. Chem. Soc. Dalton Trans. **1973** 1329/32). − [5] D.L. Reger, C. Coleman (J. Organometal. Chem. **131** [1977] 153/62).

1.5.2.3.5 Compounds with Fe–O Bonds

1.5.2.3.5.1 Compounds of the $C_5H_5Fe(CO)_2OS(O)R$, $C_5H_5Fe(CO)_2OS(O)_2R$, and $C_5H_5Fe(CO)_2OS(O)_2OR$ Type

The **sulfinates FpOS(O)R** listed in Table 32 (Nos. 1 to 5) have been identified only as intermediates in the reactions of FpR (R = alkyl, aryl) with liquid SO_2. They are reasonably stable in the presence of SO_2, the stability depending on R, decreasing in the order $R = CH_3 > C_6H_4CH_3-4 > CH_2C_6H_5$. However, after 1 to 5 h in refluxing SO_2 or on attempted isolation the corresponding sulfones $FpS(O)_2R$ (see 1.5.2.3.6.5, p. 327) are formed. The formulation of Fe–O bonded sulfinates in preference to FpS(O)OR, ionic compounds, e.g., $[FpSO_2]RSO_2$, or the conversion to $FpS(O)_2R$

is supported by ¹H NMR and IR data, and electrical conductivity measurements as well as chemical evidence. Spectroscopic evidence has also been obtained for the presence of $FpOS(O)C(CH_3)_3$ (No. 2) in the reaction of $FpCH(CH_3)_2$ with SO_2 in organic solvents. Unlike most of the alkyls examined under comparable conditions, $FpCH(CH_3)_2$ undergoes insertion at a conveniently measurable rate. The stability of $FpOS(O)CH(CH_3)_2$ is strongly dependent on the solvent. Thus, in $(CH_3)_2CHOH$ it is still detectable after 2 h. However, in benzene only the CO stretching frequencies of $FpS(O)_2CH(CH_3)_2$ can be discerned. In contrast to the conversion product $FpS(O)_2R$, $FpOS(O)R$ reacts in refluxing SO_2 with KI or KBr to yield FpX (X=I, Br). The replacement of sulfinate with iodide can be also effected in CH_2Cl_2·or benzene [5, 8].

The reported $FpOS(O)C(R)=C=CH_2$ compounds $(R=CH_3, C_6H_5)$ [2] were later shown to be the sultines I [3, 4, 6, 7, 10].

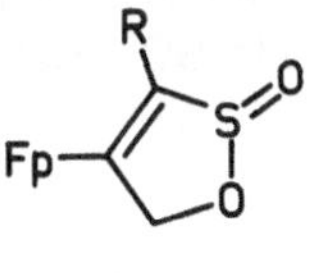

I

Explanations for Table 32: The chemical shifts given in [8] are accurate to ~0.03 ppm; the temperatures for the ¹H NMR, to ±7 °C.

Table 32
Compounds of the $C_5H_5Fe(CO)_2OS(O)R$, $C_5H_5Fe(CO)_2OS(O)_2R$, and $C_5H_5Fe(CO)_2OS(O)_2OR$ Type.
For abbreviations and dimensions, see p. 375.

No.	compound	properties and remarks explanations see above	Ref.
FpOS(O)R			
1	$FpOS(O)CH_3$	¹H NMR $(SO_2, -37°)$: 2.15 (s, CH_3), 5.25 (s, C_5H_5) IR $(SO_2, -60°, ε)$: 2012 (1740), 2061 (1830) (CO)	[8]
2	$FpOS(O)CH(CH_3)_2$	IR: 2016, 2069 (CO), ν(S–O) 828, ν(S=O) 1118 in $CHCl_3/SO_2$, 2020, 2065 (CO) in i-C_3H_7OH/SO_2	[8]
3	$FpOS(O)CH_2Si(CH_3)_3$	only reaction with I⁻ is mentioned	[8]
4	$FpOS(O)CH_2C_6H_5$	¹H NMR $(SO_2, -18°)$: 3.43, 3.59 (AB q, CH_2, $J_{AB}=12.6$), 5.09 (s, C_5H_5), 7.35 (m, C_6H_5) IR $(SO_2, -30°)$: 2012, 2062 (CO)	[5, 8]
5	$FpOS(O)C_6H_4CH_3$-4	IR $(SO_2, -30°)$: 2057, the other band is masked by unreacted alkyl	[8]

References on p. 314

Table 32 [continued]

No.	compound	properties and remarks explanations on p. 313	Ref.

FpOS(O)$_2$R

| 6 | FpOS(O)$_2$C$_6$H$_4$CH$_3$-4 | prepared from Fp$_2$ and AgOS(O)$_2$C$_6$H$_4$CH$_3$-4 in acetone at room temperature, contained paramagnetic impurities, m.p. 94° ^{1}H NMR (CD$_3$COCD$_3$): 2.35 (s, CH$_3$), 5.30 (s, C$_5$H$_5$), 7.3 (m, C$_6$H$_4$) IR (CHCl$_3$): 2028, 2076 (CO) | [9] |

FpOS(O)$_2$OR

| 7 | FpOS(O)$_2$OH | formed by treatment of FpX (X = Cl, CH$_3$) with 98% H$_2$SO$_4$, red | [1] |

References:

[1] A. Davison, W.M. McFarlane, L. Pratt, G. Wilkinson (J. Chem. Soc. **1962** 3653/66). – [2] J.L. Roustan, C. Charrier (Compt. Rend. C **268** [1969] 2113/6). – [3] M.R. Churchill, J. Wormald, D.A. Ross, J.E. Thomasson, A. Wojcicki (J. Am. Chem. Soc. **92** [1970] 1795/6). – [4] M.R. Churchill, J. Wormald (J. Am. Chem. Soc. **93** [1971] 354/9). – [5] S.E. Jacobson, P. Reich-Rohrwig, A. Wojcicki (Chem. Commun. **1971** 1526/7).

[6] D.A. Ross (Diss. Ohio State Univ. 1970; Diss. Abstr. Intern. B **31** [1971] 3905/6). – [7] J.E. Thomasson, P.W. Robinson, D.A. Ross, A. Wojcicki (Inorg. Chem. **10** [1971] 2130/7). – [8] S.E. Jacobson, P. Reich-Rohrwig, A. Wojcicki (Inorg. Chem. **12** [1973] 717/23). – [9] W.E. Williams, F.J. Lalor (J. Chem. Soc. Dalton Trans. **1973** 1329/32). – [10] L.S. Chen, S.R. Su, A. Wojcicki (Inorg. Chim. Acta **27** [1978] 79/89).

1.5.2.3.5.2   The Compound C$_5$H$_5$Fe(CO)$_2$OP(O)(OC$_6$H$_5$)$_2$

C$_5$H$_5$Fe(CO)$_2$OP(O)(OC$_6$H$_5$)$_2$ is prepared from Fp$_2$ and AgOP(O)(OC$_6$H$_5$)$_2$ in acetone at room temperature. It was unstable, could not be purified, and was identified only by its IR spectrum, ν(CO) 2010 and 2062 cm^{-1} in CHCl$_3$, W.E. Williams, F.J. Lalor (J. Chem. Soc. Dalton Trans. **1973** 1329/32).

1.5.2.3.5.3   C$_5$H$_5$Fe(CO)$_2$OOCR Compounds with R = H, Alkyl, and Aryl

The compounds summarized in Table 33 have been prepared by the following procedures:

Method I: Treatment of FpSi(CH$_3$)$_2$CH=CH$_2$ with RCOOH in CH$_2$Cl$_2$ at 25 °C for 2 to 4 d [6, 7].

Method II: Oxidation of Fp$_2$ with oxygen in acetone/aqueous HBF$_4$/HCOONa or CHCl$_3$/CCl$_3$COOH [4].

Method III: Oxidation of Fp$_2$ with AgOOCR in acetone at room temperature [5].

Method IV: Reaction of $FpCH_2C_6H_5$ with 95% CF_3COOH at $\sim 20\,°C$ for 15 min and subsequent addition to a saturated Na_2CO_3 solution [3].

Method V: Treatment of FpI with a 5% excess of AgOOCR in CH_2Cl_2 at room temperature. For No. 4, C_6H_6 may also be used as solvent, whereas no reaction occurs in coordinating solvents like THF [1].

Attempts to prepare $FpOOCC_2F_5$ from FpX (X = Cl or I) and $AgOOCC_2F_5$ (Method V) have led to brown uncrystallizable tars [1].

The perfluorocarboxylate complexes No. 4 and 5 are crystalline solids of solubility, volatility, and air–stability similar to those of FpI. The colors are lighter, owing to the lower polarizability of the perfluorocarboxylate groups [1].

Table 33
Compounds of the $C_5H_5Fe(CO)_2OOCR$ Type.
Further information on numbers preceded by an asterisk is given at the end of the table.
For abbreviations and dimensions, see p. 375.

No.	R in FpOOCR method of preparation (yield in %)	properties and remarks	Ref.
*1	H II (17.5)	red ^1H NMR $(CDCl_3)$: 5.10 (s, C_5H_5), 7.62 (s, CH) IR: 2008, 2054 (CO) in $CHCl_3$, 2016, 2055 (CO), 1293, 1620 (C=O) in C_6H_{14}	[4, 13, 14]
2	$CHCl_2$	only kinetics of FpR′ + $CHCl_2COOH$ → $FpOOCCHCl_2$ + R′H (R′ = alkyl, aryl)	[11]
3	CCl_3 I (55.7) II (49)	dark red, orange red, m.p. 92 to 94° ^1H NMR $(CDCl_3)$: 5.14 (s, C_5H_5) IR $(CHCl_3)$: 2012, 2055 or 2019, 2063 (CO)	[4, 6, 7]
*4	CF_3 I (71) III (87) IV (52) V (93)	red brown, m.p. 71 to 73°, subl. at 70 to 80°/0.1 mm ^1H NMR $(CDCl_3)$: 5.08 (s, C_5H_5) ^{19}F NMR $(CDCl_3)$: 74.2 (s, CF_3) IR: 1680 (C=O), 2019, 2059 (CO) in C_6H_{12}, 3120 (CH) in KBr	[1, 3, 5, 7 to 9]

5	$n\text{-}C_3F_7$
	V (60)
		red orange, m.p. 59 to 60°, subl. at 50°/0.1 mm
		^1H NMR $(CDCl_3)$: 5.02 (s, C_5H_5)
		^{19}F NMR $(CDCl_3)$: 116.7 (q, $\alpha\text{-}CF_2$), 127.7 ($\beta\text{-}CF_2$),
		   81.6 (t, $\gamma\text{-}CF_3$)
		IR: 1696 (C=O), 2020, 2071 (CO) in C_6H_{12},
		   3080 (CH) in KBr
			[1]

6	C_6H_5 III (42)	m.p. 97 to 98° ^1H NMR (CD_3COCD_3): 5.27 (s, C_5H_5), 7.0 to 8.0 (m, C_6H_5) IR $(CHCl_3)$: 1707 (C=O), 2017, 2059 (CO)	[5]

References on p. 317

*Further information:

C₅H₅Fe(CO)₂OOCH (Table **33**, No. **1**) was also obtained by stirring [C₅H₅-Fe(CO)₂(THF)]BF₄ with aqueous HCOONa at room temperature for 2.5 h. Water was removed and the residue extracted with CH₂Cl₂. Reduction of the volume, addition of hexane, and cooling to −10 °C afforded a red-brown product [13, 14].

A structure determination was carried out on a single crystal grown from CH₂Cl₂/hexane. The complex crystallizes in the monoclinic system with a=6.799(2), b=12.325(4), c=10.440(3) Å, and β=106.84(2)°; space group P2₁/n − C₂ₕ⁵. Z=4 gives D_c=1.76 g·cm⁻³. The structure consists of a discrete racemic mixture of molecules in the monoclinic unit cell. The formate ligand is monodentate and orientated such as to place its uncoordinated oxygen atom in the direction of the carbonyl ligands. The molecular structure is shown in **Fig. 22**; no unusually short intermolecular contacts are found [14].

Fig. 22

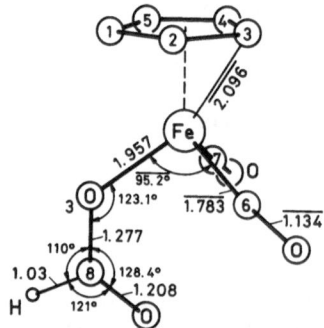

Molecular structure of C₅H₅Fe(CO)₂OOCH [14].

The complex undergoes extensive decomposition in the absence of CO. When placed in a ¹³CO-saturated heptane solution, a facil intermolecular CO exchange is noted as shown by IR spectral traces in the ν(CO) region. Upon replacement of the ¹³CO atmosphere with ¹²CO all original bands reappear with an attendant decrease in bands assignable to the ¹³CO-enriched species. After longer reaction periods at 50 °C bands appear according to decomposition to give C₅H₅Fe(CO)₂H, [C₅H₅Fe(CO)₂]₂, CO₂, and H₂. The CO ligand exchange is found to occur at a faster rate than decarboxylation which is proposed to proceed via a β elimination of a hydrogen atom to the metal center [13].

C₅H₅Fe(CO)₂OOCCF₃ (Table **33**, No. **4**) was also obtained by Method I using FpSiCl₂CH=CH₂ in CH₃OH (18.7% yield) [7]. Other cleavage reactions with CF₃COOH involved FpC₅H₄Mn(CO)₃ (92% yield) [9] and various FpR compounds (R=alkyl, aryl, cis- and trans-4-methylcyclohexyl) [11, 12]. A 57% yield was obtained from [Fp(CH₂=CHC₅H₄FeC₅H₅)]BF₄ and NaOOCCF₃ in acetone [8], but only small amounts were isolated from a similar treatment of [FpCH₂C(OC₂H₅)C₅H₄FeC₅H₅]BF₄ [10].

The fragment ions of the mass spectrum and their relative intensities are completely listed in [7]. UV irradiation with equivalent amounts of P(C₆H₅)₃ or P(OC₆H₅)₃ (²D) in benzene/hexane at room temperature gave C₅H₅Fe(CO)(²D)OOCCF₃ compounds, see 1.5.2.2.6.1.2, p. 92, and 1.5.2.2.6.1.4, p. 119. However, disubstitution occurred with $P(OCH_3)_3$ to give $C_5H_5Fe(P(OCH_3)_3)_2OOCCF_3$ and with $^2D-^2D$ ligands to give $C_5H_5Fe(^2D-^2D)OOCCF_3$ compounds with $^2D-^2D = P(CH_3)_2CH_2CH_2P(CH_3)_2$, $P(C_6H_5)_2CH_2CH_2P(C_6H_5)_2$, and cis-$P(C_6H_5)_2CH=CHP(C_6H_5)_2$, see 1.5.1.3, p. 16 [2].

References:

[1] R.B. King, R.N. Kapoor (J. Organometal. Chem. **15** [1968] 457/69). − [2] R.B. King, R.N. Kapoor (J. Inorg. Nucl. Chem. **31** [1969] 2169/77). − [3] A.N. Nesmeyanov, S.S. Churanov, I.Sh. Guzman, E.G. Perevalova (Izv. Akad. Nauk SSSR Ser. Khim. **1972** 570/1; Bull. Acad. Sci. USSR Div. Chem. Sci. **1972** 526/7). − [4] B.D. Dombek, R.J. Angelici (Inorg. Chim. Acta **7** [1973] 345/7). − [5] W.E. Williams, F.J. Lalor (J. Chem. Soc. Dalton Trans. **1973** 1329/32).

[6] W. Malisch, P. Panster (J. Organometal. Chem. **64** [1974] C5/C9). − [7] W. Malisch, P. Panster (Chem. Ber. **108** [1975] 2554/73). − [8] A.N. Nesmeyanov, E.G. Perevalova, L.I. Leont'eva (Izv. Akad. Nauk SSSR Ser. Khim. **1977** 2582/4; Bull. Acad. Sci. USSR Div. Chem. Sci. **1977** 2391/3). − [9] A.N. Nesmeyanov, E.G. Perevalova, L.I. Leont'eva, E.V. Shumilina (Izv. Akad. Nauk SSSR Ser. Khim. **1977** 1142/6; Bull. Acad. Sci. USSR Div. Chem. Sci. **1977** 1048/52). − [10] A.N. Nesmeyanov, E.G. Perevalova, L.I. Leont'eva, S.A. Eremin (Izv. Akad. Nauk SSSR Ser. Khim. **1978** 2121/4; Bull. Acad. Sci. USSR Div. Chem. Sci. **1978** 1873/6).

[11] N. De Luca (Diss. Ohio State Univ. 1978; Diss. Abstr. Intern. B **39** [1979] 3830/1). − [12] W.N. Rogers, M.C. Baird (J. Organometal. Chem. **182** [1979] C65/C68). − [13] D.J. Darensbourg, M.B. Fischer, R.E. Schmidt Jr., B.J. Baldwin (J. Am. Chem. Soc. **103** [1981] 1297/8). − [14] D.J. Darensbourg, C.S. Day, M.B. Fischer (Inorg. Chem. **20** [1981] 3577/9).

1.5.2.3.6 Compounds with Fe-S Bonds

1.5.2.3.6.1 Compounds of the $C_5H_5Fe(CO)_2SR$ Type

The compounds are listed in Table 34. They are prepared as follows.

Method I: Treatment of NaSR, generally formed from RSH and NaH in ether or THF, with (a) FpCl [16], (b) FpBr [3, 11], or (c) FpI [11]; (d) indicates reactions of RSH with NaH in mineral oil/CS_2/THF and subsequently with FpBr to yield FpSC(S)SR in addition to FpSR [4].

Method II: Refluxing Fp_2 in benzene with (a) RSH [6, 19], and (b) RSSR [6, 14].

The sulfides FpSR No. 1, 3, and 5 may be recrystallized from ether/pentane. They are air stable in the solid state, but not in solution. Their thermal stability decreases in the order $FpSC_6H_5 \gg FpSC_2H_5 > FpSCH_3$. FpSR is more stable than FpR partly because of the availability of vacant orbitals in the S atom suitable for back-donation of electrons from the Fe atom [3].

Compounds No. 1, 2, 3, 5, 6, 9, and 10 readily undergo decarbonylation to give the S-bridged dimers $(C_5H_5Fe(CO)\mu-SR)_2$. The dimerization of $FpSCH_3$ (No. 1) at room temperature proceeds rather slowly, either in ether or THF, but affords both isomers of $(C_5H_5Fe(CO)\mu-SCH_3)_2$ after 12 h, the low melting, less stable derivative being the main product. This derivative is also obtained as the exclusive product when an ether solution of $FpSCH_3$ is UV irradiated at room temperature for 6 h, while only the stable, higher melting isomer is formed in refluxing benzene after 6 h. Parallel reactions of $FpSC_2H_5$ (No. 3) give similar results. In contrast, $FpSC_6H_5$ (No. 5) gives both isomers in all cases [3]. Also the photolysis of compounds No. 2 in hexane for 30 h [18] and No. 6, 9, and 10 in benzene for 6 h [11] produces both derivatives.

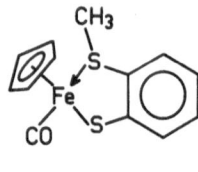

I

UV irradiation of No. 11 in CH_2Cl_2 at 20 °C for 10 min or in THF at −20 °C for 24 h forms the chelate I and the dimer $(C_5H_5FeS_2C_6H_4)_2$, respectively. In the dimer the two C_5H_5Fe groupings are probably bridged by two o-dithiolate ligands [26]. Deviating behavior is also observed on pyrolysis of compounds No. 1, 7, and 8. While $(C_5H_5\text{-}Fe(CO)\mu\text{-}SCH_3)_2$ is formed by heating No. 1 to 70 °C for 18 h at atmospheric pressure [1, 2], attempts to dimerize Nos. 7 and 8 have been unsuccessful, obviously because of the electron-withdrawing effect of C_6F_5 and C_6Cl_5 [6, 19]. For No. 8 only traces of ferrocene and Fp_2 have been isolated [6].

The dimerization is reversible. Addition of CO to $(C_5H_5Fe(CO)\mu\text{-}SR)_2$ in THF, but not in benzene, regenerates FpSR [20]. $FpSCF_3$ (No. 2) forms without addition of CO in a solution of the dimer in hexane at room temperature [18].

Compounds No. 1 and 5 react with a suspension of $Fe_2(CO)_9$ in benzene to form the unstable sulfido derivatives $C_5H_5Fe_2(SR)(CO)_6$, identified only by IR, see C3, p. 82. Such treatment converts No. 4 into $C_5H_5(CO)Fe(\mu\text{-}CO)(\mu\text{-}SC_4H_9\text{-}t)Fe(CO)_3$, see C3, p. 87 [13, 16].

Compounds No. 3 and 4 react with $(C_8H_{12}RhCl)_2$ in CH_3OH to yield $[(C_5H_5FeCO)_2(\mu\text{-}CO)(\mu\text{-}SR)]^+$ [22], with $[C_8H_{12}Rh(C_4H_8O)_2]SbF_6$ in THF [15], or with $[C_8H_{12}Rh(C_2H_5OH)_x]^+$ in C_2H_5OH [24] to yield $[Rh\{(\mu\text{-}SR)(\mu\text{-}CO)\text{-}Fe(CO)C_5H_5\}_2]^+$, see C4, pp. 27/8, 221. In contrast, No. 5 reacts with $[C_8H_{12}Rh\text{-}(C_2H_5OH)_x]^+$ in C_2H_5OH to give $[C_8H_{12}Rh\{(\mu\text{-}SR)Fp\}_2]^+$ [24].

Further reported reactions of Nos. 3 and 4 in acetone involve $[FpOC(CH_3)_2]SbF_6$, where the coordinated solvent is replaced by one FpSR molecule, giving $[Fp(\mu\text{-}SR)Fp]SbF_6$ [22], see C3, 2.5.2.2.10, p. 157.

Insertion of $CF_3C{\equiv}CCF_3$ in the Fe-S bond of Nos. 2 and 7 with formation of $(Z)\text{-}FpC(CF_3)=C(CF_3)SR$ is possible in pentane at ∼75 °C [21].

Substituent effects in $FpSC_6H_4R$ compounds were studied [8]. The Hammett σ functions can be correlated with ^1H NMR chemical shifts, CO stretching frequencies, and Mössbauer parameters.

Explanations for Table 34: CO force constants k and CO,CO interaction constants k' obtained from ν(CO) by the Cotton-Kraihanzel method are given in mdyn·$Å^{-1}$.

Table 34
Compounds of the $C_5H_5Fe(CO)_2SR$ Type.
Further information on numbers preceded by an asterisk is given at the end of the table.
For abbreviations and dimensions, see p. 375.

No.	R in FpSR method of preparation (yield in %)	properties and remarks explanations on p. 318	Ref.
*1	CH_3 Ib (75)	light reddish brown leaflets, m.p. 67 to 70, 72, 78°, subl. at 20°/10^{-2} Torr ^1H NMR: 1.91 (CH_3), 4.13 (C_5H_5) in C_6H_6, 1.58, 4.87 in CS_2 IR: 1992, 2030 (CO) in C_6H_{12}, 1981, 2029 (CO) in CCl_4, 1985, 2032 (CO) in mulls, $v(CH)$ 2860, 2920, 3010 to 3060 in KBr $k=16.28$, $k'=0.38$	[2, 3, 4, 9, 17]
2	CF_3	prepared from FpI and $AgSCF_3$ in acetone at 20 °C for 24 h (48%) red–brown liquid, m.p. $\sim$15°, air sensitive ^{19}F NMR (CH_2Cl_2): 26.1 IR (C_6H_{12}): 2000, 2044 (CO), $v(CF)$ 1083, 1109	[12]
*3	C_2H_5 Ib (57.4)	dark brown leaflets, m.p. 77.5 to 80° ^1H NMR (CS_2): 1.1 (t, CH_3), 1.97 (q, CH_2, $J=7.2$), 4.84 (C_5H_5) IR ($CHCl_3$): 1983, 2028 (CO), similar in CH_2Cl_2 $k=16.23$, $k'=0.36$	[3, 9]
4	C_4H_9-t Ia	reacts with $Fe_2(CO)_9$ in benzene at 20° to give $C_5H_5Fe(CO)(\mu\text{-}CO)(\mu\text{-}SC_4H_9\text{-}t)Fe(CO)_3$	[16]
5	C_6H_5 Ib (85) Id (56 to 85) IIb	vermilon, m.p. 80.5° ^1H NMR (CS_2): 4.83 (C_5H_5), 7.05 (C_6H_5) IR: 1987, 2030 (CO) in CCl_4, 1990, 2033 (CO) in C_6H_{12} $k=16.28$, $k'=0.35$	[3, 4, 9, 14]
6	C_6H_4F-4 Ic (ca. 25)	dark maroon, m.p. 63° ^1H NMR: 4.81 (C_5H_5), 6.65, 7.33 (t+q, C_6H_4) IR (CS_2): 1982, 2028 (CO)	[11]
*7	C_6F_5 IIa (28) IIb (38)	purple, m.p. 93° ^1H NMR ($CDCl_3$?): 5.1 ^{19}F NMR ($CHCl_3$): $\delta=132.2$ (F-2,6), 164.5 (F-3,5), 159.3 (F-4), $J_{2,3}=25.0$, $J_{2,4}<1$, $J_{2,5}=8.2$, $J_{2,6}=1.4$, $J_{3,4}=21.2$, $J_{3,5}=1.4$ IR (C_6H_{12}): 1997, 2040 (CO)	[6]
8	C_6Cl_5 IIa (60)	brown, m.p. 144 to 145°	[19]

References on p. 321

Table 34 [continued]

No.	R in FpSR method of preparation (yield in %)	properties and remarks explanations on p. 318	Ref.
9	$C_6H_4CH_3$-4 Ib (92)	forms from the isomers of $(C_5H_5Fe(CO)SC_6H_4CH_3$-4$)_2$ and CO refluxing THF (60%) green black, m.p. 84 to 85° ^{1}H NMR: 2.21 (CH_3), 4.79 (C_5H_5), 7.02 (C_6H_4) in CS_2, 2.11 (CH_3), 4.17 (C_5H_5) in C_6D_6 IR (CCl_4): 1987, 2026 (CO)	[11, 23]
10	$C_6H_4OCH_3$-4	preparation not published m.p. 55, 61°	[11, 27]
11	$C_6H_4SCH_3$-2	prepared from FpCl and $LiSC_6H_4SCH_3$-2 in THF (79%) large blue–black plates, soluble in ether, THF, CH_2Cl_2 ^{1}H NMR (CD_2Cl_2): 2.38 (SCH_3), 4.98 (C_5H_5), 6.98, 7.39 (C_6H_4) IR (CH_2Cl_2): 1983, 2030 (CO) mass spectrum: $[C_5H_5Fe(CO)_nSC_6H_4SCH_3]^+$ (n=0 to 2), $[(C_5H_5)_nFe]^+$, $[(C_5H_5)_nFeSC_6H_4S]^+$ (n=0, 1), $[C_{11}H_7FeS]^+$	[26]

*Further information:

$C_5H_5Fe(CO)_2SCH_3$ (Table 34, No. 1) is also prepared from FpCl and $(CH_3)_3SnSCH_3$ in C_6H_6 (25% yield) [17], or from Na[Fp] and $(CH_3)_3CCl$ (i.e., FpH in situ) by reaction with CH_3SSCH_3 in THF at -78 °C and stirring at room temperature (13%) [2]. It is formed on UV irradiation of Fp_2 and CH_3SSCH_3 (22%) [2] or of $Fp(CH_2)_2SCH_3$ [1] in C_6H_6.

The compound is readily soluble in most organic solvents but insoluble in petroleum ether and water [2, 3]. The solutions are sensitive to light [3].

The mass spectrum exhibits not only ions from $FpSCH_3$ but also from the pyrolysis products $C_5H_5FeC_5H_4SCH_3$, $C_5H_5FeC_5H_4CH_3$, and ferrocene. The stepwise loss of the CO groups forms the $[C_5H_5FeSCH_3]^+$ ion, which appears to undergo dehydrogenation to $C_6H_6SFe^+$. A complete list of the fragments and their relative abundance is given [10].

Polarography and controlled-potential electrolysis in $(CH_3OCH_2-)_2/[N(C_4H_9$-n$)_4]-ClO_4$ at 22 °C show that oxidation gives the radical cation $[C_5H_5Fe(CO)_2SCH_3]^+$, the process being reversible, while reduction apparently produces a dianion irreversibly. The halfwave potentials (vs. $Ag/AgClO_4$, 10^{-3} M) are $E_{1/2} = -0.4$ and -2.1 V, respectively. Multiple triangular-sweep studies have been carried out to establish reversibility in the systems [5], also see [7]. The radical cation exhibits a single sharp line in the ESR spectrum, g=1.9978 [5].

Attempts to prepare the sulfoxide or sulfone by treatment with active manganese dioxide or nickel peroxide in pentane have led to rapid decomposition [2]. Reaction with FpCl in benzene yields [Fp(μ-SCH$_3$)Fp]Cl [17]. CH$_3$I in ether immediately forms [FpS(CH$_3$)$_2$]I [1, 2].

Reaction with HgCl$_2$ in 95% ethanol causes immediate precipitation of **C$_5$H$_5$-Fe(CO)$_2$SCH$_3$·HgCl$_2$** (35% yield), orange crystals, m.p. 136 to 137 °C (dec.). The IR spectrum shows v(CO) frequencies at 1995, 2011, and 2047 cm^{-1} (mull) and v(CH) at 3060 cm^{-1} (KBr) [2].

C$_5$H$_5$Fe(CO)$_2$SC$_2$H$_5$ (Table **34**, No. **3**) crystallizes in the orthorhombic space group Pnma $-$ D$_{2h}^{16}$ (No. 62) with a = 14.118(5), b = 9.822(5), and c = 7.443(5) Å. Z = 4 gives D$_c$ = 1.53 g·cm^{-3}, while D$_m$ = 1.51 g·cm^{-3}. The molecule, shown in **Fig. 23**, is essentially octahedral with the C$_5$H$_5$ ring occupying three coordination sites. It is bisected by a crystallographic mirror plane which contains the iron and sulfur atoms and the two carbons of the ethyl group. The C$_5$H$_5$ group is disordered with the two halfrings comprising it being essentially coplanar regular pentagons [25].

Fig. 23

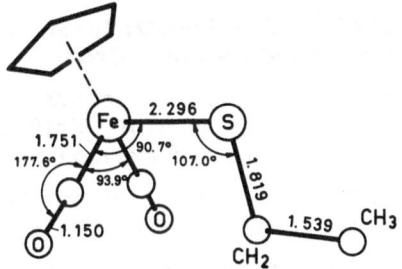

Molecular structure of C$_5$H$_5$Fe(CO)$_2$SC$_2$H$_5$ [25].

C$_5$H$_5$Fe(CO)$_2$SC$_6$F$_5$ (Table **34**, No. **7**). The mass spectrum shows the parent ion and fragments corresponding to the stepwise loss of the two CO groups [6]. Refluxing with P(OCH$_3$)$_3$ in hexane for 18 h forms C$_5$H$_5$Fe(CO)(P(OCH$_3$)$_3$)SC$_6$F$_5$ [6].

References:

[1] R.B. King, M.B. Bisnette (J. Am. Chem. Soc. **86** [1964] 1267/8). − [2] R.B. King, M.B. Bisnette (Inorg. Chem. **4** [1965] 482/5). − [3] M. Ahmad, R. Bruce, G.R. Knox (J. Organometal. Chem. **6** [1966] 1/10). − [4] R. Bruce, G.R. Knox (J. Organometal. Chem. **6** [1966] 67/75). − [5] R.E. Dessy, F.E. Stary, R.B. King, M. Waldrop (J. Am. Chem. Soc. **88** [1966] 471/6).

[6] J. Cooke, M. Green, F.G.A. Stone (J. Chem. Soc. A **1968** 170/3). − [7] R.E. Dessy, R. Kornmann, C. Smith, R. Haytor (J. Am. Chem. Soc. **90** [1968] 2001/4). − [8] R. Havlin, G.R. Knox, R. Greatrex, N.N. Greenwood (Joint Ann. Meeting Chem. Soc.-Roy. Inst. Chem. Ireland, Dublin 1968, Abstr. 3.7). − [9] R.B. King (Inorg. Chim. Acta **2** [1968] 454). − [10] R.B. King (J. Am. Chem. Soc. **90** [1968] 1429/37).

[11] M. Dekker, G.R. Knox, C.G. Robertson (J. Organometal. Chem. **18** [1969] 161/7). − [12] R.B. King, N. Welcman (Inorg. Chem. **8** [1969] 2540/3). − [13] R.J. Haines, C.R. Nolte, R. Greatrex, N.N. Greenwood (J. Organometal. Chem. **26** [1971] C45/C48). − [14] E.D. Schermer, W.H. Baddley (J. Organometal. Chem. **27** [1971] 83/8). − [15] R.J. Haines, R. Mason, J.A. Zubieta, C.R. Nolte (J. Chem. Soc. Chem. Commun. **1972** 990/1).

[16] R.J. Haines, C.R. Nolte (J. Organometal. Chem. **36** [1972] 163/75). – [17] W. Ehrl, H. Vahrenkamp (Chem. Ber. **105** [1972] 1471/85). – [18] J.L. Davidson, D.W.A. Sharp (J. Chem. Soc. Dalton Trans. **1973** 1957/60). – [19] R.D.W. Kemmitt, G.D. Rimmer (J. Inorg. Nucl. Chem. **35** [1973] 3155/9). – [20] D.D. Watkins (Diss. Univ. Nebraska 1973; Diss. Abstr. Intern. B **34** [1974] 5894).

[21] J.L. Davidson, D.W.A. Sharp (J. Chem. Soc. Dalton Trans. **1975** 2283/7). – [22] R.B. English, R.J. Haines, C.R. Nolte (J. Chem. Soc. Dalton Trans. **1975** 1030/3). – [23] D.D. Watkins, T.A. George (J. Organometal. Chem. **102** [1975] 71/7). – [24] R.J. Haines, J.C. Burckett-St. Laurent, C.R. Nolte (J. Organometal. Chem. **104** [1976] C27/C30). – [25] R.B. English, L.R. Nassimbeni, R.J. Haines (J. Chem. Soc. Dalton Trans. **1978** 1379/85).

[26] D. Sellmann, E. Unger (Z. Naturforsch. **33b** [1978] 1438/42). – [27] E.S. Bolton, R. Havlin, G.R. Knox, J. Lyon, R. Greatrex, N.N. Greenwood (unpublished from [11]).

1.5.2.3.6.2 Compounds of the $C_5H_5Fe(CO)_2(\mu\text{-SR})M$ Type with M = Transition Metal Carbonyl Group

The SR bridged heterodinuclear compounds collected in Table 35 have been prepared from equimolar amounts of FpCl and $(CH_3)_3Sn(SR)M$ (R = CH_3, C_2H_5; M = $Cr(CO)_5$, $W(CO)_5$, $Mn(CO)_2C_5H_5$) in benzene at room temperature [1, 2, 3].

The IR $v(CO)$ region of Nos. 1, 4, and 5 show more FeCO bands than expected, thus revealing rotamers resulting from hindered rotation about the Fe-S bond [3].

Compounds No. 1, 4, and 5 are moderately soluble in hydrocarbons and well soluble in other solvents [3]. They are thermally stable below ~ 100 °C. Solid-state Nos. 1 and 4 are also stable in air. No. 5 can be handled in air only for a short time, and in solution it is immediately decomposed, the color initially changing to violet. Pyrolysis of Nos. 1 and 4 or boiling their solutions forms $M(CO)_6$ (M = Cr, W) and unclean mercapto derivatives of iron [3].

Table 35
$C_5H_5Fe(CO)_2(\mu\text{-SR})M$ Compounds, M = Transition Metal Carbonyl Group.
Further information on numbers preceded by an asterisk is given at the end of the table.
For abbreviations and dimensions, see p. 375.

No.	R	M (yield in %)	properties and remarks	Ref.
*1	CH_3	$Cr(CO)_5$ (89)	red–brown crystals, m.p. 80 to 82° ^{1}H NMR (C_6H_6): 1.70 (CH_3), 4.02 (C_5H_5) IR (C_6H_{12}): 1998, 2010, 2041 (all FeCO), 1920 A_1(1), 1926 E, 1935 E, 1988 B_1, 2065 A_1(2) (all CrCO)	[1, 3]
2	C_2H_5	$Cr(CO)_5$ (32)	purple needles, m.p. 79°	[1]

Table 35 [continued]

No.	R	M (yield in %)	properties and remarks	Ref.
3	C_6F_5	$Cr(CO)_5$	no parent ion in mass spectrum, $[FpSC_6F_5]^+$ heaviest fragment, even at low temperatures	[4]
4	CH_3 (59)	$W(CO)_5$	purple crystals, m.p. 92° ^1H NMR (C_6H_6): 1.96 (CH_3), 3.93 (C_5H_5) IR (C_6H_{12}): 2001, 2008, 2048 (all FeCO), 1914 $A_1(1)$, 1925 (E), 1932 E, 1961 B_1, 2077 $A_1(2)$ (all WCO)	[3]
5	CH_3 (70)	$Mn(CO)_2C_5H_5$	dark green crystals, m.p. 88° (dec.) ^1H NMR (C_6H_6): 1.92 (CH_3), 4.17 (MnC_5H_5), 4.34 (FeC_5H_5) IR (C_6H_{12}): 1971, 1989, 2038, 2047 (all FeCO), 1830, 1854, 1858, 1914, 1923, 1947 (all MnCO)	[3]

*Further information:

$C_5H_5Fe(CO)_2S(CH_3)Cr(CO)_5$ (Table **35**, No. **1**) was also obtained from the reaction of FpCl with $LiS(CH)_3Cr(CO)_5$ in ether/benzene or with a suspension of $BrMgS(CH_3)Cr(CO)_5$ in benzene for 2 h, 46 to 47% yield [1].

The mass spectrum shows the parent ion and fragments resulting from the successive loss of the seven CO groups and then CH_3 [4].

References:

[1] W. Ehrl, H. Vahrenkamp (Chem. Ber. **103** [1970] 3563/79). — [2] H. Vahrenkamp, W. Ehrl (5th Intern. Conf. Organometal. Chem., Moscow 1971, Vol. 1, Abstr. No. 31). — [3] W. Ehrl, H. Vahrenkamp (Chem. Ber. **105** [1972] 1471/85). — [4] W. Ehrl, H. Vahrenkamp (Z. Naturforsch. **28b** [1973] 365/6).

1.5.2.3.6.3 Compounds of the $C_5H_5Fe(CO)_2SC(E)(R,X)$ Type with E=O and S

The methods used to prepare the compounds in Table 36 are described below. The table does not include the complexes FpSC(S)R, where R = CN, $NHNH_2$, NC_5H_{10} (piperidin-1-yl), NC_5H_8 (1,2,3,6-tetrahydropyridin-1-yl), and NC_4H_8O (morpholin-4-yl) because their formations, from FpCl and $[RCS_2]^-$, are given without details [7].

Method I: Reaction of FpCOR (R = CH_3, C_6H_5) and P_4S_{10} in a 1:2 mole ratio or B_2S_3 in dry ether, stirred for 4 d [9].

Method II: NaSC(E)(R,X) is reacted with FpX (X = Cl, Br, or I) in THF, benzene, or acetone at room temperature [2, 6] or refluxed [8, 10] for several hours.

Like other cyclopentadienylmetal carbonyl derivatives, the complexes FpSC(S)SR have a nearly linear relationship between C_5H_5 ^1H NMR chemical shifts and CO stretching force constants [4].

The solids No. 5 to 7 are stable in air. In solution at ambient temperature they decompose slowly, producing the chelated monocarbonyl I, see 1.5.2.2.7, p. 136.

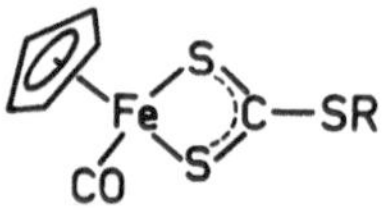

I

This product is also obtained by UV irradiation of Nos. 5 to 7 in benzene at room temperature, and it is formed as an intermediate when the compounds are refluxed in toluene. In the case of FpSC(S)SR (R=CH$_3$, C$_2$H$_5$), the final product of the last reaction is the stable isomer of the dimer, (C$_5$H$_5$Fe(CO)μ-SR)$_2$ [2]. While Nos. 3 and 4 also decompose slowly in solution, giving the chelated C$_5$H$_5$Fe(CO)S$_2$CR [9], neither refluxing of No. 2 in toluene for 24 h nor UV irradiation in benzene for 30 min leads to a similar product. Under extreme conditions decomposition is total [10].

Explanations for Table 36: Force constants (in mdyn·cm^{-1}) are designated as k for k(CO) and k' for k(CO,CO).

Table 36
Compounds of the C$_5$H$_5$Fe(CO)$_2$SC(E)(R,X) Type.
Further information on numbers preceded by an asterisk is given at the end of the table.
For abbreviations and dimensions, see p. 375.

No.	compound method of preparation (yield in %)	properties and remarks explanations see above	Ref.
*1	FpSC(O)C$_6$H$_5$	orange crystals; m.p. 125 to 126° IR (KBr): 1568, 1590 (C=O), 1990, 2030 (CO), 3060 (CH)	[1]

2	FpSC(O)NC$_4$H$_8$
	    (pyrrolidin-1-yl)
	    II
		red crystals, stable under ambient conditions,
		    soluble in CH$_3$COCH$_3$, CHCl$_3$, and C$_6$H$_6$
			[10]

*3	FpSC(S)CH$_3$
	    I (50)
		red, m.p. 98 to 100°
		^{1}H NMR (CDCl$_3$): 3.10 (CH$_3$), 5.11 (C$_5$H$_5$)
		IR: 1998, 2040 (CO) in C$_6$H$_{14}$, ν(CS) 1160 in
		    Nujol
			[9]

4	FpSC(S)C$_6$H$_5$
	    I (50)
		red, m.p. 110 to 112°
		^{1}H NMR (CDCl$_3$): 5.17 (C$_5$H$_5$), 7.25 to 8.25
		    (C$_6$H$_5$)
		IR (C$_6$H$_{14}$): 2000, 2039 (CO)
			[9]

5	FpSC(S)SCH$_3$
	    II (77)
		orange red, m.p. 104.5°
		^{1}H NMR (CS$_2$): 2.60 (CH$_3$), 5.08 (C$_5$H$_5$)
		IR (CCl$_4$): 2002, 2044 (CO)
		    k=16.53, k'=0.34
			[2, 4]

References on p. 326

Table 36 [continued]

No.	compound method of prepa- ration (yield in %)	properties and remarks explanations on p. 324	Ref.
6	FpSC(S)SC$_2$H$_5$ II (74)	orange–red platelets, m.p. 82.5° ^{1}H NMR (CS$_2$): 1.25 (t, CH$_3$, J=7.5), 3.20 (q, CH$_2$), 5.08 (C$_5$H$_5$) IR (CCl$_4$): 1998, 2040 (CO) k=16.46, k'=0.34	[2, 4]
7	FpSC(S)SC$_6$H$_5$ II (35)	orange red, m.p. 106.5° ^{1}H NMR (CS$_2$): 5.04 (C$_5$H$_5$), 7.39 (C$_6$H$_5$) IR (CCl$_4$): 2001, 2048 (CO) k=16.54, k'=0.38	[2, 4]
8	FpSC(S)N(CH$_3$)$_2$ II (~70)	brown orange, m.p. 102 to 104° ^{1}H NMR: 3.50 (CH$_3$), 5.02 (C$_5$H$_5$) IR (Nujol or mull): 1990, 2030 (CO),1480 (CN)	[6]
9	FpSC(S)NHCH(CH$_3$)$_2$ II (34)	IR (KBr): 1995, 2020 (CO),1480 (CN)	[8]

*Further information:

C$_5$H$_5$Fe(CO)$_2$SC(O)C$_6$H$_5$ (Table **36**, No. 1) is prepared by refluxing a mixture of Fp$_2$, mercapto benzoic acid, and methylcyclohexane for 6 h, 26% yield [1].

A mass spectrum exhibits ions from both FpSC(O)C$_6$H$_5$ and the pyrolysis products C$_5$H$_5$FeC$_5$H$_4$C(O)C$_6$H$_5$, ferrocene, and biphenyl. The ion [C$_5$H$_5$FeSC(O)C$_6$H$_5$]$^+$ formed by stepwise loss of the CO groups from the parent ion appears to fragment by elimination of CS or CO to give [C$_5$H$_5$FeOC$_6$H$_5$]$^+$ and [C$_5$H$_5$FeSC$_6$H$_5$]$^+$. The latter further loses two acetylene fragments successively affording [C$_9$H$_8$SFe]$^+$ and [C$_7$H$_6$SFe]$^+$. The most abundant metal-free ion is C$_6$H$_5$CO$^+$. A complete list of the fragments and their relative abundances is given [5].

Fig. 24

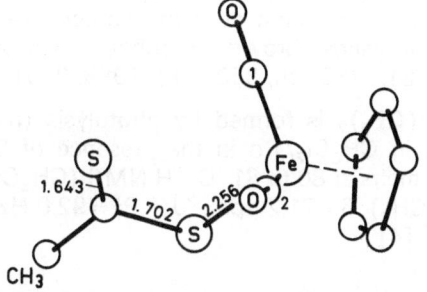

Molecular structure of C$_5$H$_5$Fe(CO)$_2$SC(S)CH$_3$ [9].

Bond angles (°):

C(1)–Fe–S   95.0(1)   C(1)–Fe–C(2)   93.7(1)
C(2)–Fe–S   96.8(1)

Polarographic reduction of the compound in dimethoxyethane/$[N(C_4H_9\text{-}n)_4]ClO_4$ at 22 °C proceeds in two steps: $E_{1/2} = -1.8$ (irreversible, n=1) and -2.2 V (referred to Ag/AgClO$_4$ 10^{-3} M in the same solvent) [3].

$C_5H_5Fe(CO)_2SC(S)CH_3$ (Table **36**, No. **3**) crystallizes in the monoclinic system with a=7.395, b=9.578, c=15.799 Å, and $\beta=101.38°$; space group $P2_1/c - C_{2h}^5$ (No. 14), Z=4. The molecular structure is shown in **Fig. 24**, p. 325.

The S-C bond lengths, which are between those of single and double bonds, indicate charge delocalization in the SCS system, probably due to charge transfer from the –S– lone pair to the lowest unfilled antibonding C=S orbital [9].

References:

[1] R.B. King (J. Am. Chem. Soc. **85** [1963] 1918/22). – [2] R. Bruce, G.R. Knox (J. Organometal. Chem. **6** [1966] 67/75). – [3] R.E. Dessy, F.E. Stary, R.B. King, M. Waldrop (J. Am. Chem. Soc. **88** [1966] 471/6). – [4] R.B. King (Inorg. Chim. Acta **2** [1968] 454). – [5] R.B. King (J. Am. Chem. Soc. **90** [1968] 1429/37).

[6] C. O'Connor, J.D. Gilbert, G. Wilkinson (J. Chem. Soc. A **1969** 84/7). – [7] L.W. Houk, L.J. Epley (Abstr. Papers 165th Natl. Meeting Am. Chem. Soc., Dallas, Texas, 1973, INOR 80). – [8] H. Brunner, T. Burgemeister, J. Wachter (Chem. Ber. **108** [1975] 3349/54). – [9] L. Busetto, A. Palazzi, E. Foresti Serantoni, L. Riva Di Sanseverino (J. Organometal. Chem. **129** [1977] C55/C58). – [10] K.R.M. Springsteen, D.L. Greene, B.J. McCormick (Inorg. Chim. Acta **23** [1977] 13/8).

1.5.2.3.6.4 Compounds of the $C_5H_5Fe(CO)_2SPR_2$ and $C_5H_5Fe(CO)_2SP(S)(R,X)_2$ Type

$C_5H_5Fe(CO)_2SP(CF_3)_2$ is prepared by photolysis (medium pressure Hg arc lamp) of FpP(S)(CF$_3$)$_2$ (see 1.5.2.3.9, p. 356, Table 43, No. 24) in CH$_3$COCH$_3$ for 3 h, 9% yield. Orange-brown crystals, m.p. 58 to 59 °C. ^{1}H NMR spectrum (CH$_3$COCH$_3$): $\delta = 5.30$ ppm (s, C$_5$H$_5$), no coupling to P. ^{19}F NMR spectrum (CH$_3$COCH$_3$): $\delta = 59.5$ ppm, $^2J(P,F) = 71.0$ Hz. IR spectrum (CH$_2$Cl$_2$): ν(CO) 2002, 2048 cm^{-1} [4].

$C_5H_5Fe(CO)_2SP(S)F_2$ is prepared from equimolar amounts of FpCl and CsPS$_2$F$_2$ in CH$_3$COCH$_3$ for 48 h (40% yield) [1, 2] or by distillation of excess HPS$_2$F$_2$ onto FpCl at -196 °C and further reaction at room temperature for 48 h and 50 °C for 18 h (60% yield) [2]. The yellow-brown amorphous solid decomposes at 95 °C. IR spectrum (KBr): 706 (P=S), ~830 sh, 850 (PF), 1998, 2051 (CO) cm^{-1} [1, 2].

$C_5H_5Fe(CO)_2SP(S)(CF_3)_2$ is formed by photolysis (medium pressure Hg arc lamp) of FpP(S)(CF$_3$)$_2$ in CH$_2$Cl$_2$/1 d in the presence of S [4] or in CH$_3$COCH$_3$ for 3 h (traces) [3, 4]. It melts at 80 to 81 °C. ^{1}H NMR (CH$_3$COCH$_3$): $\delta = 5.54$ (C$_5$H$_5$) ppm. ^{19}F NMR (CH$_3$COCH$_3$): $\delta = 71.2$ ppm, $^2J(F,P) = 92.0$ Hz. IR spectrum (CH$_2$Cl$_2$): ν(CO) 2016, 2059 cm^{-1} [4].

References:

[1] M. Lustig, L.W. Houk (Inorg. Nucl. Chem. Letters **5** [1969] 851/3). – [2] L.W. Houk, M. Lustig (Inorg. Chem. **9** [1970] 2462/5). – [3] R.C. Dobbie, P.R. Mason, R.J. Porter (J. Chem. Soc. Chem. Commun. **1972** 612/3). – [4] R.C. Dobbie, P.R. Mason (J. Chem. Soc. Dalton Trans. **1973** 1124/8).

1.5.2.3.6.5 Compounds of the C₅H₅Fe(CO)₂S(O)₂R Type with R=Alkyl, Cycloalkyl, and Aryl

The FpSO₂R compounds in Table 37 can be prepared by the following methods:

Method I: SO₂ insertion into the Fe–C bond of FpR.

a. Reaction with refluxing SO₂ for several hours [10, 12, 19, 21, 23, 31], or with SO₂ condensed onto FpR at −40 to −60 °C until all solid is dissolved [1, 2, 12, 15, 31].

b. In pentane by bubbling SO₂ at 25 °C for several hours to 4 d [2, 12]. Also in CHCl₃ [19], hexane [31], THF [24], CH₂Cl₂ at −70 °C for 12 h [35], or in HCON(CH₃)₂, CHCl₃, CH₃OH, and pentane at ∼3 bar for 8 h. (The yields decrease in the given order of solvents [23].)

Treatment of FpCH₂CR=CR′R″ by Method Ia and Ib yields FpSO₂CH₂CR=CR′R″ and the isomeric FpSO₂CR′R″CR=CH₂ in ratios strongly dependent on the reaction conditions. Isomerization is promoted by small R′ and R″, low temperatures in Method Ia, and nonpolar solvents in Method Ib [31].

Method II: From RSO₂Cl and

a. Fp₂ in THF at 27 °C for several hours [2, 11, 31] or ether at −30 °C [7].

b. [Fp]⁻ in THF at −80 to +27 °C for several hours [1, 2, 7]. RSO₂F reacts in the same way [2].

Method III: A solution of FpCl and NaSO₂R in methanol is stirred at 27 °C for 12 h [2].

Method IV: Reaction of RX (X=Cl, Br, I) at room temperature with Na[FpSO₂] prepared in situ from Na[Fp] and SO₂ in THF at −78 °C [8, 15, 18, 31, 33].

The ¹H NMR spectra of FpSO₂R show C₅H₅ downfield shifts of 0.41 to 0.82 ppm in respect to the corresponding FpR, demonstrating the strong electron–withdrawing effect of RSO₂ [2]. A linear relationship is found between ¹H NMR chemical shifts of C₅H₅ and the force constants k(CO) of FpSO₂R and other cyclopentadienyl metal carbonyls [6]. For the relationship between Mössbauer isomer shifts and ν(CO), see [22].

The IR spectra exhibit two intense absorptions in the 1035 to 1205 cm⁻¹ range. These are assigned to the asymmetric and symmetric SO stretching frequencies. They are displaced by 100 to 120 cm⁻¹ to lower wave numbers from the values reported for organic sulfones, while the ν(CO) bands occur at wave numbers about 40 to 60 cm⁻¹ higher than in FpR. Both effects are attributed to strong π interaction between iron and sulfur. With the exception of the CF₃ group the absorptions are rather insensitive to the nature of R, suggesting that the oxygens almost exclusively determine the degree of π bonding of the sulfur atom [2, 3, 4, 7, 10]. The increased Fe–S dₚ–dₚ interaction is also expressed in the shift of the ν(CS) absorptions to longer wavelengths, while the δ(OSO) vibrations are not affected [4, 7]. As may be seen in Table 37 several sulfinates show four ν(CO) frequencies in their IR spectra. This phenomenon may be attributed to the presence of rotamers owing to restricted rotation about the Fe–S bond [21].   [continued on p. 334]

Table 37
Compounds of the FpS(O)$_2$R Type, R = Alkyl, Cycloalkyl, and Aryl.
Further information on numbers preceded by an asterisk is given at the end of the table.
For abbreviations and dimensions, see p. 375.

No.	R in FpSO$_2$R method of preparation (yield in %)	properties and remarks explanations on p. 335	Ref.
1	CCl$_3$ IIa (31)	golden, m.p. 141.5 to 142° IR (CH$_2$Cl$_2$): 2025, 2065 (CO), ν(SO$_2$) 1068 sh	[31]
*2	CF$_3$ IIa (11.4) IIb (14)	yellow, dec. 150° IR: 2035 (B$_2$), 2077 (A$_1$, CO), δ(OSO) rocking 526, δ(OSO) bending 575, ν(CF$_3$) 1157, 1170 sh, 1191 in CH$_2$Cl$_2$, 1987, 2016 (B$_2$), 2059 and 2076 (A$_1$), 2125 (CO), δ(CF$_3$) 457, 611, δ(FeCO) 546, ν(CS) 748, ν_s(SO$_2$) 1070, ν_{as}(SO$_2$) 1232, ν(CF$_3$) 1163, 1178, 1222 sh in KBr	[7]
*3	CH$_3$ Ia (95) Ib (95) IIa (90) IIb (2) IV (46)	yellow, m.p. 135° ^{1}H NMR (CDCl$_3$): 3.15 (s, CH$_3$), 5.25 (s, C$_5$H$_5$) ^{13}C NMR (CDCl$_3$): 60.6 (CH$_3$), 87.6 (C$_5$H$_5$), 210.1 (CO) ^{57}Fe-γ: δ = 0.29, Δ = 1.68 IR: 2009, 2015, 2053, 2063 (CO) in CHCl$_3$ or CH$_2$Cl$_2$, ν(SO) 1052, 1061 (sh), 1181 (sh), 1194 in Nujol k = 16.72, k′ = 0.38	[1, 2, 6, 18, 22, 29, 33, 34]
4	C$_2$H$_5$ Ia (91) IIa (88) IIb (1.9)	yellow, m.p. 165° ^{1}H NMR (CDCl$_3$): 1.51 (t, CH$_3$, J ∼ 8), 3.30 (q, CH$_2$, J ∼ 8), 5.42 (s, C$_5$H$_5$) IR: 2011, 2016, 2055, 2061 (CO) in CHCl$_3$, ν(SO) 1051, 1055 sh, 1173 sh, 1185 in Nujol k = 16.72, k′ = 0.37	[1, 2, 6]
5	CH$_2$OCH$_3$ Ia (16)	m.p. 82.5° ^{1}H NMR (CDCl$_3$): 3.79 (s, CH$_3$), 4.20 (s, CH$_2$), 5.25 (s, C$_5$H$_5$) IR: 2013, 2060 (CO) in CHCl$_3$, ν(SO) 1046, 1185, 1200 in Nujol	[21]
6	CH$_2$SCH$_3$ Ia (7)	IR: 2021, 2062 (CO) in CHCl$_3$, ν(SO) 1045, 1187 in Nujol	[21]

References on p. 336

Table 37 [continued]

No.	R in FpSO$_2$R method of preparation (yield in %)	properties and remarks explanations on p. 335	Ref.
7	CH$_2$CH$_2$CN Ia (20)	m.p. 114 to 116° ¹H NMR (CDCl$_3$): 2.83 to 3.55 (m, CH$_2$CH$_2$), 5.26 (s, C$_5$H$_5$) IR: 2027, 2067 (CO) in CHCl$_3$, ν(SO) 1047, 1186 in Nujol	[21]
*8	CH$_2$CH=CH$_2$ Ia	yellow golden, m.p. 77°; subl. at 95 to 105°/∼0.1 Torr ¹H NMR (CDCl$_3$): 3.76 (d, CH$_2$, J=7), 5.22 (s, C$_5$H$_5$), 5.17 to 6.42 (m, CH= CH$_2$) IR: 2008, 2064 (CO) in CHCl$_3$, π(CH of C$_5$H$_5$) 864, ν(SO$_2$) 1021 sh, 1037, 1164 sh, 1174, ν(C=C) 1640, 1650 in Nujol	[17, 30, 31]
9	CH(CH$_3$)$_2$ Ia (57) Ib (69)	yellow, m.p. 100 to 103° IR: 2003, 2013, 2051, 2060 (CO), ν(SO) 1034, 1054, 1174 in CHCl$_3$, 2013, 2059 (CO) in i-C$_3$H$_7$OH, 1999 sh, 2008, 2055 (CO) in C$_6$H$_6$	[19]
10	CH$_2$C≡CCH$_3$ IV (17) not Ia, Ib	yellow ¹H NMR (CDCl$_3$): 1.87 (t, CH$_3$, J=2.5), 3.68 (br, CH$_2$), 5.16 (s, C$_5$H$_5$) IR: 2016, 2062 (CO) in CHCl$_3$, ν(SO) 1050, 1200 in KBr	[15]
11	CH$_2$CH$_2$CH=CH$_2$	prepared from I (see p. 335) at 125 to 130° (94%) ¹H NMR: 5.30 (C$_5$H$_5$) IR (KBr): 1980, 2059 (CO)	[13]
12	CH(CH$_3$)CH=CH$_2$ Ia (from FpCH$_2$CH=CHCH$_3$ with 100% isomerization)	yellow, m.p. 110 to 111°, subl. at 100 to 105°/∼0.1 Torr with dec. ¹H NMR (CDCl$_3$): 1.42 (d, CH$_3$, J=7), 3.33 to 4.16 (m, CH), 5.25 (s, C$_5$H$_5$), 4.85 to 6.58 (m, CH=CH$_2$) IR: 2005, 2062 [31], 1992, 2040 [10] (CO) in CHCl$_3$, π(CH of C$_5$H$_5$) 855, ν(SO$_2$) 1045, 1178, ν(C=C) 1630 in Nujol, ν(SO) 1042, 1175 in KBr	[10, 31]
13	CH$_2$C(CH$_3$)=CH$_2$ Ia	m.p. 153° (dec.) ¹H NMR (CDCl$_3$): 2.00 (s, CH$_3$), 3.73 (s, CH$_2$), 5.10 (s, =CH$_2$), 5.19 (s, C$_5$H$_5$) IR (KBr): 2000, 2045 (CO), ν(SO$_2$) 1044, 1178	[30, 31, 32]

References on p. 336

Table 37 [continued]

No.	R in FpSO$_2$R method of preparation (yield in %)	properties and remarks explanations on p. 335	Ref.
14	CH$_2$CH(CH$_3$)$_2$ Ia (56)	m.p. 103 to 105° ^{1}H NMR (CDCl$_3$): 1.07 (d, CH$_3$, J=6), 2.1 to 2.8 (m, CH), 3.03 (d, CH$_2$, J=6), 5.15 (s, C$_5$H$_5$) IR: 2001, 2012, 2049, 2060 (CO) in CHCl$_3$, ν(SO) 1050, 1172 sh, 1181 in Nujol	[21]
15	C(CH$_3$)$_3$ Ia (10 to 11)	^{1}H NMR (CDCl$_3$): 1.27 (s, CH$_3$), 5.21 (s, C$_5$H$_5$) IR: 2000, 2012, 2049, 2059 (CO) in CHCl$_3$, ν(SO) 1033, 1178 in Nujol	[21]
16	CH$_2$Si(CH$_3$)$_3$ Ia (78 to 80)	yellow, m.p. 74, 112° ^{1}H NMR (cyclohexane): −0.02 (s, CH$_3$), 2.9 (s, CH$_2$), 4.95 (s, C$_5$H$_5$) IR (CH$_2$Cl$_2$): 2003, 2049 (CO)	[19, 27]
17	CH$_2$CH$_2$C≡CCH$_3$ Ia (40)	orange, m.p. 144 to 145° ^{1}H NMR (CDCl$_3$): 1.92 (t, CH$_3$, J=2.5), 2.92 (br, CCH$_2$C), 3.32 (t, SCH$_2$, J=J=8), 5.34 (s, C$_5$H$_5$) IR: 2016, 2069 (CO) in CHCl$_3$, ν(SO) 1040, 1180 in Nujol	[15]
18	C(CH$_3$)$_2$CH=CH$_2$ see No. 19	not fully separated from No. 19, yellow IR: 2017, 2061 (CO) in CHCl$_3$, π(CH of C$_5$H$_5$) 857, ν(SO$_2$) 1028 sh, 1037, 1164 sh, 1178, ν(C=C) 1620 in Nujol no interconversion No. 18 ⇌ No. 19 in CH$_3$CN, CH$_3$CN/SO$_2$, or CH$_3$OH/SO$_2$ at 25° or in refluxing SO$_2$, some reaction No. 18 → No. 19 in refluxing C$_6$H$_6$	[30, 31]
19	CH$_2$CH=C(CH$_3$)$_2$ Ia and Ib give mixtures with 25 to 93%, No. 18 IV (13)	yellow, m.p. 125 to 126°, see also No. 18 ^{1}H NMR (CDCl$_3$): 1.78 (s), 1.83 (s) (2 CH$_3$), 3.73 (d, CH$_2$, J=8), 5.22 (s, C$_5$H$_5$), 5.44 (t, CH, J=8) IR: 2003, 2069 in CHCl$_3$, π(CH of C$_5$H$_5$) 846; ν(SO$_2$) 1026 sh, 1041, 1179, ν(C=C) 1670 in Nujol	[30, 31]
20	CH$_2$C(CH$_3$)$_3$ Ia (22)	m.p. 145 to 146° ^{1}H NMR (CDCl$_3$): 1.17 (s, CH$_3$), 3.13 (s, CH$_2$), 5.15 (s, C$_5$H$_5$) IR: 2001 to 2009 br, 2048, 2058 (CO) in CHCl$_3$, ν(SO) 1048, 1179 in Nujol	[21]

References on p. 336

Table 37 [continued]

No.	R in FpSO$_2$R method of prepa- ration (yield in %)	properties and remarks explanations on p. 335	Ref.
*21	C$_6$F$_5$ IIa (40 to >70) IIb	bright yellow-orange blocks, m.p. 165 to 170° (dec.) ^{19}F NMR (THF, 35°): 140.5 (F-2,6), 153.4 (F-4), 161.6 (F-3,5), $\pm J_{23}=$ 28.05; $\vert J_{24}\vert=1.7$; $\mp J_{25,36}=9.25$; $\vert J_{26}\vert$, $\vert J_{35}\vert=0, 3.0$; $\pm J_{34}=25.2$ IR (CHCl$_3$): 2023, 2070 (CO), v(SO) 990, 1076, 1095 in CHCl$_3$, 2025, 2068 (CO) in CH$_2$Cl$_2$, 1064 sh, 1078, 1230 (SO$_2$) in Nujol	[5, 11, 25, 31]
22	C$_6$H$_4$F-4 II (70)	yellow, m.p. 174 to 175° ^{1}H NMR (CDCl$_3$): 5.06 (s, C$_5$H$_5$), 7.38 (m, C$_6$H$_4$) IR (CH$_2$Cl$_2$): 2011, 2057 (CO), v(SO$_2$) 1047, 1205	[31]
23	C$_6$H$_5$ Ia (>46) Ib (89) IIa (89) IIb (2.3) III (73)	yellow, m.p. 137° ^{1}H NMR (CDCl$_3$): 5.30 (s, C$_5$H$_5$), 7.5 to 8.0 (m, C$_6$H$_5$) IR: 2013, 2060 (CO) in CHCl$_3$, v(SO) 1039, 1164 sh, 1182 sh, 1195 in Nujol k=16.74, k'=0.39 reaction with NaBH$_4$ in THF at 0° yields a brown solid not showing v(CO) fre- quencies	[1, 2, 6]
24	CH$_2$CH$_2$C(CH$_3$)$_3$ Ia (60)	m.p. 185 to 203° (dec.) ^{1}H NMR (CDCl$_3$): 0.93 (s, CH$_3$), 1.54 to 1.91 (m, CCH$_2$C), 2.88 to 3.23 (t, SCH$_2$), 5.15 (s, C$_5$H$_5$) IR: 2005, 2014, 2050, 2059 (CO) in CHCl$_3$, 2007, 2057 (CO), v(SO) 1049, 1175 sh, 1188 in Nujol	[16, 21]
	CHDCHDC(CH$_3$)$_3$-threo Ia (45) and Ib (54 to 94) with inversion	yellow, m.p. 174° (dec.), stable at −20° ^{1}H NMR (CHCl$_3$): 0.86 (s, CH$_3$), 1.66 (d, CHD), 2.95 (d, CHD), ^{3}J(CHDCHD)=4.3	[16, 23]
	CHDCHDC(CH$_3$)$_3$-erythro Ia (55) with inversion	^{1}H NMR (CHCl$_3$): ^{3}J(CHDCHD)=13.0	[16]
25	CH$_2$Si(CH$_3$)$_2$Si(CH$_3$)$_3$ Ib	^{1}H NMR (CS$_2$): 0.11 (Si(CH$_3$)$_3$), 0.17 (Si(CH$_3$)$_2$), 3.09 (CH$_2$), 5.08 (C$_5$H$_5$) IR (C$_6$H$_{12}$): 2000, 2021 (CO)	[24]

References on p. 336

Table 37 [continued]

No.	R in FpSO$_2$R method of preparation (yield in %)	properties and remarks explanations on p. 335	Ref.
26	CH$_2$C$_6$H$_4$F-4	^{57}Fe-γ: $\delta = 0.21$, $\Delta = 1.60$ IR (CH$_2$Cl$_2$): 1990, 2033	[22]
27	CH$_2$C$_6$H$_5$ Ia (>40)	m.p. 117° ^{1}H NMR (CDCl$_3$): 4.39 (s, CH$_2$), 5.10 (s, C$_5$H$_5$), 7.6 (s, br, C$_6$H$_5$) ^{13}C NMR (CDCl$_3$): 78.0 (CH$_2$), 87.2 (C$_5$H$_5$), 209.7 (CO) ^{57}Fe-γ: $\delta = 0.23$, $\Delta = 1.64$, indicates greater electron density at Fe than in FpCH$_2$C$_6$H$_5$ IR: 2010, 2060 (CO) in CHCl$_3$, ν(SO) 1038, 1045, 1050, 1205 in Nujol k = 16.72, k' = 0.41	[2, 6, 22, 34]
28	C$_6$H$_4$CH$_3$-2 Ia (60)	m.p. 149.5° ^{1}H NMR (CDCl$_3$): 2.81 (s, CH$_3$), 5.01 (s, C$_5$H$_5$), 7.13 to 7.44 (m, 3, C$_6$H$_4$), 7.63 to 8.01 (m, 1, C$_6$H$_4$) IR: 2011, 2056 (CO) in CHCl$_3$, ν(SO) 1040, 1192 in Nujol	[21]
29	C$_6$H$_4$CH$_3$-3 Ia (54)	m.p. 143° ^{1}H NMR (CDCl$_3$): 2.46 (s, CH$_3$), 5.06 (s, C$_5$H$_5$), 7.25 to 7.75 (m, C$_6$H$_4$) IR: 2011, 2060 (CO) in CHCl$_3$, ν(SO) 1046, 1191 in Nujol	[21]
*30	C$_6$H$_4$CH$_3$-4 Ia (75) III (90)	ochrerous, m.p. 212° ^{1}H NMR (CDCl$_3$): 2.42 (s, CH$_3$), 5.09 (s, C$_5$H$_5$), 7.18 to 7.50 (m, C$_6$H$_4$), 7.56 to 7.89 (m, C$_6$H$_4$) IR: 2015, 2058 (CO), δ(FeCO) 558 sh in CHCl$_3$, δ(OSO) rocking 503, δ(OSO) bending 572, ν(CS) 613 in CH$_2$Cl$_2$, 1978, 2000, 2029, 2060 (CO), ν_s(SO$_2$) 1040, ν_{as}(SO$_2$) 1192, ν(CH) 2930, 3085 in KBr, ν(SO) 1040, 1178 sh, 1194 in Nujol k = 16.74, k' = 0.39	[2, 4, 6, 21]
31	C$_6$H$_4$OCH$_3$-4 Ia (90)	m.p. 163 to 164° (dec.) ^{1}H NMR (CDCl$_3$): 3.99 (s, CH$_3$), 5.18 (s, C$_5$H$_5$), 6.90 to 7.28 (m, C$_6$H$_4$), 7.70 to 7.94 (m, C$_6$H$_4$) IR: 2010, 2059 (CO) in CHCl$_3$, ν(SO) 1038 br, 1193 in Nujol	[21]

References on p. 336

Table 37 [continued]

No.	R in FpSO$_2$R method of preparation (yield in %)	properties and remarks explanations on p. 335	Ref.
32	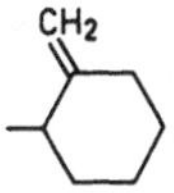	prepared from II on p. 335 and SO$_2$ m.p. 149° (dec.) ^{1}H NMR (C$_2$D$_6$SO): 1.1 to 2.3 (m, 4CH$_2$), 3.55 (m, CH), 5.13 (m, =CH$_2$), 5.33 (s, C$_5$H$_5$) IR (KBr): 2026, 2060 (CO), 1042, 1175 (SO$_2$)	[17]
33	CH(CH$_3$)C$_6$H$_5$ Ia (66) IV (small) with racemization	yellow, m.p. 180 to 182° (dec.), no decomposition in organic solvents ^{1}H NMR: 4.89 (s, C$_5$H$_5$), 7.0 (br, C$_6$H$_5$) in (CH$_3$)$_2$SO, 2.30 (d, CH$_3$, J=6.5), 3.6 (br, CH) in C$_6$H$_5$CH(CF$_3$)OH IR (CH$_2$Cl$_2$): 2015, 2064 (CO), ν(SO$_2$) 1052, 1203	[12]
	(−)$_{546}$ –compound Ia (25) Ib (ca. 60)	m.p. 172 to 173° prepared from (+)$_{546}$-FpCH(CH$_3$)C$_6$H$_5$, optical purity depends on the conditions, maximum [α]$_{546}^{27}$ = −186° (CHCl$_3$), some racemization in the solid state above 0°	[8, 12]
34	CH$_2$C$_6$H$_4$OCH$_3$-4 Ia (85)	m.p. 167 to 168° ^{1}H NMR (CDCl$_3$): 3.81 (s, CH$_3$), 4.14 (s, CH$_2$), 4.93 (s, C$_5$H$_5$), 6.71 to 7.06 (m, 2, C$_6$H$_4$), 7.19 to 7.53 (m, 2, C$_6$H$_4$) IR: 2008, 2061 (CO) in CHCl$_3$, ν(SO) 1046, 1206 in Nujol	[21]
35	(E?)-CH$_2$CH=CHC$_6$H$_5$ Ia gives Nos. 35 and 36 in a 20:80 ratio	separated from No. 36 by chromatography on Al$_2$O$_3$ yellow, m.p. 132 to 133° ^{1}H NMR (CDCl$_3$): 3.92 (d, CH$_2$, J=6.5), 5.15 (s, C$_5$H$_5$), 6.07 to 7.00 (m, CH=CH), 7.40 (m, C$_6$H$_5$) IR: 2011, 2062 (CO) in CHCl$_3$, π(CH of C$_5$H$_5$) 847, ν(SO$_2$) 1036, 1042 sh, 1177 sh, 1186, ν(C=C) 1635 in Nujol, ν(SO) 1040, 1175 in KBr	[10, 31]
36	CH(C$_6$H$_5$)CH=CH$_2$ see No. 35	could not be separated from No. 35	[31]
37	CH$_2$Si(CH$_3$)$_2$C$_6$H$_5$ Ia (75)	^{1}H NMR (C$_6$H$_{12}$): 0.62 (s, SiCH$_3$), 3.33 (s, SiCH$_2$), 5.15 (s, C$_5$H$_5$), 7.6 (C$_6$H$_5$) IR (CH$_2$Cl$_2$): 2007, 2053	[27]

References on p. 336

Table 37 [continued]

No.	R in FpSO₂R method of preparation (yield in %)	properties and remarks explanations on p. 335	Ref.

| 38 | Ib (in CH₂Cl₂ at −70°) | IR spectrum (KBr) of the brown solid indicates ionic cyclopropenylium structure [(C₆H₅)₃C₃][FpSO₂], good solubilities in toluene and ether/CH₂Cl₂ suggests equilibrium with No. 38 | [35] |

supplement

| *39 | CHDCHDC₆H₅-erythro Ia (75), Ib | yellow ¹H NMR (CDCl₃): 3.24 (CH-β), 3.40 (CH-α), ³J(H,H) = 12.4, 5.16 (C₅H₅) IR (Nujol): 1980, 2040 (CO), ν(SO) 1035, 1180 | [36, 37] |

Compounds No. 3, 4, 23, 27, and 30 are soluble in benzene, CHCl₃, CH₃OH, C₂H₅OH, acetone, water, and 6 M HCl, from which they can be recovered unchanged, but insoluble in hexane and CS₂ [2]. They appear to be stable in air for at least several months; however, CHCl₃ solutions show signs of decomposition after about 24 h of exposure to air [1, 2, 3].

Compounds No. 8, 12, 13, 18, 19, 35, and 36 are stable in air at room temperature. They are readily soluble in polar organic solvents such as CH₂Cl₂, CHCl₃, and acetone but only moderately soluble in benzene [31].

No products of desulfination could be isolated from photolysis of No. 1 in toluene for 1 h and from heating in THF at 65 °C for 5.5 h, in benzene at 80 °C for 1 h, or in toluene at 110 °C for 0.3 to 1 h. Refluxing No. 22 in toluene for 2.5 h gives only decomposition products and unreacted starting material. However, refluxing No. 21 in toluene for 1 h or photolysis in toluene for 1 h yields FpC₆F₅. There is no observable desulfination for up to 5 h in refluxing benzene or up to 3 h in dioxane or heptane [31]. No. 27 is stable in refluxing dioxane [2].

However, UV irradiation of Nos. 3, 23, and 27 in benzene at ∼30 °C for 72 h results in considerable decomposition and gives trace amounts of ferrocene [2].

No reaction has been observed with nucleophiles such as tertiary phosphines [3]. Treatment of Nos. 3, 23, and 27 in benzene with Cl₂ leads to a complete loss of both CO and SO₂ and formation of decomposition products. The reaction of Nos. 3 and 23 with Br₂ proceeds similarly, while No. 3 is not changed by treatment with I₂ at 80 °C for 1 h [2]. Treatment of Nos. 3 and 27 with Rh[P(C₆H₅)₃]₃Cl abstracts CO and forms apparently unstable iron compounds. The expected complex with a terminal SO₂ ligand was not obtained [8, 20].

References on p. 336

Explanations for Table 37: ^{57}Fe Mössbauer spectra were measured at room temperature. The terms k and k' are used for CO stretching and CO,CO interactions force constants, respectively. They were calculated by the Cotton–Kraihanzel method.

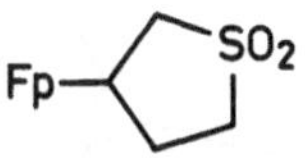

I    II

*Further information:

$C_5H_5Fe(CO)_2SO_2CF_3$ (Table **37**, No. **2**) is readily soluble in hydrocarbon halides and THF, but insoluble in ether and nonpolar solvents [7]. It is less stable than No. 30, decomposing at 150 °C. Heating in the absence of air eliminates CO but not SO_2. The lesser stability may also be related to the higher ν(CO) frequencies (in respect to No. 40), revealing a loose $d_\pi - p_\pi$ Fe–C bond [7].

The following IR bands from C_5H_5Fe vibrations were assigned on the basis of a C_{5v} local symmetry for the C_5H_5 ligand: 873 γ(CH, E_1), ν(Fe–ring, A_1), δ(Fe–ring, E_1), 1016 δ(CH, E_1), γ(CH, A_1), 1130 sh ν(CC, A_1), 1421, 1429 ν(CC, E_1), 3125, 3135 ν(CH, E_1) [7].

$C_5H_5Fe(CO)_2SO_2CH_3$ (Table **37**, No. **3**) is obtained from a solution of Fp(μ-SO_2)Fp in CH_3I at 80 to 85 °C for 3 h, 26% yield [33], by cleavage of Fp(μ-SO_2)Fp or Fp(μ-SO_2SO_2)Fp with CH_3I in refluxing THF [26, 28, 33], and from the reaction of $K[C_5H_5Fe(CO)_2SO_2] \cdot 0.5 SO_2$ with CH_3I, 9% yield [29].

The observed shifts of the ^1H resonances after addition of Eu(fod)$_3$ (Formula II on p. 104) indicates an interaction [14].

$C_5H_5Fe(CO)_2SO_2CH_2CH=CH_2$ (Table **37**, No. **8**) is formed in small amount by treatment of FpC_3H_5-cyclo with SO_2 in benzene [17]. UV irradiation with $P(C_6H_5)_3$ in benzene causes exchange of one CO group [9].

$C_5H_5Fe(CO)_2SO_2C_6F_5$ (Table **37**, No. **21**) crystallizes in the monoclinic system with a = 10.97(2), b = 12.12(2), c = 12.06(2) Å, and β = 117.6(2)°; space group $P2_1/c - C_{2h}^5$. Z = 4 gives D_c = 1.91, D_m = 1.90 g·cm^{-3}. The molecular structure is shown in **Fig. 25**, p. 336. While the Fe–S bond distance is considerably shorter than the estimated Fe–S single-bond length (2.38 Å), the S–C distance is close to the single-bond length (1.78 Å) calculated from the covalent radii of sulfur and sp^2 carbon. Both lengths indicate strong π interaction between the iron and sulfur d orbitals, but minimal interaction between the π system of the phenyl group and the sulfur d orbitals [11, 25]. This interpretation is supported by the ^{19}F NMR [5, 11, 25].

The complex is insoluble in nonpolar solvents, but it may be recrystallized from THF or aqueous acetone. It is air stable and nonvolatile. Only ions of complex decomposition products are observed by mass spectroscopy [11]. Desulfination to FpC_6F_5 occurs on photolysis in toluene or on thermolysis in boiling toluene, but not boiling benzene [31], also see [11].

$C_5H_5Fe(CO)_2SO_2C_6H_4CH_3$-4 (Table **37**, No. **30**) was also prepared from Na[Fp] and a twofold excess of $(4-CH_3C_6H_4SO_2)_2O$ in THF at −80 °C, 55% yield [4].

The complex is readily soluble in hydrocarbon halides and THF but insoluble in water, ether, and nonpolar solvents [4, 7]. It is quite stable towards air and water. It decomposes when heated above 200 °C [4].

References on p. 336

Fig. 25

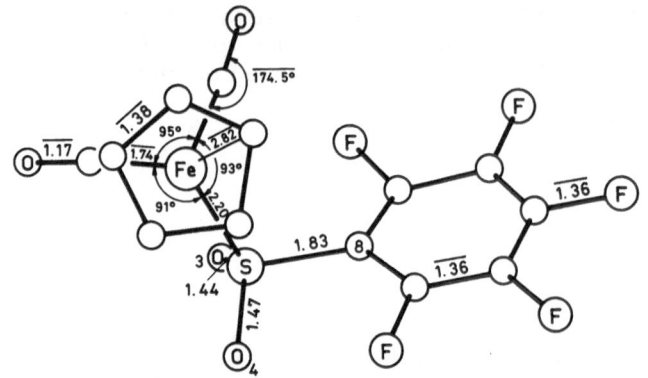

Molecular structure of $FpSO_2C_6F_5$ [11, 25].

Other bond angles (°):

O(3)-S-O(4)	115(1)	Fe-S-O(3)	112(1)
O(3)-S-C(8)	103(1)	Fe-S-O(4)	113(1)
O(4)-S-C(8)	105(1)	Fe-S-C(8)	107(1)

$C_5H_5Fe(CO)_2SO_2CHDCHDC_6H_5$-erythro (Table **37**, No. **39**). The erythro diastereomer forms exclusively from threo-$C_5H_5Fe(CO)_2CHDCHDC_6H_5$ both by Method Ia (refluxing SO_2) [37] and Method Ib (in $CHCl_3$) [36]. Thus the SO_2 insertion involves an inversion process [37].

References:

[1] J.P. Bibler, A. Wojcicki (J. Am. Chem. Soc. **86** [1964] 5051/3). – [2] J.P. Bibler, A. Wojcicki (J. Am. Chem. Soc. **88** [1966] 4862/70). – [3] A. Wojcicki, J.P. Bibler, F.A. Hartman (Proc. 9th Intern. Conf. Coord. Chem., St. Moritz, Switzerland, 1966, p. 175). – [4] E. Lindner, H. Weber (Z. Naturforsch. **22b** [1967] 1243/7). – [5] M.I. Bruce (J. Chem. Soc. A **1968** 1459/64).

[6] R.B. King (Inorg. Chim. Acta **2** [1968] 454). – [7] E. Lindner, H. Weber, G. Vitzhum (J. Organometal. Chem. **13** [1968] 431/41). – [8] A. Wojcicki, J.J. Alexander, M. Graziani, J.E. Thomasson, F.A. Hartman (New Aspects Chem. Metal Carbonyls Deriv. 1st Intern. Symp. Proc., Venice 1968, Abstr. C6, pp. 1/6). – [9] M. Graziani, A. Wojcicki (Inorg. Chim. Acta **4** [1970] 347/50). – [10] J.Y. Merour (Compt. Rend. C **271** [1970] 1397/9).

[11] M.I. Bruce, A.D. Redhouse (J. Organometal. Chem. **30** [1971] C78/C80). – [12] J.J. Alexander, A. Wojcicki (Inorg. Chim. Acta **5** [1971] 655/8). – [13] W.P. Giering, M. Rosenblum (J. Am. Chem. Soc. **93** [1971] 5299/301). – [14] T.J. Marks, J.S. Kristoff, A. Alich, D.F. Shriver (J. Organometal. Chem. **33** [1971] C35/C37). – [15] J.E. Thomasson, P.W. Robinson, D.A. Ross, A. Wojcicki (Inorg. Chem. **10** [1971] 2130/7).

[16] G.M. Whitesides, D.J. Boschetto (J. Am. Chem. Soc. **93** [1971] 1529/31). – [17] A. Cutler, R.W. Fish, W.P. Giering, M. Rosenblum (J. Am. Chem. Soc. **94** [1972] 4354/5). – [18] M.R. Churchill, B.G. DeBoer, K.L. Karla, P. Reich-Rohrwig, A. Woj-

cicki (J. Chem. Soc. Chem. Commun. **1972** 981/2). − [19] S.E. Jacobson, P. Reich-Rohrwig, A. Wojcicki (Inorg. Chem. **12** [1973] 717/23). − [20] J.J. Alexander, A. Wojcicki (Inorg. Chem. **12** [1973] 74/6).

[21] S.E. Jacobson, A. Wojcicki (J. Am. Chem. Soc. **95** [1973] 6962/70). − [22] G. Ingletto, E. Tondello, L. Di Sipio, G. Carturan, M. Graziani (J. Organometal. Chem. **56** [1973] 335/7). − [23] P.L. Bock, D.J. Boschetto, J.R. Rasmussen, J.P. Demers, G.M. Whitesides (J. Am. Chem. Soc. **96** [1974] 2814/25). − [24] K.H. Pannell, J.R. Rice (J. Organometal. Chem. **78** [1974] C35/C39). − [25] A.D. Redhouse (J. Chem. Soc. Dalton Trans. **1974** 1106/8).

[26] N.H. Tennent, S.R. Su, C.A. Poffenberger, A. Wojcicki (J. Organometal. Chem. **102** [1975] C46/C48). − [27] K.H. Pannell (Transition Metal Chem. [Weinheim] **1** [1975/76] 36/40). − [28] N.H. Tennent, S.R. Su, C.A. Poffenberger, A. Wojcicki (Proc. 17th Intern. Conf. Coord. Chem., Hamburg 1976, p. 130). − [29] C.R. Jablonski (J. Organometal. Chem. **142** [1977] C25/C30). − [30] L.S. Chen, S.R. Su, A. Wojcicki (Inorg. Chim. Acta **27** [1978] 79/89).

[31] R.L. Downs, A. Wojcicki (Inorg. Chim. Acta **27** [1978] 91/103). − [32] D.W. Lichtenberg (unpublished results from [31]). − [33] P. Reich-Rohrwig, A.C. Clark, R.L. Downs, A. Wojcicki (J. Organometal. Chem. **145** [1978] 57/68). − [34] R.G. Severson, A. Wojcicki (J. Organometal. Chem. **149** [1978] C66/C70). − [35] R. Gompper, E. Bartman, H. Nöth (Chem. Ber. **112** [1979] 218/33).

[36] K. Stanley, M.C. Baird (J. Am. Chem. Soc. **99** [1977] 1808/12). − [37] D. Dong, D.A. Slack, M.C. Baird (J. Organometal. Chem. **153** [1978] 219/28).

1.5.2.3.6.6   Further Compounds with $FpSO_n$ Units (n = 1 to 3)

Compounds of the $FpS(O)_2R$ type (R = alkyl, cycloalkyl, aryl) are treated in 1.5.2.3.6.5, p. 327. Other $FpSO_n$ compounds with Fe-O bonds are listed in Table 32, p. 313. The compounds of this section are listed in Table 38. Their preparation is either briefly given in the table or described under further information.

The **complexes** $FpSO_3R$ No. 12 to 15 are air-stable solids. In the mass spectra (70 eV, 75 to 80 °C) of Nos. 12 and 13 the molecular ion, $[M]^+$, is absent or quite weak; however, peaks corresponding to $[M-CO]^+$ and $[M-2CO]^+$ are readily discernible [5]. Compounds No. 12 and 13 readily interconvert in the appropriate refluxing alcohol, i.e., C_2H_5OH and CH_3OH, leading to $FpS(O)_2OC_2H_5$ in a 65% yield and $FpS(O)_2OCH_3$ in a 53% yield. These conversions also occur, in higher yields (55 to 85%), at ambient temperatures in the presence of small amounts of HBF_4. No. 12 was reacted with optically active octan-2-ol in the presence of HBF_4 to give No. 16. This is then converted by methanol and HBF_4 back to $FpS(O)_2OCH_3$ and octan-2-ol, which exhibits an essentially unchanged rotation. A mechanism involving $[FpSO_2]^+$ cations was proposed [5].

Table 38
Further Compounds with $FpSO_n$ Units, n=1 to 3.
Further information on numbers preceded by an asterisk is given at the end of the table.
For abbreviations and dimensions, see p. 375.

No.	compound	properties and remarks	Ref.
*1	FpS(O)(CH₃)=N- S(O)₂CH₃	yellow–orange solid, not isolated pure ¹H NMR (CDCl₃): 3.00 (O₂SCH₃), 3.40 (OSCH₃), 5.30 (C₅H₅)	[4]
*2	FpS(O)(CH₃)=N- S(O)₂C₆H₄CH₃-4	yellow–orange solid ¹H NMR (CDCl₃): 2.33 (CCH₃), 3.35 (SCH₃), 5.31 (C₅H₅) ¹³C NMR (CDCl₃): 21.4 (CCH₃), 59.4 (SCH₃), 87.9 (C₅H₅), 208.0, 208.8 (CO) IR: 2022, 2070 (CO) in CHCl₃, bands at 1010, 1025, 1088, 1115, 1145, 1280, 1295 in Nujol are assigned to ν(SN), ν(SO), ν(SO₂)	[4]
*3	FpS(O)(CH₂C₆H₅)=N- S(O)₂CH₃	yellow–orange solid ¹H NMR (CDCl₃): 3.01 (CH₃), 4.67 (CH₂), 5.08 (C₅H₅) ¹³C NMR (CDCl₃): 45.6 (CH₃), 75.9 (CH₂), 87.6 (C₅H₅), CO signal too weak for accu- rate measurement IR (Nujol): 2005, 2065 (CO), bands at 1030, 1105, 1130, 1275 are assigned to ν(SN), ν(SO), ν(SO₂)	[4]
*4	[FpSO₂]⁻ (?)	–	[3]
*5	K[FpSO₂]·0.5SO₂	orange glass, dec. above 60° under N₂	[2]
6	FpS(O)OCH₃	preparation: see further information for No. 5 m.p. above 145° (dec. under N₂) ¹H NMR (CDCl₃): 3.57 (CH₃), 5.20 (C₅H₅) IR (KBr): 2002, 2059 (CO), bands at 682, 972, 1096, and 1218 are assigned to ν(S–O), ν(C–O), and ν(S=O)	[2]
*7	FpS(O)₂Ge(CH₃)₃	yellow, m.p. 148 to 149° ¹H NMR (CHCl₃?): 0.5 (s, CH₃), 5.0 (s, C₅H₅) IR: 2000, 2060 (CO), ν(FeC) 440, νₛ(GeC) 540, δ(FeCO) 570, νₐₛ(GeC) 585, rocking (GeCH₃) 790, νₛ(S=O) 1075, νₐₛ(S=O) 1205, ν(CH) 2920, 2990, 3120	[1]
8	FpS(O)OSn(CH₃)₃ ·0.5SO₂	prepared as No. 7, apparently a polymer with intermolecular bonding from SnO to addi- tional SO₂ brown glass, dec. at 80° with gas evolution NMR: ²J(¹¹⁹Sn,H) = 64 (48 for FpSn(CH₃)₃) References on p. 341	[1]

Table 38 [continued]

No.	compound	properties and remarks	Ref.
		IR: 2000, 2050 (CO), ν(FeC) 440, ν(SnC) 550, δ(FeCO) 565, rocking (SnCH$_3$) 780, ν(SO) 950 to 990, ν_{as}(S=O) 1220, ν(CH) 2900, 2960, 3110	
9	FpS(O)$_2$OH	preparation: from Nos. 12 or 13 by hydrolysis at 25°/16 h (95%) yellow powder, m.p. 104° ^{1}H NMR (CD$_3$SOCD$_3$): 5.30 (s, C$_5$H$_5$), 11.16 (s, OH) IR (KBr): 2018, 2062 (CO), ν(S–O) 811, ν(S=O) 1038, 1184, ν(OH) 2940, ν(OD) 2230 for FpS(O)$_2$OD strong acid, reactions with NaOH/H$_2$O or ROH (R = CH$_3$, C$_2$H$_5$) give Nos. 10, 12, and 13	[5]
10	Na[FpSO$_3$]	prepared from No. 9 yellow solid, stable in air ^{1}H NMR (D$_2$O/CD$_3$COCD$_3$): 5.05 (s, C$_5$H$_5$) IR (KBr): 1994, 2050 (CO), ν(SO) 984, 1072, 1084, 1107, 1129, consistent with local C$_{3v}$ symmetry of SO$_3$ bonded through S, 1:1 electrolyte in CH$_3$OH, no reaction with CH$_3$I in refluxing CH$_3$COCH$_3$/C$_2$H$_5$OH	[5]
11	[R$_2$NH$_2$][FpSO$_3$] R = CH$_3$, C$_2$H$_5$	prepared from Nos. 12 or 13 in excess R$_2$NH at 25°, yellow precipitates, identified by ^{1}H NMR and IR	[5]
12	FpS(O)$_2$OCH$_3$	preparation: from [FpH$_2$O]BF$_4$ and Na[SO$_3$R] in CH$_3$OH at 25° ($\sim$50% yield), also see Nos. 9 and 11 yellow, m.p. 156° ^{1}H NMR (CDCl$_3$): 3.59 (s, CH$_3$), 5.21 (s, C$_5$H$_5$) IR (KBr): 2007, 2064 (CO), ν(SOC) 636, ν(SO) 971, ν(SO$_2$) 1061, 1095, 1214	[5]
13	FpS(O)$_2$OC$_2$H$_5$	prepared like No. 12 in C$_2$H$_5$OH ($\sim$50%), also see Nos. 9 and 11 yellow, m.p. 96° ^{1}H NMR (CDCl$_3$): 1.23 (t, CH$_3$, J = 7), 4.01 (q, CH$_2$, J = 7), 5.40 (s, C$_5$H$_5$) IR (KBr): 2000, 2062 (CO), ν(SO) 971, ν(SO$_2$) 1020, 1088, 1214	[5]
14	FpS(O)$_2$OC$_3$H$_7$-n	prepared from [FpH$_2$O]BF$_4$ and Na[SO$_3$C$_2$H$_5$] in excess n-C$_3$H$_7$OH at 25° yellow, m.p. 94 to 96°	[5]

References on p. 341

Table 38 [continued]

No.	compound	properties and remarks	Ref.
		^{1}H NMR (CDCl$_3$): 0.92 (t, CH$_3$, J'=6), 1.86 (m, CH$_2$CH$_3$, J=6, J'=6), 3.86 (t, SO$_3$CH$_2$, J=6), 5.32 (s, C$_5$H$_5$) IR (KBr): 2014, 2068 (CO), ν(SO) 973, ν(SO$_2$) 1087, 1211	
15	FpS(O)$_2$OC$_3$H$_7$-i	prepared like No. 14 in i-C$_3$H$_7$OH yellow, m.p. 164 to 165° ^{1}H NMR (CDCl$_3$): 1.75 (d, CH$_3$, J=6.1), 4.90 (sept, CH, J=6.1), 5.17 (s, C$_5$H$_5$) IR (KBr): 2010, 2070 (CO), ν(SOC) 640, ν(SO) 918, ν(SO$_2$) 1065, 1084, 1112, 1211	[5]
16	FpS(O)$_2$OCH(CH$_3$)- C$_6$H$_{13}$-n (+)$_{589}$ isomer	$[\alpha]_{589}^{22.5} = +23.8°$, m.p. 118° ^{1}H NMR (CDCl$_3$): 0.86 to 1.30 (m, CH$_2$, CH$_3$), 4.73 (br, CH), 5.21 (s, C$_5$H$_5$) IR (KBr): 2010, 2065 (CO), ν(SOC) 634, ν(SO) 996, ν(SO$_2$) 1088, 1212	[5]

*Further information:

C$_5$H$_5$Fe(CO)$_2$S(O)(R)=NS(O)$_2$R' (R=R'=CH$_3$; R=CH$_3$, R'=C$_6$H$_4$CH$_3$-4; R= CH$_2$C$_6$H$_5$, R'=CH$_3$, Table **38**, Nos. **1** to **3**) are formed on storage of FpN(S(O)R)-S(O)$_2$R' (see 1.5.2.3.8.2, Table 41, Nos. 5 to 7) as a solid or in solution. Their structures are derived from ^{1}H NMR and ^{13}C NMR [4].

[C$_5$H$_5$Fe(CO)$_2$SO$_2$]$^-$ (?) (Table **38**, No. **4**). The possible existence in solution, which is obtained from Na[Fp] and SO$_2$ in THF at −78 °C, is discussed in [3]. Attempts to isolate the anion by treatment of the solution with [N(P(C$_6$H$_5$)$_3$)$_2$]Cl at −78 °C and above have been unsuccessful. Reaction with CH$_3$I affords FpS(O)$_2$CH$_3$ and FpI. A possible mechanism of the latter reaction, which has been conducted under various conditions, is discussed. Similar reactions with other alkylating agents, such as 4-CH$_3$C$_6$H$_4$SO$_3$CH$_3$, CH$_3$Br, or C$_6$H$_5$CH$_2$Cl, also form FpS(O)$_2$R, but there is no IR evidence for the presence of the halides FpX, as is the case with CH$_3$I. Depending on the ratio of SO$_2$ to Na[Fp], the following products were isolated after warming to room temperature and chromatography on Florisil: Fp$_2$, Fp$_2$SO$_2$, [C$_5$H$_5$Fe(CO)]$_2$(CO)SO$_2$, and (FpSO$_2$)$_2$ [3].

K[C$_5$H$_5$Fe(CO)$_2$SO$_2$]·0.5SO$_2$ (Table **38**, No. **5**) precipitates on addition of excess or equimolar quantities of SO$_2$ to an THF solution of K[Fp] at −78 °C, 85% yield. Attempts to obtain crystals by exchange of large cations for K$^+$ have not been successful [2].

The IR spectrum in Nujol shows ν(CO) absorptions at 1984, 2020, and 2041 cm^{-1}, while in CH$_3$NO$_2$ these bands are extensively split due to ion pair formation. Removal of ion-paired species by addition of dibenzo-18-crown-6 simplifies the spectrum in CH$_2$Cl$_2$, giving frequencies at 2018 and 2052 cm^{-1} [2].

The compound is very soluble in water and methanol, moderately soluble in CH$_3$NO$_2$, and sparingly soluble in THF, CH$_2$Cl$_2$, and CHCl$_3$. In aqueous solution it is a strong electrolyte, $\Lambda = 120\ \Omega^{-1}\cdot cm^2\cdot mol^{-1}$ (10^{-3} M) [2].

The properties of the compound are expressed in Formulas I and II. These are reflected in the products obtained from alkylation with CH_3I and CH_3OSO_2F. The former results in S-alkylation to produce a 9% yield of $FpS(O)_2CH_3$ as well as a 33% yield of FpI, while the latter leads to O-alkylation, giving a 20% yield of $FpS(O)OCH_3$ (No. 6) as well as a 3% yield of $(FpSO_2)_2$ [2].

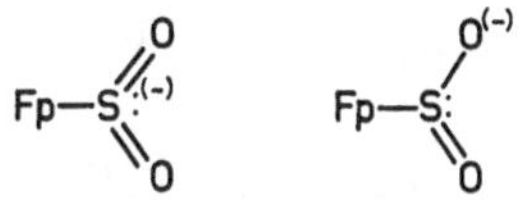

$C_5H_5Fe(CO)_2S(O)_2Ge(CH_3)_3$ (Table **38**, No. **7**) was prepared by shaking a mixture of SO_2 and $FpGe(CH_3)_3$ in a sealed tube for 5 h and isolated as a crystalline solid in an 80% yield [1].

References:

[1] R.E.J. Bichler, H.C. Clark (J. Organometal. Chem. **23** [1970] 427/30). – [2] C.R. Jablonski (J. Organometal. Chem. **142** [1977] C25/C30). – [3] P. Reich-Rohrwig, A.C. Clark, R.L. Downs, A. Wojcicki (J. Organometal. Chem. **145** [1978] 57/68). – [4] R.G. Severson, A. Wojcicki (J. Organometal. Chem. **149** [1978] C66/C70). – [5] C.A. Poffenberger, A. Wojcicki (J. Organometal. Chem. **165** [1979] C5/C9).

1.5.2.3.7 Compounds with Fe-Se and Fe-Te Bonds

The $C_5H_5Fe(CO)_2ER$ compounds (E = Se, Te) in Table 39, Nos. 1 to 4 and 7, were obtained from Fp_2 and REER by refluxing in C_6H_6 for 2 to 3 h (Nos. 1 and 7) [1, 2] or by heating in methylcyclohexane at 45 to 55 °C for 3 d (Nos. 2 to 4) [3].

Compounds No. 2 and 3 are described as malodorous substances that decompose rapidly above 70 to 80 °C [3].

Compounds No. 1 and 7 lose CO when refluxed, or more rapidly when irradiated with visible or IR light in benzene for several hours, to give the two isomers of $(C_5H_5Fe(CO)\mu-EC_6H_5)_2$. The tendency of the mononuclear complexes to dimerize, losing CO, decreases in the order S > Se > Te, paralleling the nucleophilicity of E [1, 2].

Table 39
Compounds with Fe-Se and Fe-Te Bonds.
Further information on compound No. 5 is given at the end of the table.
For abbreviations and dimensions, see p. 375.

No.	compound (yield in %)	properties and remarks	Ref.
1	FpSeC$_6$H$_5$ (53)	dark crystals, m.p. 50 to 52° ^{1}H NMR (CS$_2$): 4.79 (C$_5$H$_5$) IR (C$_6$H$_{12}$): 1984, 2026 (CO)	[1, 2]
2	FpSeCF$_3$ (33)	brown, m.p. 28° (from pentane) IR (CHCl$_3$ or KBr): 1990, 2050 (CO), ν(CF) 1060, 1090	[3]

Table 39 [continued]

No.	compound (yield in %)	properties and remarks	Ref.
3	FpSeC$_2$F$_5$ (42)	red, feathery crystals, m.p. 39 to 40° IR (CHCl$_3$ or KBr): 1995, 2045 (CO), ν(CF) 1115, 1195, 1320	[3]
4	FpSeC$_3$F$_7$-n (30)	red crystals at −78°, liquid at 20° IR (CHCl$_3$ or KBr): 2010, 2055 (CO), ν(CF) 1105, 1175, 1335	[3]
5	FpSe(O)$_2$CH$_3$	brown, m.p. 140° ^{1}H NMR (CD$_3$OD): 3.02 (s, CH$_3$), 5.05 (s, C$_5$H$_5$) IR: 2009, 2056 (CO) in CH$_2$Cl$_2$, ν(SeC) 570, ν$_s$(SeO) 728, ν$_{as}$(SeO) 859 in KBr	[5]
6	FpSeP(CF$_3$)$_2$	prepared by photolysis of FpP(Se)(CF$_3$)$_2$ in CH$_3$COCH$_3$ for 3 h m.p. 68 to 69° ^{1}H NMR (CH$_3$COCH$_3$): 5.36 (s, C$_5$H$_5$) ^{19}F NMR (CH$_3$COCH$_3$): 56.6, ^{2}J(F,P) = 66.0 IR (CH$_2$Cl$_2$): 1997, 2042 (CO)	[4]
7	FpTeC$_6$H$_5$	green, m.p. 66° ^{1}H NMR (CS$_2$): 4.77 (C$_5$H$_5$) IR (C$_6$H$_{12}$): 1976, 2018 (CO)	[1, 2]

*Further information:

C$_5$H$_5$Fe(CO)$_2$Se(O)$_2$CH$_3$ (Table **39**, No. **5**) is prepared from FpCH$_3$ and SeO$_2$ in C$_6$H$_6$ at 50 °C for 5 h, brown crystals, 12% yield, or from FpCl and NaSeO$_2$CH$_3$ in CH$_2$Cl$_2$ or CH$_3$OH, red–brown needles. It is stable in air for several days. It cannot be sublimed in high vacuum up to 130 °C. The mass spectrum (70 eV, 120 °C) shows the molecular ion, fragments resulting from successive loss of the two CO ligands and the CH$_3$SeO$_2$ group (with fragmentation), and [Fp$_2$]$^+$. In the not altered state FpSe(O)$_2$CH$_3$ is somewhat soluble in CH$_2$Cl$_2$ and CH$_3$OH, in other solvents, however, insoluble [5].

References:

[1] E.D. Schermer (Diss. Louisiana State Univ. 1971; Diss. Abstr. Intern. B **32** [1971] 807). − [2] E.D. Schermer, W.H. Baddley (J. Organometal. Chem. **27** [1971] 83/8). − [3] P. Rosenbuch, N. Welcman (J. Chem. Soc. Dalton Trans. **1972** 1963/6). − [4] R.C. Dobbie, P.R. Mason (J. Chem. Soc. Dalton Trans. **1973** 1124/8). − [5] I.P. Lorenz (Angew. Chem. **90** [1978] 60/1; Angew. Chem. Intern. Ed. Engl. **17** [1978] 53).

1.5.2.3.8 Compounds with Fe-N Bonds

1.5.2.3.8.1 Acyclic Amine and Ketenimine Ligands

The compounds of this section are listed in Table 40. The preparation of No. 2 is described under the further information. The ketenimines of the FpN=C=C(CN)-

C(CN)$_2$R type (Nos. 3 to 7) are prepared from FpR compounds and a slight excess of (CN)$_2$C=C(CN)$_2$ (1:1 mole ratio for No. 6) in CH$_2$Cl$_2$ at 25 °C for 9 to 72 h. This gives FpC(CN)$_2$-C(CN)$_2$R and the isomeric FpN=C=C(CN)-C(CN)$_2$R, which are separated by chromatography on Al$_2$O$_3$ with CHCl$_3$ eluent, followed by CH$_2$Cl$_2$ [4 to 6]. According to an earlier publication [3], reaction of FpCH$_3$ with (CN)$_2$C=C(CN)$_2$ yields only FpC(CN)$_2$C(CN)$_2$CH$_3$.

In the IR spectra of Nos. 3 to 7, the CO stretching frequencies are 11 to 23 cm^{-1} higher than those of FpC(CN)$_2$C(CN)$_2$R. This shift is supposed to be a manifestation of a substantial polarity of the Fe–N linkage [3]. The compounds are soluble in polar organic solvents but insoluble in nonpolar organic solvents. They are stable in CH$_2$Cl$_2$ solution under nitrogen for about 12 h, but after 24 to 36 h a 5 to 10% yield of FpCN can be isolated. In air, decomposition to an insoluble brown material is evident after 8 to 12 h. Storage of solutions of Nos. 3, 4, and 6 in CH$_2$Cl$_2$ at 25 °C for 3 to 6 d does not yield any detectable FpC(CN)$_2$C(CN)$_2$R [4, 6].

Table 40
Acyclic Amines and Ketenimines.
Further information on numbers preceded by an asterisk is given at the end of the table.
For abbreviations and dimensions, see p. 375.

No.	compound (yield in %)	properties and remarks	Ref.
1	FpN(C$_6$H$_5$)$_2$	no preparation given, decomposes at 150°/190 atm N$_2$ to noncarbonyl materials, heating under CO pressure yields only Fp$_2$, and no insertion product	[7]
*2	FpN=C=C(CN)$_2$	light brown leaflets, dec. p. 110° IR (Nujol): 1990, 2036 (CO), 2185 (CN), v(C–C) 1248, 1264	[1, 2]
3	FpN=C=C(CN)C(CN)$_2$CH$_3$ (5)	red needles at low temperature ^{1}H NMR (CDCl$_3$): 1.89 (s, CH$_3$), 5.32 (s, C$_5$H$_5$) IR: 2034, 2079 (CO), 2212 (CN) in CH$_2$Cl$_2$, v(N=C=C) 1295 (sym), 2140 sh (asym), 2162, neat oil	[6]
4	FpN=C=C(CN)C(CN)$_2$C$_2$H$_5$ (25)	red oil ^{1}H NMR (CDCl$_3$): 1.17 (t, CH$_3$, J=7), 2.00 (q, CH$_2$, J=7), 5.21 (s, C$_5$H$_5$) IR: 2034, 2079 (CO), 2206 (CN) in CH$_2$Cl$_2$, v(N=C=C) 1302 (sym), 2145 (asym) in mull	[6]

Table 40 [continued]

No.	compound (yield in %)	properties and remarks	Ref.
5	FpN=C=C(CN)C(CN)$_2$C$_3$H$_7$-n (15)	red oil ^{1}H NMR (CDCl$_3$): 1.0 to 2.0 (m, C$_3$H$_7$), 5.28 (s, C$_5$H$_5$) IR: 2034, 2073 (CO), 2201 (CN) in CH$_2$Cl$_2$, ν(N=C=C) 1306 (sym), 2145 (asym), neat oil	[6]
6	FpN=C=C(CN)C(CN)$_2$-CH$_2$C$_6$H$_5$ (45)	orange red, m.p. 131 to 132° ^{1}H NMR (CDCl$_3$): 3.44 (s, CH$_2$), 5.32 (s, C$_5$H$_5$), 7.59 (s, C$_6$H$_5$) IR: 2034, 2073 (CO), 2206 (CN) in CH$_2$Cl$_2$, ν(N=C=C) 1296 (sym), 2151 (asym) in Nujol	[4, 6]
*7	FpN=C=C(CN)C(CN)$_2$-CH(CH$_3$)C$_6$H$_5$ (25)	red oil ^{1}H NMR (CDCl$_3$): 1.65 (d, CH$_3$, J=7), 3.28 (q, CH, J=7), 5.08 (s, C$_5$H$_5$), 7.31 to 7.44 (m, C$_6$H$_5$) IR: 2029, 2074 (CO), 2212 (CN) in CH$_2$Cl$_2$, ν(N=C=C) 1305 (sym), 2151 (asym), neat oil	[6]

*Further information:

C$_5$H$_5$Fe(CO)$_2$N=C=C(CN)$_2$ (Table 40, No. 2) was prepared by shaking FpI and a 2.5-fold excess of Ag[C(CN)$_3$] in methanol at 20 °C for 2 h [1, 2]. While the compound was initially formulated as FpC(CN)$_3$ [1], later an Fe-N bond was assumed although the expected three CN stretching vibrations (2A$_1$ + B$_2$) due to a local symmetry C$_{2v}$ were not observed in the IR spectra of solutions [2]. The compound, which is scarcely air-sensitive, is quite soluble in THF, ether, acetone, and alcohol. It decomposes in water. Its molar conductance in acetone (18.3 mg in 20 ml) is 10.4 $\Omega^{-1} \cdot$ cm$^2 \cdot$ mol^{-1} [2].

C$_5$H$_5$Fe(CO)$_2$N=C=C(CN)C(CN)$_2$CH(CH$_3$)C$_6$H$_5$ (Table 40, No. 7). The optically active compound was prepared from FpCH(CH$_3$)C$_6$H$_5$, [α]$_{566}$ = +78.5°, and (CN)$_2$C=C(CN)$_2$. Its specific rotation could not be measured in CHCl$_3$ because of turbidity [6].

References:

[1] W. Beck, R.E. Nitzschmann, G. Neumair (Angew. Chem. 76 [1964] 346; Angew. Chem. Intern. Ed. Engl. 3 [1964] 380/1). – [2] W. Beck, R.E. Nitzschmann, H.S. Smedal (J. Organometal. Chem. 8 [1967] 547/50). – [3] A. Wojcicki, S.R. Su, J.A. Hanna (Proc. 4th Intern. Conf. Organometal. Chem., Bristol 1969, Abstr. O 5). – [4] S.R. Su, J.A. Hanna, A. Wojcicki (J. Organometal. Chem. 21 [1970] P21/P22). – [5] S.R. Su (Diss. Ohio State Univ. 1971; Diss. Abstr. Intern. B 32 [1972] 6283).

[6] S.R. Su, A. Wojcicki (Inorg. Chem. 14 [1975] 89/98). – [7] T. Inglis, M. Kilner (J. Chem. Soc. Dalton Trans. 1976 562/4).

1.5.2.3.8.2 Compounds with $C_5H_5Fe(CO)_2NSO_2$ Groups

The compounds in Table 41 are prepared by the following methods.

Method I: Reaction of FpR with a slight excess of $R'S(O)_2N=S=O$ in toluene or $CHCl_3$ at room temperature for 1 h. Excess $R'S(O)_2N=S=O$ is destroyed with 10% water in acetone, the solvent is removed, and the residue extracted with $CHCl_3$. Since the products resist crystallization, they are generally isolated by evaporation of the solvent [4].

Method II: Oxidation of Nos. 5 to 7 in CH_2Cl_2 by a slight excess of $3\text{-}ClC_6H_4\text{-}C(O)OOH$ during 1 h, neutralization with $NaHCO_3$ in methanol, solvent removal, extraction of the residue with CH_2Cl_2, and evaporation to dryness. Recrystallization from CH_2Cl_2/hexane (1:1) at low temperature [4].

Method III: Reaction of FpR with excess $CH_3S(O)_2N=S=NS(O)_2CH_3$ in $CHCl_3$, removal of unreacted sulfur diimide by filtration, and crystallization from $CHCl_3$ at low temperature [4].

Compounds No. 5 to 7 rearrange on storage as solids or in solution to yellow-orange substances, which appear to be of the $FpS(O)(R)=NSO_2R'$ type, see 1.5.2.3.6.6, p. 337. They are oxidized by $3\text{-}ClC_6H_4C(O)OOH$ (preparative Method II) to Nos. 9 to 11 [4].

Table 41
Compounds with $C_5H_5Fe(CO)_2NSO_2$ Groups.
Further information on numbers preceded by an asterisk is given at the end of the table.
For abbreviations and dimensions, see p. 375.

No.	compound method of preparation (yield in %)	properties and remarks	Ref.
*1	$FpN(SO_2F)_2$	prepared from FpI and $AgN(SO_2F)_2$ in CH_2Cl_2 (47.5% yield) red–brown crystals, dec. p. 103 to 105° ^{19}F NMR: -53.8 IR (Nujol and mull): 2033, 2068 (CO), $v_s(SO)$ 1184, 1221, $v_{as}(SO)$ 1398, 1418	[3]
*2	$FpN(SO_2Cl)COCH_2\text{-}C(CH_3)=CH_2$	prepared from $FpCH_2C(CH_3)=CH_2$ and $ClSO_2NCO$ in CH_2Cl_2 at $-40°$ (47% yield) orange red, m.p. 89 to 91° (dec.) ^1H NMR $(CDCl_3)$: 1.77 (s, CH_3), 3.61 (s, $-CH_2$), 4.77 to 4.93 (m, $=CH_2$), 5.13 (s, C_5H_5) IR: 2028, 2075 (CO) in $CHCl_3$, 1133, 1353 (SO), 1670 (C=O) in KBr	[1, 2]

References on p. 348

Table 41 [continued]

No.	compound method of preparation (yield in %)	properties and remarks	Ref.
3	FpN(SO$_2$NHC$_6$H$_5$)COCH$_2$-C(CH$_3$)=CH$_2$ (see No. 2)	orange red, m.p. 102 to 103° (dec.) ^{1}H NMR (CDCl$_3$): 1.77 (s, CH$_3$), 3.54 (s, -CH$_2$), ~4.75 (m, =CH$_2$), 4.92 (s, C$_5$H$_5$), 6.75 to 7.4 (m, C$_6$H$_5$, NH) IR: 2023, 2065 (CO) in CHCl$_3$, 1140, 1315 (SO), 1630 (C=O), 3310 (NH) in KBr	[1, 2]
4	FpN(SO$_2$NHC$_6$H$_4$CH$_3$-4)-COCH$_2$C(CH$_3$)=CH$_2$ (see No. 2)	orange red, m.p. 86 to 88° (dec.) ^{1}H NMR (CDCl$_3$): 1.78 (s, CH$_3$C=), 2.31 (s, CH$_3$C$_6$), 3.52 (s, -CH$_2$), 4.73 to 4.85 (m, =CH$_2$), 4.92 (s, C$_5$H$_5$), 6.75 to 7.25 (m, C$_6$H$_4$, NH) IR (KBr): 1997, 2055 (CO), 1140, 1316 (SO), 1628 (C=O), 3305 (NH)	[2]
5	FpN(SO$_2$CH$_3$)SOCH$_3$ I	orange–red glass ^{1}H NMR (CDCl$_3$): 2.51 (OSCH$_3$), 2.92 (O$_2$SCH$_3$), 5.23 (C$_5$H$_5$) ^{13}C NMR (CDCl$_3$): 41.9 (OSCH$_3$), 45.1 (O$_2$SCH$_3$), 85.8 (C$_5$H$_5$), 211.5, 212.2 (CO) IR: 2000, 2050 (CO) in CH$_2$Cl$_2$, 1080, 1135, 1300 (SN, SO) in Nujol	[4]
6	FpN(SO$_2$CH$_3$)-SOCH$_2$C$_6$H$_5$ I	orange–red glass ^{1}H NMR (CDCl$_3$): 2.77 (CH$_3$), 3.86, 4.28 (CH$_2$, J$_{AB}$=12.4), 5.06 (C$_5$H$_5$) ^{13}C NMR (CDCl$_3$): 43.1 (CH$_3$), 65.1 (CH$_2$), 86.0 (C$_5$H$_5$), 211.8, 212.2 (CO) IR: 2010, 2055 (CO) in CH$_2$Cl$_2$, 1080, 1140, 1300 (SN, SO) in Nujol	[4]
7	FpN(SO$_2$C$_6$H$_4$CH$_3$-4)-SOCH$_3$ I	orange–red glass ^{1}H NMR: 2.40 (2 CH$_3$), 5.17 (C$_5$H$_5$) in CDCl$_3$, 2.19, 2.58 (CH$_3$), 5.28 (C$_5$H$_5$) in C$_6$F$_6$ IR: 2015, 2060 (CO) in CHCl$_3$, 1080, 1146, 1300 (SN, SO) in Nujol	[4]
8	FpN(SO$_2$C$_6$H$_4$CH$_3$-4)-SOCH$_2$C$_6$H$_5$ I	orange–red glass	[4]
9	FpN(SO$_2$CH$_3$)$_2$ II	^{1}H NMR (CDCl$_3$): 3.06 (CH$_3$), 5.16 (C$_5$H$_5$) ^{13}C NMR (CDCl$_3$): 42.7 (CH$_3$), 85.6 (C$_5$H$_5$), 210.9 (CO) IR: 2020, 2070 (CO) in CH$_2$Cl$_2$, 1135, 1295, 1315, 1321 (SN, SO) in Nujol	[4]

References on p. 348

Table 41 [continued]

No.	compound method of prepa- ration (yield in %)	properties and remarks	Ref.
10	FpN(SO$_2$CH$_3$)- SO$_2$CH$_2$C$_6$H$_5$ II	^{1}H NMR (CDCl$_3$): 3.05 (CH$_3$), 4.64 (CH$_2$), 4.71 (C$_5$H$_5$) ^{13}C NMR (CDCl$_3$): 42.1 (CH$_3$), 60.2 (CH$_2$), 86.0 (C$_5$H$_5$), 210.9 (CO) IR: 2012, 2056 (CO) in CHCl$_3$, 1130, 1298, 1325 (SN, SO) in Nujol	[4]
11	FpN(SO$_2$CH$_3$)- SO$_2$C$_6$H$_4$CH$_3$-4 II	^{1}H NMR (CDCl$_3$): 2.44 (CCH$_3$), 3.10 (SCH$_3$), 4.92 (C$_5$H$_5$) IR: 2014, 2059 (CO) in CHCl$_3$, 1140, 1150, 1320 (SN, SO) in Nujol	[4]
12	FpN(SO$_2$CH$_3$)- S(CH$_3$)=NSO$_2$CH$_3$ III	^{1}H NMR (CDCl$_3$): 2.61, 2.96, 3.09 (CH$_3$), 5.30 (C$_5$H$_5$) ^{13}C NMR (CDCl$_3$): 39.1 (N$_2$SCH$_3$), 42.4, 43.2 (O$_2$SCH$_3$), 85.8 (C$_5$H$_5$) IR (Nujol): 1989, 2003, 2042 (CO), 1012, 1128, 1280, 1300 (SN, SO)	[4]
13	FpN(SO$_2$CH$_3$)- S(CH$_2$C$_6$H$_5$)=NSO$_2$CH$_3$ III	^{1}H NMR (CDCl$_3$): 2.36 (CH$_3$), 3.10 (CH$_3$), 4.09, 4.21 (CH$_2$, J$_{AB}$=12.1), 5.27 (C$_5$H$_5$) ^{13}C NMR (CDCl$_3$): 42.0, 42.8 (CH$_3$), 59.9 (CH$_2$), 85.8 (C$_5$H$_5$), 210.7, 212.0 (CO) IR (Nujol): 2005, 2050 (CO), 1010, 1020, 1140, 1150, 1288, 1295, 1310 (SN, SO)	[4]

*Further information:

C$_5$H$_5$Fe(CO)$_2$N(SO$_2$F)$_2$ (Table 41, No. 1) is stable at room temperature and to-wards air. The mass spectrum shows the following fragments (relative intensities): molecular ion [M]$^+$ (0.38), [M−CO]$^+$ (7.7), [M−2CO]$^+$ (57.5), [C$_5$H$_5$FeSO$_2$]$^+$ (10), [N(SO$_2$F)$_2$]$^+$ (1.5), [C$_5$H$_5$FeF]$^+$ (100), [C$_5$H$_5$Fe]$^+$ (6.1), [SO$_2$F]$^+$ (39.2), [FeF]$^+$ (34.6), and others [3].

Reaction with ^{2}D=P(C$_6$H$_5$)$_3$ or CH$_3$CN in CH$_2$Cl$_2$ produces cationic complexes [Fp^{2}D][N(SO$_2$F)$_2$] [3].

C$_5$H$_5$Fe(CO)$_2$N(SO$_2$Cl)COCH$_2$C(CH$_3$)=CH$_2$ (Table 41, No. 2) (in 2.4-fold excess) reacts with C$_6$H$_5$NH$_2$ or 4-CH$_3$C$_6$H$_4$NH$_2$ in CH$_2$Cl$_2$ at 0 °C for 1.5 h to give Nos. 3 (46%) and 4 (39%), respectively [1, 2]. However, only unreacted complex was obtained after treatment with a twofold excess of 4-NO$_2$C$_6$H$_4$NH$_2$ in CH$_2$Cl$_2$/benzene at 25 °C [2]. Fp$_2$ is formed on reaction with an excess of N(C$_2$H$_5$)$_3$ in metha-nol at ∼60 °C for 4.5 h [2] or in refluxing CH$_2$Cl$_2$ [1] but not in CH$_2$Cl$_2$ at 25 °C [2]. Treatment with excess NaOH in ethanol at 25 °C for 1 h also yields Fp$_2$ [2]. From the reaction with gaseous HCl in CHCl$_3$ at 25 °C, FpCl and CH$_2$=C(CH$_3$)CH$_2$CONH$_2$ have been isolated [1, 2].

References on p. 348

References:

[1] Y. Yamamoto, A. Wojcicki (J. Chem. Soc. Chem. Commun. **1972** 1088/9). – [2] Y. Yamamoto, A. Wojcicki (Inorg. Chem. **2** [1973] 1779/88). – [3] R. Fröböse, R. Mews, O. Glemser (Z. Naturforsch. **31** b [1976] 1497/500). – [4] R.G. Severson, A. Wojcicki (J. Organometal. Chem. **149** [1978] C66/C70).

1.5.2.3.8.3 Compounds with N-bound Heterocycles

The compounds listed in Table 42 have been prepared by the following methods:

Method I: Reaction of FpX (X=Cl, Br, I) with the Na or K salt of the heterocycle HA in CH_3CN [7, 8], C_6H_6, C_6H_6/THF [1], or THF [3, 4, 6, 8] at 45 to 60 °C for 2 to 5 h or in liquid NH_3 [2] yields FpA. Generally FpI is the starting material, but better results have been obtained with FpCl [2] and FpBr [8].

If isomers are formed, their relative yields depend on the conditions.

Method II: Reactions of 1-bound 1,2,3-triazoles FpA with H_2SO_4 or oxalic acid in THF give salts of type I [6, 8]. Similar reactions of 2-bound 1,2,3-triazoles initially give 2-bound protonated salts [6, 8], which rearrange readily on heating in THF at 45 to 60 °C to I [8].

Method III: Reaction of FpA (prepared by Method I) in H_2O/THF or H_2O/$(CH_3)_2CO$:

a. with $Na[B(C_6H_5)_4]$ gives solvent complexes of the $[FpAH^2D]$-$[B(C_6H_5)_4]$ type II, $^2D = (CH_3)_2CO$ or H_2O [7],

b. with $Na[NCB(C_6H_5)_3]$ gives ion pairs $[FpAH][NCB(C_6H_5)_3]$ of type III, in which the protonated heterocycle $[FpAH]^+$ forms a hydrogen bond with the anion $[NCB(C_6H_5)_3]^-$ [7]. Yields given for Method III are based on FpX.

Method IV: Refluxing of $[FpAH^2D][B(C_6H_5)_4]$ in THF for 1.5 h gives the corresponding donor-acceptor complexes $FpAB(C_6H_5)_3$ with $B(C_6H_5)_3$ acting as a Lewis acid towards the N-3 atom of the imidazole or 1,2,3-triazole derivative FpA [7].

In the case of triazoles the salts have been obtained only from 1-N-Fp derivatives. No NH resonance could be detected in the ^1H NMR spectra. Unlike the acetone complexes, the aqua complexes show an intense broad OH band at 3230 to 3310 cm^{-1} from coordinated water. The salts are soluble in some polar solvents (THF, acetone, methanol). The mechanism of the thermal decomposition in Method IV was discussed [7].

The triphenylcyanoborates $[FpAH][NCB(C_6H_5)_3]$ (Formula III) are formulated as contact hydrogen-bonded ion pairs in which the NH group of the heterocyclic ligand is blocked by hydrogen bonding with the nitrile group of an outer-sphere anion.

The ^1H NMR spectra confirm the hydrogen bond formation by showing broad N-H signals in a downfield region. Present below -70 °C, these signals disappear at higher temperatures, probably owing to quadrupole relaxation. In the case of No. 13b the abnormally great downfield shift of the NH resonance is thought to be due to the effect of a neighbouring phenyl group in the heterocycle located planarly. The complex pattern of the N-H vibrations in the IR spectra in the region 2500 to 2900 cm^{-1} is specific of hydrogen-bond systems [7].

The 1,2,3-triazole derivatives are soluble in THF [7], $(CH_3)_2CO$, H_2O [4], and C_6H_6 [8].

Nos. 15 and 16 are comparatively stable in the solid state but decompose in ether [4]. The mass spectra show the parent ions $[M]^+$ and fragments resulting from successive loss of the CO groups and further elimination of nitrogen and HCN molecules. As usual, in the spectra of compounds containing the Fp moiety, a rearrangement ion, $[(C_5H_5)_2Fe]^+$, is observed. The presence of $[C_5H_5Fe(CO)_2]^+$ and $[C_5H_5FeCO]^+$ ions indicates elimination of the heterocyclic ligand but retention of the carbonyl and C_5H_5 groups. With the exception of the peak $[C_5H_5FeC_6H_4N]^+$, which is of considerably lower intensity in the spectrum of compound No. 16, the mass spectra of the two isomers are practically identical. The difference may result from a No. 16 to No. 15 rearrangement before nitrogen is eliminated [4]. Similar fragmentation patterns have been observed and discussed in detail for compounds No. 10, 11, 13, and 14 [8].

The symmetric triazoles No. 11 and 16 rearrange irreversibly on heating in neutral media (C_6H_6 or THF) into the 1-bound isomers No. 10 and 15, respectively [4, 8]. However, 1-bound No. 13 forms 2-bound No. 14 on heating in THF [8]. All triazole derivatives are easily protonated at the vacant imino groups by mineral and organic acids and even by water. Dibasic acids like H_2SO_4 and oxalic acid react with 1-bound triazoles to form salts of type I, which are insoluble in THF. All 2-bound triazoles rearrange easily in H_2SO_4 and oxalic acid with formation of the salts I, which are stabilized by chelate hydrogen bonding to both free imino groups of the triazole [8].

The IR spectra of the salts I contain no stretching modes of free OH and NH groups but exhibit diffuse absorption typical of hydrogen bonds. Hindered prototropic rearrangement between N-2 and N-3 at low temperature is deduced from the temperature dependence of the ^1H and ^{13}C NMR spectra of solutions of compound No. 15a in methanol [8].

In Table 42 the heterocycles are arranged by the number of N atoms. $B(C_6H_5)_3$ complexes and salts are placed with the parent heterocycles.

The pyrrole derivative No. 1 converts on refluxing in benzene for 1 h [1] or at 100 °C in toluene for 1 h [2] into a mixture of compounds containing azaferrocene as main product, while No. 4 [1], 6, 7, and 10 [2] do not convert. An unstable π complex derived from No. 3 has not been characterized fully [1].

Imidazole and 1,2,3-triazole derivatives form with $B(C_6H_5)_3$ the donor-acceptor complexes No. 7a, 8a, 9a, 10b, and 15c (see Method IV), which are soluble in THF and $(CH_3)_2CO$ and stable in air but decompose without melting above 100 to 120 °C [7].

The solvent complexes $[FpAH^2D][B(C_6H_5)_4]$ of type II are composed of ion pairs separated by a solvent molecule (acetone or water) which is retained in an outer

References on p. 355

coordination sphere by a hydrogen bond with the heterocyclic amino group at N–3. The anions are held via nonspecific electrostatic interactions. This formulation of the salts was confirmed by X-ray analysis of No. 8c.

Table 42
Compounds FpX with N–Bound Heterocycles.
Further information on compound No. 8c is given at the end of the table.
For abbreviations and dimensions, see p. 375.

No.	compound method of preparation (yield in %)	properties and remarks	Ref.
1	Fp—N (pyrrole) I (19)	brown, m.p. 91°; subl. 60 to 70°/0.01 Torr ^{1}H NMR (CS$_2$): 4.93 (s, C$_5$H$_5$), 5.85 (d, β–H), 6.03 (m, α–H) IR: 1975, 2020 (CO)	[1]
2	COCH$_3$ Fp—N I (28)	yellow, m.p. 110 to 111°; subl. 80 to 90°/0.01 Torr ^{1}H NMR (CS$_2$): 2.11 (s, CH$_3$CO), 4.75 (s, C$_5$H$_5$), 6.05, 6.55 (m, β–H), 6.85 (m, α–H) IR: 1990, 2020 (CO)	[1]
3	Fp—N (indole) I (31.5)	red, m.p. 114 to 114.5° ^{1}H NMR (CS$_2$): 4.94 (s, C$_5$H$_5$), 6.7 (m, 6H, aromatic) IR: 1975, 2020 (CO)	[1]
4	Fp—N (carbazole) I (25)	red, m.p. 150 to 155° (dec.) ^{1}H NMR (CS$_2$): 5.05 (s, C$_5$H$_5$), 7.2, 7.75 (m's, 6+2H, aromatic), (CDCl$_3$): 5.1, 7.3, 8.1 IR: 1978, 2030 (CO)	[1]
5	Fp—N (carbazole) I (15)	m.p. 150° (dec.) IR: 1985, 2020 (CO)	[1]
6	Fp—N (pyrazole) I	unstable oil	[2]

Table 42 [continued]

No.	compound method of preparation (yield in %)	properties and remarks	Ref.
7	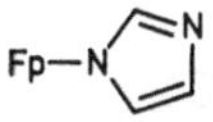 Fp—N (imidazole ring)	unstable oil	[2, 7]
7a	$FpC_3H_3N_2B(C_6H_5)_3$ IV (54.9)	yellow ^1H NMR (THF): 5.33 (s, C_5H_5), 6.71 (s, 2H), 7.11 (m, C_6H_5) IR (KBr): 2004, 2056 (CO)	[7]
7b	$[FpC_3H_3N_2H]$-$[NCB(C_6H_5)_3]$ IIIb (25.7)	^1H NMR (CH_2Cl_2, $-88°$): 5.30 (s, C_5H_5), 7.10 (m, C_6H_5), 12.15 (NH) IR (KBr): 2015, 2064 (CO), 2192 (CN), 2590 to 3210 (NH)	[7]
7c	$[FpC_3H_3N_2H(H_2O)]$-$[B(C_6H_5)_4]$ IIIa (32.5)	yellow IR (KBr): 2013, 2059 (CO), 3311 (NH) recrystallization in acetone gives No. 7d	[7]
7d	$[FpC_3H_3N_2H(OC(CH_3)_2)]$-$[B(C_6H_5)_4]$ see No. 7c	yellow ^1H NMR (CH_3COCH_3, $34°$): 5.50 (s, C_5H_5), 7.01, 7.43 (m's, C_6H_5), 8.11 (s, 2H)	[7]
8	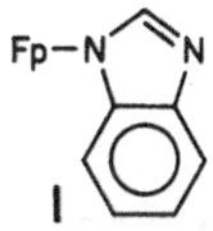 Fp—N (benzimidazole ring)	no properties reported	[7]
8a	$FpC_7H_5N_2B(C_6H_5)_3$ IV (57.3)	yellow ^1H NMR (THF?): 5.45 (s, C_5H_5), 6.87 (s, 2H), 7.11 (m, C_6H_5), 7.23 (m, heterocycle) IR (KBr): 2008, 2055 (CO)	[7]
8b	$[FpC_7H_5N_2H]$-$[NCB(C_6H_5)_3]$ IIIb	^1H NMR (THF, $34°$): 5.29 (s, C_5H_5), 7.01, 7.57 (m's, C_6H_5), 8.02 (s, 2H) IR (KBr): 2002, 2060 (CO), 2176 (CN), 2670 to 3140 (NH)	[7]
8c	$[FpC_7H_5N_2H(OC(CH_3)_2)]$-$[B(C_6H_5)_4]$ IIIa (28)	yellow ^1H NMR (CH_2COCH_3, $34°$): 5.49 (s, C_5H_5), 6.93, 7.34 (m's, C_6H_5), 8.50 (s, 2H) IR (KBr): 2009, 2050 (CO), 3300 (NH)	[7]
9	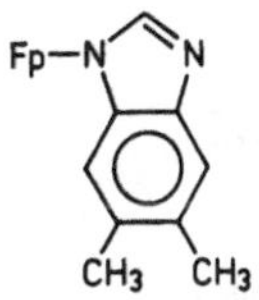 Fp—N (dimethylbenzimidazole ring) CH_3 CH_3	no properties reported	[7]

References on p. 355

Table 42 [continued]

No.	compound method of preparation (yield in %)	properties and remarks	Ref.
9a	$FpC_9H_9N_2B(C_6H_5)_3$ IV (31.8)	yellow ^1H NMR $(CD_3COCD_3, 34°)$: 2.01, 2.34, 7.55, 7.62 (s's, $C_9H_9N_2$), 5.55 (s, C_5H_5), 6.68 (s, 2H), 7.12 (m, C_6H_5) IR (KBr): 2003, 2053 (CO)	[7]
9b	$[FpC_9H_9N_2H]$- $[NCB(C_6H_5)_3]$ IIIb (45.8)	^1H NMR (THF, $-100°$): 2.31, 2.42 (s's, $C_9H_9N_2$), 5.20 (s, C_5H_5), 6.99, 7.48 (m's, C_6H_5), 8.25 (s, 2H), 13.34 (NH) IR (KBr): 2012, 2058 (CO), 2181 (CN), 2660 to 3150 (NH)	[7]
9c	$[FpC_9H_9N_2H(H_2O)]$- $[B(C_6H_5)_4]$ IIIa (30.5)	yellow IR (KBr): 2002, 2050 (CO), 3296 (NH)	[7]
9d	$[FpC_9H_9N_2H(OC(CD_3)_2)]$- $[B(C_6H_5)_4]$ from No. 9c in CD_3COCD_3	^1H NMR $(CD_3COCD_3, 34°)$: 2.36, 2.47 (s's, $C_9H_9N_2$), 5.56 (s, C_5H_5), 7.05, 7.52 (m's, C_6H_5), 8.51 (s, 2H)	[7]
10	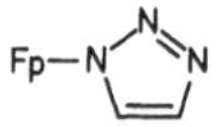 Fp—N (triazole) I (26.5) see also No. 11	oil ^1H NMR (THF): 5.30 (s, C_5H_5), 7.19, 7.38 (s's, triazole) IR (THF): 2013, 2063 (CO)	[2, 8]
10a	$[FpC_2H_2N_3H]$- $[HSO_4]$ II	^1H NMR (CH_3OH): 5.62 (s, C_5H_5), 8.27, 8.35 (s's, triazole), 17.01 (at $-90°$, s, NH) IR (KBr): 2024, 2069 (CO)	[8]
10b	$FpC_2H_2N_3B(C_6H_5)_3$ IV	yellow ^1H NMR $(CD_3COCD_3, 34°)$: 5.43 (s, C_5H_5), 7.15 (m, C_6H_5), 7.34, 7.72 (s's, triazole) IR (KBr): 2009, 2055 (CO)	[7]
11	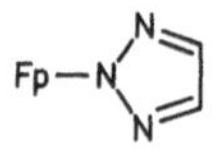 Fp—N (triazole) I (4.5)	^1H NMR (THF): 5.32 (s, C_5H_5), 7.39 (s, triazole) IR (THF): 2014, 2062 (CO) refluxing in THF gives No. 10 (32.9%)	[8]

References on p. 355

Table 42 [continued]

No.	compound method of preparation (yield in %)	properties and remarks	Ref.
12	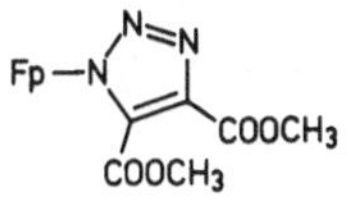 Fp—N triazole with COOCH₃ and COOCH₃ substituents	prepared from FpN_3 and $CH_3OOCC\equiv C-COOCH_3$ in boiling C_6H_6 (37%) yellow orange, m.p. 155.5 to 156.5° (gas evolution) ^1H NMR ($CDCl_3$): 3.88 (s, CH_3), 5.24 (s, C_5H_5) IR (KBr): 2015, 2075 (CO), 1730, 1750 ($COOCH_3$)	[5]
13	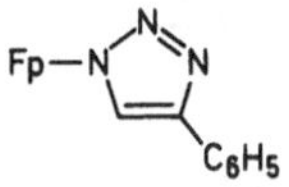 Fp—N triazole with C₆H₅ substituent I	^1H NMR (THF): 5.18 (s, C_5H_5), 6.85 (m), 7.26 (s), 7.34 (m), all triazole IR (THF): 2010, 2060 (CO) Method I in CH_3CN gives Nos. 13 and 14, at 25°/1 h (7.9 and 3.4% yield), at 45°/2.5 h (6.1 and 36.5% yield), respectively; isomerizes in THF at 60° to give No. 14 in 59% yield	[7, 8]
13a	[$FpC_8H_6N_3H$]-[HSO_4]·C_4H_8O·$0.5H_2O$ II, see also No. 14	golden scaly crystals ^1H NMR (CH_3OH): 5.61 (s, C_5H_5), 7.58, 7.87 (m's, triazole, 2.5+2.5 H), 8.57 (s, triazole, 1 H), 17.10 (at −90°, s, NH) IR (KBr): 2019, 2068 (CO)	[8]
13b	[$FpC_8H_6N_3H$]-[$NCB(C_6H_5)_3$] IIIb (29.8)	^1H NMR (THF): 5.38 (s, C_5H_5), 7.35, 7.60 (m's, BC_6H_5), 7.91 (m), 8.61 (s, $C_8H_6N_3$), 16.09 (NH) at −70°; 5.22 (s, C_5H_5), 6.95, 7.40 (m's, BC_6H_5), 7.78 (m, $C_8H_6N_3$), 8.29 (s, $C_8H_6N_3$) at +34° IR (KBr): 2028, 2072 (CO), 2196 (CN), 2370 to 2980 (NH)	[7]
13c	[$FpC_8H_6N_3H(H_2O)$]-[$B(C_6H_5)_4$] IIIa	yellow IR (KBr): 2009, 2062 (CO), 3233 (NH)	[7]
13d	[$FpC_8H_6N_3H(OC(CH_3)_2)$]-[$B(C_6H_5)_4$] IIIa	yellow ^1H NMR (CH_3COCH_3): 5.64 (s, C_5H_5), 7.10, 7.33 (m's, BC_6H_5), 7.71 (m, $C_8H_6N_3$), 8.26 (s, $C_8H_6N_3$)	[7]
14	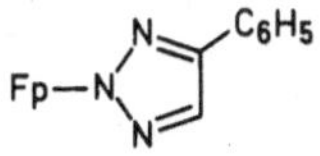 Fp—N triazole with C₆H₅ substituent I, see No. 13	^1H NMR (CH_3COCH_3): 5.26 (s, C_5H_5), 6.97, 7.41 (m's, $C_8H_6N_3$), 7.49 (s, $C_8H_6N_3$) IR (THF): 2010, 2060 (CO) reaction with H_2SO_4 in THF at 60° gives No. 13a	[8]

References on p. 355

Table 42 [continued]

No.	compound method of prepa- ration (yield in %)	properties and remarks	Ref.

15	Fp—N (triazole fused benzene ring structure)
	
	I (33), see
	also No. 16
		yellow
		^{1}H NMR (THF, 34°): 5.51 (s, C_5H_5), 7.18,
		   7.78 (m's, $C_6H_4N_3$)
		IR (THF): 2004, 2054 (CO)
		method I in THF gives Nos. 15 and 16, at
		   20°/1 h (0.02 and 10% yield), at 45°/2.5 h
		   (33 and 4% yield), respectively
			[4, 6, 8]

15a	[$FpC_6H_4N_3H$]-
	[HSO_4]
	II
		yellow
		^{1}H NMR (CH_3OH): 5.54 (s, C_5H_5), 7.82,
		   8.07 (m's, $C_6H_4N_3$), 17.29 (at $-90°$,
		   s, NH)
		^{13}C NMR (CH_3OH, 10°): 64.00, 68.16,
		   78.49, 80.18, 85.77, 98.37 (also given at
		   -85 and $-116°$)
		IR: 2020, 2066 (CO) in KBr, 2029, 2075
		   (CO) in CH_3OH, 2015, 2064 (CO) in
		   CH_3SOCH_3
			[8]

15b	[$FpC_6H_4N_3H$]- [HC_2O_4] II	yellow IR (KBr): 2018, 2069 (CO)	[8]
15c	$FpC_6H_4N_3B(C_6H_5)_3$ IV (92.1 from benzotriazole)	yellow ^{1}H NMR (CD_3COCD_3, 34°): 5.61 (s, C_5H_5), 7.20 (m, C_6H_5) IR (KBr): 2006, 2062 (CO)	[7]
15d	[$FpC_6H_4N_3H(H_2O)$]- [$B(C_6H_5)_4$] IIIa	yellow IR (KBr): 2007, 2061 (CO), 3240 (NH)	[7]
15e	[$FpC_6H_4N_3H(OC(CD_3)_2)$]- [$B(C_6H_5)_4$] IIIa	^{1}H NMR (CD_3COCD_3, 34°): 5.70 (s, C_5H_5), 7.40, 7.95 (m's, C_6H_5)	[7]
16	Fp—N (triazole fused benzene ring structure) I (see No. 15)	dark yellow ^{1}H NMR (THF, 34°): 5.48 (s, C_5H_5), 7.04, 7.66 (m's, $C_6H_4N_3$) IR (THF): 2020, 2060 (CO) refluxing in C_6H_6 for 1 h gives No. 15 (30%) $+H_2SO_4$ (THF, 45°) → No. 15a $+H_2C_2O_4$ (THF, 45°) → No. 15b	[4, 6, 8]
16a	[$FpC_6H_4N_3H$][HSO_4] II	instable, rearranges to No. 15a ^{1}H NMR (THF): 5.90 (s, C_5H_5), 8.20, 8.62 (m's, $C_6H_4N_3$), 14.00 (s, NH)	[8]

*Further information:

[$C_5H_5Fe(CO)_2C_7H_5N_2H(OC(CH_3)_2)$][$B(C_6H_5)_4$] (Table **42**, No. **8c**) crystallizes in the monoclinic system with a = 13.626(1), b = 25.410(7), c = 10.088(1) Å, and
β = 92.65(1)°; space group P2$_1$/n — C$_{2h}^5$. Z = 4 gives D$_c$ = 1.285 and D$_m$ = 1.28 g·cm^{-3}.
The crystal is composed of discrete cations [$FpC_7H_5N_2H$]$^+$, anions [$B(C_6H_5)_4$]$^-$ and
acetone molecules combined with the unsubstituted imidazole N atom via N–H···O
hydrogen bonds, as shown in **Fig. 26**. The dihedral angle between the mean planes
of C_5H_5 and $C_7H_5N_2H$ ligands is 32.0°. The dihedral angle between $C_7H_5N_2H$ and
$OC(CH_3)_2$ is 51.3°. The [$B(C_6H_5)_4$]$^-$ anion is distorted [7].

Fig. 26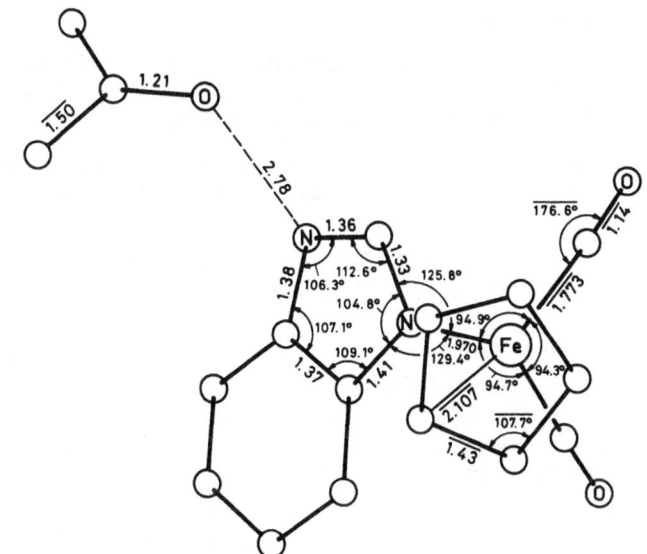

Structure of the [$C_5H_5Fe(CO)_2C_7H_5N_2H(OC(CH_3)_2)$]$^+$ cation [7].

References:

[1] P.L. Pauson, A.R. Qazi (J. Organometal. Chem. **7** [1967] 321/4). – [2] F.
Seel, V. Sperber (J. Organometal. Chem. **14** [1968] 405/10). – [3] V.N. Babin, L.A.
Fedorov, N.S. Kochetkova, Yu.A. Belousov (5th Intern. Conf. Organometal. Chem.,
Moscow 1971, Vol. 2, Abstr. No. 390). – [4] A.N. Nesmeyanov, V.N. Babin, N.S. Kochetkova, E.I. Mysov, Yu.A. Belousov, L.A. Fedorov (Dokl. Akad. Nauk SSSR **200**
[1971] 1112/5; Dokl. Chem. Proc. Acad. Sci. USSR **196/201** [1971] 838/41). –
[5] A. Rosan, M. Rosenblum (J. Organometal. Chem. **80** [1974] 103/7).

[6] A.N. Nesmeyanov, Yu.A. Belousov, V.N. Babin, E.B. Zavelovich, N.S. Kochetkova (Tezisy Dokl. 12th Vses. Chugaevskoe Soveshch. Khim. Kompleksn. Soedin.,
Novosibirsk 1975, Vol. 3, pp. 476/7; C.A. **86** [1977] No. 5590). – [7] A.N. Nesmeyanov, Yu.A. Belousov, V.N. Babin, G.G. Aleksandrov, Yu.T. Struchkov, N.S. Kochetkova (Inorg. Chim. Acta **23** [1977] 155/62). – [8] A.N. Nesmeyanov, Yu.A. Belousov, V.N. Babin, N.S. Kochetkova, S.Yu. Sil'vestrova, E.I. Mysov (Inorg. Chim. Acta
23 [1977] 173/9).

1.5.2.3.9   Compounds with Fe-E Bonds, where E = P, As, Sb, and Bi

The compounds in this section are collected in Table 43. All elements given in the heading form complexes of the Fp-E(X,R)$_2$ type (Nos. 1 to 5, 28 to 34, 42 to 48, and 55 to 57). For phosphorus, there are also Fp-P(E')E(X,R)$_2$ compounds containing pentavalent P with E' = O, S, and Se (Nos. 14 to 27). Fp-E(X,R)$_2$ compounds, except for FpBiX$_2$, can act as donors towards other transition metal carbonyls to give Fp-E(X,R)$_2$ →M(CO)$_n$L complexes (Nos. 6 to 13, 35 to 41, and 49 to 54). Some complexes with phosphorus- and arsenic-borane ligands are included at the end of Table 43.

The most common methods of preparation are given below.

Method I: Reactions of Na[Fp]

a. with ER$_2$X in nonpolar solvents or THF at −65 to +25 °C to yield FpER$_2$ for E = P, As, Sb and X = Cl [2, 7, 13, 23, 40] and Br [31]. EBr$_3$ compounds are reacted at 65 °C [31].

b. with Cl(CH$_3$)$_2$E→M compounds, where M stands for various transition metal carbonyl groups. Reactions in THF at room temperature yield Fp(μ-E(CH$_3$)$_2$)M with E = P and As [16, 17, 37].

c. with P(E')R$_2$X in THF at room temperature to yield FpP(E')R$_2$ with E' = O, S, Se and X = Cl, Br [33].

Method II: Reactions of Fp$_2$

a. with EX$_3$ in CH$_2$Cl$_2$ or C$_6$H$_6$ at room temperature to yield FpEX$_2$, E = As, Sb, Bi and X = Cl, Br, I. For As, [Fp$_2$AsX$_2$][FeX$_4$] is formed as byproduct [5, 6, 15]. SbCl$_3$ and SbBr$_3$ react only in C$_6$H$_6$ [5], for mechanisms see [5, 15].

b. with R$_2$E-ER$_2$ without solvent by refluxing for 3 h [2], in CFCl$_3$ at 65 to 70 °C for 4 h [11, 12, 18], or in toluene for 16 h [1] to yield FpER$_2$ with E = P and As.

c. with R$_2$P-E'-PR$_2$ (E = O, S, Se, mole ratio ~1:2) in CFCl$_3$ at 20 °C for 6 h to give 1:1 mixtures of FpPR$_2$ and FpP(E')R$_2$ for E' = S or Se, while for E' = O the main product is Fp(μ-PR$_2$)Fe(C$_5$H$_5$)(CO)-P(O)R$_2$ (see C3, 2.5.2.2.2, p. 108) [21].

Method III: Reactions of FpX (X = Cl [13] and I [4, 19, 42]) with R$_2$EM (M = Li [13, 39], Na [19], and K [4, 42]) in C$_6$H$_6$, THF, (CH$_3$OCH$_2$-)$_2$, or cyclopentane at room temperature to give FpER$_2$.

Method IV: Reaction of FpCl with excess P(OR)$_3$ in C$_6$H$_6$ at 25 °C to give the main product FpPO(OR)$_2$ along with C$_5$H$_5$Fe(CO)(P(OR)$_3$)Cl and small amounts of C$_5$H$_5$Fe(CO)(P(OR)$_3$)PO(OR)$_2$ [3, 8]. Similar reactions were used with P(OCH$_2$CH=CH$_2$)$_2$C$_6$H$_5$, P(OCH$_2$CH=CH$_2$)(C$_6$H$_5$)$_2$ [3, 8], and P(OCH$_2$CH$_2$Cl)(CF$_3$)$_2$ [41]. For mechanisms see [3, 8].

Method V: Reaction of [FpC$_6$H$_{10}$-cyclo]PF$_6$ with the [N(CH$_3$)$_4$]$^+$ salts of [B$_{10}$H$_{12}$P]$^-$, [B$_{10}$H$_{12}$As]$^-$, or [7,8-B$_9$H$_{10}$As$_2$]$^-$ or with K[7,8-B$_9$H$_{10}$CHP] in acetone at room temperature for 2 h and subsequent refluxing for 3 h [47].

References on p. 373

Phosphides, Arsenides, and Antimonides. Solutions of compounds No. 28, 34, and 43 in suitable solvents (Table 43) exhibit four $\nu(CO)$ absorption bands instead of the two predicted. This $\nu(CO)$ doubling is attributed to conformational isomerism owing to hindered rotation about the Fe-As and Fe-Sb bonds [31, 40]. Assignment of the CO stretching frequencies to the separate rotamers has been accomplished for No. 34 by use of the ^{13}CO satellite bands [40] but was unsuccessful for Nos. 28 and 43 because of their low solubility [31, 40]. The favored conformations of No. 34 are I and II (also for No. 28, with two Cl atoms at the oxygen positions in I and II). Steric factors are responsible for the stabilities [40].

Isomer I (C_s)   Isomer II (C_1)

⬤ Fe
◯ As

I   II

In contrast, the low-temperature ^1H NMR spectra of Nos. 34 ($-60\,°C$) and 43 ($-80\,°C$) do not show any sign of isomerism resulting from hindered rotation about the Fe-As or Fe-Sb bonds [31, 40]. However, the spectra of No. 34 ($-60\,°C$ in $C_6D_5CD_3$) show a clear broadening of the CH_3 peaks both in ^1H and ^{13}C NMR, indicating the beginning of ring-inversion freezing [40].

The mass spectra of Nos. 3, 32, 43, and 45 all show the parent ion $[M]^+$ and the fragments $[M-nCO]^+$ with $n=1$ to 2 [2, 31]. There are also strong peaks corresponding to the species $[(C_6F_5)_2E]^+$, $[C_6F_5E]^+$ ($E=P$, As), and $[C_5H_5Fe]^+$ [2].

Transition Metal Carbonyls Substituted by Phosphides, Arsenides, and Antimonides. In the ^1H NMR spectra of Nos. 35 to 40 the CH_3 signals are at lower field than those of $FpAs(CH_3)_2$. This was ascribed to a more positive As atom due to coordination. This inductive effect influences the chemical shifts of the FeC_5H_5 protons only to a small degree so that their positions are generally quite similar to those of the free base [35]. In the case of Nos. 51 to 54 the signals are also shifted somewhat to low field [31]. The C_5H_5 resonance of Nos. 7, 8, and 10 (nothing reported about No. 9) is a closely spaced doublet on account of P,H coupling [28]. The ^{19}F NMR spectra of Nos. 7 to 10 also consist of a simple doublet in every case. The range of the ^{19}F NMR shifts is narrow, emphasizing the similar magnetic environment of the CF_3 groups, always to high field of those for $FpP(CF_3)_2$. The equivalence of the CF_3 ligands and the detection of only one signal for each complex is attributed to rotation about metal-phosphorus bonds that is rapid on the NMR time scale. Neither No. 8 nor 9 shows changes in the ^{19}F NMR spectrum when cooled to $-60\,°C$ [28].

The coordination shifts the $\nu(CO)$ absorptions in the IR spectra of the bridged compounds to higher frequencies than those found for the free $FpE(R,X)_2$ complexes [28, 31, 35, 40]. The bands of nearly equal intensity in the spectrum of each of the compounds No. 7 to 9, at ~2020 and $\sim2055\ cm^{-1}$, are assigned to the $\nu(CO)$ vibrations of the Fp grouping. In the case of No. 10 the use of the more polar acetone slightly lowers the wave number [28]. The observed doubling of the $\nu(CO)$ bands of the Fp unit in Nos. 37 [35], 52, and 53 [31] in cyclohexane has been ascribed to conformational isomerism owing to hindered rotation about the Fe-E bond, while the additional $\nu(CO)$ absorptions in the IR spectra of Nos. 8 and 9 are attributed to restricted rotation about the P-Co and P-Mn bonds, respectively [28]. Similar explanations are given for the additional v(NO) vibrations of Nos. 8, 9 [28], and 36 [35]. In solvents other than cyclohexane, such as THF, CH_2Cl_2, or C_6H_6, the critical bands are only broadened by solvent effects [31]. The results of variable-temperature IR experiments (20 to $-40\,°C$) on the relative intensities of the v(CO) and v(NO) absorptions of No. 8 were found to be also consistent with rotational isomerism [28]. The absence of v(CO) doubling in Nos. 41 [40] and 51 [31] is supposed to indicate the presence of only one particularly stable conformer. The importance of steric and electronic factors in causing isomerism is discussed in [31].

Compounds No. 8 to 10 are reported to be stable in air for short periods, but solutions are markedly air sensitive [28]. Compounds No. 35 to 40 are air stable in the solid state, and, with the exception of No. 35, they are also stable in solution [35].

The mass spectra of Nos. 51 to 54 show the parent ion $[M]^+$. The dominant fragmentation process is the loss of the CO groups. Sb-M cleavage tends to produce the starting $M(CO)_6$ complexes (M = Cr, W) and $C_5H_5Mn(CO)_3$ [31].

While photolysis in benzene at 20 °C does not affect Nos. 6 [16], 35, and 36 [35], Nos. 37 to 40 decompose to give Fp_2, $M(CO)_6$, and $C_5H_5Mn(CO)_3$, generally as main products in addition to smaller amounts of $(M(CO)_4)(\mu-As(CH_3)_2)_2$ (M = Mo, W), $(C_5H_5Fe(CO)\mu-As(CH_3)_2)_2$, and unidentified substances [35]. Photolysis of No. 38 in cyclohexane forms a flaky precipitate, from which $Cr(CO)_6$ was sublimed, while the residue, insoluble in cyclohexane and benzene, could not be identified [17].

Electrochemical oxidation and reduction of Nos. 36, 37, and 40 in C_6H_5CN at Pt electrodes gave the following half-wave potentials (vs. Ag/AgCl):

No.		$E_{1/2}$ (in V) oxidation	reduction
36	$FpAs(CH_3)_2Co(CO)_2NO$	0.760	-1.480
37	$FpAs(CH_3)_2Mn(CO)_2C_5H_5$	0.480	-1.610
40	$FpAs(CH_3)_2W(CO)_5$	0.1090	-1.430

Only the oxidation of No. 37 was reversible. In all cases $[Fp]^-$ could be identified as reduction product. The results are discussed in connection with other arsenic-bridged dinuclear metal carbonyls [48].

$C_5H_5Fe(CO)_2P(E')(X,R)_2$ Compounds with Pentavalent P. The structures of the products have been established in most cases from their spectra. In general, there is coupling between the C_5H_5 protons and the phosphorus, revealing the presence of an Fe-P rather than an Fe-E' bond [18, 33]. The apparent absence of coupling in some cases has been related to the broadness of the peaks [8, 18]. While in No. 23 the phosphorus coupling of the C_5H_5 resonance seemed beyond the resolution of the spectrometer, in the alkylated derivatives the doublet was indeed observed [33]. Also the ^{19}F NMR chemical shifts of Nos. 14, 24, and 26, which appear as doublets at higher field from the $FpP(CF_3)_2$ resonance, and the enlarged $^2J(F,P)$ coupling constants are consistent with pentavalent P [11, 18]. A comparison of the NMR of No. 22 with that obtained for P^{III} and P^V derivatives as well as with analogous complexes containing metal-P bonds is reported [10] and confirms this type of bonding.

References on p. 373

The IR spectra of FpP(E')R$_2$ (E'=S, Se) show that the v(CO) frequencies increase with the electron-withdrawing power of R; thus, the complexes where R=CF$_3$ show the highest values and the complex containing CH$_3$ groups shows the lowest v(CO) values. This suggests a higher electron-density donation $d_\pi \rightarrow p_\pi^*$ from metal to CO in the CH$_3$ derivative than in the CF$_3$ derivative. The lowest δ(C$_5$H$_5$) value in the ^{1}H NMR spectrum is observed when R=CH$_3$ [33]. The almost identical CO stretching frequencies in the complexes FpP(E')(CF$_3$)$_2$ (E'=O, S, Se) suggest that the Fe-P bonding is quite similar in all these complexes [20].

The decrease in the electronegativity along the series may be compensated by a decrease in the competitive E'$\rightarrow$Pπ bonding [20].

Compounds No. 14, 24, and 26 are air stable, and most of the solutions are stable for short periods in the presence of air and light. The parent ion is observed in the mass spectrum, and fragmentation involves loss of the CO groups and fluorocarbon fragments [18].

UV irradiation of Nos. 24 and 26 in acetone gives the isomeric complexes FpE'P(CF$_3$)$_2$ (E'=S, Se, see 1.5.2.3.6.4, p. 326, and 1.5.2.3.7, p. 341), whereas photolysis of No. 14 in CH$_2$Cl$_2$ produces a complex mixture of products, from which ferrocene, (C$_5$H$_5$Fe(CO)P(CF$_3$)$_2$)$_2$ (see C4, 2.5.2.3.1.4, p. 17) in small quantities, and (C$_5$H$_5$)$_2$Fe$_3$(CO)$_2$(OP(CF$_3$)$_2$)$_4$ as the major product have been isolated. When a sample of No. 24 slightly contaminated with sulfur is irradiated in CH$_2$Cl$_2$ for one day, the product formed is FpSP(S)(CF$_3$)$_2$, see 1.5.2.3.6.4, p. 326 [11, 18].

Protonation of Nos. 14, 24, and 26 yielding the cations [Fp{P(CF$_3$)$_2$(E'H)}]$^+$ is effected by treating the complexes with a mixture of HCl and SnCl$_4$ in CH$_3$NO$_2$. The protonation of No. 26 is always accompanied by precipitation of red selenium and formation of [Fp(P(CF$_3$)$_2$H)]$^+$. The formulation of the products is mainly based on the NMR spectra. While No. 26 deposits only selenium in 98% H$_2$SO$_4$, the NMR spectra of Nos. 14 and 24 in H$_2$SO$_4$ have shown almost identical results to those obtained with HCl-SnCl$_4$ [22].

Compounds No. 23, 25, and 27 react with CH$_3$I or [O(CH$_3$)$_3$][BF$_4$] in CH$_2$Cl$_2$ at room temperature by S or Se methylation to afford the ionic complexes [FpPR$_2$(E'CH$_3$)]X (X=I, BF$_4$). But alkylation with C$_2$H$_5$I or [O(C$_2$H$_5$)$_3$][BF$_4$] is only possible with No. 23 [33].

On treating Nos. 23, 25, and 27 with an excess of P(C$_6$H$_5$)$_3$ or P(CH$_3$)$_2$C$_6$H$_5$ (=^{2}D) in refluxing ligroin one CO group is replaced by the ^{2}D ligand to give C$_5$H$_5$-Fe(CO)(^{2}D)P(E')R$_2$ [33].

Compounds with Heteroatom Borane Ligands. On the basis of the NMR data it is proposed that in Nos. 58 to 61 the borane is σ-bonded to iron via phosphorus or arsenic as shown in **Fig. 27** for No. 58.

Fig. 27

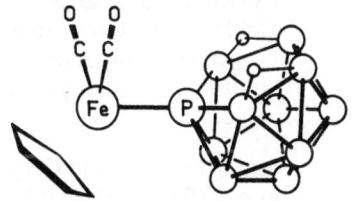

Proposed structure for C$_5$H$_5$Fe(CO)$_2$PB$_{10}$H$_{12}$ [47].

References on p. 373   24*

In the ^{11}B NMR spectra all signals are doublets with $J(B,H) = 150 \pm 15$ Hz. All complexes change the color irreversibly at ~ 200 °C [47].

Table 43
Compounds with Fe–E Bonds, where E = P, As, Sb, and Bi.
Further information on numbers preceded by an asterisk is given at the end of the table.
For abbreviations and dimensions, see p. 375.

No.	compound method of preparation (yield in %)	properties and remarks IR assigned to $\nu(CO)$ if not stated otherwise	Ref.
	compounds with trivalent P		
*1	FpP(CF$_3$)$_2$ IIb (63), IIc, IV (small)	orange brown, m.p. 56 to 58° ^{1}H NMR (CH$_3$COCH$_3$): 5.4 (d, C$_5$H$_5$, ^{3}J(P,H) = 2.3) ^{19}F NMR (CH$_3$COCH$_3$): 45.5 (^{2}J(P,F) = 54.3) IR: 2000, 2046 in CH$_2$Cl$_2$, 2008, 2048 in C$_6$H$_{14}$, 2008, 2049 in CCl$_4$ prepared also from Fp$_2$ + HP(CF$_3$)$_2$ at 100° for 7 h (42%)	[12, 18, 21, 28, 41]
*2	FpP(C$_6$H$_5$)$_2$ Ia, III	IR (C$_6$H$_{12}$): 1966, 2015	[7, 13, 19]
3	FpP(C$_6$F$_5$)$_2$ Ia (25), IIb and III failed	red, m.p. 132° ^{1}H NMR (CDCl$_3$): 5.02 (C$_5$H$_5$, ^{3}J(P,H) = 2.8) ^{19}F NMR (CHCl$_3$): 127.1 (F-2,6), 154.4 (F-4), 162.0 (F-3,5) IR (C$_2$Cl$_4$): 1989, 2034 air stable, UV irradiation yields Fp$_2$	[2]
*4	Ia (51), III (21)	yellow orange, brown, m.p. 195°, 200 to 210° (dec.) ^{1}H NMR: $\sim$3.9 (s, C$_5$H$_5$), 6.7 to 7.3 (m, C$_6$H$_5$) IR (KBr): 1972, 2020 stable in refluxing mesitylene for 1.5 h and on sublimation at 180° in vacuo solutions are readily oxidized by air to No. 21	[4, 43]
4a	FpPC$_4$(C$_6$H$_5$)$_4$·CH$_2$Cl$_2$	forms from No. 4 in CH$_2$Cl$_2$ m.p. 175 to 180° (dec.) IR (KBr): 1965, 2008	[4]

References on p. 373

Table 43 [continued]

No.	compound method of preparation (yield in %)	properties and remarks IR assigned to ν(CO) if not stated otherwise	Ref.
5	FpP(Si(CH₃)₃)₂ III	heating the solid or a solution yields P(Si(CH₃)₃)₃ but no phosphido-bridged dimer reaction with Ni(CO)₄ or Fe₂(CO)₉ yields No. 13 and FpP(Si(CH₃)₃)₂-Fe(CO)₄	[39]

metal carbonyl complexes with FpPR₂ compounds

6	FpP(CH₃)₂Ni(CO)₃ Ib (49)	orange yellow, m.p. 18 to 20° ¹H NMR (C₆H₆): 1.39 (CH₃, ²J(P,H) = 4.8), 4.06 (C₅H₅, ³J(P,H) = 1.9) IR (C₆H₁₂): 1985, 2027 (FeCO), 1976, 2054, 2067 (NiCO) mass spectrum (20°): [M]⁺, loss of all CO, followed by loss of both CH₃; above 80° [C₅H₅Fe(CO)P(CH₃)₂]₂⁺ forms decomposes slowly in air	[16]
7	FpP(CF₃)₂Ni(CO)₃ see No. 1	chrome yellow, m.p. 65 to 70° ¹H NMR (CDCl₃): 5.20 (d, C₅H₅, ³J(P,H) = 2.0) ¹⁹F NMR (CDCl₃): 53.0 (d, CF₃, ²J(P,F) = 54) IR (C₆H₁₄): 2020, 2054 (FeCO), 1996, 2079 (NiCO)	[28]
8	FpP(CF₃)₂Co(CO)₂NO see No. 1	red, m.p. 113 to 116° ¹H NMR (CDCl₃): 5.20 (d, C₅H₅, ³J(P,H) = 2.0) ¹⁹F NMR (CDCl₃): 53.5 (d, CF₃, ²J(P,F) = 53) IR (C₆H₁₄): ν(NO) and ν(CO) at 1752, 1782 sh, 1976, 1998, 2016, 2036 sh, 2046, and 2059	[28]
9	FpP(CF₃)₂Mn(CO)₃NO see No. 1	dark red, m.p. 119 to 121° ¹H NMR (CH₃COCH₃): 5.55 (C₅H₅) ¹⁹F NMR (CH₃COCH₃): 52.4 (d, CF₃, ²J(P,F) = 49) IR (C₆H₁₄): ν(NO) and ν(CO) at 1704, 1748, 1921, 1932, 1969, 1991, 2022, 2033 sh, 2039, and 2060	[28]

References on p. 373

Table 43 [continued]

No.	compound method of prepa- ration (yield in %)	properties and remarks IR assigned to $\nu(CO)$ if not stated otherwise	Ref.
10	$FpP(CF_3)_2Mn(CO)_4H$ see No. 1	dark yellow, m.p. 71 to 74° ^1H NMR (CH_3COCH_3): -7.62 (MnH, $^2J(P,H)=43$), 5.60 (d, C_5H_5, $^3J(P,H)=2.0$) ^{19}F NMR (CH_3COCH_3): 53.0 (d, CF_3, $^2J(P,F)=45$) IR (CH_3COCH_3): 1970 br, 2000, 2042, 2062 IR and NMR results show trans structure around Mn	[28]
*11	$FpP(C_6H_5)_2Ni(CO)_3$	yellow orange, m.p. $\sim$105° (dec.) ^1H NMR: 3.93 (d, C_5H_5, $^3J(P,H)=1.2$) in C_6D_6, 4.73 (s, br, C_5H_5) in $CDCl_3$ IR (CH_2Cl_2): 1982, 2006, 2043	[9]
12	$[FpP(C_6H_5)_2Ru(CO)_2-$ $C_5H_5][B(C_6H_5)_4]$ see No. 2	yellow ^1H NMR $(CD_2Cl_2, 38°)$: 4.45 (d, C_5H_5, $^3J(P,H)=1.5$), 4.78 (s, C_5H_5) IR (CH_2Cl_2): 2004, 2039, 2053 not completely free of $[(C_5H_5M(CO)_2)_2P(C_6H_5)_2][B(C_6H_5)_4]$ (M = Fe, Ru) 1:1 electrolyte in acetone irradiation in THF gives Fp_2 and $(C_5H_5Ru(CO)_2)_2$	[19]
13	$FpP(Si(CH_3)_3)_2Ni(CO)_3$ see No. 5	crystals stable at room temperature cleavage of P–Si bonds in CH_3OH yields complexes with bridging PH_2 groups	[39]
compounds with pentavalent P			
*14	$FpPO(CF_3)_2$ IIc see No. 1	yellow needles, m.p. 131° ^1H NMR (CH_3COCH_3): 5.64 (s, br, C_5H_5) ^{19}F NMR (CH_3COCH_3): 70.5 (d, $^2J(P,F)=71.8$) IR (CH_2Cl_2): 2019, 2062, $\nu(P=O)$ 1210	[11, 18, 20, 21, 25]
15	$FpPO(C_6H_5)_2$ IV (45)	^1H NMR $(CDCl_3, 38°)$: 4.94 (s, C_5H_5) IR: 1975, 2031 in C_6H_{12}, $\nu(P=O)$ 1126 in CS_2	[8]
16	$FpPO(OCH_3)_2$ IV	obtained in low yields from nucleophilic attack of $[C_5H_5Mo(CO)_3]^-$ on $[FpP(OCH_3)_3]^+$ and identified only by IR (CH_2Cl_2): 1989, 2043	[3, 8]

References on p. 373

Table 43 [continued]

No.	compound method of prepa- ration (yield in %)	properties and remarks IR assigned to $\nu(CO)$ if not stated otherwise	Ref.
17	FpPO(OC$_2$H$_5$)$_2$ IIc (30), IV (40)	yellow ^{1}H NMR (CDCl$_3$, 38°): 1.29 (t, CH$_3$, ^3J(H,H) = 7.0), 4.04 (m, CH$_2$), 5.09 (s, C$_5$H$_5$) IR: 1993, 2045 in C$_6$H$_{12}$, ν(P=O) 1180 in CS$_2$ also prepared from FpCl and NaOP(OC$_2$H$_5$)$_2$ in THF	[3, 8, 45]
18	FpPO(OCH$_2$CH=CH$_2$)$_2$ IV (35)	oil ^{1}H NMR (CDCl$_3$, 38°): 5.10 (s, C$_5$H$_5$) IR: 1994, 2044 in C$_6$H$_{12}$, ν(PO) 1179 in CS$_2$	[8]
19	FpPO(OCH$_2$CH=CH$_2$)C$_6$H$_5$ IV (35)	oil ^{1}H NMR (CDCl$_3$, 38°): 4.91 (s, C$_5$H$_5$) IR: 1986, 2039 in C$_6$H$_{12}$, ν(PO) 1167 in CS$_2$	[8]
20	FpPO(OC$_4$H$_9$-n)$_2$ IV	identified only by IR (CH$_2$Cl$_2$): 1988, 2039	[8]
21	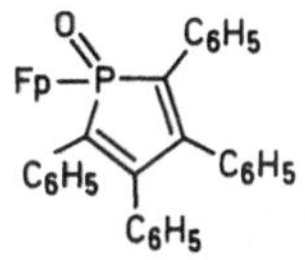 see No. 4	yellow–orange crystals, dec. 245 to 260° structure based on IR, which is almost identical with that of No. 4 except ν(PO) 1117	[4]
22	FpP(S)F$_2$ Ic (15, from SPF$_2$Br)	orange brown ^{1}H NMR: 5.90 (C$_5$H$_5$, ^3J(P,H) = 1.5) ^{19}F NMR: −17.25 (d, ^2J(P,F) = 1236) IR (CH$_3$COOC$_2$H$_5$): 2020, 2055	[10]
23	FpPS(CH$_3$)$_2$ Ic (32)	yellow crystals ^{1}H NMR (CD$_3$COCD$_3$): 1.90 (d, CH$_3$, ^2J(P,H) = 10.0), 5.14 (d, C$_5$H$_5$) IR (CH$_2$Cl$_2$): 1978, 2028	[33]
24	FpPS(CF$_3$)$_2$ IIc (96) see also No. 1	also from FpCl + HSP(CF$_3$)$_2$ in CFCl$_3$ at 50°/4 d in the dark (28% after sublima- tion at 70°) yellow, m.p. 185 to 188° ^{1}H NMR (CH$_3$COCH$_3$): 5.60 (d, C$_5$H$_5$, ^3J(P,H) = 1.5) ^{19}F NMR (CH$_3$COCH$_3$): 67.0 (d, ^2J(P,F) = 71.0); same in CCl$_4$ or CFCl$_3$ IR (CH$_2$Cl$_2$ or CCl$_4$): 2021, 2062	[18, 21]

References on p. 373

Table 43 [continued]

No.	compound method of preparation (yield in %)	properties and remarks IR assigned to v(CO) if not stated otherwise	Ref.
25	FpPS(OC$_2$H$_5$)$_2$ Ic (57)	yellow–brown powder ^{1}H NMR (CD$_3$COCD$_3$): 1.37 (t, C$_2$H$_5$), 4.17 (m, C$_2$H$_5$), 5.12 (d, C$_5$H$_5$, ^{3}J(P,H) = 1.0) IR (CH$_2$Cl$_2$): 1996, 2045	[33]
26	FpPSe(CF$_3$)$_2$ IIc (98) see also No. 1	yellow crystals, m.p. 199 to 201° ^{1}H NMR (CH$_3$COCH$_3$): 5.48 (d, C$_5$H$_5$, ^{3}J(H,P) = 1.8) ^{19}F NMR (CH$_3$COCH$_3$): 65.4 (d, ^{2}J(P,F) = 67.0), same in CCl$_4$ and CFCl$_3$ IR (CH$_2$Cl$_2$ or CCl$_4$): 2021, 2062	[18, 21]
27	FpPSe(C$_6$H$_5$)$_2$ Ic (53)	^{1}H NMR (CD$_3$COCD$_3$): 5.11 (d, C$_5$H$_5$, ^{3}J(P,H) = 1.5) IR (CH$_2$Cl$_2$): 1985, 2035	[33]

compounds with As

No.	compound method of preparation (yield in %)	properties and remarks IR assigned to v(CO) if not stated otherwise	Ref.
28	FpAsCl$_2$ IIa see also No. 34	orange-yellow needles, m.p. 106 to 108° (dec.) ^{1}H NMR (C$_6$D$_6$): 3.98 (C$_5$H$_5$) IR: 1979, 2023 (isomer I), 1993, 2035 (isomer II) in C$_6$H$_{12}$, 1976, 2022 (isomer I), 1989, 2034 (isomer II) in CS$_2$, for the isomers, see p. 357 may be handled briefly in air, at 100° or in THF at 25° rapid, quantitative formation of FpCl	[5, 15, 40]
29	FpAsBr$_2$ IIa	orange crystals IR (CH$_2$Cl$_2$): 1990, 2036	[5, 15]
*30	FpAs(CH$_3$)$_2$ Ia	red liquid ^{1}H NMR (C$_6$H$_6$): 1.28 (CH$_3$), 4.00 (C$_5$H$_5$) IR (C$_6$H$_{12}$): 1950, 1997	[23, 35]
31	FpAs(CF$_3$)$_2$ IIb (65)	volatile orange crystals, m.p. 48 to 50° ^{1}H NMR (CFCl$_3$): 4.92 (s, C$_5$H$_5$) ^{19}F NMR (CFCl$_3$): 42.16 (s, CF$_3$) IR: 1995, 2038 in CS$_2$, 2000, 2044 in mull decomposes slowly in air, very soluble, yields (C$_5$H$_5$Fe(CO)μ-As(CF$_3$)$_2$)$_2$ on UV irradiation in methylcyclohexane	[1]

References on p. 373

Table 43 [continued]

No.	compound method of preparation (yield in %)	properties and remarks IR assignéd to ν(CO) if not stated otherwise	Ref.
32	$FpAs(C_6F_5)_2$ Ia (39), IIb (20)	yellow, m.p. 138° ^{1}H NMR $(CDCl_3)$: 4.98 ^{19}F NMR $(CHCl_3)$: 124.2 (F-2,6), 153.9 (F-4), 161.6 (F-3,5) IR (C_2Cl_4): 1984, 2030 UV irradiation in methylcyclohexane yields brown insoluble crystals, proposedly polymeric $(C_5H_5Fe(CO)-As(C_6F_5)_2)_n$	[2]
33	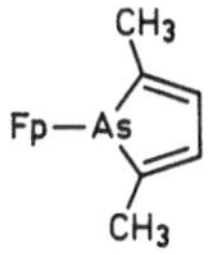 CH$_3$ Fp—As CH$_3$ III	red, only characterized by IR $(C_{16}H_{34})$: 1965, 2010 gives 2,5-dimethylarsaferrocene and $(C_5H_5Fe(CO)AsC_4H_2(CH_3)_2)_2$ in boiling xylene	[42]
33a	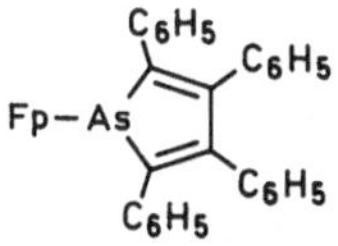 C$_6$H$_5$ C$_6$H$_5$ Fp—As C$_6$H$_5$ C$_6$H$_5$ Ia (72), III (30)	brown, m.p. 209 to 211° (dec.) ^{1}H NMR: ~3.9 (s, C_5H_5), 6.7 to 7.3 (m, C_6H_5) IR (C_6H_{14}): 1968, 2012 no $[M]^+$ in the mass spectrum but $[M-2CO]^+$ 60-h refluxing in $C_6H_5CH_3$ gives 2,3,4,5-tetraphenyl-1-arsaferrocene cleavage by X_2 (X=Cl, Br) forms $XAsC_4(C_6H_5)_4$ and FpX, also see No. 4	[43, 44, 46]
*34	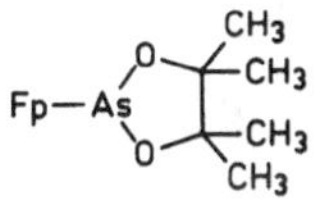 O CH$_3$ CH$_3$ Fp—As CH$_3$ O CH$_3$ Ia (62.5)	golden-yellow leaflets, m.p. 124 to 126° (dec.) ^{1}H NMR (C_6D_6): 1.03 (s, CH_3, trans to As), 1.10 (s, CH_3, cis to As), 4.05 (C_5H_5) ^{13}C NMR $(C_6D_5CD_3)$: 25.1 (CH_3), 83.4 (C ring), 85.5 (C_5H_5), 213.4 (CO) IR (C_6H_{12}): 1961, 2010 (isomer I), 1970, 2013.5 (isomer II), for the isomers, see p. 357	[40]

metal carbonyl complexes with FpAsR$_2$ compounds

| 35 | $FpAs(CH_3)_2Ni(CO)_3$ see No. 30 | impure crystals ^{1}H NMR (C_6H_6): 1.31 (CH_3), 4.00 (C_5H_5) IR (C_6H_{12}): 1974, 2016 (FeCO), 1985, 1994, 2062 (NiCO) | [35] |

References on p. 373

Table 43 [continued]

No.	compound method of preparation (yield in %)	properties and remarks IR assigned to $\nu(CO)$ if not stated otherwise	Ref.
		solutions are stable at room temperature only in the presence of $Ni(CO)_4$, otherwise, Ni, $Ni(CO)_4$, and No. 30 form	
36	$FpAs(CH_3)_2Co(CO)_2NO$ see No. 30	red, m.p. 32° ^1H NMR (C_6H_6): 1.33 (CH_3), 4.00 (C_5H_5) IR (C_6H_{12}): 1976, 2015 (FeCO), 1753 sh, 1793, 1966, 2034 (NO and CoCO)	[35]
37	$FpAs(CH_3)_2Mn(CO)_2C_5H_5$ Ib (69) see No. 30	purple, m.p. 136° ^1H NMR (C_6H_6): 1.55 (CH_3), 4.13 (FeC_5H_5), 4.20 (MnC_5H_5) IR (C_6H_{12}): 1961, 1971 sh, 2010, 2020 sh (FeCO), 1863, 1924 (MnCO)	[35, 37]
38	$FpAs(CH_3)_2Cr(CO)_5$ Ib (32) see No. 30	small red–brown plates, m.p. 50 to 52° ^1H NMR (C_6H_6): 1.38 (CH_3), 4.03 (C_5H_5) IR (C_6H_{12}): 1980, 2022 (FeCO), 1925, 1938, 1976 sh, 2054 (CrCO) [35], 1976, 2031 (FeCO), 1922, 1935, 2068 (CrCO) [17] stable to air and moisture	[17, 35]
39	$FpAs(CH_3)_2Mo(CO)_5$ see No. 30	yellow brown, m.p. 58° ^1H NMR (C_6H_6): 1.38 (CH_3), 4.00 (C_5H_5) IR (C_6H_{12}): 1980, 2021 (FeCO), 1933, 1943, 2064 (MoCO)	[35]
40	$FpAs(CH_3)_2W(CO)_5$ see No. 30	ochreous, m.p. 68° ^1H NMR (C_6H_6): 1.53 (CH_3), 4.03 (C_5H_5) IR (C_6H_{12}): 1981, 2024 (FeCO), 1924, 1936, 1971 sh, 2065 (WCO)	[35]
41	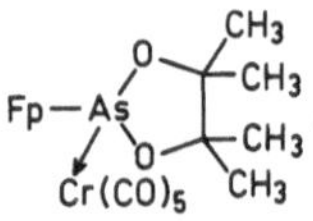 see No. 34	light yellow, m.p. 101 to 103° ^1H NMR (C_6D_6): 1.18 (s, CH_3, trans to As), 1.35 (s, CH_3, cis to As), 4.13 (C_5H_5) IR (C_6H_{12}): 1989, 2030 (FeCO), 1931, 1950, 2060 (CrCO), symmetric conformation I (see p. 357) assumed soluble in organic solvents	[40]

References on p. 373

Table 43 [continued]

No.	compound method of preparation (yield in %)	properties and remarks IR assigned to $\nu(CO)$ if not stated otherwise	Ref.
compounds with Sb			
42	$FpSbCl_2$ IIa	no properties reported	[5]
*43	$FpSbBr_2$ Ia (40.8), IIa	orange-yellow needles, m.p. 144 to 145° (dec.) ^1H NMR (C_6D_6): 3.98 (s, C_5H_5) IR: 1972, 1984, 2016, 2027 in CS_2, 1980, 2025 in CH_2Cl_2, see p. 357	[5, 31]
44	$FpSbI_2$ IIa	red crystals	[5, 15]
*45	$FpSb(CH_3)_2$ Ia (51.5)	red oil, m.p. 12 to 14° ^1H NMR (C_6D_6): 1.21 (s, CH_3), 4.15 (s, C_5H_5) IR (C_6H_{12}): 1949, 1997	[24, 31]
46	$FpSbBrCH_3$	IR (C_6H_{12}): 1966, 2010 prepared from Na[Fp] and CH_3SbBr_2 reaction with $Na[C_5H_5Mo(CO)_3]$ yields Fp_2SbCH_3 (see C3, 2.5.2.2.2, p. 108) and $(C_5H_5Mo(CO)_3)_2SbCH_3$	[26]
46a	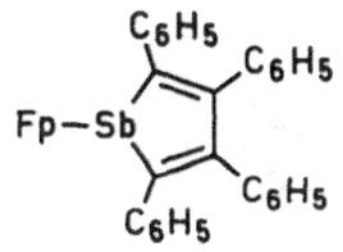 Ia (55), III (25)	brown, m.p. 127 to 129° (dec.) ^1H NMR: ~3.9 (s, C_5H_5), 6.7 to 7.3 (m, C_6H_5) IR (C_6H_{14}): 1960, 2005 no useful mass spectrum obtained at 180° inlet temperature cleavage of the Fp-Sb bond by Cl_2 and Br_2, also see No. 4	[43]
47	$FpSbBrMo(CO)_3C_5H_5$ see No. 43	black-brown leaflets, m.p. 121 to 123° (dec.) ^1H NMR (C_6D_6): 4.35 (s, FeC_5H_5), 5.02 (s, MoC_5H_5) IR (C_6H_{12}): 1910, 1938 sh, 1944, 1968, 1977, 1992, 1996, 2022, 2031	[27]
48	$FpSbBrW(CO)_3C_5H_5$ see No. 43	black-brown leaflets; m.p. 138 to 140° (dec.) ^1H NMR (C_6D_6): 4.37 (s, FeC_5H_5), 5.02 (s, WC_5H_5) IR (C_6H_{12}): 1907, 1932 sh, 1939, 1965, 1971, 1991, 1995, 2021, 2030	[27]

References on p. 373

Table 43 [continued]

No.	compound method of preparation (yield in %)	properties and remarks IR assigned to ν(CO) if not stated otherwise	Ref.

metal carbonyl complexes with FpSb(X,R)$_2$ compounds

49	FpSbF$_2$Cr(CO)$_5$ see No. 52	orange–yellow needles, m.p. 162 to 164° ^{1}H NMR (C$_6$D$_6$): 3.97 (s, C$_5$H$_5$) ^{19}F NMR (THF): 103.4 (s) IR (THF): 1954 br, 1996, 2039, 2071 with Na[Fp] the F atoms can be stepwise substituted by Fp up to Fp$_3$SbCr(CO)$_5$	[30, 32]
50	FpSbF$_2$W(CO)$_5$ see No. 53	brown–yellow needles, m.p. 163 to 165° ^{1}H NMR (C$_6$D$_6$): 3.97 (s, C$_5$H$_5$) ^{19}F NMR (THF): 108.0 (s) IR (THF): 1953 br, 1994, 2038, 2079 reacts with Na[Fp] like No. 49	[30, 32]
51	FpSbBr$_2$Mn(CO)$_2$C$_5$H$_5$ see No. 43	deep black crystals, dec. at 147° ^{1}H NMR (C$_6$D$_6$): 4.30 (s, MnC$_5$H$_5$), 4.40 (s, FeC$_5$H$_5$) IR: 2001, 2041 (FeCO), 1888, 1942 (MnCO) in THF, 2005, 2044 (FeCO), 1887, 1941 (MnCO) in CS$_2$ Cotton–Kraihanzel force constant k(CO in Fp) = 14.81 mdyn·Å^{-1} soluble in THF to give a green solution, stable in air for some time	[31]
52	FpSbBr$_2$Cr(CO)$_5$ see No. 43	red, m.p. 88 to 90° (dec.) ^{1}H NMR (C$_6$D$_6$): 4.15 (s, C$_5$H$_5$) IR (C$_6$H$_{12}$): 2006, 2013, 2041, 2049 (FeCO), 1941, 1953, 1966, 1976, 2073 (CrCO) reaction with AgBF$_4$ in C$_6$H$_6$ at 60° for 2 h gives No. 49 (90.8%), Na[Fp] in cyclo- hexane at room temperature gives Fp$_2$Sb- BrCr(CO)$_5$ (see C3, 2.5.2.2.2, p. 108)	[31, 32]
53	FpSbBr$_2$W(CO)$_5$ see No. 43	brown yellow; m.p. 91 to 93° (dec.) ^{1}H NMR (C$_6$D$_6$): 4.03 (s, C$_5$H$_5$) IR (C$_6$H$_{12}$): 2002, 2009, 2038, 2046 (FeCO), 1939, 1951, 1963, 1967, 2079 (WCO) reaction with AgBF$_4$ in C$_6$H$_6$ at 60° for 2 h gives No. 50 (95.4%)	[31, 32]
54	FpSb(CH$_3$)$_2$Cr(CO)$_5$ see No. 45	light brown, m.p. 59 to 61° ^{1}H NMR (C$_6$D$_6$): 1.35 (s, CH$_3$), 4.22 (s, C$_5$H$_5$) IR (C$_6$H$_{12}$): 1975, 2014 (FeCO), 1929, 1945, 1963, 2050 (CrCO)	[31]

References on p. 373

Table 43 [continued]

No.	compound method of preparation (yield in %)	properties and remarks IR assigned to ν(CO) if not stated otherwise	Ref.

compounds with Bi

55	FpBiCl$_2$ IIa	dark red sparkling crystals IR (CH$_3$COCH$_3$): 1970, 2016, 2048 Λ = 4 Ω^{-1}·cm^2·mol^{-1} (in CH$_3$COCH$_3$, ~0.002 M)	[5, 15]
56	FpBiBr$_2$ IIa	no properties reported	[5]
57	FpBiI$_2$ IIa	no properties reported	[5]

compounds with heteroatom borane ligands

58	FpPB$_{10}$H$_{12}$ V	yellow ^{1}H NMR (CD$_3$COCD$_3$): −4.1 (s, br, bridging BH), 5.61 (d, C$_5$H$_5$, ^{3}J(P,H) = 2.8) ^{11}B NMR (CH$_2$Cl$_2$): −3.4, +8.3, +10.5, +16.5, +24.1 ^{13}C NMR (CH$_2$Cl$_2$): 86.4 (s, C$_5$H$_5$), 209.7 (d, CO, ^{2}J(P,C) = 21.8) IR (KBr): 704, 864, 952 sh, 1020, 1094, 1422, 1991, 2045, 2564, 3103 for a proposed structure, see p. 359	[47]
59	FpPCHB$_9$H$_{10}$ V	yellow ^{1}H NMR (CD$_3$COCD$_3$): 1.59 (d, carborane CH, J(P,H) = 20), 5.72 (d, C$_5$H$_5$, ^{3}J(P,H) = 2.9) ^{11}B NMR (CH$_2$Cl$_2$): −1.0, +6.9, +9.0, +12.0, +14.9, +25.6, +32.5 IR (KBr): 734, 748, 819, 866, 989, 1013, 1030, 1420, 1612, 1995, 2045, 2523, 2833, 2880, 3064	[47]
60	FpAsB$_{10}$H$_{12}$ V (45)	yellow needles ^{1}H NMR (CD$_3$COCD$_3$): −4.2 (s, br, bridging BH), 5.54 (s, C$_5$H$_5$) ^{11}B NMR (CH$_3$COCH$_3$): −4.8, +6.5, +14.6, +22.6 IR (KBr): 800, 861, 980, 1017, 1083, 1411, 1613, 1986, 2040, 2503 sh, 2519, 3117 for a proposed structure, see p. 359	[47]

References on p. 373

Table 43 [continued]

No.	compound method of preparation (yield in %)	properties and remarks IR assigned to ν(CO) if not stated otherwise	Ref.
61	FpAs$_2$B$_9$H$_{10}$ V	yellow ^{1}H NMR (CD$_3$COCD$_3$): -4.2 (bridging BH), 5.56 (s, C$_5$H$_5$) ^{11}B NMR (CH$_2$Cl$_2$): -10.0, -8.1, $+4.4$, $+6.1$, $+8.0$, $+16.4$, $+28.7$ IR (KBr): 731, 872, 983, 1004, 1072, 1419, 1615, 1990, 2040, 2533, 2967, 3107	[47]

*Further information:

C$_5$H$_5$Fe(CO)$_2$P(CF$_3$)$_2$ (Table **43**, No. **1**) crystallizes in the monoclinic system with a$=$8.602(7), b$=$11.924(9), c$=$12.859(9) Å, and β$=$112.75(9)°; space group P2$_1$/c$-$C$_{2h}^5$. Z$=$4 gives D$_c$=1.89 g·cm^{-3} [20, 25]. The molecular structure is shown in **Fig. 28**. The crystals are of irregular shape [25]. For comparison with FpPO(CF$_3$)$_2$, see No. 14.

Fig. 28

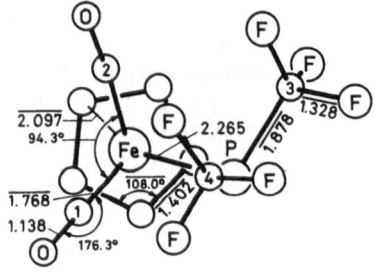

Molecular structure of C$_5$H$_5$Fe(CO)$_2$P(CF$_3$)$_2$ [20, 25].

Other bond angles (°):

| P–Fe–C(1) | 91.3(2) | Fe–P–C(3,4) | 107.4 (av) |
| P–Fe–C(2) | 95.8(2) | C(3)–P–C(4) | 94.5(3) |

UV irradiation in cyclohexane for 4.5 h or heating of the solution causes dimerization to trans-C$_5$H$_5$Fe(CO)(μ-P(CF$_3$)$_2$)$_2$(CO)FeC$_5$H$_5$ (see C4, 2.5.2.3.1.4, p. 17) [11, 18]. The complex does not react with anhydrous HCl or HI at room temperature in either polar or nonpolar solvents; at higher temperature, the Fe–P bond is cleaved and P(CF$_3$)$_2$H is liberated [22]. A mixture of SnCl$_4$ and HCl protonates the compound in CH$_2$Cl$_2$ to give [FpP(CF$_3$)$_2$H][SnCl$_5$]. Action of excess of X$_2$ or ICl in CFCl$_3$ affords the salts [FpP(CF$_3$)$_2$X][X$_3$] (X$=$Cl, Br, I) quantitatively. Reaction with [FpP(CF$_3$)$_2$Br][Br$_3$] in CH$_3$CN gives FpBr and P(CF$_3$)$_2$Br. Treatment with SnCl$_4$ and Cl$_2$ in CH$_3$NO$_2$ forms [FpP(CF$_3$)$_2$Cl][SnCl$_5$] [22].

FpP(CF$_3$)$_2$ acts as Lewis base towards metal carbonyls and nitrosyls such as Ni(CO)$_4$, Co(CO)$_3$NO, Fe(CO)$_2$(NO)$_2$, Mn(CO)$_4$(NO), and Mn(CO)$_5$H in CFCl$_3$, CHCl$_3$, or n-C$_5$H$_{12}$. Reactions require several days to give No. 7 (81%), No. 8 (78%),

No. 9 (55%), and No. 10 (85%) [28]. Treatment with $Fe_2(CO)_9$ in CH_2Cl_2 at 20 °C for 1 d gives $Fp(\mu-P(CF_3)_2)Fe(CO)_4$ (see C3, 2.5.1.1.1, p. 82), and reaction with $Co_2(CO)_8$ in $CFCl_3$ at 20 °C for 5 min gives $(FpP(CF_3)_2)_2Co_2(CO)_6$ [28].

With NO in $CFCl_3$ No. 14 is prepared almost quantitatively. Reactions in CS_2 with excess sulfur at 60 °C for 3 d in a sealed tube or excess P_4S_{10} at room temperature for 3 d and red selenium at room temperature for 1 d, then at 50 °C for 1 h give Nos. 24 (up to 58%) and 26 (70%) [11, 18].

$C_5H_5Fe(CO)_2P(C_6H_5)_2$ (Table 43, No. 2) can react as donor in C_6H_6, displacing the Cl atom in $C_5H_5M(CO)_2Cl$ (M = Fe, Ru) to give $[Fp(\mu-P(C_6H_5)_2)Fp]^+$ cations (see C3, 2.5.2.2.10, p. 157) and No. 12 ($\sim$45% yield in the presence of $Na[B(C_6H_5)_4]$) [7, 19]. With $Fe_2(CO)_9$ in C_6H_6 at room temperature $Fp(\mu-P(C_6H_5)_2)Fe(CO)_4$ (see C3, 2.5.1.1.1, p. 82) is obtained [7, 13]. Complexes of the types $[(C_5H_5FeCO)_2M-(CO)_2(P(C_6H_5)_2)_2]X$ (M = Rh, Ir) and $[(C_5H_5FeCO)_2Rh(P(C_6H_5)_3)(CO)_2-(P(C_6H_5)_2)_2]X$ were isolated from reactions with $RhCl_3 \cdot 3H_2O$, $(Rh(CO)_2Cl)_2$, or $(Rh(C_8H_{12})Cl)_2$ in CH_3OH, with $[Ir(C_8H_{12})(C_4H_8O)_2]SbF_6$ in THF, and with $Rh(C_8H_{12})(P(C_6H_5)_3)Cl$ in C_2H_5OH, respectively [14] (see C4, 2.5.3.2.6.2, p. 219). $(FpP(C_6H_5)_2)_2Rh(CO)Cl$ is formed with $(Rh(CO)_2Cl)_2$ in C_6H_6 [34]. Treatment with $[L_2Rh(solvent)_x]SbF_6$ (L = $P(OC_6H_5)_3$ or L_2 = C_8H_{12}) yields complexes of the $[Fp(P(C_6H_5)_2)RhL_2]SbF_6$ type [34].

$C_5H_5Fe(CO)_2PC_4(C_6H_5)_4$ (Table 43, No. 4). No molecular ion was observed in the mass spectrum since inlet temperatures of 180 to 200 °C were necessary. The heaviest fragment is $[M-2CO]^+$, which presumably has the ferrocene-like structure III [43]. This ferrocene analogue, 2,3,4,5-tetraphenyl-1-phosphaferrocene, is formed in a 35% yield when No. 4 is heated in refluxing $C_6H_5CH_3$ for 60 h [44]. Cleavage of the Fe-P bond with Cl_2 or Br_2 (X_2) in C_6H_{14}/CCl_4 at -10 to +20 °C gives FpX and $XPC_4(C_6H_5)_4$ [43].

$R = C_6H_5$

III

$C_5H_5Fe(CO)_2P(C_6H_5)_2Ni(CO)_3$ (Table 43, No. 11) was prepared from FpCl and $HP(C_6H_5)_2Ni(CO)_3$ (1:1 mole ratio) by treatment with excess $(C_2H_5)_2NH$ in benzene/ether at room temperature for 2 d, 77% yield [9].

The compound shows X-ray emission bands corresponding to Ni-K_α at 36°12', 43°64', and 51°34', to Fe-K_α at 42°32', 51°36', and 61°08', and to P-K_α at 26°38' and 54°54' [9].

The substance is unstable when heated but is stable to UV radiation [9].

$C_5H_5Fe(CO)_2PO(CF_3)_2$ (Table 43, No. 14) is also formed on treatment of $[FpP(CF_3)_2Br][Br_3]$ with water [22].

The compound crystallizes in the monoclinic system with a = 11.938(8), b = 7.603(6), c = 13.818(9) Å, and β = 100.97(8)°; space group $P2_1/c-C_{2h}^5$. Z = 4 gives

References on p. 373

$D_c = 1.95$ g·cm^{-3} [20, 25]. The molecular structure, shown in **Fig. 29**, is almost identical with that of FpP(CF$_3$)$_2$ (No. 1). The main structural change occurring on oxidation at the phosphorus atom is a shortening of the Fe-P bond. This has been interpreted in terms of increased Fe→P $d_\pi - d_\pi$ back bonding, an assumption supported by the shift in the ν(CO) bands to higher frequencies, indicating reduced Fe→COπ bonding, and the downfield ^{1}H NMR shift of the C$_5$H$_5$ resonance. The lengthening of the Fe-CO and the shortening of the C-O bonds, both effects are small and within the limits of experimental error. The similar P-C bond lengths in Nos. 1 and 14 are also consistent with increased back-donation to phosphorus [20, 25].

Fig. 29

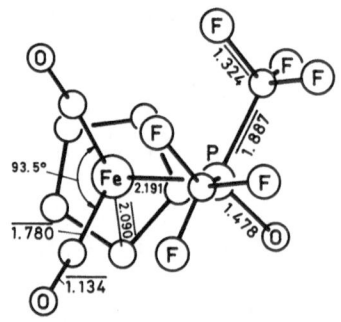

Molecular structure of C$_5$H$_5$Fe(CO)$_2$PO(CF$_3$)$_2$ [20, 25].

Bond angles (°) at the P atom:

| Fe-P-C | 111.6(15) (av) | C-P-C | 96.4(3) |
| Fe-P-O | 121.6(2) | C-P-O | 106.4(4) (av) |

C$_5$H$_5$Fe(CO)$_2$As(CH$_3$)$_2$ (Table **43**, No. **30**) is formed by exchange from FpSi(CH$_3$)$_3$ and As(CH$_3$)$_2$Cl in nonpolar solvents at 25 °C for 12 h [23].

The substance is stable in common organic solvents for a long time. It reacts with various halides RX to give the ionic complexes [FpAs(CH$_3$)$_2$R]X [23]. Reactions with excess of Ni(CO)$_4$, Co(CO)$_3$NO, Fe$_2$(CO)$_9$, or Fe(CO)$_2$(NO)$_2$ in cyclohexane or n-hexane at 20 °C yield No. 35 (35%), No. 36 (67%), FpAs(CH$_3$)$_2$Fe(CO)$_4$, and FpAs(CH$_3$)$_2$Fe(CO)(NO)$_2$, respectively [35]. Reactions with C$_5$H$_5$Mn(CO)$_2$OC$_4$H$_8$, Cr(CO)$_5$OC$_4$H$_8$ and W(CO)$_5$OC$_4$H$_8$ in THF (formed by UV irradiation of carbonyls in THF) as well as with Mo(CO)$_5$HNC$_5$H$_{10}$ in hexane give the arsinide-substituted carbonyls No. 37 (24%), No. 38 (57%), No. 40 (21%), and No. 39 (23%), respectively [35]. However, excess C$_5$H$_5$Co(CO)$_2$ in cyclohexane at 85 °C gives (C$_5$H$_5$-Fe(CO)As(C$_6$H$_5$)$_2$)$_2$ and (C$_5$H$_5$Fe)$_2$C$_5$H$_5$Co(CO)$_3$(As(CH$_3$)$_2$)$_2$, see C4, 2.5.3.2.6.1, p. 215. Even under mild conditions no FpAs(CH$_3$)$_2$Co(CO)C$_5$H$_5$ has been obtained [35, 38]. With ^{4}L M(CO)$_4$ compounds (M = Cr, Mo, W; ^{4}L = norbornadiene) in cyclohexane or THF, complexes of the (FpAs(CH$_3$)$_2$)$_2$M(CO)$_4$ type are formed, whereas treatment with Fe$_2$(CO)$_6$(SCH$_3$)$_2$ in C$_6$H$_6$ leads to FpAs(CH$_3$)$_2$Fe$_2$(CO)$_5$(SCH$_3$)$_2$ [36].

C$_5$H$_5$Fe(CO)$_2$AsO$_2$C$_2$(CH$_3$)$_4$ (Table **43**, No. **34**) is assumed to prefer an axial orientation of Fp in respect to the nonplanar dioxarsolane ring on the basis of ^{1}H NMR results [40], also see remarks on conformation on p. 357.

The compound is extraordinarily soluble in hydrocarbons. Decomposition with formation of Fp$_2$ is observed on melting or on storage in solution, decomposition favored by polarity of the solvent. Reaction with CH_3COCl gives No. 28 (56.2%). Treatment with photochemically formed $Cr(CO)_5OC_4H_8$ in THF yields No. 41 (65.7%) [40].

$C_5H_5Fe(CO)_2SbBr_2$ (Table **43**, No. **43**) reacts with Na[Fp] (1:1 mole ratio) in cyclohexane at room temperature to give Fp_2SbBr [29 to 31]. Reactions with $Na[C_5H_5M(CO)_3]$ (M = Mo, W) in cyclohexane at 60 to 70 °C yield No. 47 (39.9%) and No. 48 (39.2%) [27]. However, with $C_5H_5Mn(CO)_2{}^2L$ (^2L = cycloheptene) or with photochemically formed $M(CO)_5OC_4H_8$ (M = Cr, W) in THF the stibine-substituted carbonyls No. 51 (46.7%), No. 52 (69.8%), and No. 53 (58.0%) can be prepared [31].

$C_5H_5Fe(CO)_2Sb(CH_3)_2$ (Table **43**, No. **45**). The pure solid is almost indefinitely stable. According to [24] even in refluxing THF for 48 h it shows no change, whereas in [31] mention is made of thermal and solvolytic instability. Reaction with the photochemically formed $Cr(CO)_5OC_4H_8$ in THF leads to No. 54 (46.2%) [31].

References:

[1] W.R. Cullen, R.G. Hayter (J. Am. Chem. Soc. **86** [1964] 1030/2). – [2] M. Cooke, M. Green, D. Kirkpatrick (J. Chem. Soc. A **1968** 1507/10). – [3] R.J. Haines, A.L. du Preez, I.L. Marais (J. Organometal. Chem. **24** [1970] C26/C28). – [4] E.H. Braye, K.K. Joshi (Bull. Soc. Chim. Belges **80** [1971] 651/3). – [5] W.R. Cullen, D.J. Patmore, J.R. Sams, M.J. Newlands, L.K. Thompson (Chem. Commun. **1971** 952/3).

[6] E. Eisner, M.J. Newlands, L.K. Thompson (5th Intern. Conf. Organometal. Chem., Moscow 1971, Vol. 2, Abstr. No. 303). – [7] R.J. Haines, C.R. Nolte, R. Greatrex, N.N. Greenwood (J. Organometal. Chem. **26** [1971] C45/C48). – [8] R.J. Haines, A.L. du Preez, I.L. Marais (J. Organometal. Chem. **28** [1971] 405/13). – [9] K. Yasufuku, H. Yamazaki (J. Organometal. Chem. **28** [1971] 415/21). – [10] C.B. Colburn, W.E. Hill, D.W.A. Sharp (Inorg. Nucl. Chem. Letters **8** [1972] 625/7).

[11] R.C. Dobbie, P.R. Mason, R.J. Porter (J. Chem. Soc. Chem. Commun. **1972** 612/3). – [12] R.C. Dobbie, M.J. Hopkinson, D. Whittaker (J. Chem. Soc. Dalton Trans. **1972** 1030/4). – [13] R.J. Haines, C.R. Nolte (J. Organometal. Chem. **36** [1972] 163/75). – [14] R.J. Haines, R. Mason, J.A. Zubieta, C.R. Nolte (J. Chem. Soc. Chem. Commun. **1972** 990/1). – [15] W.R. Cullen, D.J. Patmore, J.R. Sams (Inorg. Chem. **12** [1973] 867/72).

[16] W. Ehrl, H. Vahrenkamp (J. Organometal. Chem. **63** [1973] 389/98). – [17] W. Ehrl, H. Vahrenkamp (Chem. Ber. **106** [1973] 2550/5). – [18] R.C. Dobbie, P.R. Mason (J. Chem. Soc. Dalton Trans. **1973** 1124/8). – [19] R.J. Haines, A.L. du Preez, C.R. Nolte (J. Organometal. Chem. **55** [1973] 199/203). – [20] M.J. Barrow, G.A. Sim, R.C. Dobbie, P.R. Mason (J. Organometal. Chem. **69** [1974] C4/C6).

[21] R.C. Dobbie, M.J. Hopkinson (J. Chem. Soc. Dalton Trans. **1974** 1290/3). – [22] R.C. Dobbie, P.R. Mason (J. Chem. Soc. Dalton Trans. **1974** 2439/42). – [23] W. Malisch, M. Kuhn (Angew. Chem. **86** [1974] 51/2). – [24] W. Malisch, P. Panster (J. Organometal. Chem. **76** [1974] C7/C10). – [25] M.J. Barrow, G.A. Sim (J. Chem. Soc. Dalton Trans. **1975** 291/5).

[26] W. Malisch, P. Panster (Chem. Ber. **108** [1975] 700/15). – [27] W. Malisch, P. Panster (Z. Naturforsch. **30b** [1975] 229/34). – [28] R.C. Dobbie, P.R. Mason

(J. Chem. Soc. Dalton Trans. **1976** 189/91). − [29] W. Malisch, P. Panster (Proc. 17th Intern. Conf. Coord. Chem., Hamburg 1976, p. 352). − [30] W. Malisch, P. Panster (Angew. Chem. **88** [1976] 680/1).

[31] P. Panster, W. Malisch (Chem. Ber. **109** [1976] 692/704). − [32] P. Panster, W. Malisch (Chem. Ber. **109** [1976] 2112/20). − [33] P. Piraino, F. Faraone, M.C. Aversa (J. Chem. Soc. Dalton Trans. **1976** 610/3). − [34] R.J. Haines, J.C. Burckett-St. Laurent, C.R. Nolte (J. Organometal. Chem. **104** [1976] C27/C30). − [35] R. Müller, H. Vahrenkamp (Chem. Ber. **110** [1977] 3910/9).

[36] F. Richter, H. Vahrenkamp (J. Chem. Res. S **1977** 155). − [37] U. Richter, H. Vahrenkamp (J. Chem. Res. S **1977** 156/7). − [38] E. Röttinger, R. Müller, H. Vahrenkamp (Angew. Chem. **89** [1977] 341/2). − [39] H. Schäfer (8th Intern. Conf. Organometal. Chem., Kyoto, Japan, 1977, Abstr. 2 A14). − [40] W. Malisch, P. Panster (Z. Naturforsch. **33b** [1978] 1405/9).

[41] W. Clegg, S. Morton (J. Chem. Soc. Dalton Trans. **1978** 1452/4). − [42] G. Thiollet, F. Mathey, R. Poilblanc (Inorg. Chim. Acta **32** [1979] L67/L68). − [43] E.W. Abel, C. Towers (J. Chem. Soc. Dalton Trans. **1979** 814/9). − [44] E.W. Abel, N. Clark, C. Towers (J. Chem. Soc. Dalton Trans. **1979** 1552/6). − [45] A.L. du Preez, I.L. Marais, R.J. Haines, A. Pidcock, M. Safari (J. Chem. Soc. Dalton Trans. **1981** 1918/23).

[46] E. Abel, C. Towers (5th Intern. Conf. Organometal. Chem., Moscow 1971, Vol. 1, Abstr. No. 63). − [47] T. Yamato, L.J. Todd (J. Organometal. Chem. **67** [1974] 75/80). − [48] T. Madach, H. Vahrenkamp (Z. Naturforsch. **34b** [1979] 1195/8).

Remarks on Abbreviations and Dimensions

Many compounds in this volume are presented in tables in which numerous abbreviations are used and the dimensions are omitted for the sake of conciseness. This necessitates the following clarification.

Temperatures are given in °C, otherwise K stands for Kelvin. Abbreviations used with temperatures are m.p. for melting point and dec. for decomposition.

NMR represents **nuclear magnetic resonance**. Chemical shifts are given as δ values in ppm; reference substances and signs for δ are shown in the scheme below:

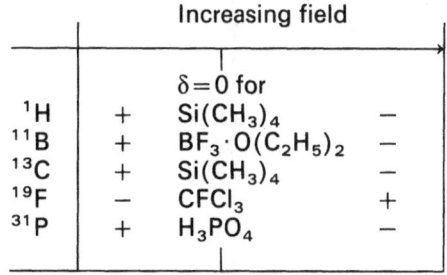

coupling constants J in Hz given as J(A, B) or as J(1,3) referring to labelled structural formulas

Multiplicities of the signals are abbreviated as s, d, t, q (singlett to quartet), quint, sext, sept (quintet to septet), and m (multiplet); terms like dd (double doublet) and t's (triplets) are also used. Assignments referring to labelled structural formulas are given in the form C-4, H-3,5.

Mössbauer spectra are represented by ^{57}Fe-γ and ^{119}Sn-γ: both the isomer shift δ (vs. $Na_2[Fe(CN)_5NO]$ and $BaSnO_3$ at room temperature) and the quadrupole splitting Δ are given in $mm \cdot s^{-1}$; the experimental error has generally been omitted. Other reference substances for δ are indicated after the numerical value, e.g., $\delta = 0.23$ (Fe).

Optical spectra are labelled as IR (infrared), R (Raman), and UV (electronic spectrum including the visible region). IR bands and Raman lines are given in cm^{-1}; the assigned bands are usually labelled with the symbols v for stretching vibration and δ for deformation vibration whereas unlabelled bands belong to CO stretching vibrations. Intensities occur in parentheses either in the common qualitative terms (s, m, w, vs, etc.) or as numerical relative intensities. The UV absorption maxima, λ_{max}, are given in nm followed by the extinction coefficient ε ($L \cdot cm^{-1} \cdot mol^{-1}$) or $\log \varepsilon$ in parentheses; sh means shoulder.

Solvents or the **physical state** of the sample and the temperature (in °C or K) are given in parentheses immediately after the spectral symbol, e.g., R (solid), ^{13}C NMR (C_6D_6, 50 °C). Common solvents are given by their formula (C_6H_{12}=cyclohexane) except THF, which represents tetrahydrofuran.

Figures give only selected parameters. Barred bond lengths (in Å) or angles are mean values for parameters of the same type. In the lists of data below the figures, mean values are labelled as (av). C_5H_5(c) indicates the center of the C_5H_5 ring.

Empirical Formula Index

In the following index the compounds are listed in the order of increasing carbon content. Empirical formulas of ionic compounds are given in brackets; ions as well as components of solvates and adducts are separated by a period.

Page references are printed in ordinary types, table numbers in bold-face, and compound numbers of the tables in italics.

$C_{17}H_{21}FeO_2P$	206, **21**, 4
$C_{17}H_{24}As_2FeO$	228, **25**, 4
$C_{17}H_{25}FeOPSn$	112, **9**, 6
$C_{17}H_{26}FeN_2O$	238, **26**, 3
$C_{17}H_{26}FeP_2$	55, **4**, 16
$[C_{17}H_{28}FeP_2S]^+ \cdot [BF_4]^-$	41, **3**, 2
$C_{17}H_{29}F_4FeN_3P_2$	20, **2**, 26
$C_{17}H_{31}FeIN_2$	18, **2**, 12
$C_{17}H_{35}Cl_3FeO_6P_2Sn$	20, **2**, 31
$\mathbf{C_{18}H_5CrF_5FeO_7S}$	323, **35**, 3
$C_{18}H_{13}Cl_2FeN_3O$	139, **13**, 1
$[C_{18}H_{13}Cl_2FeN_4O_2]^+ \cdot [PF_6]^-$	146
$C_{18}H_{14}FeN_2O$	240, **26**, 16
$C_{18}H_{14}FeOS_2$	142, **13**, 31
$C_{18}H_{16}BrFeOP$	112, **9**, 8
$C_{18}H_{17}AsCrFeO_9$	366, **43**, 41
$C_{18}H_{18}FeN_2O_5S$	346, **41**, 3
$[C_{18}H_{18}FeN_3O_3]^+ \cdot [B(C_6H_5)_4]^-$	353, **42**, 13d
$C_{18}H_{20}ClFeO_3P$	128/9
$C_{18}H_{23}FeO_2P$	206, **21**, 5
$C_{18}H_{24}FFeOP$	155, **15**, 4
$C_{18}H_{25}Cl_2FeOPSn$	89, **7**, 10
$C_{18}H_{32}BrFeOP$	89, **7**, 13
$C_{18}H_{32}ClFeOP$	89, **7**, 12
$C_{18}H_{32}FeIOP$	89, **7**, 14
$\mathbf{C_{19}H_5AsF_{10}FeO_2}$	365, **43**, 32
$C_{19}H_5F_{10}FeO_2P$	360, **43**, 3
$C_{19}H_{12}FeO_9Ru_3$	49/50
$C_{19}H_{13}FeNO_2$	350, **42**, 4
$[C_{19}H_{15}D_6FeN_2O_3]^+ \cdot [B(C_6H_5)_4]^-$	352, **42**, 9d
$C_{19}H_{15}FeNO$	68
$C_{19}H_{15}FeNOS_2$	141, **13**, 22
$C_{19}H_{15}FeNO_2$	343, **40**, 1
$C_{19}H_{15}FeO_2P$	360, **43**, 2
$C_{19}H_{15}FeO_2PSe$	364, **43**, 27
$C_{19}H_{15}FeO_3P$	362, **43**, 15
$C_{19}H_{15}FeO_5P$	314
$C_{19}H_{16}ClFeN_3O$	139, **13**, 2
$C_{19}H_{17}FeNO_2$	350, **42**, 5
$C_{19}H_{18}AsFeIO$	131, **11**, 1
$C_{19}H_{18}FeIOP$	112, **9**, 9
$C_{19}H_{20}FeN_2O_5S$	346, **41**, 4
$C_{19}H_{21}FeOP$	206, **21**, 6
$C_{19}H_{23}FeIN_2$	18, **2**, 15
$C_{19}H_{28}FeN_2O$	238, **26**, 4
$[C_{19}H_{29}FeO_7P_2]^+ \cdot [BF_4]^-$	82, **6**, 31
$[C_{19}H_{30}FeP_3]^+ \cdot I^-$	8, **1**, 41

Ligand Formula Index

Ligands containing carbon atoms (except CO, CN, CNO, and CNS) are used in the following Ligand Formula Index to locate a compound in the order of increasing carbon content of the respective ligand. The number of identical ligands in a compound and the nature of bonding are not taken into consideration, so that several compounds may be listed at one position, if need be. On the other hand, compounds having two or more different types of carbon–containing ligands occur at two or more positions, respectively. The following examples illustrate the arrangement.

$C_5H_5Fe(CO)_2Cl$:

| C_5H_5 | — | — | CO | Cl |

$[C_5H_5Fe(P(CH_3)_3)_3]PF_6$:

| C_3H_9P | C_5H_5 | — | — | — |
| C_5H_5 | C_3H_9P | — | — | — |

$C_5H_5Fe(CO)(P(OC_6H_5)_3)CH_3$:

CH_3	C_5H_5	$C_{18}H_{15}O_3P$	CO	—
C_5H_5	CH_3	$C_{18}H_{15}O_3P$	CO	—
$C_{18}H_{15}O_3P$	CH_3	C_5H_5	CO	—

In view of the very large number of carbonyl compounds, CO is not included in the first column; it is given in column 4 along with other non-organic ligands, such as H, halogen, NO, etc., in columns 3 and 5.

Page references are printed in ordinary types, table numbers in bold-face, and compound numbers of the tables in italics.

$CAlCl_3N$	C_5H_5	—	CO	—	307, **31**, *13*
$CBBr_3N$	C_5H_5	—	CO	—	306, **31**, *10*
$CBCl_3N$	C_5H_5	—	CO	—	306, **31**, *9*
CBF_3N	C_5H_5	—	CO	—	306, **31**, *8*
CCl_3GaN	C_5H_5	—	CO	—	307, **31**, *15*
CCl_3O_2S	C_5H_5	—	CO	—	328, **37**, *1*
CD_2NO	C_5H_5	$C_{18}H_{15}P$	CO	—	173
CF_3	C_5H_5	$C_{18}H_{15}P$	CO	—	178, **18**, *5*
CF_3O_2S	C_5H_5	—	CO	—	328, **37**, *2*
CF_3S	C_5H_5	—	CO	—	319, **34**, *2*
CF_3Se	C_5H_5	—	CO	—	341, **39**, *2*
CHO	C_2H_3O	C_5H_5	CO	—	251, **27**, *1*
CHO	C_5H_5	C_7H_5O	CO	—	252, **27**, *2*
CHO	C_5H_5	$C_8H_7O_2$	CO	—	252, **27**, *3*
CHO_2	C_5H_5	—	CO	—	315, **33**, *1*
CHO_2	C_5H_5	$C_{18}H_{15}P$	CO	—	169
CHS_2	C_5H_5	$C_{26}H_{24}P_2$	—	—	24, **2**, *64*
CH_2Br	C_5H_5	$C_{18}H_{15}P$	CO	—	177, **18**, *3*
CH_2Cl	C_5H_5	$C_{18}H_{15}P$	CO	—	177, **18**, *2*
CH_2I	C_5H_5	$C_{18}H_{15}P$	CO	—	177, **18**, *4*
CH_2NO	C_5H_5	$C_6H_{15}P$	CO	—	155, **15**, *3*

CH₂NO	C₅H₅	C₁₈H₁₅P	CO	–	173
CH₂NO	C₁₈H₁₅	C₁₈H₁₅P	CO	–	229
CH₃	C₅H₅	–	–	NO	49
CH₃	C₃H₉O₃P	C₅H₅	–	–	53, **4**, *2*
CH₃	C₃H₉O₃P	C₅H₅	CO	–	212, **22**, *1*
CH₃	C₃H₉P	C₅H₅	–	–	53, **4**, *1*
CH₃	C₃H₉P	C₅H₅	CO	–	155, **15**, *1*
CH₃	C₄H₁₀S	C₅H₅	CO	–	228, **25**, *7*
CH₃	C₅H₅	C₅H₅N	CO	–	154
CH₃	C₅H₅	C₁₀H₁₆As₂	CO	–	228, **25**, *4*
CH₃	C₅H₅	C₁₂H₂₇O₃P	CO	–	213, **22**, *6*
CH₃	C₅H₅	C₁₂H₂₇P	CO	–	155, **15**, *5*
CH₃	C₅H₅	C₁₇H₁₅P	CO	–	208, **21**, *25*
CH₃	C₅H₅	C₁₈H₁₅As	CO	–	228, **25**, *1*
CH₃	C₅H₅	C₁₈H₁₅O₃P	CO	–	213, **22**, *8*
CH₃	C₅H₅	C₁₈H₁₅P	CO	–	177, **18**, *1*
CH₃	C₅H₅	C₁₈H₁₅Sb	CO	–	228, **25**, *5*
CH₃	C₅H₅	C₂₁H₂₂NP	CO	–	222, **23**, *11*
CH₃	C₅H₅	C₂₂H₂₃P	CO	–	208, **21**, *21*
CH₃	C₅H₅	C₂₆H₂₄P₂	–	–	53, **4**, *6*; 67/8
CH₃	C₁₂H₁₁	C₁₈H₁₅P	CO	–	229/30
CH₃	C₁₈H₁₅P	C₂₄H₁₉	CO	–	232
CH₃BN	C₅H₅	–	CO	–	70; 306, **31**, *7*
CH₃BrSb	C₅H₅	–	CO	–	367, **43**, *46*
CH₃F₂Si	C₅H₅	–	CO	H	71, **5**, *2*
CH₃F₄NP₂	C₅H₅	–	–	Cl	18, **2**, *6*
CH₃F₄NP₂	C₅H₅	–	CO	Cl	130
CH₃F₄NP₂	C₄H₄N	C₅H₅	–	–	19, **2**, *24*
CH₃O₂S					
S(O)OCH₃	C₅H₅	–	CO	–	338, **38**, *9*
OS(O)CH₃	C₅H₅	–	CO	–	313, **32**, *1*
OS(O)CH₃	C₅H₅	C₁₈H₁₅P	CO	–	97, **8**, *9*
SO₂CH₃	C₅H₅	–	CO	–	328, **37**, *3*
SO₂CH₃	C₃H₉O₃P	C₅H₅	CO	–	120, **10**, *5*
SO₂CH₃	C₅H₅	C₁₂H₂₇O₃P	CO	–	121, **10**, *18*
SO₂CH₃	C₅H₅	C₁₂H₂₇P	CO	–	89, **7**, *15*
SO₂CH₃	C₅H₅	C₁₈H₁₅O₃P	CO	–	122, **10**, *25*
SO₂CH₃	C₅H₅	C₁₈H₁₅P	CO	–	97, **8**, *11*
SO₂CH₃	C₅H₅	C₂₆H₂₄P₂	–	–	24, **2**, *65*
SO₂CH₃	C₁₂H₁₁	C₁₈H₁₅P	CO	–	135/6
CH₃O₂Se	C₅H₅	–	CO	–	342, **39**, *5*
CH₃O₃S	C₅H₅	–	CO	–	339, **38**, *12*
CH₃S	C₅H₅	–	CO	–	67; 319, **34**, *1*
CH₄F₂NP	C₅H₅	–	CO	Cl	129/30
CH₁₁B₉P	C₅H₅	–	CO	–	369, **43**, *59*
C₂AsF₆	C₅H₅	–	CO	–	364, **43**, *31*

C_2H_3O	C_5H_5	$C_{14}H_{11}O$	CO	—	252, **27**, *8, 9*
C_2H_3O	C_5H_5	$C_{18}H_{15}As$	CO	—	228, **25**, *2*
C_2H_3O	C_5H_5	$C_{18}H_{15}OP$	CO	—	221, **21**, *2*
C_2H_3O	C_5H_5	$C_{18}H_{15}O_2P$	CO	—	221, **23**, *4*
C_2H_3O	C_5H_5	$C_{18}H_{15}O_3P$	CO	—	214, **22**, *13*
C_2H_3O	C_5H_5	$C_{18}H_{15}P$	CO	—	159, **16**, *1*
C_2H_3O	C_5H_5	$C_{18}H_{15}Sb$	CO	—	228, **25**, *6*
C_2H_3O	C_5H_5	$C_{18}H_{33}P$	CO	—	156, **15**, *11*
C_2H_3O	C_5H_5	$C_{19}H_{17}P$	CO	—	208, **21**, *20*
C_2H_3O	C_5H_5	$C_{20}H_{20}NP$	CO	—	221, **23**, *7*
C_2H_3O	C_5H_5	$C_{21}H_{21}P$	CO	—	208, **21**, *24*
C_2H_3O	C_5H_5	$C_{21}H_{22}NP$	CO	—	222, **23**, *12*
C_2H_3O	C_5H_5	$C_{22}H_{23}P$	CO	—	208, **21**, *22*
C_2H_3O	C_5H_5	$C_{22}H_{24}NP$	CO	—	222, **23**, *13*
C_2H_3O	C_5H_5	$C_{27}H_{26}NP$	CO	—	222, **24**, *14*
C_2H_3O	C_5H_5	$C_{34}H_{33}As_2P$	CO	—	223, **23**, *22*
C_2H_3O	C_5H_5	$C_{34}H_{33}P_3$	CO	—	223, **23**, *18*
C_2H_3O	C_5H_5	$C_{42}H_{42}P_4$	CO	—	223, **23**, *21*
C_2H_3O	$C_{12}H_{11}$	$C_{18}H_{15}P$	CO	—	230/1
C_2H_3O	$C_{18}H_{15}P$	$C_{24}H_{19}$	CO	—	232
$C_2H_3O_2$	C_5H_5	$C_{18}H_{15}P$	CO	—	169/70
$C_2H_3S_2$	C_5H_5	—	CO	—	140, **13**, *8*; 324, **36**, *3*
$C_2H_3S_3$	C_5H_5	—	CO	—	142, **13**, *25*; 324, **36**, *5*
C_2H_4NO	C_5H_5	$C_{18}H_{15}P$	CO	—	173
$C_2H_4NS_2$	C_5H_5	—	CO	—	140, **13**, *10*
C_2H_5	$C_3H_9O_3P$	C_5H_5	—	—	58, **4**, *35*
C_2H_5	C_3H_9P	C_5H_5	—	—	58, **4**, *34*
C_2H_5	C_5H_5	$C_8H_{11}P$	CO	—	205, **21**, *1*
C_2H_5	C_5H_5	$C_{18}H_{15}OP$	CO	—	221, **23**, *1*
C_2H_5	C_5H_5	$C_{18}H_{15}O_2P$	CO	—	221, **23**, *3*
C_2H_5	C_5H_5	$C_{18}H_{15}O_3P$	—	—	53, **4**, *3*
C_2H_5	C_5H_5	$C_{18}H_{15}O_3P$	CO	—	213, **22**, *9*
C_2H_5	C_5H_5	$C_{18}H_{15}P$	CO	—	178, **18**, *5*
C_2H_5	C_5H_5	$C_{18}H_{33}P$	CO	—	156, **15**, *10*
C_2H_5	C_5H_5	$C_{19}H_{17}P$	CO	—	208, **21**, *19*
C_2H_5	C_5H_5	$C_{21}H_{21}P$	CO	—	208, **21**, *23*
C_2H_5O	C_5H_5	$C_{18}H_{15}P$	CO	—	182, **18**, *29*
C_2H_5O	C_5H_5	$C_{26}H_{24}P_2$	—	—	56, **4**, *23*
$C_2H_5O_2S$	C_5H_5	—	CO	—	328, **37**, *4*
$C_2H_5O_2S$	C_5H_5	$C_{18}H_{15}O_3P$	CO	—	122, **10**, *26*
$C_2H_5O_2S$	C_5H_5	$C_{18}H_{15}P$	CO	—	97, **8**, *12*
$C_2H_5O_3S$					
$\quad S(O)_2OC_2H_5$	C_5H_5	—	CO	—	339, **38**, *13*
$\quad SO_2CH_2OCH_3$	C_5H_5	—	CO	—	328, **37**, *5*
$\quad CH_2SO_2OCH_3$	C_5H_5	$C_{18}H_{15}P$	CO	—	184, **18**, *41*
C_2H_5S	C_5H_5	—	CO	—	319, **34**, *3*
C_2H_6As	C_3H_9P	C_5H_5	CO	—	88, **7**, *4*
C_2H_6As	C_5H_5	—	CO	—	364, **43**, *30*
$C_2H_6F_2NP$	C_2H_3O	C_5H_5	CO	—	222, **23**, *15*

C₂H₆F₂NP	C₄H₄N	C₅H₅	–	–	19, **2**, 25
C₂H₆NO₃S₂	C₅H₅	–	CO	–	338, **38**, 1; 346, **41**, 5
C₂H₆NO₄S₂	C₅H₅	–	CO	–	346, **41**, 9
C₂H₆OS	C₅H₅	C₆H₁₁O	CO	–	228, **25**, 8
C₂H₆OS	C₅H₅	C₈H₁₃O	CO	–	228, **25**, 9
C₂H₆O₃P	C₃H₉O₃P	C₅H₅	CO	–	120, **10**, 6
C₂H₆O₃P	C₅H₅	–	CO	–	362, **43**, 16
C₂H₆PS	C₅H₅	–	CO	–	363, **43**, 23
C₂H₆Sb	C₅H₅	–	CO	–	367, **43**, 45
C₃F₇	C₅H₅	C₁₃H₁₃P	CO	–	206, **21**, 8
C₃F₇	C₅H₅	C₁₈H₁₅P	CO	–	178, **18**, 15
C₃F₇	C₅H₅	C₂₆H₂₂P₂	–	–	56, **4**, 21
C₃F₇	C₅H₅	C₂₆H₂₄P₂	–	–	56, **4**, 20
C₃F₇	C₅H₅	C₂₆H₂₄P₂	CO	–	223, **23**, 20
C₃F₇Se	C₅H₅	–	CO	–	342, **39**, 4
C₃H₃	C₅H₅	C₂₆H₂₄P₂	–	–	55, **4**, 15
C₃H₃N	C₅H₅	C₁₈H₁₅P	CO	–	79, **6**, 9
C₃H₃NO₂	C₅H₅	C₁₈H₁₅O₃P	–	–	6, **1**, 24
C₃H₃N₂	C₅H₅	–	CO	–	350, **42**, 6; 351, **42**, 7
C₃H₄ClN	C₅H₅	C₁₈H₁₅O₃P	–	–	6, **1**, 21
C₃H₄N					
CH(CH₃)CN	C₅H₅	C₁₈H₁₅P	CO	–	184, **18**, 43
CH₂CH₂CN	C₅H₅	C₁₈H₁₅O₃P	–	–	53, **4**, 5
CH₂CH₂CN	C₅H₅	C₁₈H₁₅P	CO	–	178, **18**, 14
C₃H₄NO₂S	C₅H₅	–	CO	–	329, **37**, 7
C₃H₄N₂	C₅H₅	–	CO	–	351, **42**, 7b
C₃H₄N₂·H₂O	C₅H₅	–	CO	–	351, **42**, 7c
C₃H₄N₂·OC(CH₃)₂	C₅H₅	–	CO	–	351, **42**, 7d
C₃H₄N₂O	C₅H₅	C₁₈H₁₅O₃P	–	–	6, **1**, 23
C₃H₅					
CH₂CH=CH₂	C₅H₅	C₁₈H₁₅P	CO	–	179, **18**, 13
C(CH₃)=CH₃	C₅H₅	C₁₈H₁₅P	CO	–	185, **18**, 45
C₃H₅N	C₅H₅	C₁₈H₁₅O₃P	–	–	6, **1**, 19
C₃H₅O					
COC₂H₅	C₃H₉O₃P	C₅H₅	CO	–	212, **22**, 4
COC₂H₅	C₅H₅	–	CO	CN	149, **14**, 4, 5
COC₂H₅	C₅H₅	C₈H₁₁P	CO	–	206, **21**, 4
COC₂H₅	C₅H₅	C₁₂H₂₇P	CO	–	155, **15**, 9
COC₂H₅	C₅H₅	C₁₃H₁₃P	CO	–	207, **21**, 9
COC₂H₅	C₅H₅	C₁₈H₁₅O₃P	CO	–	215, **22**, 14
COC₂H₅	C₅H₅	C₁₈H₁₅P	CO	–	160, **16**, 3
COC₂H₅	C₅H₅	C₂₀H₂₀NP	CO	–	221, **23**, 8
C(OCH₃)=CH₂	C₅H₅	C₁₈H₁₅P	CO	–	182, **18**, 30
C₃H₅O₂	C₅H₅	C₁₈H₁₅P	CO	–	172
C₃H₅O₂S	C₅H₅	–	CO	–	329, **37**, 8
C₃H₅O₂S	C₅H₅	C₁₈H₁₅P	CO	–	97, **8**, 14
C₃H₅O₂S₂	C₅H₅	–	CO	–	328, **37**, 6

C$_3$H$_5$S$_3$	C$_5$H$_5$	—	CO	—	142, **13**, *26*; 325, **36**, *6*
C$_3$H$_6$NO	C$_5$H$_5$	C$_{18}$H$_{15}$P	CO	—	173
C$_3$H$_6$NSSe	C$_5$H$_5$	—	CO	—	142, **13**, *28*
C$_3$H$_6$NS$_2$	C$_5$H$_5$	—	CO	—	140, **13**, *12*; 325, **36**, *8*
C$_3$H$_6$NSe$_2$	C$_5$H$_5$	—	CO	—	142, **13**, *29*
C$_3$H$_6$N$_2$	C$_5$H$_5$	C$_{18}$H$_{15}$O$_3$P	—	—	6, **1**, *22*
C$_3$H$_6$O	C$_5$H$_5$	C$_6$H$_{16}$P$_2$	—	—	7, **1**, *31*
C$_3$H$_6$O	C$_5$H$_5$	C$_{26}$H$_{24}$P$_2$	—	—	7, **1**, *34, 35*
C$_3$H$_7$	C$_5$H$_5$	C$_{18}$H$_{15}$P	CO	—	179, **18**, *12*; 184, **18**, *44*
C$_3$H$_7$O	C$_5$H$_5$	C$_{18}$H$_{15}$P	CO	—	182, **18**, *31*
C$_3$H$_7$O	C$_5$H$_5$	C$_{26}$H$_{24}$P$_2$	—	—	56, **4**, *24*
C$_3$H$_7$O$_2$S					
OS(O)CH(CH$_3$)$_2$	C$_5$H$_5$	—	CO	—	313, **32**, *2*
SO$_2$CH$_2$CH$_2$CH$_3$	C$_5$H$_5$	C$_{18}$H$_{15}$P	CO	—	102, **8**, *41*
SO$_2$CH(CH$_3$)$_2$	C$_5$H$_5$	—	CO	—	329, **37**, *9*
C$_3$H$_7$O$_3$S	C$_5$H$_5$	—	CO	—	339, **38**, *14*; 340, **38**, *15*
C$_3$H$_8$N	C$_5$H$_5$	—	CO	—	237, **26**, *1*
C$_3$H$_9$GeO$_2$S	C$_5$H$_5$	—	CO	—	338, **38**, *7*
C$_3$H$_9$N$_2$O$_4$S$_3$	C$_5$H$_5$	—	CO	—	347, **41**, *12*
C$_3$H$_9$O$_2$SSn	C$_5$H$_5$	—	CO	—	338, **38**, *8*
C$_3$H$_9$O$_3$P	CH$_3$	C$_5$H$_5$	—	—	53, **4**, *2*
C$_3$H$_9$O$_3$P	CH$_3$	C$_5$H$_5$	CO	—	212, **22**, *1*
C$_3$H$_9$O$_3$P	CH$_3$O$_2$S	C$_5$H$_5$	CO	—	120, **10**, *5*
C$_3$H$_9$O$_3$P	C$_2$F$_2$O$_2$	C$_5$H$_5$	—	—	19, **2**, *22*
C$_3$H$_9$O$_3$P	C$_2$H$_5$	C$_5$H$_5$	—	—	58, **4**, *35*
C$_3$H$_9$O$_3$P	C$_2$H$_6$O$_3$P	C$_5$H$_5$	CO	—	120, **10**, *6*
C$_3$H$_9$O$_3$P	C$_3$H$_5$O	C$_5$H$_5$	CO	—	212, **22**, *4*
C$_3$H$_9$O$_3$P	C$_4$H$_{11}$Si	C$_5$H$_5$	CO	—	213, **22**, *5*
C$_3$H$_9$O$_3$P	C$_5$H$_5$	—	—	—	4, **1**, *5*
C$_3$H$_9$O$_3$P	C$_5$H$_5$	—	—	Br	18, **2**, *9*
C$_3$H$_9$O$_3$P	C$_5$H$_5$	—	—	H	18, **2**, *4*
C$_3$H$_9$O$_3$P	C$_5$H$_5$	—	—	I	19, **2**, *19*
C$_3$H$_9$O$_3$P	C$_5$H$_5$	—	—	SnCl$_3$	20, **2**, *30*
C$_3$H$_9$O$_3$P	C$_5$H$_5$	—	CO	Br	120, **10**, *3*
C$_3$H$_9$O$_3$P	C$_5$H$_5$	—	CO	Cl	120, **10**, *2*
C$_3$H$_9$O$_3$P	C$_5$H$_5$	—	CO	H	120, **10**, *1*
C$_3$H$_9$O$_3$P	C$_5$H$_5$	—	CO	I	120, **10**, *4*
C$_3$H$_9$O$_3$P	C$_5$H$_5$	—	CO	SnCl$_3$	120, **10**, *7*
C$_3$H$_9$O$_3$P	C$_5$H$_5$	C$_7$H$_7$	CO	—	212, **22**, *2*
C$_3$H$_9$O$_3$P	C$_5$H$_5$	C$_{11}$H$_9$	CO	—	212, **22**, *3*
C$_3$H$_9$O$_3$P	C$_5$H$_5$	C$_{12}$H$_8$Cl$_2$N$_3$	—	—	25, **2**, *76*
C$_3$H$_9$O$_3$P	C$_5$H$_5$	C$_{14}$H$_{14}$N$_3$	—	—	25, **2**, *9*
C$_3$H$_9$O$_3$P	C$_6$H$_7$	—	CO	Cl	133, **12**, *3*
C$_3$H$_9$O$_3$P	C$_6$H$_7$	—	CO	I	133, **12**, *4*
C$_3$H$_9$P	CH$_3$	—	—	—	53, **4**, *1*
C$_3$H$_9$P	CH$_3$	C$_5$H$_5$	CO	—	155, **15**, *1*
C$_3$H$_9$P	C$_2$D$_6$OS	C$_5$H$_5$	—	—	5, **1**, *10, 11*

C_3H_9P	C_2H_3N	C_5H_5	–	–	5, **1**, 12/14
C_3H_9P	C_2H_5	C_5H_5	–	–	58, **4**, 34
C_3H_9P	C_2H_6As	C_5H_5	CO	–	88, **7**, 4
C_3H_9P	C_3H_9Si	C_5H_5	–	–	20, **2**, 27
C_3H_9P	C_3H_9Sn	C_5H_5	–	–	21, **2**, 37
C_3H_9P	C_5H_5	–	–	–	3; 4, **1**, 1/4
C_3H_9P	C_5H_5	–	–	CN	26, **2**, 81
C_3H_9P	C_5H_5	–	–	H	18, **2**, 3
C_3H_9P	C_5H_5	–	–	I	19, **2**, 18
C_3H_9P	C_5H_5	–	CO	–	78, **6**, 1/3
C_3H_9P	C_5H_5	–	CO	Cl	88, **7**, 2
C_3H_9P	C_5H_5	–	CO	CN	88, **7**, 6
C_3H_9P	C_5H_5	–	CO	H	88, **7**, 1
C_3H_9P	C_5H_5	–	CO	I	88, **7**, 3
C_3H_9P	C_5H_5	C_6H_5S	–	–	41, **3**, 2
C_3H_9P	C_5H_5	$C_{11}H_{15}O$	CO	–	155, **15**, 2
C_3H_9P	C_5H_5	$C_{14}H_{17}As\text{-}CoMnO_5$	–	CO	88, **7**, 5
C_3H_9P	C_5H_5	$C_{18}H_{15}P$	–	–	5, **1**, 15
C_3H_9P	C_5H_5	$C_{18}H_{15}Sn$	–	–	21, **2**, 39; 41, **3**, 3
C_3H_9Si	C_3H_9P	C_5H_5	–	–	20, **2**, 27
C_3H_9Si	C_5H_5	$C_{18}H_{15}O_3P$	CO	–	132, **10**, 32
C_3H_9Si	C_5H_5	$C_{18}H_{15}P$	CO	–	100, **8**, 30
C_3H_9Si	C_5H_5	$C_{26}H_{22}P_2$	–	–	24, **2**, 67
C_3H_9Si	C_5H_5	$C_{26}H_{24}P_2$	–	–	24, **2**, 66
C_3H_9Sn	C_3H_9P	C_5H_5	–	–	21, **2**, 37
C_3H_9Sn	C_5H_5	$C_8H_{11}P$	CO	–	112, **9**, 6
C_3H_9Sn	C_5H_5	$C_{13}H_{10}AsF_3$	CO	–	131, **11**, 2
C_3H_9Sn	C_5H_5	$C_{13}H_{10}F_3P$	CO	–	113, **9**, 14
C_3H_9Sn	C_5H_5	$C_{13}H_{13}P$	CO	–	113, **9**, 12
C_3H_9Sn	C_5H_5	$C_{18}H_{15}As$	CO	–	131, **11**, 7
C_3H_9Sn	C_5H_5	$C_{18}H_{15}O_3P$	CO	–	123, **10**, 33
C_3H_9Sn	C_5H_5	$C_{18}H_{15}P$	CO	–	101, **8**, 35
C_3H_9Sn	C_5H_5	$C_{18}H_{15}Sb$	–	–	21, **2**, 38
C_3H_9Sn	C_5H_5	$C_{18}H_{15}Sb$	CO	–	132, **11**, 11
C_3H_9Sn	C_5H_5	$C_{26}H_{22}P_2$	–	–	25, **2**, 73
C_3H_9Sn	C_5H_5	$C_{26}H_{24}P_2$	–	–	24, **2**, 72; 41, **3**, 11
C_3H_9Sn	C_5H_5	$C_{29}H_{20}F_6P_2$	CO	–	114, **9**, 19
C_4ClF_4	C_5H_5	$C_{18}H_{15}P$	–	CO	201, **20**, 1
$C_4CoF_6NO_3P$	C_5H_5	–	–	CO	361, **43**, 8
$C_4F_6S_2$	C_5H_5	–	–	–	43
$C_4F_7O_2$	C_5H_5	–	–	CO	315, **33**, 5
C_4HF_6	C_5H_5	$C_{26}H_{24}P_2$	–	–	56, **4**, 22
$C_4HF_6N_2$	C_5H_5	–	–	CO	240, **26**, 15
$C_4HF_6N_2$	C_5H_5	$C_8H_{11}P$	–	–	57, **4**, 28
$C_4HF_6N_2$	C_5H_5	$C_{18}H_{15}O_3P$	–	–	57, **4**, 30

405

$C_4HF_6N_2$	C_5H_5	$C_{18}H_{15}P$	—	—	57, **4**, *29*
C_4H_3O	C_5H_5	$C_{18}H_{15}P$	—	CO	202, **20**, *7*
C_4H_3S	C_5H_5	$C_{18}H_{15}P$	—	CO	202, **20**, *10*
$C_4H_4F_3S$	C_5H_5	—	—	CO	238, **26**, *8*
C_4H_4N	$CH_3F_4NP_2$	C_5H_5	—	—	19, **2**, *24*
C_4H_4N	$C_2H_6F_2NP$	C_5H_5	—	—	19, **2**, *25*
C_4H_4N	$C_4H_{10}F_2NP$	C_5H_5	—	—	20, **2**, *26*
C_4H_4N	C_5H_5	—	—	PF_3	19, **2**, *23*
C_4H_4N	C_5H_5	—	—	CO	350, **42**, *1*
C_4H_5N	C_5H_5	$C_{18}H_{15}P$	—	CO	79, **6**, *10*
C_4H_5O					
$\quad CH=CHCOCH_3$	C_5H_5	$C_{18}H_{15}P$	—	CO	163, **16**, *26*
$\quad COC_3H_5$-cyclo	C_5H_5	$C_{18}H_{15}P$	—	CO	163, **16**, *28*
$C_4H_5O_2S$	C_5H_5	—	—	CO	329, **37**, *10*
$C_4H_6AsCoNO_3$	C_5H_5	—	—	CO	366, **43**, *36*
C_4H_7					
$\quad CH=C(CH_3)_2$	C_5H_5	$C_{26}H_{24}P_2$	—	—	54, **4**, *12*
$\quad C(CH_3)=CHCH_3$	C_5H_5	$C_{18}H_{15}P$	CO	—	185, **18**, *46*
$\quad CH_2C_3H_5$-cyclo	C_5H_5	$C_{18}H_{15}P$	CO	—	181, **18**, *22*
$C_4H_7D_2$					
$\quad CD_2CH_2CH_2CH_3$	C_5H_5	$C_{18}H_{15}P$	CO	—	180, **18**, *17*
$\quad CD_2CH(CH_3)_2$	C_5H_5	$C_{18}H_{15}P$	CO	—	181, **18**, *21*
$\quad CH_2CD_2CH_2CH_3$	C_5H_5	$C_{18}H_{15}P$	CO	—	180, **18**, *18*
C_4H_7O					
$\quad COCH(CH_3)_2$	C_2H_3O	C_5H_5	CO	—	251, **27**, *6*; 252, **27**, *7*
$\quad COCH(CH_3)_2$	C_5H_5	$C_8H_{11}P$	CO	—	206, **21**, *5*
$\quad COCH(CH_3)_2$	C_5H_5	$C_{18}H_{15}P$	CO	—	161, **16**, *10*
$\quad COCH(CH_3)_2$	C_5H_5	$C_{20}H_{20}NP$	CO	—	222, **23**, *9*
$\quad C(OC_2H_5)=CH_2$	C_5H_5	$C_{18}H_{15}P$	CO	—	183, **18**, *34*
C_4H_7OS	C_5H_5	—	CO	—	241, **26**, *23*
$C_4H_7O_2$	C_5H_5	$C_{18}H_{15}P$	CO	—	170
$C_4H_7O_2S$					
$\quad SO_2CH_2CH_2CH=CH_2$	C_5H_5	—	CO	—	329, **37**, *11*
$\quad SO_2CH_2CH_2CH=CH_2$	C_5H_5	$C_{18}H_{15}P$	CO	—	98, **8**, *16*
$\quad SO_2CH_2C(CH_3)=CH_2$	C_5H_5	—	CO	—	329, **37**, *13*
$\quad SO_2CH(CH_3)CH=CH_2$	C_5H_5	—	CO	—	329, **37**, *12*
C_4H_8DO	C_5H_5	—	CO	—	183, **18**, *33*
C_4H_8NO	C_5H_5	$C_{18}H_{15}P$	CO	—	173
$C_4H_8NO_2P$	C_5H_5	C_6H_5	CO	—	224, **23**, *24*
$C_4H_8NS_2$	C_5H_5	—	CO	—	140, **13**, *11*; 325, **36**, *9*
C_4H_8O	C_5H_5	$C_{18}H_{15}O_3P$	CO	—	78, **6**, *7*
C_4H_8O	C_5H_5	$C_{18}H_{15}P$	CO	—	78, **6**, *6*
C_4H_8O	C_5H_5	$C_{26}H_{24}P_2$	—	—	8, **1**, *36*
C_4H_9					
$\quad C_4H_9$-n	C_5H_5	$C_{18}H_{15}O_3P$	CO	—	214, **22**, *10*
$\quad C_4H_9$-n	C_5H_5	$C_{18}H_{15}P$	CO	—	180, **18**, *16*
$\quad C_4H_9$-n	C_5H_5	$C_{26}H_{24}P_2$	—	—	54, **4**, *9*
$\quad CH(CH_3)CH_2CH_3$	C_5H_5	$C_{18}H_{15}O_3P$	CO	—	214, **22**, *11*
$\quad CH(CH_3)CH_2CH_3$	C_5H_5	$C_{18}H_{15}P$	CO	—	180, **18**, *19*

C_4H_9					
$\quad$ CH(CH$_3$)CH$_2$CH$_3$	C_5H_5	$C_{26}H_{24}P_2$	–	–	54, **4**, *10*
$\quad$ CH$_2$CH(CH$_3$)$_2$	C_5H_5	$C_{18}H_{15}P$	CO	–	180, **18**, *20*
C_4H_9AlN	C_5H_5	–	CO	–	307, **31**, *14*
C_4H_9BN	C_5H_5	–	CO	–	306, **31**, *11*
C_4H_9GaN	C_5H_5	–	CO	–	307, **31**, *16*
C_4H_9NO	C_5H_5	–	–	I	18, **2**, *14*
C_4H_9O	C_5H_5	$C_{18}H_{15}P$	CO	–	183, **18**, *32*
$C_4H_9O_2S$					
$\quad$ SO$_2$CH$_2$CH(CH$_3$)$_2$	C_5H_5	–	CO	–	330, **37**, *14*
$\quad$ SO$_2$CH$_2$CH(CH$_3$)$_2$	C_5H_5	$C_{18}H_{15}P$	CO	–	97, **8**, *15*
$\quad$ SO$_2$C(CH$_3$)$_3$	C_5H_5	–	CO	–	330, **37**, *15*
C_4H_9S	C_5H_5	–	CO	–	319, **34**, *4*
C_4H_9Si	C_5H_5	$C_{13}H_{13}P$	–	–	20, **2**, *28*
C_4H_9Si	C_5H_5	$C_{13}H_{13}P$	CO	–	112, **9**, *11*
$C_4H_{10}F_2NP$	C_2H_3O	C_5H_5	CO	–	223, **23**, *16*
$C_4H_{10}F_2NP$	C_5H_5	–	–	I	18, **2**, *16*
$C_4H_{10}F_2NP$	C_5H_5	–	CO	I	130
$C_4H_{10}O_2PS$	C_5H_5	–	CO	–	364, **43**, *25*
$C_4H_{10}O_3P$	C_5H_5	–	CO	–	363, **43**, *17*
$C_4H_{10}O_3P$	C_5H_5	$C_6H_{15}O_3P$	CO	–	121, **10**, *10*
$C_4H_{10}O_3P$	C_5H_5	$C_{18}H_{15}P$	CO	–	183, **18**, *37*
$C_4H_{10}S$	C_5H_5	–	CO	–	228, **25**, *7*
$C_4H_{11}N$	C_5H_5	–	–	I	18, **2**, *11*
$C_4H_{11}O_2SSi$					
$\quad$ OS(O)CH$_2$Si(CH$_3$)$_3$	C_5H_5	–	CO	–	313, **32**, *3*
$\quad$ SO$_2$CH$_2$Si(CH$_3$)$_3$	C_5H_5	–	CO	–	330, **37**, *16*
$C_4H_{11}Si$	$C_3H_9O_3P$	C_5H_5	CO	–	213, **22**, *5*
$C_4H_{11}Si$	C_5H_5	$C_{13}H_{13}P$	CO	–	207, **21**, *10*
$C_4H_{11}Si$	C_5H_5	$C_{14}H_{15}P$	CO	–	208, **21**, *17*
$C_4H_{11}Si$	C_5H_5	$C_{18}H_{15}P$	CO	–	184, **18**, *38*
C_4N_3	C_5H_5	–	CO	–	343, **40**, *2*
C$_5$Br$_2$CrO$_5$Sb	C_5H_5	–	CO	–	368, **43**, *52*
$C_5Br_2O_5SbW$	C_5H_5	–	CO	–	368, **43**, *53*
$C_5CrF_2O_5Sb$	C_5H_5	–	CO	–	368, **43**, *49*
$C_5F_2O_5SbW$	C_5H_5	–	CO	–	368, **43**, *50*
$C_5F_6MnNO_4P$	C_5H_5	–	CO	–	361, **43**, *9*
$C_5F_6NiO_3P$	C_5H_5	–	CO	–	361, **43**, *7*
C_5F_9S	C_5H_5	–	CO	–	239, **26**, *10*
$C_5H_3F_6S$	C_5H_5	–	CO	–	239, **26**, *9*
C_5H_3OS	C_5H_5	$C_{18}H_{15}P$	CO	–	163, **16**, *25*
$C_5H_3O_2$	C_5H_5	$C_{18}H_{15}P$	CO	–	161, **16**, *12*
$C_5H_4F_3OS$	C_5H_5	–	CO	–	241, **26**, *24*
C_5H_5	–	–	–	–	1
C_5H_5	–	–	–	H	1
C_5H_5	–	–	–	B$_5$H$_{10}$	47
C_5H_5	–	–	–	B$_{10}$H$_{10}$	47/8
C_5H_5	–	–	–	B$_{10}$H$_{15}$	47
C_5H_5	–	–	–	NO	48

C_5H_5	–	I	–	NO	48
C_5H_5	–	–	–	PF_3	1/2
C_5H_5	–	D	–	PF_3	17, **2**, 2
C_5H_5	–	H	–	PF_3	17, **2**, 1
C_5H_5	–	–	CO	$AsB_{10}H_{12}$	369, **43**, 60
C_5H_5	–	–	CO	$AsBr_2$	364, **42**, 29
C_5H_5	–	–	CO	$AsCl_2$	364, **43**, 28
C_5H_5	–	–	CO	$As_2B_9H_{10}$	370, **43**, 61
C_5H_5	–	–	CO	B_3H_8	147
C_5H_5	–	–	CO	$B_{10}H_{12}P$	369, **43**, 58
C_5H_5	–	–	CO	$BiBr_2$	369, **43**, 56
C_5H_5	–	–	CO	$BiCl_2$	369, **43**, 55
C_5H_5	–	–	CO	BiI_2	369, **43**, 57
C_5H_5	–	–	CO	Br	257/93; 293, **30**, 4, 5, 16, 17
C_5H_5	–	–	CO	Cl	257/93; 293, **30**, 1/3, 7/14
			–	–	294, **30**, 15
C_5H_5	–	–	CO	CN	69/70; 305, **31**, 1
C_5H_5	–	–	CO	F	257/93
C_5H_5	–	–	CO	D	256
C_5H_5	–	–	CO	H	255/6
C_5H_5	–	–	CO	I	257/93; 293, **30**, 6; 294, **30**, 18/20
C_5H_5	–	–	CO	N_3	306, **31**, 6
C_5H_5	–	–	CO	NCO	305, **31**, 2
C_5H_5	–	–	CO	NCS	305, **31**, 3
C_5H_5	–	–	CO	NO	146
C_5H_5	–	–	CO	NO_2	312
C_5H_5	–	–	CO	NO_3	312
C_5H_5	–	–	CO	$N(SO_2F)_2$	345, **41**, 1
C_5H_5	–	–	CO	$OS(O_2)OH$	314, **32**, 7
C_5H_5	–	I	CO	PF_3	129
C_5H_5	–	–	CO	$P(S)F_2$	363, **43**, 22
C_5H_5	–	–	CO	$SP(S)F_2$	326
C_5H_5	–	–	CO	SCN	305, **31**, 4
C_5H_5	–	–	CO	SO_2	338, **38**, 4/5; 339, **38**, 10/11
C_5H_5	–	–	CO	$S(O)_2OH$	339, **38**, 9
C_5H_5	–	–	CO	$SbBr_2$	367, **43**, 43
C_5H_5	–	–	CO	$SbCl_2$	367, **43**, 42
C_5H_5	–	–	CO	SbI_2	367, **43**, 47
C_5H_5	–	–	CO	SeCN	306, **31**, 5
C_5H_5	–	–	CO	$SiCl_3$	72, **5**, 5/7
C_5H_5	–	H	CO	$SiCl_3$	71, **5**, 1
C_5H_5	$CAlCl_3N$	–	CO	–	307, **31**, 13
C_5H_5	$CBBr_3N$	–	CO	–	306, **31**, 10

C_5H_5	$CBCl_3N$	–	CO	–	306, **31**, *9*
C_5H_5	CBF_3N	–	CO	–	306, **31**, *8*
C_5H_5	CCl_3GaN	–	CO	–	307, **31**, *15*
C_5H_5	CCl_3O_2S	–	CO	–	328, **37**, *1*
C_5H_5	CD_2NO	$C_{18}H_{15}P$	CO	–	173
C_5H_5	CF_3	$C_{18}H_{15}P$	CO	–	178, **18**, *5*
C_5H_5	CF_3O_2S	–	CO	–	328, **37**, *2*
C_5H_5	CF_3S	–	CO	–	319, **34**, *2*
C_5H_5	CF_3Se	–	CO	–	341, **39**, *2*
C_5H_5	CHO	C_2H_3O	CO	–	251, **27**, *1*
C_5H_5	CHO	C_7H_5O	CO	–	251, **27**, *2*
C_5H_5	CHO	C_8H_7O	CO	–	251, **27**, *3*
C_5H_5	CHO_2	–	CO	–	315, **33**, *1*
C_5H_5	CHO_2	$C_{18}H_{15}P$	CO	–	169
C_5H_5	CHS_2	$C_{26}H_{24}P_2$	–	–	24, **2**, *64*
C_5H_5	CH_2Br	$C_{18}H_{15}P$	CO	–	177, **18**, *3*
C_5H_5	CH_2Cl	$C_{18}H_{15}P$	CO	–	177, **18**, *2*
C_5H_5	CH_2I	$C_{18}H_{15}P$	CO	–	177, **18**, *4*
C_5H_5	CH_2NO	$C_6H_{15}P$	CO	–	155, **15**, *3*
C_5H_5	CH_2NO	$C_{18}H_{15}P$	CO	–	173
C_5H_5	CH_3	–	–	NO	49
C_5H_5	CH_3	$C_3H_9O_3P$	–	–	53, **4**, *2*
C_5H_5	CH_3	$C_3H_9O_3P$	CO	–	212, **22**, *1*
C_5H_5	CH_3	C_3H_9P	–	–	53, **4**, *1*
C_5H_5	CH_3	C_3H_9P	CO	–	155, **15**, *1*
C_5H_5	CH_3	$C_4H_{10}S$	CO	–	228, **25**, *7*
C_5H_5	CH_3	C_5H_5N	CO	–	154
C_5H_5	CH_3	$C_{10}H_{16}As_2$	CO	–	228, **25**, *4*
C_5H_5	CH_3	$C_{12}H_{27}P$	CO	–	155, **15**, *5*
C_5H_5	CH_3	$C_{17}H_{15}P$	CO	–	208, **21**, *25*
C_5H_5	CH_3	$C_{18}H_{15}As$	CO	–	228, **25**, *1*
C_5H_5	CH_3	$C_{18}H_{15}O_3P$	CO	–	213, **22**, *8*
C_5H_5	CH_3	$C_{18}H_{15}P$	CO	–	177, **18**, *1*
C_5H_5	CH_3	$C_{18}H_{15}Sb$	CO	–	228, **25**, *5*
C_5H_5	CH_3	$C_{21}H_{22}NP$	CO	–	222, **23**, *11*
C_5H_5	CH_3	$C_{22}H_{23}P$	CO	–	208, **21**, *21*
C_5H_5	CH_3	$C_{26}H_{24}P_2$	–	–	53, **4**, *6*; 67/8
C_5H_5	CH_3BN	–	CO	–	306, **31**, *7*; 70
C_5H_5	CH_3BrSb	–	CO	–	367, **43**, *46*
C_5H_5	CH_3F_2Si	–	CO	H	71, **5**, *2*
C_5H_5	$CH_3F_4NP_2$	–	–	Cl	18, **2**, *6*
C_5H_5	$CH_3F_4NP_2$	–	CO	Cl	130
C_5H_5	$CH_3F_4NP_2$	C_4H_4N	–	–	19, **2**, *24*
C_5H_5	CH_3O_2S	–	CO	–	313, **32**, *1*; 328, **37**, *3*; 338, **38**, *6*
C_5H_5	CH_3O_2S	$C_3H_9O_3P$	CO	–	120, **10**, *5*
C_5H_5	CH_3O_2S	$C_{12}H_{27}O_3P$	CO	–	121, **10**, *18*
C_5H_5	CH_3O_2S	$C_{12}H_{27}P$	CO	–	89, **7**, *15*

C_5H_5	CH_3O_2S	$C_{18}H_{15}O_3P$	CO	–	122, **10**, *25*
C_5H_5	CH_3O_2S	$C_{18}H_{15}P$	CO	–	97, **8**, *9, 11*
C_5H_5	CH_3O_2S	$C_{26}H_{24}P_2$	–	–	24, **2**, *65*
C_5H_5	CH_3O_2Se	–	CO	–	342, **39**, *5*
C_5H_5	CH_3O_3S	–	CO	–	339, **38**, *12*
C_5H_5	CH_3S	–	CO	–	67;
					319, **34**, *1*
C_5H_5	CH_4F_2NP	–	CO	Cl	129/30
C_5H_5	$CH_{11}B_9P$	–	CO	–	369, **43**, *59*
C_5H_5	C_2AsF_6	–	CO	–	364, **43**, *31*
C_5H_5	$C_2Cl_3O_2$	–	CO	–	315, **33**, *3*
C_5H_5	C_2D_6OS	C_3H_9P	–	–	5, **1**, *10, 11*
C_5H_5	C_2F_3	$C_{18}H_{15}P$	CO	–	179, **18**, *11*
C_5H_5	C_2F_3O	$C_{18}H_{15}P$	CO	–	159, **16**, *2*
C_5H_5	$C_2F_3O_2$	–	CO	–	315, **33**, *4*
C_5H_5	$C_2F_3O_2$	$C_3H_9O_3P$	–	–	19, **2**, *22*
C_5H_5	$C_2F_3O_2$	$C_6H_{16}P_2$	CO	–	23, **2**, *60*
C_5H_5	$C_2F_3O_2$	$C_{18}H_{15}O_3P$	CO	–	122, **10**, *24*
C_5H_5	$C_2F_3O_2$	$C_{18}H_{15}P$	CO	–	96, **8**, *8*
C_5H_5	$C_2F_3O_2$	$C_{26}H_{22}P_2$	–	–	23, **2**, *62*
C_5H_5	$C_2F_3O_2$	$C_{26}H_{24}P_2$	–	–	23, **2**, *61*
C_5H_5	C_2F_5	$C_{18}H_{15}P$	CO	–	178, **18**, *9*
C_5H_5	C_2F_5	$C_{26}H_{22}P_2$	–	–	55, **4**, *19*
C_5H_5	C_2F_5	$C_{26}H_{24}P_2$	–	–	55, **4**, *18*
C_5H_5	C_2F_5	$C_{26}H_{24}P_2$	CO	–	223, **23**, *19*
C_5H_5	C_2F_5Se	–	CO	–	342, **39**, *3*
C_5H_5	C_2F_6OP	–	CO	–	362, **43**, *14*
C_5H_5	C_2F_6P	–	CO	–	360, **43**, *1*
C_5H_5	C_2F_6PS	–	CO	–	326;
					363, **43**, *24*
C_5H_5	$C_2F_6PS_2$	–	CO	–	326
C_5H_5	C_2F_6PSe	–	CO	–	342, **39**, *6*;
					364, **43**, *26*
C_5H_5	C_2H	$C_{26}H_{24}P_2$	–	–	55, **4**, *14*
C_5H_5	$C_2HCl_2O_2$	–	CO	–	315, **33**, *2*
C_5H_5	C_2HF_4	$C_{18}H_{15}P$	CO	–	179, **18**, *10*
C_5H_5	C_2H_2ClN	$C_{18}H_{15}O_3P$	–	–	6, **1**, *20*
C_5H_5	$C_2H_2D_3O_2S$	$C_{18}H_{15}P$	CO	–	97, **8**, *13*
C_5H_5	C_2H_2N	$C_{18}H_{15}P$	CO	–	178, **18**, *8*
C_5H_5	$C_2H_2N_3$	–	CO	–	352, **42**, *10, 11*
C_5H_5	$C_2H_3AlBr_3O$	–	CO	–	233/4
C_5H_5	$C_2H_3D_2$	$C_{18}H_{15}P$	CO	–	178, **18**, *7*
C_5H_5	C_2H_3N	C_3H_9P	–	–	5, **1**, *12, 13, 14*
C_5H_5	C_2H_3N	$C_{18}H_{15}O_3P$	–	–	5, **1**, *8*
C_5H_5	C_2H_3N	$C_{18}H_{15}P$	CO	–	178, **7**, *8*
C_5H_5	C_2H_3N	$C_{26}H_{24}P_2$	–	–	8, **1**, *39, 40*
C_5H_5	$C_2H_3N_3$	–	CO	–	352, **42**, *10a*
C_5H_5	C_2H_3O	–	CO	–	147/8;
					251, **27**, *4, 5*
C_5H_5	C_2H_3O	–	CO	CN	149, **14**, *1/3, 7*
C_5H_5	C_2H_3O	–	CO	NH_3	153/4

C_5H_5	C_2H_3O	$C_2H_6F_2NP$	CO	–	222, **23**, *15*
C_5H_5	C_2H_3O	C_4H_7O	CO	–	252, **27**, *6, 7*
C_5H_5	C_2H_3O	$C_4H_{10}F_2NP$	CO	–	223, **23**, *16*
C_5H_5	C_2H_3O	$C_5H_{10}F_2NP$	CO	–	223, **23**, *17*
C_5H_5	C_2H_3O	$C_8H_{11}P$	CO	–	206, **21**, *3*
C_5H_5	C_2H_3O	$C_{12}H_{23}O_3P$	CO	–	221, **23**, *6*
C_5H_5	C_2H_3O	$C_{12}H_{27}O_3P$	CO	–	213, **22**, *7*
C_5H_5	C_2H_3O	$C_{12}H_{27}P$	CO	–	155, **15**, *8*
C_5H_5	C_2H_3O	$C_{14}H_{11}O$	CO	–	252, **27**, *8, 9*
C_5H_5	C_2H_3O	$C_{18}H_{15}As$	CO	–	228, **25**, *2*
C_5H_5	C_2H_3O	$C_{18}H_{15}OP$	CO	–	221, **23**, *2*
C_5H_5	C_2H_3O	$C_{18}H_{15}O_2P$	CO	–	221, **23**, *4*
C_5H_5	C_2H_3O	$C_{18}H_{15}O_3P$	CO	–	214, **22**, *13*
C_5H_5	C_2H_3O	$C_{18}H_{15}P$	CO	–	159, **16**, *1*
C_5H_5	C_2H_3O	$C_{18}H_{15}Sb$	CO	–	228, **25**, *6*
C_5H_5	C_2H_3O	$C_{18}H_{33}P$	CO	–	156, **15**, *11*
C_5H_5	C_2H_3O	$C_{19}H_{17}P$	CO	–	208, **21**, *20*
C_5H_5	C_2H_3O	$C_{20}H_{20}NP$	CO	–	221, **23**, *7*
C_5H_5	C_2H_3O	$C_{21}H_{21}P$	CO	–	208, **21**, *24*
C_5H_5	C_2H_3O	$C_{21}H_{22}NP$	CO	–	222, **23**, *12*
C_5H_5	C_2H_3O	$C_{22}H_{23}P$	CO	–	208, **21**, *22*
C_5H_5	C_2H_3O	$C_{22}H_{24}NP$	CO	–	222, **23**, *13*
C_5H_5	C_2H_3O	$C_{29}H_{26}NP$	CO	–	222, **23**, *14*
C_5H_5	C_2H_3O	$C_{34}H_{33}As_2P$	CO	–	223, **23**, *22*
C_5H_5	C_2H_3O	$C_{34}H_{33}P_3$	CO	–	223, **23**, *18*
C_5H_5	C_2H_3O	$C_{42}H_{42}P_4$	CO	–	223, **23**, *21*
C_5H_5	$C_2H_3O_2$	$C_{18}H_{15}P$	CO	–	169/70
C_5H_5	$C_2H_3S_2$	–	CO	–	140, **13**, *8*; 324, **36**, *3*
C_5H_5	$C_2H_3S_3$	–	CO	–	142, **13**, *25*; 324, **36**, *5*
C_5H_5	C_2H_4NO	$C_{18}H_{15}P$	CO	–	173
C_5H_5	$C_2H_4NS_2$	–	CO	–	140, **13**, *10*
C_5H_5	C_2H_5	$C_3H_9O_3P$	–	–	58, **4**, *35*
C_5H_5	C_2H_5	C_3H_9P	–	–	58, **4**, *34*
C_5H_5	C_2H_5	$C_8H_{11}P$	CO	–	205, **21**, *1*
C_5H_5	C_2H_5	$C_{18}H_{15}OP$	CO	–	221, **23**, *1*
C_5H_5	C_2H_5	$C_{18}H_{15}O_2P$	CO	–	221, **23**, *3*
C_5H_5	C_2H_5	$C_{18}H_{15}O_3P$	–	–	53, **4**, *3*
C_5H_5	C_2H_5	$C_{18}H_{15}O_3P$	CO	–	213, **22**, *9*
C_5H_5	C_2H_5	$C_{18}H_{15}P$	CO	–	178, **18**, *6*
C_5H_5	C_2H_5	$C_{18}H_{33}P$	CO	–	156, **15**, *10*
C_5H_5	C_2H_5	$C_{19}H_{17}P$	CO	–	208, **21**, *19*
C_5H_5	C_2H_5	$C_{21}H_{21}P$	CO	–	208, **21**, *23*
C_5H_5	C_2H_5O	$C_{18}H_{15}P$	CO	–	182, **18**, *29*
C_5H_5	C_2H_5O	$C_{26}H_{24}P_2$	–	–	56, **4**, *23*
C_5H_5	$C_2H_5O_2S$	–	CO	–	328, **37**, *4*
C_5H_5	$C_2H_5O_2S$	$C_{18}H_{15}O_3P$	CO	–	122, **10**, *26*
C_5H_5	$C_2H_5O_2S$	$C_{18}H_{15}P$	CO	–	97, **8**, *12*
C_5H_5	$C_2H_5O_3S$	–	CO	–	328, **37**, *5*; 339, **38**, *13*

C_5H_5	$C_2H_5O_3S$	$C_{18}H_{15}P$	CO	–	184, **18**, *41*
C_5H_5	C_2H_5S	–	CO	–	319, **34**, *3*
C_5H_5	C_2H_6As	–	CO	–	364, **43**, *30*
C_5H_5	C_2H_6As	C_3H_9P	CO	–	88, **7**, *4*
C_5H_5	$C_2H_6F_2NP$	C_4H_4N	–	–	19, **2**, *25*
C_5H_5	$C_2H_6NO_3S_2$	–	CO	–	338, **38**, *1;* 346, **41**, *5*
C_5H_5	$C_2H_6NO_4S_2$	–	CO	–	346, **41**, *9*
C_5H_5	C_2H_6OS	$C_6H_{11}O$	CO	–	228, **20**, *8*
C_5H_5	C_2H_6OS	$C_8H_{13}O$	CO	–	228, **25**, *9*
C_5H_5	$C_2H_6O_3P$	–	CO	–	362, **43**, *16*
C_5H_5	$C_2H_6O_3P$	$C_3H_9O_3P$	CO	–	120, **10**, *6*
C_5H_5	C_2H_6PS	–	CO	–	363, **43**, *23*
C_5H_5	C_2H_6Sb	–	CO	–	367, **43**, *45*
C_5H_5	C_3F_7	$C_{13}H_{13}P$	CO	–	206, **21**, *8*
C_5H_5	C_3F_7	$C_{18}H_{15}P$	CO	–	168, **18**, *15*
C_5H_5	C_3F_7	$C_{26}H_{22}P_2$	–	–	56, **4**, *21*
C_5H_5	C_3F_7	$C_{26}H_{24}P_2$	–	–	56, **4**, *20*
C_5H_5	C_3F_7	$C_{26}H_{24}P_2$	CO	–	223, **23**, *20*
C_5H_5	C_3F_7Se	–	CO	–	342, **39**, *4*
C_5H_5	C_3H_3	$C_{26}H_{24}P_2$	–	–	55, **4**, *15*
C_5H_5	C_3H_3N	$C_{18}H_{15}P$	CO	–	79, **6**, *9*
C_5H_5	$C_3H_3NO_3$	$C_{18}H_{15}O_3P$	–	–	6, **1**, *24*
C_5H_5	$C_3H_3N_2$	–	CO	–	350, **42**, *6, 7*
C_5H_5	C_3H_4ClN	$C_{18}H_{15}O_3P$	–	–	6, **1**, *21*
C_5H_5	C_3H_4N	$C_{18}H_{15}O_3P$	–	–	53, **4**, *5*
C_5H_5	C_3H_4N	$C_{18}H_{15}P$	CO	–	179, **18**, *14;* 184, **18**, *43*
C_5H_5	$C_3H_4NO_2S$	–	CO	–	329, **37**, *7*
C_5H_5	$C_3H_4N_2$	–	CO	–	351, **42**, *7b*
C_5H_5	$C_3H_4N_2 \cdot H_2O$	–	CO	–	351, **41**, *7c*
C_5H_5	$C_3H_4N_2 \cdot$ $OC(CH_3)_2$	–	CO	–	351, **41**, *7d*
C_5H_5	$C_3H_4N_2O$	$C_{18}H_{15}O_3P$	–	–	6, **1**, *23*
C_5H_5	C_3H_5	$C_{18}H_{15}P$	CO	–	179, **18**, *13;* 185, **18**, *45*
C_5H_5	C_3H_5N	$C_{18}H_{15}O_3P$	–	–	6, **1**, *19*
C_5H_5	C_3H_5O	–	CO	CN	149, **14**, *4, 5*
C_5H_5	C_3H_5O	$C_3H_9O_3P$	CO	–	212, **22**, *4*
C_5H_5	C_3H_5O	$C_8H_{11}P$	CO	–	206, **21**, *4*
C_5H_5	C_3H_5O	$C_{12}H_{27}P$	CO	–	155, **15**, *9*
C_5H_5	C_3H_5O	$C_{13}H_{13}P$	CO	–	207, **21**, *9*
C_5H_5	C_3H_5O	$C_{18}H_{15}O_3P$	CO	–	215, **22**, *14*
C_5H_5	C_3H_5O	$C_{18}H_{15}P$	CO	–	160, **16**, *3;* 182, **18**, *30*
C_5H_5	C_3H_5O	$C_{20}H_{20}NP$	CO	–	221, **23**, *8*
C_5H_5	$C_3H_5O_2$	$C_{18}H_{15}P$	CO	–	172
C_5H_5	$C_3H_5O_2S$	–	CO	–	329, **37**, *8*
C_5H_5	$C_3H_5O_2S$	$C_{18}H_{15}P$	CO	–	97, **8**, *14*
C_5H_5	$C_3H_5O_2S_2$	–	CO	–	328, **37**, *6*
C_5H_5	$C_3H_5S_3$	–	CO	–	142, **13**, *26;* 325, **36**, *6*

C_5H_5	C_3H_6NO	$C_{18}H_{15}P$	CO	–	173
C_5H_5	C_3H_6NSSe	–	CO	–	142, **13**, *28*
C_5H_5	$C_3H_6NS_2$	–	CO	–	140, **13**, *12*; 325, **36**, *8*
C_5H_5	$C_3H_6NSe_2$	–	CO	–	142, **13**, *29*
C_5H_5	$C_3H_6N_2$	$C_{18}H_{15}O_3P$	–	–	6, **1**, *22*
C_5H_5	C_3H_6O	$C_6H_{16}P_2$	–	–	7, **1**, *31*
C_5H_5	C_3H_6O	$C_{26}H_{24}P_2$	–	–	7, **1**, *34, 35*
C_5H_5	C_3H_7	$C_{18}H_{15}P$	CO	–	179, **18**, *12*; 184, **18**, *44*
C_5H_5	C_3H_7O	$C_{18}H_{15}P$	CO	–	182, **18**, *31*
C_5H_5	C_3H_7O	$C_{26}H_{24}P_2$	–	–	56, **4**, *24*
C_5H_5	$C_3H_7O_2S$	–	CO	–	313, **32**, *2*; 329, **37**, *9*
C_5H_5	$C_3H_7O_2S$	$C_{18}H_{15}P$	CO	–	102, **8**, *41*
C_5H_5	$C_3H_7O_3S$	–	CO	–	339, **38**, *14*; 340, **38**, *15*
C_5H_5	C_3H_8N	–	CO	–	237, **26**, *1*
C_5H_5	$C_3H_9GeO_2S$	–	CO	–	338, **38**, *7*
C_5H_5	$C_3H_9N_2O_4S_3$	–	CO	–	347, **41**, *12*
C_5H_5	$C_3H_9O_2SSn$	–	CO	–	338, **38**, *8*
C_5H_5	$C_3H_9O_3P$	–	–	–	4, **1**, *5*
C_5H_5	$C_3H_9O_3P$	–	–	Br	18, **2**, *9*
C_5H_5	$C_3H_9O_3P$	–	–	H	18, **2**, *4*
C_5H_5	$C_3H_9O_3P$	–	–	I	19, **2**, *19*
C_5H_5	$C_3H_9O_3P$	–	–	$SnCl_3$	20, **2**, *30*
C_5H_5	$C_3H_9O_3P$	–	CO	Br	120, **10**, *3*
C_5H_5	$C_3H_9O_3P$	–	CO	Cl	120, **10**, *2*
C_5H_5	$C_3H_9O_3P$	–	CO	H	120, **10**, *1*
C_5H_5	$C_3H_9O_3P$	–	CO	I	120, **10**, *4*
C_5H_5	$C_3H_9O_3P$	–	CO	$SnCl_3$	120, **10**, *7*
C_5H_5	$C_3H_9O_3P$	$C_4H_{11}Si$	CO	–	213, **22**, *5*
C_5H_5	$C_3H_9O_3P$	C_7H_7	CO	–	212, **22**, *2*
C_5H_5	$C_3H_9O_3P$	$C_{11}H_9$	CO	–	212, **22**, *3*
C_5H_5	$C_3H_9O_3P$	$C_{12}H_8Cl_2N_3$	–	–	25, **2**, *76*
C_5H_5	$C_3H_9O_3P$	$C_{14}H_{14}N_3$	–	–	25, **2**, *79*
C_5H_5	C_3H_9P	–	–	–	3; 4, **1**, *1/4*
C_5H_5	C_3H_9P	–	–	CN	26, **2**, *81*
C_5H_5	C_3H_9P	–	–	H	18, **2**, *3*
C_5H_5	C_3H_9P	–	–	I	19, **2**, *18*
C_5H_5	C_3H_9P	–	CO	–	78, **6**, *1/3*
C_5H_5	C_3H_9P	–	CO	Cl	88, **7**, *2*
C_5H_5	C_3H_9P	–	CO	CN	88, **7**, *6*
C_5H_5	C_3H_9P	–	CO	H	88, **7**, *1*
C_5H_5	C_3H_9P	–	CO	I	88, **7**, *3*
C_5H_5	C_3H_9P	C_3H_9Si	–	–	20, **2**, *27*
C_5H_5	C_3H_9P	C_3H_9Sn	–	–	21, **2**, *37*
C_5H_5	C_3H_9P	C_6H_5S	–	–	41, **3**, *2*
C_5H_5	C_3H_9P	$C_{11}H_{15}O$	CO	–	155, **15**, *2*
C_5H_5	C_3H_9P	$C_{14}H_{17}As_2-$ $CoMnO_5$	CO	–	88, **7**, *5*

C_5H_5	C_3H_9P	$C_{18}H_{15}P$	–	–	5, **1**, *15*
C_5H_5	C_3H_9P	$C_{18}H_{15}Sn$	–	–	21, **2**, *39*; 41, **3**, *3*
C_5H_5	C_3H_9Si	$C_{18}H_{15}O_3P$	CO	–	123, **10**, *33*
C_5H_5	C_3H_9Si	$C_{18}H_{15}P$	CO	–	100, **8**, *30*
C_5H_5	C_3H_9Si	$C_{26}H_{22}P_2$	–	–	24, **2**, *67*
C_5H_5	C_3H_9Si	$C_{26}H_{24}P_2$	–	–	24, **2**, *66*
C_5H_5	C_3H_9Sn	$C_8H_{11}P$	CO	–	112, **9**, *6*
C_5H_5	C_3H_9Sn	$C_{13}H_{10}AsF_3$	CO	–	131, **11**, *2*
C_5H_5	C_3H_9Sn	$C_{13}H_{10}F_3P$	CO	–	113, **9**, *14*
C_5H_5	C_3H_9Sn	$C_{13}H_{13}P$	CO	–	113, **9**, *12*
C_5H_5	C_3H_9Sn	$C_{18}H_{15}As$	CO	–	131, **11**, *7*
C_5H_5	C_3H_9Sn	$C_{18}H_{15}O_3P$	CO	–	123, **10**, *33*
C_5H_5	C_3H_9Sn	$C_{18}H_{15}P$	CO	–	101, **8**, *35*
C_5H_5	C_3H_9Sn	$C_{18}H_{15}Sb$	–	–	21, **2**, *38*
C_5H_5	C_3H_9Sn	$C_{18}H_{15}Sb$	CO	–	132, **11**, *11*
C_5H_5	C_3H_9Sn	$C_{26}H_{22}P_2$	–	–	25, **2**, *73*
C_5H_5	C_3H_9Sn	$C_{26}H_{24}P_2$	–	–	24, **2**, *72*; 41, **3**, *11*
C_5H_5	C_3H_9Sn	$C_{29}H_{20}F_6P_2$	CO	–	114, **9**, *19*
C_5H_5	C_4ClF_4	$C_{18}H_{15}P$	CO	–	201, **20**, *1*
C_5H_5	$C_4CoF_6NO_3P$	–	CO	–	361, **43**, *8*
C_5H_5	$C_4F_6S_2$	–	–	–	43
C_5H_5	$C_4F_7O_2$	–	CO	–	315, **33**, *5*
C_5H_5	C_4HF_6	$C_{26}H_{24}P_2$	–	–	56, **4**, *22*
C_5H_5	$C_4HF_6N_2$	–	CO	–	240, **26**, *15*
C_5H_5	$C_4HF_6N_2$	$C_8H_{11}P$	–	–	57, **4**, *28*
C_5H_5	$C_4HF_6N_2$	$C_{18}H_{15}O_3P$	–	–	57, **4**, *30*
C_5H_5	$C_4HF_6N_2$	$C_{18}H_{15}P$	–	–	57, **4**, *29*
C_5H_5	C_4H_3O	$C_{18}H_{15}P$	CO	–	202, **20**, *7*
C_5H_5	C_4H_3S	$C_{18}H_{15}P$	CO	–	202, **20**, *10*
C_5H_5	$C_4H_4F_3S$	–	CO	–	238, **26**, *8*
C_5H_5	C_4H_4N	–	–	PF_3	19, **2**, *23*
C_5H_5	C_4H_4N	–	CO	–	350, **42**, *1*
C_5H_5	C_4H_4N	$C_4H_{10}F_2NP$	–	–	20, **2**, *26*
C_5H_5	C_4H_5N	$C_{18}H_{15}P$	CO	–	79, **6**, *10*
C_5H_5	C_4H_5O	$C_{18}H_{15}P$	CO	–	163, **16**, *26, 28*
C_5H_5	$C_4H_5O_2S$	–	CO	–	329, **37**, *10*
C_5H_5	$C_4H_6AsCoNO_3$	–	CO	–	366, **43**, *36*
C_5H_5	C_4H_7	$C_{18}H_{15}P$	CO	–	181, **18**, *22*; 185, **18**, *46*
C_5H_5	C_4H_7	$C_{26}H_{24}P_2$	–	–	54, **4**, *12*
C_5H_5	$C_4H_7D_2$	$C_{18}H_{15}P$	CO	–	180, **18**, *17, 18*; 181, **18**, *21*
C_5H_5	C_4H_7O	$C_8H_{11}P$	CO	–	206, **21**, *5*
C_5H_5	C_4H_7O	$C_{18}H_{15}P$	CO	–	161, **16**, *10*; 183, **18**, *34*
C_5H_5	C_4H_7O	$C_{20}H_{20}NP$	CO	–	222, **23**, *9*
C_5H_5	C_4H_7OS	–	CO	–	241, **26**, *23*
C_5H_5	$C_4H_7O_2$	$C_{18}H_{15}P$	CO	–	170
C_5H_5	$C_4H_7O_2S$	–	CO	–	329, **37**, *11/13*

C_5H_5	$C_4H_7O_2S$	$C_{18}H_{15}P$	CO	—	98, **8**, 16
C_5H_5	C_4H_8DO	$C_{18}H_{15}P$	CO	—	183, **18**, 33
C_5H_5	C_4H_8NO	$C_{18}H_{15}P$	CO	—	173
C_5H_5	$C_4H_8NO_2P$	C_6H_5	CO	—	224, **23**, 24
C_5H_5	$C_4H_8NS_2$	—	CO	—	140, **13**, 11; 325, **36**, 9
C_5H_5	C_4H_8O	$C_{18}H_{15}O_3P$	CO	—	78, **6**, 7
C_5H_5	C_4H_8O	$C_{26}H_{24}P_2$	—	—	8, **1**, 36
C_5H_5	C_4H_9	$C_{18}H_{15}O_3P$	CO	—	214, **22**, 10, 11
C_5H_5	C_4H_9	$C_{18}H_{15}P$	CO	—	180, **18**, 16, 19, 20
C_5H_5	C_4H_9	$C_{26}H_{24}P_2$	—	—	54, **4**, 9, 10
C_5H_5	C_4H_9AlN	—	CO	—	307, **31**, 14
C_5H_5	C_4H_9BN	—	CO	—	306, **31**, 11
C_5H_5	C_4H_9GaN	—	CO	—	307, **31**, 16
C_5H_5	C_4H_9NO	—	—	I	18, **2**, 14
C_5H_5	C_4H_9O	$C_{18}H_{15}P$	CO	—	183, **18**, 32
C_5H_5	$C_4H_9O_2S$	—	CO	—	330, **37**, 14, 15
C_5H_5	$C_4H_9O_2S$	$C_{18}H_{15}P$	CO	—	97, **8**, 15
C_5H_5	C_4H_9S	—	CO	—	319, **34**, 4
C_5H_5	C_4H_9Si	$C_{13}H_{13}P$	—	—	20, **2**, 28
C_5H_5	C_4H_9Si	$C_{13}H_{13}P$	CO	—	112, **9**, 11
C_5H_5	$C_4H_{10}F_2NP$	—	—	I	18, **2**, 16
C_5H_5	$C_4H_{10}F_2NP$	—	CO	I	130
C_5H_5	$C_4H_{10}O_2PS$	—	CO	—	364, **43**, 25
C_5H_5	$C_4H_{10}O_3P$	—	CO	—	363, **43**, 17
C_5H_5	$C_4H_{10}O_3P$	$C_6H_{15}O_3P$	CO	—	121, **10**, 10
C_5H_5	$C_4H_{10}O_3P$	$C_{18}H_{15}P$	CO	—	183, **18**, 37
C_5H_5	$C_4H_{11}N$	—	—	I	18, **2**, 11
C_5H_5	$C_4H_{11}O_2Si$	—	CO	—	313, **32**, 3; 330, **37**, 16
C_5H_5	$C_4H_{11}Si$	$C_{13}H_{13}P$	CO	—	207, **21**, 10
C_5H_5	$C_4H_{11}Si$	$C_{14}H_{15}P$	CO	—	208, **21**, 17
C_5H_5	$C_4H_{11}Si$	$C_{18}H_{15}P$	CO	—	184, **18**, 38
C_5H_5	C_4N_3	—	CO	—	343, **40**, 2
C_5H_5	$C_5Br_2CrO_5Sb$	—	CO	—	368, **43**, 52
C_5H_5	$C_5Br_2O_5SbW$	—	CO	—	368, **43**, 53
C_5H_5	$C_5CrF_2O_5Sb$	—	CO	—	368, **43**, 49
C_5H_5	$C_5F_2O_5SbW$	—	CO	—	368, **43**, 50
C_5H_5	$C_5F_6MnNO_4P$	—	CO	—	361, **43**, 9
C_5H_5	$C_5F_6NiO_3P$	—	CO	—	361, **43**, 7
C_5H_5	C_5F_9S	—	CO	—	239, **26**, 10
C_5H_5	$C_5H_3F_6S$	—	CO	—	239, **26**, 9
C_5H_5	C_5H_3OS	$C_{18}H_{15}P$	CO	—	163, **16**, 25
C_5H_5	$C_5H_3O_2$	$C_{18}H_{15}P$	CO	—	161, **16**, 12
C_5H_5	$C_5H_4F_3OS$	—	CO	—	241, **26**, 24
C_5H_5	C_5H_5O	$C_{18}H_{15}P$	CO	—	202, **20**, 8
C_5H_5	$C_5H_6AsNiO_3$	—	CO	—	365, **43**, 35
C_5H_5	C_5H_6N	$C_{18}H_{15}P$	CO	—	185, **18**, 47
C_5H_5	$C_5H_6NiO_3P$	—	CO	—	361, **53**, 6
C_5H_5	$C_5H_7ClNO_3S$	—	CO	—	345, **41**, 2

C_5H_5	C_5H_7N	$C_{18}H_{15}P$	CO	–	79, **6**, *11*
C_5H_5	C_5H_7O	$C_{18}H_{15}P$	CO	–	163, **16**, *30*
C_5H_5	$C_5H_7O_2$	–	–	–	43
C_5H_5	$C_5H_7O_2S$	–	CO	–	330, **37**, *17*
C_5H_5	C_5H_8NOS	–	CO	–	324, **36**, *2*
C_5H_5	$C_5H_8NOS_2$	–	CO	–	142, **13**, *24*
C_5H_5	C_5H_9	$C_{18}H_{15}P$	CO	–	185, **18**, *47*
C_5H_5	C_5H_9O	$C_{18}H_{15}P$	CO	–	161, **16**, *11*; 163, **16**, *29*
C_5H_5	$C_5H_9O_2S$	–	CO	–	330, **37**, *18, 19*
C_5H_5	$C_5H_9O_3P$	–	CO	Br	124, **10**, *41*
C_5H_5	$C_5H_9O_3P$	–	CO	Cl	124, **10**, *40*
C_5H_5	$C_5H_9O_3P$	–	CO	I	124, **10**, *42*
C_5H_5	$C_5H_{10}NF_2P$	–	–	Br	18, **2**, *7*
C_5H_5	$C_5H_{10}NF_2P$	–	–	I	19, **2**, *17*
C_5H_5	$C_5H_{10}NO$	–	CO	–	239, **26**, *12*
C_5H_5	$C_5H_{10}NS_2$	–	CO	–	141, **13**, *16*
C_5H_5	C_5H_{11}	$C_{18}H_{15}P$	CO	–	181, **18**, *23, 24*
C_5H_5	C_5H_{11}	$C_{26}H_{24}P_2$	–	–	54, **4**, *11*
C_5H_5	$C_5H_{11}N$	–	–	I	18, **2**, *13*
C_5H_5	$C_5H_{11}OSi$	$C_{13}H_{13}P$	CO	–	207, **21**, *11*
C_5H_5	$C_5H_{11}OSi$	$C_{14}H_{15}P$	CO	–	208, **21**, *18*
C_5H_5	$C_5H_{11}OSi$	$C_{18}H_{15}P$	CO	–	160, **16**, *4*
C_5H_5	$C_5H_{11}O_2S$	–	CO	–	330, **37**, *20*
C_5H_5	C_6Cl_5S	–	CO	–	319, **34**, *8*
C_5H_5	C_6F_5	$C_{18}H_{15}O_3P$	CO	–	216, **22**, *21*
C_5H_5	C_6F_5	$C_{18}H_{15}P$	CO	–	198, **19**, *10*
C_5H_5	$C_6F_5O_2S$	–	CO	–	331, **37**, *21*
C_5H_5	C_6F_5S	–	CO	–	319, **34**, *7*
C_5H_5	$C_6HF_6MnO_4P$	–	CO	–	362, **43**, *10*
C_5H_5	$C_6H_3CrO_5S$	–	CO	–	322, **35**, *1*
C_5H_5	$C_6H_3F_6OS$	–	CO	–	241, **26**, *25*
C_5H_5	$C_6H_3O_5W$	–	CO	–	323, **35**, *4*
C_5H_5	C_6H_4Cl	$C_{18}H_{15}O_3P$	CO	–	216, **22**, *20*
C_5H_5	C_6H_4Cl	$C_{18}H_{15}P$	CO	–	198, **19**, *7*
C_5H_5	C_6H_4F	$C_6H_{15}P$	CO	–	155, **15**, *4*
C_5H_5	C_6H_4F	$C_{10}H_{15}P$	CO	–	207, **21**, *13*
C_5H_5	C_6H_4F	$C_{14}H_{15}P$	CO	–	207, **21**, *16*
C_5H_5	C_6H_4F	$C_{18}H_{15}O_3P$	CO	–	215, **22**, *18*; 216, **22**, *19*
C_5H_5	C_6H_4F	$C_{18}H_{15}P$	CO	–	198, **19**, *5, 6*
C_5H_5	C_6H_4F	$C_{21}H_{21}P$	CO	–	204
C_5H_5	C_6H_4F	$C_{30}H_{27}Fe_3P$	CO	–	224, **23**, *23*
C_5H_5	$C_6H_4FO_2S$	–	CO	–	331, **37**, *22*
C_5H_5	C_6H_4FS	–	CO	–	319, **34**, *6*
C_5H_5	$C_6H_4N_3$	–	CO	–	354, **42**, *15, 16*
C_5H_5	C_6H_5	$C_6H_{16}P_2$	–	–	55, **4**, *16*
C_5H_5	C_6H_5	$C_8H_9O_2P$	CO	–	221, **23**, *5*
C_5H_5	C_6H_5	$C_8H_{11}P$	–	–	53, **4**, *4*
C_5H_5	C_6H_5	$C_{10}H_{15}P$	CO	–	207, **21**, *12*
C_5H_5	C_6H_5	$C_{14}H_{15}P$	CO	–	207, **21**, *15*

C_5H_5	C_6H_5	$C_{18}H_{15}O_3P$	CO	–	215, **22**, *17*
C_5H_5	C_6H_5	$C_{18}H_{15}P$	CO	–	197, **19**, *1*
C_5H_5	C_6H_5	$C_{21}H_{21}P$	CO	–	204
C_5H_5	C_6H_5	$C_{26}H_{24}P_2$	–	–	55, **4**, *17*
C_5H_5	$C_6H_5Cl_2Sn$	$C_6H_{15}P$	CO	–	89, **7**, *10*
C_5H_5	$C_6H_5Cl_2Sn$	$C_{18}H_{15}O_3P$	–	–	21, **2**, *35*
C_5H_5	$C_6H_5Cl_2Sn$	$C_{18}H_{15}O_3P$	CO	–	124, **10**, *36*
C_5H_5	$C_6H_5Cl_2Sn$	$C_{18}H_{15}P$	CO	–	102, **8**, *39*
C_5H_5	$C_6H_5F_6O$	$C_{18}H_{15}P$	CO	–	201, **20**, *4*
C_5H_5	$C_6H_5N_3$	–	CO	–	354, **42**, *15a,* *15b, 16a*
C_5H_5	$C_6H_5N_3 \cdot H_2O$	–	CO	–	354, **42**, *15d*
C_5H_5	$C_6H_5N_3 \cdot$ $OC(CD_3)_2$	–	CO	–	354, **42**, *15e*
C_5H_5	$C_6H_5O_2$	$C_{18}H_{15}P$	CO	–	161, **16**, *13*
C_5H_5	$C_6H_5O_2S$	–	CO	–	331, **37**, *23*
C_5H_5	C_6H_5S	–	CO	–	319, **34**, *5*
C_5H_5	C_6H_5S	$C_{26}H_{24}P_2$	–	–	23, **2**, *63;* 41, **3**, *9, 10*
C_5H_5	C_6H_5Se	–	CO	–	341, **39**, *1*
C_5H_5	C_6H_5Te	–	CO	–	342, **39**, *7*
C_5H_5	C_6H_6NO	–	CO	–	350, **42**, *2*
C_5H_5	$C_6H_6N_3O_4$	–	CO	–	353, **42**, *12*
C_5H_5	C_6H_7	$C_{18}H_{15}P$	CO	–	185, **18**, *47*
C_5H_5	C_6H_7P	–	CO	I	111, **9**, *1*
C_5H_5	$C_6H_7O_4$	$C_{26}H_{24}P_2$	–	–	57, **4**, *27*
C_5H_5	C_6H_8As	–	CO	–	365, **43**, *33*
C_5H_5	C_6H_9	$C_{18}H_{15}P$	CO	–	196
C_5H_5	C_6H_9N	$C_{18}H_{15}P$	CO	–	79, **6**, *12*
C_5H_5	$C_6H_{10}NS_2$	–	CO	–	141, **13**, *23*
C_5H_5	$C_6H_{10}O_3P$	–	CO	–	363, **43**, *18*
C_5H_5	$C_6H_{10}O_3P$	$C_9H_{15}O_3P$	CO	–	121, **10**, *17*
C_5H_5	C_6H_{11}	$C_{18}H_{15}O_3P$	CO	–	214, **22**, *12*
C_5H_5	C_6H_{11}	$C_{18}H_{15}P$	CO	–	185, **18**, *48*
C_5H_5	C_6H_{11}	$C_{26}H_{24}P_2$	–	–	54, **4**, *7;* 55, **4**, *13*
C_5H_5	$C_6H_{11}D_2O_2S$	–	CO	–	331, **37**, *24*
C_5H_5	$C_6H_{11}O$	$C_{18}H_{15}P$	CO	–	185, **18**, *47*
C_5H_5	$C_6H_{11}O_3P$	–	CO	Br	125, **10**, *44*
C_5H_5	$C_6H_{11}O_3P$	–	CO	Cl	124, **10**, *43*
C_5H_5	$C_6H_{11}O_3P$	–	CO	I	125, **10**, *45*
C_5H_5	$C_6H_{12}AsO_2$	–	CO	–	365, **43**, *34*
C_5H_5	$C_6H_{13}N$	–	–	I	18, **2**, *12*
C_5H_5	$C_6H_{13}O_2S$	–	CO	–	331, **37**, *24*
C_5H_5	$C_6H_{15}O_3P$	–	–	–	4, **1**, *6*
C_5H_5	$C_6H_{15}O_3P$	–	–	$SnCl_3$	20, **2**, *31*
C_5H_5	$C_6H_{15}O_3P$	–	CO	Cl	121, **10**, *8*
C_5H_5	$C_6H_{15}O_3P$	–	CO	I	121, **10**, *9*
C_5H_5	$C_6H_{15}O_3P$	–	CO	$SnCl_3$	121, **10**, *14*
C_5H_5	$C_6H_{15}O_3P$	$C_{17}H_{15}Si$	CO	–	121, **10**, *12*
C_5H_5	$C_6H_{15}O_3P$	$C_{18}H_{15}Si$	CO	–	121, **10**, *11*

C_5H_5	$C_6H_{15}O_3P$	$C_{18}H_{15}Sn$	CO	–	121, **10**, *13*
C_5H_5	$C_6H_{15}P$	–	CO	I	88, **7**, *7*
C_5H_5	$C_6H_{15}P$	–	CO	$SnCl_3$	89, **7**, *11*
C_5H_5	$C_6H_{15}P$	$C_{12}H_{10}ClSn$	CO	–	89, **7**, *9*
C_5H_5	$C_6H_{15}P$	$C_{18}H_{15}O_3P$	–	–	5, **1**, *16*
C_5H_5	$C_6H_{15}P$	$C_{18}H_{15}Sn$	CO	–	88, **7**, *8*
C_5H_5	$C_6H_{16}P_2$	–	–	Br	22, **2**, *52*
C_5H_5	$C_6H_{16}P_2$	–	–	Cl	22, **2**, *49*
C_5H_5	$C_6H_{16}P_2$	–	–	H	22, **2**, *44*
C_5H_5	$C_6H_{16}P_2$	–	–	I	23, **2**, *55*
C_5H_5	$C_6H_{16}P_2$	–	CO	–	80, **6**, *17/19*
C_5H_5	$C_6H_{16}P_2$	$C_{18}H_{15}O_3P$	–	–	7, **1**, *32, 33*
C_5H_5	$C_6H_{16}P_2$	$C_{43}H_{38}^-$ MoN_2P_2	–	–	10, **1**, *50*
C_5H_5	$C_6H_{17}O_2SSi_2$	–	CO	–	331, **37**, *25*
C_5H_5	$C_6H_{17}Si_2$	$C_{18}H_{15}P$	CO	–	100, **8**, *31*
C_5H_5	$C_6H_{18}N_3P$	–	CO	Cl	91, **7**, *26*
C_5H_5	$C_6H_{18}N_3P$	–	CO	H	91, **7**, *25*
C_5H_5	$C_6H_{18}PSi_2$	–	CO	–	361, **43**, *5*
C_5H_5	$C_7H_3N_4$	–	CO	–	343, **40**, *3*
C_5H_5	$C_7H_3N_4$	$C_{18}H_{15}O_3P$	CO	–	123, **10**, *30*
C_5H_5	$C_7H_3N_4$	$C_{18}H_{15}P$	CO	–	99, **8**, *25*
C_5H_5	C_7H_4ClO	$C_{18}H_{15}P$	CO	–	162, **16**, *19*
C_5H_5	C_7H_4FO	$C_{18}H_{15}P$	CO	–	162, **16**, *18*
C_5H_5	$C_7H_5Br_2^-$ MnO_2Sb	–	CO	–	368, **43**, *51*
C_5H_5	$C_7H_5CrO_5S$	–	CO	–	322, **35**, *2*
C_5H_5	$C_7H_5F_6S$	–	CO	–	242, **26**, *27;* 253
C_5H_5	$C_7H_5N_2$	–	CO	–	351, **42**, *8*
C_5H_5	C_7H_5O	–	CO	CN	149, **14**, *6, 8*
C_5H_5	C_7H_5O	$C_{10}H_{15}P$	CO	–	207, **21**, *14*
C_5H_5	C_7H_5O	$C_{18}H_{15}P$	CO	–	161, **16**, *15;* 198, **19**, *3*
C_5H_5	C_7H_5O	$C_{26}H_{24}P_2$	–	–	56, **4**, *26*
C_5H_5	C_7H_5OS	–	CO	–	324, **36**, *1*
C_5H_5	$C_7H_5O_2$	–	CO	–	315, **33**, *6*
C_5H_5	$C_7H_5O_2$	$C_{18}H_{15}P$	CO	–	162, **16**, *17*
C_5H_5	$C_7H_5S_2$	–	CO	–	140, **13**, *9;* 324, **36**, *4*
C_5H_5	$C_7H_5S_3$	–	CO	–	142, **13**, *27;* 325, **36**, *7*
C_5H_5	$C_7H_6AsCrO_5$	–	CO	–	366, **43**, *38*
C_5H_5	$C_7H_6AsMoO_5$	–	CO	–	366, **43**, *39*
C_5H_5	$C_7H_6AsO_5W$	–	CO	–	366, **43**, *40*
C_5H_5	$C_7H_6CrO_5Sb$	–	CO	–	368, **43**, *54*
C_5H_5	$C_7H_6FO_2S$	–	CO	–	332, **37**, *26*
C_5H_5	$C_7H_6N_2$	–	CO	–	351, **42**, *8b*
C_5H_5	$C_7H_6N_2$ $OC(CH_3)_2$	–	CO	–	351, **42**, *8c*
C_5H_5	C_7H_7	$C_8H_{11}P$	CO	–	205, **21**, *2*

C_5H_5	C_7H_7	$C_{12}H_{27}P$	CO	–	155, **15**, *7*
C_5H_5	C_7H_7	$C_{13}H_{13}P$	CO	–	206, **21**, *7*
C_5H_5	C_7H_7	$C_{18}H_{15}As$	CO	–	228, **25**, *3*
C_5H_5	C_7H_7	$C_{18}H_{15}P$	CO	–	182, **18**, *25*; 198, **19**, *2*
C_5H_5	C_7H_7	$C_{26}H_{24}P_2$	–	–	54, **4**, *8*
C_5H_5	C_7H_7O	$C_{18}H_{15}P$	CO	–	198, **19**, *4, 8*
C_5H_5	C_7H_7OS	–	CO	–	320, **34**, *10*
C_5H_5	$C_7H_7O_2S$	–	CO	–	313, **32**, *5*; 332, **37**, *27/30*
C_5H_5	$C_7H_7O_2S$	$C_{18}H_{15}As$	CO	–	131, **11**, *4*
C_5H_5	$C_7H_7O_2S$	$C_{18}H_{15}P$	CO	–	97, **8**, *10*; 98, **8**, *17*
C_5H_5	$C_7H_7O_3S$	–	CO	–	314, **32**, *6*; 332, **37**, *31*
C_5H_5	C_7H_7S	–	CO	–	320, **34**, *9*
C_5H_5	$C_7H_7S_2$	–	CO	–	320, **34**, *11*
C_5H_5	C_7H_9N	–	–	I	18, **2**, *15*
C_5H_5	C_7H_9O	$C_{12}H_{27}P$	CO	–	155, **15**, *6*
C_5H_5	$C_7H_{10}NO$	–	CO	–	239, **26**, *13*
C_5H_5	$C_7H_{11}O$	$C_{18}H_{15}P$	CO	–	162, **16**, *21*
C_5H_5	$C_7H_{11}D_2O$	$C_{18}H_{15}P$	CO	–	160, **16**, *7*
C_5H_5	$C_7H_{11}O_2S$	–	CO	–	333, **37**, *32*
C_5H_5	$C_7H_{12}NO$	–	CO	–	239, **26**, *14*
C_5H_5	$C_7H_{13}O_3P$	–	CO	Br	125, **10**, *47*
C_5H_5	$C_7H_{13}O_3P$	–	CO	Cl	125, **10**, *46*
C_5H_5	$C_7H_{13}O_3P$	–	CO	I	125, **10**, *48*
C_5H_5	$C_7H_{17}O_2Si_2$	$C_{18}H_{15}P$	CO	–	160, **16**, *5*
C_5H_5	$C_8H_3Co_2O_6$	–	CO	–	250
C_5H_5	$C_8H_4F_3O_2$	–	CO	–	44
C_5H_5	$C_8H_4F_9S$	–	CO	–	242, **26**, *28*
C_5H_5	$C_8H_4MnO_3$	$C_{18}H_{15}P$	CO	–	201, **20**, *3*
C_5H_5	C_8H_5	$C_{18}H_{15}P$	CO	–	196
C_5H_5	$C_8H_5BrMo-O_3Sb$	–	CO	–	367, **43**, *47*
C_5H_5	$C_8H_5BrO_3SbW$	–	CO	–	367, **43**, *48*
C_5H_5	$C_8H_5N_4$	–	CO	–	343, **40**, *4*
C_5H_5	$C_8H_5N_4$	$C_{18}H_{15}P$	CO	–	99, **8**, *26*
C_5H_5	C_8H_5O	$C_{18}H_{15}P$	CO	–	202, **20**, *9*
C_5H_5	C_8H_6N	–	CO	–	350, **42**, *3*
C_5H_5	$C_8H_6N_3$	–	CO	–	353, **42**, *14, 15*
C_5H_5	$C_8H_7D_2O_2S$	–	CO	–	334, **37**, *39*
C_5H_5	$C_8H_7N_3$	–	CO	–	353, **42**, *13a, 13b*
C_5H_5	$C_8H_7N_3 \cdot H_2O$	–	CO	–	353, **42**, *13c*
C_5H_5	$C_8H_7N_3 \cdot OC-(CH_3)_2$	–	CO	–	353, **42**, *13d*
C_5H_5	C_8H_7O	$C_{18}H_{15}P$	CO	–	161, **16**, *9*; 162, **16**, *20*
C_5H_5	C_8H_7O	$C_{20}H_{20}NP$	CO	–	222, **23**, *10*
C_5H_5	$C_8H_7O_2$	$C_{18}H_{15}P$	CO	–	162, **16**, *16*

C_5H_5	$C_8H_8MnO_2S$	—	CO	—	323, **35**, *5*
C_5H_5	C_8H_8NS	—	CO	—	139, **13**, *4*
C_5H_5	$C_8H_8NS_2$	—	CO	—	140, **13**, *14*
C_5H_5	$C_8H_8O_3P$	—	CO	—	240, **26**, *21*
C_5H_5	C_8H_9O	$C_{26}H_{24}P_2$	—	—	56, **4**, *25*
C_5H_5	C_8H_9O	$C_{18}H_{15}P$	CO	—	183, **18**, *35*
C_5H_5	$C_8H_9O_2S$	—	CO	—	313, **32**, *4*; 333, **37**, *33*
C_5H_5	$C_8H_9O_3S$	—	CO	—	333, **37**, *34*
C_5H_5	$C_8H_{10}NO_3S_2$	—	CO	—	338, **38**, *2, 3*; 346, **41**, *6, 7*
C_5H_5	$C_8H_{10}NO_4S_2$	—	CO	—	347, **41**, *10, 11*
C_5H_5	$C_8H_{11}P$	—	—	Br	18, **2**, *8*
C_5H_5	$C_8H_{11}P$	—	CO	—	206, **21**, *6*
C_5H_5	$C_8H_{11}P$	—	CO	Cl	11, **9**, *3*
C_5H_5	$C_8H_{11}P$	—	CO	CN	111, **9**, *5*
C_5H_5	$C_8H_{11}P$	—	CO	H	111, **9**, *2*
C_5H_5	$C_8H_{11}P$	—	CO	I	111, **9**, *4*
C_5H_5	$C_8H_{11}P$	$C_{18}H_{15}Sn$	—	—	21, **2**, *40*
C_5H_5	$C_8H_{11}P$	$C_{18}H_{15}Sn$	CO	—	112, **9**, *7*
C_5H_5	$C_8H_{11}Si$	—	CO	H	72, **5**, *3*
C_5H_5	$C_8H_{11}Sn$	$C_{18}H_{15}P$	CO	—	101, **8**, *36*
C_5H_5	$C_8H_{13}O$	$C_{18}H_{15}P$	CO	—	160, **16**, *8*
C_5H_5	$C_8H_{15}S$	$C_{18}H_{15}P$	CO	—	185, **18**, *47*
C_5H_5	$C_8H_{17}O_3S$	—	CO	—	340, **38**, *16*
C_5H_5	$C_8H_{18}O_3P$	—	CO	—	363, **43**, *20*
C_5H_5	$C_8H_{18}O_3P$	$C_{12}H_{27}P$	CO	—	122, **10**, *19*
C_5H_5	$C_9H_4MnO_4$	$C_{18}H_{15}P$	CO	—	161, **16**, *14*
C_5H_5	$C_9H_5O_2$	$C_{18}H_{15}P$	CO	—	163, **16**, *24*
C_5H_5	$C_9H_7N_4$	—	CO	—	344, **40**, *5*
C_5H_5	$C_9H_7N_4$	$C_{18}H_{15}P$	CO	—	99, **8**, *27*
C_5H_5	C_9H_7O	$C_{18}H_{15}P$	CO	—	163, **16**, *27*
C_5H_5	C_9H_8NOS	—	CO	—	241, **26**, *26*
C_5H_5	$C_9H_9N_2$	—	CO	—	351, **42**, *9*
C_5H_5	$C_9H_9O_2$	$C_{18}H_{15}P$	CO	—	198, **19**, *9*
C_5H_5	$C_9H_9O_2S$	—	CO	—	333, **37**, *35, 36*
C_5H_5	$C_9H_{10}NS_2$	—	CO	—	140, **13**, *13*; 141, **13**, *15, 18*
C_5H_5	$C_9H_{10}N_2$	—	CO	—	342, **42**, *9b*
C_5H_5	$C_9H_{10}N_2 \cdot H_2O$	—	CO	—	352, **42**, *9c*
C_5H_5	$C_9H_{10}N_2 \cdot OC(CH_3)_2$	—	CO	—	352, **42**, *9d*
C_5H_5	$C_9H_{10}O_2P$	—	CO	—	363, **43**, *19*
C_5H_5	$C_9H_{10}O_2P$	$C_{12}H_{12}O_2P$	CO	—	128/9
C_5H_5	C_9H_{11}	$C_{18}H_{15}P$	CO	—	182, **18**, *26*
C_5H_5	$C_9H_{11}AsMnO_2$	—	CO	—	366, **43**, *37*
C_5H_5	$C_9H_{11}O_2S$	$C_{18}H_{15}P$	CO	—	98, **8**, *18*
C_5H_5	$C_9H_{13}N_2O_4S_3$	—	CO	—	347, **41**, *13*
C_5H_5	$C_9H_{13}O_2SSi$	—	CO	—	333, **37**, *37*
C_5H_5	$C_9H_{13}Si$	$C_{18}H_{15}P$	CO	—	184, **18**, *39*
C_5H_5	$C_9H_{15}O_3P$	—	CO	Cl	121, **10**, *16*

C$_5$H$_5$	C$_{12}$H$_{10}$O$_3$P	–	CO	–	314
C$_5$H$_5$	C$_{12}$H$_{10}$P	–	CO	–	360, **43**, 2
C$_5$H$_5$	C$_{12}$H$_{10}$PSe	–	CO	–	364, **43**, 27
C$_5$H$_5$	C$_{12}$H$_{11}$P	–	CO	Br	112, **9**, 8
C$_5$H$_5$	C$_{12}$H$_{12}$N	–	CO	–	350, **42**, 5
C$_5$H$_5$	C$_{12}$H$_{15}$N$_2$O$_3$S	–	CO	–	346, **41**, 4
C$_5$H$_5$	C$_{12}$H$_{15}$O$_2$P	–	CO	Cl	128/9
C$_5$H$_5$	C$_{12}$H$_{21}$O$_2$	C$_{18}$H$_{15}$P	CO	–	171/2
C$_5$H$_5$	C$_{12}$H$_{21}$O$_3$S	C$_{18}$H$_{15}$P	CO	–	98, **8**, 19
C$_5$H$_5$	C$_{12}$H$_{27}$P	–	CO	Br	89, **7**, 13
C$_5$H$_5$	C$_{12}$H$_{27}$P	–	CO	Cl	89, **7**, 12
C$_5$H$_5$	C$_{12}$H$_{27}$P	–	CO	I	89, **7**, 14
C$_5$H$_5$	C$_{12}$H$_{27}$P	–	CO	NCO	89, **7**, 17
C$_5$H$_5$	C$_{12}$H$_{27}$P	C$_{18}$H$_{15}$Si	CO	–	89, **7**, 16
C$_5$H$_5$	C$_{13}$H$_7$N$_4$	–	CO	–	344, **40**, 6
C$_5$H$_5$	C$_{13}$H$_7$N$_4$	C$_{18}$H$_{15}$P	CO	–	99, **8**, 28
C$_5$H$_5$	C$_{13}$H$_{10}$F$_3$P	C$_{18}$H$_{15}$Sn	CO	–	113, **9**, 15
C$_5$H$_5$	C$_{13}$H$_{10}$N	–	CO	–	68
C$_5$H$_5$	C$_{13}$H$_{10}$NS$_2$	–	CO	–	141, **13**, 22
C$_5$H$_5$	C$_{13}$H$_{11}$ClN$_3$	–	CO	–	139, **13**, 2
C$_5$H$_5$	C$_{13}$H$_{13}$As	–	CO	I	131, **11**, 1
C$_5$H$_5$	C$_{13}$H$_{13}$P	–	CO	I	112, **9**, 9
C$_5$H$_5$	C$_{13}$H$_{13}$P	–	CO	CN	112, **9**, 10
C$_5$H$_5$	C$_{13}$H$_{13}$P	C$_{18}$H$_{15}$Sn	–	–	21, **2**, 41
C$_5$H$_5$	C$_{13}$H$_{13}$P	C$_{18}$H$_{15}$Sn	CO	–	113, **9**, 15
C$_5$H$_5$	C$_{13}$H$_{13}$Si	C$_{18}$H$_{15}$P	CO	–	101, **8**, 33
C$_5$H$_5$	C$_{13}$H$_{13}$Si	C$_{18}$H$_{33}$P	CO	–	90, **7**, 21
C$_5$H$_5$	C$_{13}$H$_{23}$N$_2$	–	CO	–	238, **26**, 4
C$_5$H$_5$	C$_{14}$H$_5$N$_4$	C$_{18}$H$_{15}$P	CO	–	182, **18**, 27
C$_5$H$_5$	C$_{14}$H$_7$O$_9$Ru$_3$	–	–	–	49/50
C$_5$H$_5$	C$_{14}$H$_9$N$_4$	–	CO	–	344, **40**, 7
C$_5$H$_5$	C$_{14}$H$_{12}$NS$_2$	–	CO	–	141, **13**, 21
C$_5$H$_5$	C$_{14}$H$_{13}$NPS	–	CO	–	139, **13**, 5
C$_5$H$_5$	C$_{14}$H$_{13}$NPS$_2$	–	CO	–	143, **13**, 33
C$_5$H$_5$	C$_{14}$H$_{13}$S$_2$	–	CO	–	142, **13**, 32
C$_5$H$_5$	C$_{14}$H$_{14}$NO$_3$S$_2$	–	CO	–	346, **41**, 8
C$_5$H$_5$	C$_{14}$H$_{14}$N$_3$	–	CO	–	139, **13**, 3
C$_5$H$_5$	C$_{14}$H$_{14}$N$_3$	–	CO	NO	146
C$_5$H$_5$	C$_{14}$H$_{14}$N$_3$	C$_{18}$H$_{15}$O$_3$P	–	–	26, **2**, 80
C$_5$H$_5$	C$_{14}$H$_{14}$N$_3$	C$_{18}$H$_{15}$P	–	–	25, **2**, 78
C$_5$H$_5$	C$_{14}$H$_{14}$PS	–	CO	–	142, **13**, 30
C$_5$H$_5$	C$_{14}$H$_{25}$P$_3$	–	–	–	8, **1**, 41
C$_5$H$_5$	C$_{15}$H$_{10}$NiO$_3$P	–	CO	–	362, **43**, 11
C$_5$H$_5$	C$_{15}$H$_{11}$O$_2$	–	–	–	43
C$_5$H$_5$	C$_{15}$H$_{14}$NS$_2$	–	CO	–	141, **13**, 20
C$_5$H$_5$	C$_{15}$H$_{15}$OP	–	CO	Cl	129
C$_5$H$_5$	C$_{15}$H$_{15}$NPS	–	CO	–	139, **13**, 6
C$_5$H$_5$	C$_{15}$H$_{16}$O$_2$PS	–	CO	–	143, **13**, 35
C$_5$H$_5$	C$_{15}$H$_{16}$P	–	CO	–	240, **26**, 17
C$_5$H$_5$	C$_{15}$H$_{25}$N$_2$	–	CO	–	238, **26**, 5
C$_5$H$_5$	C$_{16}$H$_{13}$O	–	CO	–	241, **26**, 22

C_5H_5	$C_{16}H_{14}N_2O$	–	–	–	44
C_5H_5	$C_{17}H_{14}P$	–	CO	–	240, **26**, *18*
C_5H_5	$C_{17}H_{15}Si$	$C_{18}H_{15}P$	CO	–	101, **8**, *34*
C_5H_5	$C_{17}H_{15}Si$	$C_{18}H_{33}P$	CO	–	90, **7**, *22*
C_5H_5	$C_{18}H_{14}OP$	–	CO	–	240, **26**, *20*
C_5H_5	$C_{18}H_{14}O_3P$	$C_{18}H_{15}O_3P$	–	–	57, **4**, *32*
C_5H_5	$C_{18}H_{15}As$	–	CO	NCS	131, **11**, *5*
C_5H_5	$C_{18}H_{15}As$	–	CO	I	131, **11**, *3*
C_5H_5	$C_{18}H_{15}As$	–	CO	$SnCl_3$	131, **11**, *6*
C_5H_5	$C_{18}H_{15}As$	–	–	NO	49
C_5H_5	$C_{18}H_{15}As$	I	–	NO	48
C_5H_5	$C_{18}H_{15}As$	$C_{18}H_{15}Sn$	CO	–	131, **11**, *8*
C_5H_5	$C_{18}H_{15}Ge$	$C_{26}H_{24}P_2$	–	–	24, **2**, *68*
C_5H_5	$C_{18}H_{15}OP$	–	CO	H	129
C_5H_5	$C_{18}H_{15}O_2P$	–	CO	H	129
C_5H_5	$C_{18}H_{15}O_3P$	–	–	–	4, **1**, *7, 8*; 5, **1**, *9*
C_5H_5	$C_{18}H_{15}O_3P$	–	–	Br	18, **2**, *10*
C_5H_5	$C_{18}H_{15}O_3P$	–	–	H	18, **2**, *5*
C_5H_5	$C_{18}H_{15}O_3P$	–	–	I	19, **2**, *20*; 41, **3**, *1*
C_5H_5	$C_{18}H_{15}O_3P$	I	–	NO	48
C_5H_5	$C_{18}H_{15}O_3P$	–	–	SCN	19, **2**, *21*
C_5H_5	$C_{18}H_{15}O_3P$	–	–	$SnBr_3$	20, **2**, *33*
C_5H_5	$C_{18}H_{15}O_3P$	–	–	$SnCl_3$	20, **2**, *32*
C_5H_5	$C_{18}H_{15}O_3P$	–	–	SnF_3	20, **2**, *29*
C_5H_5	$C_{18}H_{15}O_3P$	–	–	SnI_3	21, **2**, *34*
C_5H_5	$C_{18}H_{15}O_3P$	–	–	SO_2	5, **1**, *17*
C_5H_5	$C_{18}H_{15}O_3P$	–	CO	–	75; 78, **6**, *5*; 215, **22**, *16*
C_5H_5	$C_{18}H_{15}O_3P$	–	CO	Br	122, **10**, *22*
C_5H_5	$C_{18}H_{15}O_3P$	–	CO	Cl	122, **10**, *21*
C_5H_5	$C_{18}H_{15}O_3P$	–	CO	CN	123, **10**, *31*
C_5H_5	$C_{18}H_{15}O_3P$	–	CO	H	122, **10**, *20*
C_5H_5	$C_{18}H_{15}O_3P$	–	CO	I	122, **10**, *23*
C_5H_5	$C_{18}H_{15}O_3P$	–	CO	NCO	123, **10**, *29*
C_5H_5	$C_{18}H_{15}O_3P$	–	CO	SeCN	123, **10**, *28*
C_5H_5	$C_{18}H_{15}O_3P$	–	CO	$SnBr_2I$	124, **10**, *39*
C_5H_5	$C_{18}H_{15}O_3P$	–	CO	$SnBr_3$	134, **10**, *38*
C_5H_5	$C_{18}H_{15}O_3P$	–	CO	$SnCl_3$	124, **10**, *37*
C_5H_5	$C_{18}H_{15}O_3P$	$C_{18}H_{15}Sn$	–	–	21, **2**, *42*
C_5H_5	$C_{18}H_{15}O_3P$	$C_{18}H_{15}Sn$	CO	–	123, **10**, *34*
C_5H_5	$C_{18}H_{15}P$	–	–	NO	49
C_5H_5	$C_{18}H_{15}P$	I	–	NO	48
C_5H_5	$C_{18}H_{15}P$	–	CO	–	75; 78, **6**, *4*
C_5H_5	$C_{18}H_{15}P$	–	CO	Br	95, **8**, *5*
C_5H_5	$C_{18}H_{15}P$	–	CO	D	95, **8**, *2*
C_5H_5	$C_{18}H_{15}P$	–	CO	Cl	95, **8**, *4*
C_5H_5	$C_{18}H_{15}P$	–	CO	CN	100, **8**, *29*

C_5H_5	$C_{18}H_{15}P$	–	CO	F	95, **8**, *3*
C_5H_5	$C_{18}H_{15}P$	–	CO	H	95, **8**, *1*
C_5H_5	$C_{18}H_{15}P$	–	CO	I	96, **8**, *6*
C_5H_5	$C_{18}H_{15}P$	–	CO	I·(SCN)$_2$	96, **8**, *7*
C_5H_5	$C_{18}H_{15}P$	–	CO	NCO	98, **8**, *22*
C_5H_5	$C_{18}H_{15}P$	–	CO	NCS	99, **8**, *23*
C_5H_5	$C_{18}H_{15}P$	–	CO	NCSe	99, **8**, *24*
C_5H_5	$C_{18}H_{15}P$	–	CO	SCN	98, **8**, *20*
C_5H_5	$C_{18}H_{15}P$	–	CO	SeCN	98, **8**, *21*
C_5H_5	$C_{18}H_{15}P$	–	CO	SnCl$_3$	102, **8**, *40*
C_5H_5	$C_{18}H_{15}P$	$C_{18}H_{15}Si$	CO	–	101, **8**, *32*
C_5H_5	$C_{18}H_{15}P$	$C_{18}H_{15}Sn$	CO	–	101, **8**, *37*
C_5H_5	$C_{18}H_{15}P$	$C_{19}H_{17}P$	CO	–	233
C_5H_5	$C_{18}H_{15}P$	$C_{21}H_{21}P$	CO	–	233
C_5H_5	$C_{18}H_{15}P$	$C_{23}H_{18}CoO_2$	CO	–	201, **20**, *2*
C_5H_5	$C_{18}H_{15}Sb$	–	–	NO	49
C_5H_5	$C_{18}H_{15}Sb$	I	–	–	48; 132, **11**, *9*
C_5H_5	$C_{18}H_{15}Sb$	–	CO	SnCl$_3$	132, **11**, *10*
C_5H_5	$C_{18}H_{15}Sb$	$C_{18}H_{15}Sn$	CO	–	132, **11**, *12*
C_5H_5	$C_{18}H_{15}Si$	$C_{18}H_{33}P$	CO	–	90, **7**, *20*
C_5H_5	$C_{18}H_{33}P$	–	CO	H	90, **7**, *19*
C_5H_5	$C_{19}H_{15}BN$	–	CO	–	71; 306, **31**, *12*
C_5H_5	$C_{19}H_{15}N$	–	CO	–	237, **26**, *2*
C_5H_5	$C_{19}H_{15}NPS$	–	CO	–	140, **13**, *7*
C_5H_5	$C_{19}H_{15}NPS_2$	–	CO	–	143, **13**, *34*
C_5H_5	$C_{19}H_{15}PO$	–	CO	–	232
C_5H_5	$C_{19}H_{15}O_2PRu$	–	CO	–	362, **43**, *12*
C_5H_5	$C_{19}H_{17}P$	–	CO	H	113, **9**, *16*
C_5H_5	$C_{20}H_{15}P$	–	CO	Br	114, **9**, *21*
C_5H_5	$C_{20}H_{17}BN_3$	–	CO	–	352, **42**, *10b*
C_5H_5	$C_{21}H_{15}O_2S$	–	CO	–	334, **37**, *38*
C_5H_5	$C_{21}H_{18}BN_2$	–	CO	–	351, **42**, *7a*
C_5H_5	$C_{21}H_{21}NP$	–	CO	–	238, **26**, *6*
C_5H_5	$C_{21}H_{21}P$	–	CO	H	90, **7**, *23*
C_5H_5	$C_{21}H_{22}NP$	–	CO	Br	114, **9**, *23*
C_5H_5	$C_{21}H_{22}NP$	–	CO	Cl	114, **9**, *22*
C_5H_5	$C_{21}H_{22}NP$	–	CO	I	114, **9**, *24*
C_5H_5	$C_{21}H_{22}OPRh$	–	CO	–	250
C_5H_5	$C_{22}H_{15}OP$	–	CO	–	363, **43**, *21*
C_5H_5	$C_{22}H_{24}NP$	–	CO	I	114, **9**, *25*
C_5H_5	$C_{22}H_{29}P$	–	CO	CN	113, **9**, *18*
C_5H_5	$C_{22}H_{29}P$	–	CO	I	113, **9**, *17*
C_5H_5	$C_{24}H_{19}BN_3$	–	CO	–	354, **42**, *15c*
C_5H_5	$C_{24}H_{29}P_3$	–	–	–	9, **1**, *42*
C_5H_5	$C_{25}H_{20}BN_2$	–	CO	–	351, **42**, *8a*
C_5H_5	$C_{25}H_{22}P_2$	–	CO	–	80, **6**, *15, 16*
C_5H_5	$C_{26}H_{18}P$	–	CO	–	240, **26**, *19*
C_5H_5	$C_{26}H_{22}P_2$	–	–	Br	23, **2**, *54*
C_5H_5	$C_{26}H_{22}P_2$	–	–	Cl	22, **2**, *51*

C_5H_5	$C_{26}H_{22}P_2$	—	CO	—	81, **6**, 27
C_5H_5	$C_{26}H_{23}NP$	—	CO	—	238, **26**, 7
C_5H_5	$C_{26}H_{24}P_2$	—	—	Br	23, **2**, 53; 41, **3**, 5
C_5H_5	$C_{26}H_{24}P_2$	—	—	Cl	22, **2**, 50; 41, **3**, 4
C_5H_5	$C_{26}H_{24}P_2$	—	—	CN	23, **2**, 58; 41, **3**, 7
C_5H_5	$C_{26}H_{24}P_2$	—	—	D	22, **2**, 46
C_5H_5	$C_{26}H_{24}P_2$	—	—	H	22, **2**, 45
C_5H_5	$C_{26}H_{24}P_2$	—	—	I	23, **2**, 56; 41, **3**, 6
C_5H_5	$C_{26}H_{24}P_2$	—	—	MgBr	25, **2**, 74
C_5H_5	$C_{26}H_{24}P_2$	—	—	NCS	41, **3**, 8
C_5H_5	$C_{26}H_{24}P_2$	—	—	NH_3	8, **1**, 37
C_5H_5	$C_{26}H_{24}P_2$	—	—	N_2H_4	8, **1**, 38
C_5H_5	$C_{26}H_{24}P_2$	—	—	NO	49
C_5H_5	$C_{26}H_{24}P_2$	—	—	SCN	23, **2**, 59
C_5H_5	$C_{26}H_{24}P_2$	—	—	S_2O_3	26, **2**, 82
C_5H_5	$C_{26}H_{24}P_2$	—	—	$SnBr_3$	24, **2**, 70
C_5H_5	$C_{26}H_{24}P_2$	—	—	$SnCl_3$	24, **2**, 69
C_5H_5	$C_{26}H_{24}P_2$	—	—	SnI_3	24, **2**, 71
C_5H_5	$C_{26}H_{24}P_2$	—	CO	—	80, **6**, 20; 81, **6**, 21/26
C_5H_5	$C_{27}H_{24}BN_2$	—	CO	—	352, **42**, 9a
C_5H_5	$C_{28}H_{20}As$	—	CO	—	365, **43**, 33a
C_5H_5	$C_{28}H_{20}P$	—	CO	—	360, **43**, 4, 4a
C_5H_5	$C_{28}H_{20}Sb$	—	CO	—	367, **43**, 46a
C_5H_5	$C_{29}H_{20}F_6P_2$	$C_{18}H_{15}Sn$	CO	—	114, **9**, 20
C_5H_5	$C_{30}H_{29}P_2$	—	—	—	58, **4**, 33
C_5H_5	$C_{31}H_{28}P_2$	—	CO	—	81, **6**, 29; 82, **6**, 30
C_5H_5	$C_{31}H_{30}EuF_{21}$-NO_6	—	CO	—	307, **31**, 17
C_5H_5	$C_{31}H_{30}F_{21}$-$HoNO_6$	—	CO	—	307, **31**, 17
C_5H_5	$C_{31}H_{30}F_{21}$-NO_6Pr	—	CO	—	307, **31**, 19
C_5H_5	$C_{31}H_{30}F_{21}$-NO_6Yb	—	CO	—	302, **31**, 17
C_5H_5	$C_{34}H_{33}P_3$	—	—	—	9, **1**, 43, 44
C_5H_5	$C_{34}H_{33}P_3$	—	—	D	22, **2**, 48
C_5H_5	$C_{34}H_{33}P_3$	—	—	H	22, **2**, 47
C_5H_5	$C_{36}H_{37}O_2P_2$	—	—	I	23, **2**, 57
C_5H_5	$C_{41}H_{39}P_3$	—	—	—	9, **1**, 45/47
C_5H_5	$C_{42}H_{42}P_4$	—	—	—	9, **1**, 48, 49
C_5H_5	$C_{42}H_{42}P_4$	—	CO	—	81, **6**, 28
C_5H_5	$C_{42}H_{42}P_4$	—	CO	I	90, **7**, 24
C_5H_5	$C_{49}H_{40}O_7P_3Rh$	—	CO	—	250
C_5H_5	$C_{61}H_{60}Eu_2F_{42}$-NO_{12}	—	CO	—	307, **31**, 18

C_5H_5	$C_{61}H_{60}F_{42}Ho_2\text{-}NO_{12}$	–	CO	–	307, **31**, *18*
C_5H_5	$C_{61}H_{60}F_{42}\text{-}NO_{12}Yb$	–	CO	–	307, **31**, *18*
C_5H_5	$C_{61}H_{60}F_{42}\text{-}NO_{12}Pr_2$	–	CO	–	307, **31**, *18*
C_5H_5N	CH_3	C_5H_5	CO	–	154
C_5H_5O	C_5H_5	$C_{18}H_{15}P$	CO	–	202, **20**, *8*
$C_5H_6AsNiO_3$	C_5H_5	–	CO	–	365, **43**, *35*
C_5H_6N	C_5H_5	$C_{18}H_{15}P$	CO	–	185, **18**, *47*
$C_5H_6NiO_3P$	C_5H_5	–	CO	–	361, **43**, *6*
$C_5H_7ClNO_3S$	C_5H_5	–	CO	–	345, **41**, *2*
C_5H_7H	C_5H_5	$C_{18}H_{15}P$	CO	–	79, **6**, *11*
C_5H_7O	C_5H_5	$C_{18}H_{15}P$	CO	–	163, **16**, *30*
$C_5H_7O_2$	C_5H_5	–	–	–	43
$C_5H_7O_2S$	C_5H_5	–	CO	–	330, **37**, *17*
C_5H_8NOS	C_5H_5	–	CO	–	324, **36**, *2*
$C_5H_8NOS_2$	C_5H_5	–	CO	–	142, **13**, *24*
C_5H_9	C_5H_5	$C_{18}H_{15}P$	CO	–	185, **18**, *47*
C_5H_9O					
$\quad COC(CH_3)_3$	C_5H_5	$C_{18}H_{15}P$	CO	–	161, **16**, *11*
$\quad COCH(CH_3)\text{-}CH_2CH_3$	C_5H_5	$C_{18}H_{15}P$	CO	–	161, **16**, *29*
$C_5H_9O_2S$					
$\quad SO_2C(CH_3)_2\text{-}CH=CH_2$	C_5H_5	–	CO	–	330, **37**, *18*
$\quad SO_2CH_2\text{-}CH=C(CH_3)_2$	C_5H_5	–	CO	–	330, **37**, *19*
$C_5H_9O_3P$	C_5H_5	–	CO	Br	124, **10**, *41*
$C_5H_9O_3P$	C_5H_5	–	CO	Cl	124, **10**, *40*
$C_5H_9O_3P$	C_5H_5	–	CO	I	124, **10**, *42*
$C_5H_9O_3P$	C_6H_7	–	CO	I	133, **12**, *6*
$C_5H_{10}F_2NP$	C_2H_3O	C_5H_5	CO	–	223, **23**, *17*
$C_5H_{10}F_2NP$	C_5H_5	–	–	Br	18, **2**, *7*
$C_5H_{10}F_2NP$	C_5H_5	–	–	I	19, **2**, *17*
$C_5H_{10}NO$	C_5H_5	–	CO	–	239, **26**, *12*
$C_5H_{10}NS_2$	C_5H_5	–	CO	–	141, **13**, *16*
$C_5H_{11}\text{-}n$	C_5H_5	$C_{26}H_{24}P_2$	–	–	54, **4**, *11*
C_5H_{11}					
$\quad CH_2CH(CH_3)\text{-}CH_2CH_3$	C_5H_5	$C_{18}H_{15}P$	CO	–	181, **18**, *23*
$\quad CH_2CH_2\text{-}CH(CH_3)_2$	C_5H_5	$C_{18}H_{15}P$	CO	–	181, **18**, *24*
$C_5H_{11}N$	C_5H_5	–	–	I	18, **2**, *13*
$C_5H_{11}OSi$	C_5H_5	$C_{13}H_{13}P$	CO	–	207, **21**, *11*
$C_5H_{11}OSi$	C_5H_5	$C_{14}H_{15}P$	CO	–	208, **21**, *18*
$C_5H_{11}OSi$	C_5H_5	$C_{18}H_{15}P$	CO	–	160, **16**, *4*
$C_5H_{11}O_2S$	C_5H_5	–	CO	–	330, **37**, *20*
C_6Cl_5S	C_5H_5	–	CO	–	319, **34**, *8*
C_6F_5	C_5H_5	$C_{18}H_{15}O_3P$	CO	–	216, **22**, *21*

C_6F_5	C_5H_5	$C_{18}H_{15}P$	CO	—	198, **19**, *10*
$C_6F_5O_2S$	C_5H_5	—	CO	—	331, **37**, *21*
C_6F_5S	C_5H_5	—	CO	—	319, **34**, *7*
$C_6HF_6MnO_4P$	C_5H_5	—	CO	—	362, **43**, *10*
$C_6H_3CrO_5S$	C_5H_5	—	CO	—	322, **35**, *1*
$C_6H_3F_6OS$	C_5H_5	—	CO	—	241, **26**, *25*
$C_6H_3O_5SW$	C_5H_5	—	CO	—	323, **35**, *4*
C_6H_4Cl-4	C_5H_5	$C_{18}H_{15}O_3P$	CO	—	216, **22**, *20*
C_6H_4Cl-4	C_5H_5	$C_{18}H_{15}P$	CO	—	198, **19**, *7*
C_6H_4F					
$\quad C_6H_4F$-3	C_5H_5	$C_{18}H_{15}O_3P$	CO	—	216, **22**, *19*
$\quad C_6H_4F$-3	C_5H_5	$C_{18}H_{15}P$	CO	—	198, **19**, *6*
$\quad C_6H_4F$-4	C_5H_5	$C_6H_{15}P$	CO	—	155, **15**, *4*
$\quad C_6H_4F$-4	C_5H_5	$C_{10}H_{15}P$	CO	—	207, **21**, *13*
$\quad C_6H_4F$-4	C_5H_5	$C_{14}H_{15}P$	CO	—	207, **21**, *16*
$\quad C_6H_4F$-4	C_5H_5	$C_{18}H_{15}O_3P$	CO	—	215, **22**, *18*
$\quad C_6H_4F$-4	C_5H_5	$C_{18}H_{15}P$	CO	—	198, **19**, *5*
$\quad C_6H_4F$-4	C_5H_5	$C_{21}H_{21}P$	CO	—	204
$\quad C_6H_4F$-4	C_5H_5	$C_{30}H_{27}Fe_3P$	CO	—	224, **23**, *23*
$C_6H_4FO_2S$	C_5H_5	—	CO	—	331, **37**, *22*
C_6H_4FS	C_5H_5	—	CO	—	319, **34**, *6*
$C_6H_4N_3$	C_5H_5	—	CO	—	354, **42**, *15, 16*
C_6H_5	$C_4H_8NO_2P$	C_5H_5	CO	—	224, **23**, *24*
C_6H_5	C_5H_5	$C_6H_{16}P_2$	—	—	55, **4**, *16*
C_6H_5	C_5H_5	$C_8H_9O_2P$	CO	—	221, **23**, *5*
C_6H_5	C_5H_5	$C_8H_{11}P$	—	—	53, **4**, *4*
C_6H_5	C_5H_5	$C_{10}H_{15}P$	CO	—	207, **21**, *12*
C_6H_5	C_5H_5	$C_{14}H_{15}P$	CO	—	207, **21**, *15*
C_6H_5	C_5H_5	$C_{18}H_{15}O_3P$	CO	—	215, **22**, *17*
C_6H_5	C_5H_5	$C_{18}H_{15}P$	CO	—	197, **19**, *1*
C_6H_5	C_5H_5	$C_{21}H_{21}P$	CO	—	204
C_6H_5	C_5H_5	$C_{26}H_{24}P_2$	—	—	55, **4**, *17*
$C_6H_5Cl_2Sn$	C_5H_5	$C_6H_{15}P$	CO	—	89, **7**, *10*
$C_6H_5Cl_2Sn$	C_5H_5	$C_{18}H_{15}O_3P$	—	—	21, **2**, *35*
$C_6H_5Cl_2Sn$	C_5H_5	$C_{18}H_{15}O_3P$	CO	—	124, **10**, *36*
$C_6H_5Cl_2Sn$	C_5H_5	$C_{18}H_{15}P$	CO	—	102, **8**, *39*
$C_6H_5F_6O$	C_5H_5	$C_{18}H_{15}P$	CO	—	201, **20**, *4*
$C_6H_5N_3$	C_5H_5	—	CO	—	354, **42**, *15a, 15b, 16a*
$C_6H_5N_3 \cdot H_2O$	C_5H_5	—	CO	—	354, **42**, *15d*
$C_6H_5N_3 \cdot OC(CH_3)_2$	C_5H_5	—	CO	—	354, **42**, *15e*
$C_6H_5O_2$	C_5H_5	$C_{18}H_{15}P$	CO	—	161, **16**, *13*
$C_6H_5O_2S$	—	—	CO	—	331, **37**, *23*
C_6H_5S	C_3H_9P	C_5H_5	—	—	41, **3**, *2*
C_6H_5S	C_5H_5	—	CO	—	319, **34**, *5*
C_6H_5S	C_5H_5	$C_{26}H_{24}P_2$	—	—	23, **2**, *63;* 41, **3**, *9, 10*
C_6H_5Se	C_5H_5	—	CO	—	341, **39**, *1*
C_6H_5Te	C_5H_5	—	CO	—	342, **39**, *7*
C_6H_6NO	C_5H_5	—	CO	—	350, **42**, *2*
$C_6H_6N_3O_4$	C_5H_5	—	CO	—	353, **42**, *12*

C_6H_7

$C(CH_3)=C(CH_3)-$ $C\equiv CH$	C_5H_5	—	CO	—	185, **18**, *47*
$CH_3C_5H_4$	—	—	CO	$SiCl_3$	72, **5**, *8*
$CH_3C_5H_4$	—	H	CO	$SiCl_3$	72, **5**, *4*
$CH_3C_5H_4$	$C_3H_9O_3P$	—	CO	Cl	133, **12**, *3*
$CH_3C_5H_4$	$C_3H_9O_3P$	—	CO	I	133, **12**, *4*
$CH_3C_5H_4$	$C_5H_9O_3P$	—	CO	I	133, **12**, *6*
$CH_3C_5H_4$	$C_6H_{15}O_3P$	—	—	—	2
$CH_3C_5H_4$	$C_9H_{21}O_3P$	—	—	I	133, **12**, *5*
$CH_3C_5H_4$	$C_{12}H_{27}P$	—	CO	I	133, **12**, *1*
$CH_3C_5H_4$	$C_{18}H_{15}As$	—	CO	I	133, **12**, *10*
$CH_3C_5H_4$	$C_{18}H_{15}O_3P$	—	CO	Br	133, **12**, *8*
$CH_3C_5H_4$	$C_{18}H_{15}O_3P$	—	CO	Cl	133, **12**, *7*
$CH_3C_5H_4$	$C_{18}H_{15}O_3P$	—	CO	I	133, **12**, *9*
$CH_3C_5H_4$	$C_{18}H_{15}P$	—	CO	I	133, **12**, *2*
C_6H_7-cyclo	$C_6H_{11}O_3P$	—	CO	—	82, **6**, *31*
$C_6H_7O_4$	C_5H_5	$C_{26}H_{24}P_2$	—	—	57, **4**, *27*
C_6H_7P	C_5H_5	—	CO	I	111, **9**, *1*
C_6H_8As	C_5H_5	—	CO	—	365, **43**, *33*
C_6H_9	C_5H_5	$C_{18}H_{15}P$	CO	—	196
C_6H_9N	C_5H_5	$C_{18}H_{15}P$	CO	—	79, **6**, *12*
$C_6H_{10}NS_2$	C_5H_5	—	CO	—	141, **13**, *23*
$C_6H_{10}O_3P$	C_5H_5	—	CO	—	363, **43**, *18*
$C_6H_{10}O_3P$	C_5H_5	$C_9H_{15}O_3P$	CO	—	121, **10**, *17*

C_6H_{11}

$C(C_2H_5)=CHC_2H_5$	C_5H_5	$C_{18}H_{15}O_3P$	CO	—	214, **22**, *12*
$C(C_2H_5)=CHC_2H_5$	C_5H_5	$C_{18}H_{15}P$	CO	—	185, **18**, *48*
$CH_2-C_5H_9$-cyclo	C_5H_5	$C_{26}H_{24}P_2$	—	—	54, **4**, *7*
$CH_2(CH_2)_3-$ $CH=CH_2$	C_5H_5	$C_{26}H_{24}P_2$	—	—	55, **4**, *13*
$C_6H_{11}D_2O_2S$	C_5H_5	—	CO	—	331, **37**, *24*
$C_6H_{11}O$	C_5H_5	$C_{18}H_{15}P$	CO	—	185, **18**, *47*
$C_6H_{11}O_3P$	C_5H_5	—	CO	Br	125, **10**, *44*
$C_6H_{11}O_3P$	C_5H_5	—	CO	Cl	124, **10**, *43*
$C_6H_{11}O_3P$	C_5H_5	—	CO	I	125, **10**, *45*
$C_6H_{11}O_3P$	C_6H_7	—	CO	—	82, **6**, *31*
$C_6H_{11}O_3P$	C_7H_9	—	CO	—	82, **6**, *32*
$C_6H_{12}AsO_2$	C_5H_5	—	CO	—	365, **43**, *34*
$C_6H_{13}N$	C_5H_5	—	—	I	18, **2**, *12*
$C_6H_{13}O_2S$	C_5H_5	—	CO	—	331, **37**, *24*
$C_6H_{15}O_3P$	$C_4H_{10}O_3P$	C_5H_5	CO	—	121, **10**, *10*
$C_6H_{15}O_3P$	C_5H_5	—	—	—	4, **1**, *6*
$C_6H_{15}O_3P$	C_5H_5	—	—	$SnCl_3$	20, **2**, *31*
$C_6H_{15}O_3P$	C_5H_5	—	CO	Cl	121, **10**, *8*
$C_6H_{15}O_3P$	C_5H_5	—	CO	I	121, **10**, *9*
$C_6H_{15}O_3P$	C_5H_5	—	CO	$SnCl_3$	121, **10**, *4*
$C_6H_{15}O_3P$	C_5H_5	$C_{17}H_{15}Si$	CO	—	121, **10**, *12*
$C_6H_{15}O_3P$	C_5H_5	$C_{18}H_{15}Si$	CO	—	121, **10**, *11*
$C_6H_{15}O_3P$	C_5H_5	$C_{18}H_{15}Sn$	CO	—	121, **10**, *13*
$C_6H_{15}O_3P$	C_6H_7	—	—	—	2

$C_7H_6AsMoO_5$	C_5H_5	–	CO	–	366, **43**, *39*
$C_7H_6AsO_5W$	C_5H_5	–	CO	–	366, **43**, *40*
$C_7H_6CrO_5Sb$	C_5H_5	–	CO	–	368, **43**, *54*
$C_7H_6FO_2S$	C_5H_5	–	CO	–	332, **37**, *26*
$C_7H_6N_2$	C_5H_5	–	CO	–	351, **42**, *8b*
$C_7H_6N_2 \cdot OC(CH_3)_2$	C_5H_5	–	CO	–	351, **42**, *8c*
C_7H_7					
$CH_2C_6H_5$	$C_3H_9O_3P$	C_5H_5	CO	–	212, **22**, *2*
$CH_2C_6H_5$	C_5H_5	$C_8H_{11}P$	CO	–	205, **21**, *2*
$CH_2C_6H_5$	C_5H_5	$C_{12}H_{27}P$	CO	–	155, **15**, *7*
$CH_2C_6H_5$	C_5H_5	$C_{13}H_{13}P$	CO	–	206, **21**, *7*
$CH_2C_6H_5$	C_5H_5	$C_{18}H_{15}As$	CO	–	228, **25**, *3*
$CH_2C_6H_5$	C_5H_5	$C_{18}H_{15}P$	CO	–	182, **18**, *25*
$CH_2C_6H_5$	C_5H_5	$C_{26}H_{24}P_2$	–	–	54, **4**, *8*
$CH_2C_6H_5$	$C_{10}H_9$	$C_{18}H_{15}P$	CO	–	232
$C_6H_4CH_3-4$	C_5H_5	$C_{18}H_{15}P$	CO	–	198, **19**, *2*
C_7H_7O					
$C_6H_4CH_2OH-4$	C_5H_5	$C_{18}H_{15}P$	CO	–	198, **19**, *4*
$C_6H_4OCH_3-4$	C_5H_5	$C_{18}H_{15}P$	CO	–	198, **19**, *8*
C_7H_7OS	C_5H_5	–	CO	–	320, **34**, *10*
$C_7H_7O_2S$					
$OS(O)CH_2C_6H_5$	C_5H_5	$C_{18}H_{15}P$	CO	–	97, **8**, *10*
$OS(O)C_6H_4CH_3-4$	C_5H_5	–	CO	–	313, **32**, *5*
$SO_2CH_2C_6H_5$	C_5H_5	–	CO	–	332, **37**, *27*
$SO_2CH_2C_6H_5$	C_5H_5	$C_{18}H_{15}As$	CO	–	131, **11**, *4*
$SO_2CH_2C_6H_5$	C_5H_5	$C_{18}H_{15}P$	CO	–	98, **8**, *17*
$SO_2C_6H_4CH_3-2$	C_5H_5	–	CO	–	332, **37**, *28*
$SO_2C_6H_4CH_3-3$	C_5H_5	–	CO	–	332, **37**, *29*
$SO_2C_6H_4CH_3-4$	C_5H_5	–	CO	–	332, **37**, *30*
$C_7H_7O_3S$					
$OS(O)_2C_6H_4$-CH_3-4	C_5H_5	–	CO	–	314, **32**, *6*
$SO_2C_6H_4OCH_3-4$	C_5H_5	–	CO	–	332, **37**, *31*
C_7H_7S	C_5H_5	–	CO	–	320, **34**, *9*
$C_7H_7S_2$	C_5H_5	–	CO	–	320, **34**, *11*
C_7H_9-cyclo	C_5H_5	$C_6H_{11}O_3P$	CO	–	82, **6**, *32*
C_7H_9N	C_5H_5	–	–	I	18, **2**, *15*
C_7H_9O	C_5H_5	$C_{12}H_{27}P$	CO	–	155, **15**, *6*
$C_7H_{10}NO$	C_5H_5	–	CO	–	239, **26**, *13*
$C_7H_{11}D_2O$	C_5H_5	$C_{18}H_{15}P$	CO	–	160, **16**, *7*
$C_7H_{11}O$	C_2H_6OS	C_5H_5	CO	–	228, **25**, *8*
$C_7H_{11}O$	C_5H_5	$C_{18}H_{15}P$	CO	–	162, **16**, *21*
$C_7H_{11}O_2S$	C_5H_5	–	CO	–	333, **37**, *32*
$C_7H_{12}NO$	C_5H_5	–	CO	–	239, **26**, *14*
$C_7H_{13}O_3P$	C_5H_5	–	CO	Br	125, **10**, *47*
$C_7H_{13}O_3P$	C_5H_5	–	CO	Cl	125, **10**, *46*
$C_7H_{13}O_3P$	C_5H_5	–	CO	I	125, **10**, *48*
$C_7H_{17}OSi_2$	C_5H_5	$C_{18}H_{15}P$	CO	–	160, **16**, *5*
$C_8H_3CoO_6$	C_5H_5	–	CO	–	250
$C_8H_4F_3O_2$	C_5H_5	–	–	–	44

$C_8H_4F_9S$	C_5H_5	—	CO	—	242, **26**, *28*
$C_8H_4MnO_3$	C_5H_5	$C_{18}H_{15}P$	CO	—	201, **20**, *3*
C_8H_5	C_5H_5	$C_{18}H_{15}P$	CO	—	196
$C_8H_5BrMoO_3Sb$	C_5H_5	—	CO	—	367, **43**, *47*
$C_8H_5BrO_3SbW$	C_5H_5	—	CO	—	367, **43**, *48*
$C_8H_5N_4$	C_5H_5	—	CO	—	343, **40**, *4*
$C_8H_5N_4$	C_5H_5	$C_{18}H_{15}P$	CO	—	99, **8**, *26*
C_8H_6N	C_5H_5	—	CO	—	350, **42**, *3*
$C_8H_6N_3$	C_5H_5	—	CO	—	353, **42**, *13, 14*
$C_8H_7D_2O_2S$	C_5H_5	—	CO	—	334, **37**, *39*
$C_8H_7N_3$	C_5H_5	—	CO	—	353, **42**, *13a, 13b*
$C_8H_7N_3 \cdot H_2O$	C_5H_5	—	CO	—	353, **42**, *13c*
$C_8H_7N_3 \cdot OC(CH_3)_2$	C_5H_5	—	CO	—	353, **42**, *13d*
C_8H_7O					
$\quad COCH_2C_6H_5$	C_5H_5	$C_{18}H_{15}P$	CO	—	161, **16**, *9*
$\quad COCH_2C_6H_5$	C_5H_5	$C_{20}H_{20}NP$	CO	—	222, **23**, *10*
$\quad COCH_2C_6H_5$	$C_{10}H_9$	$C_{18}H_{15}P$	CO	—	232
$\quad COC_6H_4CH_3-4$	CHO	C_5H_5	CO	—	252, **27**, *3*
$\quad COC_6H_4CH_3-4$	C_5H_5	$C_{18}H_{15}P$	CO	—	162, **16**, *20*
$C_8H_7O_2$	C_5H_5	$C_{18}H_{15}P$	CO	—	162, **16**, *16*
$C_8H_8MnO_2S$	C_5H_5	—	CO	—	323, **35**, *5*
C_8H_8NS	C_5H_5	—	CO	—	139, **13**, *4*
$C_8H_8NS_2$	C_5H_5	—	CO	—	140, **13**, *14*
$C_8H_8O_3P$	C_5H_5	—	CO	—	240, **26**, *21*
C_8H_9O	C_5H_5	$C_{18}H_{15}P$	CO	—	183, **18**, *35*
C_8H_9O	C_5H_5	$C_{26}H_{24}P_2$	—	—	56, **4**, *25*
$C_8H_9O_2P$	C_5H_5	C_6H_5	CO	—	221, **23**, *5*
$C_8H_9O_2S$					
$\quad OS(O)CH_2CH_2-C_6H_5$	C_5H_5	—	CO	—	313, **32**, *4*
$\quad SO_2CH(CH_3)C_6H_5$	C_5H_5	—	CO	—	333, **37**, *33*
$C_8H_9O_3S$	C_5H_5	—	CO	—	333, **37**, *34*
$C_8H_{10}NO_3S_2$					
$\quad S(O)(CH_3)=N-S(O)_2-C_6H_4-CH_3-4$	C_5H_5	—	CO	—	338, **38**, *2*
$\quad S(O)-(CH_2C_6H_5)=N-S(O)_2CH_3$	C_5H_5	—	CO	—	338, **38**, *3*
$\quad N(SO_2CH_3)-SOCH_2C_6H_5$	C_5H_5	—	CO	—	346, **41**, *6*
$\quad N(SO_2C_6H_4-CH_3-4)SOCH_3$	C_5H_5	—	CO	—	346, **41**, *7*
$C_8H_{10}NO_4S_2$					
$\quad N(SO_2CH_3)-SO_2CH_2C_6H_5$	C_5H_5	—	CO	—	347, **41**, *10*
$\quad N(SO_2CH_3)-SO_2C_6H_4CH_3-4$	C_5H_5	—	CO	—	347, **41**, *11*
$C_8H_{11}P$	C_2H_3O	C_5H_5	CO	—	206, **21**, *3*
$C_8H_{11}P$	C_2H_5	C_5H_5	CO	—	205, **21**, *1*

$C_8H_{11}P$	C_3H_5O	C_5H_5	CO	–	206, **21**, *4*
$C_8H_{11}P$	C_3H_9Sn	C_5H_5	CO	–	112, **9**, *6*
$C_8H_{11}P$	$C_4HF_6N_2$	C_5H_5	–	–	57, **4**, *28*
$C_8H_{11}P$	C_4H_7O	C_5H_5	CO	–	206, **21**, *5*
$C_8H_{11}P$	C_5H_5	–	–	Br	18, **2**, *8*
$C_8H_{11}P$	C_5H_5	–	CO	–	206, **21**, *6*
$C_8H_{11}P$	C_5H_5	–	CO	Cl	111, **9**, *3*
$C_8H_{11}P$	C_5H_5	–	CO	CN	111, **9**, *5*
$C_8H_{11}P$	C_5H_5	–	CO	H	111, **9**, *2*
$C_8H_{11}P$	C_5H_5	–	CO	I	111, **9**, *4*
$C_8H_{11}P$	C_5H_5	C_6H_5	–	–	53, **4**, *4*
$C_8H_{11}P$	C_5H_5	C_7H_7	CO	–	205, **21**, *2*
$C_8H_{11}P$	C_5H_5	$C_{18}H_{15}Sn$	–	–	21, **2**, *40*
$C_8H_{11}P$	C_5H_5	$C_{18}H_{15}Sn$	CO	–	112, **9**, *7*
$C_8H_{11}Si$	C_5H_5	–	CO	H	72, **5**, *3*
$C_8H_{11}Sn$	C_5H_5	$C_{18}H_{15}P$	CO	–	101, **8**, *36*
$C_8H_{13}O$	C_2H_6OS	C_5H_5	CO	–	228, **25**, *9*
$C_8H_{13}O$	C_5H_5	$C_{18}H_{15}P$	CO	–	160, **16**, *8*
$C_8H_{15}S$	C_5H_5	$C_{18}H_{15}P$	CO	–	185, **18**, *47*
$C_8H_{17}O_3S$	C_5H_5	–	CO	–	340, **38**, *16*
$C_8H_{18}O_3P$	C_5H_5	–	CO	–	363, **43**, *20*
$C_8H_{18}O_3P$	C_5H_5	$C_{12}H_{27}O_3P$	CO	–	122, **10**, *19*
$\mathbf{C_9}H_4MnO_4$	C_5H_5	$C_{18}H_{15}P$	CO	–	161, **16**, *14*
$C_9H_5O_2$	C_5H_5	$C_{18}H_{15}P$	CO	–	163, **16**, *24*
$C_9H_7N_4$	C_5H_5	–	CO	–	344, **40**, *5*
$C_9H_7N_4$	C_5H_5	$C_{18}H_{15}P$	CO	–	99, **8**, *27*
C_9H_7O	C_5H_5	$C_{18}H_{15}P$	CO	–	163, **16**, *27*
C_9H_8NOS	C_5H_5	–	CO	–	241, **26**, *26*
$C_9H_9N_2$	C_5H_5	–	CO	–	351, **42**, *9*
$C_9H_9O_2$	C_5H_5	$C_{18}H_{15}P$	CO	–	198, **19**, *9*
$C_9H_9O_2S$					
$\quad SO_2CH(C_6H_5)-$ $\quad\quad CH{=}CH_2$	C_5H_5	–	CO	–	333, **37**, *36*
$\quad SO_2CH_2-$ $\quad\quad CH{=}CHC_6H_5$	C_5H_5	–	CO	–	333, **37**, *35*
$C_9H_{10}NS_2$					
$\quad S_2CN(CH_3)CH_2-$ $\quad\quad C_6H_5$	C_5H_5	–	CO	–	140, **13**, *13*
$\quad S_2CN(CH_3)-$ $\quad\quad C_6H_4CH_3{-}4$	C_5H_5	–	CO	–	141, **13**, *15*
$\quad S_2CN(C_2H_5)C_6H_5$	C_5H_5	–	CO	–	141, **13**, *18*
$C_9H_{10}N_2$	C_5H_5	–	CO	–	352, **42**, *9b*
$C_9H_{10}N_2{\cdot}H_2O$	C_5H_5	–	CO	Br	352, **42**, *9c*
$C_9H_{10}N_2{\cdot}OC(CH_3)_2$	C_5H_5	–	CO	–	352, **42**, *9d*
$C_9H_{10}O_2P$	C_5H_5	–	CO	–	363, **43**, *19*
$C_9H_{10}O_2P$	C_5H_5	$C_{12}H_{15}O_2P$	CO	–	128/9
C_9H_{11}	C_5H_5	$C_{18}H_{15}P$	CO	–	182, **18**, *26*
$C_9H_{11}AsMnO_2$	C_5H_5	–	CO	–	366, **43**, *37*
$C_9H_{11}O_2S$	C_5H_5	$C_{18}H_{15}P$	CO	–	98, **8**, *18*

$C_9H_{13}N_2O_4S_3$	C_5H_5	—	CO	—	347, **41**, *13*
$C_9H_{13}O_2SSi$	C_5H_5	—	CO	—	333, **37**, *37*
$C_9H_{13}Si$	C_5H_5	$C_{18}H_{15}P$	CO	—	184, **18**, *39*
$C_9H_{15}O_3P$	C_5H_5	—	CO	Cl	121, **10**, *16*
$C_9H_{15}O_3P$	C_5H_5	$C_6H_{10}O_3P$	CO	—	121, **10**, *17*
$C_9H_{18}N$	C_5H_5	—	CO	—	68
$C_9H_{18}NS_2$	C_5H_5	—	CO	—	141, **13**, *19*
$C_9H_{18}NiO_3PSi_2$	C_5H_5	—	CO	—	362, **43**, *13*
$C_9H_{21}O_3P$	C_5H_5	—	CO	I	121, **10**, *15*
$C_9H_{21}O_3P$	C_6H_7	—	CO	I	133, **12**, *5*
$C_9H_{27}Sn_4$	C_5H_5	$C_{18}H_{15}O_3P$	—	—	22, **3**, *43*
$C_{10}F_{11}S$	C_5H_5	—	CO	—	239, **26**, *11*
$C_{10}H_8N_2$	C_5H_5	—	CO	—	80, **6**, *13*
$C_{10}H_8N_2$	C_5H_5	$C_{18}H_{15}As$	—	—	7, **1**, *26*
$C_{10}H_8N_2$	C_5H_5	$C_{18}H_{15}P$	—	—	6, **1**, *25*
$C_{10}H_8N_2$	C_5H_5	$C_{18}H_{15}Sb$	—	—	7, **1**, *27*
$C_{10}H_9$	C_7H_7	$C_{18}H_{15}P$	CO	—	232
$C_{10}H_9$	C_8H_7O	$C_{18}H_{15}P$	CO	—	232
$C_{10}H_{10}P$	C_5H_5	—	CO	—	360, **43**, *3*
$C_{10}H_{11}$	C_5H_5	$C_{18}H_{15}P$	CO	—	185, **18**, *47*
$C_{10}H_{11}O$	C_5H_5	$C_{18}H_{15}P$	CO	—	160, **16**, *6*
$C_{10}H_{11}S$	C_5H_5	$C_{18}H_{15}P$	CO	—	185, **18**, *47*
$C_{10}H_{12}NS_2$	C_5H_5	—	CO	—	141, **13**, *17*
$C_{10}H_{14}NO_2P$	C_5H_5	—	CO	—	80, **6**, *14*
$C_{10}H_{15}O_2SSi$	C_5H_5	$C_{18}H_{15}O_3P$	CO	—	122, **10**, *27*
$C_{10}H_{15}P$	C_5H_5	C_6H_4F	CO	—	207, **21**, *13*
$C_{10}H_{15}P$	C_5H_5	C_6H_5	CO	—	207, **21**, *12*
$C_{10}H_{15}P$	C_5H_5	C_7H_5O	CO	—	207, **21**, *14*
$C_{10}H_{15}Si$	C_5H_5	$C_{18}H_{15}O_3P$	CO	—	215, **22**, *15*
$C_{10}H_{15}Si$	C_5H_5	$C_{18}H_{15}P$	CO	—	184, **18**, *40*
$C_{10}H_{16}As_2$	CH_3	C_5H_5	CO	—	228, **25**, *4*
$C_{11}CrF_5O_5S$	C_5H_5	—	CO	—	323, **35**, *3*
$C_{11}H_6F_5$	C_5H_5	$C_{18}H_{15}O_3P$	CO	—	216, **22**, *22*
$C_{11}H_9$	$C_3H_9O_3P$	C_5H_5	CO	—	212, **22**, *3*
$C_{11}H_9$	C_5H_5	$C_{18}H_{15}P$	CO	—	182, **18**, *28*
$C_{11}H_9$	$C_{18}H_{15}O_3P$	—	—	—	3
$C_{11}H_9$	$C_{18}H_{15}O_3P$	—	—	I	16
$C_{11}H_{12}AsCrO_7$	C_5H_5	—	CO	—	366, **43**, *41*
$C_{11}H_{13}N_2O_3S$	C_5H_5	—	CO	—	346, **41**, *3*
$C_{11}H_{15}O$	C_3H_9P	C_5H_5	CO	—	155, **15**, *2*
$C_{11}H_{15}O$	C_5H_5	$C_{18}H_{15}P$	CO	—	162, **16**, *22, 23*
$C_{11}H_{17}O_4$	C_5H_5	$C_{18}H_{15}P$	CO	—	185, **18**, *47*
$C_{11}H_{19}O_2$	C_5H_5	$C_{18}H_{15}P$	CO	—	170/1
$C_{11}H_{21}N_2$	C_5H_5	—	CO	—	238, **26**, *3*
$C_{11}H_{21}O$	C_5H_5	$C_{18}H_{15}P$	CO	—	183, **18**, *36*
$C_{11}H_{21}O_3S$	C_5H_5	$C_{18}H_{15}P$	CO	—	184, **18**, *42*

$C_{12}AsF_{10}$	C_5H_5	—	CO	—	365, **43**, *32*
$C_{12}H_7F_6O$	C_5H_5	$C_{18}H_{15}P$	CO	—	202, **20**, *6*
$C_{12}H_8Cl_2N_3$	$C_3H_9O_3P$	C_5H_5	—	—	25, **2**, *76*
$C_{12}H_8Cl_2N_3$	C_5H_5	—	CO	—	139, **13**, *1*
$C_{12}H_8Cl_2N_3$	C_5H_5	—	CO	NO	146
$C_{12}H_8Cl_2N_3$	C_5H_5	$C_{18}H_{15}O_3P$	—	—	25, **2**, *77*
$C_{12}H_8Cl_2N_3$	C_5H_5	$C_{18}H_{15}P$	—	—	25, **2**, *75*
$C_{12}H_8N$	C_5H_5	—	CO	—	350, **42**, *4*
$C_{12}H_8N_2$	C_5H_5	$C_{18}H_{15}As$	—	—	7, **1**, *29*
$C_{12}H_8N_2$	C_5H_5	$C_{18}H_{15}P$	—	—	6, **1**, *28*
$C_{12}H_8N_2$	C_5H_5	$C_{18}H_{15}Sb$	—	—	7, **1**, *30*
$C_{12}H_9F_6O$	C_5H_5	$C_{18}H_{15}P$	CO	—	202, **20**, *5*
$C_{12}H_9N_2$	C_5H_5	—	CO	—	240, **26**, *16*
$C_{12}H_9N_2$	C_5H_5	$C_{18}H_{15}P$	—	—	57, **4**, *31*
$C_{12}H_9S_2$	C_5H_5	—	CO	—	142, **13**, *31*
$C_{12}H_{10}ClSn$	C_5H_5	$C_6H_{15}P$	CO	—	89, **7**, *9*
$C_{12}H_{10}ClSn$	C_5H_5	$C_{18}H_{15}O_3P$	—	—	21, **2**, *36*
$C_{12}H_{10}ClSn$	C_5H_5	$C_{18}H_{15}O_3P$	CO	—	123, **10**, *35*
$C_{12}H_{10}ClSn$	C_5H_5	$C_{18}H_{15}P$	CO	—	102, **8**, *38*
$C_{12}H_{10}N$	C_5H_5	—	CO	—	343, **40**, *1*
$C_{12}H_{10}OP$	C_5H_5	—	CO	—	362, **43**, *15*
$C_{12}H_{10}OP$	C_5H_5	$C_{15}H_{15}OP$	CO	—	129
$C_{12}H_{10}O_3P$	C_5H_5	—	CO	—	314
$C_{12}H_{10}P$	C_5H_5	—	CO	—	360, **43**, *2*
$C_{12}H_{10}PSe$	C_5H_5	—	CO	—	364, **43**, *27*
$C_{12}H_{11}$	CH_3	$C_{18}H_{15}P$	CO	—	229/30
$C_{12}H_{11}$	CH_3O_2S	$C_{18}H_{15}P$	CO	—	135/6
$C_{12}H_{11}$	C_2H_3O	$C_{18}H_{15}P$	CO	—	230/1
$C_{12}H_{11}$	$C_{18}H_{15}P$	—	CO	I	134/5
$C_{12}H_{11}P$	C_5H_5	—	CO	Br	112, **9**, *8*
$C_{12}H_{12}N$	C_5H_5	—	CO	—	350, **42**, *5*
$C_{12}H_{12}O_2P$	C_5H_5	$C_9H_{10}O_2P$	CO	—	128/9
$C_{12}H_{15}N_2O_3S$	C_5H_5	—	CO	—	346, **41**, *4*
$C_{12}H_{15}O_2P$	C_5H_5	—	CO	Cl	128/9
$C_{12}H_{21}O_2$	C_5H_5	$C_{18}H_{15}P$	CO	—	171/2
$C_{12}H_{21}O_3S$	C_5H_5	$C_{18}H_{15}P$	CO	—	98, **8**, *19*
$C_{12}H_{23}O_3P$	C_2H_3O	C_5H_5	CO	—	221, **23**, *6*
$C_{12}H_{27}O_3P$	CH_3	C_5H_5	CO	—	213, **22**, *6*
$C_{12}H_{27}O_3P$	CH_3O_2S	C_5H_5	CO	—	121, **10**, *18*
$C_{12}H_{27}O_3P$	C_2H_3O	C_5H_5	CO	—	213, **22**, *7*
$C_{12}H_{27}O_3P$	C_5H_5	$C_8H_{18}O_3P$	CO	—	122, **10**, *19*
$C_{12}H_{27}P$	CH_3	C_5H_5	CO	—	155, **15**, *5*
$C_{12}H_{27}P$	CH_3O_2S	C_5H_5	CO	—	89, **7**, *15*
$C_{12}H_{27}P$	C_2H_3O	C_5H_5	CO	—	155, **15**, *8*
$C_{12}H_{27}P$	C_3H_5O	C_5H_5	CO	—	155, **15**, *9*
$C_{12}H_{27}P$	C_5H_5	—	CO	Br	89, **7**, *13*
$C_{12}H_{27}P$	C_5H_5	—	CO	Cl	89, **7**, *12*
$C_{12}H_{27}P$	C_5H_5	—	CO	I	89, **7**, *14*
$C_{12}H_{27}P$	C_5H_5	—	CO	NCO	89, **7**, *17*
$C_{12}H_{27}P$	C_5H_5	C_7H_7	CO	—	155, **15**, *7*
$C_{12}H_{27}P$	C_5H_5	C_7H_9O	CO	—	155, **15**, *6*

$C_{12}H_{27}P$	C_5H_5	$C_{13}H_7N_4$	CO	–	90, **7**, *18*
$C_{12}H_{27}P$	C_5H_5	$C_{18}H_{15}Si$	CO	–	89, **7**, *16*
$C_{12}H_{27}P$	C_6H_7	–	CO	I	133, **12**, *1*
$C_{13}H_7N_4$	C_5H_5	–	CO	–	344, **40**, *6*
$C_{13}H_7N_4$	C_5H_5	$C_{12}H_{27}P$	CO	–	90, **7**, *18*
$C_{13}H_7N_4$	C_5H_5	$C_{18}H_{15}P$	CO	–	99, **8**, *28*
$C_{13}H_{10}AsF_3$	C_3H_9Sn	C_5H_5	CO	–	131, **11**, *2*
$C_{13}H_{10}F_3P$	C_3H_9Sn	C_5H_5	CO	–	113, **9**, *14*
$C_{13}H_{10}F_3P$	C_5H_5	$C_{18}H_{15}Sn$	CO	–	113, **9**, *15*
$C_{13}H_{10}N$	C_5H_5	–	CO	–	68
$C_{13}H_{10}NS_2$	C_5H_5	–	CO	–	141, **13**, *22*
$C_{13}H_{11}ClN_3$	C_5H_5	–	CO	–	139, **13**, *2*
$C_{13}H_{13}As$	C_5H_5	–	CO	I	131, **11**, *1*
$C_{13}H_{13}P$	C_3F_7	C_5H_5	CO	–	206, **21**, *8*
$C_{13}H_{13}P$	C_3H_5O	C_5H_5	CO	–	207, **21**, *9*
$C_{13}H_{13}P$	C_3H_9Sn	C_5H_5	CO	–	113, **9**, *12*
$C_{13}H_{13}P$	C_4H_9Si	C_5H_5	–	–	20, **2**, *28*
$C_{13}H_{13}P$	C_4H_9Si	C_5H_5	CO	–	112, **9**, *11*
$C_{13}H_{13}P$	$C_4H_{11}Si$	C_5H_5	CO	–	207, **21**, *10*
$C_{13}H_{13}P$	C_5H_5	–	CO	CN	112, **9**, *10*
$C_{13}H_{13}P$	C_5H_5	–	CO	I	112, **9**, *9*
$C_{13}H_{13}P$	C_5H_5	$C_5H_{11}OSi$	CO	–	207, **21**, *11*
$C_{13}H_{13}P$	C_5H_5	C_7H_7	CO	–	206, **21**, *7*
$C_{13}H_{13}P$	C_5H_5	$C_{18}H_{15}Sn$	–	–	21, **2**, *41*
$C_{13}H_{13}P$	C_5H_5	$C_{18}H_{15}Sn$	CO	–	113, **9**, *13*
$C_{13}H_{13}Si$	C_5H_5	$C_{18}H_{15}P$	CO	–	101, **8**, *33*
$C_{13}H_{13}Si$	C_5H_5	$C_{18}H_{33}P$	CO	–	90, **7**, *21*
$C_{13}H_{23}N_2$	C_5H_5	–	CO	–	238, **26**, *4*
$C_{14}H_5N_4$	C_5H_5	$C_{18}H_{15}P$	CO	–	182, **18**, *27*
$C_{14}H_7O_9Ru_3$	C_5H_5	–	–	–	49/50
$C_{14}H_9N_4$	C_5H_5	–	CO	–	344, **40**, *7*
$C_{14}H_{11}O$	C_2H_3O	C_5H_5	CO	–	252, **27**, *8, 9*
$C_{14}H_{12}NS_2$	C_5H_5	–	CO	–	141, **13**, *21*
$C_{14}H_{13}NPS$	C_5H_5	–	CO	–	139, **13**, *5*
$C_{14}H_{13}NPS_2$	C_5H_5	–	CO	–	143, **13**, *33*
$C_{14}H_{13}S_2$	C_5H_5	–	CO	–	142, **13**, *32*
$C_{14}H_{14}NO_3S_2$	C_5H_5	–	CO	–	346, **41**, *8*
$C_{14}H_{14}N_3$	$C_3H_9O_3P$	C_5H_5	–	–	25, **2**, *79*
$C_{14}H_{14}N_3$	C_5H_5	–	CO	–	139, **13**, *3*
$C_{14}H_{14}N_3$	C_5H_5	–	CO	NO	146
$C_{14}H_{14}N_3$	C_5H_5	$C_{18}H_{15}O_3P$	–	–	26, **2**, *80*
$C_{14}H_{14}N_3$	C_5H_5	$C_{18}H_{15}P$	–	–	25, **2**, *78*
$C_{14}H_{14}PS$	C_5H_5	–	CO	–	142, **13**, *30*
$C_{14}H_{15}P$	$C_4H_{11}Si$	C_5H_5	CO	–	208, **21**, *17*
$C_{14}H_{15}P$	C_5H_5	$C_5H_{11}OSi$	CO	–	208, **21**, *18*
$C_{14}H_{15}P$	C_5H_5	C_6H_4F	CO	–	207, **21**, *16*
$C_{14}H_{15}P$	C_5H_5	C_6H_5	CO	–	207, **21**, *15*

$C_{14}H_{17}As_2CoMnO_5$	C_3H_9P	C_5H_5	CO	–	88, **7**, *5*
$C_{14}H_{25}P_3$	C_5H_5	–	–	–	8, **1**, *41*
$C_{15}H_{10}NiO_3P$	C_5H_5	–	CO	–	362, **43**, *11*
$C_{15}H_{11}O_2$	C_5H_5	–	–	–	43
$C_{15}H_{14}NS_2$	C_5H_5	–	CO	–	141, **13**, *20*
$C_{15}H_{15}NPS$	C_5H_5	–	CO	–	139, **13**, *6*
$C_{15}H_{15}OP$	C_5H_5	$C_{12}H_{10}OP$	CO	–	129
$C_{15}H_{15}OP$	C_5H_5	–	CO	Cl	129
$C_{15}H_{16}O_2PS$	C_5H_5	–	CO	–	143, **13**, *35*
$C_{15}H_{16}P$	C_5H_5	–	CO	–	240, **26**, *17*
$C_{15}H_{25}N_2$	C_5H_5	–	CO	–	238, **26**, *5*
$C_{16}H_{13}O$	C_5H_5	–	CO	–	241, **26**, *22*
$C_{16}H_{14}N_2O_2$	C_5H_5	–	–	–	44
$C_{17}H_{14}P$	C_5H_5	–	CO	–	240, **26**, *18*
$C_{17}H_{15}P$	CH_3	C_5H_5	CO	–	208, **21**, *25*
$C_{17}H_{15}Si$	C_5H_5	$C_6H_{15}O_3P$	CO	–	121, **10**, *12*
$C_{17}H_{15}Si$	C_5H_5	$C_{18}H_{15}P$	CO	–	101, **8**, *34*
$C_{17}H_{15}Si$	C_5H_5	$C_{18}H_{33}$	CO	–	90, **7**, *22*
$C_{18}H_{14}OP$	C_5H_5	–	CO	–	240, **26**, *20*
$C_{18}H_{14}O_3P$	C_5H_5	$C_{18}H_{15}O_3P$	–	–	57, **4**, *32*
$C_{18}H_{15}$	CH_2NO	$C_{18}H_{15}P$	CO	–	229
$C_{18}H_{15}$	C_5H_5	$C_{18}H_{15}P$	CO	Cl	134
$C_{18}H_{15}As$	CH_3	C_5H_5	CO	–	228, **25**, *1*
$C_{18}H_{15}As$	C_2H_3O	C_5H_5	CO	–	228, **25**, *2*
$C_{18}H_{15}As$	C_3H_9Sn	C_5H_5	CO	–	131, **11**, *7*
$C_{18}H_{15}As$	C_5H_5	–	CO	I	131, **11**, *3*
$C_{18}H_{15}As$	C_5H_5	–	CO	NCS	131, **11**, *5*
$C_{18}H_{15}As$	C_5H_5	–	–	NO	49
$C_{18}H_{15}As$	C_5H_5	I	–	NO	48
$C_{18}H_{15}As$	C_5H_5	–	CO	$SnCl_3$	131, **11**, *6*
$C_{18}H_{15}As$	C_5H_5	C_7H_7	CO	–	228, **25**, *3*
$C_{18}H_{15}As$	C_5H_5	$C_7H_7O_2S$	CO	–	131, **11**, *4*
$C_{18}H_{15}As$	C_5H_5	$C_{10}H_8N_2$	–	–	7, **1**, *26*
$C_{18}H_{15}As$	C_5H_5	$C_{18}H_{15}Sn$	CO	–	131, **11**, *8*
$C_{18}H_{15}As$	C_6H_7	–	CO	I	133, **12**, *10*
$C_{18}H_{15}Ge$	C_5H_5	$C_{26}H_{24}P_2$	–	–	24, **2**, *68*
$C_{18}H_{15}OP$	C_2H_3O	C_5H_5	CO	–	221, **23**, *2*
$C_{18}H_{15}OP$	C_2H_5	C_5H_5	CO	–	221, **23**, *1*
$C_{18}H_{15}OP$	C_5H_5	–	CO	H	129
$C_{18}H_{15}O_2P$	C_2H_3O	C_5H_5	CO	–	221, **23**, *4*
$C_{18}H_{15}O_2P$	C_2H_5	C_5H_5	CO	–	221, **23**, *3*
$C_{18}H_{15}O_2P$	C_5H_5	–	CO	H	129
$C_{18}H_{15}O_3P$	CH_3	C_5H_5	CO	–	213, **22**, *8*

$C_{18}H_{15}O_3P$	CH_3O_2S	C_5H_5	CO	–	122, **10**, *25*
$C_{18}H_{15}O_3P$	$C_2F_3O_2$	C_5H_5	CO	–	122, **10**, *24*
$C_{18}H_{15}O_3P$	C_2H_2ClN	C_5H_5	–	–	6, **1**, *20*
$C_{18}H_{15}O_3P$	C_2H_3N	C_5H_5	–	–	5, **1**, *18*
$C_{18}H_{15}O_3P$	C_2H_3O	C_5H_5	CO	–	214, **22**, *13*
$C_{18}H_{15}O_3P$	C_2H_5	C_5H_5	–	–	53, **4**, *3*
$C_{18}H_{15}O_3P$	C_2H_5	C_5H_5	CO	–	213, **22**, *9*
$C_{18}H_{15}O_3P$	$C_2H_5O_2S$	C_5H_5	CO	–	122, **10**, *26*
$C_{18}H_{15}O_3P$	$C_3H_3NO_2$	C_5H_5	–	–	6, **1**, *24*
$C_{18}H_{15}O_3P$	C_3H_4ClN	C_5H_5	–	–	6, **1**, *21*
$C_{18}H_{15}O_3P$	C_3H_4N	C_5H_5	–	–	53, **4**, *5*
$C_{18}H_{15}O_3P$	$C_3H_4N_2O$	C_5H_5	–	–	6, **1**, *23*
$C_{18}H_{15}O_3P$	C_3H_5N	C_5H_5	–	–	6, **1**, *9*
$C_{18}H_{15}O_3P$	C_3H_5O	C_5H_5	CO	–	215, **22**, *14*
$C_{18}H_{15}O_3P$	$C_3H_6N_2$	C_5H_5	–	–	6, **1**, *22*
$C_{18}H_{15}O_3P$	C_3H_9Si	C_5H_5	CO	–	123, **10**, *32*
$C_{18}H_{15}O_3P$	C_3H_9Sn	C_5H_5	CO	–	123, **10**, *33*
$C_{18}H_{15}O_3P$	$C_4HF_6N_2$	C_5H_5	–	–	57, **4**, *30*
$C_{18}H_{15}O_3P$	C_4H_8O	C_5H_5	CO	–	78, **6**, *7*
$C_{18}H_{15}O_3P$	C_4H_9	C_5H_5	CO	–	214, **22**, *10, 11*
$C_{18}H_{15}O_3P$	C_5H_5	–	–	–	4, **1**, *7, 9;* 5, **1**, *9*
$C_{18}H_{15}O_3P$	C_5H_5	–	–	Br	18, **2**, *10*
$C_{18}H_{15}O_3P$	C_5H_5	–	–	H	18, **2**, *5*
$C_{18}H_{15}O_3P$	C_5H_5	–	–	I	19, **2**, *20;* 41, **3**, *1*
$C_{18}H_{15}O_3P$	C_5H_5	I	–	NO	48
$C_{18}H_{15}O_3P$	C_5H_5	–	–	SCN	19, **2**, *21*
$C_{18}H_{15}O_3P$	C_5H_5	–	–	SO_2	5, **1**, *17*
$C_{18}H_{15}O_3P$	C_5H_5	–	–	$SnBr_3$	20, **2**, *33*
$C_{18}H_{15}O_3P$	C_5H_5	–	–	$SnCl_3$	20, **2**, *32*
$C_{18}H_{15}O_3P$	C_5H_5	–	–	SnF_3	20, **2**, *29*
$C_{18}H_{15}O_3P$	C_5H_5	–	–	SnI_3	21, **2**, *34*
$C_{18}H_{15}O_3P$	C_5H_5	–	CO	–	75; 78, **6**, *5;* 215, **22**, *16*
$C_{18}H_{15}O_3P$	C_5H_5	–	CO	Br	122, **10**, *22*
$C_{18}H_{15}O_3P$	C_5H_5	–	CO	Cl	122, **10**, *21*
$C_{18}H_{15}O_3P$	C_5H_5	–	CO	CN	123, **10**, *31*
$C_{18}H_{15}O_3P$	C_5H_5	–	CO	H	122, **10**, *20*
$C_{18}H_{15}O_3P$	C_5H_5	–	CO	I	122, **10**, *23*
$C_{18}H_{15}O_3P$	C_5H_5	–	CO	NCO	123, **10**, *29*
$C_{18}H_{15}O_3P$	C_5H_5	–	CO	SeCN	123, **10**, *28*
$C_{18}H_{15}O_3P$	C_5H_5	–	CO	$SnBr_2I$	124, **10**, *39*
$C_{18}H_{15}O_3P$	C_5H_5	–	CO	$SnBr_3$	124, **10**, *38*
$C_{18}H_{15}O_3P$	C_5H_5	–	CO	$SnCl_3$	124, **10**, *37*
$C_{18}H_{15}O_3P$	C_5H_5	C_6F_5	CO	–	216, **22**, *21*
$C_{18}H_{15}O_3P$	C_5H_5	C_6H_4Cl	CO	–	216, **22**, *20*
$C_{18}H_{15}O_3P$	C_5H_5	C_6H_4F	CO	–	215, **22**, *18*
$C_{18}H_{15}O_3P$	C_5H_5	–	–	–	216, **22**, *19*
$C_{18}H_{15}O_3P$	C_5H_5	C_6H_5	CO	–	215, **22**, *17*

$C_{18}H_{15}O_3P$	C_5H_5	$C_6H_5Cl_2Sn$	–	–	21, **2**, *35*
$C_{18}H_{15}O_3P$	C_5H_5	$C_6H_5Cl_2Sn$	CO	–	124, **10**, *36*
$C_{18}H_{15}O_3P$	C_5H_5	C_6H_{11}	CO	–	214, **22**, *12*
$C_{18}H_{15}O_3P$	C_5H_5	$C_6H_{15}P$	–	–	5, **1**, *16*
$C_{18}H_{15}O_3P$	C_5H_5	$C_6H_{16}P_2$	–	–	7, **1**, *32, 33*
$C_{18}H_{15}O_3P$	C_5H_5	$C_7H_3N_4$	CO	–	123, **10**, *30*
$C_{18}H_{15}O_3P$	C_5H_5	$C_{10}H_{15}$-O_2SSi	CO	–	122, **10**, *27*
$C_{18}H_{15}O_3P$	C_5H_5	$C_{10}H_{15}Si$	CO	–	215, **22**, *15*
$C_{18}H_{15}O_3P$	C_5H_5	$C_{11}H_6F_5$	CO	–	216, **22**, *22*
$C_{18}H_{15}O_3P$	C_5H_5	$C_{12}H_8Cl_2N_3$	–	–	25, **2**, *77*
$C_{18}H_{15}O_3P$	C_5H_5	$C_{12}H_{10}ClSn$	–	–	21, **2**, *36*
$C_{18}H_{15}O_3P$	C_5H_5	$C_{12}H_{10}ClSn$	CO	–	123, **10**, *35*
$C_{18}H_{15}O_3P$	C_5H_5	$C_{14}H_{14}N_3$	–	–	26, **2**, *80*
$C_{18}H_{15}O_3P$	C_5H_5	$C_{18}H_{14}O_3P$	–	–	54, **4**, *32*
$C_{18}H_{15}O_3P$	C_5H_5	$C_{18}H_{15}Sn$	–	–	21, **2**, *42*
$C_{18}H_{15}O_3P$	C_5H_5	$C_{18}H_{15}Sn$	CO	–	123, **10**, *34*
$C_{18}H_{15}O_3P$	C_6H_7	–	CO	Br	133, **12**, *8*
$C_{18}H_{15}O_3P$	C_6H_7	–	CO	Cl	133, **12**, *7*
$C_{18}H_{15}O_3P$	C_6H_7	–	CO	I	133, **12**, *9*
$C_{18}H_{15}O_3P$	$C_{11}H_9$	–	–	–	3; 16
$C_{18}H_{15}O_3P$	$C_{23}H_{18}O_3P$	–	–	I	44/6
$C_{18}H_{15}P$	CD_2NO	C_5H_5	CO	–	173
$C_{18}H_{15}P$	CF_3	C_5H_5	CO	–	178, **18**, *5*
$C_{18}H_{15}P$	CHO_2	C_5H_5	CO	–	169
$C_{18}H_{15}P$	CH_2Br	C_5H_5	CO	–	177, **18**, *3*
$C_{18}H_{15}P$	CH_2Cl	C_5H_5	CO	–	177, **18**, *2*
$C_{18}H_{15}P$	CH_2I	C_5H_5	CO	–	177, **18**, *4*
$C_{18}H_{15}P$	CH_2NO	C_5H_5	CO	–	173
$C_{18}H_{15}P$	CH_2NO	$C_{18}H_{15}$	CO	–	229
$C_{18}H_{15}P$	CH_3	C_5H_5	CO	–	177, **18**, *1*
$C_{18}H_{15}P$	CH_3	$C_{12}H_{11}$	CO	–	229/30
$C_{18}H_{15}P$	CH_3	$C_{24}H_{19}$	CO	–	232
$C_{18}H_{15}P$	CH_3O_2S	C_5H_5	CO	–	97, **8**, *9, 11*
$C_{18}H_{15}P$	CH_3O_2S	$C_{12}H_{11}$	CO	–	135/6
$C_{18}H_{15}P$	C_2F_3	C_5H_5	CO	–	179, **18**, *11*
$C_{18}H_{15}P$	C_2F_3O	C_5H_5	CO	–	159, **16**, *2*
$C_{18}H_{15}P$	$C_2F_3O_2$	C_5H_5	CO	–	96, **8**, *8*
$C_{18}H_{15}P$	C_2F_5	C_5H_5	CO	–	178, **18**, *9*
$C_{18}H_{15}P$	C_2HF_4	C_5H_5	CO	–	179, **18**, *10*
$C_{18}H_{15}P$	$C_2H_2D_3O_2S$	C_5H_5	CO	–	97, **8**, *13*
$C_{18}H_{15}P$	C_2H_2N	C_5H_5	CO	–	178, **18**, *8*
$C_{18}H_{15}P$	$C_2H_3D_2$	C_5H_5	CO	–	178, **18**, *7*
$C_{18}H_{15}P$	C_2H_3N	C_5H_5	CO	–	78, **6**, *8*
$C_{18}H_{15}P$	C_2H_3O	C_5H_5	CO	–	159, **16**, *1*
$C_{18}H_{15}P$	C_2H_3O	$C_{12}H_{11}$	CO	–	230/1
$C_{18}H_{15}P$	C_2H_3O	$C_{24}H_{19}$	CO	–	232
$C_{18}H_{15}P$	$C_2H_3O_2$	C_5H_5	CO	–	169/70
$C_{18}H_{15}P$	C_2H_4NO	C_5H_5	CO	–	173
$C_{18}H_{15}P$	C_2H_5	C_5H_5	CO	–	178, **18**, *5*
$C_{18}H_{15}P$	C_2H_5O	C_5H_5	CO	–	182, **18**, *29*

$C_{18}H_{15}P$	$C_2H_5O_2S$	C_5H_5	CO	–	97, **8**, *12*
$C_{18}H_{15}P$	$C_2H_5O_3S$	C_5H_5	CO	–	184, **18**, *41*
$C_{18}H_{15}P$	C_3F_7	C_5H_5	CO	–	178, **18**, *15*
$C_{18}H_{15}P$	C_3H_3N	C_5H_5	CO	–	79, **6**, *9*
$C_{18}H_{15}P$	C_3H_4N	C_5H_5	CO	–	178, **18**, *14*; 184, **18**, *43*
$C_{18}H_{15}P$	C_3H_5	C_5H_5	CO	–	179, **18**, *13*; 185, **18**, *45*
$C_{18}H_{15}P$	C_3H_5O	C_5H_5	CO	–	160, **16**, *3*; 182, **18**, *30*
$C_{18}H_{15}P$	$C_3H_5O_2$	C_5H_5	CO	–	172
$C_{18}H_{15}P$	$C_3H_5O_2S$	C_5H_5	CO	–	97, **8**, *14*
$C_{18}H_{15}P$	C_3H_6NO	C_5H_5	CO	–	173
$C_{18}H_{15}P$	C_3H_7	C_5H_5	CO	–	178, **18**, *12*; 184, **18**, *44*
$C_{18}H_{15}P$	C_3H_7O	C_5H_5	CO	–	182, **18**, *31*
$C_{18}H_{15}P$	$C_3H_7O_2S$	C_5H_5	CO	–	102, **8**, *41*
$C_{18}H_{15}P$	C_3H_9P	C_5H_5	–	–	4, **1**, *15*
$C_{18}H_{15}P$	C_3H_9Si	C_5H_5	CO	–	100, **8**, *30*
$C_{18}H_{15}P$	C_3H_9Sn	C_5H_5	CO	–	101, **8**, *35*
$C_{18}H_{15}P$	C_4ClF_4	C_5H_5	CO	–	201, **20**, *1*
$C_{18}H_{15}P$	$C_4HF_6N_2$	C_5H_5	–	–	57, **4**, *29*
$C_{18}H_{15}P$	C_4H_3O	C_5H_5	CO	–	202, **20**, *7*
$C_{18}H_{15}P$	C_4H_3S	C_5H_5	CO	–	202, **20**, *10*
$C_{18}H_{15}P$	C_4H_5N	C_5H_5	CO	–	79, **6**, *10*
$C_{18}H_{15}P$	C_4H_5O	C_5H_5	CO	–	163, **16**, *26, 28*
$C_{18}H_{15}P$	C_4H_7	C_5H_5	CO	–	181, **18**, *22*; 185, **18**, *46*
$C_{18}H_{15}P$	$C_4H_7D_2$	C_5H_5	CO	–	180, **18**, *17, 18*; 181, **18**, *21*
$C_{18}H_{15}P$	C_4H_7O	C_5H_5	CO	–	161, **16**, *10*; 183, **18**, *34*
$C_{18}H_{15}P$	$C_4H_7O_2$	C_5H_5	CO	–	170
$C_{18}H_{15}P$	$C_4H_7O_2S$	C_5H_5	CO	–	98, **8**, *16*
$C_{18}H_{15}P$	C_4H_8DO	C_5H_5	CO	–	183, **18**, *33*
$C_{18}H_{15}P$	C_4H_8NO	C_5H_5	CO	–	173
$C_{18}H_{15}P$	C_4H_8O	C_5H_5	CO	–	78, **6**, *6*
$C_{18}H_{15}P$	C_4H_9	$\cdot C_5H_5$	CO	–	180, **18**, *16, 19, 20*
$C_{18}H_{15}P$	C_4H_9O	C_5H_5	CO	–	183, **18**, *32*
$C_{18}H_{15}P$	$C_4H_9O_2S$	C_5H_5	CO	–	97, **8**, *15*
$C_{18}H_{15}P$	$C_4H_{10}O_3P$	C_5H_5	CO	–	183, **18**, *37*
$C_{18}H_{15}P$	$C_4H_{11}Si$	C_5H_5	CO	–	184, **18**, *38*
$C_{18}H_{15}P$	C_5H_3OS	C_5H_5	CO	–	163, **16**, *25*
$C_{18}H_{15}P$	$C_5H_3O_2$	C_5H_5	CO	–	161, **16**, *12*
$C_{18}H_{15}P$	C_5H_5	–	–	NO	49
$C_{18}H_{15}P$	C_5H_5	I	–	NO	48
$C_{18}H_{15}P$	C_5H_5	–	CO	–	75; 78, **6**, *4*
$C_{18}H_{15}P$	C_5H_5	–	CO	Br	95, **8**, *5*
$C_{18}H_{15}P$	C_5H_5	–	CO	Cl	95, **8**, *4*

$C_{18}H_{15}P$	C_5H_5	—	CO	CN	100, **8**, *29*
$C_{18}H_{15}P$	C_5H_5	—	CO	D	95, **8**, *2*
$C_{18}H_{15}P$	C_5H_5	—	CO	F	95, **8**, *3*
$C_{18}H_{15}P$	C_5H_5	—	CO	H	95, **8**, *1*
$C_{18}H_{15}P$	C_5H_5	—	CO	I	96, **8**, *6*
$C_{18}H_{15}P$	C_5H_5	—	CO	$I \cdot (SCN)_2$	96, **8**, *7*
$C_{18}H_{15}P$	C_5H_5	—	CO	NCO	98, **8**, *22*
$C_{18}H_{15}P$	C_5H_5	—	CO	NCS	98, **8**, *23*
$C_{18}H_{15}P$	C_5H_5	—	CO	NCSe	99, **8**, *24*
$C_{18}H_{15}P$	C_5H_5	—	CO	SCN	98, **8**, *20*
$C_{18}H_{15}P$	C_5H_5	—	CO	SeCN	98, **8**, *21*
$C_{18}H_{15}P$	C_5H_5	—	CO	$SnCl_3$	102, **8**, *40*
$C_{18}H_{15}P$	C_5H_5	C_5H_5O	CO	—	202, **20**, *8*
$C_{18}H_{15}P$	C_5H_5	C_5H_6N	CO	—	185, **18**, *47*
$C_{18}H_{15}P$	C_5H_5	C_5H_7N	CO	—	97, **6**, *11*
$C_{18}H_{15}P$	C_5H_5	C_5H_7O	CO	—	163, **16**, *30*
$C_{18}H_{15}P$	C_5H_5	C_5H_9	CO	—	185, **18**, *47*
$C_{18}H_{15}P$	C_5H_5	C_5H_9O	CO	—	161, **16**, *11*; 163, **16**, *29*
$C_{18}H_{15}P$	C_5H_5	C_5H_{11}	CO	—	181, **18**, *23, 24*
$C_{18}H_{15}P$	C_5H_5	$C_5H_{11}OSi$	CO	—	160, **16**, *4*
$C_{18}H_{15}P$	C_5H_5	C_6F_5	CO	—	198, **19**, *10*
$C_{18}H_{15}P$	C_5H_5	C_6H_4Cl	CO	—	198, **19**, *7*
$C_{18}H_{15}P$	C_5H_5	C_6H_4F	CO	—	198, **19**, *5, 6*
$C_{18}H_{15}P$	C_5H_5	C_6H_5	CO	—	197, **19**, *1*
$C_{18}H_{15}P$	C_5H_5	$C_6H_5Cl_2Sn$	CO	—	102, **8**, *39*
$C_{18}H_{15}P$	C_5H_5	$C_6H_5F_6O$	CO	—	201, **20**, *4*
$C_{18}H_{15}P$	C_5H_5	$C_6H_5O_2$	CO	—	161, **16**, *13*
$C_{18}H_{15}P$	C_5H_5	C_6H_7	CO	—	185, **18**, *47*
$C_{18}H_{15}P$	C_5H_5	C_6H_9	CO	—	196
$C_{18}H_{15}P$	C_5H_5	C_6H_9N	CO	—	79, **6**, *12*
$C_{18}H_{15}P$	C_5H_5	C_6H_{11}	CO	—	185, **18**, *48*
$C_{18}H_{15}P$	C_5H_5	$C_6H_{11}O$	CO	—	185, **18**, *47*
$C_{18}H_{15}P$	C_5H_5	$C_6H_{17}Si_2$	CO	—	100, **8**, *31*
$C_{18}H_{15}P$	C_5H_5	$C_7H_3N_4$	CO	—	99, **8**, *25*
$C_{18}H_{15}P$	C_5H_5	C_7H_4ClO	CO	—	162, **16**, *19*
$C_{18}H_{15}P$	C_5H_5	C_7H_4FO	CO	—	162, **16**, *18*
$C_{18}H_{15}P$	C_5H_5	C_7H_5O	CO	—	161, **16**, *15*; 198, **19**, *3*
$C_{18}H_{15}P$	C_5H_5	$C_7H_5O_2$	CO	—	162, **16**, *17*
$C_{18}H_{15}P$	C_5H_5	C_7H_7	CO	—	182, **18**, *25*; 198, **19**, *2*
$C_{18}H_{15}P$	C_5H_5	C_7H_7O	CO	—	198, **19**, *4, 8*
$C_{18}H_{15}P$	C_5H_5	$C_7H_7O_2S$	CO	—	97, **8**, *10, 17*
$C_{18}H_{15}P$	C_5H_5	$C_7H_{11}D_2O$	CO	—	160, **16**, *7*
$C_{18}H_{15}P$	C_5H_5	$C_7H_{11}O$	CO	—	162, **16**, *21*
$C_{18}H_{15}P$	C_5H_5	$C_7H_{17}OSi_2$	CO	—	160, **16**, *5*
$C_{18}H_{15}P$	C_5H_5	$C_8H_4MnO_3$	CO	—	201, **20**, *3*
$C_{18}H_{15}P$	C_5H_5	C_8H_5	CO	—	196
$C_{18}H_{15}P$	C_5H_5	$C_8H_5N_4$	CO	—	99, **8**, *26*
$C_{18}H_{15}P$	C_5H_5	C_8H_6O	CO	—	202, **20**, *9*

$C_{18}H_{15}P$	C_5H_5	C_8H_7O	CO	–	161, **16**, 9; 162, **16**, 20
$C_{18}H_{15}P$	C_5H_5	$C_8H_7O_2$	CO	–	162, **16**, 16
$C_{18}H_{15}P$	C_5H_5	C_8H_9O	CO	–	183, **18**, 35
$C_{18}H_{15}P$	C_5H_5	$C_8H_{11}Sn$	CO	–	101, **8**, 36
$C_{18}H_{15}P$	C_5H_5	$C_8H_{13}O$	CO	–	160, **16**, 8
$C_{18}H_{15}P$	C_5H_5	$C_8H_{15}S$	CO	–	185, **18**, 47
$C_{18}H_{15}P$	C_5H_5	$C_9H_4MnO_4$	CO	–	161, **16**, 14
$C_{18}H_{15}P$	C_5H_5	$C_9H_5O_2$	CO	–	163, **16**, 24
$C_{18}H_{15}P$	C_5H_5	$C_9H_7N_4$	CO	–	99, **8**, 27
$C_{18}H_{15}P$	C_5H_5	C_9H_7O	CO	–	163, **16**, 27
$C_{18}H_{15}P$	C_5H_5	$C_9H_9O_2$	CO	–	198, **19**, 9
$C_{18}H_{15}P$	C_5H_5	C_9H_{11}	CO	–	182, **18**, 26
$C_{18}H_{15}P$	C_5H_5	$C_9H_{11}O_2S$	CO	–	98, **8**, 18
$C_{18}H_{15}P$	C_5H_5	$C_9H_{13}Si$	CO	–	184, **18**, 39
$C_{18}H_{15}P$	C_5H_5	$C_{10}H_8N_2$	–	–	6, **1**, 25
$C_{18}H_{15}P$	C_5H_5	$C_{10}H_{11}$	CO	–	185, **18**, 47
$C_{18}H_{15}P$	C_5H_5	$C_{10}H_{11}O$	CO	–	160, **16**, 6
$C_{18}H_{15}P$	C_5H_5	$C_{10}H_{11}S$	CO	–	185, **18**, 47
$C_{18}H_{15}P$	C_5H_5	$C_{10}H_{15}Si$	CO	–	184, **18**, 40
$C_{18}H_{15}P$	C_5H_5	$C_{11}H_9$	CO	–	182, **18**, 28
$C_{18}H_{15}P$	C_5H_5	$C_{11}H_{15}O$	CO	–	162, **16**, 22, 23
$C_{18}H_{15}P$	C_5H_5	$C_{11}H_{17}O_4$	CO	–	185, **18**, 47
$C_{18}H_{15}P$	C_5H_5	$C_{11}H_{19}O_2$	CO	–	170/1
$C_{18}H_{15}P$	C_5H_5	$C_{11}H_{21}O$	CO	–	183, **18**, 36
$C_{18}H_{15}P$	C_5H_5	$C_{11}H_{21}O_3S$	CO	–	184, **18**, 42
$C_{18}H_{15}P$	C_5H_5	$C_{12}H_7F_6O$	CO	–	202, **20**, 6
$C_{18}H_{15}P$	C_5H_5	$C_{12}H_8Cl_2N_3$	–	–	25, **2**, 75
$C_{18}H_{15}P$	C_5H_5	$C_{12}H_8N_2$	–	–	6, **1**, 28
$C_{18}H_{15}P$	C_5H_5	$C_{12}H_9F_6O$	CO	–	202, **20**, 5
$C_{18}H_{15}P$	C_5H_5	$C_{12}H_9N_2$	–	–	57, **4**, 31
$C_{18}H_{15}P$	C_5H_5	$C_{12}H_{10}ClSn$	CO	–	102, **8**, 38
$C_{18}H_{15}P$	C_5H_5	$C_{12}H_{21}O_2$	CO	–	171/2
$C_{18}H_{15}P$	C_5H_5	$C_{12}H_{21}O_3S$	CO	–	98, **8**, 19
$C_{18}H_{15}P$	C_5H_5	$C_{13}H_7N_4$	CO	–	99, **8**, 28
$C_{18}H_{15}P$	C_5H_5	$C_{13}H_{13}Si$	CO	–	101, **8**, 33
$C_{18}H_{15}P$	C_5H_5	$C_{14}H_5N_4$	CO	–	182, **18**, 27
$C_{18}H_{15}P$	C_5H_5	$C_{14}H_{14}N_3$	–	–	25, **2**, 78
$C_{18}H_{15}P$	C_5H_5	$C_{17}H_{15}Si$	CO	–	101, **8**, 34
$C_{18}H_{15}P$	C_5H_5	$C_{18}H_{15}$	CO	Cl	134
$C_{18}H_{15}P$	C_5H_5	$C_{18}H_{15}Si$	CO	–	101, **8**, 32
$C_{18}H_{15}P$	C_5H_5	$C_{18}H_{15}Sn$	CO	–	101, **8**, 37
$C_{18}H_{15}P$	C_5H_5	$C_{19}H_{17}P$	CO	–	233
$C_{18}H_{15}P$	C_5H_5	$C_{21}H_{21}P$	CO	–	233
$C_{18}H_{15}P$	C_5H_5	$C_{23}H_{18}CoO_2$	CO	–	201, **20**, 2
$C_{18}H_{15}P$	C_6H_7	–	CO	I	133, **12**, 2
$C_{18}H_{15}P$	C_7H_7	$C_{10}H_9$	CO	–	232
$C_{18}H_{15}P$	C_8H_7O	$C_{10}H_9$	CO	–	232
$C_{18}H_{15}P$	$C_{12}H_{11}$	–	CO	I	134/5
$C_{18}H_{15}Sb$	CH_3	C_5H_5	CO	–	228, **25**, 5
$C_{18}H_{15}Sb$	C_2H_3O	C_5H_5	CO	–	228, **25**, 6

$C_{18}H_{15}Sb$	C_3H_9Sn	C_5H_5	—	—	21, **2**, *38*
$C_{18}H_{15}Sb$	C_3H_9Sn	C_5H_5	CO	—	132, **11**, *11*
$C_{18}H_{15}Sb$	C_5H_5	—	CO	I	132, **11**, *9*
$C_{18}H_{15}Sb$	C_5H_5	—	—	NO	49
$C_{18}H_{15}Sb$	C_5H_5	I	—	NO	48
$C_{18}H_{15}Sb$	C_5H_5	—	CO	$SnCl_3$	132, **11**, *10*
$C_{18}H_{15}Sb$	C_5H_5	$C_{10}H_8N_2$	—	—	7, **1**, *27*
$C_{18}H_{15}Sb$	C_5H_5	$C_{12}H_8N_2$	—	—	7, **1**, *30*
$C_{18}H_{15}Sb$	C_5H_5	$C_{18}H_{15}Sn$	CO	—	132, **11**, *12*
$C_{18}H_{15}Si$	C_5H_5	$C_6H_{15}O_3P$	CO	—	121, **10**, *11*
$C_{18}H_{15}Si$	C_5H_5	$C_{12}H_{27}P$	CO	—	89, **7**, *16*
$C_{18}H_{15}Si$	C_5H_5	$C_{18}H_{15}P$	CO	—	101, **8**, *32*
$C_{18}H_{15}Si$	C_5H_5	$C_{18}H_{33}P$	CO	—	90, **7**, *20*
$C_{18}H_{15}Sn$	C_3H_9P	C_5H_5	—	—	21, **2**, *39*; 41, **3**, *3*
$C_{18}H_{15}Sn$	C_5H_5	$C_6H_{15}O_3P$	CO	—	121, **10**, *13*
$C_{18}H_{15}Sn$	C_5H_5	$C_6H_{15}P$	CO	—	88, **7**, *8*
$C_{18}H_{15}Sn$	C_5H_5	$C_8H_{11}P$	—	—	21, **2**, *40*
$C_{18}H_{15}Sn$	C_5H_5	$C_8H_{11}P$	CO	—	112, **9**, *7*
$C_{18}H_{15}Sn$	C_5H_5	$C_{13}H_{10}F_3P$	CO	—	113, **9**, *15*
$C_{18}H_{15}Sn$	C_5H_5	$C_{13}H_{13}P$	—	—	21, **2**, *41*
$C_{18}H_{15}Sn$	C_5H_5	$C_{13}H_{13}P$	CO	—	113, **9**, *13*
$C_{18}H_{15}Sn$	C_5H_5	$C_{18}H_{15}As$	CO	—	131, **11**, *8*
$C_{18}H_{15}Sn$	C_5H_5	$C_{18}H_{15}O_3P$	—	—	21, **2**, *42*
$C_{18}H_{15}Sn$	C_5H_5	$C_{18}H_{15}O_3P$	CO	—	123, **10**, *34*
$C_{18}H_{15}Sn$	C_5H_5	$C_{18}H_{15}P$	CO	—	101, **8**, *37*
$C_{18}H_{15}Sn$	C_5H_5	$C_{18}H_{15}Sb$	CO	—	132, **11**, *12*
$C_{18}H_{15}Sn$	C_5H_5	$C_{29}H_{20}F_6P_2$	CO	—	114, **9**, *20*
$C_{18}H_{33}P$	C_2H_3O	C_5H_5	CO	—	156, **15**, *11*
$C_{18}H_{33}P$	C_2H_5	C_5H_5	CO	—	156, **15**, *10*
$C_{18}H_{33}P$	C_5H_5	—	CO	H	90, **7**, *19*
$C_{18}H_{33}P$	C_5H_5	$C_{13}H_{13}Si$	CO	—	90, **7**, *21*
$C_{18}H_{33}P$	C_5H_5	$C_{17}H_{15}Si$	CO	—	90, **7**, *22*
$C_{18}H_{33}P$	C_5H_5	$C_{18}H_{15}Si$	CO	—	90, **7**, *20*
$C_{19}H_{15}BN$	C_5H_5	—	CO	—	71; 306, **31**, *12*
$C_{19}H_{15}N$	C_5H_5	—	CO	—	237, **26**, *2*
$C_{19}H_{15}NPS$	C_5H_5	—	CO	—	140, **13**, *7*
$C_{19}H_{15}NPS_2$	C_5H_5	—	CO	—	143, **13**, *34*
$C_{19}H_{15}O_2PRu$	C_5H_5	—	CO	—	362, **43**, *12*
$C_{19}H_{15}PO$	C_5H_5	—	CO	—	232
$C_{19}H_{17}P$	C_2H_3O	C_5H_5	CO	—	208, **21**, *20*
$C_{19}H_{17}P$	C_2H_5	—	CO	—	208, **21**, *19*
$C_{19}H_{17}P$	C_5H_5	—	CO	H	113, **9**, *16*
$C_{19}H_{17}P$	C_5H_5	$C_{18}H_{15}P$	CO	—	233
$C_{20}H_{15}P$	C_5H_5	—	CO	Br	114, **9**, *21*
$C_{20}H_{17}BN_3$	C_5H_5	—	CO	—	352, **42**, *10b*

$C_{20}H_{20}NP$	C_2H_3O	C_5H_5	CO	–	221, **23**, *7*
$C_{20}H_{20}NP$	C_3H_5O	C_5H_5	CO	–	221, **23**, *8*
$C_{20}H_{20}NP$	C_4H_7O	C_5H_5	CO	–	222, **23**, *9*
$C_{20}H_{20}NP$	C_5H_5	C_8H_7O	CO	–	222, **23**, *10*
$C_{21}H_{15}O_2S$	C_5H_5	–	CO	–	334, **37**, *38*
$C_{21}H_{18}BN_2$	C_5H_5	–	CO	–	351, **42**, *7a*
$C_{21}H_{21}NP$	C_5H_5	–	CO	–	238, **26**, *6*
$C_{21}H_{21}P$					
$CH_2CH_2CH_2$-$P(C_6H_5)_3$	C_5H_5	$C_{18}H_{15}P$	CO	–	233
$P(CH_2C_6H_5)_3$	C_2H_3O	C_5H_5	CO	–	208, **21**, *24*
$P(CH_2C_6H_5)_3$	C_2H_5	C_5H_5	CO	–	208, **21**, *23*
$P(CH_2C_6H_5)_3$	C_5H_5	–	CO	H	90, **7**, *23*
$P(C_6H_4CH_3-4)_3$	C_5H_5	C_6H_4F	CO	–	204
$P(C_6H_4CH_3-4)_3$	C_5H_5	C_6H_5	CO	–	204
$C_{21}H_{22}NP$	CH_3	C_5H_5	CO	–	222, **23**, *11*
$C_{21}H_{22}NP$	C_2H_3O	C_5H_5	CO	–	222, **23**, *12*
$C_{21}H_{22}NP$	C_5H_5	–	CO	Br	114, **9**, *23*
$C_{21}H_{22}NP$	C_5H_5	–	CO	Cl	114, **9**, *22*
$C_{21}H_{22}NP$	C_5H_5	–	CO	I	114, **9**, *24*
$C_{21}H_{22}OPRh$	C_5H_5	–	CO	–	250
$C_{22}H_{15}OP$	C_5H_5	–	CO	–	363, **43**, *21*
$C_{22}H_{23}P$	CH_3	C_5H_5	CO	–	208, **21**, *21*
$C_{22}H_{23}P$	C_2H_3O	C_5H_5	CO	–	208, **21**, *22*
$C_{22}H_{24}NP$	C_2H_3O	C_5H_5	CO	–	222, **23**, *13*
$C_{22}H_{24}NP$	C_5H_5	–	CO	I	114, **9**, *25*
$C_{22}H_{29}P$	C_5H_5	–	CO	CN	113, **91**, *18*
$C_{22}H_{29}P$	C_5H_5	–	CO	I	113, **9**, *17*
$C_{23}H_{18}CoO_2$	C_5H_5	$C_{18}H_{15}P$	CO	–	201, **20**, *2*
$C_{23}H_{18}O_3P$	$C_{18}H_{15}O_3P$	–	–	I	45/6
$C_{24}H_{19}$	CH_3	$C_{18}H_{15}P$	CO	–	232
$C_{24}H_{19}$	C_2H_3O	$C_{18}H_{15}P$	CO	–	232
$C_{24}H_{19}BN_3$	C_5H_5	–	CO	–	352, **42**, *15c*
$C_{24}H_{29}P_3$	C_5H_5	–	–	–	9, **1**, *42*
$C_{25}H_{20}BN_2$	C_5H_5	–	CO	–	351, **42**, *8a*
$C_{25}H_{22}P_2$	C_5H_5	–	CO	–	80, **6**, *15, 16*
$C_{26}H_{18}P$	C_5H_5	–	CO	–	240, **26**, *19*
$C_{26}H_{22}P_2$	$C_2F_3O_2$	C_5H_5	–	–	23, **2**, *62*
$C_{26}H_{22}P_2$	C_2F_5	C_5H_5	–	–	55, **4**, *19*

$C_{26}H_{22}P_2$	C_5H_5	—	—	Br	23, **2**, *54*
$C_{26}H_{22}P_2$	C_5H_5	—	—	Cl	22, **2**, *51*
$C_{26}H_{22}P_2$	C_5H_5	—	CO	—	81, **6**, *27*
$C_{26}H_{23}NP$	C_5H_5	—	CO	—	238, **26**, *7*
$C_{26}H_{24}P_2$	CHS_2	C_5H_5	—	—	24, **2**, *64*
$C_{26}H_{24}P_2$	CH_3	C_5H_5	—	—	53, **4**, *6*; 67/8
$C_{26}H_{24}P_2$	CH_3O_2S	C_5H_5	—	—	24, **2**, *65*
$C_{26}H_{24}P_2$	$C_2F_3O_2$	C_5H_5	—	—	23, **2**, *61*
$C_{26}H_{24}P_2$	C_2F_5	C_5H_5	—	—	55, **4**, *18*
$C_{26}H_{24}P_2$	C_2F_5	C_5H_5	CO	—	223, **23**, *19*
$C_{26}H_{24}P_2$	C_2H	C_5H_5	—	—	55, **4**, *14*
$C_{26}H_{24}P_2$	C_2H_3N	C_5H_5	—	—	8, **1**, *39, 40*
$C_{26}H_{24}P_2$	C_2H_5O	C_5H_5	—	—	56, **4**, *23*
$C_{26}H_{24}P_2$	C_3F_7	C_5H_5	—	—	56, **4**, *20, 21*
$C_{26}H_{24}P_2$	C_3F_7	C_5H_5	CO	—	223, **23**, *20*
$C_{26}H_{24}P_2$	C_3H_3	C_5H_5	—	—	55, **4**, *15*
$C_{26}H_{24}P_2$	C_3H_6O	C_5H_5	—	—	7, **1**, *34, 35*
$C_{26}H_{24}P_2$	C_3H_7O	C_5H_5	—	—	56, **4**, *24*
$C_{26}H_{24}P_2$	C_3H_9Si	C_5H_5	—	—	24, **2**, *66, 67*
$C_{26}H_{24}P_2$	C_3H_9Sn	C_5H_5	—	—	24, **2**, *72*; 25, **2**, *73*; 41, **3**, *11*
$C_{26}H_{24}P_2$	C_4HF_6	C_5H_5	—	—	56, **4**, *22*
$C_{26}H_{24}P_2$	C_4H_7	C_5H_5	—	—	54, **4**, *12*
$C_{26}H_{24}P_2$	C_4H_8O	C_5H_5	—	—	8, **1**, *36*
$C_{26}H_{24}P_2$	C_4H_9	C_5H_5	—	—	54, **4**, *9, 10*
$C_{26}H_{24}P_2$	C_5H_5	—	—	Br	23, **2**, *53*; 41, **3**, *5*
$C_{26}H_{24}P_2$	C_5H_5	—	—	Cl	22, **2**, *50*; 41, **3**, *4*
$C_{26}H_{24}P_2$	C_5H_5	—	—	CN	23, **2**, *58*; 41, **3**, *7*
$C_{26}H_{24}P_2$	C_5H_5	—	—	D	22, **2**, *46*
$C_{26}H_{24}P_2$	C_5H_5	—	—	H	22, **2**, *45*
$C_{26}H_{24}P_2$	C_5H_5	—	—	I	23, **2**, *56*; 41, **3**, *6*
$C_{26}H_{24}P_2$	C_5H_5	—	—	MgBr	25, **2**, *74*
$C_{26}H_{24}P_2$	C_5H_5	—	—	NH_3	8, **1**, *37*
$C_{26}H_{24}P_2$	C_5H_5	—	—	N_2H_4	8, **1**, *38*
$C_{26}H_{24}P_2$	C_5H_5	—	—	NCS	41, **3**, *8*
$C_{26}H_{24}P_2$	C_5H_5	—	—	NO	49
$C_{26}H_{24}P_2$	C_5H_5	—	—	SCN	23, **2**, *59*
$C_{26}H_{24}P_2$	C_5H_5	—	—	S_2O_3	26, **2**, *82*
$C_{26}H_{24}P_2$	C_5H_5	—	—	$SnBr_3$	24, **2**, *70*
$C_{26}H_{24}P_2$	C_5H_5	—	—	$SnCl_3$	24, **2**, *69*
$C_{26}H_{24}P_2$	C_5H_5	—	—	SnI_3	24, **2**, *71*
$C_{26}H_{24}P_2$	C_5H_5	—	CO	—	80, **6**, *20*; 81, **6**, *21/26*
$C_{26}H_{24}P_2$	C_5H_5	C_5H_{11}	—	—	54, **4**, *11*
$C_{26}H_{24}P_2$	C_5H_5	C_6H_5	—	—	55, **4**, *17*

Formula					Reference
$C_{26}H_{24}P_2$	C_5H_5	C_6H_5S	–	–	23, **2**, *63*; 41, **3**, *9, 10*
$C_{26}H_{24}P_2$	C_5H_5	$C_6H_7O_4$	–	–	57, **4**, *27*
$C_{26}H_{24}P_2$	C_5H_5	C_6H_{11}	–	–	54, **4**, *7*; 55, **4**, *13*
$C_{26}H_{24}P_2$	C_5H_5	C_7H_7	–	–	54, **4**, *8*
$C_{26}H_{24}P_2$	C_5H_5	C_7H_5O	–	–	56, **4**, *26*
$C_{26}H_{24}P_2$	C_5H_5	C_8H_9O	–	–	56, **4**, *25*
$C_{26}H_{24}P_2$	C_5H_5	$C_{18}H_{15}Ge$	–	–	24, **2**, *68*
$\mathbf{C_{27}H_{24}BN_2}$	C_5H_5	–	CO	–	352, **42**, *9a*
$\mathbf{C_{28}H_{20}As}$	C_5H_5	–	CO	–	365, **43**, *33a*
$C_{28}H_{20}P$	C_5H_5	–	CO	–	360, **43**, *4, 4a*
$C_{28}H_{20}Sb$	C_5H_5	–	CO	–	367, **43**, *46a*
$\mathbf{C_{29}H_{20}F_6P_2}$	C_3H_9Sn	C_5H_5	CO	–	114, **9**, *19*
$C_{29}H_{20}F_6P_2$	C_5H_5	$C_{18}H_{15}Sn$	CO	–	114, **9**, *20*
$C_{29}H_{26}NP$	C_2H_3O	C_5H_5	CO	–	222, **23**, *14*
$\mathbf{C_{30}H_{27}Fe_3P}$	C_5H_5	C_6H_4F	CO	–	224, **23**, *23*
$C_{30}H_{29}P_2$	C_5H_5	–	–	–	58, **4**, *33*
$\mathbf{C_{31}H_{28}P_2}$	C_5H_5	–	CO	–	81, **6**, *29*; 82, **6**, *30*
$C_{31}H_{30}EuF_{21}NO_6$	C_5H_5	–	CO	–	307, **31**, *17*
$C_{31}H_{30}F_{21}HoNO_6$	C_5H_5	–	CO	–	307, **31**, *17*
$C_{31}H_{30}F_{21}NO_6Pr$	C_5H_5	–	CO	–	307, **31**, *17, 19*
$C_{31}H_{30}F_{21}NO_6Yb$	C_5H_5	–	CO	–	307, **31**, *17*
$\mathbf{C_{34}H_{33}As_2P}$	C_2H_3O	C_5H_5	CO	–	223, **23**, *22*
$C_{34}H_{33}P_3$	C_5H_5	–	–	–	9, **1**, *43, 44*
$C_{34}H_{33}P_3$	C_5H_5	–	–	D	22, **2**, *48*
$C_{34}H_{33}P_3$	C_5H_5	–	–	H	22, **2**, *47*
$C_{34}H_{33}P_3$	C_2H_3O	C_5H_5	CO	–	223, **23**, *18*
$\mathbf{C_{36}H_{37}O_2P_2}$	C_5H_5	–	–	I	23, **2**, *57*
$\mathbf{C_{41}H_{39}P_3}$	C_5H_5	–	–	–	9, **1**, *45/47*
$\mathbf{C_{42}H_{42}P_4}$	C_2H_3O	C_5H_5	CO	–	223, **23**, *21*

$C_{42}H_{42}P_4$	C_5H_5	–	–	–	9, **1**, *48, 49*
$C_{42}H_{42}P_4$	C_5H_5	–	CO	–	81, **6**, *28;*
					90, **7**, *24*
$C_{43}H_{37}MoN_2P_2$	C_5H_5	$C_6H_{16}P_2$	–	–	10, **1**, *50*
$C_{49}H_{40}O_7P_3Rh$	C_5H_5	–	CO	–	250
$C_{61}H_{60}Eu_2F_{42}NO_{12}$	C_5H_5	–	CO	–	307, **31**, *18*
$C_{61}H_{60}F_{42}Ho_2NO_{12}$	C_5H_5	–	CO	–	307, **31**, *18*
$C_{61}H_{60}F_{42}NO_{12}Pr_2$	C_5H_5	–	CO	–	307, **31**, *18*
$C_{61}H_{60}F_{42}NO_{12}Yb_2$	C_5H_5	–	CO	–	307, **31**, *18*

Table of Conversion Factors

Following the notation in Landolt-Börnstein [7], values which have been fixed by convention are indicated by a bold-face last digit. The conversion factor between calorie and Joule that is given here is based on the thermochemical calorie, $cal_{th\,ch}$, and is defined as 4.1840 J/cal. However, for the conversion of the „Internationale Tafelkalorie", cal_{IT}, into Joule, the factor 4.1868 J/cal is to be used [1, p. 147]. For the conversion factor for the British thermal unit, the Steam Table Btu, BTU_{ST}, is used [1, p. 95].

Force	N	dyn	kp
1 N (Newton)	1	10^5	0.1019716
1 dyn	10^{-5}	1	1.019716×10^{-6}
1 kp	9.80665	9.80665×10^5	1

Pressure	Pa	bar	kp/m²	at	atm	Torr	lb/in²
1 Pa (Pascal) = 1 N/m²	1	10^{-5}	1.019716×10^{-1}	1.019716×10^{-5}	0.986923×10^{-5}	0.750062×10^{-2}	145.0378×10^{-6}
1 bar = 10^6 dyn/cm²	10^5	1	10.19716×10^3	1.019716	0.986923	750.062	14.50378
1 kp/m² = 1 mm H₂O	9.80665	0.980665×10^{-4}	1	10^{-4}	0.967841×10^{-4}	0.735559×10^{-1}	1.422335×10^{-3}
1 at = 1 kp/cm²	0.980665×10^5	0.980665	10^4	1	0.967841	735.559	14.22335
1 atm = 760 Torr	1.01325×10^5	1.01325	1.033227×10^4	1.033227	1	760	14.69595
1 Torr = 1 mm Hg	133.3224	1.333224×10^{-3}	13.59510	1.359510×10^{-3}	1.315789×10^{-3}	1	19.33678×10^{-3}
1 lb/in² = 1 psi	6.89476×10^3	68.9476×10^{-3}	703.069	70.3069×10^{-3}	68.0460×10^{-3}	51.7149	1

Work, Energy, Heat	J	kWh	kcal	Btu	MeV
1 J (Joule) = 1 Ws = 1 Nm = 10^7 erg	1	2.778×10^{-7}	2.39006×10^{-4}	9.4781×10^{-4}	6.242×10^{12}
1 kWh	3.6×10^6	1	860.4	3412.14	2.247×10^{19}
1 kcal	4184.0	1.1622×10^{-3}	1	3.96566	2.6117×10^{16}
1 Btu (British thermal unit)	1055.06	2.93071×10^{-4}	0.25164	1	6.5858×10^{15}
1 MeV	1.602×10^{-13}	4.450×10^{-20}	3.8289×10^{-17}	1.51840×10^{-16}	1

Power	kW	h. p. (PS)	kp m/s	kcal/s
1 kW = 10^{10} erg/s	1	1.35962	101.972	0.239006
1 h. p. (PS)	0.73550	1	75	0.17579
1 kp m/s	9.80665×10^{-3}	0.01333	1	2.34384×10^{-3}
1 kcal/s	4.1840	5.6886	426.650	1

References:
[1] A. Sacklowski, Die neuen SI-Einheiten, Goldmann, München 1979. (Conversion tables in an appendix.)
[2] International Union of Pure and Applied Chemistry, Manual of Symbols and Terminology for Physicochemical Quantities and Units, Pergamon, London 1979; Pure Appl. Chem. 51 [1979] 1/41.
[3] The International System of Units (SI), National Bureau of Standards Specl. Publ. 330, [1972].
[4] H. Ebert, Physikalisches Taschenbuch, 5th Ed., Vieweg, Wiesbaden 1976.
[5] Kraftwerk Union Information, Technical and Economic Data on Power Engineering, Mülheim/Ruhr 1978.
[6] E. Padelt, H. Laporte, Einheiten und Größenarten der Naturwissenschaften, 3rd Ed., VEB Fachbuchverlag, Leipzig 1976.
[7] Landolt-Börnstein, 6. Aufl., II. Bd., 1. Tl., 1971, S. 1/14.

Key to the Gmelin System
of Elements and Compounds

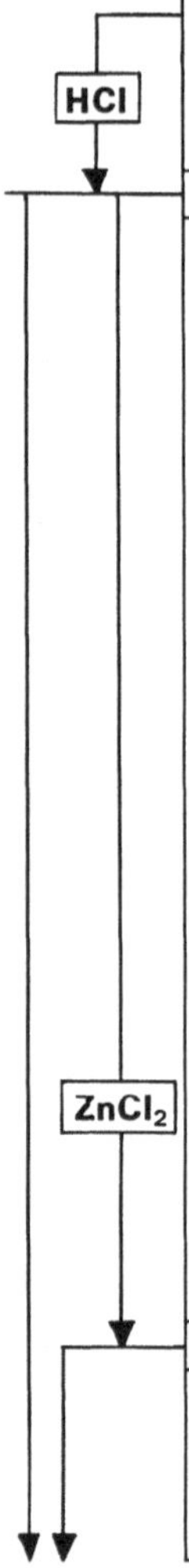

System Number	Symbol	Element
1		Noble Gases
2	H	Hydrogen
3	O	Oxygen
4	N	Nitrogen
5	F	Fluorine
6	**Cl**	**Chlorine**
7	Br	Bromine
8	I	Iodine
	At	Astatine
9	S	Sulfur
10	Se	Selenium
11	Te	Tellurium
12	Po	Polonium
13	B	Boron
14	C	Carbon
15	Si	Silicon
16	P	Phosphorus
17	As	Arsenic
18	Sb	Antimony
19	Bi	Bismuth
20	Li	Lithium
21	Na	Sodium
22	K	Potassium
23	NH_4	Ammonium
24	Rb	Rubidium
25	Cs	Caesium
	Fr	Francium
26	Be	Beryllium
27	Mg	Magnesium
28	Ca	Calcium
29	Sr	Strontium
30	Ba	Barium
31	Ra	Radium
32	**Zn**	**Zinc**
33	Cd	Cadmium
34	Hg	Mercury
35	Al	Aluminium
36	Ga	Gallium

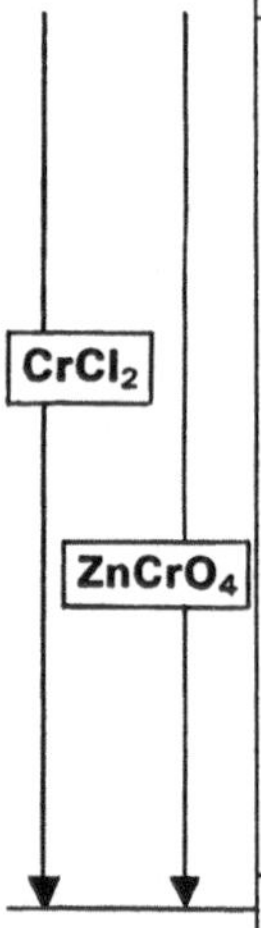

System Number	Symbol	Element
37	In	Indium
38	Tl	Thallium
39	Sc, Y La—Lu	Rare Earth Elements
40	Ac	Actinium
41	Ti	Titanium
42	Zr	Zirconium
43	Hf	Hafnium
44	Th	Thorium
45	Ge	Germanium
46	Sn	Tin
47	Pb	Lead
48	V	Vanadium
49	Nb	Niobium
50	Ta	Tantalum
51	Pa	Protactinium
52	**Cr**	**Chromium**
53	Mo	Molybdenum
54	W	Tungsten
55	U	Uranium
56	Mn	Manganese
57	Ni	Nickel
58	Co	Cobalt
59	Fe	Iron
60	Cu	Copper
61	Ag	Silver
62	Au	Gold
63	Ru	Ruthenium
64	Rh	Rhodium
65	Pd	Palladium
66	Os	Osmium
67	Ir	Iridium
68	Pt	Platinum
69	Tc	Technetium[1]
70	Re	Rhenium
71	Np,Pu...	Transuranium Elements

HCl

$CrCl_2$

$ZnCrO_4$

$ZnCl_2$

Material presented under each Gmelin System Number includes all information concerning the element(s) listed for that number plus the compounds with elements of lower System Number.

For example, zinc (System Number 32) as well as all zinc compounds with elements numbered from 1 to 31 are classified under number 32.

[1] A Gmelin volume titled "Masurium" was published with this System Number in 1941.

A Periodic Table of the Elements with the Gmelin System Numbers is given on the Inside Front Cover